Eighth International Congress of Pesticide Chemistry

Eighth International Congress of Pesticide Chemistry

Options 2000

EDITED BY

Nancy N. Ragsdale
U.S. Department of Agriculture

Philip C. Kearney
U.S. Department of Agriculture

Jack R. Plimmer
ABC Laboratories

Proceedings of a conference organized by the American Chemical Society and sponsored by the International Union of Pure and Applied Chemistry
Washington, DC, July 4–9, 1994

American Chemical Society, Washington, DC, 1995

Library of Congress Cataloging-in-Publication Data

International IUPAC Congress of Pesticide Chemistry (8th: 1994: Washington, D.C.)

Eighth International Congress of Pesticide Chemistry: options 2000: proceedings of a conference / organized by the American Chemical Society and sponsored by the International Union of Pure and Applied Chemistry, Washington, DC, July 4–9, 1994: edited by Nancy N. Ragsdale, Philip C. Kearney, Jack R. Plimmer.

p. cm.—(ACS Conference Proceedings Series, ISSN 1054–7487).

Includes bibliographical references and index.

ISBN 0–8412–2995–3

1. Pesticides—Congresses.

I. Ragsdale, Nancy N., 1938– . II. Kearney, P. C. (Philip C.), 1932– . III. Plimmer, Jack R., 1927– . IV. American Chemical Society. V. International Union of Pure and Applied Chemistry. VI. Title. VII. Series: Conference proceedings series (American Chemical Society).

SB950.93.I57 1994
668'65—dc20 94–42675
CIP

The paper used in this publication meets the minimum requirements of American National Standard for Information Sciences—Permanence of Paper for Printed Library Materials, ANSI Z39.48–1984. ♾

PRINTED IN THE UNITED STATES OF AMERICA

1994 Advisory Board

ACS Conference Proceedings Series

M. Joan Comstock, *Series Editor*

Eighth International Congress of Pesticide Chemistry

Executive Committee

President
Philip C. Kearney
U.S. Department of Agriculture

Chairman, Organizing Committee
Jack R. Plimmer
ABC Laboratories

Chairman, Scientific Program Committee
Nancy N. Ragsdale
U.S. Department of Agriculture

Finance Committee

Chairman
John McCarthy
American Crop Protection Association

Stan Cath
Agricultural Research Institute

Richard Herrett
Agricultural Research Institute

Organizing Committee

Budget Coordinator
John Bourke
Cornell University

Social Coordinator
Willa Garner
Garndal Associates

Scheduling and Facilities Coordinator
Paul Hedin
U.S. Department of Agriculture

ACS Coordinator
Dianne Ruddy
American Chemical Society

Publications Coordinator
Wayne Ivie
U.S. Department of Agriculture

Public Relations Coordinator
Marguerite Leng
Leng Associates

Poster Coordinator
Judd Nelson
University of Maryland

Bursar
Joel Coats
Iowa State University

Exposition Coordinator
John Parfet
ABC Laboratories

Scientific Program Committee

Vice Chairman
Elmo Beyer
DuPont

Synthesis
Joseph Fenyes
Buckman Laboratories

Delivery
Barrington Cross
American Cyanamid

Environmental Fate and Behavior
Robert Menzer
U.S. Environmental Protection Agency

Residues
Willis Wheeler
University of Florida

Biotechnology
Robert Hollingworth
Michigan State University

Metabolism
Gaylord Paulson
U.S. Department of Agriculture

Mode of Action
Julius Menn
U.S. Department of Agriculture

Resistance
Ronald Kuhr
North Carolina State University

Regulatory
John Gardiner
DuPont

Risk Assessment
Perry Gehring
DowElanco

Contents

RISK ASSESSMENT

Preface

THE SCOPE OF PESTICIDE CHEMISTRY embraces a wide variety of disciplines and extends far beyond chemistry. Responding to the challenge of providing new molecules to control pests was initially a straightforward task with high rates of scientific success and considerable commercial reward. The history of outstanding successes in controlling disease vectors and crop pests continued for many years. Dramatic advances in new chemistry and methods of delivery continue; however, pest control practitioners must today confront the problem of providing suitable control techniques for pest species in which the appearance of resistance to chemical controls may rapidly negate years of effort. In addition, public concerns about the safety of materials that are released into the environment in large quantities demand that chemical approaches not adversely affect people and the environment.

As questions of safety have become a major issue, it has become extremely important to detect and measure trace chemicals in many different matrices and to investigate their fate and effects in a variety of organisms and ecological systems. Such studies are mandated by regulatory agencies, whose activities are intensifying globally. The questions of safety assessment and the burden of the increasing costs of mandatory tests have affected the pesticide industry worldwide. The search for more effective and economical techniques to satisfy the regulatory issues is a major effort. Biotechnology offers some promise to deliver alternative approaches to conventional pesticide chemicals and new approaches to pest control, but these approaches are not acceptable without a more complete understanding of the potential consequences of their use for human health and the environment.

The International Union of Pure and Applied Chemistry (IUPAC) has sponsored congresses in pesticide chemistry at four-year intervals over the past 32 years. The Eighth International Congress of Pesticide Chemistry, organized by the American Chemical Society, was held in Washington, DC, in July 1994. The theme was the options for pest control chemistry as we enter the 21st century. This was the first time that the Congress had been held in the United States, and the breadth of international representation, with participants from more than 50 countries, was extremely gratifying. The Congress provided an opportunity for scientists from around the world to meet and discuss the dramatic advances and the rapid changes that are taking place in pesticide chemistry.

The organizers were faced with the difficulty of integrating the many components and disciplines that now comprise pesticide chemistry. The Program Committee selected 10 major topic areas, and these areas became the basis for the Congress symposia. At each symposium four invited speakers gave presentations; this proceedings volume is based on these presentations. An introductory overview was added to this volume. The Congress also included more than 900 poster presentations and more than 30 workshops as well as a satellite symposium on pesticide use in the developing countries. These proceedings reflect the current thoughts and opinions of internationally recognized scientists in each of the 10 topic areas of the Congress.

Acknowledgments

The organizers of the Eighth International Congress of Pesticide Chemistry gratefully acknowledge the generous financial support of the following companies and organizations:

BASF Corporation
DuPont
American Crop Protection Association
American Cyanamid Company
DowElanco
Ciba Plant Protection
Miles Inc.
Monsanto Corporation
ZENECA, Inc.
Merck & Co., Inc.
Valent U.S.A. Corporation

We thank the speakers, participants, and all those who worked diligently to make the Congress a success.

NANCY N. RAGSDALE
U.S. Department of Agriculture
National Agricultural Pesticide Impact
  Assessment Program
Administration Building, Room 321–A
Washington, DC 20250

PHILIP C. KEARNEY
U.S. Department of Agriculture
Natural Resources Institute
Room 208, Building 003, BARC-West
Beltsville, MD 20705

JACK R. PLIMMER
ABC Laboratories
7200 ABC Lane, P.O. Box 1097
Columbia, MO 65205

November 1, 1994

We also thank the following U.S. government agencies for their support and cooperation:

U.S. Department of Agriculture
  Agricultural Research Service
  Cooperative State Research Service
U.S. Environmental Protection Agency
  Office of Prevention, Pesticides,
    and Toxic Substances
U.S. Food and Drug Administration
  Regulatory Affairs
  Center for Food Safety and Applied
    Nutrition

World Food Security up to 2010 and the Global Pesticide Situation

W. Klassen, Joint FAO/IAEA Division of Nuclear Techniques in Food and Agriculture, International Atomic Energy Agency, Vienna, Austria

This paper describes progress over the past three decades and prospects to 2010 toward improved global food security and nutrition, adequacy of the natural resource base for sustained development and deceleration of demographic growth. Since population growth rates recede as people overcome poverty, it is important that in the decades immediately ahead significant progress be made; otherwise the insidious processes threatening sustainability will gain overwhelming strength. Improvements in food security are expected to correlate closely with per caput GDP[2] growth. The latter is projected to vary from 5.7% in East Asia to 0.7% in Sub-Saharan Africa. Between 1990 and 2010 the population of developing countries is expected to increase by 52%, while overall agricultural production is projected to increase by 62%. Two-thirds of the latter is projected to come from yield increases, one fifth from expanding land use and one-eighth from intensified cropping[3]. Cropped area per caput will sink to o.17 ha in Near East-N. Africa, 0.14 in E. Asia (excl. China) and 0.12 in S.Asia; water shortages will be acute in some regions. The ability of the agrochemicals industry to meet the needs of developing countries may be enhanced through the efforts of the OECD Pesticide Project to harmonize registration and re-registration requirements, by wide adoption of the Code of Conduct on the Distribution and Use of Pesticides, and by brisk economic growth in Asia and Latin America. GATT may assure that residues approved by the Codex Alimentarius Commission will move freely in international trade.

Introduction: There is considerable anxiety that before the end of the 21st Century the human population will have become so large that science and technology will not be able to prevent irreversible damage to the natural resource base and to the environment and that humanity generally will sink into abject poverty and chronic hunger (Brown et al., 1993). Thus, Pimentel et al. (1994) recently wrote an essay on "Natural Resources and an Optimum Human Population", in which they asked: "Does human society want 10 to 15 billion humans living in poverty and malnourishment or 1 or 2 billion living with abundant resources and a quality environment?"

Questions of how to largely overcome undernutrition, how to lessen adverse environmental impacts and how the world population may make the transition to very low demographic growth are of considerable concern to the Food and Agriculture, Organization (Alexandratos, 1988) and to the International Atomic Energy Agency. Therefore, I am grateful for this opportunity to share with you data and analyses which these organizations have assembled. Many of the data presented here were assembled by the FAO Global Perspective Studies Unit led by Dr. N. Alexandratos, and presented in November 1993 to the FAO Conference in the draft document: "Agriculture: Towards 2010" (FAO, 1993a). "Agriculture: Towards 2010" presents what is likely to happen by 2010 when the human population is expected to have increased to about 7.2 billion. Unfortunately, at the time when this document

[1]Countries comprising some FAO Regions: (a) **S. Asia**: Bangladesh, India, Nepal, Pakistan & Sri Lanka; (b) **E. Asia**: Cambodia, China, Indonesia, DPRK, ROK, Laos, Malaysia, Myanmar, Philippines, Thailand, & Viet Nam; (c) **Near East/N. Africa**: Afghanistan, Algeria, Egypt, Iran, Iraq, Jordan, Lebanon, Libya, Morocco, Saudi Arabia, Syria, Tunisia, Turkey & Yemen. [2]GDP or Gross Domestic Product is the annual value of goods and services produced in a country. [3]Cropping intensity is an index of the frequency per year of use of a plot of arable land. If two crops are harvested per year from the plot, the index is 200%. If two crops are harvested in each of two years followed by one of fallow, the index is 66%.

1054–7487/95/0001$15.00/0

was developed, there were still deficiencies in the databases on natural resources in China . Consequently, a number of projections for China will not be available until later this year.

This presentation is relevant to the following three questions with special reference to developing countries for the time period ending 2010:

1. How good are the prospects for progress towards the elimination of undernutrition and food insecurity?

2. How good are the prospects for safeguarding the productive potential and the broader environmental functions of agricultural resources for future generations (the essence of sustainability), while satisfying food and other needs?

3. How good are the prospects that the agrochemicals industry will continue to make significant contributions to the elimination of food insecurity and to the sustainability of agricultural production?

Nutritional needs and progress in reducing malnutrition and hunger: How much food does a person need? Generally, a 25 year old woman when pregnant requires 2510 calories per day, while a 25 year old male engaged in heavy activity requires 3320 calories per day (FAO, 1992a). However, on average a person should receive 2700 calories per day in order to work productively (see James and Schofield, 1990; World Health Organization, 1985). Undernourishment exists when the minimum caloric requirement of 1.54 times the basal metabolic rate is not met (FAO, 1992a, 1993a).

Globally, since about 1960 significant but still unsatisfactory progress has been made toward improved nutrition and food security (Fig. 1). During the past three decades, world gross agricultural production has grown more rapidly than the world population, and this trend is projected to continue through 2010. Thus, during 1988/90 world per caput food supplies were 18 per cent higher than 30 years ago (FAO, 1993a). Yet (Fig. 2) progress has varied from highly impressive as in E. Asia, and N. East/N. Africa to highly unsatisfactory as in Sub-Saharan Africa and S. Asia.

With the significant exception of Sub-Saharan Africa (Fig. 3), there has also been a significant steady decline in the percentage of the population which suffers from undernutrition, albeit the decline in the total number of undernourished has been much less rapid. The number of chronically undernourished people in developing countries now stands at 790 million or one out of five. Continuing significant

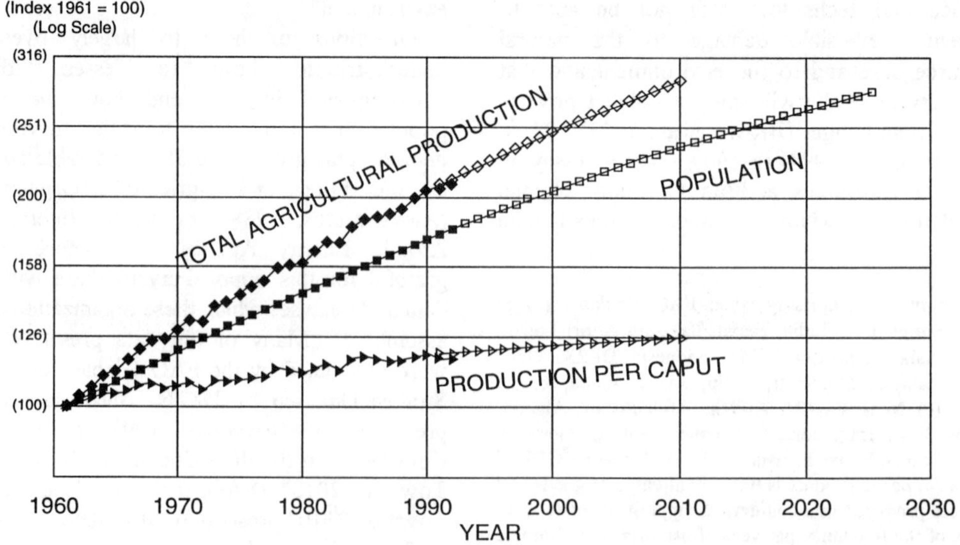

Fig. 1. Trends in world gross agricultural production, population and production per caput: historical to 1992 and projected beyond.

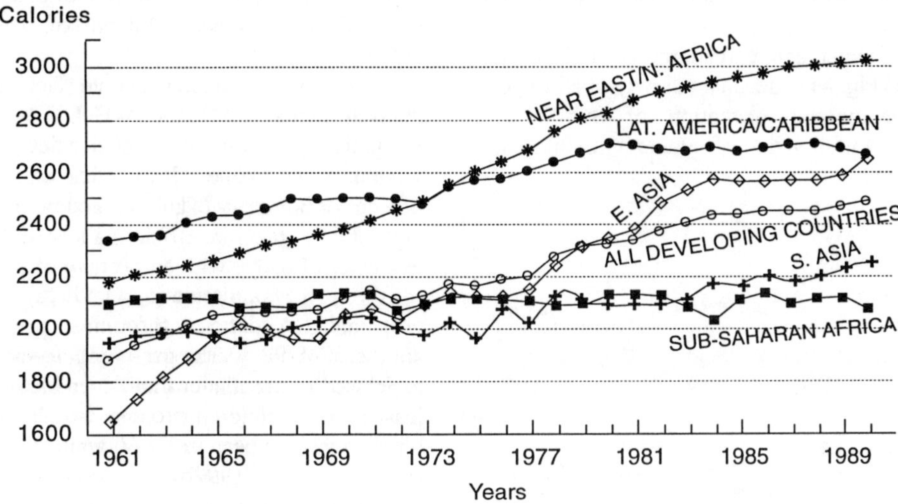

Fig. 2. Trends in food supplies per caput (calories/day) in developing country regions. Data obtained with permission from FAO (1993c).

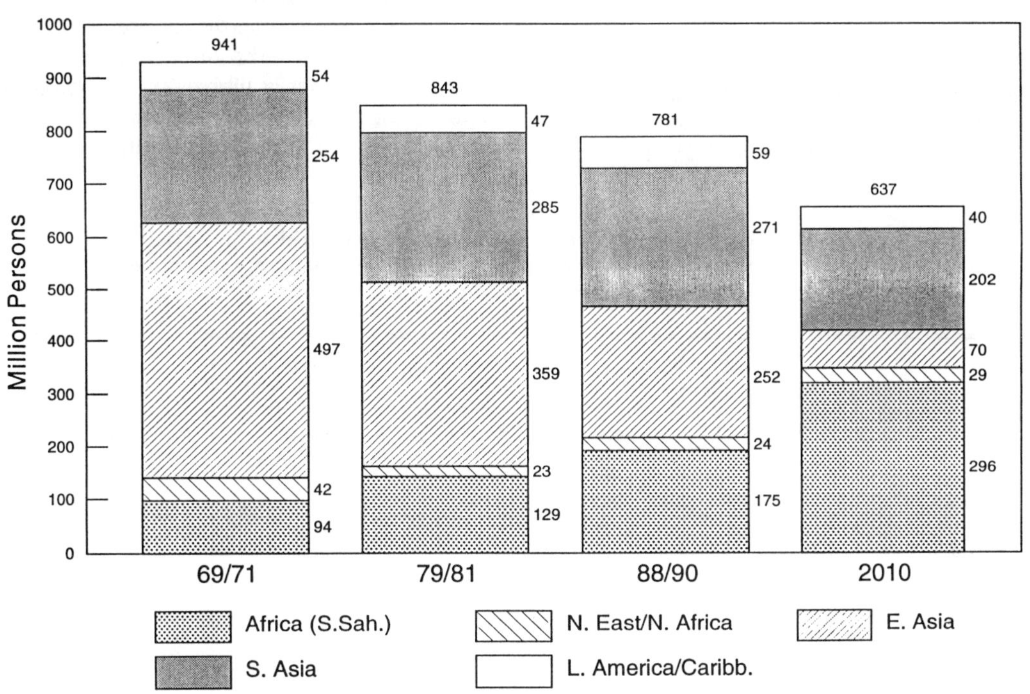

Fig. 3. Numbers (millions) of undernourished people developing country regions: historical and projected to 2010.

progress in overcoming chronic undernutrition is projected for E. Asia, slow progress for S. Asia and progressive worsening for Sub-Saharan Africa.

The number and percentage of underweight children (Fig. 4) is declining slowly in all regions with the notable exception of Sub-Saharan Africa. Mortality rate of infants and children is closely correlated with the percentage that is underweight. Indeed, in the least developed countries the death rate of infants and children under 5 years exceeds one out of five (FAO, 1993b).

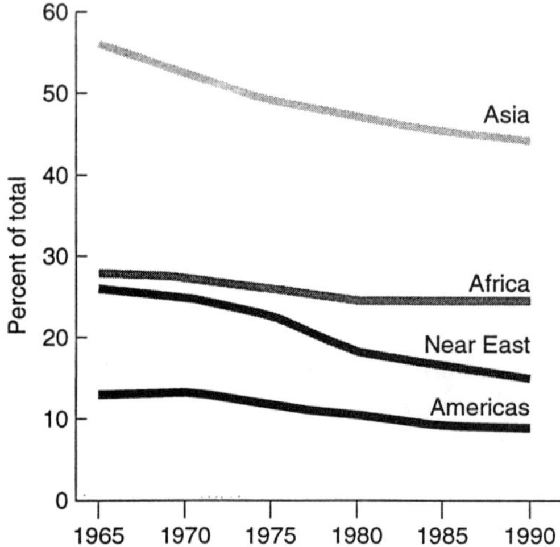

Fig. 4. Incidence of underweight children in regions of developing countries. Data were obtained with permission from FAO (1933a).

More than 2 billion people lack essential micronutrients (FAO and WHO, 1992) . Of particular concern are deficiencies in iron, vitamin A and iodine . Iron deficiency affects 30 percent of humanity (Table 1), and ranges from a high of 43 percent in Africa to a low of 7 percent in western Europe with an average of 29 percent for the world population (FAO 1992a).

Vitamin A deficiency is most prevalent in southeast Asia. Vitamin A deficiency may be expressed as xerothalmia and even blindness. At least 500,000 children become partially or totally blind each year because of Vitamin A deficiency (FAO, 1988, 1992b). Iodine deficiency disorders afflict about one billion people, and are most prevalent in southeast Asia (FAO, 1992a).

Trends: Total Population, Population Economically Active in Agriculture and GDP: The world population grew from 2.5 billion in 1950 to 5.3 billion in 1990; more than doubling in 40 years (Table 2). Although the rate of growth is slowing, the annual increment (Table 3) is approaching its expected peak of 94

Table 1. Populations at risk of and affected by micronutrient malnutrition, by WHO region, 1991 (population in millions). Reproduced with permission from International Conference on Nutrition: Nutrition and Development - a global assessment - 1992.

REGION	Iodine deficiency disorders		Vitamin A deficiency		Iron-deficient or anaemic
	At risk	Affected (Goitre)	At risk*	Affected* (xerophthalmia)	
Africa	150	39	18	1.3	206
Americas	55	30	2	0.1	94
South-East Asia	280	100	138	10.0	616
Europe	82	14	-	-	27
Eastern Mediterranean	33	12	13	1.0	149
Wesern Pacific	405	30	19	1.4	1058
Total	1005	225	190	13.8	2150

* Pre-school children only

Table 2. World population size and growth, medium variant. Reproduced with permission from United Nations (1993).World Population Prospects: The 1992 Revision (United Nations publication , Sales No. E. 93.XIII.7, Copyright 1993 United Nations.)

Year	Population (billions)
1950	2.5
1990	5.3
1992	5.5
2000	6.2
2010	7.2
2020	8.1
2025	8.5

Table 3. World population growth and annual increments, medium variant projections, 1950 - 2025. Reproduced with permission from United Nations (1993). World Population Prospects (The 1992 Revision, United Nations publication, Sales No. E.93.XIII.7. Copyright 1993 United Nations.)

Period	Annual increment (millions)	Annual growth rate (percentage)
1950 - 1955	47	1.8
1965 - 1970	72	2.1
1975 - 1980	74	1.7
1985 - 1990	88	1.7
1990 - 1995	93	1.7
1995 - 2000	94	1.6
2020 - 2025	85	1.0

million additional people per year (United Nations, 1991, 1993).

In the 20 years between 1990 and 2010, population growth rates (Table 4) are projected to decline in both developing and developed countries. Yet by 2010, the world population is projected to have increased by 1.9 billion. Ninety-four percent of the increase would occur in developing countries.

Population growth rates would be highest in Sub-Saharan Africa which, together with only modest economic growth, would preclude a significant improvement in per caput GDP.

By 2010, the GDP per caput is projected to grow annually at 3.4 percent for the developing countries taken as a whole (World Bank., 1993). The GDP growth rates per caput for the various regions are projected to vary from highest to lowest in the following order: E. Asia (5.7%), S. Asia (3.0%), L. America and Caribbean (2.3%), N. East/N. Africa (1.9%), Sub-Saharan Africa (0.7%) It seems likely that the rates of improvement in food security will correlate fairly closely with these GDP per caput growth rates.

By 2010 the total number of people economically active in agriculture (Table 5) in developing countries as a whole is expected to increase slightly, but the percentage of the total population engaged in agriculture is projected to decline from 59.6 percent to 46.9 percent. By 2010 the percentages of the populations economically engaged in agriculture are projected to vary from highest to lowest in the following order: Sub-Saharan Africa (59.7%), S. Asia (56.5%), Mainland China (51.7%), E. Asia excluding China (36.9%), N. East/N. Africa (23.5%), L. America/Caribbean (16.9%), and all developed countries (3.6%).

The high percentage of people engaged in agriculture in Sub-Saharan Africa and in S. Asia, means, of course, that the majority of the poor in these countries depend on agriculture both for employment and incomes. So long as this dependence remains high, the growth of food production and of agricultural productivity will continue to be the principal means for alleviating their poverty, improving their nutrition, and for eventually reducing birth rates. In Sub-Saharan Africa, and to a large extent in S. Asia, the food security problem cannot be alleviated significantly through international trade, because sufficient cash needed to pay for food imports is

Table 4. Population projections and GDP growth assumptions. Adapted with permission from FAO (1993a).

	Population						GDP Growth Rates	
	1989	2010	70-80	80-90	90-2000	2000-10	1989 - 2010	
	million						Total	Per caput
	million		growth rates percent per annum				%/annum	
World	5 205	7 209	1.9	1.8	1.7	1.4		
All Developing Countries	3 960	5 835	2.2	2.1	2.0	1.7		
93 Developing Countries	3 905	5 758	2.2	2.1	2.0	1.7	5.3	3.4
Sub-Saharan Africa	473	915	2.9	3.2	3.3	3.1	3.9	0.7
Near East/North Africa	297	493	2.7	2.8	2.6	2.2	4.4	1.9
East Asia	1 588	2 001	1.9	1.5	1.5	0.9	7.0	5.7
South Asia	1 144	1 728	2.3	2.4	2.2	1.8	5.1	3.0
Latin America + Caribbean	433	622	2.4	2.2	1.9	1.6	4.0	2.3
Developed Countries	1 244	1 373	0.8	0.7	0.5	0.4	2.6	2.1
Western Europe	399	410	0.4	0.3	0.2	0.0	2.7	2.5
North America	274	311	1.0	0.9	0.6	0.5	2.2	1.6
Others	182	214	1.4	1.0	0.9	0.7	4.2	3.4
E. Europe and former USSR	387	435	0.9	0.8	0.6	0.5	0.5	0.0

not being generated. Rather, the solution must begin locally with the marked improvement in the productivity of agriculture.

Experience has shown that the capability of people to be effective economic agents is often a more important factor in economic development than natural resources and man-made physical capital (FAO, 1993a). The mounting requirements in developing countries to intensify and modernize agriculture and to provide supporting services requires that illiteracy rates be reduced, so that people can be educated and trained to handle properly modern technologies such as pesticides, as well as to find off-farm employment. Rural illiteracy rates (Fig. 5) are especially high in N. East/N. Africa and in Sub-Saharan Africa; being higher invariably for females than for males.

Reduction of rural illiteracy is crucial in efforts to alleviate poverty, to improve nutrition, to reduce the mortality rates of infants and children and to reduce the birthrate.

Dietary Sources of Daily Energy and Nutrient Supply: Globally, the average daily energy supply of the average person is provided as follows: cereals, 51%; roots and tubers 5.3%; animal products, 13.5%; fruits, pulses, vegetables and nuts, 8.2%, and oils and fats sugars and honey, 19.1% (FAO, 1992b). However, people in developed countries rely less on cereals and more on animal products and fruits and vegetables (Fig. 6). In Sub-Saharan Africa, roots and tubers are major sources of calories.

Pathways to Increased Crop Production: Between 1990 and 2010, crop production in developing countries, excluding China, is projected to increase at an annual rate of 2.4 percent or

Table 5. Population economically active in agriculture. Adapted with permission from FAO (1993a).

	Population Economically Active In Agriculture					
	Million People			Percent of Total Economically Active Population		
	1990	2000	2010	1990	2000	2010
World	1 101.5	1 165.5	1 214.9	46.6	42.1	37.8
Developed Countries	50.1	34.8	23.8	8.3	5.5	3.6
All Developing Countries	1 051.4	1 130.7	1 191.2	59.6	53.0	46.7
93 Developing Countries	1 039.0	1 118.2	1 178.8	59.8	53.3	46.9
Sub-Saharan Africa	140.0	168.8	205.8	71.2	65.6	59.7
Lat. Amer./Caribbean	41.1	41.5	40.4	26.3	21.2	16.9
Near East/North Africa	34.6	37.6	39.4	37.2	29.9	23.5
East Asia minus China	98.6	103.8	103.8	50.2	43.3	36.9
China (mainland)	405.3	448.4	423.9	67.5	59.8	51.7
South Asia	274.5	318.1	365.5	64.7	60.7	56.5

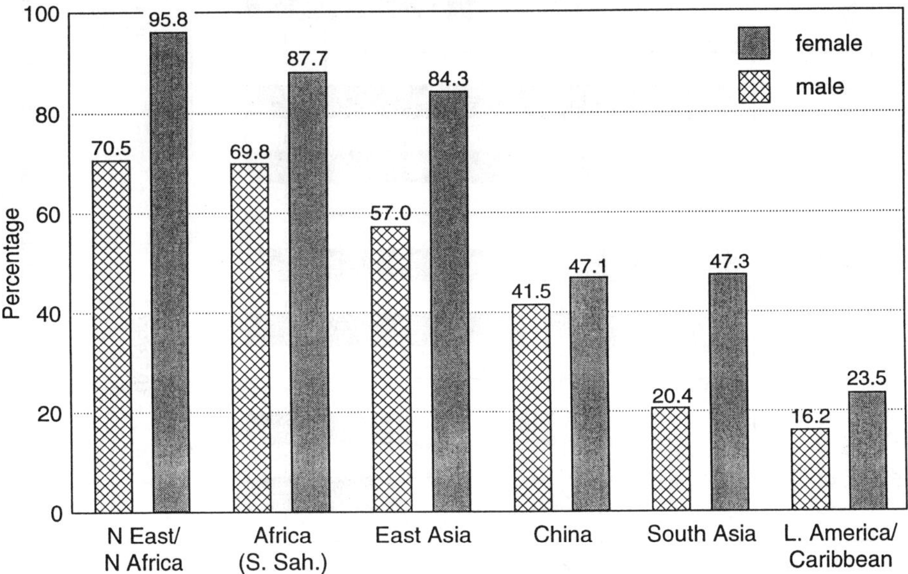

Fig. 5. Rural illiteracy rates by gender and developing country region. Data aggregated by FAO (1993c) from UNESCO, Statistical Yearbook, 1992, Paris, for those countries having data . Data do not necessarily correspond to the same year for each country.

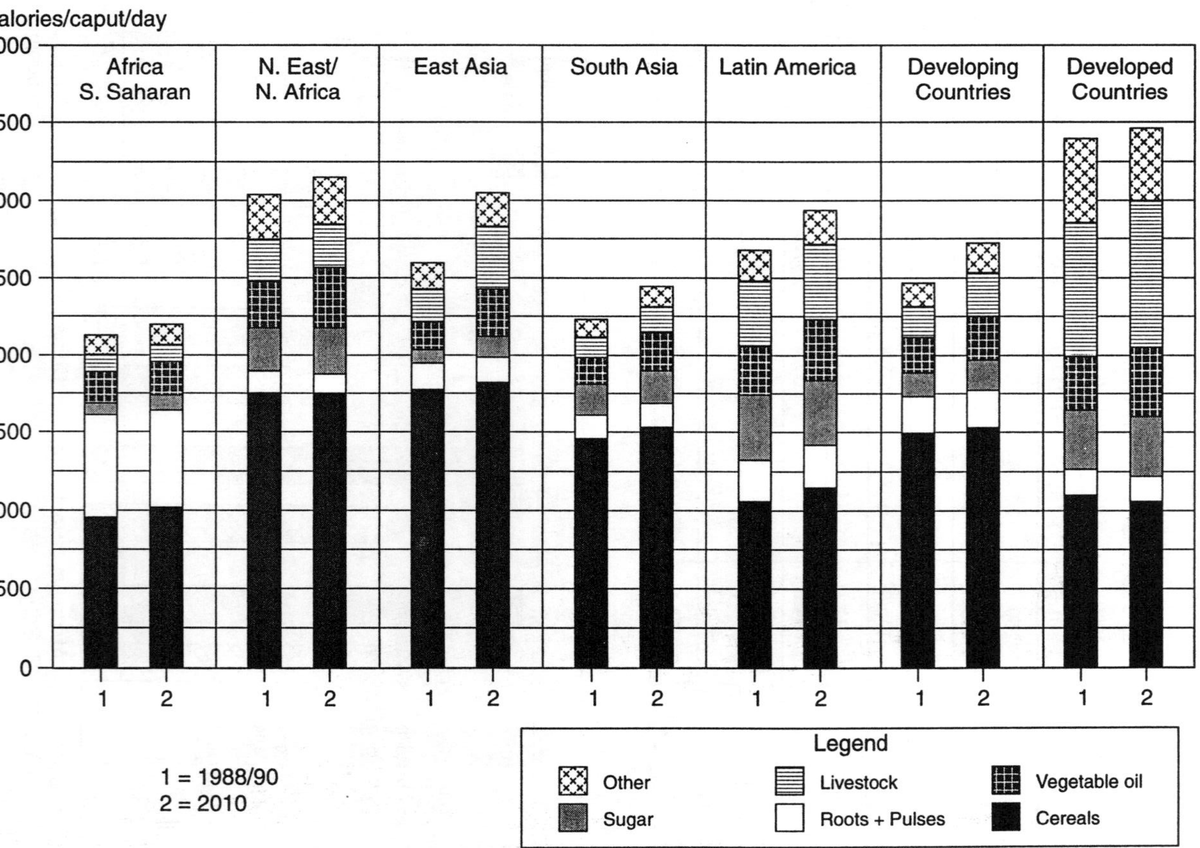

Fig. 6. Total calories/caput/day by major food groups in developing country regions, 1988/90 and projected for 2010.

overall by 66 percent; whereas the population is projected to increase by 52 percent (Fig. 7). During the previous two decades, the corresponding annual rate of increase in crop production was 2.9 percent for an overall increase of 69 percent. On a global basis (Table 6), about two-thirds of the increase must come from yield increases, one-fifth from bringing more land into production and one eighth from more intensive use of the land such as growing multiple crops per season and by reducing fallow periods. Cropping intensity is expected to increase from 79 percent in 1999/90 to 85 percent by 2010. Wide variations between regions in the relative importance of these three pathways to production will occur.

Urbanization is expected to spur the commercialization of agriculture, and to intensify the demand for animal products. Further mounting population pressure on a shrinking land base will force intensification of production and foster mixed

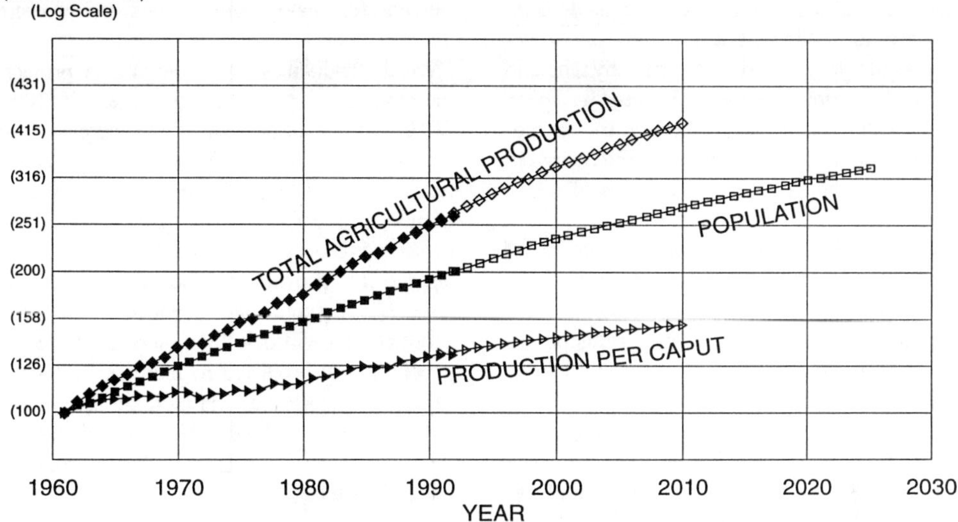

Fig. 7. Trends in developing countries (excluding China) gross agricultural production, population and production per caput: historical to 1992 and projected beyond.

Table 6. Sources of Growth in Crop Production in Developing Countries excluding China. Adapted with permission from FAO (1993a).

	Sources of growth over 1988/90 to 2010		
	Increased Yields	More Land	Intensified Cropping
	Percent Contribution		
93 Developing Countries	66	21	13
Sub-Saharan Africa	53	30	17
Near East/N. Africa	71	9	20
East Asia	64	27	9
South Asia	80	7	13
Latin Amer./Carib.	53	28	19

crop-livestock systems as an efficient and sustainable strategy of food production (Winrock International, 1992).

Food Production Projected to 2010: Production, area planted and yields of major crops in developing countries, excluding China, for 1988/90 and 2010 are projected to increase substantially (Table 7). Percent increases in production projected to be reached by 2010 are: rice, 51; wheat, 55; maize, 75; groundnut, 88; seed cotton, 100 and soybean 108. Steady increases in yields of the major cereals achieved in the past two decades are expected to continue, but at somewhat slower rates per annum (Fig. 8). The progress by China in raising yields during the past 15 years has been remarkable, and was achieved through the greater use of inputs, through the development of hybrid rice and through use of better genotypes of maize and wheat.

Growth rates (Fig. 9) in the world production of cereals began to decline about 15 years ago because of efforts by exporting countries to stem over production. Consequently, during this period world, cereal production (Fig. 10) failed to keep pace with population growth (Alexandratos, 1994).

The dependency of the developing countries on imports of cereals (Fig. 11) is projected to grow from the current net import level of about 90 million tons/year to 160 million tons with most of this increase to be supplied by N. America and Oceania. However, the former USSR and eastern European countries probably will no longer import significant quantities, and could turn into modest net exporters.

During the past two decades, the production of soybean in the western hemisphere and palm oil in Asia increased dramatically. By 2010 vegetable oil production by developing countries (excluding China) is expected to increase two fold from 36.7 million tons in 1988/90 to 72 million tons. In the next two decades, palm oil and palmkernel oil production are likely to grow by 130 percent; and their share of the vegetable oil market is projected to increase from 32 percent in 1988/90 to 38 percent by 2010. The market share of soybean oil is expected to increase from 19 percent in 1988/90 to 20 percent in 2010. Nevertheless, soybean production is projected to increase by 108 percent; half of which will come from area expansion and half from yield increases. Developing countries are expected to increase their exports of both vegetable oils and oilseed meal for livestock feed. Cotton production is projected to double in developing countries excluding China. Moreover, developing countries are expected to import significant quantities of cotton to meet the demand of their textile mills. Their exports of cotton clothing and of natural rubber will likely increase.

Production of fruit and vegetables in developing countries expanded very strongly during the past two decades, and increases of about 60 to 90 percent for major commodities are projected by 2010 (Table 8).

Meat production in developing is projected to increase by 120 percent between 1988/90 and 2010. Forty six percent of the increase (Fig. 12) would come from pigs and 29 percent from poultry. The annual rate of growth of cattle numbers in Sub-Saharan Africa and S. Asia is projected to be about 1 percent, and far less than is needed to provide to the number of draught animals required to achieve efficient crop production (Winrock International, 1992). Milk production is projected to increase by 68 percent. Developing countries will increase their imports of hides and skins and their exports of leather goods.

Only modest increases in global production of fish are projected (Table 9).

Adequacy of the Resource Base Projected to 2010: Overall the resource base is adequate both for increasing food production and for performing other vital functions, however severe problems exist at the regional and country levels. The overall percentage of total rainfed land that has some potential as cropland (Classes AT1 through AT7) is 40 percent (Table 10), but varies from 6% in N. East/N. Africa to 52% in L. America/Caribbean. Further, the crop production potential of substantial portions of such land is constrained by unfavorable conditions of terrain and soils (Table 11). Low natural soil fertility is the dominant constraint in Sub-Saharan Africa and in L. America, while in the N. East/N. Africa the steep slopes of the mountain ranges are the dominant constraint. In E. Asia, low natural soil fertility and poor soil drainage are the dominant constraints, and those in S. Asia are steep slopes, poor drainage and

Table 7. Production, Area Planted, and Yields for Major Crops in Developing Countries Excluding China. Numbers in brackets are the percent increase of the 2010 values over the 1988/90 values.

Commodity	Year	Area Harvested million hectares	Yield tons/ha	Total Production million tons
Wheat	1988/90	70	1.9	132
	2010	77 (10%)	2.7 (42%)	205 (55%)
Rice (paddy)	1988/90	109	2.8	303
	2010	120 (10%)	3.8 (36%)	459 (51%)
Maize	1988/90	63	1.8	112
	2010	80 (27%)	2.5 (39%)	196 (75%)
Barley	1988/90	17	1.3	22
	2010	19 (12%)	1.8 (38%)	35 (59%)
Millet	1988/90	32	0.7	22
	2010	38 (19%)	0.8 (14%)	32 (45%)
Sorghum	1988/90	37	1.0	37
	2010	50 (35%)	1.2 (20%)	62 (68%)
Total Cereals	1988/90	331	1.9	631
	2010	389 (18%)	2.6 (37%)	995 (58%)
Cassava	1988/90	15	10.1	153
	2010	18 (20%)	12.2 (21%)	223 (46%)
Sugarcane	1988/90	15	59.6	882
	2010	18 (20%)	75.4 (27%)	1 365 (55%)
Pulses	1988/90	52	0.6	30
	2010	61 (17%)	0.8 (33%)	48 (60%
Soybean	1988/90	22	1.7	38
	2010	33 (50%)	2.4 (41%)	79 (108%)
Groundnut	1988/90	17	1.0	16
	2010	21 (24%)	1.4 (40%)	30 (88%)
Coffee	1988/90	11	0.5	6
	2010	12 (9%)	0.7 (40%)	8 (33%)
Seed Cotton*	1988/90	19	1.1	21
	2010	22 (16%)	1.9 (73%)	42 (100%)

*Additional 1990 data for cotton: China, 5.6 million ha and production of 13.5 million tons; developed countries, 8.7 million ha and production of 19.1 million tons; and world, 32.8 million ha and production of 53.7 million tons.

SOURCE: Adapted with permission from FAO (1933a).

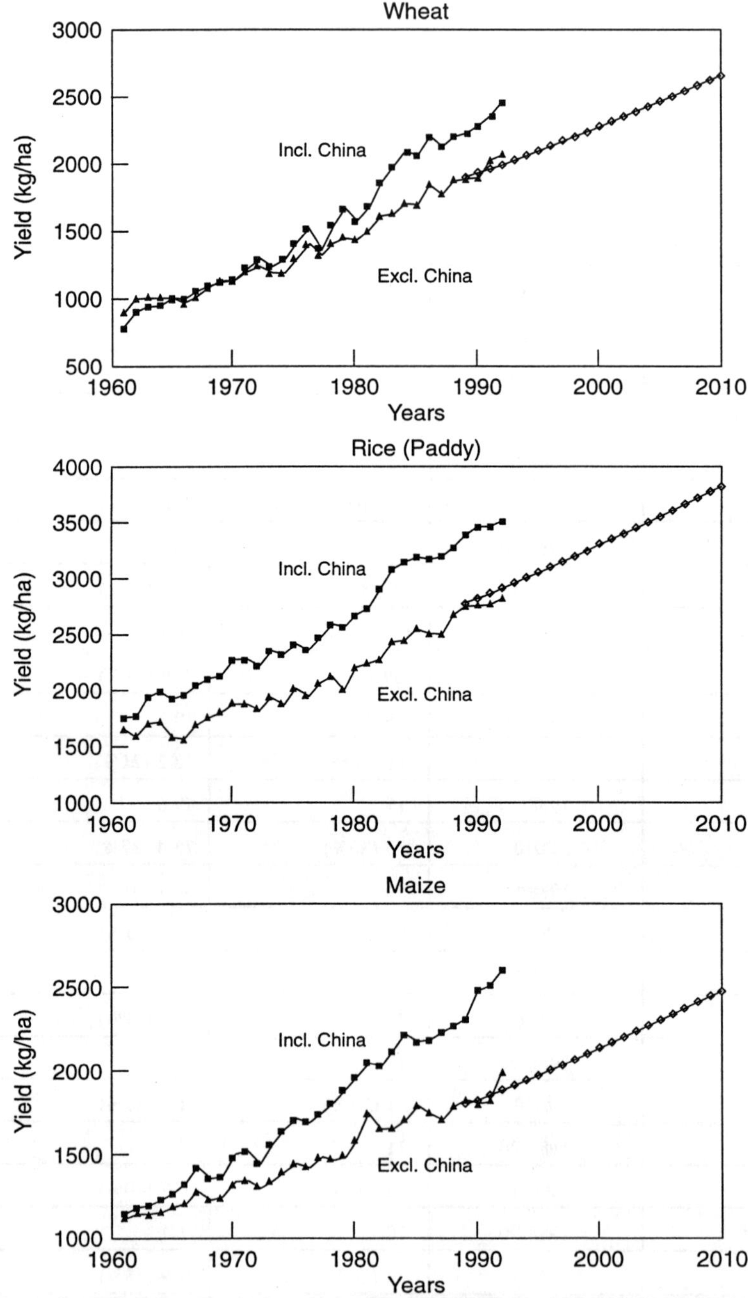

Fig. 8. Trends in yields of wheat, rice and maize in developing countries.

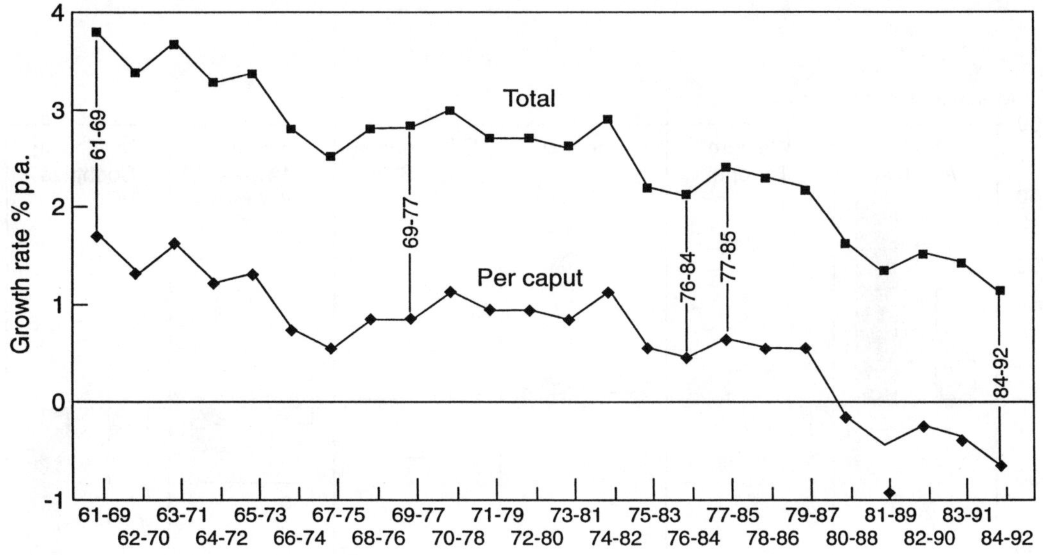

Note: After USSR cereals data conversion from bunker to clean weight.

Fig. 9. Trends in rates of growth of world cereal production based on averages of all 8-year periods in 1961-1992.

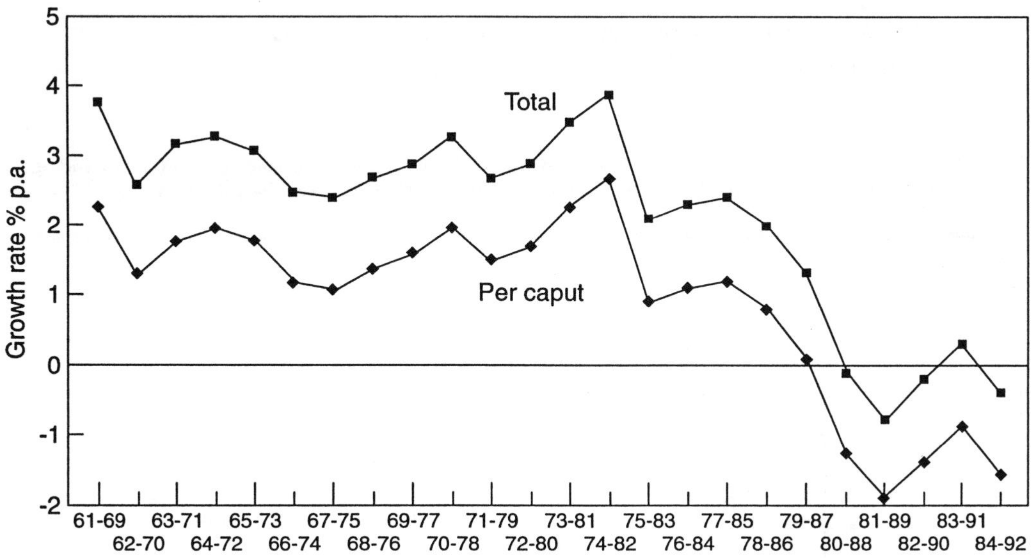

Fig. 10. Trends in rates of growth of world cereal production in net cereals exporting countries based on averages of all 8-year periods in 1961-1992 . Countries are USA, Canada, Australia, New Zealand, Argentina, Uruguay, EC-12, Austria, Sweden, Hungary, Turkey, Guyana, Malawi, Nepal, Myanmar, Pakistan, Zimbabwe, South Africa, Suriname, Thailand, and former Yugoslavia.

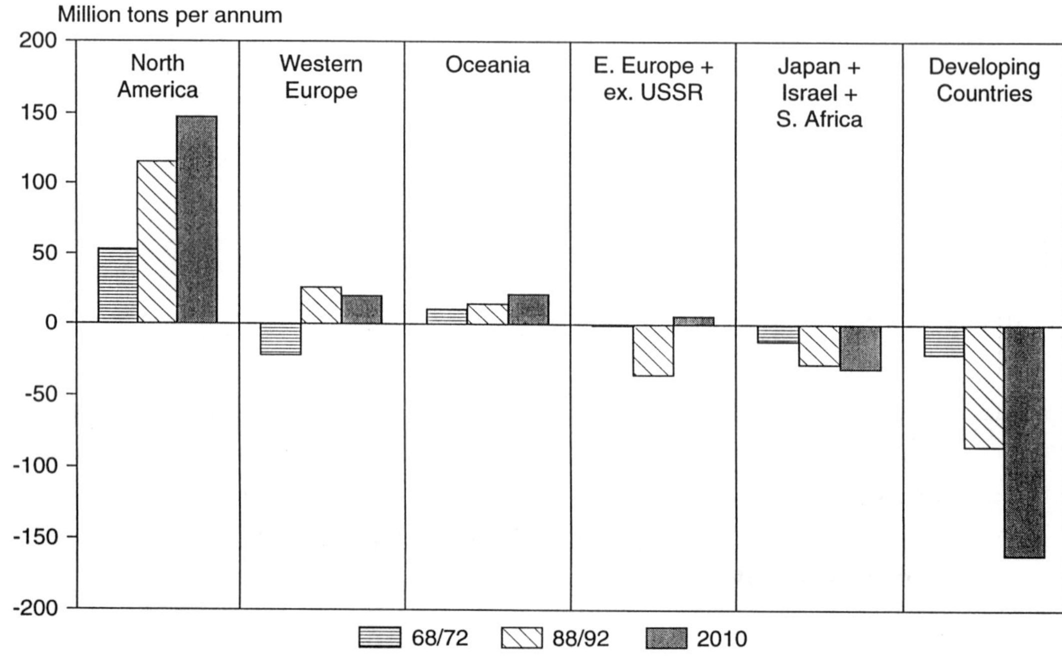

Fig. 11. Trends in positions in net trade in total cereals by major regions with projections to 2010.

Table 8. Fruit and vegetables: urea harvested, yield and total production in developing countries excluding China. Numbers in brackets are the percent increase of the 2010 values over the 1988/90 values.

Commodity	Year	Area Harvested ('000 hectares)	Yield (tons/ha)	Total Production ('000 tons)
Vegetables	1988/90	20 200	7.7	156 100
	2010	28 500 (41%)	10.0 (30%)	284 500 (82%)
Plantains	1988/90	3 900	6.5	25 700
	2010	4 900 (26%)	8.5 (31%)	41 200 (60%)
Bananas	1988/90	3 600	11.4	40 700
	2010	4 500 (25%)	15.5 (35%)	70 200 (73%)
Citrus	1988/90	3 500	11.9	41 300
	2010	4 500 (29%)	14.4 (21%)	65 300 (58%)
Other Fruit	1988/90	12 700	8.2	104 600
	2010	18 900 (49%)	10.5 (28%)	198 700 (90%)

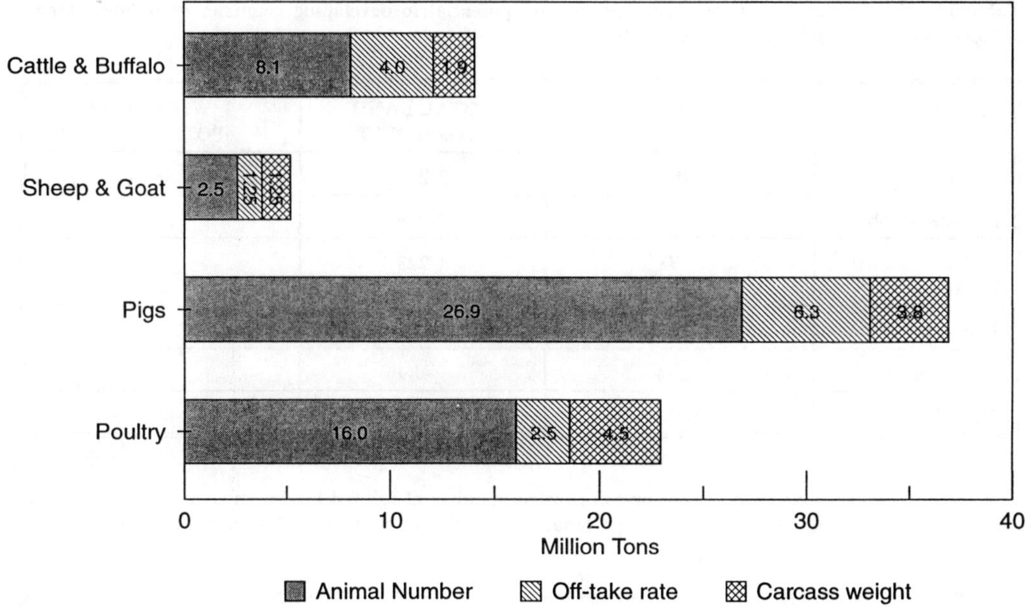

Fig. 12. Sources of growth in meat production in developing countries projected for 1988/90 to 2010. Data obtained with permission from FAO(1993c).

Table 9. Present and possible future production levels of fish world-wide. Adapted with permission from FAO (1993a).

(million tons, liveweight)

Type	1989/91	2010
Capture	86	90 - 110
Marine	79	
Inland	7	
Culture	12	15 - 20
Marine	4.5	
Inland	7.5	
TOTAL	98	105 - 130

sandy and stony soils. Overall in the developing countries, the dominant constraint is low natural fertility of the soil.

During 1988/90 757 million ha. in developing countries excluding China were in use for crop production out of a total potential base of 2.5 billion ha., mostly with severe constraints. By 2010 land in use may rise by 93 million to 850 million ha. Currently 123 million ha are irrigated of which 36 million is irrigated desert (Table 12). By 2010 146 million ha are projected to be irrigated, including an additional 2 million ha of desert (Fig. 13).

The amount of land currently available per person for producing food and fibre (Table 13) varies from a high of 0.45 hectare in Sub-Saharan Africa to a low of 0.17 in E. Asia, excluding China, for an average of 0.27 in these developing countries. By 2010, land available to feed and clothe each person is projected to vary from a high of 0.35 hectare in L. America and Caribbean to a low of 0.12 in S. Asia with a mean of 0.19.

More important is the amount of crop land per person economically active in agriculture (Table 13). As economic development proceeds, large efficiencies in production must be relaized so that one farm worker can grow food for many workers engaged in other sectors of the economy. Currently the number of hectares per farm worker varies from 4.6. in L. America/Caribbean to a low of 0.73 in S. Asia. Unfortunately, by 2010 this deplorable situation is projected to improve slightly only in L. America/Caribbean and E. Asia. This means that

Table 10. Rainfed land with some crop production potential in developing countries excluding China. Adapted with permisssion from FAO (1993a).

REGION	AT1 to AT 7 Hectares X 10^6	TOTAL LAND Hectares X 10^6	% WITH CROP POTENTIAL
Sub-S. Africa	1 008	2 214	46
L. Amer/Carib	1 054	2 038	52
N. East N. Africa	78	1 223	6
East Asia	153	380	38
South Asia	244	489	50
TOTAL	2 538	6 390	40

Table 11. Share of land with terrain or soil constraints in total rainfed land with some crop production potential in developing countries excluding China. Adapted with permission from FAO (1993a).

(percent)

Constraints	Regions					
	SSA	LA&C	NENA	EA	SA	Total
Steep slopes 16 - 45%	11	6	24	13	19	10
Shallow soils <50 cm	1	10	4	1	1	1
Low natural fertility	42	47	1	28	4	38
Poor soil drainage	15	28	2	26	11	20
Sandy and stony soils	36	15	17	11	11	23
Soil chemical constraints*	1	2	3	1	2	1
Percent of total AT1 to AT7 land with constraints	72	72	43	63	42	67

* Salinity, Sodicity and Gypsum. ** Individual constraints are non-additive, i.e. they usually overlap.

the majority of people in Africa and S. Asia will continue to have no option other than to rely on agriculture both for income and employment. Thus in 2010 poverty, chronic undernutrition and high birthrates would continue to prevail.

__Land Degradation__: Water and wind erosion, nutrient mining, salinization of soils and contamination of water are extensive (Fig. 14), and are caused principally by mismanagement of arable land, overgrazing and deforestation. Non-sustainable practices are of special concern with regard to Sub-Saharan Africa, where almost one-half of all additional rainfed land brought into production by 2010 will be located. During the next twenty years shifting cultivation (slash and burn, chitemene, etc.) will continue to be practiced widely in Sub-Saharan Africa. Moreover, the trend towards less productive and unsustainable systems may gain momentum. The slash and burn chitemene system cannot be practiced without woodland (the renewable resource), because the productivity of the soil is restored only through giving the nitrogen-fixing trees a long enough fallow period to regenerate to their original status. However, because of progressively increasing population pressure, the

Table 12. Total rainfed land with crop production potential, total area in production, and amounts of irrigated rainfed land and irrigated desert in production (1988/90) and projected for 2010 in developing countries excluding China. Adapted with permission from (FAO 1993a).

Region	Total Rainfed Land with Crop Potential (million hectares)	Total Area in Production (million hectares)		Irrigated Rainfed Land in Use (million hectares)		Irrigated Arid and Hyperarid Area (million hectares)	
		1988/90	2010	1988/90	2010	1988/90	2010
Sub-Sah. Africa	1 008	212	255	4.6	6.2	0.7	0.8
Lat. Amer. - Carib.	1 054	190	217	9.9	13.2	5.1	5.1
N. East - N. Africa	78	77	81	5.3	6.5	14.8	16.2
E. Asia minus China	153	87	103	19.3	21.5	0.0	0.0
South Asia	244	191	195	48.1	60.5	15.3	15.8
Total	2 538	757	850	87.1	108.0	35.9	37.9

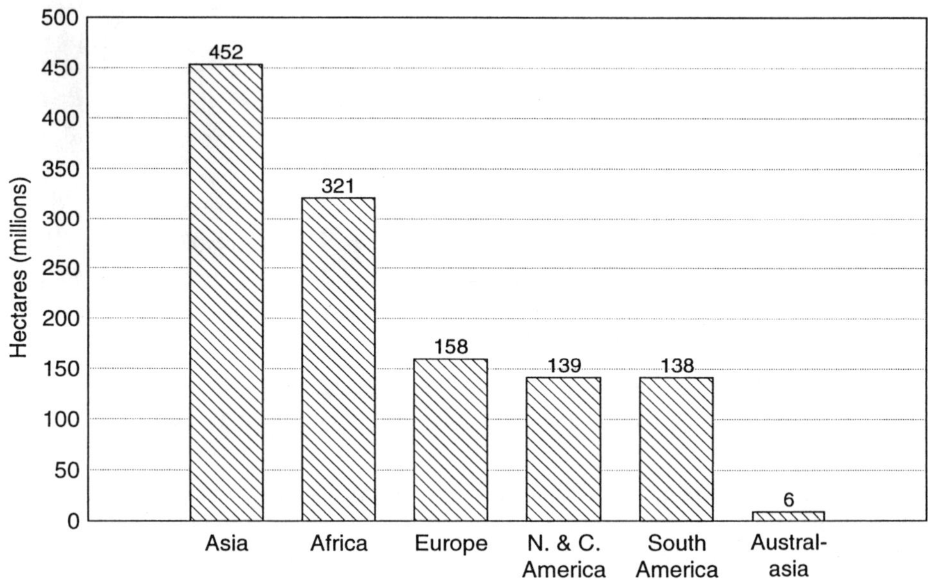

Fig. 13. Amount of cropland irrigated in developing countries excluding China.

Table 13. Hectares per caput of cropland in use 1990 and projected for 2010 for developing countries excluding China. The numbers in brackets are the number of hectares per caput economically active in agriculture. Adapted with permission from FAO (1993a).

Region	1990		2010	
	Millions of Hectares of Rainfed and Irrigated Land in Use	Hectares Per Caput	Millions of Hectares of Rainfed and Irrigated Land in Use	Hectares Per Caput
Sub-Saharan Africa	212	0.45 (1.5)	254	0.28 (1.2)
Lat. Amer.-Carib.	190	0.44 (4.6)	217	0.35 (5.4)
N.East-N.Africa	77	0.26 (2.2)	82	0.17 (2.1)
East Asia excl. China	77	0.17 (0.78)	88	0.14 (0.85)
South Asia	201	0.18 (0.73)	210	0.12 (0.57)
Average	757	0.27 (1.3)	851	0.19 (1.1)

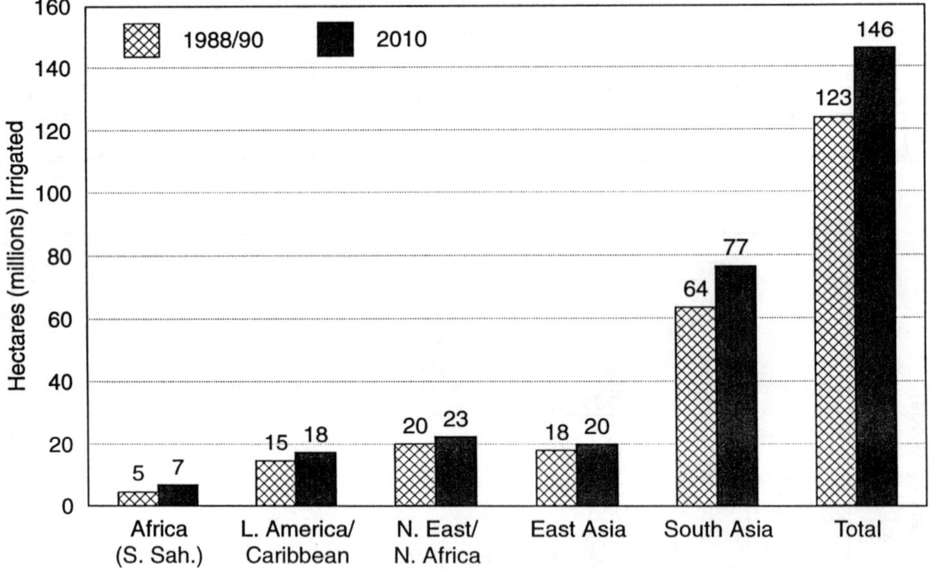

Fig. 14. Extent of moderate to severe soil degradation, i.e. area affected by water and wind erosion and by chemical and physical degradation.

fallow period is being reduced with the consequence that trees are disappearing and grass is taking over. Thus, farmers are being forced to adopt less productive systems such as the "fundikira" in Zambia, i.e. crops planted on soil that has been spread over a layer of decomposing grass residue. In such areas, the soil has become infertile and erosion has become extensive.

<u>**Water**</u>: Agriculture uses nearly 70 percent of managed water resources, industry uses 21 percent and 9 percent is used domestically.

The amount of managed water per caput (Table 14) has been declining steadily. Many countries are closer to their water limits than they are to their land limits. Already about one-half of the production of cereals in developing countries is accomplished through irrigation. In

Table 14. Per Caput Water Availability By Region, 1950 - 2000 . Adapted with permission from (FAO 1993a).

Region	1950	1960	1970	1980	2000
	'000 m^3				
Africa	20.6	16.5	12.7	9.4	5.1
Asia	9.6	7.9	6.1	5.1	3.3
Latin. America	105.0	80.2	61.7	48.8	28.3
Europe	5.9	5.4	4.9	4.4	4.1
North America	37.2	30.2	25.2	21.3	17.5

Original source: Ayibotele, N.B. 1992 . The world's water: Assessing the resource. Keynote paper at the International Conference on Water and the Environment, Dublin, Ireland, 1992.

1988/90, 37 percent of agricultural products were grown on irrigated land and by 2010 this is projected to increase to 42 percent. Thus, irrigation will become progressively more essential in meeting the food needs of developing countries.

International rivers whose waters may lead to disputes include the Euphrates, Ganges, Jordan, Mekong and the Nile (FAO, 1993d). In the Middle East, disputes over water, ethnic rivalries and associated territorial aspirations interact to endanger the slow path towards peace (Oman and Roudi, 1994).

Deforestation: The retention of natural forests, especially in the tropics (Table 15), is of great importance globally in relation to safeguarding biological diversity and the vital roles of forests in the dynamics of weather and climate; not the least of which is to serve as a carbon sink to counteract the possibility of global warming. In developing countries, forests are vitally necessary for providing fuel wood and other essentials. About 26 percent of the world's land area is forested for a total of 3.4 billion ha. of which one-half is tropical forest in developing countries (FAO, 1993a, 1993b). Annual deforestation is estimated at 15.4 million ha. per year (0.8%) up from 11.4 million ha. per year in 1980. Deforestation is caused mainly by land clearing for agriculture, necessitated by growth of the impoverished rural population (FAO, 1993a).

In many developing countries, high population growth rates, in combination with limited employment opportunities, persistent poverty, inequality of access to land and insecurity of the food supply, mean that the only option for subsistence is migration, often to forest areas, to find land for agriculture or pasture and shifting cultivation. On average some 60 percent of the net population increase in developing countries is absorbed migration to the urban areas, the rest swells the rural population - more in sub-Saharan Africa and S. Asia, less in L. America and the N. East/N. Africa (FAO, 1993a). Deforestation is exacerbated strongly by migration induced by war.

Clearly over the next 20 years the need to produce more food for growing populations will create progressively higher pressures on land, water, forests and fisheries. However FAO (1993a) notes that:"More than the need to produce more food for growing populations, the

Table 15. Tropical forest cover area and deforestation by geographical region. Percent of forested land is shown in brackets. Adapted with permission from FAO (1993a, 1993b).

Region	Forest Cover in 1990 (Million Hectares)	Annual Deforestation 1981 - 1990	
		Million Hectares	Percent Per Annum
Africa	527.6 (23.6%)	4.1	0.7
Asia and Pacific	310.6 (34.8%)	3.9	1.2
Latin America/Carib.	918.1 (55.6%)	7.4	0.8
Total	1 756.3 (39.2%)	15.4	0.8

pressures threatening sustainability are likely to be those emanating from growth of rural poverty, as more and more people attempt to extract a living out of dwindling resources. When these processes occur in an environment of poor limited resources and when the circumstances for introducing sustainable technologies and practices are not propitious, the risk grows that a vicious circle of poverty and resource degradation will set in. Poverty related environmental pressures are, however, only part of the story.

"Agricultural practices, consumption patterns and policies on the part of the rich also contribute to the problem. Responding to environmental pressures from this origin will depend on changes in policies to remove incentives for environmentally damaging practices and indeed to introduce disincentives for controlling them."

Production Inputs:

a. *Genetically improved crops.* Uninterrupted genetic improvement of crops is necessary to expand both the yield and the area planted to various crops. Hybrid maize selected for harsh environments is displacing sorghum and millet . High yielding semi-dwarf cereals adapted to local conditions will have a significant impact as will hybrid rice which was pioneered in China and which is already spreading throughout Asia.

Asexually reproducing and vegetatively propagated crops, such as banana, cassava, sweet potato and yams, are being improved through combinations of mutation breeding and in vitro culture techniques. Also in vitro techniques are used to free propagules from viruses. Genetically engineered crops are likely to begin to contribute significantly by the end of the next decade.

b. *Fertilizers.* Fertilizer use per ha. (Fig. 15) in 1988/90 and projected for 2010 is far too low in Sub-Saharan Africa and in individual countries in other regions. Growth in fertilizer consumption (Fig. 16) is projected to be substantial in L. America and S. Asia.

c. *Agrochemicals.* Protecting progressively higher yielding high yielding crops will require superb technology and skillful management of ecosystems. A significant concern is whether the agrochemicals industry will be able to meet the needs of developing countries for very safe and environmentally innocuous pesticide technologies. This need is much greater in developing than in developed countries because

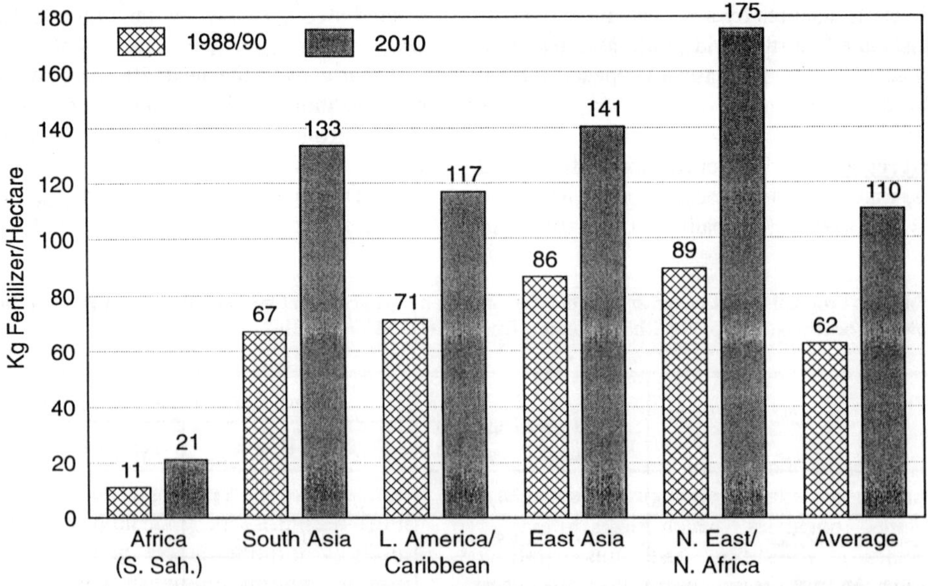

Fig. 15. Intensity of fertilizer (N,P,K) use in developing countries excluding China: 1988/90 and projected for 2010.

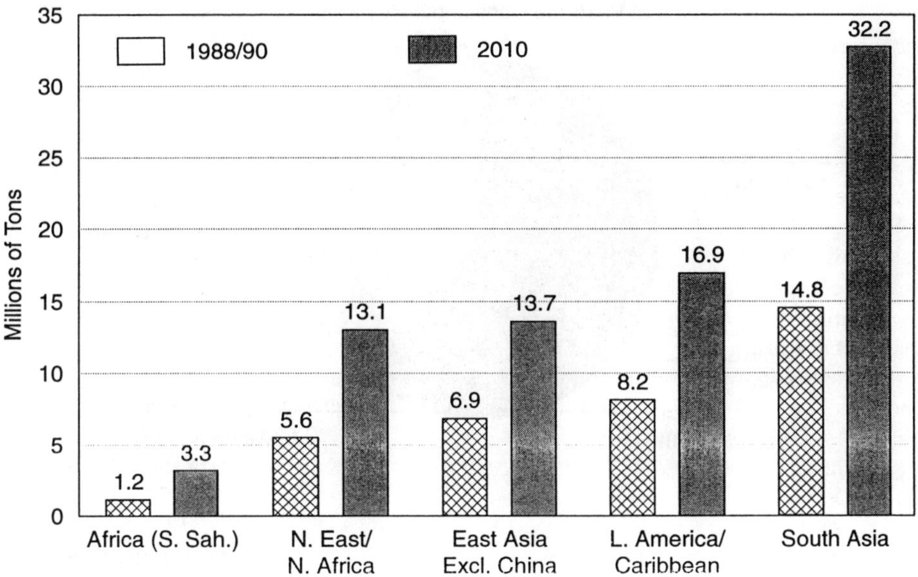

Fig. 16. Annual consumption of fertilizer (N,P,K) in developing countries excluding China: 1988/90 and projected for 2010.

many developing counties have serious deficiencies in training, use of protective clothing, proper disposal of pesticide containers, etc. (Waibel, 1993; Fleischer and Waibel, 1994).

Developing countries farm 54 percent of available land, yet they spend scarcely more than one-fifth ($5.7 billion) of the $25 - $26 billion (Table 16) spent world wide for pesticides (Aspelin et al., 1992; Klassen, 1993; McDougall, 1994). Consumption of pesticides by region (Fig. 17) is currently as follows: E. Asia including China, 38 percent; L. America, 30 percent; N. East/N. Africa, 15 percent; S. Asia 13 percent and Sub-Saharan Africa, 4 percent. Significant growth in pesticide consumption is likely to occur in S. Asia, E. Asia and L. America (FAO 1993a, Waibel 1994). Developing countries account for about 50 percent of world use of insecticides, 20 percent of fungicides, and 10 percent of herbicides.

Professor Zadoks of Wageningen University has argued that developing countries should by-pass the approach to plant protection used in developed countries, which relies heavily on pesticides. Pesticide use has perceived externalities that are unacceptable to the general public and, hence, to politicians (Kendrick and Stimmann, 1985; Waibel 1993; Zadoks, 1992).

Table 16. Approximate world market for chemical pesticides, 1991. Reproduced from Klassen (1993) Copyright 1993 American Chemical Society.

Class	Percent	US Dollars (billions)
Herbicides	44	11.44
Insecticides and acaricides	29	7.54
Fungicides	21	5.46
Nematicides and plant growth regulators	6	1.56
Total	100	26

Thus, the European Union has established a maximum of 0.1 ppb of any pesticide in drinking level regardless of toxicological hazard. The levels of pesticide use per ha. (Fig. 18) are quite high in several European countries, and the concern is so great that their governments have legislated initiatives to reduce pesticide usage by

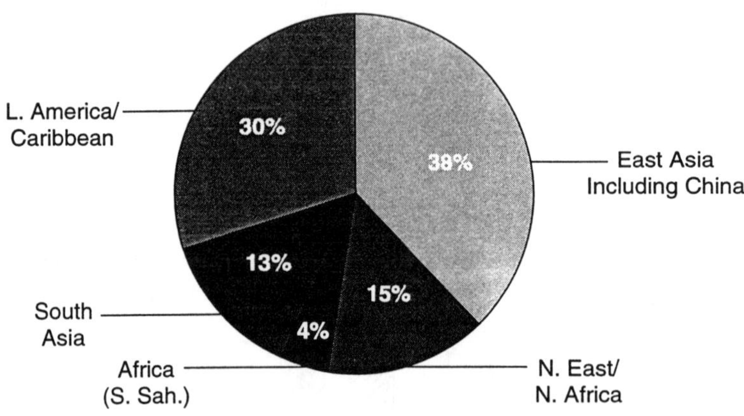

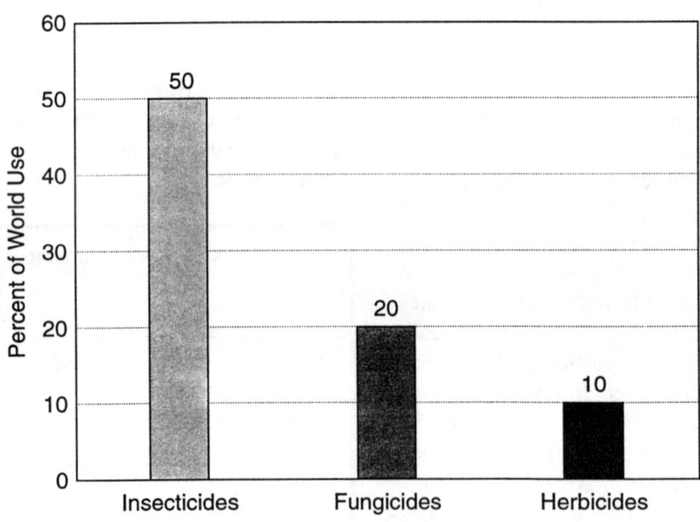

Fig. 17. Pesticide consumption in developing countries by region and class.

50 percent or more by 2000 (Neale and Sutton, 1993; Bellinder et al. in press, Zadoks, 1992; Netherlands Min. Agriculture, 1990). Thus, Zadoks foresees a radical departure from the use of methods perceived to have high externalities to new practices which meet the criteria of sustainability. According to this scenario, much use of pesticides would be replaced by much more extensive use of pest resistant crop cultivars and biological controls, better seed technology, precision application of better pesticide formulations, greater reliance on forecasting of populations of damaging entities, and the application of information and knowledge in decision support systems.

I believe the likelihood that Zadoks' vision

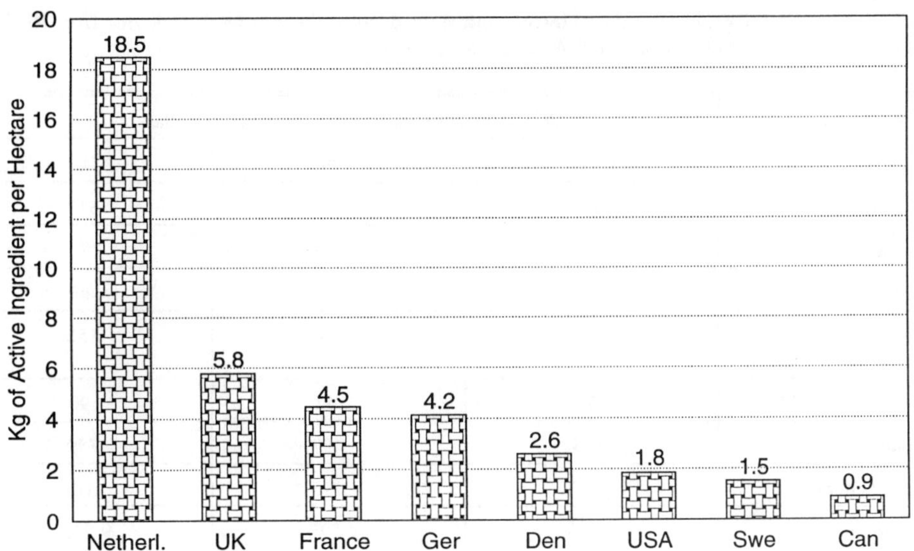

Fig. 18. Intensity of pesticide use in selected developed countries in 1987. Data from Bellinder et al. (1994).

may be realized eventually are the greatest for management of insect pest problems, intermediate for plant pathogens and largely unrealizable for cropland weeds. With regard to the use of alternatives (Table 17) to conventional insecticides, Ridgway et al. (1993) found that the use of biological controls, behavior modifying chemicals and specialty chemicals is growing fairly rapidly, but that by 2001 they will have captured less than 10 percent of the world market for insect control products. Clearly, then, even coping with insect pests will continue to require the extensive use of conventional chemical insecticides for decades to come.

In an attempt to establish a framework for forecasting roughly future trends in the use of pesticides and alternative technologies based on principles of economics and natural resource management, Waibel (1994) has proposed that the main factors which determine levels of use of pesticides in developing countries are as follows:

1. Major technological changes in agriculture, such as replacement of small mixed farming operations with monocultures of uniform varieties, increased field size, use of irrigation and fertilizers, use of resistant cultivars, biological controls, etc.

2. Substitution of family labour with external inputs as a result of economic growth which creates jobs in sectors other than agriculture.

3. Government policies concerning food security which tend to foster excessive use of pesticides.

4. Principle of substitution, whereby the relative prices of chemical versus non-chemical methods determine the farmer's decisions. Examples include the displacement of hand or mechanical weeding by herbicides based on relative costs; and the replacement of chemicals with information obtained by scouts and monitoring equipment as in certain integrated pest management programmes.

5. Principle of marginality, i.e. additional returns from an intervention must exceed the additional costs thereof . In the case of high value commodities, economic thresholds are exceeded at fairly low pest densities, which trigger interventions. This is especially relevant to high value crops which are highly susceptible to pest damage such as fruit, vegetables and cotton, and whose acreage will increase strongly to meet the demands of the growing population.

6. Improvements in pest management technology and programmes. More effective and safer active ingredients or formulations stimulate the use of pesticides, whereas advances in resistant cultivars, biological controls, etc. militate against their use. For example the eradication of the boll weevil from the southeastern U.S.A. has reduced insecticide use

Table 17. Estimated and projected sales of insect control products at the user level. Adapted from Rigway et al. 1992. Copyright 1993 International Atomic Energy Agency.

Product group	World market (millions of constant US dollars)		
	1 991	1 996	2 001
Microbial agents[a,b]	157	219	318
Nematodes[c]	4	30	70
Arthropods[c,d]	35	47	60
Behaviour modifying chemicals[b]	60	80	158
Botanicals[b]	70	81	90
Subtotal	326	457	759
Specialty chemicals[b,e]	100	116	181
Conventional insecticides[b]	9 358	9 597	9 212
Total	9 784	10 170	10 152

[a]Bacteria, viruses and fungi, and toxins produced by these organisms. [b]Estimates from SRI International, Menlo Park, CA.
[c]Estimates from Wall Street Journal and an anonymous source. [d]Includes primarily predators and parsitoids; some pollinators for use in glasshouses are also included. [e]Includes primarily insect growth regulators.

on cotton by 50 to 70 percent (Carlson et al., 1989).

7. The political economy of the agrochemicals industry, such as efforts by the industry to meet the groundswell of public resistance to use of pesticides (see Beyer, 1991), and the tendency to shift manufacturing facilities to Asian countries. Such shifts may change the interests of developing countries as they become exporters of pesticides.

Moreover it is essential fundamentally to recognize that since almost 80 percent of pesticide sales are in developed countries, the capacity of the agrochemicals industry in the near term to provide chemicals and application technology well suited to the needs of developing countries will depend primarily on how well the industry fares in industrialized countries. Over half of world pesticide sales (Figure 19) are in Western Europe and the United States (Neale and Sutton, 1993; Waibel, 1994). Since the early 1970s, when pesticide registration was made subject to substantially more stringent environmental safety and toxicological criteria, the R&D activities of the agrochemicals industry have focussed narrowly on the major commodities in these two markets, namely small grain cereals, maize, soybeans and cotton. Active ingredients registered on these commodities tend to become readily available for corresponding markets in developing countries (Table 18). On the other hand various other commodities such as beverages, fruit and vegetables, ornamentals, spices, pulses, livestock, etc. constitute neglected minor markets. For the past two decades, the industry has almost completely stopped introducing new active ingredients for many minor uses or to re-register older compounds for them. However these minor uses correspond directly to the most important export commodities of developing countries. Also during the past two decades the agrochemicals industry has undergone profound structural changes. These were caused, in part, by the weak demand for pesticides since about 1980 to the present (Fig. 20), when western pesticide markets approached saturation, and when economic growth and world trade decelerated. During this time agricultural production exceeded demand in the OECD countries; with unwanted production being subsidized. Concurrently, western Europe

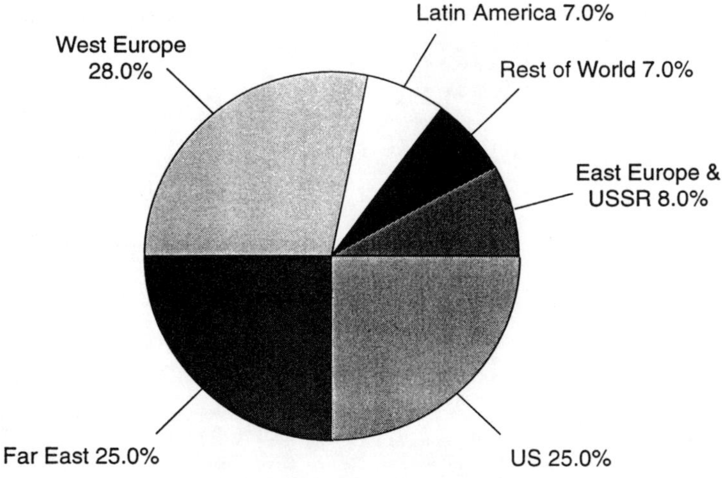

Fig. 19. World pesticide market. Reproduced with permission from Sutton and Neale (1993). Copyright British Crop Protection Council.

Table 18. World Pesticide Use in 1990 by Crop Category. Adapted with permission from FAO (1993a).

Crops	Herbicides (%)	Insecticides (%)	Fungicides (%)	Total Pesticides (%)
Fruit & Vegetables	16	27	43	26
Rice	11	17	16	14
Maize	18	8	1	11
Cotton	5	25	2	11
Wheat	14	2	13	10
Soybean	17	3	2	9
Sugar beet	6	3	2	4
Others	13	15	21	15
Total	100	100	100	100

became a significant exporter of cereals, and the limited global export market shrank as drastic economic and political changes in the Eastern Bloc countries forced them to reduce their grain imports. OECD countries adjusted by reducing the area planted and commodity support prices. In 1983, the United States introduced its "Payment in Kind" programme (PIK), and by 1987 more than 17 million hectares had been taken out of production in developed countries (Finney, 1988). Prior to PIK, the agrochemical market had been expanding at an average annual rate of 6.3 percent; but in 1983 PIK caused a real decline of 2.9 percent. Subsequently, overall growth of the agrochemicals market has proceeded sluggishly and even declined in the first few years of the 1990s (Finney, 1988, Waibel, 1994).

The industry's difficulties were exacerbated by the progressive ballooning of R&D costs (Fig. 21), especially those relating to increasingly more stringent standards of toxicology and environmental safety required for registration and re-registration. In the United States, these requirements were tightened by several amendments of the Federal Insecticide Fungicide

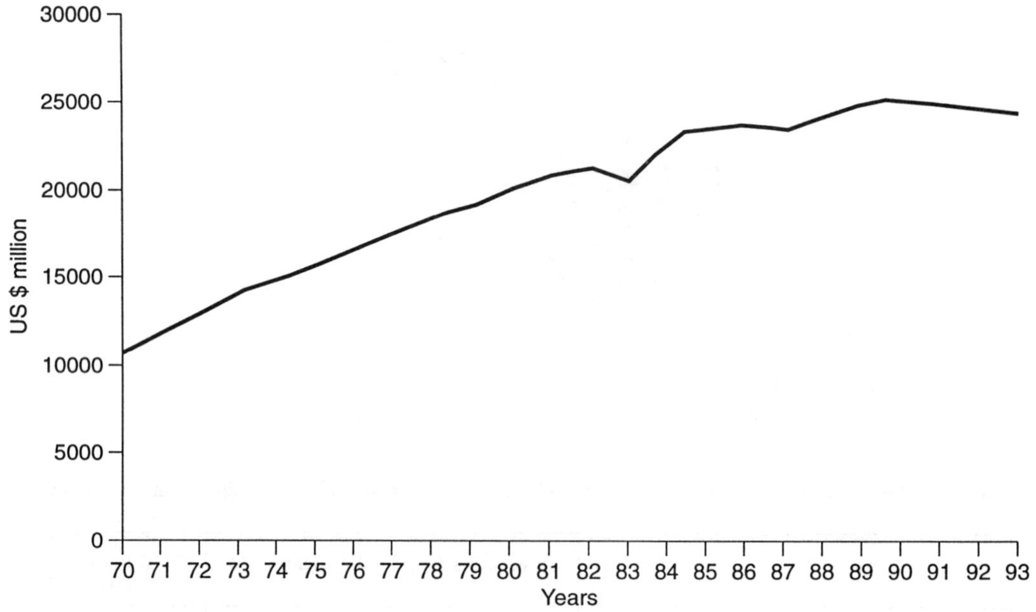

Fig. 20. Trends in global pesticide sales in 1992 US$. Data provided by Allan Woodburn Associates and by Wood McKenzie Consultants; adapted from Waibel (1994).

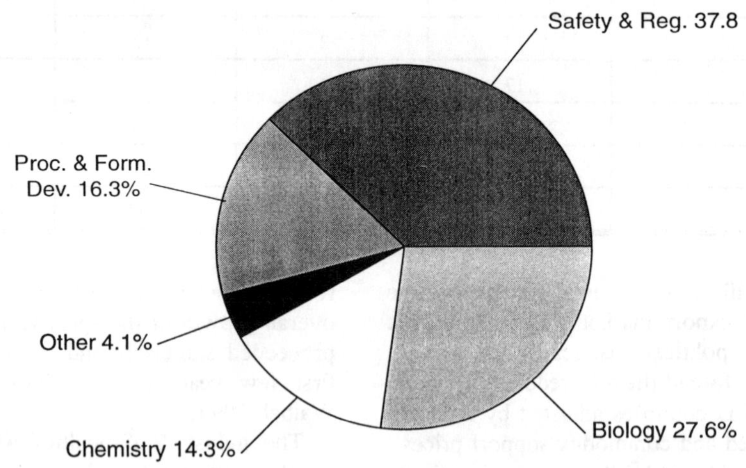

Fig. 21. Pesticide R&D costs in 1992 . Reproduced with permission from Sutton and Neale (1993). Copyright British Crop Protection Council.

and Rodenticide Act. In Europe, the broad stringent requirements for environmental fate and ecotoxicity data have been elaborated in Council Directive 91/44/EEC (Riley, 1993), also shown as the EC Registration Directive (McMinn and Thomas, 1991) and in annexes thereto (Thomas, 1993).

According to GIFAP, the costs of developing a new active ingredient are running between US$ 80-100 million not including the hardware costs in production (Gardiner, personal communication). The minimum time from discovery of a new active ingredient to its practical deployment (Fig. 22) is around 8 years (Neale and Sutton, 1993), and well over half of the period of patent protection may have been exhausted before the cumulative discounted cash flow reaches the break-even point (Finney, 1988; Menn and Henrick, 1985). Re-registration costs per active ingredient vary between US$2 million and US$10 million, depending on data requirements (Nesheim, 1992). Consequently, only very large companies are able to support the R&D needed to innovate and to support already existing products. For example, in Germany, between 1987 and 1990, the number of registered active ingredients declined from 308 to 216 or - 24.5 percent, while the number of formulated products declined from 1695 to 958 or - 29.6 percent (Finney, 1990). Whereas more than 600 fungicidal, herbicidal, insecticidal and nematicidal active ingredients (Worthing, 1979) were brought to the market between 1950 and 1980 for an average of about 20 per year, the number between 1980 and 1992 was 119 for an average of about 9 per year (Fig. 23). Prior to 1980, there were more than fifty profitable pesticide firms with significant R&D programmes. By 1989 there were about one dozen significant firms and about 75 percent of the world market was dominated by 10 firms (Finney, 1990). There have been a number of major acquisitions and mergers since 1984 (Table 19).

These structural changes are of very great concern, because industry's pesticide research effort will likely be roughly proportional to the number of pesticide producers (Zadoks, 1992). Clearly the search for new active ingredients, new formulations and better application technology needed to meet sustainability requirements would benefit from greater competition.

Moreover, during the past two decades national governments have retreated unremittingly from funding research related to pest problems (Kinney, 1985; Klassen, 1991; Wolf, 1985), with the effect that the industry, government, academic coalition for research on principles and practices of crop protection has been weakened significantly in North America and Western Europe.

Collectively these changes exacerbate the difficulties faced by developing countries in producing and exporting high value fruit, vegetable and specialty crops - all of which are

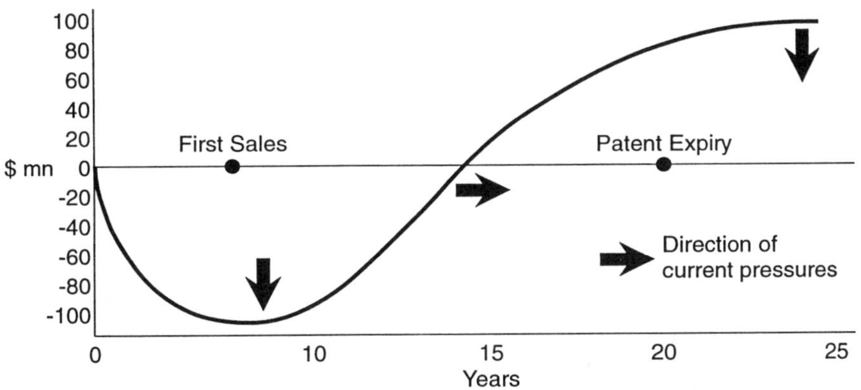

Fig. 22. Typical cumulative discounted cash flow for a successful new product . Adapted with permission from Finney (1988). Copyright British Crop Protection Council.

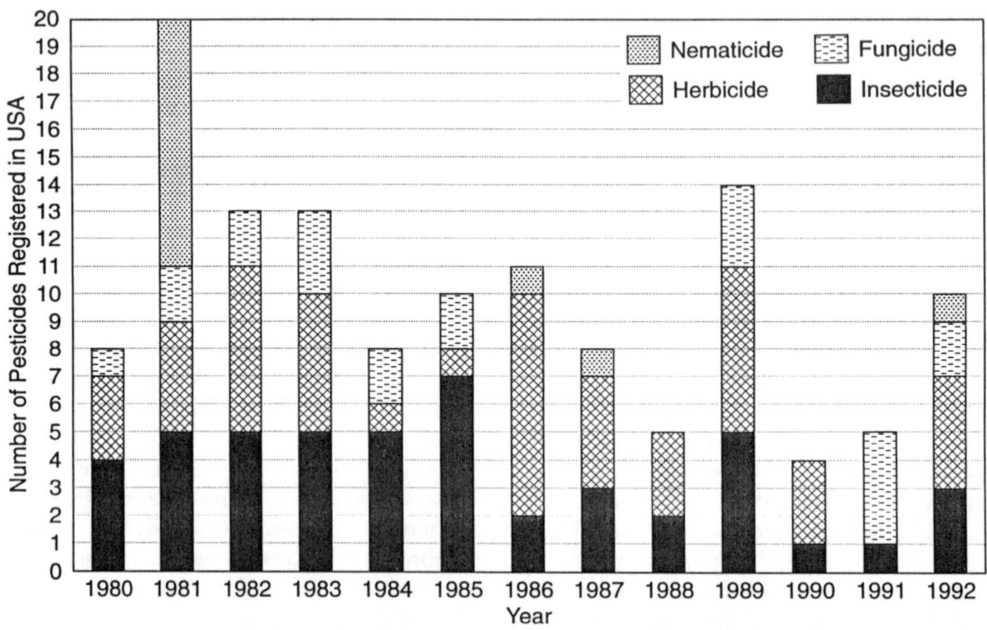

Fig. 23. Trends in the introduction of new active ingredients into sales each year in U.S.A. Data provided by the Environmental Protection Agency, Washington, D.C.

Table 19. Major acquisitions and mergers:1984 - 1994.

ICI + Stauffer
Rhone Poulenc + Union Carbide
DuPont + Shell (US)
Sandoz + Velsicol + SDS Biotech
Shell + Celamerck
Sumitomo + Chevron
American Cynamid +Shell
Hoechst + Schering form AgrEvo

"minor uses", as well as the difficulties in reducing pesticide-related illnesses, deaths and environmental impacts. The current slow pace of innovation in pesticide technology (including formulations and applications technology) means that the developing countries will tend to be stuck for at least the next decade with much antiquated technology. That a great potential exists to discover and develop a new pesticides with apparently excellent safety to and environmental innocuity is exemplified by the introduction of sulfonyl ureas Beyer (1991) and imidazolinone herbicides (Stinson, 1994). However the short-term outlook for the timely and substantial realization of this potential seems poor.

Nevertheless several developments and initiatives are underway which have some potential to revitalize innovation. Firstly, the OECD Pesticide Project is working toward the harmonization of approaches and data requirements of registration and re-registration among all Member States. OECD's long term objective is to reach a level of mutual confidence and agreement whereby a registration package accepted by one Member Government would be accepted by all (OECD, 1993; Visser, pers. comm.). Doubtless this would provide considerable savings to governments and to the industry. Secondly, developing countries which are parties to Agenda 21 have pledged to adopt the International Code of Conduct on the Distribution and Use of Pesticides by 2000 (FAO, 1990). It is strongly in the interests of all developing countries to implement fully the Code of Conduct which, among other things, requires governments to introduce legislation and regulatory programmes needed to minimize adverse consequences from use of pesticides. Adoption of the Code is being encouraged by some OECD governments, which also require industrial firms to comply with the Prior Informed Consent article of the Code (Hanson, 1994). It is important that initiatives of OECD governments do not have the net effect of depriving growers in developing countries of high quality products.

Most importantly, the continuing economic boom in Asia and resumption of economic growth in N. America and L. America, and will enable farmers in these regions to purchase more inputs and provide resources for industrial R&D.

I venture to make the following summary predictions regarding the future of pesticides and the agrochemicals industry:

1. During the next decade, pesticide sales are likely to grow substantially in a number of developing countries such as Brazil, China, Colombia, India, and S. Korea, to increase modestly in N. America and Japan, but to remain static in western Europe.

2. The substantial growth in pesticide use foreseen for Asia, and Latin America (see Waibel 1994) will induce more R&D on pesticide technology to be conducted in these regions by new companies and by subsidiaries of established companies. Concurrently biologically based products will increasingly be used especially to meet needs which require neither rapid action nor nearly total protection against pest damage, as well as in other niche markets.

3. High regulatory standards of pesticide safety and environmental innocuity will be required and enforced globally, although the pressure to harmonize them will result in significant modifications of specific requirements in national legislation.

4. The GATT requires that importation of a food commodity deemed acceptable by the Codex Alimentarius Commission or by other international bodies/conventions cannot be prevented by a participating country. Therefore, "circle of poison" and related laws which prohibit the importation of commodities with residues of pesticides not registered in the importing country will have to be waived unless the importing country can prove scientifically that the residues in question are unsafe.

The agrochemicals industry has an extremely important role to play in meeting food, sustainability and environmental needs. This challenge is difficult, but one that the industry cannot avoid, and which is both a great commercial opportunity, and humanitarian service. Chemists have especially important roles in efforts develop sustainable systems for safeguarding the global environment and for intensified food production.

Since population growth rates recede as people overcome poverty (United Nations, 1987), and since increasing food production is the principal means of combatting poverty in most developing countries, major improvements in pest management, including better agrochemical technologies are urgently needed. Policies which favour agriculture in developing countries create a sequence of employment and income multiplier effects throughout a nation's economy (Bautista and Valdes, 1993; von Braun et al. 1993), which are likely to bring about reductions in adverse environmental impacts and reduced demographic growth. It is important that, during the next two decades, very significant progress be made in combatting poverty in developing countries in order to stave off major problems which would surely arise if the world's population of rural poor were to double during the next century. However, if we fail to take timely and strong remedial actions, the insidious processes threatening sustainability will gain strength rapidly and ruin the productive potential of vast areas of arable and forest land.

Acknowledgments: N. Alexandratos, Chief, Golbal Perspective Studies Unit, FAO, Rome provided most of the data in this paper except those pertaining to agrochemicals. Without his counsel, this paper could not have been written. I am very grateful to R. Platzer for invaluable assistance in preparing figures. Valuable help was received from L.R. Batra, H. de Haen, R. Gardiner, R. Hance, A. Hassan, A. Marcoux, J. McCarthy, L. Shawa, B. Sigurbjörnsson, J.W. Snow and H. Waibel.

References

Alexandratos, N. (editor). *World agriculture Toward 2000*. Food and Agriculture

Organization and Belhaven Press, London, U.K., **1988**, 338 pp.

Alexandratos, N. IFPRI Roundtable:"Population and food in the 21st century: Meeting future needs of an increasing world population." Talk given during roundtable at Washington, D.C., 14-16 February 1994. Unpublished manuscript, **1994**, 41 pp.

Aspelin, A.L., A.H. Grunke and R. Tola. "Pesticides industry sales and usage: 1990 and 1991 market estimates." Econ. Anal. Br. Ofc. Pesticide Prog., Environ. Prot. Agency, Washington, D.C., **1992**, 37 pp.

Bautista, R.M.; Valdez A. (editors). *The bias against agriculture: trade and macroeconomic policies in developing countries.* Institute for Contemporary Studies Press, San Francisco, **1993**, 339 pp.

Bellinger, R.R.; Gummesson, G.; Karlsson, K."Percentage-driven government mandates for pesticide reduction: The Swedish Model."Weed Technology, **1994,** *8(1)*: 1-10.

Beyer, E.M. "Crop protection: Meeting the challenge." *Proceedings Brighton Crop Protection Conference - Weeds - 1991*, **1991**, 1, 3-22.

Brown, L.R.; Flavin, C.; Postel, S.; Starke, L.*The state of the world - 1993*. W.W. Norton & Co. New York, **1993**, 268 pp.

Carlson, G.A., G. Sappie, and M. Hammig.1989. Economic returns to boll weevil eradication. Econ. Res. Serv. Rept No. 621. **1989**, 31 pp.

Finney, J.R. "World crop protection prospects:Demisting the crystal ball." *Proceedings Brighton Crop Protection Conference - Pests and Diseases*, **1988**, 1, 3-14.

Finney, J.R. "World Crop Protection Prospects:Where do we stand? Where do we go?" Presentation at 7th International Conference on Pesticide Chemistry, 5-10 August 1990, Hamburg, Germany. ICI Agrochemicals, Bracknell, Berks., U.K., **1990**, 27 pp.

Fleischer, G. and H. Waibel "Survey onnational pesticide and pest management policies, **1994**, FAO, Rome, 28 pp.

Food and Agriculture Organization. Requirements of vitamin A, iron, folate and vitamin B_{12}. Report of a Joint FAO/WHO Expert Consultation. FAO Food and Nutrition, **1988**, Series No. 23. 107 pp.

Food and Agriculture Organization. *International Code of Conduct on the Distribution and Use of Pesticides,* FAO, Rome, **1990**, 34 pp.

Food and Agriculture Organization. *World foodsupplies and prevalence of chronic undernutrition in developing regions as assessed in 1992.* **1992a**, Document ESS/MISC/1/92. 25 pp.

Food and Agriculture Organization. *Food and Nutrition at the Turn of the Millenium.* **1992b**, Document U8259/E/1/10.92/10.500. 1 sheet.

Food and Agriculture Organization. Agriculture: Towards 2010. C 93/24 Document of 27th Session of FAO Conference, 6-25 November 1993, Rome, Italy, **1993a**, 362 pp.

Food and Agriculture Organization."RDSpecial: Nutrition." *Rural Development*, **1993b**, *14*, 10-16.

Food and Agriculture Organization. "RDTrends: Tropical Forests." *Rural Development*, **1993c**, *14*, 8.

Food and Agriculture Organization. *The State of Food and Agriculture*, Rome, Italy, **1993d**, 306 pp.

FAO and WHO. "Major issues for nutrition strategies: Preventing specific micronutrient deficiencies." Theme paper No. 6 for International Conference on Nutrition, PREP/COM/ICN/92/INF/11, **1992**, 40 pp.

Gardiner, G.R. Letter of 14 February **1994** toW. Klassen.

Hanson, D.J. "Administration seeks tightercurbs on exports of unregistered pesticides." *C&EN*, **1994**, *72(7)*, 16-17, February 14, **1994**.

International Bank for Reconstruction and Development. Global Economic Prospects and the Developing Countries. World Bank, Washington, D.C., **1993**, 72 pp.

James, W.P.T.; Schofield, E.C. *Human energy requirements: a manual for planners and nutritionists.* Oxford University Press, Oxford, **1990**, 172 pp.

Kendrick, J.B.; Stimmann, M.W. "Impacts of chemicals on agricultural production and the environment." p. 37-42. in J.L. Hilton (ed.)

Agricultual Chemicals of the Future. Roman & Allanheld, Totowa, New Jersey, **1985**, 464 pp.

Kinney, T.B. "The future of chemicals in agriculture." p. 3-7. in J.L. Hilton (ed.) Agricultual Chemicals of the Future. Roman & Allanheld, Totowa, New Jersey, **1985**, 464 pp.

Klassen, W. "Pest management and biologically based technologies: A look to the future." p. 410-422. in R.D. Lumsden and J.L. Vaughn (editors), Pest Management: Biologically Based Technologies, *Proc. of Beltsville Symposium XVIII*, May 2-6, 1993. ACS Conference Proceedings Series, American Chemical Society, Washington, D.C., **1993**, 435 pp.

McDougall, J. "Guarded optimism in agrochemicals." *Chemistry and Industry*, **1994**, *7*, 264, 4 April 1994.

McMinn, A.L.; Thomas, B. "The impact of community regulations on the future development of new compounds: An industry view." *Proceedings Brighton Crop Protection Conference - Weeds - 1991*, **1991**, *2*, 563-570.

Menn, J.J.; Henrick. "Newer chemicals for insect control." p. 247-265. in J.L. Hilton (ed.) Agricultual Chemicals of the Future. Roman & Allanheld, Totowa, New Jersey, **1985**, 464 pp.

Neale, M.C.; Sutton, A. "European view of horticultural crop protection." p. 79-88 in D. Tyson (editor) Crop Protection: Crisis for UK Horticulture? British Crop Protection Council, Farnham. 98. **1993**.

Nesheim, O.N. Impact of current U.S.pesticide issues on availability of pesticides for tropical fruits, *Proc. fourth Intern. Mango Symp.*, 5-11 July 1992, Miami, Acta Horticulturae, **1992**, 341, 545-549.

Organization for Economic Co-operation and Development. "Proposed OECD activity on pesticides for 1994-1996." 1993, ENV/MC/CHEM(93)14/REV1, 8 pp.

Omran, A.; Roudi, F. "The Middle East Population Puzzle." *Population Bulletin*, 48(1), 1-40.

Pimentel, D.; Harman, R.; Pacenza, M.;Pecarsky, J.; Pimentel, M. "Natural resources and optimum human population." *Population and Environment: A Journal of Interdisciplinary Studies*, **1994**, *15*, 15(5): 347-369.

Ridgway, R.L.; Inscoe, M.N.; Thorpe, K.W. "Advances and trends in managing insect pests", p. 3-15 in International Atomic Energy Agency, *Management of Insect Pests: Nuclear and related molecular and genetic techniques*, Proceedings of a Symposium, Vienna, Austria, 19 - 23 October 1992, jointly organized by IAEA and FAO. Vienna, Austria, **1993**, 669 pp.

Riley, D. "Consequences for environment data/risk assessment requirements". *Proceedings Brighton Crop Protection Conference - Weeds - 1993*. **1993**, *3*, 1081-1086.

Stimson, S. Perkin Medal awarded to Cyanamid's Marinus Los. *C&EN*, . **1994**, *72*(10), 29.

The Netherlands Ministry of Agriculture,Nature Management and Fisheries. Multi-year crop protection plan. The Hague, **1990**, 13 pp.

Thomas, B. "An industrial epilogue." *Proceedings Brighton Crop Protection Conference - Weeds - 1993*, **1993**, *3*, 1097-1104.

United Nations Department of International Economic and Social Affairs. "Fertility Behaviour in the Context of Development." Population Studies No. 100, New York, **1987**, 383 pp.

United Nations Department of International Economic and Social Affairs. "World population prospects 1990." Population Studies No. 120, New York, **1991**, 607 pp.

United Nations. "World population prospects."The 1992 Revision, New York, **1993**, 677 pp.

Visser, R. Letter of 14 March **1994** to W. Klassen.

Von Braun, J.; Hopkins, R.F.; Puetz, D.; Pandya-Lorch, R. "Aid to agriculture: reversing the decline." International Food Policy Research Institute, Washington, D.C., **1993**, 18 pp.

Waibel, H. "Government intervention in crop protection in developing countries." p. 76-93, in Cadwick, D. and J. Marsh (eds.), CIBA Foundation Symp. 177, Wiley & Sons, Chichester, **1993**, 285 pp.

Waibel, H. "Global pesticide markets and futureprospects for pesticides." Presented at 16th Session of FAO/UNEP Panel of Experts on IPM, Rome 25-29 April **1994**, 15 pp.

Winrock International. *Animal agriculture inSub-Saharan Africa*. Winrock International Institute for Agricultural Development, Morrilton, Arkansas, USA, **1992**, 125 pp.

Wolf, E.E. "Future trends of agriculturalchemicals research in industry and the public sector." p.9-14 in J.L. Hilton (ed.) *Agricultural Chemicals of the Future*. Roman & Allanheld, Totowa, New Jersey, **1985**, 464 pp.

World Health Organization. "Energy and protein requirements." Report of the Joint FAO/WHO/UNU Expert Consultation. Technical Report Series 724, **1985**, 206 pp.

Worthing, C.R. *The Pesticide Manual. A World Compendium,* 6th Ed. British Crop Protection Council, **1979**, 655 pp.

Zadoks, J.C. "The costs of change in plant protection." *Jour. Pl. Prot. in the Tropics.* **1992**, *9(2)*: 151-159.

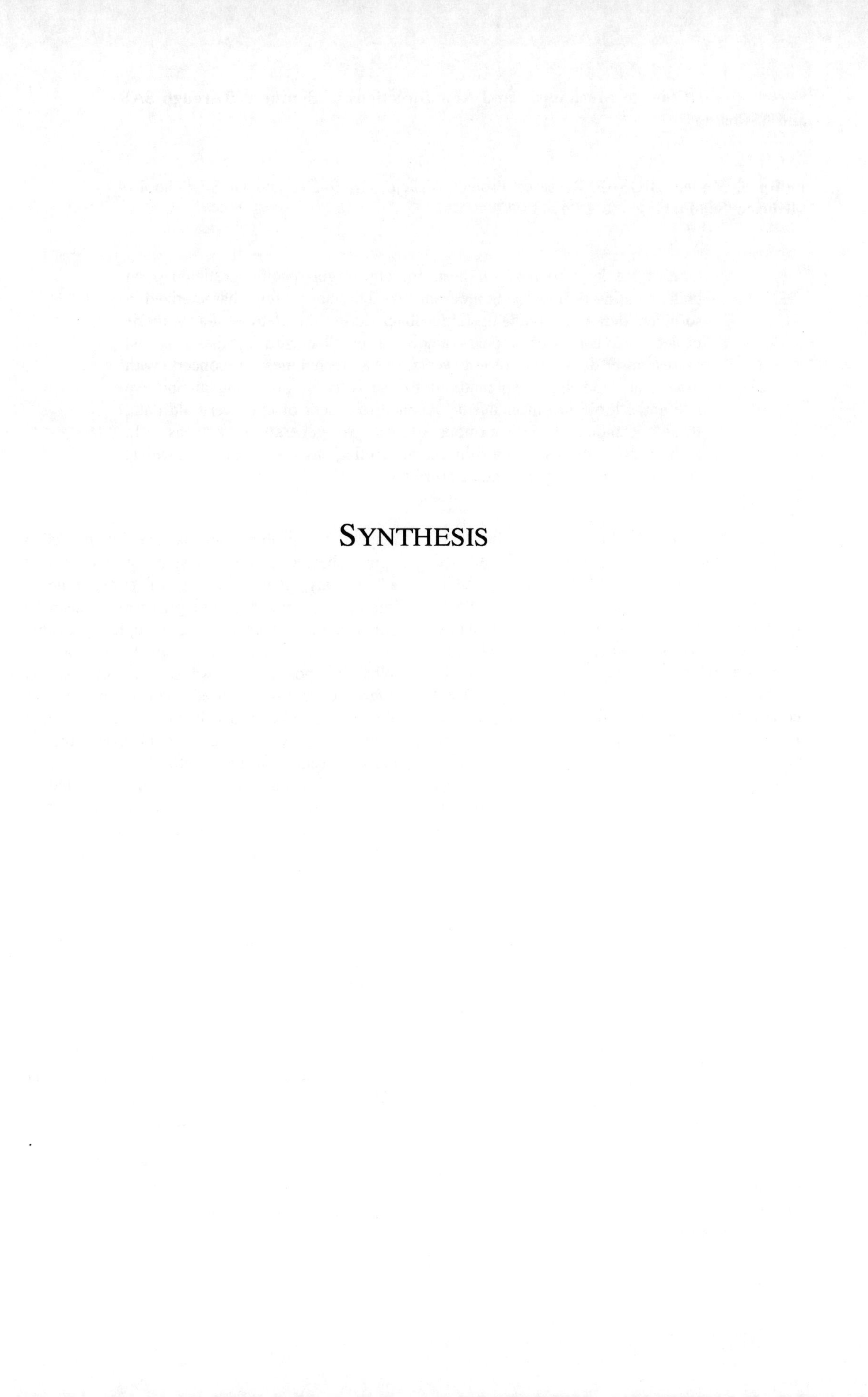

SYNTHESIS

Searching for Clues to Mechanism and New Directions in Synthesis Through SAR and Modeling

Philip S. Magee, BIOSAR Research Project, Vallejo, CA 94591 and UCSF School of Medicine, San Francisco, CA 94143

One of the keys to new synthesis for any target-specific pesticide is an understanding of underlying mechanisms. These are rarely characterized in sufficient detail to provide useful feedback during the lifetime of a synthesis project. What does appear rapidly in parallel with synthesis is the bioresponse data. In recent years, SAR techniques in concert with molecular modeling have matured to the level of providing an incisive inferential look into mechanism. Applied to recent or even very old data, these techniques have enormous capacity to generate new ideas. In addition to synthesis, it is also necessary to flag environmental and medical concerns to allow the setting of priorities.

While synthesis is the primary thrust in pesticide development, the field has broadened into environmental and medical issues of public concern where mechanism also plays a role in problem solving. In these toxicological domains, the literature is overflowing with "old data", the analysis of which has immediate applications to current problems. This paper will address modern approaches to both recent and old data (as old as 1901) with an emphasis on mechanistic insight. While the examples presented all depend on historical data, the real message concerns the use of inferential studies (SAR, modeling) in concert with on-going or recent synthesis to generate new ideas and at least flag environmental and medical concerns to allow the setting of priorities.

Pesticide Development

Mathematical (QSAR) and molecular (classical and quantum) modeling are increasing in vigor and provide many opportunities to assist in synthetic design. The two general fields are complementary as one depends principally on molecular data, while the other relies on molecular structure. Ultimately, they focus on the same goal and both have enormous capacity to build on experience. As this paper will show, they can be supplementary in problem solving as well as independent (Magee, 1989). With respect to probing mechanism by modeling approaches, one point must be stressed above all, namely, that modeling of either type is inferential. One does not prove mechanism by either quantum mechanics or multiple regression, but only through well-designed experiments. What is possible, however, is very strong inference leading to a useful working hypothesis. The inventive synthesis chemist needs little more than this to generate his next idea, and therein lies the strength of these methods.

The following examples are not intended to provide new synthetic ideas, but are selected to illustrate the power of the methods and to stress the continued usefulness of historical data in understanding biomolecular interactions.

Phenoxyacetic acids. Those of us who remember the beginning of modern QSAR or the Hansch Approach (Van Valkenburg, 1972) will appreciate this example (Hansch *et al.*, 1963). Plant growth regulation activity was measured by the Avena cylinder straight growth method for 25 bioactive phenoxyacetic acids with measured LogP(o/w) values. Correlation of the data was curious in requiring the use of para-sigmas to describe the electronic effect of meta-substituents. In fact, the authors postulated a two-point attachment theory developed earlier (Muir and Hansch, 1955) where both the side chain and a ring position were involved in binding.

1054–7487/95/0034$12.00/0

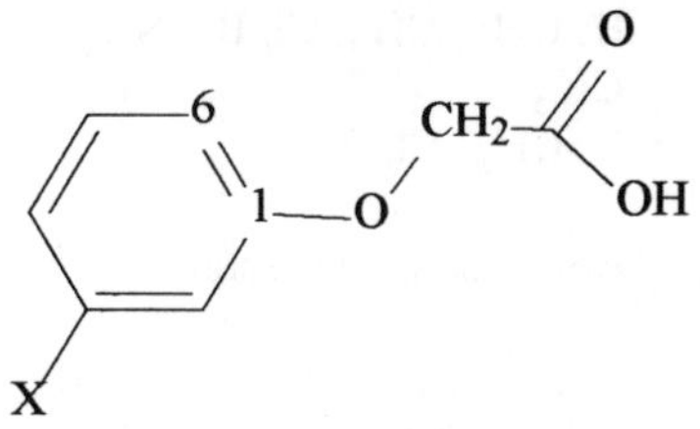

$$\text{Log}(1/C) = 3.19\,\text{Pi} - 1.80\,\text{Pi}^2 + 1.72\,\text{Sig}(*) + 4.09$$
T = 6.60 5.10 4.96

n = 19 r = 0.898 F = 20.74

(*) - Sigma Para for meta groups

To provide support for this concept, a semi-empirical study was made of the sigma charge induced on the C1 and C6 carbons of the ring using AM1 (Stewart, 1990). The C6 carbon is para to the activating meta group. To facilitate minimization, the COOH group was replaced by the identically inductive methoxy group (sigmaI = 0.30). Replacing the reversed sigma constants with the AM1 charges reveals explicity that both positions are involved in the binding mechanism.

$$\text{Log}(1/C) = 2.82\,\text{Pi} - 1.70\,\text{Pi}^2 + 27.0\,\text{CHG1} + 30.8\,\text{CHG6} + 6.53$$
T = 4.27 3.36 1.96 2.87

n = 19 r = 0.823 F = 7.37

As readily seen, electropositive character supports higher activity at both C1 and C6. The correlation is somewhat weaker as AM1 charges developed on gas-phase analogs cannot provide the nuances of a solution measured pKa (Sigma). While of no current interest, it is clear that these insights would be valuable in designing superior auxins.

N-Methyl meta-substituted phenyl carbamates. An old set of data indeed, collected by T.R. Fukuto and R.L.Metcalf between 1960-1967. The entire collection of ortho-, meta- and para-carbamates (n = 269) has been analyzed (Goldblum, Yoshimoto and Hansch, 1981) by regression to provide a QSAR equation with satifactory statistics (r = 0.892, F = 15.0). Later, a subset of the meta-substituted carbamates was analyzed by a hypermolecule approach which used 11 positions to characterize all of the meta-groups (Magee, 1990). The alpha-atom at the 3-position was the only non-carbon atom in the hypermolecule and, as such, was parameterized for electronic (Chi, electronegativity), London forces (MR), solvation (LogP fragment values) and steric effects (van der Waals radii, Rv). All other positions were dichotomous (1.0/0.0) for the presence or absence of a hydrocarbyl group. What proved so unusual in this case was the dependance of the pI50 data for housefly AChE on only 3 of the 11 positions. The alpha-atom was described strongly by electronegativity with a negative coefficient. As Chi correlates negatively with sigma charge, this implied that electropositive character at the alpha-atom was associated with strong binding at an anionic site.

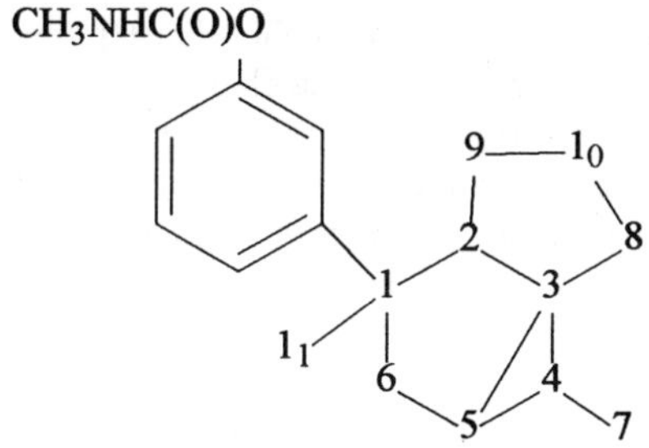

P1 = C, F, Cl, Br, I, O, N, P, S, Si
P2-P11 = C, CH, CH_2, CH_3

$$\text{pI50} = -0.835\,\text{X1} + 0.693\,\text{P2} + 0.299\,\text{P5} + 7.18 \qquad X = \text{Chi}$$
T = 6.09 3.51 2.04

n = 36 r = 0.841 F = 25.82

To establish the actual charge dependance at the alpha-atom, a series of meta-substituted MeX-phenyl acetates were modeled semi-empirically by PM3 (Stewart, 1990). Using the acetate in place of the N-methylcarbamate group and simplifying the RX-groups to MeX was critical in achieving reasonable convergence times without sacrificing too much in accuracy. The induced sigma charge on the alpha-atom, X, was then used for each of the RX-substituents regardless of actual chain length. Considering the approximations introduced, the following correlation is quite satisfactory in confirming induced charge on the meta-alpha-atom as a major factor in carbamate pI50 activity.

$$pI50 = \underset{T = 2.18}{1.19\,PM3CHG1} + \underset{2.85}{0.773\,P2} + \underset{2.31}{0.456\,P5} + 4.66$$

$$n = 36 \qquad r = 0.672 \qquad F = 8.74$$

While the difference may seem trivial and is clearly weaker in correlation, it is a step forward to confirm induced charge as a contributing factor, rather than proposing it on the basis of a correlation with element electronegativity. The electropositive order for the alpha-atoms from the PM3 sigma charge is O, F, C, Br, I, Cl, S, Si, P, providing a direct recipe for new syntheses. Of singular interest is the fact that H, which falls between S and Si in sigma charge, is an outlier in both correlations. The experimental pI50 is an order of magnitude below that predicted by either equation. This strongly suggests that bulk in the form of polarizable volume (MR) is a requirement of the meta-substituent.

Ethoxychlor analogs. Ethoxychlor was still in substantial use when these data were developed (Brown, Metcalf, Sternberg & Coats, 1981). I was attracted to this data set by the simple variations of the alkyl group because of recent success in analyzing the anesthetic activity of simple C1-C3 haloalkanes (Magee, 1991). Positional analysis of a hypermolecule gave exceptional results for the anesthetics and also seemed appropriate for the ethoxychlor analogs. Basically, there are five occupied positions, though P2 and P4 are weakly populated with respect to P1, P3 and P5. The descriptors explored were MR (polarizable volume), F (solvation), X (Chi, electronegativity), and ICH = 1.0 for chirality at the central carbon. Positions were assigned by size in the order: 1>3>5. Chirality proved unimportant, as did Chi, while variations in F and MR for positions 1, 3 and 5 dominated the correlation.

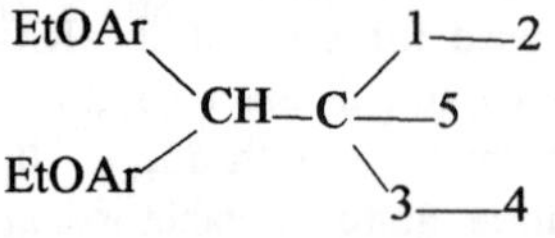

```
EtOAr            1——2
     \          /
      CH—C——5
     /          \
EtOAr            3——4
```

P1, P3 - H, CH_2, CH_3, Cl, Br, NO_2
P2, P4 - CH_3
P5 - H, CH_3, Cl, Br

ED50 = American cockroach crural nerve

$$Log(1/ED50) = \underset{T = 10.12}{0.183\,MR3} - \underset{3.44}{0.0602\,MR5} - \underset{3.71}{0.272\,F1} - \underset{3.89}{0.455\,F3} + 4.75$$

$$n = 20 \qquad r = 0.953 \qquad F = 36.65$$

This very strong equation provides a virtual roadmap to synthesis optimisation. As in the case of the anesthetics, the dependance of activity on most, if not all, the positions indicates that binding for this group is non-specific.

Environmental Issues

Environmental thought depends so much on the concept of distribution and the apparently simple shake-flask descriptor, LogP(octanol/w), that good judgement suggests rocking the boat quite gently. As the science of solvation remains incompletely resolved, it follows that nothing about differential solvation or LogP can possibly be simple, other than the mental construct of gentle non-mechanistic passage from one phase to another. In theory, 2.303RTLogP represents the difference in Free Energy for a solute to form a minimum energy cavity in n-octanol and water. In actuality, there is water in the n-octanol and vice-versa, which destroys the thermodynamic neatness of the partition. Nor is it a gentle, non-mechanistic passage as shown by an activation energy of 10.3 kcal/mol for the transfer of methyl nicotinate from water to octanol (Guy and Honda, 1984). The increasing complexity and decreasing accuracy of calculated LogP's is reflected in the examples at the back of a recent book on drug lipophilicity (Rekker and Mannhold, 1992). It is clear that this additive-constitutive descriptor becomes more and more constitutive with increasing molecular complexity.

It can be shown that LogP is a function of several intermolecular forces, including MR, HBA, HBD and electronic descriptors (Charton and Charton, 1982). The major correlator is MR (polarizable volume or London forces) and this implies something interesting. If LogP is applied to a problem where the partitioning media is very different from n-octanol, then the addition of MR to LogP may be required for optimum correlation, in effect, adjusting the underlying MR scale. This was observed nicely in the inhibition of bioluminescence of *Photobacterium phosphoreum* by nonreactive organics (Hermens, Busser, Leeuwangh and Musch, 1985). My own contribution to adjusting LogP was to reason that partitioning into a phase dissimilar to octanol would result in selective interaction with either lipophilic or hydrophilic substructures. To accomplish this insight, LogP was simply factored into PL and PH by using ordinary LogP calculation methods (Magee, 1991). It was soon discovered that this procedure often required MR as an adjunct descriptor and sometimes the explicit inclusion of H-bonding descriptors. The following examples illustrate the power of the technique in refining LogP correlations with greater mechanistic insight.

Partitioning into hydrocarbon oils. A revealing publication that previewed the many problems of LogP correlations was an attempt to reconcile the differences between partitioning from water into n-octanol and various hydrocarbons (Seiler, 1974). For a set of 230 values, Seiler found 35 outliers and required 23 descriptors to correlate the remaining 195 in an equation relating LogP(o/w) to LogP(HC/w). This astounding equation was an early warning of troubles to come for "simple" LogP relations, of which the literature contains thousands. If you can't do hydrocarbons, then all of the one-on-one relations for soil and aquatic species come under suspicion of over-simplification.

Seiler's data set contained some compounds of uncertain structure and some that were so complex that factoring into PL and PH was not appropriate. Nineteen of his structures were discarded and the remaining 211 contained 22 of his 35 outliers. In addition to factoring LogP, MR and all types of H-bonding descriptors were included for a total of 6 descriptors. There were 19 intramolecular H-bonds in the set (IHB=1.0). All 6 descriptors were important. There were no outliers and the residuals closely match a computer generated normal distribution, indicating no remaining information other than experimental error. The relation is instructive.

LogP(HC/w) = - 0.0380 MR + 1.45 PL
T = 4.76 15.11

+ 1.29 PH - 1.90 HBD + 0.494 HBA
15.62 14.17 3.23

+ 0.633 IHB - 0.025
2.86

n = 211 r = 0.903 F = 149.89

Note that the coefficient of PL is larger than that of PH, indicating selective solvation and that the contribution of MR is negative. This is the only case I have found where preferential solvation of lipophilic structures occurs, perhaps not surprising, as there are few solvents more lipophilic than n-octanol (LogP = 2.93) other than hydrocarbons or chlorocarbons. It is now profitable to reexamine some important cases of soil/water and fish/water partitioning.

Binding of pesticides to soil. Soil absorption data are well measured and corrected for organic content, KOM = organic/water partition. Most data sets give satisfactory to excellent LogP correlations, but analysis with PL, PH, MR and H-bonding descriptors is superior and provides more information of mechanistic interest. Of the available data, soil absorption of a large set of polar compounds is of special interest as many pesticides and synthetic intermediates are represented (Sabljic, 1987). The LogP expression is shown first for comparison.

LogKOM = 0.565 LogP + 0.687
T = 20.01

n = 132 r = 0.869 F = 400.4

LogKOM = 0.220 PL + 0.379 PH
T = 3.72 8.39

$$+\ 0.0272\ \underset{5.87}{MR}\ -\ 0.379\ \underset{4.20}{HBD}\ +\ 0.596$$

$$n = 132 \qquad r = 0.908 \qquad F = 148.9$$

The residuals are nearly perfect, suggesting that all relevant information has been extracted. Note the predominance of PH and the positive coefficient of MR. This reflects partitioning into a more polar medium than n-octanol as one would expect complex soil matter to be. The dependance on H-bond donors suggests that reversible binding is part of the partitioning process.

Of fishes and frogs. Like soil organic matter, proteinaceous meat and organs of aquatic species are also far from the polarity of octanol. To illustrate the effect of LogP factoring, we select two cases of interest. The first set includes both non-polar and polar compounds, many of which would be described as typical pesticide intermediates (Protic and Sabljic, 1989). Measured for aquatic toxicity to fathead minnows, it was selected for special study because of exceptional molecular variation and polarity range (LogP covers >6.0 orders of magnitude). The second set is of singular interest in dating from the very beginning of bioactivity studies. It concerns the narcosis of frog tadpoles in water by a broadly diverse set of simple aliphatic and aromatic compounds (Overton, 1901). It serves the dual purpose of supporting our secondary theme that well measured data are timeless and need to be periodically reviewed. It is also interesting in providing a connection between narcosis, anesthesia and death.

Protic and Sabljic (1989) have provided a major set of aquatic toxicity data on fathead minnows for 179 compounds (96 hour LC50). It correlates moderately well with LogP and only somewhat better by factoring. H-bonding donors are borderline correlators and were not included. In addition, the residuals are slightly skewed, leading one to suspect that unknown factors are at work. In brief, this is not a perfect set, but perhaps it is more typical experimentally than those that are. The LogP correlation is shown first for comparison. One outlier, the reactive phenacyl bromide, was identified and deleted.

$$\mathrm{Log(1/LC50)} = 0.744\ \mathrm{LogP} - 0.865$$
$$T = 20.22$$

$$n = 178 \qquad r = 0.836 \qquad F = 408.8$$

$$\mathrm{Log(1/LC50)} = 0.279\ \underset{4.62}{PL} + 0.737\ \underset{11.26}{PH}$$
$$+\ 0.0498\ \underset{7.79}{MR}\ -\ 1.31$$

$$n = 178 \qquad r = 0.882 \qquad F = 203.7$$

Despite a less than perfect correlation, this relation supports the general hypothesis underlying the factoring of LogP, namely the effect of partitioning into tissue of greater polarity than n-octanol. Note the extreme selection for hydrophilic substructures and the very strong positive contribution of MR to adjust the underlying scale. This is one of the strongest examples of selective substructure solvation so far.

Data more than 90 years old certainly qualify as historical data and some might question the wisdom of taking it seriously. Those who do are forgetting that German technology measured indices of refraction to six decimal places and produced conductometrically measured rate constants and equilibria of unsurpassed quality. Precision measurements were in full bloom a century ago. And so, with great pleasure, we revisit some of the first seriously measured bioresponse data of the Victorian era (Overton, 1901). How good an experimentalist was Overton? What can his data on frog tadpole narcosis tell us today? First of all, narcosis is defined as the minimum concentration in water to induce a state of coma-like immobility in a subject, which condition is rapidly reversed on restoring the animal to pure water. One can imagine the difficulties of defining such a number for a test compound. Narcosis has been related to anesthesia (Hermens, 1986) and the latter effect has been linked to death by anesthesia (Magee, 1991). Neural impairment is implicated in all three responses.

If correlation with LogP alone were to stand the test of time, then Overton would be considered a very competent experimentalist. Viewing his data by factoring, however, reveals a biochemist of superior performance. The LogP correlation is shown first for comparison.

$$\text{Log}(1/C) = 0.843\ \text{LogP} + 0.988$$
$$T = 17.35$$

$$n = 52 \qquad r = 0.926 \qquad F = 301.0$$

$$\text{Log}(1/C) = 0.200\ \text{PL} + 0.672\ \text{PH}$$
$$T = 2.74 \qquad 12.58$$

$$+\ 0.0695\ \text{MR} + 0.529$$
$$9.65$$

$$n = 52 \qquad r = 0.976 \qquad F = 315.5$$

It is rare indeed to find bioresponse data of this quality. There are no outliers and the residuals are normally distributed. Moreover, the underlying mechanism of the partition is clearly supported by the much later data of Protic and Sabljic on minnows. Tadpoles and minnows are indeed parallel in response and display similar selectivity with respect to partitioning.

Medical Issues

One generally thinks of oral and dermal toxicity as a key medical risk in handling pesticides. Most of these data on various animals and by various routes of administration are not only historical, but come from many different laboratories where animal sources, strain, gender, age, vehicle and technique are all hidden variables. Despite the obstacles, this type of interlaboratory data was successfully analyzed in mechanistic terms by allowing 8-10 compounds per variable to achieve statistical strength (Magee, 1989). It was possible to show that mechanism varied between species and between routes of administration. Successful correlations were obtained for phenols, anilines, pyridines, N-methyl carbamates, organophosphates and diarylamine rodenticides.

More recently, another problem related to the handling of pesticides, namely contact dermatitis, has been resolved despite even greater problems in assembling data reliable enough to build a model (Magee, Hostynek and Maibach, 1994). Allergic contact dermatitis is an immune response with highly variable laboratory and clinical results depending on the animal or human test and on the experience of the investigator. Over a three year period, we were able to assemble enough reliable data on allergens and non-allergens to build a stable model that proved to predict new candidates with about the same success as clinical testing. The technique employed was two-value (1.0/0.0) regression, a rarely used method that I feel has enormous potential for resolving dichotomous problems in agriculture and medicine. The only disadvantage is that each class needs to be equally populated; in this case by 36 allergens and 36 non-allergens (one outlier). Descriptors for the model were a combination of mechanistic factors related to transport and binding, and a number of key substructure descriptors that flag the potential for either hapten or prohapten activity. Of singular interest was the failure of LogP to correlate, while the lipophilic contribution (PL) was among the strongest factors. The reader is referred to the original publication for a full interpretation of the following expression, which is too lengthy to include here.

	Coeff	T Value
CLASS =	0.00974 MR	2.10
	-0.153 PL	3.21
	0.0468 HBA	2.52
	-0.154 HBD	3.13
	-0.251 ICOOR	3.09
	0.127 IX	3.40
	0.215 IOH	2.79
	0.564 IPOS	6.70
	0.465 IQUIN	6.01
	0.203 (constant)	

$$n = 71 \qquad r = 0.823 \qquad F = 14.19$$

In addition to drugs and environmental chemicals, this expression is currently being applied to over 150 marketed pesticides in an effort to establish probable risk.

Conclusions

In modeling the growth regulation activity of the phenoxyacetic acids, an attempt was made to directly support the contention of the 1963 authors that binding of the ring involved dual sites. This was implied in the original correlation by a sigma dependance that pointed to the ring position para to the meta-substituent. By using sigma charges derived from semi-empirical AM1 modeling, the 1- and 6-ring positions were directly implicated as binding sites. While less precise than solution measured sigma's, the modeled charges are far more incisive in assessing factors underlying mechanism. In addition, they provide a direct path between modeling and synthesis.

In a similar fashion, the increase of HFAChE pI50 with electropositive behavior at the meta-alpha atom is directly confirmed by the PM3 charge study, while only implied in the earlier use of electronegativity. This concept in combination with the significance of P2 and P5 is all that a synthesis chemist needs to generate new ideas. In this study also, the charges were less well correlated than Chi, but the result is far more meaningful.

The positional study of the ethoxychlor analogs is not very revealing in a mechanistic sense as it is difficult to decide if the interactions are specific or non-specific since neither steric nor electronic factors are involved. However, the quality of the correlation and the positional factors provide a virtual recipe for synthetic optimization. These positional descriptors are far more powerful for refinement than whole molecule or substituent descriptors like LogP, MR, sigma and steric measures.

In refining the analysis of environmental issues, an attempt was made to retain the valuable concept of simple LogP mediated partitioning. The value of this concept for clarifying complex problems to the public and to funding agencies is beyond question. Nevertheless, neither LogP itself nor any of the LogP mediated mechanisms are as simple as they often appear and we have a responsibility to examine these critical distributions in more detail. The method described here may not be the most incisive in the long run, but it has the value of doing little damage to the general concept of passive partitioning of chemicals in nature. There are now thousands of fair to excellent LogP correlations in the literature and those of greatest importance need to be reexamined in more detail, either by this method of factoring or by another which recognizes the independance of intermolecular forces.

Medical issues will continue to expand in scope as pesticide use goes into the future and faces increasing scrutiny. The first published SAR model of allergic contact dermatitis was presented here to illustrate the need for modeling adverse human effects. The overlap between pesticide science and medical science is growing as more issues come before the growers and the public. Public distrust of pesticides and concerns about the need for animal testing is constantly increasing the obstacles we face in solving many problems. One thing, however, is certainly true. The literature is swamped with unanalyzed historical toxicology data and at some point, we must begin to explore the massive amounts of information buried in these numbers. It is a future responsibility to begin the modeling and extraction of mechanistic information from our current supply of well measured data while concurrently devising safer and less invasive tests to extract new information when animal testing cannot be avoided. Such efforts are already in progress in programs such as "Alternatives to Animal Testing", supported by Johns Hopkins University.

Lastly, the point needs to be made that well measured data will stand the test of time by definition. What fades into the past are old techniques and old interpretations. As new methods of statistical, quantum chemical or neural networks come onto the scene, one should always search the literature first to see if the needed data are not already there. I may be the last person in this century to examine some of these older sets, but others in coming centuries will do it again, if for no other reason than to test a new technique against reliable data. As mechanisms clarify, feedback into synthesis becomes more direct and predictable. And, as computers and programs improve, the availability of advanced techniques for SAR and molecular modeling become available to all chemists for experimentation and insight.

References

Brown, D.D.; Metcalf, R.L.; Sternberg, J.G.; Coats, J.R. *Pest. Biochem. Physiol.* **1981**, *15*, 43.

Charton, M.; Charton, B.I. *J. Theoret. Biol.* **1982**, *99*, 629.

Goldblum, A.; Yoshimoto, M.; Hansch, C. *J. Agric. Food Chem.* **1981**, *29*, 277.

Guy, R.H.; Honda, D.H. *Int. J. Pharm.* **1984**, *19*, 129.

Hansch, C.; Muir, R.M.; Fujita, T.; Maloney, P.P.; Geiger, F.; Streich, M. *J. Am. Chem. Soc.* **1963**, *85*, 2817-2824.

Hermens, J.; Busser, F.; Leewangh, P.; Musch, A. *Ecotoxicol. Environ. Saf.* **1985**, *9*, 17-25.

Hermens, J.L.M. *Pest. Sci.* **1986**, *17*, 287-296.

Magee, P.S., in *Probing Bioactive Mechanisms,* ed. by Magee, P.S.; Henry, D.R.; Block, J.H., ACS Symposium Series 413; American Chemical Society: Washington, DC, 1989, 37-56.

Magee, P.S., in *Probing Bioactive Mechanisms,* ed. by Magee, P.S.; Henry, D.R.; Block, J.H., ACS Symposium Series 413; American Chemical Society: Washington, DC, 1989, 390-399.

Magee, P.S., *Quant. Struct.-Act. Relat.* **1990**, *9*, 202-215.

Magee, P.S., in *QSAR: Rational Approaches to the Design of Bioactive Compounds*, ed. by Silipo, C.; Vittoria, A.; Elsevier: Amsterdam, 1991, 549-552.

Magee, P.S. *Sci. Total Environ.* **1991**, *109/110,* 155-178.

Magee, P.S.; Hostynek, J.J.; Maibach, H.I. *Quant. Struct.-Act. Relat.* **1994**, *13*, 22-33.

Muir, R.M.; Hansch, C. *Ann. Rev. Plant Physiol.* **1955**, *6,* 157.

Overton, E. *Studien uber die Narkose*; Gustav Fischer: Jena, 1901.

Protic, M.; Sabljic, A. *Aquat. Toxicol.* **1989**, *14,* 47-64.

Rekker, R.F.; Mannhold, R. *Calculation of Drug Lipophilicity*; **VCH**: Weinheim, 1992, 85-104.

Sabljic, A. *Environ. Sci. Technol.* **1987**, *21,* 358-366.

Seiler, P. *Eur. J. Med. Chem.* **1974**, *9,* 473-479.

Stewart, J.J.P. *J. Comp.-Aided Mol. Design* **1990**, *4(1),* 1-105.

Van Valkenburg, W., ed., *Biological Correlations-The Hansch Approach*; ACS Symposium Series 114; American Chemical Society: Washington, DC, 1972.

3-Aryl-4-halo-5-substituted Pyrazoles: A New Class of Highly Active Preemergent and Postemergent Herbicides

Bruce C. Hamper, Deborah A. Mischke, Kindrick L. Leschinsky, Lisa L. McDermott and S. Douglas Prosch, Monsanto Company, The Agricultural Group, 800 North Lindbergh Blvd., St. Louis, MO 63167 (USA)

The 3-aryl-4-halo-5-haloalkylpyrazoles and 3-aryl-4-halo-5-alkylsulfonylpyrazoles represent a unique structural class of herbicides which have significant preemergent and postemergent activity in the grams per hectare range. In greenhouse and field studies, chosen compounds were highly active, providing broadspectrum control of broadleaf and narrowleaf weeds. Both structural types were prepared from acetophenone precursors by conversion to diketones or diketone equivalents followed by cyclocondensation with hydrazines. Various substituted arylpyrazoles were prepared for structure-activity and herbicide spectrum comparisons.

In the last ten years, numerous herbicidal compounds from structurally diverse areas have been reported to inhibit protoporphrin IX oxidase or PROTOX (Duke, 1991). These include commercially important nitrodiphenyl ethers (oxyfluorfen, acifluorfen and its esters), N-arylheterocycles (oxadiazon) and the newer N-phenylimides (S-23031).

We have previously reported a series of substituted phenylisoxazoles which are potent pre- and postemergent herbicides and appear to have a mode of action similar to that of known PROTOX inhibitors (Hamper, 1994). These compounds, arylisoxazoles **1** having a carbon-carbon link between the two rings, were prepared by cycloaddition of nitrile oxides to acetylenic esters (Fig. 1). Conversion of the isoxazolecarboxylate esters **1** to carboxamides **2** improved the activity of compounds in this series, affording broadspectrum weed control. The best overall activity was observed with compounds having a combination of three substitutents: a 2,4 or 2,4,5 substituted phenyl ring in the 3 position, a primary or secondary amide in the 4 position and a CF_2Cl group in the 5 position of the isoxazole ring.

A series of pyrazole phenyl ethers herbicides **3** (Moedritzer, 1992), prepared from regioisomeric hydroxypyrazoles (Hamper, 1992), have also been reported as PROTOX inhibitors. By comparison of these pyrazole phenyl ethers and the direct carbon-carbon linked isoxazoles, substituted phenylpyrazoles (**4**, $R_1 = CH_3$, $R_2 = CF_3$, SO_2CH_3; $R_3 = Cl$, Br) were proposed as potential new herbicides (Woodard, 1994; Mischke, 1992) and inhibitors of PROTOX (Fig. 2). A related series of difluoromethoxypyrazole (**4**, $R_2 = OCF_2H$) herbicides have recently been reported (Miura, 1993). The compounds in this report include trifluoromethyl- and alkylsulfonylpyrazoles (**4**, $R_2 = CF_3$, SO_2CH_3).

Synthesis

The targeted phenylpyrazoles required starting materials that would allow construction of a carbon-carbon bond between the two rings. This was accomplished by building the pyrazole ring from suitably substituted acetophenone precursors **5**. For direct comparison with some of the more active N-phenylimides (Hamada, 1989) and isoxazoles **4**, we prepared 2-fluoro-4-chloro-5-methoxyacetophenone **6** (Fig. 3). The most direct route would be acylation of anisole **7**, however, acetylation occurs meta to the fluorine group rather than ortho as desired. In the presence of strong Lewis acids, the methoxy group is hydrolysed and the initial O-acetylated product undergoes a Fries rearrangement to give a mixture of **8** and **9**. By contrast, formylation with dichloromethyl methyl ether occurs ortho to the fluorine to give the benzaldehyde in 55% yield. Treatment with methyl Grignard followed by oxidation gave acetophenone **6** with the desired 2,4,5 substitution pattern.

Acetophenones were the initial starting materials for construction of both the trifluoropyrazoles **13** and methylsulfonylpyrazoles **17** (Fig. 4 and 6). Claisen condensation of the acetophenone with trifluoroacetate gave diketone **10**. The regioisomeric pyrazoles **11** and **12** were obtained by direct cyclocondensation of **10** with methylhydrazine in 85% overall yield. Alternatively, treatment of **10** with hydrazine followed by methylation of the intermediate NH pyrazole gave mixtures of **11** and **12** in two steps. The isolated regioisomers were

1054–7487/95/0042$12.00/0

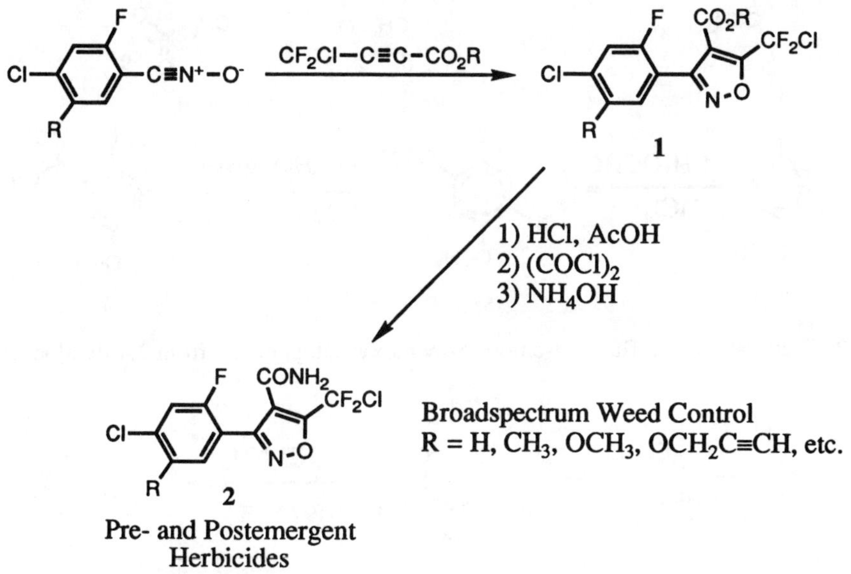

$CF_2Cl{-}C{\equiv}C{-}CO_2R$

1

1) HCl, AcOH
2) $(COCl)_2$
3) NH_4OH

2

Pre- and Postemergent Herbicides

Broadspectrum Weed Control
R = H, CH_3, OCH_3, $OCH_2C{\equiv}CH$, etc.

Fig. 1. Preparation of substituted 3-aryl-4-haloalkyl-5-isoxazolecarboxamide herbicides from cycloaddition of nitrile oxides and acetylenic esters.

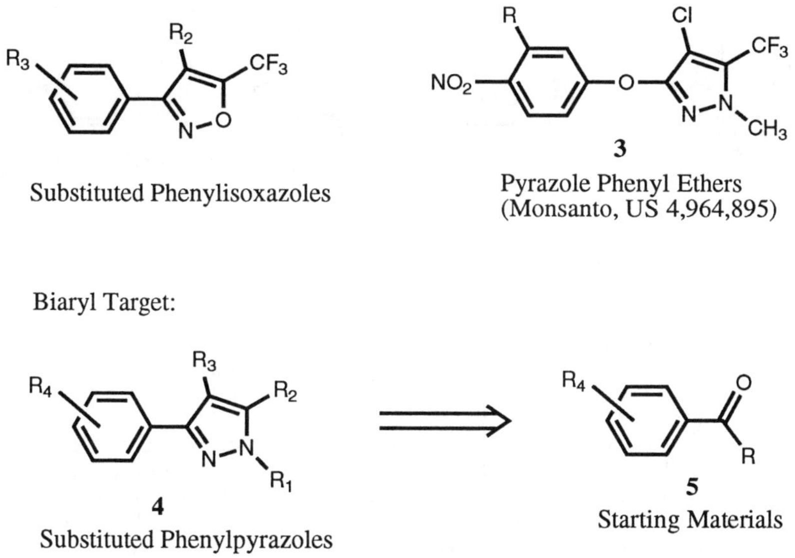

Substituted Phenylisoxazoles

3

Pyrazole Phenyl Ethers
(Monsanto, US 4,964,895)

Biaryl Target:

4

Substituted Phenylpyrazoles

5

Starting Materials

Fig. 2. Structural target for new PROTOX inhibitors.

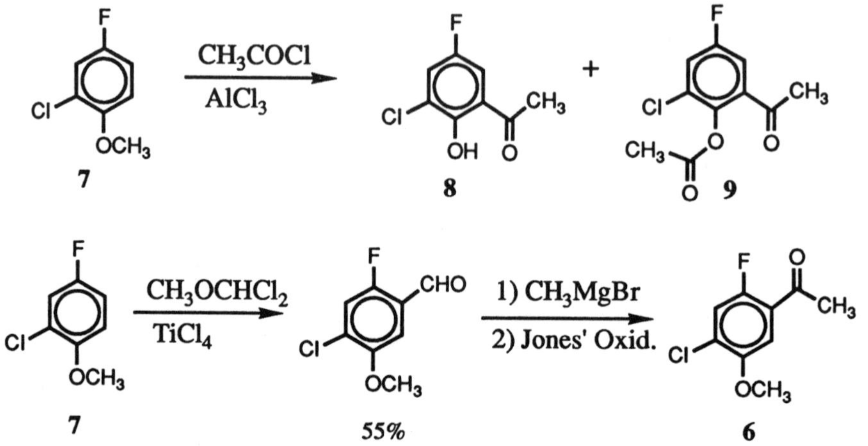

Fig. 3. Preparation of 2-fluoro-4-chloro-5-methoxyacetophenone from 2,4-dihaloanisole.

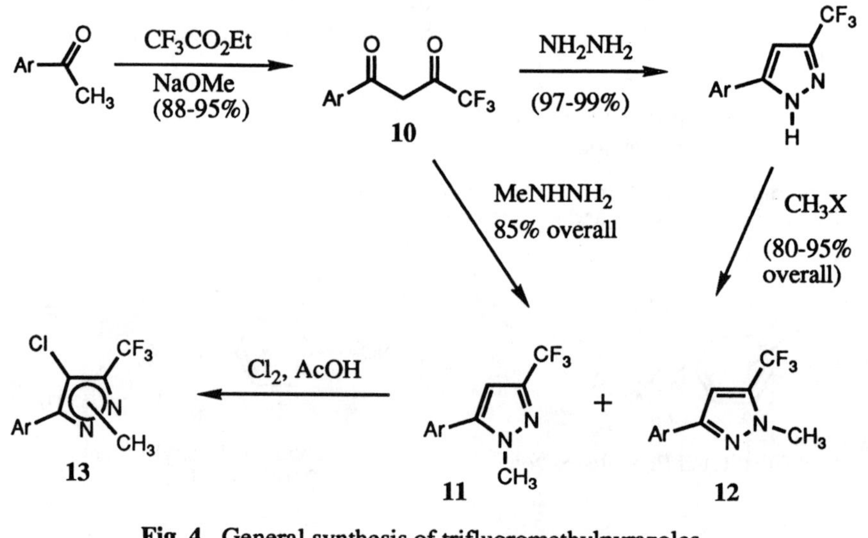

Fig. 4. General synthesis of trifluoromethylpyrazoles.

chlorinated to give the fully substituted pyrazoles **13**. Control of the regioselectivity in formation of **11** and **12** was critical for larger scale preparations of these pyrazoles in order to avoid separation problems and lower yields of the desired isomer (Fig. 5). Activity comparisons had shown that compounds derived from the 3-aryl isomers **12** were consistently more active than those from 5-aryl **11**. Direct cyclocondensation of the diketone **10** with methylhydrazine gave the 5-aryl isomer **11** as the major product. Methylation of an intermediate NH pyrazole, however, gave the desired 3-aryl isomer **12** as the major product. In the presence of a base, such a potassium carbonate, a 3 to 7 ratio was obtained of **11** and **12**, respectively. Under neutral or acidic conditions, regiospecific formation of **12** was observed with less than 4% of the 5-aryl isomer **11**.

The alkylsulfonylpyrazoles **17** were prepared by initial formation of dithioketene **15** (Fig. 6). A sodium enolate of the acetophenone was trapped with carbon disulfide and exhaustively methylated to give **15**. Addition of methylhydrazine afforded the 3-arylpyrazole **16** as the only observed regioisomer. The next two steps, oxidation and chlorination, were carried out in either of the two possible sequences, depending on which transformation was carried out first, to give 4-chloro-5-methylsulfonylpyrazole **17**.

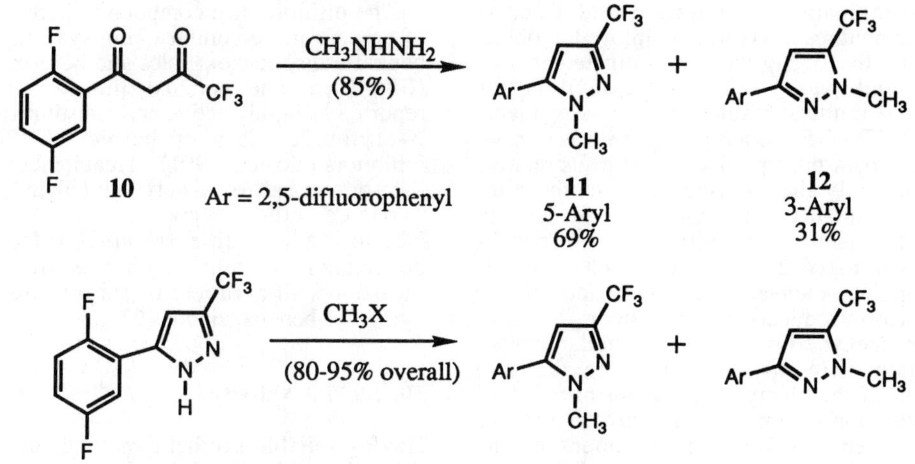

CH_3NHNH_2 (85%)

10 Ar = 2,5-difluorophenyl

11 5-Aryl 69%

12 3-Aryl 31%

CH_3X (80-95% overall)

CH_3I, K_2CO_3	30%	70%
$(CH_3)_2SO_4$, toluene, heat	4%	96%

Fig. 5. Preparation of an alkylsulfonylpyrazole from 2,5-difluoroacetophenone.

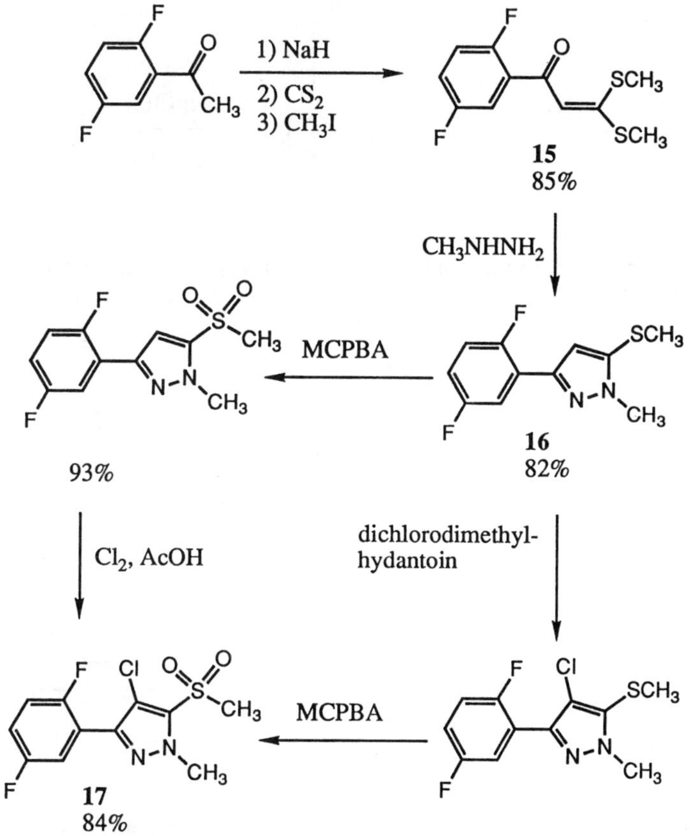

1) NaH 2) CS_2 3) CH_3I

15 85%

CH_3NHNH_2

16 82%

MCPBA

93%

Cl_2, AcOH

dichlorodimethyl-hydantoin

MCPBA

17 84%

Fig. 6. Regioselective synthesis of 3-aryl and 5-aryl-trifluoromethylphenylpyrazoles.

A combination of nitration and fluorine displacement reactions were employed to obtain many of the compounds investigated in this series and was a convenient method for obtaining multisubstituted phenyl substituents (Fig. 7). The 2,5-difluorophenyl **18** was nitrated to give a para-nitro product. Due to its relative position, only the fluorine ortho to the nitro group is suitably activated for displacement. Thus, treatment with methoxide afforded 2-fluoro-4-nitro-5-methoxy **19**. Other nucleophiles such as mercaptans and amines were also introduced in this manner. The 2,4-difluorophenyl **20** was used in a similar manner to obtain different substituents in the para position of the phenyl ring. After nitration of **20**, both fluorines are formally activated towards displacement, however, replacement of the fluorine ortho to the nitro group is observed as the major or sole product **21**.

The difluoronitro compounds were also used to prepare more complex ring systems such as benzofurans, benzoxazoles and benzoxazinones (Fig. 8). The benzoxazinones have been reported as highly active aryl substitutents in the N-arylimide class of herbicidal PROTOX inhibitors (Yosida, 1991). Treatment of **22** with the sodium salt of *n*-butyl glycolate gave the expected ether derivative in 72% yield. Reduction of this product afforded a benzoxazinone ring which was alkylated on nitrogen with a variety of substituents to give pyrazolyl-benzoxazinones **23**.

Herbicidal Activity

Having reliable synthetic methods in hand for the preparation of a variety of substituted phenylpyrazoles, compounds were prepared for

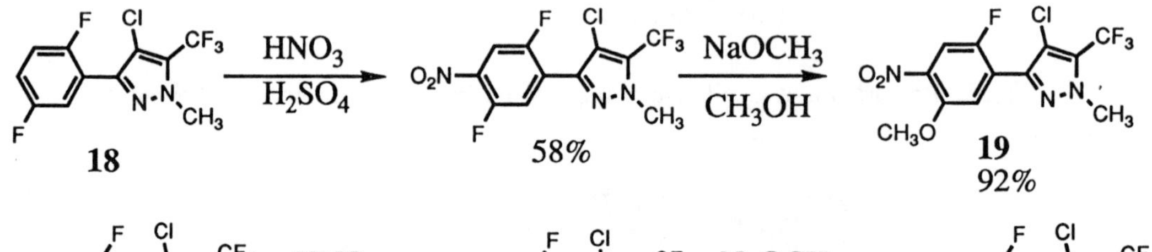

Fig. 7. Nitration and fluoride displacement as a means of preparing 2,4,5-trisubstituted phenylpyrazoles.

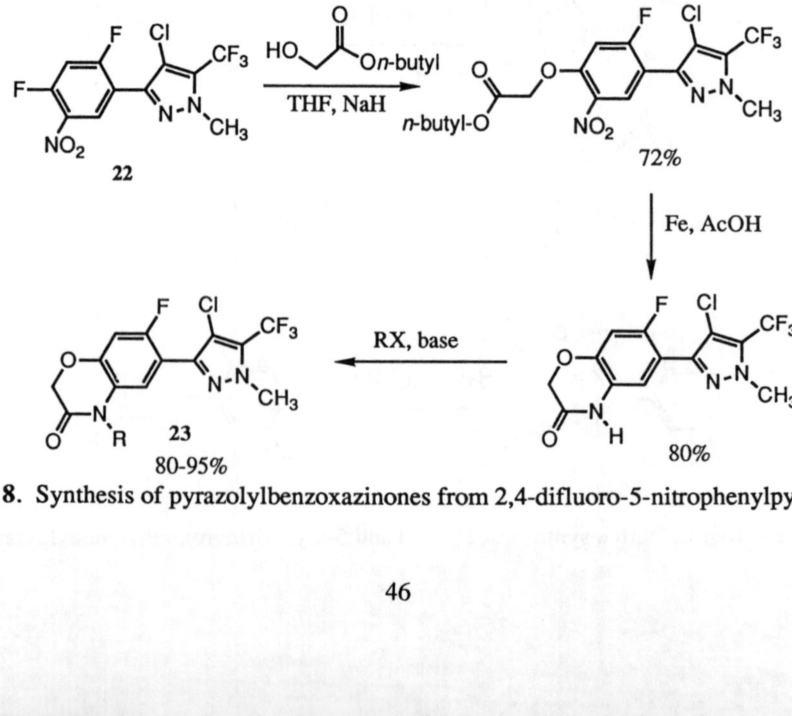

Fig. 8. Synthesis of pyrazolylbenzoxazinones from 2,4-difluoro-5-nitrophenylpyrazoles.

direct comparisons of herbicidal activity. Comparisons of compounds having phenyl substituents common to the nitrodiphenyl ether or N-phenylimide herbicides are shown in the Table. The activity of the test compounds on narrowleaf (NL) and broadleaf (BL) weeds in these whole plant studies are expressed as GR80 values or the amount of material in kg per hectare required to provide 80% control of the weed species. In these tests, the 2-fluoro-4-chloro-5-methoxyphenylpyrazoles (entries 3 and 4) are more active than either para-nitro or 2,4-dichloro compounds (entries 1 and 2). These trisubstituted compounds provide excellent control of broadleaf weeds in both pre- and postemergent test. The methylsulfonylpyrazole is generally weaker than the CF_3 analog in preemergent tests, but shows better overall post activity. Compounds lacking a chlorine substituent in the 4 position of the pyrazole ring (entry 5) are completely inactive even a rates of 5 kg per hectare. The regioisomeric 5-aryl-4-chloropyrazoles (i.e., entry 6) did exhibit activity in preemergent tests, but were in general less active than their 3-aryl counterparts. This difference was greater than 10X in a direct

Table. Relative Herbicidal Activity of Substituted Phenylpyrazoles.

entry	compound	Preemergent (kg/ha) NL	BL	Postemergent (kg/ha) NL	BL
1	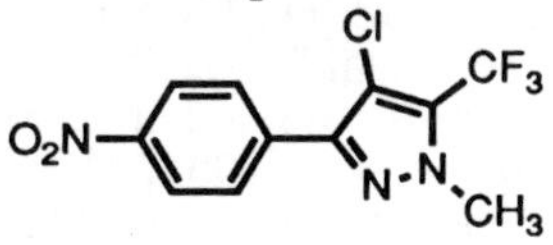 O_2N; Cl; CF_3; N-N; CH_3	4.2	5.7	9.5	6.6
2	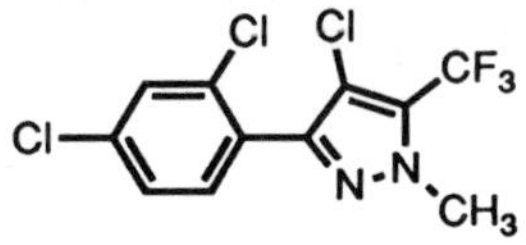 Cl; Cl; Cl; CF_3; N-N; CH_3	0.40	0.29	2.7	2.3
3	F; Cl; Cl; CH_3O; CF_3; N-N; CH_3	0.024	0.047	2.3	0.25
4	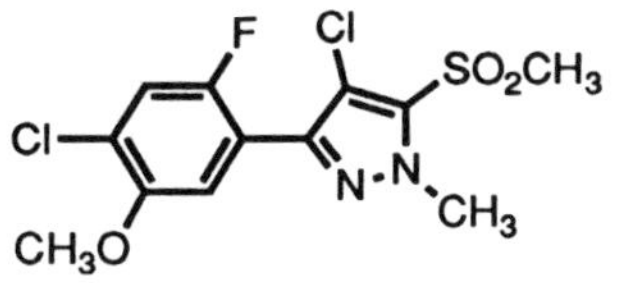 F; Cl; Cl; CH_3O; SO_2CH_3; N-N; CH_3	0.082	0.13	0.32	0.044
5	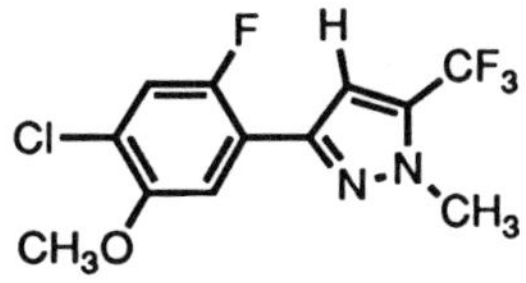 F; H; Cl; CH_3O; CF_3; N-N; CH_3	>5	>5	>5	>5
6	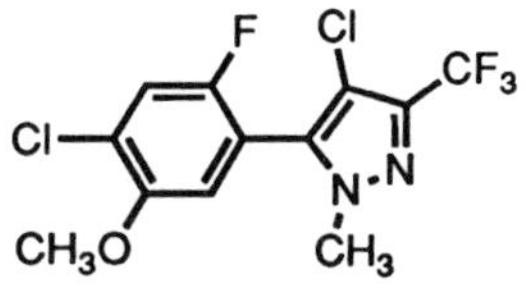 F; Cl; Cl; CH_3O; CF_3; N-N; CH_3	0.76	0.51	>5	3

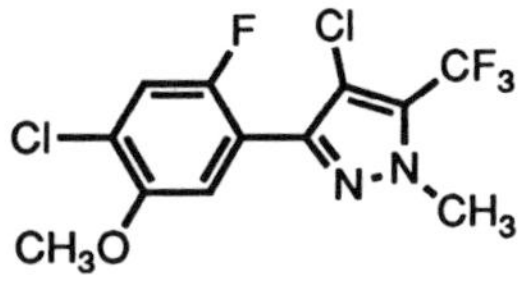

comparison of the 2-fluoro-4-chloro-5-methoxyphenylpyrazoles.

Summary

A new structural class of herbicides, substituted phenylpyrazoles, has been investigated which provide excellent weed control in both greenhouse and field studies. Herbicidal activity is dependent on the substituents of the phenyl and pyrazole rings and regiochemistry of the pyrazole.

References

Duke, S.O.; Lydon, J.; Bercerril, J.M.; Sherman, T.D.; Lehnen, L.P.; Matsumoto, H. *Weed Science* **1991**, *39*, 465-473.

Hamada, T.; Yoshida, R.; Nagano, E.; Oshio, H.; Kamoshita, K. In *Brighton Crop Protection Conference-Weeds-1989;* Proceedings Volume 1; British Crop Protection Council: Surrey, England, 1989, pg. 41-46.

Hamper, B.C.; Kurtzweil, M.L.; Beck, J.P. *J. Org. Chem.* **1992,** *57,* 5680-5685.

Hamper, B.C.; Leschinsky, K.L.; Massey, S.S.; Bell, C.L.; Brannigan, L.H.; Prosch, S.D. *J. Food Agric. Chem.* submitted for publication.

Mischke, D.A.; Hamper, B.C.; Woodard, S.S.; Moedritzer, K.; Rogers, M.D.; Dutra, G.A. WO 92/02509, **1992**.

Miura, Y.; Ohnishi, M.; Mabuchi, T.; Yanai, I. In *Brighton Crop Protection Conference-Weeds-1993*, Proceedings Volume 1; British Crop Protection Council: Surrey, England, 1993, pp 35-40.

Moedritzer, K.; Allgood, S.G.; Charumilind, P.; Clark, R.D.; Gaede, B.J.; Kurtzweil, M.L.; Mischke, D.A.; Parlow, J.J.; Rogers, M.D.; Singh, R.K.; Stikes, G.L.; Webber, R.K. In *Synthesis and Chemistry of Agrochemicals III*; Baker, D.R.; Fenyes, J.G.; Steffens, J.J., Eds.; ACS Symposium Series No. 504; American Chemical Society: Washington, DC, 1992, pp 147-160.

Woodard, S. S.; Hamper, B. C.; Moedritzer, K.; Rogers, M. D.; Mischke, D. A.; Dutra, G. A. U. S. Patent 5,281,571, **1994**.

Yoshida, R.; Sakaki, M.; Sato, R.; Haga, T.; Nagano, E.; Oshio, H.; Kamoshita, K. In *Brighton Crop Protection Conference-Weeds-1991;* Proceedings Volume 1; British Crop Protection Council: Surrey, England, 1991, pg. 69-75.

Chloronicotinyl Insecticides: Development of Imidacloprid

K. Shiokawa, S. Tsuboi and K. Moriya, Yuki Research Center, Nihon Bayer Agrochem Co., Ltd., Yuki, Ibaraki 307, Japan
S. Kagabu, Department of Chemistry,Faculty of Education, Gifu University,Gifu 501-11, Japan

Novel 1-arylmethyl-2-nitromethyleneimidazolidines whose aryl groups included heteroaromatic rings were synthesized and tested for insecticidal activity against green rice leafhoppers (*Nephotettix cincticeps*). 6-Chloro-3-pyridylmethyl- and 2-chloro-5-thiazolylmethyl were found to be highly active arylmethyl groups and the former was then used as a substituent in other related systems. From these 6-chloro-3-pyridylmethyl-2-nitroiminoimidazolidine, imidacloprid, was selected for development, founding a new class of insecticides displaying excellent wide activity coupled with systemicity and long duration of action. Synthetic aspects, structure-activity relationships and the biological profile of imidacloprid are discussed.

Many organophosphates, carbamates, synthetic pyrethroids and chitin biosynthesis inhibitors have been developed for use as insecticides in crop protection. In the meantime field resistance and crossresistance have become a problem and there is a need for a new product with a fundamentally different mechanism of action.

At the beginning of our program our attention was caught by the nitromethylene derivatives, reported by Soloway in 1979, because of their highly unusual structure and interesting biological activity. Of particular interest subsequently was their mode of action, being deferent from those of current insecticides (Schroeder, 1984; Harris, 1986; Sattelle, 1989).

We chose the green rice leafhopper (*Nephotettix cincticeps*), a major pest found in rice cultivation, as our target organism for screening and optimizing biological activity. Table 1 shows the activity of typical market insecticides against this pest.

Fenthion and propoxur, an organophosphate and a carbamate respectively, showed no practical efficacy against multi-resistant strains of green rice leafhopper. Ethofenprox and buprofezin, a synthetic pyrethroid and a chitin biosynthesis inhibitor, showed good efficacy against both susceptible and multi-resistant strains.

SKI-71 (**1**) (Soloway, 1979) also exhibited good activity against susceptible and multi-resistant strains. Other unsubstituted nitromethylene heterocycles (**2**, **3**) showed only very weak activity but one of the first *N*-benzyl derivatives we synthesized (**5**) was active at 200 ppm. We therefore initiated a synthesis program based on arylmethyl derivatives.

Synthesis

1-Arylmethyl-2-nitromethyleneimidazolidines are accessible either by arylmethylation of ethylenediamine followed by ring formation with 1,1-bis(methylthio)-2-nitroethylene or by direct

Table 1. Insecticidal activities (LC_{90})[a] of some current insecticides and nitromethylene heterocycles.

No.	Compounds	Green rice leafhopper MR[b)]	($S^{c)}$)
	fenthion	-[d)]	(200)
	propoxur	1000	(40)
	ethofenprox	8	(8)
	buprofezin	8	(8)
1	HN S CHNO$_2$	40	(40)
2	HN S CHNO$_2$	1000	
3	HN NH CHNO$_2$	1000	
4	Benzyl-N NH CHNO$_2$	200	(200)

a) LC_{90} is the lowest concentration necessary to kill over 90% of the insects tested in six stages between 1000 and 0.32 ppm. b) MR is a multi-resistant strain which is reared in the presence of organophosphorus and carbamate insecticides. c) S is a susceptible strain which is reared without chemical pressure. LC_{90} against susceptible strain of green rice leafhopper shows in parentheses. d) - represents no activity at a 1000 ppm dose.

arylmethylation of the pre-formed ring system, 2-nitromethyleneimidazolidine.

The corresponding nitroimino derivatives are also available by the same routes; direct arylmethylation of 2-nitroiminoimidazolidine or reaction of the arylmethylethylenediamine with nitroguanidine as shown in Fig. 1.

These general routes formed the basis of our synthetic program.

Structure-activity Relationships

We first focused our attention on the influence of simple substituents on the benzyl group. Some examples are shown in Table 2.

In the series of 1-(chlorophenylmethyl)-2-nitromethyleneimidazolidines shown **(5-7)**, a chlorine atom in the para-position **(7)** enhanced insecticidal activity relative to the unsubstituted compound **(4)**. The meta-chloro derivative had the some activity as the unsubstituted derivative, but moving the chlorine atom to the ortho-position **(5)** destroyed the activity completely.

A variety of heteroarylmethyl groups were introduced at the 1-position of 2-nitromethylene-imidazolidine **(8-32)**(Tables 3-7). Table 3 shows a series of furan, thiophene and pyrrole derivatives, of which the 5-bromo-2-thienylmethyl derivative **(10)** displayed the best activity. The 2-furylmethyl, 2-thienylmethyl and 2-pyrrolylmethyl derivatives **(8,9,11)** were as active as the benzyl derivative**(4)**.

We then moved to the pyridyl derivatives, which were additionally interesting because they represent part of the essential structure of nicotine, a natural antagonist of acetylcholine. Some of these **(12-25)** are shown in Tables 4-6.

The unsubstituted pyridylmethyl derivatives (Table 4) all showed improved activity relative to their corresponding chlorobenzyl partners. The 3- and 4-pyridylmethyl derivatives **(13, 14)** were

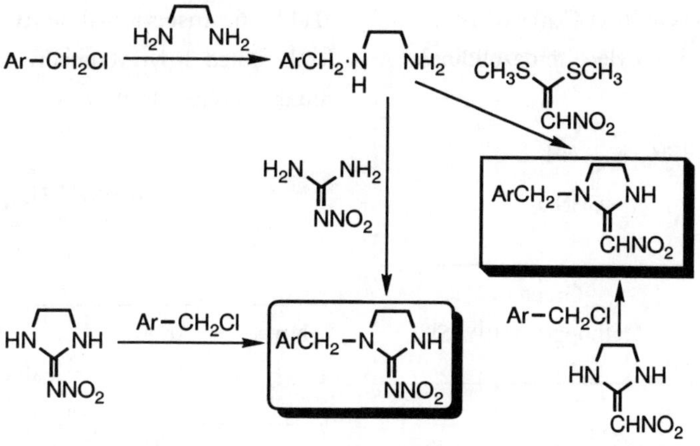

Fig. 1. Synthesis of 1-arylmethyl-2-nitromethylene- and -2-nitroiminoimidazolidines.

Table 2. Insecticidal activity (LC_{90})[a] of 1-chlorophenylmethyl-2-nitromethyleneimidazolidines.

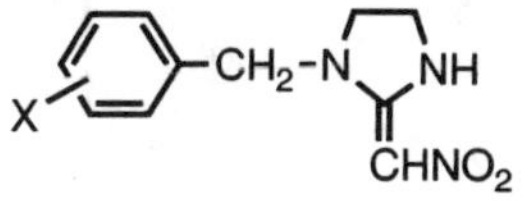

No.	X	Green rice leafhopper MR[b]	($S^{c)}$)
5	2-Cl	-[d]	
6	3-Cl	200	
7	4-Cl	40	(40)

Table 3. Insecticidal activity (LC_{90})[a] of 1-arylmethyl-2-nitromethyleneimidazolidines.

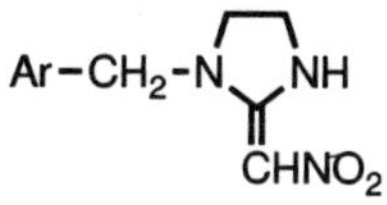

No.	Ar	Green rice leafhopper MR[b]
8	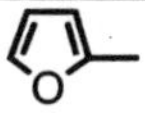 2-furyl	200
9	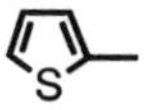 2-thienyl	200
10	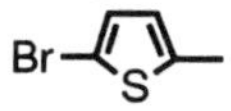 5-bromo-2-thienyl	40
11	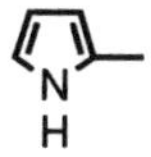 2-pyrrolyl	200

active at 8 ppm, i.e. at almost the same levels as ethofenprox and buprofezin, while the 2-pyridylmethyl derivative (**12**) was only as active as the unsubstituted benzyl derivative (**4**).

Since the introduction of a chlorine atom enhanced activity in the benzyl series, it was introduced into the pyridine ring of pyridylmethyl-2-nitromethyleneimidazolidine to obtain **15-19**.

A chlorine atom in the 5-position of the 2-pyridylmethyl derivative (**15**) caused no enhancement of activity. This was also the case for a chlorine in the 2-position of the 4-pyridylmethyl derivative (**19**). However, among the 3-pyridylmethyl derivatives the 6-chloro compound (**18**)was 25 times more active than the parent; a chlorine in the 2-position (**16**) or 5-position (**17**) diminished the activity. The 6-chloro-3-pyridylmethyl derivative (**18**) was

Table 4. Insecticidal activity (LC_{90})[a] of 1-pyridylmethyl-2-nitromethyleneimidazolidines.

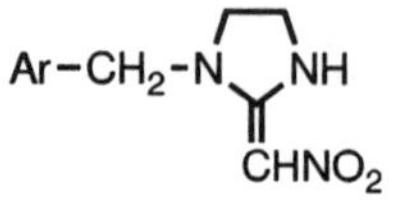

$Ar-CH_2-N$ NH, $CHNO_2$

No.	Ar	Green rice leafhopper MR[b]	(S[c])
12	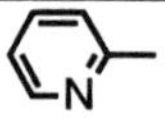 N	200	
13	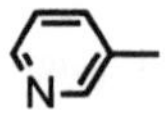 N	8	(8)
14	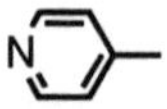 N	8	(8)

Table 5. Insecticidal activity (LC_{90})[a] of 1-(chloropyridyl)methyl-2-nitromethyleneimidazolidines.

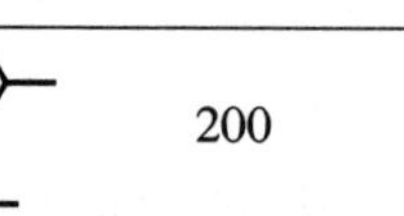

$Ar-CH_2-N$ NH, $CHNO_2$

No.	Ar	Green rice leafhopper MR[b]	(S[c])
15	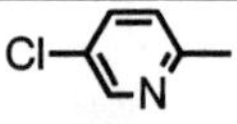 Cl, N	200	
16	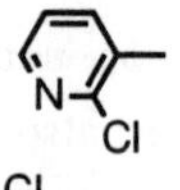 N, Cl	200	
17	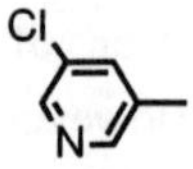 Cl, N	40	
18	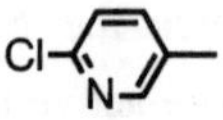 Cl, N	0.32	(0.32)
19	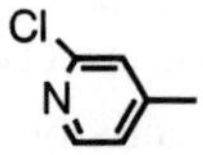 Cl, N	200	

Table 6. Insecticidal activity (LC_{90})[a] of 1-(substituted-3-pyridyl)methyl-2-nitromethylene-imidazolidines.

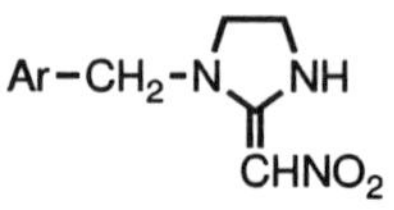

$Ar-CH_2-N$ NH, $CHNO_2$

No.	Ar	Green rice leafhopper MR[b]	(S[c])
20	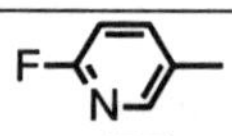 F, N	0.32	(0.32)
21	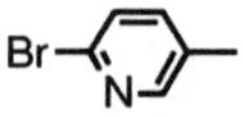 Br, N	1.6	(0.32)
22	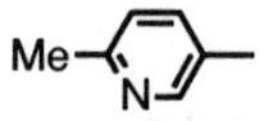 Me, N	1.6	(1.6)
23	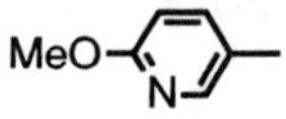 MeO, N	200	
24	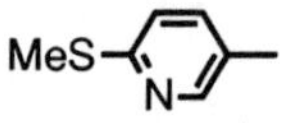 MeS, N	1000	
25	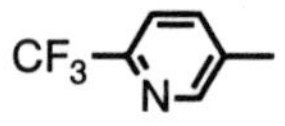 CF_3, N	8	

highly active against both strains of green rice leafhopper, the LD_{90} values being 0.32 ppm (Kagabu, 1992).

Table 6 shows the effect of replacing the chlorine atom in this derivative by other substituents.

In this series (**20-25**) the 6-halo and 6-methyl derivatives (**20-22**) were also very active. A methoxy (**23**) or methylthio (**24**) group seriously reduced activity. The trifluoromethyl derivative (**25**) is only moderately active. Other heteroarylmethyl systems were extensively studied, and many heteroaromatics bearing a halogen atom or methyl group as substituent proved to be highly active (Table 7).

Of these the 2-chloro-5-thiazolylmethyl derivative (**32**), a representative of the five-membered heteroaromatics, was as active as the 6-chloro-3-pyridylmethyl derivative (**18**), which was our best insecticide thus far in the 1-

arylmethyl-2-nitromethyleneimidazoline series (Moriya, 1993).

The structural requirements upon the arylmethyl moiety for active 2-nitromethyleneimidazolidines can thus be summarized as follows (Fig. 2);

1) Hydrogen seems to be the best substituent at the 2-position of the aryl ring.

2) An iminyl nitrogen atom should be at the 3-position.

3) A halogen atom or methyl group should be the substituent at the 4-position.

When the methylene spacer group between the 6-chloro-3-pyridyl ring and the imidazolidine moiety in the best compound (**18**) was replaced by others, activity was reduced (Table 8).

The order of activity was 6-chloro-3-pyridylmethyl (**18**) > α-(6-chloro-3-pyridyl)ethyl (**34**) > 6-chloro-3-pyridyl (**33**) > β-(6-chloro-3-pyridyl)ethyl (**35**).Ring-expansion and annulation of the imidazolidine moiety were not well tolerated (Table 9).

Cl, F > Br, Me ⟸ X–(4-position of pyridine ring, N at 3, position 2)–CH_2N imidazolidine NH, =$CHNO_2$

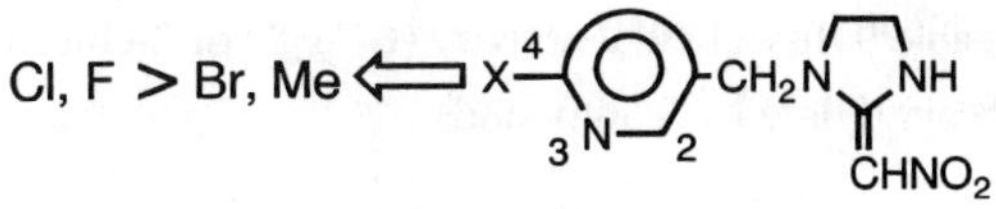

Fig. 2. Structural requirements for arylmethyl moieties of insecticidal 2-nitromethyleneimidazolidines.

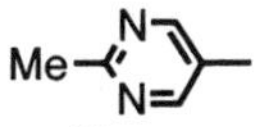

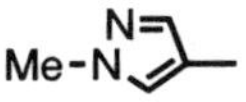

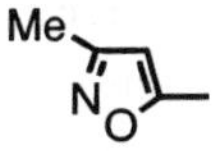

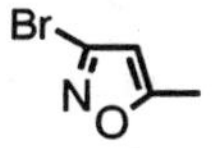

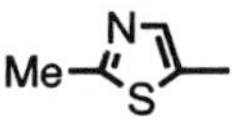

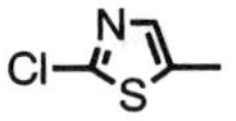

Expansion to the 6-membered ring gave an equally active derivative (**37**) but further expansion to the 7-membered ring (**38**) diminished the activity as did methylation of the other imidazolidine nitrogen of **18** to give **36**. Annulation of the imidazolidine ring (**39**) caused complete loss of activity.

The imidazolidine moiety in 6-chloro-5-pyridylmethyl-2-nitromethyleneimidazolidine (**18**) was then converted into some other heterocycles (Table 10). The pyrrolidine and oxazolidine derivatives (**40, 41**) were somewhat less active

Table 7. Insecticidal activity (LC_{90})[a] of 1-arylmethyl-2-nitromethyleneimidazolidines.

Ar–CH_2–N(imidazolidine)NH, =$CHNO_2$

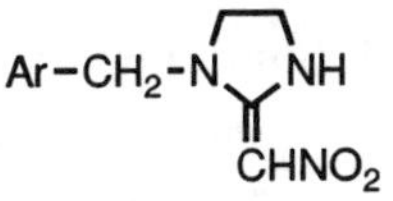

No.	Ar	Green rice leafhopper MR[b]	(S[c])
26	Me-pyrimidinyl	1.6	(1.6)
27	Me-pyrimidinyl	1.6	(8)
28	Me-N pyrazolyl	1.6	(1.6)
29	Me-isoxazolyl	1.6	(1.6)
30	Br-isoxazolyl	1.6	(8)
31	Me-thiazolyl	1.6	(1.6)
32	Cl-thiazolyl	0.32	(0.32)

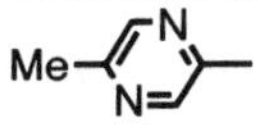

Table 8. Insecticidal activity (LC_{90})[a] of 2-chloro-5-pyridyl compounds

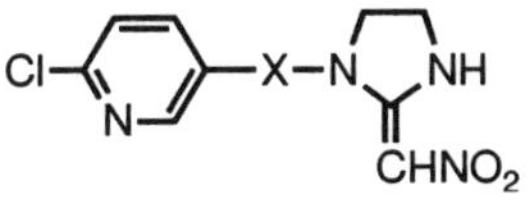

Cl–(pyridyl)–X–N(imidazolidine)NH, =$CHNO_2$

No.	X	Green rice leafhopper MR[b]
33	-	8
34	$CH(CH_3)$	1.6
35	CH_2CH_2	40

Table 9. Insecticidal activity (LC_{90})[a] of 2-chloro-5-pyridylmethyl compounds

Cl-(pyridyl)-CH_2-A

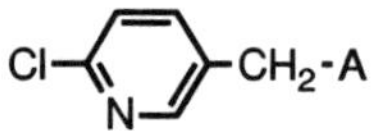

No.	A	Green rice leafhopper MR[b]	(S[c])
18	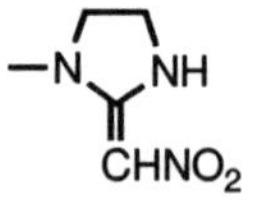 imidazolidine (–N, NH, =$CHNO_2$)	0.32	(0.32)
36	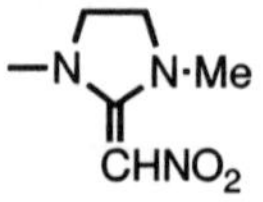 imidazolidine (–N, N·Me, =$CHNO_2$)	8	
37	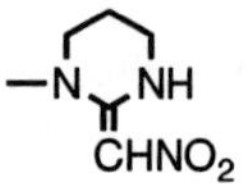 hexahydropyrimidine (–N, NH, =$CHNO_2$)	0.32	(0.32)
38	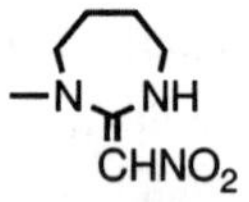 diazepane (–N, NH, =$CHNO_2$)	8	
39	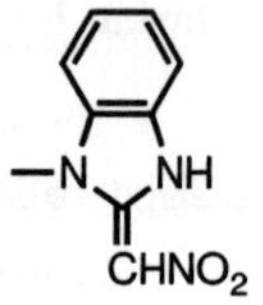 benzimidazoline (–N, NH, =$CHNO_2$)	-[d]	

Table 10. Insecticidal activity (LC_{90})[a] of 2-chloro-5-pyridylmethyl compounds

Cl-(pyridyl)-CH_2·N(ring, X)=$CHNO_2$

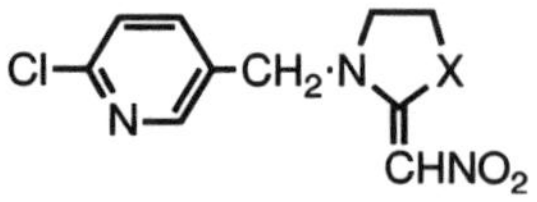

No.	X	Green rice leafhopper MR[b]	(S[c])
40	CH	1.6	
41	O	8	
42	S	0.32	(0.32)

but the thiazolidine derivative (**42**) was as active against the green rice leafhopper as the hitherto best compound (**18**) (Shiokawa, 1992).

When the nitromethylene group in our best compound (**18**) was replaced by other systems the effect was quite dramatic (Table 11). The cyanomethylene (**43**), keto (**45**) and imino (**46**) derivatives were inactive but strongly electron-withdrawing systems like dicyanomethylene (**44**), nitroimino (**48**) and cyanoimino (**49**) retained at least some activity. The nitroimino derivative (**48**) was equal to our nitromethylene compound (**18**) in this test with an LC_{90} value of 0.32 ppm.

When the imidazolidine moiety in 6-chloro-5-pyridylmethyl-2-nitroiminoimidazolidine (**48**) was replaced by the desaturated systems of imidazoline, thiazoline and 1,2-dihydropyridine, all these derivatives (**50-52**) proved to be highly active (Table 12).

The structural requirement for insecticidal activity among 6-chloro-3-pyridylmethyl derivatives can thus be summarized as shown in Fig. 3. A methylene group was the best spacer A. As a ring size, a five- or six-membered ring was best and could be saturated or unsaturated. NH and S were the best for part X. For the part Y a methylene or an imino group were nitromethylene and nitroimino.

From the 6-chloro-3-pyridylmethyl derivatives, representatives were examined in the residual efficacy test against green rice leafhoppers and green peach aphids for possible practical value. The results are shown in Table 13.

The 2-nitroiminoimidazolidine (**48**) showed long

Table 11. Insecticidal activity (LC_{90})[a)] of 6-chloro-3-pyridylmethylimidazolidine compounds

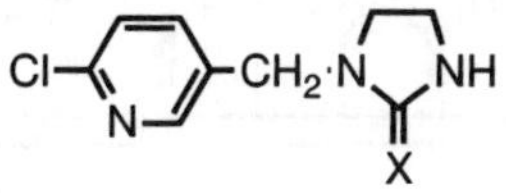

No.	X	Green rice leafhopper MR[b)]	(S[c)])
18	$CHNO_2$	0.32	(0.32)
43	CHCN	-[d)]	
44	$C(CN)_2$	200	
45	O	-[d)]	
46	NH	-[d)]	
48	NNO_2	0.32	(0.32)
49	NCN	1.6	(1.6)

Table 12. Insecticidal activity (LC_{90})[a)] of 2-chloro-5-pyridylmethyl compounds

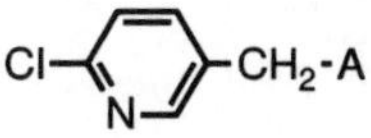

No.	A	Green rice leafhopper MR[b)]	(S[c)])
50	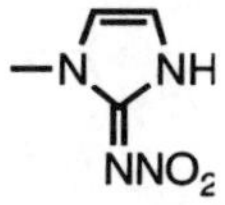 2-nitroiminoimidazoline ($-N$, NH, NNO_2)	0.32	(0.32)
51	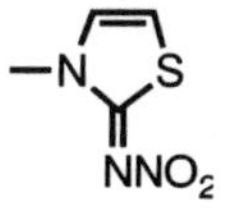 2-nitroiminothiazoline ($-N$, S, NNO_2)	0.32	(0.32)
52	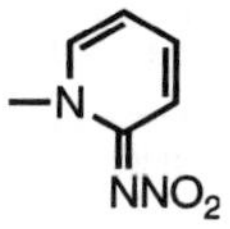 2-nitroiminopyridine ($-N$, NNO_2)	0.32	(0.32)

residual efficacy against both the leafhopper and the aphid. The 2-nitromethyleneimidazolidine (**18**) was somewhat less active in this test system. The 2-nitromethylenethiazolidine (**42**) and 1,2-dihydro-2-nitroiminopyridine (**52**) showed long residual activity only against the leafhopper.After exhaustive greenhouse and field trials 1-(6-chloro-3-pyridy)methyl-2-nitroiminoimidazolidine (**48**) was selected and developed into a practical insecticide named imidacloprid.

Photostability and Physicochemical Parameters

One major handicap of the original nitromethylene derivatives reported by Soloway was their instability when exposed to sunlight. The UV spectra of tetrahydro-2-nitromethylene-4*H*-1,3-thiazine (**1**, SKI-71), 1-(6-chloro-3-pyridyl)methyl-2-nitromethyleneimidazolidine (**18**) and imidacloprid (**48**) were measured in 10 ppm aqueous solution, as shown in Fig. 4. Their λ max values were 348, 321 and 270 nm respectively. The UV absorption maximum of imidacloprid (**48**) is considerably shifted to shorter wavelength compared with those of **1** and **18**. The shortest wavelength of sunlight to reach the earth's surface is 290 nm, and the light of wavelength between 290 and 400 nm is very important for photodegradation of pesticides in the field (Fukami, 1981). Since imidacloprid (**48**) has no absorption peak between 290 and 400 nm and only shows very slight absorption at all in this range, it is far less susceptible to photodegradation by sunlight than the other two compounds.This greatly improved photostability

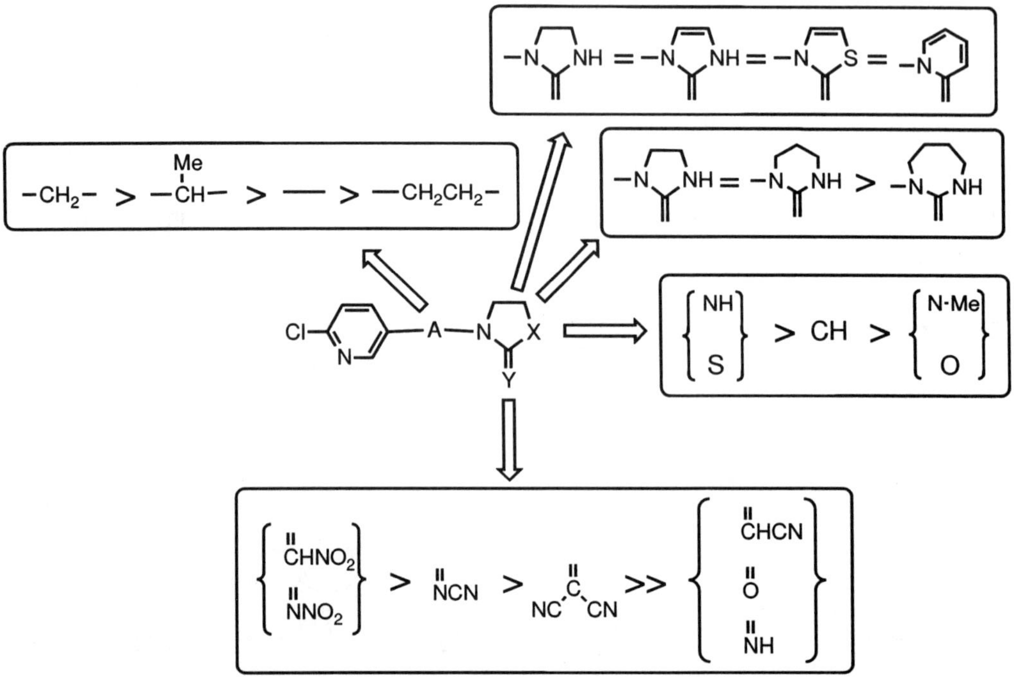

Fig. 3. Structural requirements for active 6-chloro-3-pyridyl derivatives.

Table 13. Residual efficacy of 1-(6-chloro-3-pyridyl)methyl-2-nitromethyleneimidazolidine (**18**), (6-chloro-3-pyridyl)methyl-2-nitroiminoimidazolidine (**48**), 1-(6-chloro-3-pyridyl)methyl-2-nitromethylenethiazolidine (**42**) and 1-(6-chloro-3-pyridyl)methyl-1,2-dihydro-2-nitroiminopyrid (**52**) against green rice leafhopper and green peach aphid at 200 ppm.

		Mortality (%)						
Compounds	Insects	2	7	14	21	27	35	42 DAT[e]
18	leafhopper	100	100	100	75	70	30	
	aphid	100	95	95	90	45	5	
42	leafhopper	100	100	100	100	100	95	
	aphid	95	80	0				
48	leafhopper	100	100	100	100	100	100	100
	aphid	100	100	100	100	100	40	
52	leafhopper	100	100	100	100	100	100	
	aphid	100	25	0				

e) Days after treatment.

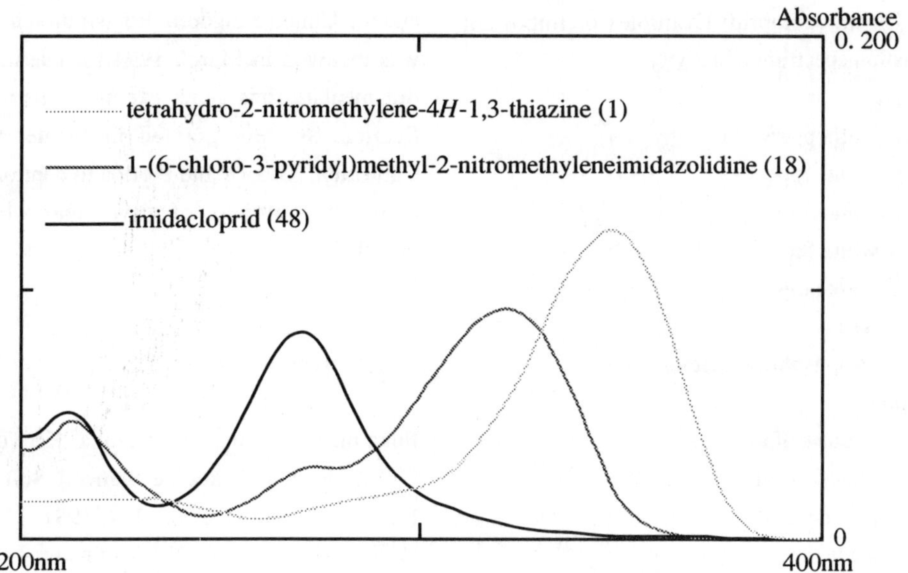

Fig. 4. UV spectrum of tetrahydro-2-nitromethylene-4*H*-1,3-thiazine (**1**), 1-(6-chloro-3-pyridyl)methyl-2-nitromethyleneimidazolidine (**18**), and imidacloprid (**48**).

of imidacloprid is combined with favorable physicochemical parameters (Table 14) and enables the product to be used in a broad range of applications and its biological activity to be exploited to the full.

Thanks to its chemical constitution as a quasibetaine, imidacloprid is a polar substance with a water solubility of ca. 500ppm and has a very low log P value. As a consequence imidacloprid has excellent systemic properties; it is rapidly distributed within the plant. It is taken up by roots and transported to upper plant parts, and also enters the phloem, which is the food source for sucking insects such as aphids.

Table 14. Imidacloprid: Physicochemical data

Cl-(pyridyl)-CH_2 N-imidazolidine-NH, =NNO_2

Appearance:	Colorless crystals	
Melting point:	143.8°C (Modification 1)	
	136.4°C (Modification 2)	
Vapor pressure:	2 x 10^{-9} hPa at 20°C	
Solubility:	Water	0.51
(g/l solvent at 20°C)	n-Hexane	<0.1
	Dichloromethane	55
	2-Propanol	1.2
	Toluene	0.68
Partition coefficient:		
(octanol/water)	log P 0.57	

Biological Profile

Table 15 shows a selection of major sucking and chewing pests susceptible to imidacloprid is predestined to be a seed treatment product. Seed treatment protects sugar beet, corn, cereals, rice, potatoes and cotton during the germination phase and well into the growing season from soil pests such as wireworms and pigmy beetle and from leaf pests such as aphids which also act as virus vectors.

Drip irrigation and seedling box treatment are also very relevant application techniques/ It is a highly effective spray treatment in vegetables,

Table 15. imidacloprid: Examples of important pests within spectrum of activity

Hemiptera
- leafhoppers
- planthoppers
- aphids
- white flies
- plant bugs

Thysanoptera
- various thrips species

Lepidoptera
- Citrus leafminer
- apple leafminers
- Peach leafminer
- tea leafrollers

Coleoptera
- Rice water weevil
- wire worms
- Colorado potato beetle
- other leaf eating beetles
- grubs

Diptera
- various leafminers
- fruit flies
- Onion fly

Isoptera
- termites

citrus and tree fruit with outstanding efficacy against resistant or difficult to control pests.

Conclusion

Imidacloprid is the first member of a new class of insecticides acting on the nicotinergic acetylcholine receptor to find wide application in practice. It is now registered in 39 countries including France (since 1992), Japan, Germany and the United Kingdom. Registration in the USA was received in March 1994 for use in turf and ornamentals. Emergency exemption permits under Section 18 were granted in Winter 93/94 in California, Arizona and Florida to control a highly damaging outbreak of silver leaf whitefly in vegetables.

References

Fukami, J.;, Uesugi, Y.; Isizuka, K.; Tomizawa, C. "Methods in Pesticide Sience," Soft Science, Inc. Tokyo, Vol. 4, 230 - 237, 1981.

Harris, M.; Price, R. N.; Robinson, J.; May, T. E. *Proceedings of the British Crop Protection Conference - Pests and Diseases* 1, 115-122, 1986.

Kagabu, S.; Moriya, K.; Shibuya, K.; Hattori, Y.; Tsuboi, S.; Shiokawa, K. *Biosci. Biotech. Biochem.*, **56**, 362-363, 1992

Moriya, K.; Shibuya, K.; Hattori, Y.; Tsuboi, S.; Shiokawa, K.; Kagabu, S. *Biosci. Biotech. Biochem.*, **57**, 127-128, 1993

Sattelle, D. B.; Buckingham, S. A.; Wafford, K. A.; Sherby, S. M.; Bakry, N. M.; Eldefrawi, A. T.; Eldefrawi, M. E.; May, T. E. *Proc. R. Sod. Lond.* B **237**, 501-514, 1989

Schroeder, M. E.; Flattum, R. F. *Pesticide Biochem. Physiol.*, **22**, 148-169, 1984.

Shiokawa, K.; Tsuboi, S.; Kagabu, S.; Moriya, K. Jp. Pat. Sho 61-267575, Nov. 27, 1986.

Shiokawa, K.; Moriya, K.; Shibuya, K.; Hattori, Y.; Tsuboi, S.; Kagabu, S. *Biosci. Biotech. Biochem.*, **56**, 1364-1365, 1992

Soloway, S. B.; Henry, A. C.; Kollmeyer, W. D.; Padgett, W. M.; Powell, K. E.; Roman, S. A.; Tieman, C. H.; Corey, R. A.; Horne, C. A. *Advances in Pesticide Science*; Part 2, Pergamon Press, Oxford, 206-217, 1979.

The Synthesis of Fungicidal β-Methoxyacrylates

John M. Clough, Vivienne M. Anthony, Paul J. de Fraine, Torquil E. M. Fraser, Christopher R. A. Godfrey, Jeremy R. Godwin and David Youle,
ZENECA Agrochemicals, Jealott's Hill Research Station, Bracknell, Berkshire, RG12 6EY, U.K.

We describe the discovery of a new systemic β-methoxyacrylate fungicide with a broad spectrum of activity and a novel mode of action; this compound has been given the code name ICIA5504. The evolution of ideas which led ultimately to the synthesis of this fungicide can be traced back to a family of naturally-occurring β-methoxyacrylates. An early objective was to prepare readily-accessible analogues of these natural products which were of sufficient stability to be used as agricultural fungicides, and this led to β-methoxyacrylates which were linked at the α-position to a variety of substituted benzenes. Building on this foundation, the synthesis of further compounds then enabled fungicidal activity, systemic movement in plants, and crop safety to be optimised. The discovery of ICIA5504 was the culmination of this work.

The Need for New Fungicides

The global market for agricultural fungicides was valued at $4.735 billion at the end-user level in 1993 (Wood Mackenzie, 1994), and we estimate that more than 150 different compounds are now in use. Nevertheless, for the following reasons, there is a continuous need for the discovery and development of new compounds. Firstly, a fungicide of a new class may have technical advantages over existing products, in terms of its breadth of spectrum, levels of activity, or safety in the environment, for example. In addition, many established fungicides become less effective as they are used year after year owing to the onset of resistance. It follows that one of the most important aspects of fungicide research is the discovery of compounds with novel modes of action: such compounds are not likely to be cross-resistant to fungicides already in use, and may have a unique technical profile. New fungicides are, of course, also important from a commercial point of view because they may be patentable in their own right, and because competitors' patents are less likely to influence decisions relating to their manufacture and sale.

The subject of this paper is the work which led to the discovery of a new broad spectrum systemic fungicide with the code name ICIA5504, a compound with a novel mode of action which is currently being developed by ZENECA Agrochemicals (Figure 1) (Godwin *et al.*, 1992).

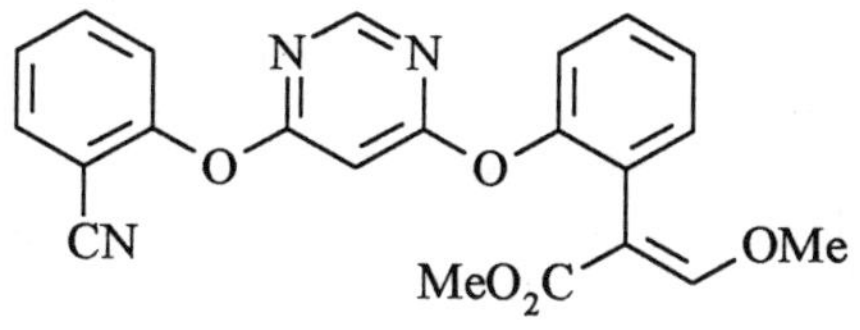

N N O O CN MeO_2C OMe

ICIA5504

Fig. 1. ICIA5504.

Origins: A Family of Natural Products

Our interest in the β-methoxyacrylates can be traced back to a family of fungicidal natural products, the strobilurins, oudemansins and myxothiazols, which are themselves all derivatives of β-methoxyacrylic acid (for a review, see Clough, 1993). Over thirty of these natural products have now been isolated and characterised, and the simplest of the strobilurins and oudemansins, together with the most abundant myxothiazol, are shown in Figure 2. The strobilurins and oudemansins are found in several genera of small fungi, such as *Strobilurus tenacellus* and *Oudemansiella*

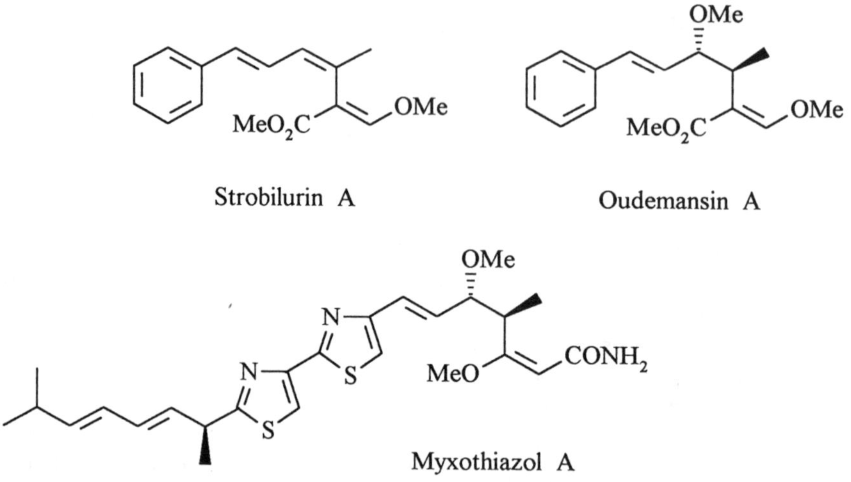

Strobilurin A

Oudemansin A

Myxothiazol A

Fig. 2. Natural fungicidal derivatives of β-methoxyacrylic acid.

mucida, which grow on rotting wood, whereas the myxothiazols have been isolated from various strains of the gliding bacterium *Myxococcus fulvus*.

The fungicidal activity of the strobilurins, oudemansins and myxothiazols is a direct result of their ability to inhibit mitochondrial respiration in fungi. More specifically, it is known that these compounds bind at the so-called Q_o-site on cytochrome b and thereby interfere with the function of the cytochrome bc_1 complex, which is located in the inner mitochondrial membrane of fungi and other eukaryotes (Figure 3). The function of this complex is to catalyse the transfer of electrons from ubiquinol to cytochrome c. As this takes place, protons are translocated across the membrane in which the complex is embedded, and this establishes a proton gradient which, in turn, drives the synthesis of ATP. In the presence of a strobilurin, oudemansin or myxothiazol, ubiquinol is still able to bind at cytochrome b in its usual way, but it is not oxidized, and it has been suggested that this is the result of a conformational distortion of cytochrome b which slightly displaces ubiquinol at its binding site (Brandt *et al.*, 1988).

A key piece of information from our point of view, already established when we first became interested in this area of chemistry in the early 1980's, was that none of the many other known inhibitors of respiration bind at the same specific site on cytochrome b as the strobilurins, oudemansins and myxothiazols, so that cross-resistance to existing fungicides would not be expected. We also realised at the outset of this work that respiration inhibitors have the potential to produce toxicity towards mammals. However, the low acute mammalian toxicity of strobilurin A and oudemansin A reassured us that toxicities towards fungi and mammals are not inextricably linked in this class of compound.

The first step was to confirm the fungicidal activity claimed for the natural products. This was important because it is our experience that many compounds which are reported in the literature to be fungicidal are, in reality, only very weakly active or have a narrow spectrum of activity. Samples of oudemansin A and myxothiazol A, generously provided by Prof. T. Anke (University of Kaiserslautern, Germany) and Prof. H. Reichenbach (Gesellschaft für Biotechnologische Forschung, Braunschweig, Germany), respectively, when applied as foliar sprays at a concentration of 33mg/l, were shown to have activity against several commercially important fungal pathogens growing on plants in the glasshouse. By contrast, a sample of strobilurin A, obtained by total synthesis

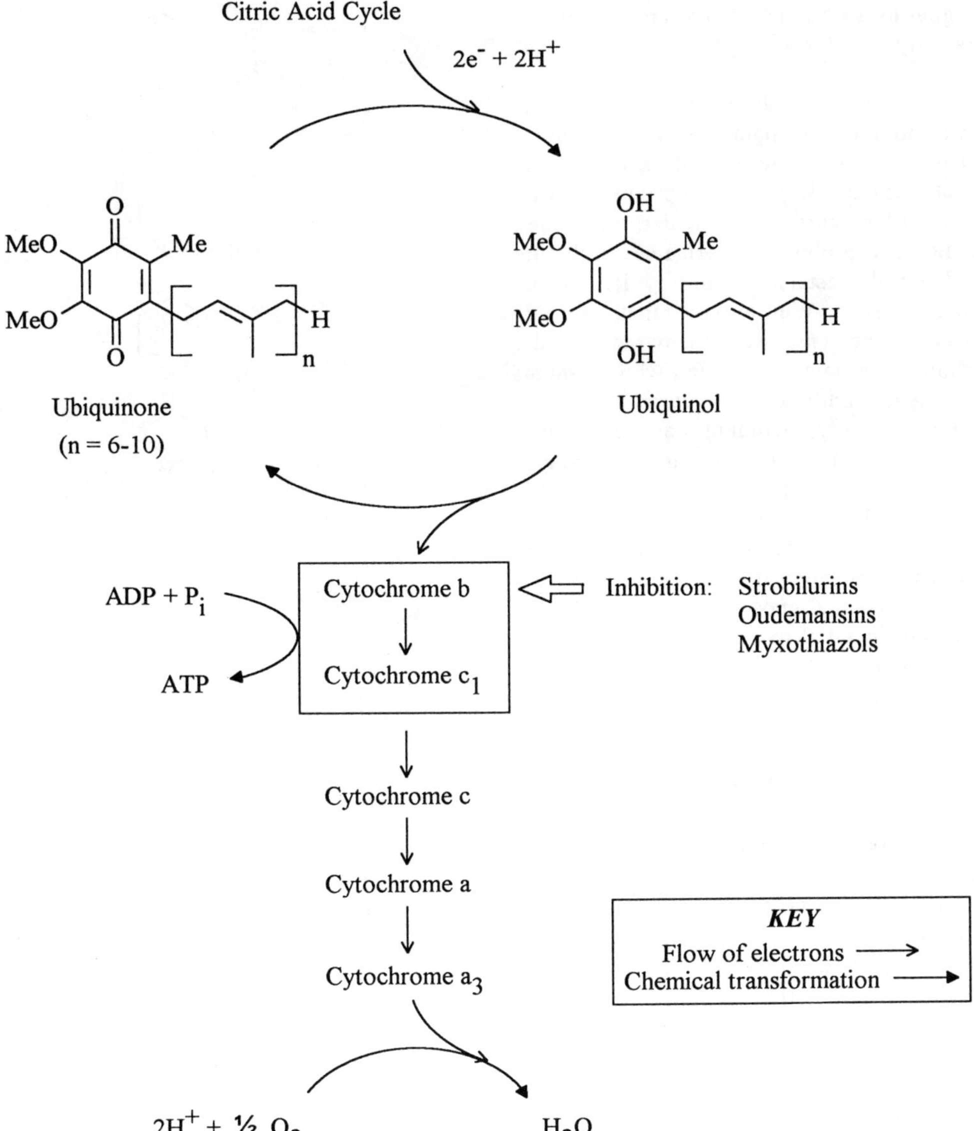

Fig. 3. Mitochondrial respiration: the electron transport chain.

(Beautement and Clough, 1987), gave quite different results: although active at 25mg/l against a variety of plant pathogenic fungi growing on agar in subdued light, the compound had no activity when applied as a foliar spray at the same rate to fungi growing on plants in the glasshouse. We were able to show that this lack of activity *in vivo* followed from the photochemical instability and relatively high volatility of strobilurin A, through which it is rapidly lost from the leaf surface.

The results which we had obtained from the natural products were therefore mixed: oudemansin A and myxothiazol A had good activity against fungi growing on plants in the glasshouse, while strobilurin A, for reasons which were now understood, was active only *in vitro*.

Analogues of the Natural Products: Solutions to Stability Problems

A programme of synthesis was initiated. Our aim was to prepare analogues of the natural products with high levels of activity and physical properties which were optimised for an agricultural fungicide. In particular, we needed to avoid the problems associated with the photochemical instability and volatility of strobilurin A. We argued that the β-methoxyacrylate unit was likely to be the toxophore, important for activity, while it might be possible to modify other structural features of the natural products, including those responsible for the photoinstability of strobilurin A, without loss of activity. A wide variety of compounds were prepared and tested, and this premise turned out to be broadly true: while many compounds containing the β-methoxyacrylate group are fungicidal, most modifications to this group result in a fall in activity. Nevertheless, certain other toxophores, each of which resembles the (*E*)-methyl β-methoxyacrylate group in shape, size and electronic character, are in fact able to confer activity when linked to an appropriate backbone. Most notable amongst these are the (*E*)-methyl methoxyiminoacetate group, the toxophore in BASF's development compound BAS 490 F (Ammermann *et al.*, 1992), and the (*E*)-*N*-methyl methoxyiminoacetamide group, found in Shionogi's SSF-126 (Hayase *et al.*, 1989; Masuko *et al.*, 1993) (Figure 4).

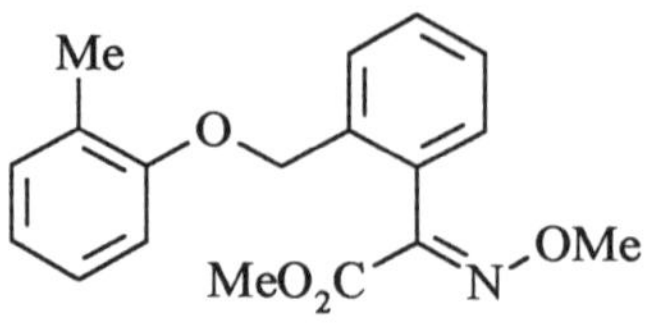

Me

O

MeO_2C N OMe

BAS 490 F

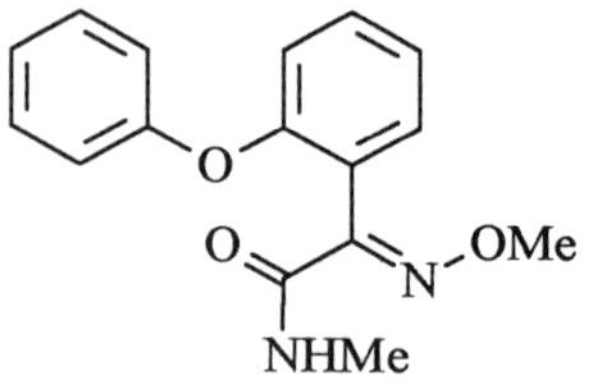

O

O N OMe

NHMe

SSF-126

Fig. 4. Alternative toxophores in BAS 490 F and SSF-126.

Of our various ideas for stabilising strobilurin A, the most fruitful was to lock the (*Z*)-olefinic bond into a benzene ring to give the stilbene **1** (Figure 5) (Bushell *et al.*, 1984). In contrast to strobilurin A, the stilbene was highly active against fungi growing on plants in the glasshouse as well as against fungi growing on agar. Nevertheless, it still had insufficient photochemical stability to express good activity in the field (Beautement *et al.*, 1991; Clough *et al.*, 1992). Interestingly, Anke and Steglich have independently prepared the stilbene **1** as well as various other related compounds (Anke and Steglich, 1989).

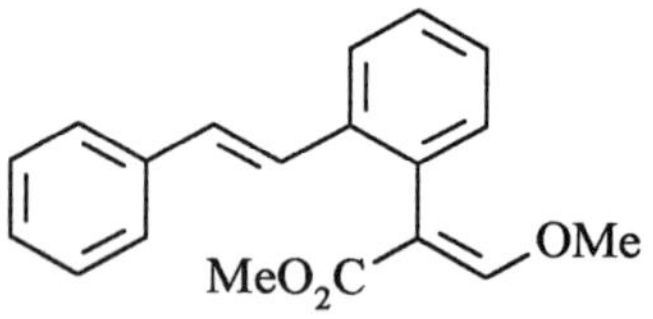

Further structural modification produced a far more useful fungicide, the diphenyl ether **2** (Figure 5) (Bushell *et al.*, 1984), which is considerably more photochemically stable than the stilbene **1**: the times taken for loss of the first 50% of the stilbene **1** and the diphenyl ether **2**, when irradiated as thin films by a xenon lamp to simulate sunlight, were three minutes and thirty *hours* respectively. The diphenyl ether **2** controlled the growth of a variety of

MeO_2C OMe

1

O

MeO_2C OMe

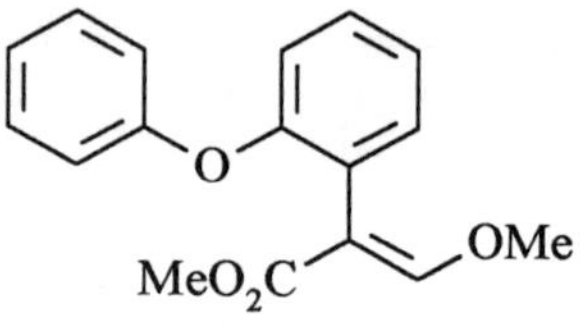

2

Fig. 5. Stilbene and diphenyl ether analogues of strobilurin A.

commercially-important fungal pathogens of wheat, barley and vines in the field. However, it produced unacceptable damage to cereals and vines in some field trials (Clough *et al.*, 1992; Clough *et al.*, 1994).

An important property of the diphenyl ether **2**, almost certainly contributing to the good activity (and, indeed, to the phytotoxicity) seen in the field, was its systemic movement in plants. Systemicity is a key attribute of modern agricultural fungicides which improves field performance through redistribution of the compound within plant tissue after spraying, and potentially allows the farmer to use the fungicide as a foliar spray, paddy-applied granule, seed or soil treatment. The first evidence for the systemicity of the diphenyl ether **2** [log P (octanol/water) = 3.25; water-solubility = 30 ppm] came from indicator tests in the glasshouse in which the compound, when applied as a root drench, protected both cereal and broad-leaved crops from infection by fungal spores which were applied as a foliar spray suspension 24 hours later. (The diphenyl ether also gave good control in parallel tests in which it was applied as a foliar spray). In further tests, solutions of **2**, applied as small spots at the base of a leaf of a growing barley plant, gave protection from infection by *Erysiphe graminis* f. sp. *hordei* over the surface of the leaf between the site of application and the leaf tip. The fact that there was no disease control between the site of application and the base of the leaf indicated that this effect was due to xylem translocation (apoplastic movement) and not phloem translocation (symplastic movement) or vapour activity. Further studies with a ^{14}C-labelled sample of **2** confirmed that the material moved from the site of application towards the tip of the leaf.

To summarise, the diphenyl ether **2** was an important lead compound. Unlike the natural products, it was easy to prepare and provided a useful springboard for further synthesis. Importantly, its much improved photochemical stability in comparison to that of strobilurin A enabled it to express broad spectrum systemic activity under field conditions. However, the application rates required for good activity were still relatively high (750g/ha on temperate cereals, for example) and, at these rates, phytotoxicity was sometimes a problem.

Optimisation of Systemic Activity

While studying the scope for the introduction of substituents in the diphenyl ether **2**, we found that highly fungicidal compounds resulted when single atoms, such as halogens, or small groups, such as methyl or methoxyl, were incorporated at the 3- or 4- (but not at the 2-) position of the phenoxy group (Clough *et al.*, 1992). By contrast, the much larger phenoxy substituent could only be accommodated at the 3-position of the phenoxy group: the tricyclic compound **3** was highly fungicidal, more active in the glasshouse than the diphenyl ether **2**, while the regioisomeric compound **4** was only a very weak fungicide (Figures 6 and 7) (Clough *et al.*, 1994). Incidentally, the three possible regioisomeric *tetracyclic* derivatives of **3** in which a further phenoxy group had been added to the terminal ring were also prepared, but each was less active than **3** (compounds **7-9**, Figure 7).

As we expected, the introduction of a greasy phenoxy substituent to **2** to give **3** resulted in a loss of systemic movement: **3** was simply too lipophilic to be mobile in plant tissue [log P (octanol/water) estimated to be about 5.1; water-solubility (measured) = 0.03ppm]. Thus, solutions of **3**, applied as a root drench to a range of cereal and broad-leaved crops, gave no protection from subsequent fungal infection of

3

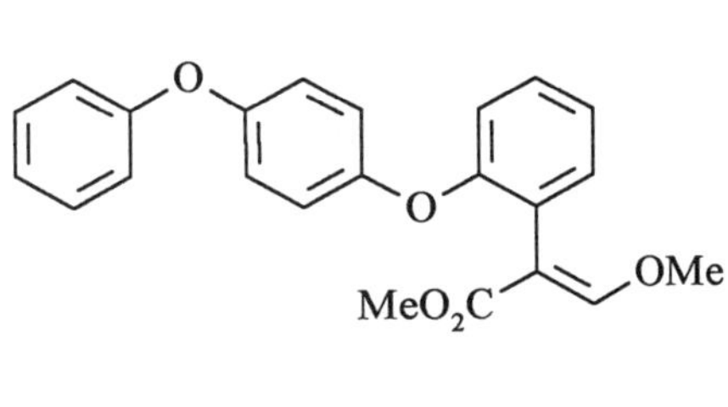

4

Fig. 6. Tricyclic acrylates.

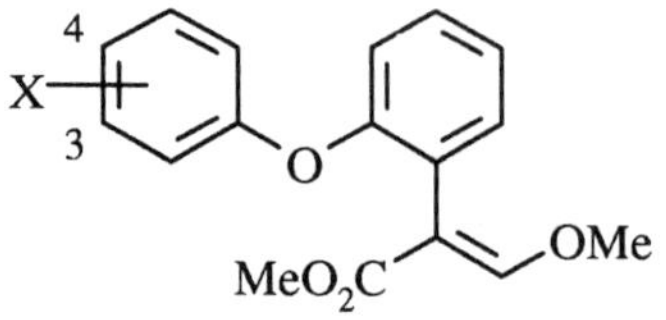

Cpd. No.	X	Log P	PUCCRT		PLASVI		VENTIN		PYRIOR	
			foliar	root	foliar	root	foliar	root	foliar	root
2	H	3.25	*	*	*	-	*	*	**	*
3	3-PhO	5.1^+	**	-	**	-	**	-	**	-
4	4-PhO	5.1^+	-	-	-	-	-	-	-	-
7	3-(2-$PhOC_6H_4O$)	6.9^+	**	-	**	-	*	-	**	-
8	3-(3-$PhOC_6H_4O$)	6.9^+	**	-	*	-	**	-	*	-
9	3-(4-$PhOC_6H_4O$)	6.9^+	-	-	**	-	*	-	-	-

KEY : - Weak activity, * Moderate activity, ** Activity ≥ standard, M = Missing.
Log P measured in octanol/water (+ estimated value); measured and estimated values are rounded to the nearest 0.05 and 0.1 respectively.

PUCCRT = *Puccinia recondita*, brown rust on wheat
PLASVI = *Plasmopara viticola*, downy mildew on vines
VENTIN = *Venturia inaequalis*, apple scab
PYRIOR = *Pyricularia oryzae*, rice blast

The following standards were used:

PUCCRT, Foliar and Root - Cyproconazole
PLASVI, Foliar - Captafol
PLASVI, Root - Metalaxyl
VENTIN, Foliar - Hexaconazole
VENTIN, Root - (5*RS*,6*RS*)-6-Hydroxy-2,2,7,7-tetramethyl-5-(1,2,4-triazol-1-yl)octan-3-one (see de Fraine and Worthington, 1986)
PYRIOR, Foliar - Tricyclazole
PYRIOR, Root - Pyroquilon

Untreated plants had at least 60% disease coverage in all tests.

Fig. 7. Bi-, tri- and tetracyclic acrylates.

the foliage, while application of the same solutions *via* foliar spray provided excellent control (Figure 7). So, while **3** clearly had an excellent shape for binding at cytochrome b, its physical properties prevented it from being redistributed effectively within plants.

As well as preparing derivatives of the diphenyl ether **2**, as outlined above, we synthesised analogues in which one or both of the benzene rings of **2** had been replaced by a heterocycle. For example, phenyl pyridinyl ethers of the types **5** (Anthony *et al.*, 1986, EP 242,070; Anthony *et al.*, 1986, EP 242,081) and **6** (Anthony *et al.*, 1986, EP 243,012) (Figure 8: X and Y = H, halogen, alkyl, *etc.*) were especially active series, while the other regioisomers, in which the ether link is at the 3- or 4-position of a pyridine ring, were, in general, less fungicidal compounds. Like the diphenyl ether **2**, many of these heterocyclic compounds had systemic properties (**5** and **6** in which X = Y = H have octanol/water log P values of 1.80 and 2.30 respectively) (Clough *et al.*, 1992).

As more β-methoxyacrylates were synthesised and tested, it became clear that systemic activity through root uptake generally occurs for compounds which have octanol/water log P's below about 3.5, whereas slightly more lipophilic compounds may still be redistributed in leaves, as indicated by leaf-spotting tests of the type described above. These values represent only rough guidelines, since systemicity is also related to water-solubility (a function of partition coefficient and melting point). Furthermore, there is evidence that the log P-threshold for systemic activity through root uptake varies from one host plant to another. A moderately active compound may be sufficiently active to be fungicidal when applied as a foliar spray, but too weak to show activity when applied as a root drench, even if it has the appropriate physical properties. This is a consequence of the fact that only a small proportion of the root-applied compound is taken up into the plant tissues.

Various ways of modifying the structure of the tricyclic compound **3**, in order to lower its partition coefficient whilst retaining its overall shape, could be envisaged. However, we calculated that only by replacing at least one of the benzene rings in **3** with a heterocycle could the partition coefficient be lowered sufficiently for systemicity to be restored (Clough *et al.*, 1994). We worked on the hypothesis that it should be possible to "fine-tune" lipophilicity and other important physical properties by the careful selection of suitable rings and substituents on them. Clearly, a large number of compounds, with different combinations of heterocycles and substituents, were possible targets for synthesis, and we did not know at the outset which modifications to the tricyclic compound **3** would be accommodated at the active site, or would be beneficial to binding. Therefore, compounds with a representative range of physical properties, especially log P and pKa, were initially chosen for synthesis. Many compounds were prepared and tested in the glasshouse, and the most promising were taken forward for field trials, where the best showed high fungicidal activity and little or no phytotoxicity. The discovery of ICIA5504 was the culmination of this work.

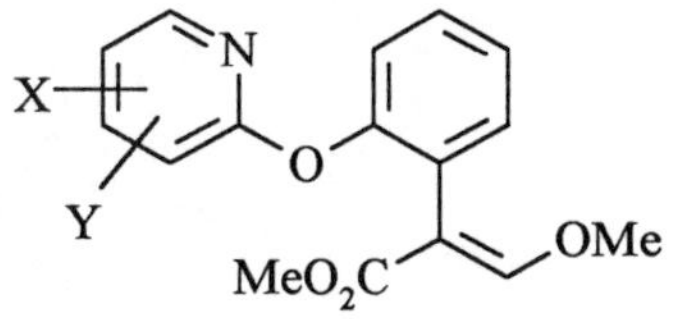

X N Y O MeO_2C OMe

5

X N Y O MeO_2C OMe

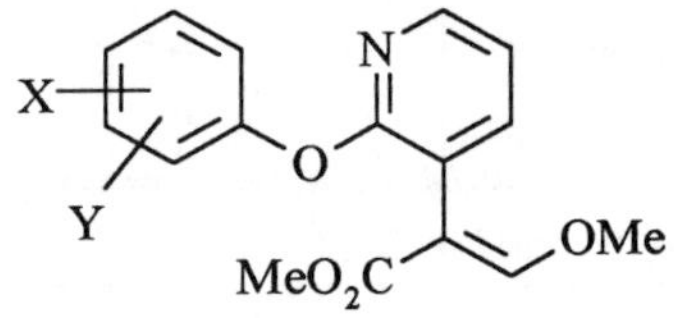

6

Fig. 8. Pyridine analogues of the diphenyl ether **2**.

Fungicidal Activity of ICIA5504 and Related Compounds

The fungicidal activity of a variety of compounds related to ICIA5504 is shown in Figures 9 to 14. All these results are from 24-hour protectant tests in the glasshouse, *i.e.*, tests in which the plants were treated first with the β-

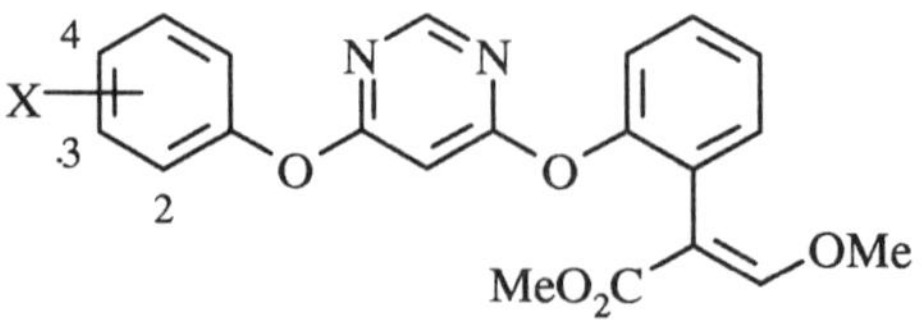

Cpd. No.	X	Log P	PUCCRT foliar	PUCCRT root	PLASVI foliar	PLASVI root	VENTIN foliar	VENTIN root	PYRIOR foliar	PYRIOR root
ICIA5504	2-CN	2.65	**	*	**	**	**	-	**	**
10	3-CN	2.55	**	-	**	-	*	-	-	-
11	4-CN	2.45	**	-	**	-	*	-	-	-
12	2-NO_2	2.90	**	*	**	*	**	*	**	-

KEY : As for Figure 7

Fig. 9. Close analogues of ICIA5504.

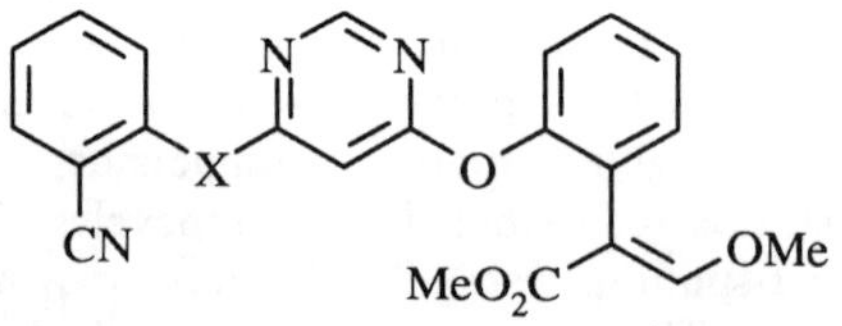

Cpd. No.	X	Log P	PUCCRT foliar	PUCCRT root	PLASVI foliar	PLASVI root	VENTIN foliar	VENTIN root	PYRIOR foliar	PYRIOR root
13	S	2.70	**	-	**	-	*	-	-	-
14	NH	2.40	-	-	**	-	*	-	-	-
15	NMe	2.20	*	-	*	-	-	-	-	-
16	CH_2	2.6^+	-	-	-	-	-	-	*	-
17	CH_2O	2.4^+	**	-	**	-	*	-	**	-
18	SO_2O	2.50	*	-	-	-	-	-	-	-

KEY : As for Figure 7

Fig. 10. Alternative linking groups between rings.

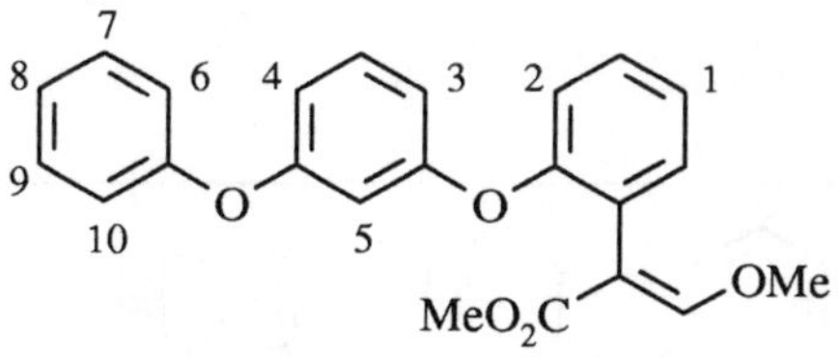

Cpd. No.	Nitrogen Positions	Log P	PUCCRT		PLASVI		VENTIN		PYRIOR	
			foliar	root	foliar	root	foliar	root	foliar	root
19	1-N, 2-N	3.90	-	-	**	-	-	-	-	-
20	3-N, 4-N	2.90	**	-	**	-	**	*	**	**
21	4-N, 5-N	2.70	*	-	*	-	*	-	*	*
22	3-N, 5-N	2.60	**	*	**	-	*	-	**	**
23	6-N, 10-N	2.65	*	-	**	-	**	-	**	**
24	6-N, 8-N	2.95	-	-	**	-	*	-	-	-
25	7-N, 9-N	3.05	-	-	**	-	-	-	*	-

KEY : As for Figure 7

Fig. 11. Regioisomeric pyrimidines.

methoxyacrylate, either as a foliar spray or as a root drench, and then, 24 hours later, with a spore suspension of the fungal pathogen as a foliar spray. The four pathogens shown were selected from those used in screening because they represent each of the main sub-divisions of eumycotic plant pathogenic fungi: *Puccinia recondita* is a Basidiomycete, *Plasmopara viticola* an Oomycete, *Venturia inaequalis* an Ascomycete and *Pyricularia oryzae* a Deuteromycete. Furthermore, the plants which these four pathogens infect - wheat, vines, apples and rice, respectively - are a representative cross-section of crops to which fungicides are applied around the world.

Regioisomers of ICIA5504 in which the cyano group has been moved to alternative positions on the terminal ring were less active than ICIA5504 itself (Figure 9). These compounds usefully illustrate a general point made earlier: the 3-cyano and 4-cyano regioisomers (**10** and **11** respectively) were fungicidal when sprayed on the foliage, but were too weak to express activity when applied as a root drench, even though they had suitable partition coefficients. Other substituents, such as nitro, could replace cyano at the 2-position of the terminal ring (Clough *et al.*, 1989, EP 382,375).

Activity fell to a greater or lesser extent when the oxygen atom between the pyrimidine and cyanophenyl rings of ICIA5504 was replaced with the alternative one- or two-atom linking groups shown in Figure 10. For example, methyleneoxy was a reasonable replacement, while a simple methylene link led to a poorly fungicidal analogue of ICIA5504 (compounds **17** and **16** respectively) (Clough *et al.*, 1990, EP 468,695).

The activity of the acrylate **20**, the analogue of ICIA5504 in which the cyano group has been removed, is compared in Figure 11 with that of six of its regioisomers. This series of compounds has the potential to be isosteric with

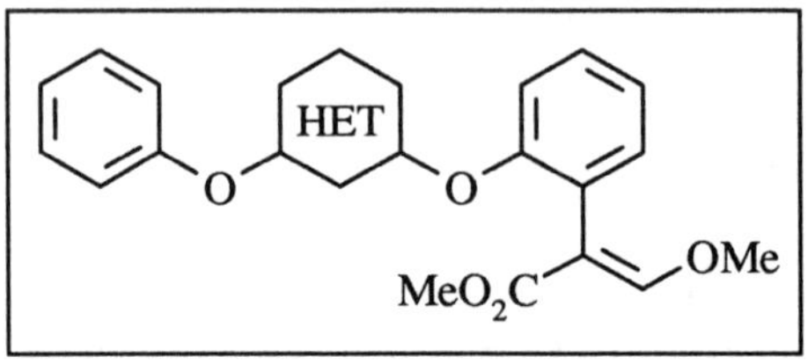

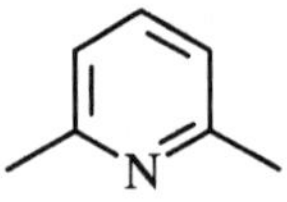

26

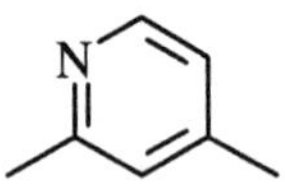

27

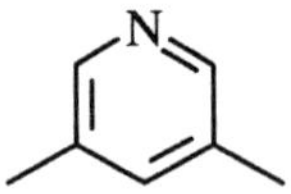

28

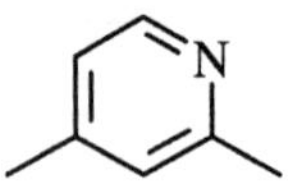

29

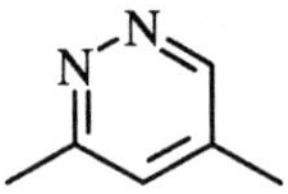

30

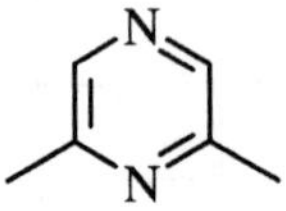

31

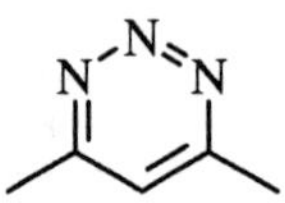

32

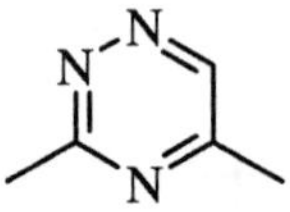

33

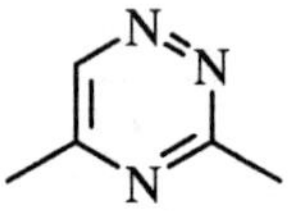

34

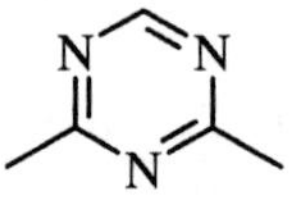

35

Cpd. No.	Log P	PUCCRT		PLASVI		VENTIN		PYRIOR	
		foliar	root	foliar	root	foliar	root	foliar	root
26	3.8^+	*	-	**	-	-	-	*	-
27	3.7^+	*	-	**	-	**	-	*	-
28	4.10	*	-	**	-	**	-	**	-
29	3.65	**	-	**	*	**	-	**	-
30	2.90	-	-	*	-	**	-	-	-
31	3.35	*	-	**	-	*	-	*	-
32	2.1^+	-	-	**	*	*	-	-	-
33	2.1^+	-	-	*	-	-	-	-	-
34	2.10	-	-	**	-	*	-	-	-
35	2.10	**	*	**	-	**	*	-	-

KEY : As for Figure 7

Fig. 12. Pyridines, diazines and triazines in the central ring position.

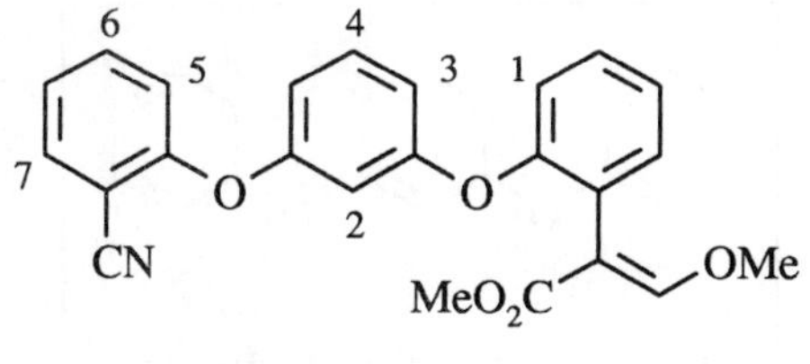

Cpd. No.	Nitrogen Positions	Log P	PUCCRT		PLASVI		VENTIN		PYRIOR	
			foliar	root	foliar	root	foliar	root	foliar	root
36	None	4.7$^+$	**	-	**	-	**	-	**	-
37	N-1	3.70	*	-	**	-	**	-	*	-
38	N-2	3.45	*	-	**	-	M	M	-	-
39	N-3	3.00	*	-	**	-	*	-	**	-
40	N-4	3.8$^+$	*	M	**	-	*	*	M	-
41	N-5	3.50	**	-	**	-	**	-	**	**
42	N-6	3.1$^+$	-	-	*	-	-	-	-	-
43	N-7	3.1$^+$	-	-	**	-	-	-	**	-

KEY : As for Figure 7

Fig. 13. Benzene and pyridine analogues of ICIA5504.

the original tricyclic compound **3** although, of course, they do not have superimposable lowest energy conformations for electronic reasons. It is clear that the binding site is very sensitive to the position and orientation of the pyrimidine ring: some of the regioisomers, such as **22** and **23**, had good activity, while others, such as **24** and **25**, did not (Anthony *et al.*, 1987; Clough *et al.*, 1989, EP 382,375).

The activity of other analogues of the acrylate **20** in which the pyrimidine ring was replaced by a pyridine (Anthony *et al.*, 1986, EP 242,081), pyrazine, pyridazine (Clough and Godfrey, 1986), 1,2,3-triazine (Worthington *et al.*, 1991), 1,2,4-triazine or 1,3,5-triazine ring (Clough *et al.*, 1989, EP 405,782) are shown in Figure 12. Again, these compounds have the potential to be isosteric with the tricyclic compound **3**. While some analogues, such as the 2,4-linked pyridine **29** and the 1,3,5-triazine **35**, were highly active, others, such as the 1,2,4-triazines **33** and **34** and the pyrazine **31**, quite unpredictably, were relatively poor compounds. The activity of the pyridines was of sufficient interest to justify the synthesis of several of the possible analogues incorporating a cyano group (Figure 13) (Anthony *et al.*, 1986, EP 242,081; Anthony *et al.*, 1987; Clough *et al.*, 1987).

Finally, Figure 14 shows results for analogues of ICIA5504 in which the β-methoxyacrylate toxophore has been replaced by an alternative, isosteric group. Compounds which incorporate the methoxyiminoacetate (**44**, Clough *et al.*, 1990, EP 468,684) or the *N*-methyl methoxyiminoacetamide toxophores (**45**, Clough *et al.*, 1991) had good activity. By contrast, analogues in which the β-methoxy group of the toxophore has been replaced with a methylthio or an ethyl group (**46** and **47** respectively) were much weaker fungicides.

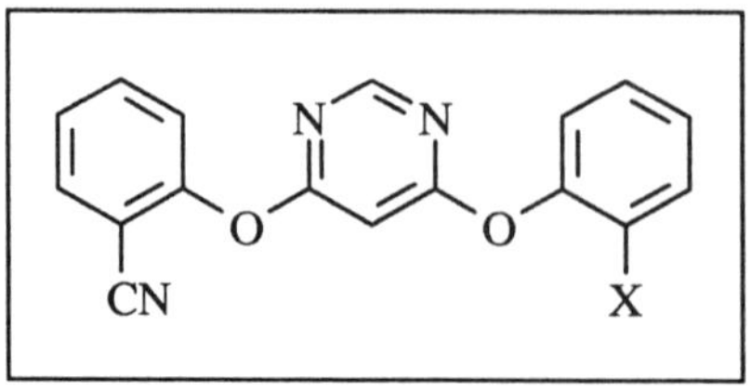

X =

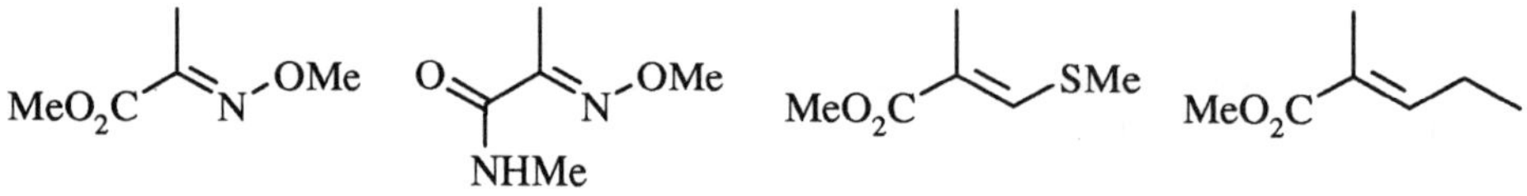

44 45 46 47

Cpd. No.	Log P	PUCCRT		PLASVI		VENTIN		PYRIOR	
		foliar	root	foliar	root	foliar	root	foliar	root
44	2.50	*	-	**	-	**	-	**	-
45	1.90	**	-	**	-	*	**	**	**
46	3.30	-	-	*	-	-	-	-	-
47	3.65	-	-	-	-	-	-	-	-

KEY : As for Figure 7

Fig. 14. Analogues of ICIA5504 with alternative toxophores.

ICIA5504

ICIA5504 was prepared in the laboratory by the steps shown in Figure 15 (Clough *et al.*, 1989, EP 382,375). This proved to be an excellent sequence, providing the divergent intermediates **48** and **49** for the synthesis of a variety of related compounds (Clough *et al.*, 1994). ICIA5504 is a white crystalline solid with a melting point of 118-9 °C. Single crystal X-ray analysis (Figure 16) showed that the molecule adopts a strongly twisted conformation in the crystal lattice, with the three rings lying in planes which are roughly orthogonal to each other, and the cyano group pointing towards the toxophore. The pronounced twist between the almost planar β-methoxyacrylate group and the ring to which it is attached (torsion angle between least squares mean planes = 63°) is a feature which we had observed for several related compounds, such as the stilbene **1** (Beautement *et al.*, 1991).

ICIA5504 has eradicant, protectant, translaminar and systemic properties, giving it the potential for use as a foliar, paddy water, seed or soil treatment (Godwin *et al.*, 1992). Its versatility and broad spectrum of activity have been demonstrated in field trials against a wide range of economically important crop diseases, including members of all four major sub-divisions of eumycotic plant pathogenic fungi. For example, as a foliar treatment on wheat and

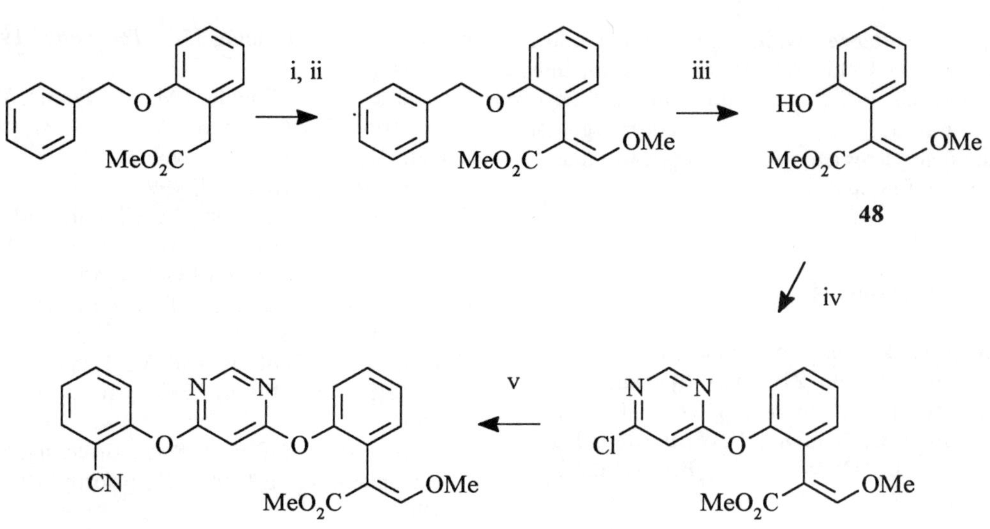

48

ICIA5504

49

Reagents: i, HCO_2Me, NaH, DMF, then H_3O^+; ii, Me_2SO_4, K_2CO_3, DMF; iii, H_2, 5% Pd-C, EtOAc; iv, 4,6-dichloropyrimidine, K_2CO_3, DMF; v, 2-cyanophenol, K_2CO_3, catalytic CuCl, DMF.

Fig. 15. Laboratory synthesis of ICIA5504.

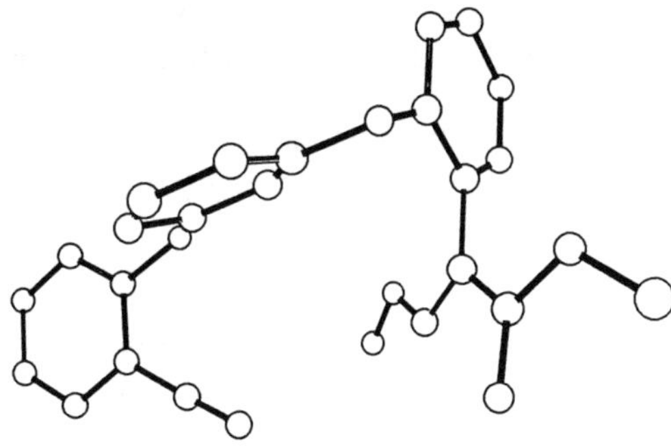

Fig. 16. Single crystal X-ray structure of ICIA5504.

barley, it has shown good control of a range of diseases which attack the stem-base, leaves and ears, and it has demonstrated an impressive persistence of fungicidal effect, with the result that there is longer retention of green leaf tissue and significant benefits in yield. ICIA5504 controls both of the important fungal pathogens of rice, namely rice blast and sheath blight, and it may be applied either as a foliar spray or as a granule to paddy water. Similarly, its systemic properties allow it to be applied as a seed treatment on barley for the control of the important foliar pathogen powdery mildew. Furthermore, ICIA5504 has shown control of a wide range of fungal diseases of vines and other fruits.

As expected on the basis of its novel mode of action, ICIA5504 controls the growth of strains of fungi which are resistant to 14-demethylase inhibitors, phenylamides, dicarboximides or benzimidazoles.

Finally, of critical importance is the low acute oral toxicity of ICIA5504 ($LD_{50} > 5000$ mg/kg in the rat, male and female).

It is anticipated that, following commercial development, ICIA5504 will make a major global impact in the control of fungal pathogens in agriculture. Research on the β-methoxyacrylate fungicides has now become of major importance within the agrochemical industry: by October 1994, twenty companies and research institutes had published a total of 215 patent applications claiming β-methoxyacrylates and related compounds, mainly for use as fungicides.

Acknowledgements

We thank our colleagues in ZENECA Agrochemicals and ZENECA Specialties who have participated in the ICIA5504 project, especially B.C. Baldwin, D.W. Bartlett, P.J.V. Cleare, P.J. Crowley, I. Ferguson, M.J. Houghton, M.G. Hutchings, G.J. Sexton and T.E. Wiggins. We are grateful to D.J. Williams, Imperial College of Science, Technology and Medicine, London, for determination of the X-ray crystal structure of ICIA5504.

References

Ammermann, E.; Lorenz, G.; Schelberger, K.; Wenderoth, B.; Sauter, H.; Rentzea, C. *Brighton Crop Prot. Conf.: Pests and Diseases* - **1992**, Vol. 1, British Crop Protection Council, Farnham, U.K., **1992**, p. 403.

Anke, T.; Steglich, W. In *Biologically Active Molecules, Identification, Characterization and Synthesis*, Schlunegger, U.P., Ed., Springer-Verlag, Berlin and Heidelberg, **1989**, p. 9.

Anthony, V.M.; Clough, J.M.; de Fraine, P.J.; Godfrey, C.R.A. EP 242,070, priority 17 April **1986**.

Anthony, V.M.; Clough, J.M.; de Fraine, P.J.; Godfrey, C.R.A.; Ferguson, I.; Crowley, P.J.; Hutchings, M.G. EP 242,081, priority 17 April **1986**.

Anthony, V.M.; Clough, J.M.; de Fraine, P.J.; Godfrey, C.R.A.; Crowley, P.J.; Anderton, K. EP 243,012, priority 17 April **1986**.

Anthony, V.M.; Heaney, S.P.; Beautement, K.; Clough, J.M.; Crowley, P.J.; Godfrey, C.R.A.; de Fraine, P.J.; Buckley, A.J.; Hutchings, M.G.; Ferguson, I. EP 307,103, priority 9 Sept. **1987**.

Beautement, K.; Clough, J.M. *Tet. Letts.*, **1987**, *28*, 475.

Beautement, K.; Clough, J.M.; de Fraine, P.J.; Godfrey, C.R.A. *Pestic. Sci.*, **1991**, *31*, 499.

Brandt, U., Schägger, H.; von Jagow, G. *Eur. J. Biochem.*, **1988**, *173*, 499.

Bushell, M.J.; Beautement, K.; Clough, J.M.; de Fraine, P.J.; Anthony, V.M.; Godfrey, C.R.A. EP 178,826, priority 19 Oct. **1984**.

Clough, J.M.; Godfrey, C.R.A. EP 260,794, priority 20 Aug. **1986**.

Clough, J.M.; Godfrey, C.R.A.; Heaney, S.P.; Anderton, K. EP 312,243, priority 15 Oct. **1987**.

Clough, J.M.; Godfrey, C.R.A.; Streeting, I.T.; Cheetham, R. EP 382,375, priority 10 Feb. **1989**.

Clough, J.M.; Streeting, I.T.; Godfrey, C.R.A. EP 405,782, priority 28 June **1989**.

Clough, J.M.; Godfrey, C.R.A.; Streeting, I.T.; Cheetham, R.; de Fraine, P.J. EP 468,684, priority 27 July **1990**.

Clough, J.M.; Godfrey, C.R.A.; Streeting, I.T.; Cheetham, R.; de Fraine, P.J.; Bartholomew, D.; Eshelby, J.J. EP 468,695, priority 27 July **1990**.

Clough, J.M.; Godfrey, C.R.A.; de Fraine, P.J. GB 2,253,624, priority 30 Jan. **1991**.

Clough, J.M.; de Fraine, P.J.; Fraser, T.E.M.; Godfrey, C.R.A. In *Synthesis and Chemistry of Agrochemicals III*, Baker, D.R.; Fenyes, J.G.; Steffens, J.J., Eds., ACS Symposium Series No. 504, American Chemical Society, Washington, DC, **1992**, p. 372.

Clough, J.M. *Nat. Prod. Reports*, **1993**, *10*, 565.

Clough, J.M.; Evans, D.A.; de Fraine, P.J.; Fraser, T.E.M.; Godfrey, C.R.A.; Youle, D. In *Natural and Engineered Pest Management Agents*, Hedin, P.A.; Menn, J. J.; Hollingworth, R.M., Eds., ACS Symposium Series No. 551, American Chemical Society, Washington, DC, **1994**, p. 37.

de Fraine, P.J.; Worthington, P.A. *Pestic. Sci.*, **1986**, *17*, 343.

Godwin, J.R.; Anthony, V.M.; Clough, J.M.; Godfrey, C.R.A. *Brighton Crop Prot. Conf.: Pests and Diseases* - **1992**, Vol. 1, British Crop Protection Council, Farnham, U.K., **1992**, p. 435.

Hayase, Y.; Kataoka, T.; Takenaka, H.; Ichinari,

M.; Masuko, M.; Takahashi, T.; Tanimoto, N. EP 398,692, priority 17 May **1989**.

Masuko, M.; Niikawa, M.; Kataoka, T.; Ichinari, M.; Takenaka, H.; Hayase, Y.; Hayashi, Y.; Takeda, R. 6th International Congress of Plant Pathol., 28 July-6 Aug. **1993**, Montréal, Canada, Abstract No. 3.7.16.

Wood Mackenzie Consultants Limited, Edinburgh and London, *Agrochemical Service*, May **1994**.

Worthington, P.A.; Clough, J.M.; Godfrey, C.R.A.; Streeting, I.T. GB 2,255,092, priority 23 April **1991**.

DELIVERY

Dispersions and Dispersible Systems

Tharwat F. Tadros, Zeneca Agrochemicals (Formerly part of the ICI group), Jealott's Hill Research Station, Bracknell, Berkshire, U.K.

This overview deals with the fundamentals of dispersible systems, their properties as well as their applications. Dispersions encompass solid/liquid dispersions (suspensions, SC's), emulsion concentrates (EW's), mixtures of SC and EW (SE's), micoemulsions (ME's), micoencapsulated systems, multiple emulsions and gels. The SC's are technically and fundamentally well advanced, whereas EW's require more research to enable one to produce stable and robust systems. The ME's offer the advantage of being thermodynamically stable and may offer enhancement in the biological efficacy of the agrochemicals. They are not yet widely used, since they still require great deal of research to formulate them. Microencapsulated systems are now widely used, since they offer many advantages such as a reduction in operating hazards, controlled release and protection of many chemicals against degradation. Recently gels have been suggested as a method of formulation of emulsifiable concentrates and aqueous solutions.

Recently, there has been considerable interest in "solid" formulations that can be easily dispersed into water. Water dispersible granules (WG's) are by far the most attractive, since they overcome the problem of dustiness encountered with wettable powders. They can also be packed in water soluble bags, thus reducing operator hazards. Research is still required to reduce the cost of production, to make them easily dispersible in spray tanks and to be able to incorporate adjuvants such as surfactants in the system. Other "solid" formulations include tablets and "dried" microcapsules.

A comparison between "liquid" and "solid" formulation is given at the end of the overview, highlighting the advantages and disadvantages of both systems.

Agrochemicals formulations cover a wide range of systems that are prepared to suit a specific application (Tadros, 1994). These formulations need to satisfy a number of criteria such as optimum biological efficacy, ease of application with minimum hazard to the operator, convenient and cheap production methods and a suitable procedure for pack disposal. These criteria put some constraints on the use of the early formulations such as wettable powders (WP's) and emulsifiable concentrates (EC's), both of which may cause problems on application such as dustiness with WP's and harmful volatile solvents with EC's. For these reasons, suspensions concentrates (SC's) and emulsion concentrates (EW's) have been introduced as alternatives. The basic principles for formulating SC's, which are well established, will be summarized in the first part of this overview. This is followed by a description of the stability/instability of EW's, which still requires fundamental research to enable one to produce adequate formulations. Recently, mixtures of SC's and EW's were introduced and these proved to be difficult to formulate due to the interactions between the particles of the SC and the droplets of the EW. The major problems encountered with SE's and possible solutions will be given in the third section. Microemulsions are also potentially useful formulations for agrochemicals. As we will see later, microemulsions are thermodynamically stable systems which offer a number of advantages such as enhanced biological efficacy. Microencapsulated products are also beneficial for controlled release and they also may be safer to the operator. Other dispersions which are potentially useful are multiple emulsions and gels and these will be briefly described in this overview. It should be mentioned that dispersions still constitute significant proportion of agrochemical formulations on the market. Apart from their popularity with the farmers due to their ease of application, such dispersions enable the formulation chemist to incorporate various adjuvants that enhance the biological

1054–7487/95/0076$12.00/0

efficacy (Tadros, 1994). However, in recent years there has been continuous debate about disperse systems for a number of reasons. Firstly, most disperse systems (with the exception of microemulsions) are thermodynamically unstable and they may have limited shelf life. Secondly, dispersions may require accurate control of the manufacturing process in order to prepare reproducible products. Thirdly, the formulation of complex dispersions such as SE's is not easy and requires a great deal of knowledge of colloid and interface science. Fourthly, the problem of pack disposal of the formulation has not been resolved. Legislation requires that the empty container should not contain a significant amount of the agrochemical (in many cases a concentration as low as 0.1% is required). Rinsing and cleaning of the empty containers is not straightforward, particularly when one is dealing with concentrated dispersions with high viscosity. Also, with many SC's the particulate solid may adhere to the walls of the container, particularly when drying occurs. This makes the cleaning process more difficult. For these reasons, many agrochemical companies have invested a great deal in research and development of "solid" systems, with the hope of overcoming many of the above mentioned problems. However, as we will see later in the section on solids, these hopes have not yet materialized, since the problems encountered with dispersions have been replaced by other problems inherent to solid formulations. For example, the perception that "solids" are inherently more stable than "liquid" formulations is not always true. Many "solids" such as water dispersible grains may undergo "compaction" and stronger particle adhesion on storage, which makes the dispersibility of these systems more difficult after prolonged storage. Although initially, it was thought that the problem of pack disposal was not an issue with solids, since these may be packed in water soluble bags, it was later realised that these bags must be supported by hard board materials, which may not be burnt due to some contamination by the chemical. One of the most crucial problems with "solid" formulations is the difficulty of incorporating the necessary adjuvants which may be liquid or semi-solid surfactants. In addition, solid formulations are generally more expensive to produce compared to dispersions and they require investment in costly capital equipments. These point will be addressed in the section on solid formulations.

Suspension Concentrates (SC's)

Suspension concentrates (SC's) are easy to prepare. The solid (powder or paste) is dispersed in an aqueous solution of a surfactant or polymer using a high speed stirrer, followed by a wet milling process (bead mills are the most commonly used). The surfactant or polymer (sometimes referred to as the dispersing agent) enables the wetting of the powder, its dispersion into single particles and finally its grinding (comminution). The dispersing agent should provide sufficient stability in the colloid sense (Tadros, 1980,1989,1990). In other words, it should prevent particle aggregation during dispersion and milling and also during storage. Several dispersing agents are used, the most common are those consisting of sulphonated naphthalene formaldehyde condensates and lignosulphonates. These dispersants are of high molecular weight and they contain ionic groups on the chain (polyelectrolyte type). They provide stability by a combination of double layer repulsion (due to the presence of charges) and steric repulsion (due to the dangling chains at the solid/liquid interface). More recently, dispersants based on block and graft nonionic polymers have been introduced, which are suitable for preparation of highly concentrated SC's (Heath et al,1984). An example of these dispersants (marketed by ICI Surfactants) is Atlox 4913, which consists of a backbone of methylmethacrylate (the "anchoring" part of the chain that is strongly adsorbed on the particle surface) and several chains of poly(ethylene oxide) (PEO), which provides the stabilising components. Since PEO is strongly solvated with the medium (water), when two particles covered with these chains approach to a separation distance that is smaller than twice the adsorbed layer thickness, overlap of the chains is prohibited as a result of the increase in osmotic pressure in the overlap region. This results in strong repulsion between the particles. This is referred to as the mixing interaction term, G_{mix}. Another repulsive term arises from the loss in configurational entropy between the layers on overlap, this is referred to as the elastic term, G_{el}. This is illustrated in Fig.1 which shows the variation of G_A (the van der Waals attraction), G_{mix}, G_{el} and G_T (total energy of interaction) with distance of separation between the particles. Since these PEO chains are not very long (few nms), highly concentrated suspensions (up to a volume fraction of ~ 0.6) can be obtained without increase in the viscosity of the suspension. These highly concentrated suspensions are required in some applications such as seed dressing. Once a suspension that is colloidally stable is prepared, a suspending agent (sometimes referred to as an antisettling agent) should be added in the continuous phase to prevent settling and/or claying of the particles (Tadros, 1980,1989,1990). The antisettling agent usually produces a "gel" network in the continuous phase. The most commonly used antisettling systems in SC's are polysaccharides (mostly xanthan gum), swellable clays (such as sodium montmorillonite), finely divided silica or mixtures of these. These "gels" have a very high viscosity at low shear rates, but a low viscosity at relatively high shear rates. The high viscosity at low shear (sometimes referred to as the residual viscosity) can reach very high values (of the order of 10^4 Pas or higher) thus preventing any sedimentation of the particles. As the shear rate is increased the viscosity decreases rapidly and it reaches reasonably low values in the high shear limit (of the order of 0.1 Pas or lower.

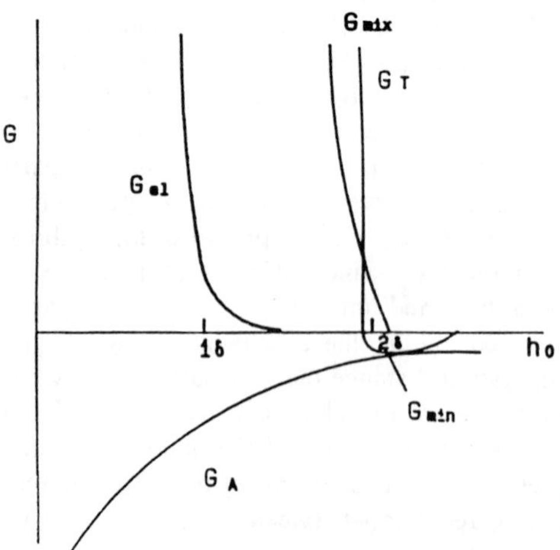

Fig. 1. Variation of G_A, G_{mix}, G_{el}, and G_T with interparticle separation.

This is precisely what is required for an SC, i.e. the system should have very high viscosity at low shear to prevent sedimentation, but once the shear rate is increased by pouring the formulation, the viscosity drops rapidly allowing easy flow and dispersion in the spray tank. The "gel" in the continuous phase should also have enough "elasticity" to prevent separation of the formulation on storage. Many SC's separate on storage leaving a clear liquid layer at the top (a phenomenon sometimes referred to as syneresis). Although technically this may not be a problem, since the suspension can be redispersed by gently shaking the container, it is not acceptable to the customer who prefers to have a homogenous system.

Another instability problem with SC's is that of Ostwald ripening or crystal growth. This results from the difference in solubility between the small and large particles (Tadros, 1980,1989,1990). Most pesticides have some solubility in the continuous phase. The smaller particles will have higher solubility (chemical potential) than the larger particles and hence on storage, the smaller particles dissolve and become deposited on the larger ones. Apart from the effect on the physical stability of the suspension (whereby settling may be enhanced), the production of larger particles may affect the biological efficacy of the agrochemical. In some cases, surfactant micelles may enhance crystal growth as a result of solubilization of the chemical (Tadros, 1980). In this case, it is preferable to use polymeric agents as dispersants. These polymers may also reduce crystal growth by strong adsorption on the particle surfaces.

The long term physical stability of the SC can be assessed using various physical techniques. Flocculation and Ostwald ripening may be assessed using particle size analysis as a function of time. This requires diluting the SC and one should be very careful in the dilution process because this may destroy the flocs which have formed. The particle size distribution may be measured using microscopy (combined with image analysis), a Coulter Counter or various light diffraction (combined with light scattering) techniques. The bulk properties of the SC can be investigated using rheological techniques (Tadros, 1980,1989,1990). Steady state shear stress-shear rate measurements allow one to obtain the yield value and the plastic viscosity.

The yield value may give an indication of the stability against settling and/or claying. The high shear viscosity provides information on the dispersion of the suspension on dilution. For more quantitative assessment of the physical stability of SC's, low deformation measurements are required. Two types of measurements may be applied. The first is constant stress (creep) measurements which allows one to obtain the residual (zero shear) viscosity and the critical stress above which flow occurs. Both may be related to the long term stability of the suspension. The second technique applies oscillation (dynamic measurements). In this case, a sinusoidal strain is applied on the system and the stress and strain sine waves are followed simultaneously. For a viscoelastic system (such as that of an SC), the stress oscillates with the same frequency as the strain, but is out of phase with it. From the amplitudes of the stress and strain, one can obtain the complex modulus G^* (which is simply the ratio of the amplitudes of stress and strain). The storage modulus G' (the elastic component of the complex modulus) is simply the product of G^* and the cosine of the phase angle shift. The loss modulus G" (the viscous component of the complex modulus) is the product of G^* and the sine of the phase angle shift. G' is a useful parameter that may be used to predict separation in SC's (Tadros,1990). By combining creep and oscillatory measurements, one may be able to predict the long term physical stability of suspensions.

Emulsion Concentrates (EW's)

As mentioned above EW's are not as well understood as SC's. These systems are prepared by emulsification of the oil (which may be the active ingredient or its mixture with an oil such as liquid paraffin or an aromatic solvent) in an aqueous solution of a surfactant (referred to as the emulsifier). For this purpose, a high speed mixer or homogenizer may be used. The emulsifier should satisfy a number of roles (Walstra,1989). Firstly, it should reduce the oil/water interfacial tension γ to reduce the energy required for emulsification and produce small droplets. Secondly, it should provide an interfacial tension gradient (Gibbs elasticity $\varepsilon = d\gamma/d\ln A$, where A is the interfacial area) to reduce coalescence during emulsification. As a result of the interfacial tension gradient, molecules will diffuse from the bulk to the interface carrying liquid with them (this effect is referred to as the Marangoni effect). The Gibbs-Marangoni effect prevents coalescence during emulsification. Thirdly, the surfactant should aid disruption of the liquid (which may produce threads) into small droplets.

Once an emulsion is prepared (usually one aims at an average droplet size of the order of 1 μm), it is essential to stabilize the system against the various possible breakdown processes (Tadros and Vincent, 1983). The latter are schematically represented in Fig.2. Creaming or sedimentation may be prevented by the addition of thickeners such as polysaccharides (swellable clays may cause problems with emulsions) in the same way as discussed for SC's. Flocculation is also prevented by applying the same principles used for SC's (electrostatic and/or steric repulsion). Ostwald ripening in emulsions may occur if the oil phase has some solubility in the continuous medium. In some cases, by proper choice of the emulsifier, Ostwald ripening may be significantly reduced. The emulsifier reduces the process by two main effects: reduction of the interfacial tension and increasing the Gibbs elasticity (Vincent, 1984). Another practical method for reducing Ostwald ripening is to incorporate a small proportion of a highly insoluble oil in the droplets (Davis and Smith,1974).

The most difficult and least understood instability process is that of coalescence. It arises from the thinning and disruption of the liquid film between the droplets. The surface forces between oil droplets in an emulsion play an important role in this process. It has also been claimed that the properties of the interfacial film such as its dilational elasticity and interfacial viscosity may play a crucial role (Tadros,1994). However, there is not sufficient experimental evidence that supports the role of interfacial rheology in stability against coalescence and research is required to explain the coalescence process, which may only occur after a long induction period. These complications prohibited the development of EW's in many cases, inspite of their obvious attraction. Apart from their suitability of

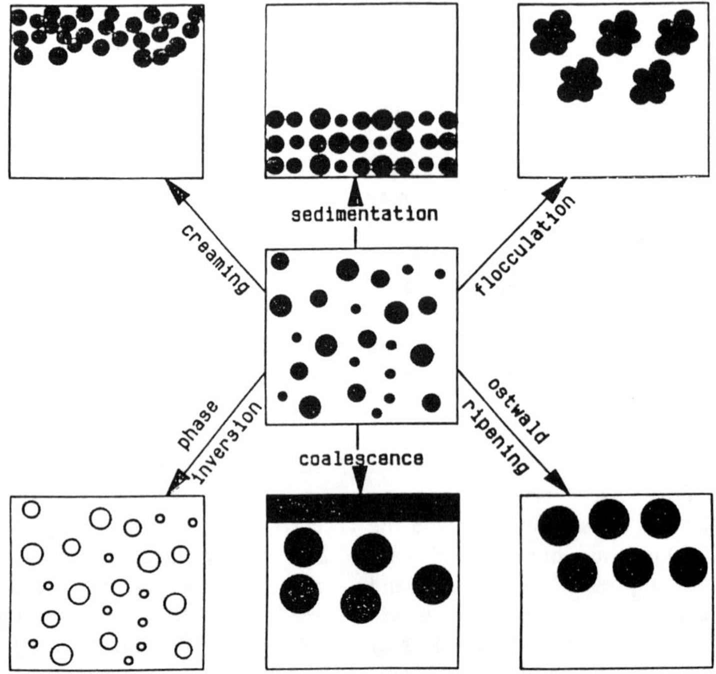

Fig. 2. Schematic representation of the breakdown processes in emulsions.

application, EW's are attractive since they have high flash points, are safer to the operator and in some cases can enhance the activity of the agrochemical.

Inspite of the lack of understanding of coalescence, a number of guide lines may be used to stabilize the emulsion. Generally speaking, mixed surfactant films (such as a mixture of Span and Tween) enhance stability against coalescence (Tadros,1986). This enhancement may be due to the increased elasticity of the film and slower diffusion of the molecules from the interface. In some cases, mixed surfactants may produce liquid crystalline phases which are more effective for stabilization of the system. Macromolecular surfactants can also be effective in reducing coalescence, again as a result of producing rigid and elastic films at the interface.

The last instability process in emulsions is that of phase inversion which may occur as a result of several factors. The most obvious reason for phase inversion is when the oil phase volume fraction exceeds a critical value. Alternatively phase inversion may occur if the properties of the surfactant change on storage. For example if an emulsion is prepared using an optimum high HLB number surfactant and the temperature of the system increases above a critical value, inversion may occur since at higher temperatures, the effective HLB number decreases and becomes unsuitable for formation of an o/w emulsion.

Various physical methods may be applied to investigate the properties of emulsions such as interfacial tension and interfacial rheology measurements, droplet size analysis as a function of time and rheological techniques (Tadros, 1994).

Suspoemulsions (SE's)

These are systems consisting of a mixture of a suspension and emulsion. At first sight these systems may appear to be easy to formulate. Only seldomly is a stable SE produced by mixing a stable suspension and a stable emulsion. This is because the interactions between the particles and the droplets lead to a number of phenomena, which may not occur with the individual SC and EW. For example, the emulsifier may displace the dispersant even though the latter is of higher molecular weight. This phenomenon is well known when polymer-surfactant mixtures are used in suspensions. The lower molecular weight surfactant can diffuse rapidly to the solid/solution interface and become adsorbed on the surface initially replacing one of the polymer segments on the surface. This process continues until most of the dispersant molecules are displaced. The adsorbed surfactant molecules may not be efficient in stabilizing the particles and flocculation of the suspension occurs. In addition, as the emulsion becomes deficient of the emulsifier and the droplets approach the particles, coalescence of these droplets occur. In some cases, one observes under the microscope a large number of particles surrounding a large coalesced drop. Alternatively heteroflocculation between the particles and droplets may occur and the flocs produced may reach a large size. If the oil of the emulsion is able to wet the particle surface, this heteroflocculation process may be enhanced and finally a heavily flocculated system results. If the solid particles have some solubility in the oil phase, these particles may enter the oil drops and recrystallize into large crystals. These phenomena have been observed with many suspeoemulsions. The actual mechanism is far from being fully understood.

A very effective method of stabilizing SE's is to use powerful dispersants and emulsifiers such that transfer between the two materials become improbable. For example, we have found (Tadros,1988) that with some suspoemulsions addition of the "comb" graft copolymer Atlox 4913 (from ICI Surfactants) is effective in preventing heteroflocculation. Presumably by strong adsorption on the particles, the latter became coated with a layer of PEO thus preventing interaction with the emulsion droplets which were also stabilized by a surfactant containing PEO chains. In some cases, this polymer also reduced crystal growth in the SE. However, it should be mentioned that these effects are specific to a particular system and in every case a suspoemulsion is formulated one has to discover the best dispersant and emulsifier.

Microemulsions (ME's)

This is a very attractive type of formulation, since the system is a single optically isotropic and thermodynamically stable dispersion of oil, water and amphiphile (one or more surfactants) (Danielsson and Lindman, 1981). The origin of thermodynamic stability arises from the low interfacial energy of the system which is outweighed by the negative entropy of dispersion. In this case the free energy of formation of the system becomes zero or negative. This ultra-low interfacial tension is usually achieved by the use of two or more emulsifiers. This can be understood from the effect of surfactant concentration, C, and nature on the interfacial tension γ between oil and water. Addition of surfactant to the aqueous or oil phase causes a gradual lowering of γ, reaching a limiting value at the critical micelle concentration (cmc). Any further increase in C above the cmc causes little or no further decrease in γ. The limiting γ reached with most surfactants is seldom lower than 0.1 mNm^{-1}, which is not low enough for microemulsion formation (that requires values of γ of 10^{-2} mNm^{-1} or lower). However, if a surfactant mixture is used with one component predominantly water soluble such as sodium dodecyl sulphate and one predominantly oil soluble such as a medium chain alcohol (usually referred to as the cosurfactant), the limiting γ value can reach values lower than 10^{-2} mNm^{-1} or even becomes transiently negative (Overbeek et al, 1984). In the latter case, the interface expands spontaneously adsorbing all surfactant molecules till a small positive γ is reached. This behaviour is schematically shown in Fig.3. As is clearly shown, addition of the cosurfactant causes a shift in the γ - log C curve of the surfactant to lower values and the cmc is reduced.

Microemulsions offer a number of advantages over EW's for the following reasons (Tadros, 1994). Once the composition of the microemulsion is identified, the system is prepared by simple mixing of all the components without the need of any appreciable shear. Due to their thermodynamic stability, these formulations undergo no separation or breakdown on storage (within a certain temperature range depending on the system).

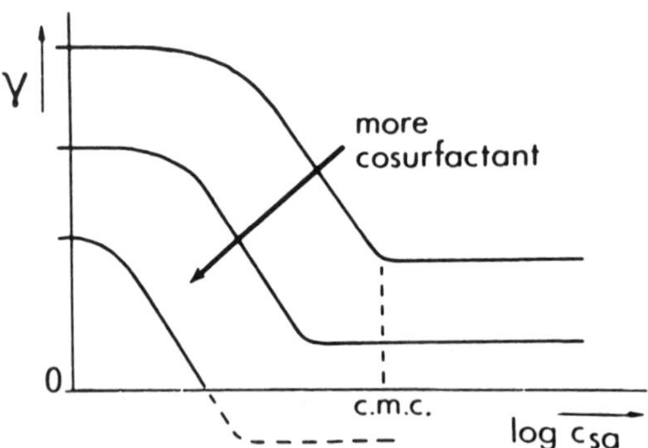

Fig. 3. Influence of addition of cosurfactant on the γ/log C curve of a surfactant.

The low viscosity of the system ensures their ease of pourability and they leave little residue in the container. Another main attraction of microemulsions is their possible enhancement of biological efficacy of many agrochemicals. This is due to solubilization of the chemical by the microemulsion droplets. Two effects may be considered which are complimentary. The first effect is due to enhanced penetration of the chemical through fine pores as a result of the low interfacial tension. The second effect is due to enhancement of the flux due to diffusion, as a result of solubilization (Tadros, 1994). At present microemulsions are prepared by simple trial and error methods, due to the lack of understanding of the various interactions that are necessary to produce the system. In addition, most microemulsion systems are produced within a narrow temperature range which may not be sufficient for agrochemical formulations.

Microencapsulated Dispersions

The most common procedure for encapsulation of agrochemicals is by application of interfacial polymerization, whereby two polyfunctional monomers A and B are chosen, one being oil soluble and the other water soluble. Together these monomers should be able to undergo polycondensation or polyaddition (Kondo,1979). If the material to be encapsulated is oil soluble, oil-dispersible or an oil itself, an o/w emulsion is prepared (using a stabilizer such as polyvinyl alcohol) and the hydrophobic monomer is dissolved in the oil phase. On the other hand, if the material to be encapsulated is water soluble, a w/o emulsion is prepared with the hydrophilic monomer dissolved in the internal water phase. Examples of oil soluble monomers are polybasic acid chloride, bis-haloformate and polyisocyanate, whereas water soluble monomers could be polyamine, or glycol polyphenol. Thus, a capsule of polyamide, polyurethane or polyurea may be formed. The ultimate capsule size is determined by the size of the dispersed droplets and this size is determined by the nature and concentration of the emulsifier. The thickness of the capsule wall can also be controlled by controlling the concentration of the reactants and the process.

The capsule dispersion may also contain an antisettling system to reduce sedimentation. The final formulation is similar in its properties to an SC. However, it offers a number of advantages such as reduced acute mammalian toxicity (which may be a problem with many insecticides), longer residual efficacy, reduced fish toxicity and reduced phytotoxicity. Additional advantages may be protection of chemical against breakdown, reduced volatility, safer for operator, improved compatibility with other formulations, reduced ground water contamination and in some cases reduced pesticide application rate.

Multiple Emulsions

Multiple emulsions (w/o/w or o/w/o) are ideal systems for application in agrochemicals for the following reasons. One may have three active ingredients in one formulation and can incorporate additives in three compartments. They can be usefully applied for sustained release by control of the breakdown process on application. The formulation can have an inherently low mammalian toxicity (Tadros,1986). The main criteria for preparation of stable w/o/w multiple emulsions are : two emulsifiers, one with low and one with high HLB numbers (Tadros, 1992). Emulsifier 1 should ideally produce a viscoelastic film to reduce transport during storage. A very stable primary emulsion is required, whereby coalescence is reduced to a minimum. In this respect polymeric surfactants such Atlox 4912 (an ABA block copolymer of polyhydroxystearic acid and polyethylene oxide) are the most preferred. An optimum osmotic balance between the internal and external phases is required. For example, if an agrochemical such as paraquat is used in the internal phase (w/o), the outside continuous phase should contain sufficient electrolyte (such as $MgCl_2$) to balance the osmotic pressure. The secondary emulsifier should produce an effective barrier to prevent flocculation and coalescence of the multiple emulsion droplets. The multiple emulsion is prepared in two steps. The w/o emulsion is prepared using high speed mixers to produce droplets in the region of 1 μm), whereas the secondary emulsion is prepared using a low speed stirrer to produce drops of 10-100 μm. The scheme for preparation of the multiple emulsion is illustrated in Fig.4. A typical photomicrograph and freeze fractured electron micrograph of a w/o/w multiple emulsion of paraquat dichloride is shown in Fig.5. The stability of the multiple emulsion can be investigated using the same techniques used for emulsions.

Gel Formulations

There has been some interest in recent years in "gel" formulations which become suitable for use by putting them in water soluble or water dispersible bags. This makes the product safer to handle and safer for the environment. A number of patents on water dispersible organic gels have recently been filed by Rhone-Poulenc (Hodakowski et al, 1992) to which the reader should refer to for detailed information. Essentially a gel is produced in the oil based formulation (such as an EC) using a nonionic, ionic or amphoteric surfactant (water soluble or water dispersible) and in some cases a polyacrylic acid polymer. Alternatively particulate gelling agents (modified clays and silicas) may also be used. The patents claim that the gels are easily dispersible in water. Clearly more research and development work is required before such formulations are introduced onto the market place.

Solid Formulations

As mentioned in the introduction, there has been considerable interest in recent years into formulating agrochemicals as solids that are easily dispersed in water. Several solid formulations have been introduced for many years, such as water dispersible grains (WG's), tablets and granules. The WG's are probably the most attractive since they offer the possibility of producing safer products (non dusty) that can also be placed in water soluble or water dispersible bags hence removing the problem of pack disposal. A number of other advantages for WG's are also obvious. For example, they allow one to obtain a higher concentration of the active

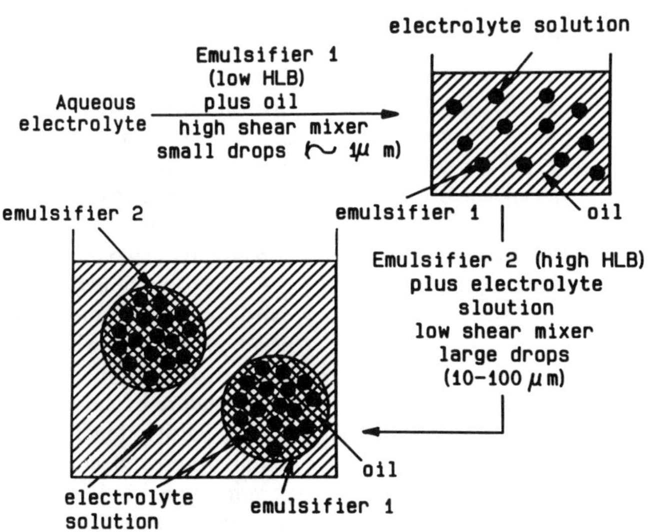

Fig. 4. Scheme for the preparation of multiple emulsions.

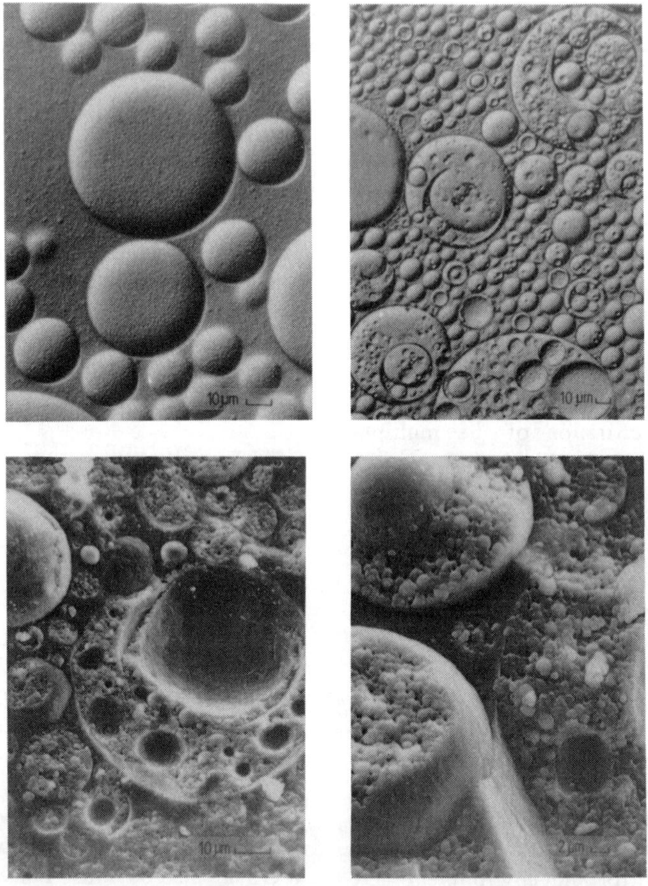

Fig. 5. Optical (top) and freeze-fractured electron (bottom) micrographs of a multiple emulsion.

ingredient (loading as high as 90% of the active ingredient is possible in principle), than those encountered with SC's and EW's. It is also claimed that WG's are physically stable over a long period of time. However, there are indications that WG's may become more difficult to disperse when the formulation is stored for a long period. This is not surprising, since with time the particles may adhere more strongly with each other, particularly when there are traces of liquids or impurities. Other advantages for WG's are simpler and safer dispensing, lack of spillage encountered with liquids, and possible packing in water soluble bags.

Several methods may be applied to produce WG's of which the following is worth mentioning. The first method (referred to as extrusion) is carried out by extruding a paste to produce cylindrical granules of ~ 0.5 - 1.0 mm diameter. The advantage of this method is that it can be carried out at "ambient" temperatures and hence it is suitable even for low m.p. compounds. Two main drawbacks of the method may be mentioned. Firstly, for adequate extrusion one requires adequate control of the paste rheology, which depends on a large number of parameters such as particle size and shape, nature of the dispersant and filler used as well as particle loading and water content. This requires a great deal of adjustment of the various parameters as well as the process. Secondly, the dispersion of extruded granules may not be sufficiently fast and it may also deteriorate on storage. An alternative procedure for making

WG's is to use spray drying. This method has the advantage of producing easily dispersible granules. Unfortunately, the method can only be applied to high m.p. compounds and the product may also be dusty. Another attractive method is fluid bed granulation, whereby a binding liquid is added to a powder fluidised by air or nitrogen. The agglomerates produced are rounded by attrition within the bed to granules of 1-2 mm diameter. A similar procedure to fluid bed granulation is mechanical agglomeration, whereby the powder is kept in motion by mechanical means rather than using air. Pan granulation has also been used for producing WG's. In this case a liquid is added to a powder as it rotates on a tilted dish.

All the above mentioned methods of making WG's require a great deal of engineering skill, accurate adjustment of the formulation parameters and high capital cost. Clearly, the answers to these problems will emerge in time when various agrochemical companies develop these methods. In addition, the problem of fast dispersion needs careful investigation of the adhesive forces between the particles in the grains and how these can be reduced and maintained low on storage.

Another solid formulation that can be used with liquid agrochemicals is that of granules. In this case, the liquid is absorbed in the pores of the filler and the granules can be applied in the soil or sprinkled on crops. One of the main problems with granules is the difficulty of releasing the chemical from the pores. This is due to the surface forces operating when a liquid is entrapped in a pore. If the o/w interfacial tension is appreciable (as is the case with many liquids), a very high pressure gradient exists within the pore. This is simply due to the Laplace pressure which is proportional to $2\gamma/r$. For a pore with two radii that are small and unequal, the pressure gradient can be very high and the liquid remains entrapped in the pore. For quick release of the liquid, one has to reduce the interfacial tension to very low values and this may not be easy in practice.

Another solid formulation that is suitable for highly active agrochemicals is tabletting. In this case, the active ingredient that may be a solid or a liquid is formulated as a tablet using inert materials such as micronized cellulose. The tablet produced should quickly disperse in the spray tank and this requires a good wetting and dispersing agent. An attraction of tablet formulations is that the chemical can be applied in specified doses and packaging becomes easy. In addition, tablets are very safe for handling. However, the cost of producing tablets may be high and the method can only be applied to very active chemicals that require only a few grams per hectare.

Another potentially useful solid formulation is that of "dried" microcapsules. The latter may contain a liquid active ingredient thus providing a convenient method of converting liquids to solids. The main problem with dried microcapsules is the possibility of deterioration of dispersion on storage. This is due to the elastic nature of the capsule wall, which may undergo deformation with time, increasing the area of contact between the capsules. This increases the adhesive forces and makes dispersion more difficult. Methods are required to overcome this problem.

Conclusions

In this overview, which is by no means exhaustive, I have highlighted the various types of available dispersions in agrochemicals. In my opinion, these dispersions are still very attractive for formulating agrochemicals and they are very popular with the farmer. There are still a number of drawbacks, which need to be resolved. The long term physical stability of these dispersions may cause problems and this requires research in the field of colloid and interface science. With my long experience in this research field, I cannot forsee problems on stability that cannot be solved. Another major problem with dispersions is the issue of pack disposal. However, if methods can be applied to clean the containers such that the remaining active ingredient residue is sufficiently low, these empty containers my be incinerated or shredded into small pieces and used as compost in the soil. Recently a number of biodegradable polymers suitable for making containers (such as Biopol manufactured by Zeneca) became available and these may be ideal for packaging of dispersions.

The above problems may be overcome if one uses solid formulations in water soluble or dispersible bags. However, the technology of making solids is not very advanced and it may be costly. Incorporation of adjuvants (which may be liquids or semi-solids) is not easy and requires further formulation work. In addition, some of the issues such as storage stability and ease of dispersion need to resolved.

My personal opinion is to continue the research on both fronts, improving on dispersions and solving the pack disposal problem, as well as carrying out research and development on solids. Both types of formulations will be applied in the future depending on the nature of the agrochemical, availability and success.

Acknowledgement

The author is grateful to Miss Julia Cutler for her critical reading of the manuscript and making improvements to the text.

References

Danielsson, I. and Lindman, B., *Colloids and Surfaces* **1981**,3,391.

Davis, S.S. and Smith, A., in *Theory and Practice of Emulsion Technology*, Ed. Smith, A.L., Academic Press, London **1974**, p.285

Heath, D., Knott, R.D., Knowles, D.A. and Tadros, Th.F., *ACS Symp. Ser.*, **1984**,254,11

Hodakowski, L.E., Couch, R.W., Couge, S.T. and Ligon, R.C., *Int. Patent. Publ.* **1994**, WO 92/01377.

Kondo, A. *Microencapsulation Process and Technology*, Marcel Dekker, New York **1979.**

Overbeek, J.Th.G., De Bruyn, P.L. and Verhoecks, F. in *Surfactants*, Ed. Tadros, Th.F., Academic Press, London **1984**, p.111

Tadros, Th.F., *Adv. Colloid Interface Sci.* **1980**,12, 141.

Tadros, Th.F. and Vincent, B., in *Encyclopedia of Emulsion Technology*, Ed. Becher, P. Marcel Dekker, N.Y.**1983**, p.139

Tadros, Th.F., *European Patent,* 88303995-4 **1986.**

Tadros, Th.F., *European Patent*, 0289356 **1988.**

Tadros, Th.F., *Proceedings of the Second World Congress on Surfactants*, Vol.III **1988**, 321-332.

Tadros, Th.F., *Pesticide Sci.* **1989**,26, 51.

Tadros, Th.F., *Adv. Colloid Interface Sci.* **1990**, 32,205.

Tadros, Th.F., *Int. J. Cosmet. Sci.* **1992**,14,93-111.

Tadros, Th.F., *Colloids and Surfaces* **1994**, in press.

Tadros, Th.F., *Surfactants in Agrochemicals*, Marcell Dekker, N.Y., **1994**, in press.

Vincent, B., in *Surfactants*, Editor Tadros, Th.F., Academic Press, London **1984.**

Walstra, P., in *Encyclopedia of Emulsion Technology*, Editor Becher, P., Vol.I. Marcel Dekker, N.Y. **1983**, Chapter 2.

Packaging and Application Technology

Robert L. Denny, Agricultural Container Research Council, 1225 I Street N.W., Washington, DC 20005

During the last five years there have been significant developments in the fields of product packaging and pesticide delivery systems. In the area of packaging technology, the idea that the container and the chemical formulation within the container are an integrated unit and are inseparable has become an accepted concept and an accepted field of research and development. In the United States this has contributed to the reduction of absorbed residues on/in containers, allowing for reuse/recycling of that packaging material. Changing practices as well as evolving package engineering have also contributed to the virtual elimination of the pesticide container material in some applications. Successful development of refillable containers of all sizes, water dissolvable films or gels, tablets, and microencapsulated pesticides substantially reduce or eliminate the need for recycling or disposal of packaging that has been exposed to pesticide residue.

In the area of pesticide delivery or application technology, again there has been significant progress in relating the science of formulation to application equipment engineering. Innovative approaches have contributed to reducing off-target damage, worker exposure, and other residue management issues.

Although significant research will be required to achieve full potential, the next conceptual step in research and development is already underway. That concept centers on the unification of the pesticide formulation, packaging, and the available delivery systems while providing multiple doses per application at a viable cost.

In the earliest days of pest control, right up until the post World War II chemical revolution, the goal of research scientists, manufacturers, and applicators alike was to find a product that 1. controlled the pest, and 2. was economical to manufacture, market and apply. There were other considerations, but the controlling forces were efficacy and price. In the United States, the earliest pesticide law, the Insecticide Act of 1910, was primarily written to provide safeguards for farmers, assuring them that the product that they paid for was indeed an efficacious insecticide. In 1947, the Federal Insecticide, Fungicide and Rodenticide Act replaced and broadened the scope of the older statute, but still, the primary focus was initially registration, simple labeling, and consumer protection.

It is no wonder, that in this regulatory and economic climate that the registrants of DDT, aldrin, dieldrin, various arsenicals and the host of other broad spectrum pesticides paid scant attention to either expanding the formulation choices or focusing on packaging options. The formulations available had already proven themselves efficacious, at least in the short run. There was every likelihood that a company could lose market advantage if reformulating or innovative packaging forced up total costs.

There were several events that forever altered these binary forces that initially shaped the form and substance of world pesticides. In the USA, a biologist by the name of Rachel Carson published the 1962 Pulitzer Prize winning book, Silent Spring, warning of dire environmental health consequences of the uncontrolled use of broad spectrum pesticides. No matter how challenged the substance of the research and text may have been by contemporary scientists, the work surely inspired development of alternative, more specific active ingredients, contributed to user demand for fewer oil based formulations,

1054–7487/95/0087$12.00/0

and concurrently signaled the inevitable removal of pesticide regulation from the labeling/efficacy orientation of the US Department of Agriculture and folding that program into a newly (1970) formed Environmental Protection Agency.

Focus on the Formulation

Since the mid 1980's, an increasing number of pesticide producers and formulators have listened to the new market forces that shape the design of crop protection materials and other pesticide products. Certainly, the historic concerns for efficacy and cost effectiveness are still mandatory, but the environmental health and safety concerns coupled with conscious planning for management or elimination of wastes are also equally important in order to remain viable competitors in today's markets. How individual companies have faced these combined challenges is an interesting exercise, but case studies are only levelers of the status quo unless the concept of the shifting, but nevertheless identified goals are kept in constant awareness.

Using the historic and other forces shaping the direction of pest management materials, then what is the ideal? The ideal is a chemical or biological material whose integrity can safely be maintained throughout its lifetime of transportation and storage, and that when applied to the target organism, controls that pest without effectively altering nontarget organisms or posing risk to any handler or applicator. The material encasing this idealize pest control material is also of no concern as either a drain of resources or a vehicle of environmental contamination. Of course, all of the elements of this ideal must be accomplished at a price that will facilitate demand and use by the applicator (value) and a reasonable profit for the producer. Effectively, what we are witnessing in the 1990's is the realization that pesticide product = formulated active ingredient + packaging + delivery/use system. Each component of the product cannot be created or judged separately, each component of the product is interdependent on the interface between every other component or function.

The directions that the international pesticide industry are taking illustrate how companies are attempting to meet the aforementioned idealized goals and provide clues for research that is still needed to attain a closer unity with those ideals.

A dialogue is underway today at the US Environmental Protection Agency in an effort to promote either "safer" pesticides or reduce the quantities of pesticides used. Traditionally, one might consider designing formulations that impact only the target organism. This is, of course, a difficult task since so many living things have similar biochemistry. Still, corporations have been successful in culturing bacteria, parasites, and engineering chemicals that are unusually specific to a few classes or even one class of organisms. Selectivity for the target pest can then be enhanced by either the application technology or timing of the application.

Another technique of the formulator is to prepare active ingredients that may be applied at ultra low volume, theoretically reducing pesticide uptake by off-target organisms. This method is dependent on parallel improvements in application technology and technique. In terms of effects on packaging, certainly materials that are efficacious in low doses are usually conservers of packaging materials as well although, as yet, the removal of residues from those packages or the use of refillable containers has not been extensively developed.

At issue in the US, as well as elsewhere, is the debate between the attributes of dry versus liquid formulations. The research, development, and ultimate user acceptance will not be settled by inherent merits of the physical state of the formulation per se, but also the other two components of an effective product: packaging and handling/application/use. There have been two significant developments in packaging, handling, and application of dry formulations. First, equipment manufactures have successfully developed equipment for ground applying large quantities of dry bulk materials from either large bulk bins or refillable bags capable of holding 1000 lb. or more of material. The down side of this technology is the concern for worker exposure unless dusts are controlled through formulation or handling technology. A second equipment innovation is the American Cyanamid-John Deere partnership resulting in the Lock-n-Load refillable container that replaces the applicator filled bins on dry application/planting equipment. This refillable container has a one-way valve and functions effectively as a "closed system" protecting the applicator/handler from filling exposures. The disadvantage of applications of dry materials is that unless techniques and formulations are developed that allow homogeneous mixtures of several different crop protection chemicals and nutrients then simultaneous applications will be prohibited, limiting users to a single pass with each formulation.

Some industry leaders interviewed for this address find a decided advantage in favor of aqueous suspensions in use as pesticide mixtures. Custom application of complex mixtures crafted for individual sites is effectively creating a new formulation for each site depending on the pests needing control, the season, the nutrients required, anticipated rainfall etc. Liquid formulations may be better adapted to dose measurement and utilization in current application equipment. Problems do arise however, when considering the packaging storage and disposition for so many components and the question of compatibility within the mix tank. It is not clear that traditional liquid mix tanks provide the uniform homogeneity for effective application and pest control.

One of the most interesting, but slow developing areas of formulation research is that of timed release formulations. Here again the product interface between the characteristics of the formulation and the packaging and handling/application may impact development and acceptance in a major way. Of course, timed release formulations can be either liquid or dry and the development of these formulations will be a major topic at this IUPAC conference. But one of the potential advantages of both liquid and dry timed release formulations is that residue contamination of packaging could potentially be drastically reduced, if not eliminated. Thus the starch/polymer coating effectively takes on the characteristics of the pesticide "containment" and any other required vessel or packaging could be viewed as a secondary vehicle for minimizing environmental/health impacts.

FOCUS ON PACKAGING

The United States is in the midst of a fairly rapid evolution in pesticide packaging spurred on by standing or proposed Federal, State, and local regulations and ordinances and customer demand.

Packaging was not at first a shaping force in the design of crop protection products or other chemicals. Most of the early pesticide producers in the West traced their roots to petroleum and chemical production and the choices for containers were the same as any other segment of the chemical industry. If a liquid product was required in quantity, a steel drum was the obvious choice, smaller quantities usually found a home in the ubiquitous 5 gallon flat top can. Dry products were contained in sewn bags.

The first force changing this accepted pattern occurred starting in the late 1970's and early 1980's. Observant sales staff noticed that at approximately 10lb/gallon, there was a marked prejudice against drums of any size where field handling equipment was scarce. Even repetitive lifting 50 or so pounds above shoulder height to load mix tanks was not only tiring, but at times unsafe. Plastic containers of 2-1/2 gallon size were easier to handle, easier to open, essentially corrosion proof and emptied well. An added attraction was that at the end of a day, the plastic jugs could be handled just at the multiwall bags had been for years: they were burned. Metal containers were harder to get rid of.

At about the same time, Midwestern growers of large tracts of corn/soybean monoculture became interested in avoiding large piles of one-way containers altogether and began working with pesticide producers and retail establishments to essentially utilize intermediate bulk containers, called mini-bulks when used in US agriculture.

These two customer driven departures from the traditional chemical industry packaging started the process of considering pesticide packaging in the context of what is contained and how it will be handled/applied in the field.

In the US, solid and hazardous waste regulations began reducing the availability of landfills placing a new emphasis on finding alternatives to traditional, disposable packaging. Although poorly enforced, prohibitions against open burning are another incentive to find alternatives to past practices. In 1988, NACA conducted a survey of agricultural chemical container usage. Already, the HDPE plastic jug had virtually vanquished the steel pail that once dominated the market just a short decade before. Plastic jugs of all sizes contained 74% of the US pesticide market, while steel pails accounted for only 6% of the liquid packaging. Dry products were still contained almost exclusively by bags of multiwall or, in some cases LDPE plastic, construction. By 1991, the most recent year where information is available, NACA reported that although the total gallons of agricultural products were down 29%, the market share packaged in HDPE actually rose 3%. By 1991, the hammer lock hold on the market enjoyed by one-way bags was showing signs of easing somewhat. Water-soluble bags were clearly the fastest growing segment of the pesticide packaging choices. There are some powerful advantages for water-dissolvable packaging. If the product is dropped into the mix tank with proper agitation, the soluble pack functions like a closed system for the mixer–loader of the pesticide. Water-dissolvable packaging (WDP) is now available in either cast or blown films and even laminates for additional strength. If films are inadequate for the task, gels and dissolvable bottles or ampoules can be used depending on the characteristics of the formulation. In all of these applications, there is one problem for the WDP family of packaging; that is, how does the manufacturer select the size? If the package is too large, there may be either too much mix prepared for the acreage, or the applicator may be tempted to slit open the package and thereby defeat the "closed-system" appeal of the packaging.

Another minor incursion into the world of one-way dry containers is the packaging of dry materials into rigid HDPE plastic. The advantage of this latter move will become apparent in the following discussion. And finally refillable FIB's or bulk bags as well as blow molded, one-way valved refillable containers that function as dry closed systems are starting to erode the market share of multiwall and other bag type, nonrefillable packaging. Why are we seeing these new innovations? There are several reasons. Emptying pesticide into application equipment can sometimes result in unacceptable exposures to the handler. Any system, whether liquid or dry that mates with the application equipment forming a closed transfer system can eliminate one of the primary routes of mixer/loader handler exposure. Another reason is that in addition to the obvious conservation measures gained by utilizing refillable or recyclable packaging, many of the traditional containers such as flat top containers of all kinds and multiwall or plastic bags retain relatively large quantities of formulation. Studies have shown that both of these packages retain on the order of 10 grams or so, while a well constructed HDPE container can retain less than 10 milligrams of formulation per container. Residue retention for refillable containers is usually not a concern unless the refillable container is utilized for a different formulation.

ACRC

One nationwide program that has evolved over the past five years is assuring that millions of crop protection chemical containers are being managed quite differently than the traditional open burning and burial of the past. The reason for this waste reduction is a mutually beneficial partnership between the agricultural chemical industry, state or local governments and the users of crop protection chemicals. After several years of trial and error, certain key ingredients are obviously contributing to the success of this return-reuse program.

As an outgrowth of a 1989 pilot partnership between the Mississippi Department of Agriculture, farmers, the National Agricultural Chemical Association and elements of the crop protection chemical industry, a program developed into what is now known as the Agricultural Containers Research Council or ACRC; a non-profit corporation dedicated to working with organization interested in collecting, consolidating, and recycling plastic crop protection containers. The ACRC is sponsored by the leading crop protection chemical producers, packagers, and distributors and is now providing the knowledge, research, training assistance, and support for collection and final disposition of millions of plastic containers. And the ACRC is accomplishing this stewardship while maintaining, and even encouraging, State and local flexibility in that effort to remove crop protection chemical containers from the environment. Currently, there are few other examples of national, cooperative efforts to retrieve packaging from the consumer or end-user.

The ACRC has identified that there are a number of components of a national collection program for agricultural chemical containers.

Components of a National Collection Program for Agricultural Chemical Containers

Personnel Training

Education for everyone involved from corporate and regulatory managers to crop protection mixers/loaders has been a key component of the Agricultural Container Research Council crop protection container recovery program. The ACRC External Affairs Committee has developed video and slide-script format materials in both English and Spanish that assist in training mixers/loaders of crop chemicals as well as the inspector charged with assuring that the clean HDPE containers are acceptable to the ACRC program. The committee has also developed printed media such as brochures and posters for promoting the return of clean chemical containers. Public agencies or educational institutions may obtain these materials by contacting the ACRC in Washington, DC.

Site Development

The ACRC recognizes that a local entity such as State/local government or an interested user-group can best select, advertise, and manage collection sites. There are two basic kinds of sites utilized by the ACRC. In the first programs, sponsors held collection return days where the consolidation equipment, inspectors, and holders of the containers converged on an advertized site on a specific day. The advantage of this approach is that there are no containers to store. The disadvantage is that it is labor intensive for everyone and not always convenient for the farmer or other pesticide user. The alternative to this approach is the supervised collection site where a trained inspector is on hand during the crop protection chemical application season and the consolidation equipment appears only at the end of the year for grinding and shipment to a storage facility. Growers like this approach, but there must obviously be trained inspectors and storage facilities available at each site.

Consolidation

The ACRC has developed the standards for reducing the volume of plastic containers by as much as 50:1. Quick and efficient volume reduction is essential for scheduling, transportation of chipped plastic, and safe and inexpensive storage.

Transportation

In the earliest days of consolidating agricultural containers, the predecessors of the ACRC used large wooden and corrugated box containers. This unit of shipment has been replaced by the reusable flexible intermediate bulk container, or FIBC. A rigid box displaces the same cubic footage whether it is empty, full or somewhere in between. An FIBC displaces a volume proportionate to its contents.

Reuse/Recycling

The ACRC Technical Committee has responsibility for acceptable uses of material.

The tasks of the Technical Committee include:

- preparing protocols
- conducting analyses
- receiving and evaluating bids and protocols from testing laboratories
- reviewing and evaluating progress and final reports from such laboratories
- preparing summaries of studies developed or supported by the Council for use by the members for publication and scientific journals or for presentation to scientific groups or regulatory agencies
- performing any other task or actions within the scope of its authority as designated by the Board of Directors

The objectives of the Technical Committee are to develop analytical methods for residues in plastic, develop data to support end uses, evaluate recycling options and to scrutinize recycling operations that are part of any Technical Committee research project.

The Technical Committee has compiled information for the ACRC paper on "White Coal," which supports use of high density polyethylene (HDPE) crop protection containers as an energy source. The Technical Committee has developed protocols and procedures for evaluating facilities that are conducting research for the ACRC. A project to determine residue levels from properly rinsed HDPE containers and to compare analytical procedures for TE and TOC (Total Extraction and Total Organic Carbon) is near completion. Two pallet projects are in final stages. These projects are evaluating plastic pallets made from collected HDPE and from virgin plastic. A hazardous waste drum project using up to 28% ACRC collected HDPE mixed with virgin plastic is also in its final evaluation stage.

Funding

After only 30 months of existence, leaders of the ACRC report that the pivotal element in organizing a nationwide collection program of this type, and perhaps any kind, is the assurance to the participants that the overall program is administered in the most cost effective manner possible, and that there is an equitable distribution of those associated costs no matter how large or how small the supporter.

Tackling the issue of paying for container recovery has never been a simple task. In the late 1980's legislation was developed in Congress by Sen. Wyche Fowler, D-GA, that proposed a fee to be assessed on each nonrefillable crop protection chemical container. This money would be collected by the Treasury and administered by the EPA to provide collection sites in each State. Estimates on the amount of money to be assessed varied widely, but most participants estimated $1 or more per container. The bill did not pass, but the concept remains as an expensive alternative if private efforts like the ACRC should falter. Thus, it became clear to many packagers and distributors of agricultural products that sharing expenses of these programs was preferable to Federal administration or to each company providing collection programs for their own containers. It was, at first, awkward to devise a procedure to share those collection and disposition expenses. Dividing expenses evenly among participants was unacceptable to smaller companies, and counting and deciding on collection site expenses at the time of return was simply too cumbersome and labor intensive. It was through this experience that the organization and current funding mechanism of the ACRC evolved and is still evolving.

The current, procedure for membership assessment within the ACRC involves estimating the upper limit of pounds of high density polyethylene that could be collected in the coming year. This figure is based on the growth rate of plastic collected for recent years relative to the number of states and collection sites involved.

A look at the growth of the number of states involved is also a measure for predicting program expenses.

States with at least one collection site:

1989: MS

1990: FL, IA, IL, ME, MN, MO, MS, NC, OR, WA

1991: FL, IA, ID, ME, MN, MS, NC, OR, VT, WA

1992: AL, AR, CA, FL, GA, IN, IA, KS, KY, LA, ME, MI, MN, MS, MO, NE, NJ, NY, NC, OH, OR, PA, TX, VA, VT, WI

1993: AL, AZ, AR, CA, CO, CT, DE, FL, GA, HI, IL, IN, IA, KS, KY, LA, ME, MD, MI, MN, MS, MO, MT, NE, NJ, NM, NY, NC, OK, OR, PA, SC, SD, TN, TX, VA, VT, WA, WI, WY

Of course, the above numbers do not accurately predict containers returned. There are still a number of States that have only small private programs or just one or two pilot projects in 1994. Nevertheless, States that have been participating since 1990, such as Florida, Iowa, Maine, Minnesota, Missouri, Oregon, and Washington have demonstrated that once a program is begun, the effort steadily increases to a truly statewide effort in just a few years. For this reason, the ACRC estimates that the amount of plastic used for budgeting purposes will be 5 million pounds in 1994.

The ACRC contracts with two primary companies for the grinding, storage, and shipment to a recycler or energy recovery facility. These two contractors are: SCT Environmental of Pasadena, Texas and Tri-Rinse, Inc. of St. Louis, Missouri. Primary contractors are free to negotiate subcontracts with other companies, but all plastic returned and the final disposition of the plastic must follow rigid contractual requirements. Using the expenses per pound of the previous (or current) year, the ACRC calculates the costs of collection, granulation, and disposition. This total operational budget, when combined with any administrative costs, is divided between the crop protection chemical and adjuvant producers and formulators according to a formula. Surveys have shown that the primary contribution to the pounds of HDPE plastic comes from the accumulation of one gallon and 2 1/2 gallon containers.

The emphasis on the program was originally focused on agricultural pesticides; however, some classes of herbicides used for right-of-way management, golf courses, forest vegetative management, commercial lawn care, etc. are understandably blurred with similar, if not identical, agricultural formulations. Also, in practice, collections in many agricultural areas were in practice accepting the almost identical HDPE containers that once held crop adjuvants. For these reasons, the ACRC decided in 1994 to expand the program to include all FIFRA regulated materials packaged in nonrefillable HDPE containers and similarly packaged agricultural adjuvants. Consumer or homeowner chemicals are not included.

Once the totals of the two sizes of containers are counted and the market share is calculated, then a company's contribution to the pounds of plastic that are projected for the coming year can be estimated. This is calculated using the average weight of the one gallon container of 0.33 lb. and the 2 1/2 gallon container weight of 0.75 lb. ACRC membership accounts for approximately 91-92% of all agricultural crop protection chemicals registered in the United States. This final figure is confidentially invoiced to the ACRC member companies.

Although the costs associated with running ACRC programs do attract attention, the question remains as to viable alternatives. As the ACRC achieves success due to expanding scope and numbers of participants, the costs for supporting this effort have rapidly increased. Yet, unless a producer ignores all stewardship responsibilities, there appear to be no less expensive options.

A logical alternative to HDPE for liquid concentrates is the mini-bulk or small-volume refillable. Industry calculations, based on average estimates of refills per season, storage costs, and maintenance indicate that refillables are in the range of $1.20 to $2.45 per delivered gallon of crop protection per season. These estimates are calculated based on container and equipment costs prior to the implementation of the EPA's proposed regulations on refillable containers. If promulgated as proposed, new regulations will eventually increase costs due to drop test data requirements, sunset provisions on plastic containers, and other valving and shielding requirements. Eventually refillable containers must also be destroyed or recycled.

Nonrefillable containers, or at least those made from HDPE, are the container itself, cap, label-booklet, carton, and a shared fraction of pallet space-cost. The National Agricultural Chemical Association estimates that these materials contribute to a price of $1.43 per gallon for one gallon containers and $1.02 for a 2-1/2 gallon size. ACRC experience pegs the current price for collection and disposition of HDPE containers at approximately $0.70 per lb. Therefore the 1994 contribution of collection and disposition costs for HDPE nonfillable containers is somewhere in the $0.20 per gallon range or less. This means that nonrefillable plastic containers are competitive in today's market to the least expensive refillable containers, even when collection-recycling costs of the ACRC are added to those costs. And importantly, the ACRC costs per gallon delivered in HDPE should decline while the costs per gallon delivered in refillable containers may increase.

The other alternative to a non-profit type of program like the ACRC's is a Federal collection program or a mosaic of State programs. Either alternative is virtually guaranteed to increase container costs far above current levels. Fortunately, the ACRC is achieving nationwide recovery of a large percentage of one of the most common forms of crop protection product containers.

In the examples of a massive collection effort sponsored by the US pesticide industry, HDPE plastic that is usually used to package liquid concentrates in the US, can be collected and reutilized. What remains to be developed is a container for dry product that empties well and there is an infrastructure for collecting that spent packaging for some useful purpose.

FOCUS ON APPLICATION TECHNOLOGY

Throughout these discussions, it has been apparent that one cannot mention the developments and progress in the one area without also referencing the associated impact on the other two areas. For instance, the mating of refillable packaging to the application equipment is the fusion of formulation, packaging, and application technology. Yet, this focus on application technology will be my briefest discussion. This is in part due to the result of a very practical consideration. If a manufacturer develops a formulation and a plug in module of packaging that can only be used by a specific piece of equipment, then that manufacturer must have enormous market share and demand for the product in order for that product to competitively survive. As an example of this illustration, there are only a handful of application principles in common usage. Whether the formulation is applied aerially, by flotation tired ground rig, or a small tractor sprayer or even backpack, the manufacturer is usually compelled to fit the packaging and formulation to the existing application equipment. And furthermore, the operational principles that characterize a commercial custom application rig can be very different to the smallest boom sprayer mounted behind a tractor. Companies that ignore the available equipment, are usually not capable of staying in the market for very long.

Still there are opportunities for technological breakthroughs in the application equipment area. One of the most exciting opportunities has been opened by the technology that was developed for the NATO military a few years ago at great expense, but is now approaching the point where one can expect peacetime applications if the cost can be made reasonable. Examples of this technology, called geographical information systems or GIS, already include the utilization of navigational satellite positioning to indicate to the applicator where the last application was made in a field. Marking foams and flaggers may not be necessary if this technology can be inexpensively applied.

The detail of remote, sensing, satellite mapping is well known to all of us. The fingerprint of the biosphere can be so well characterized that diseases and insect infestations can sometimes be detected from space orbit. If pest problems can be detected from several thousand meters in space, then imagine, if you will the potential for detection of the presence of weeds, insects, and diseases from the edge of a spray boom. Combine GIS and RS into a single piece of ground application equipment and precise metering of pesticides is possible. Costs are probably too expensive to contemplate at his time, but large scale technological integrated pest management (IPM) could result if electronic scouts triggered microbursts of crop protection chemicals that only impact target pests.

At the opposite end of the technology scale, application systems need to be engineered in developing regions of the world or in specialty pest control uses where labor is not the primary determinant of crop costs or other pest control. In these cases, the formulation constituents of the product equation are extremely important, because human proximity to the agricultural site is virtually assured. Still, there is a need for application development here. Containment in closed systems could markedly lessen accidental exposures to pesticides. Sealed plug-in modules fitted with one-way valving designed for backpack sprayers, filled by gravity or hand operated pumps could provide affordable, low risk crop protection for areas where labor is plentiful. One key is to keep the value of the containers and the application equipment as crop protection tools, not potential vessels for cooking or domestic livestock feeding.

Conclusion

Briefly, I have tried to cover a professional lifetime of observations and insights that I have formed and those around me have shared. Obviously to do an effective, comprehensive reporting of those insights in a little over a half hour is impossible. But if I leave you with one thought, let it be this one. It does not matter whether you are a scientist, and engineer, a corporate executive, or even a government regulator; if you change one component of the product equation, that is that Product = formulated activity ingredient + packaging + application technology or you forget that the product must have attainable value to the end user, then you must recognize that any component that you change will impact all other components. Pesticide formulations, packaging, or application technology must be viewed as a dynamic unity. And if they are, they will have value.

References

Agricultural Research Institute and GEOSAT Committee, Inc. *Proc. Forum on Remote Sensing in Agricultural Decision-Making.* **1992**

Carson, R. *Silent Spring;* Houghton Mifflin Company: Boston, PA, 1962.

Denny, Robert. *Chemical Packaging Review.*

Denny, Robert. *Chemical Packaging Review.*

Pesticide Container Study, A Report to Congress, U.S. Environmental Protection Agency, U.S. Government Printing Office: Washington, DC, **1992**.

State of the States Report: Pesticide Storage, Disposal and Transportation, U.S. Environmental Protection Agency, U.S. Government Printing Office: Washington, DC, **1992**.

Controlled Release Systems: Prospects for the Future

Richard M Wilkins, Department of Agricultural and Environmental Science, Newcastle University, Newcastle upon Tyne NE1 7RU, U.K.

The types of controlled release systems for effective pesticide delivery are briefly reviewed, emphasizing their real benefits, in terms of published recent results of performance, especially in soil and field conditions. Microencapsulation has shown the greatest commercial uptake due to its use in formulation concentrates for spraying. However, natural and biodegradable polymers also used in larger particle sized formulations or devices such as gels, crosslinked polymers and composites, represents an important trend in controlled release pesticides. The perceived role of the technology in terms of reducing environmental impact combined with efficacy for existing and novel pest control agents will be central in the prospects for the future.

There is a definite change in attitudes to crop protection practices and this can be seen as part of the drive towards sustainable agriculture. The definitions of this are many but include the reduction of the use and impact of pesticides on the environment. Examples of definitions include integrated systems of agricultural production less dependant on high inputs of energy and synthetic chemicals....(Francis *et al*, 1990). Thus, as a step towards sustainable pest control systems controlled release technology has a role to play, both in reducing pesticide inputs whilst maintaining control levels and facilitating the use of short-lived, biorational materials or biopesticides. Even considering the range of definitions of sustainable agriculture, which all imply a time element, then controlled release systems fit in well, especially in particular situations where, for example, leaching in soils or safety to non-target natural enemies of pest insects is concerned. Conversely, it should be recognised that there are many situations where controlled release has no benefit and no application (Wilkins, 1993).

Defining controlled release (an overall term expressing slow release, delayed release, etc.) presents some difficulties, especially in the open environment, where the conditions are ill defined and are changeable.

The scope of this paper will be that of pesticides, conventional and novel, but not including biopesticides based on micro-organisms, and will include products aimed at agriculture and related areas. Delivery mainly to soil and plants will be emphasized, thus excluding release to the atmosphere (volatile pheromones). The status of controlled release systems applied to this topic will be reviewed, according to the formulation system, and the prospects for the future based on how far the evidence can support the efficacy and benefits. In evaluating controlled release formulations the relationship with time, the kinetics of release, under practical conditions is important. Where this information has been gained then realistic assessment of such treatments is possible. Thus the correlation of lab with field needs to be established.

The kinetics of release vary according to the mechanisms controlling this release. The rate controlling step is usually determined under laboratory conditions (often with a water immersion test); the release kinetics in the field may differ. Obviously, the release behaviour under environmental situations is most relevant to pest management, but there is little published information on this performance. Depending on the formulation type release mechanisms vary. Microcapsules release their active ingredient by diffusion or cell rupture (Tsuji, 1990). Granular forms can release by diffusion (of the pesticide out or water in) combined with a pore generating system (Boehm and

Anderson, 1990). Other mechanisms include swelling of dry gel matrices, degradation of the matrix with swelling or erosion of the surface. The role of such mechanisms in controlling release depends on the microenvironment of the device and this dependence emphasizes the importance of determining kinetics under real conditions. Ideally, a mathematical relationship for the release can be used and this could be matched to the concentration-time requirements for optimum pest control and minimum evironmental impact. Clearly, good models of the behaviour of pesticides in the environment are a prerequisite for the application of this approach to efficient delivery.

Sprayable Systems

Microcapsules

For pesticide application microencapsulation was the first real commercial usage and has been the most popular approach, partly due to the ease of application by conventional spray methodology, concurrent with the importance of foliar treatment and the reduction of solvents in spray concentrates. Of the various techniques of microcapsule manufacture that of interfacial condensation polymerization has in the last 20 years become the method of choice for environmental applications (Marrs and Scher, 1990). Typically the pesticide in an organic phase with isocyanate monomers is dispersed into the aqueous phase containing an emulsifier and a protective colloid. The monomers are polymethylene-polyphenyl-isocyanate (PAPI) and toluenediisocyanate (TDI). The emulsion is heated to a temperature such that the monomers are hydrolysed at the interface to form amines which react with the monomers to form a polyurea microcapsule wall. This produces a very thin cell wall, no monomer in the aqueous phase and a high pesticide loading at low cost. Release by diffusion is influenced by particle size, wall thickness and permeability (and particle size distribution); all of which can be adjusted by altering the proportions of the component reactants (e.g. ratio of PAPI to TDI) and emulsifiers. Other mechanisms of release can be by rupture (by trampling or mastication by a pest insect) of a capsule wall (e.g. polyurethane) that is impermeable to the active agent. Diffusion controlled release may be suitable for foliar applications whilst rupture release may be used for wood or plant protection against chewing termites.

Many such microencapsulated formulations (produced as capsule suspension concentrates for spraying) have reached the market (for listing see Wilkins, 1990a; Duncan and Seymour, 1989) with applications for insect, weed and rodent control and crop regulation. As well as for foliar and soil there are applications that include pheromone and repellant sprays, wood and rubbish treatment and seed dressings. Microcapsules can also be included in dry concentrates such as water dispersable granules (WG). This illustrates the additional value of microencapsulation in converting a variable active agent (which may be liquid or experience phase changes (melts) at ambient temperatures) into a fine flowable (or suspendible) solid with consistent surface properties. A final step would be in converting microcapsules into microgranules.

Analyses of the performance of microcapsules in the field is spectacularly missing in spite of two decades of use. The biological efficacy of control of grass weeds by microencapsulated alachlor (Lasso Micro-Tech™) increased when compared to e.c. formulations. However, further evaluation of this formulation found that there were no differences in annual grass weed control between encapsulated and e.c. formulations under tilled conditions but for no-till the microcapsule gave significantly better control (Wilson *et al*, 1988). This difference was related to the release from the capsules, influenced by the wet-and-dry cycle on the surface of the straw residues in the field (Petersen and Shea, 1989). Further understanding of the behaviour of these microcapsules in soils has been provided by Capri and Walker (1993) who established that at higher temperatures and dry soil the activity was enhanced. This was further studied in lab and field conditions (Negre *et al*, 1992); the encapsulated form being relatively more persistent in moist soils.

In the case of encapsulated insecticides evidence has been accumulated that this type of formulation is safer to natural enemies and other non-target organisms than conventional spray concentrates (Wilkins, 1991). In field conditions frequent use of the encapsulated methyl parathion (Penncap-M[R]) for the control of spider mites on grapes in southern France led to higher numbers of predacious mites, when compared to the suspension concentrate formulation (Vidal, 1992). In cotton, microencapsulated sulprofos (prepared using polymers; polymethyl methacrylate, ethyl cellulose, polymethyl styrene, cellulose acetate butyrate) only slightly increased activity against larval and egg stages (Lateef *et al*, 1993). The use of microcapsules is opening up new applications and against new pests. The applications and benefits have been ably described by Tsuji (1990) especially where novel release mechanisms (such as capsule rupture through biting and trampling by insects) give control advantages. The use of these approaches against out-door pests is further exemplified in subterranean termite control by microencapsulated chlorpyrifos (Davis *et al*, 1993) and by microencapsulated permethrin which is transmitted undamaged between termites within the nest and leads to delayed toxicity and effective control (Schoknecht *et al*, 1994).

Improvement of microcapsule formulations is continually sought (Tocker, 1993). New polymers (such as poly-amides, Janssen *et al*, 1993), which are able to provide slow release in thin walls and survive the open environment are recent additions to the list of established materials.

Other Sprayable Systems

Other sprayable systems that have been developed and evaluated include emulsified high polymers and other materials such as bitumens. Such latex formulations produce a continuous film following loss of water after spraying. Whether applied in microcapsules or a film forming polymer losses of the pesticide may be reduced, especially to volatilization, leading to increased uptake into the plant. Reduction of losses during, and immediately after foliar application can have a major impact on reducing losses and efficacy. Foliar applications of chlorpyrifos using emulsified polymers also reduced losses to volatilization and increased persistence (Chen *et al*, 1994).

Spray application to soils using latex additives has been known for some time. The bio-availability of trifluralin was increased using this method and phenolic resin emulsions (Chen *et al*, 1992).

Non-sprayable Systems

Granular formulations

The larger sizes of granules for soil application provide for greater variety of release mechanisms and compounding techniques compared to the small particles suitable for spraying. Granules can be coated with a reservoir of active agent or can contain the active dispersed in a matrix or monolith. The matrix is easily prepared by blending the active with a polymer and then cohered by compression moulding, injection moulding, screw extrusion, calendering or casting. With elastomers the active may be incorporated using the usual methods of the rubber industry.

Release from matrices are normally rate declining and often follow a square root of time relationship. Release from thermoplastic polymer matrix structures can be sustained by including finely ground salts of low solubility that can be leached to generate pores. This approach was first used in 1985 with chlorpyrifos for control of root-attacking sugar cane grubs in Australia (Boehm and Anderson, 1990). The commercial development has followed with a range of such granules (suSCon[R]) that target soil insect pests otherwise difficult to control and replacing environmentally detrimental organochlorines. This method is much more efficient than the original thermoplastic matrix which often only released 10-30% of the pesticide loading. Evidence from the field has accumulated to the efficacy of this formulation type and can be exemplified in the protection of sugar cane with a "suSCon" chlorpyrifos granule (14% ai) when compared to the volatile organochlorine soil sterilant, ethylene dibromide, EDB, (at 118kg ai/ha) in the field in Tanzania. At the application rates of 5.88 and 7.84 kg ai/ha the granule was better than repeated EDB for the first two years. In the third year the original granule application was still giving control while the other treatments had complete crop failure (Saidi *et al*, 1992).

The thermoplastic granules are also safer to non-target organisms, such as birds and mammals. For chlorpyrifos, for example, the acute oral toxicity to white bob quail birds was determined as about 70 mg/kg, whereas for the 10% thermoplastic granule formulation this was increased to 1390 mg/kg (Wilkins, 1990b).

Granule or pellet (tablet) forms of herbicides in dicalcium phosphate have been developed for weed control in nurseries. These dry pressed tablets released the low water soluble napropamide over several weeks (Gorski, 1993). Granules based on kaolin with the herbicide metribuzin and crosslinked with calcium alginate and linseed oil (Pepperman and Kuan, 1993) gave a slow release. Another new granular formulation was based upon cellulose xanthide, which is crosslinked by oxidation to entrap the soil applied insecticide, carbofuran (Bote *et al*, 1993).

Biodegradable Systems

Just as the trend away from organic solvents has encouraged the emergence of microcapsule suspension concentrate formulations (e.g. Lasso Micro-Tech™) another environmental consideration, that of biodegradability, may deter them. However, the large particle size of granule formulation can employ natural biodegradable polymers which may be limited for microcapsule dimensions. Thus, an important trend in controlled release is away from synthetic polymers and towards natural, low cost materials such as starches, other polysaccharides and lignins (Lohmann, 1992).

Starch

In many ways starch is an ideal polymer for agricultural granule formulations: safe and non-toxic, readily available and of low cost, food-based technology, high molecular weight and water insoluble, easily modified chemically and physically, can be solubilized, has branched and linear forms, biodegradable and has environmentally safe metabolites.

Continued research to entrap pesticides in a starch matrix has evaluated many technologies including gelatinizing, mixing in the pesticide and then crosslinking with xanthide, calcium (and other multivalent cations), borate or calcium-borate. However recently these processes have been replaced with steam injection cooking of the starch, followed by mixing of the pesticide and natural crosslinking through retrogradation. Previously a batch method formulation can now be achieved in a continuous operation using a heated screw extruder (Wing *et al*, 1993). Encapsulation efficiencies of 94-99% (atrazine), 72-95% (alachlor) and 87-92% (metolachlor) were obtained.

Other materials can be incorporated into the starch matrix to reduce costs or modify release characteristics. Clay (attapulgite) was blended with unmodified cornstarch and processed with water and herbicides atrazine or metolachlor in a twin-screw extruder (Carr *et al*, 1994). Entrapment efficiency decreased with the clay content of the extrudate. Release rates in swirled water of the dried granules so produced showed almost complete release in 70-90 hours (atrazine) or 3-50 hours (metolachlor).

Starch entrapment technology has been continually progressed by the USDA since 1976, passing through many manufacturing processes and the performance evaluated under lab and field conditions. With pilot scale production by extrusion pearl starch formulations have undergone greater field trials to validate their use and benefits. Extensive tests (Schreiber *et al*, 1993) at 10 locations were made with atrazine, alachlor and metolachlor. Acceptable weed control and reduced leaching were obtained. However, reports of the early field performance indicated that the starch formulations (both jet cooked and borate formulations) of alachlor and metribuzin had only a limited effect on movement in the soil (Fleming *et al*, 1992).

Comparison of the performance of starch encapsulated with conventional formulations of herbicides has recently been done in the field (Gish *et al*, 1994). Atrazine and alachlor formulations were applied to no-till and tilled fields and evaluated over two years. Starch encapsulated atrazine was more persistent than conventional forms (due to less leaching) but the two formulations of alachlor showed little differences (as this released faster from the starch granules). Losses by volatilization were also measured (Wienhold and Gish, 1994b) and were reduced by starch encapsulation, especially in the case of the more volatile alachlor. This field experiment underlined the dependence of the release rates on water solubility of the active ingredient. This was further described (Wienhold and Gish, 1994a) in encapsulating five herbicides (alachlor, atrazine, EPTC, metolachlor, metribuzin) in starch; release rates (in water but with polyethylene glycol added to simulate the water potentials of soils) varied according to water solubility of the herbicide. Field experiments have been reported showing that starch granule formulations can also deliver several herbicides in maize when combined together in the same granule (Schreiber *et al*, 1994). Under a variety of agronomic conditions (tilled and non-tilled) weed control and grain yields were generally as good as normal spray applications of these herbicides as tank mixes.

Crosslinked Gels

Crosslinked gel formulations are convenient to prepare (based on mild aqueous conditions at ambient temperatures) and used in medical applications. The most popular, based on alginates, are expensive and this limits agricultural exploitation. Encapsulation of aldicarb in calcium alginate with clay increased the safety of this highly toxic pesticide and its efficiency in field experiments (Pussemier *et al*, 1992). For aquatic pesticide applications alternative effective slow release formulations have been based on gelatin in a carboxymethyl-cellulose gel crosslinked with cupric ion (Prasad and Kalyanasundaram, 1993). Laboratory release profiles of fenthion were influenced by the presence of the gelatin. Also developed mainly for aquatic applications are the superabsorbent polymers, based on polyacryl-amides or polyacrylates (Culigel[R]), that can swell in water, imbibe a pesticide and entrap it on drying (Levy *et al*, 1993). Release occurs on rehydration.

Other Matrix Methods

Matrix formulation methods based upon natural degradable polymers include those based on lignins. Lignin also is most suited as a pesticide carrier, similarly to starch, with the advantages, that, due to its aromatic nature, it protects against ultra-violet radiation and also does not degrade as fast. Pesticides can be bound to lignin, both covalently and by sorption, trapped by crosslinking or by an adhesive matrix, or by combinations of methods to provide a controlled release system (Wilkins, 1990a). A semi-commercial application of the matrix method is the 50% granule (Reforest-Aid[R]), containing 2,4-D and pine kraft lignin, for conifer protection and establishment.

Granules of a lignin-carbofuran matrix have been field trialled in tropical rice validating their efficacy against rice insects and virus diseases, and showing increased grain yields compared to conventional formulations. Further studies showed release profiles in the field very similar to those obtained in the laboratory (Wilkins, 1990a). The performance of controlled release formulations in the field, i.e. their release profiles under real conditions, has rarely been measured, but to realize the potential of these formulations this needs to be done. The release rates of carbofuran-lignin granules in water and soils has been modelled and this allows prediction of the formulation performance according to environmental conditions (Ali and Wilkins, 1990).

New Concepts

New concepts which may show how the future is developing for pesticide (or pest control agents including chemical and microbial components) delivery include target specific formulations which are activated, or release an active, at, or near, the target site. Medium molecular weight polymers designed to liberate an insecticide/acaricide, chlordimeform, under site specific conditions have been prepared (Lohmann and D'Hondt, 1988). These are designed for leaf surfaces with varying pH or microbial activity and can release the pesticide according to pH, light or cations activity. The role of light in natural activation processes and in activating the release of bio-active compounds has been extensively examined (Lohmann and Petrak, 1989).

The use of temperature to regulate release has been promoted based on side chain crystalline polymers (Intelimers[R]) which change phase at certain temperatures (Greene *et al*, 1993). Microcapsule formulations based on these temperature responsive polymers prolonged weed control and reduced leaching in the field (Greene *et al*, 1992).

Enzymatic mechanisms of release could be used, with targets within pests being considered. Such targets could include gut and detoxification enzymes. For example, starch formulations release in the presence of amylase, which method has been used for their analysis.

Entrapment of pesticides at the molecular level is increasingly being proposed to control their availability. Examples include the use of conjugates with cyclodextrin (e.g. Pyrethrin CD; Dainihon Jochuugiku, Japan) and with the new starburst polymers (Tomelia and Wilson, 1993).

Benefits and future trends

With the recent increased frequency of field evaluation of controlled release systems the benefits claimed by their proponents are gradually being realized. Possibly the earliest publication on controlled availability (Ripper *et al*, 1948) of pesticides concerned the improvement of the selectivity of the broad spectrum insecticide, DDT. When coated with cellulose DDT particles were found to have reduced toxicity to a number of natural enemies of a lepidopteran pest. This effect of increased safety to natural enemies has been repeatedly shown (Wilkins, 1991) since this date with the most recent field studies (Vidal, 1993) demonstrating increased safety of microencapsulated methyl parathion to predacious mites which attack a pest mite of grapevines.

Investigation into the processes concerned with pesticide application and dosage transfer to pests is a vital part of the development of controlled release technologies and the improvement of safe pest control (Hall, 1993). Safety to non-targets (Wilkins, 1990b) and sprayer operators is important but with some pesticide-microcapsule combinations this is not always improved. For example applications still wet did not reduce the risk for human uptake of microencapsulated lindane but when dried there was substantial reduced risk (Berisford *et al*, 1991).

Major areas of environmental concern are leaching and volatilization of pesticides. Considerable evidence is accumulating that polymer formulations do reduce leaching risk whilst maintaining weed control. An important field study for nonpoint contamination of surface and ground waters confirmed that starch formulations of atrazine significantly reduced these losses and also reduced soil concentrations of the more readily leached atrazine metabolites (Mills and Thurman, 1994). Furthermore, these field experiments demonstrate the efficacy of granular herbicides for soil treatment, which if accepted by farmers, can lead to substantial reduction of environmental contamination by eliminating spray drift.

There are two important areas where controlled release could have an impact, possibly either positively or negatively. These relate to the enhanced degradation of soil applied pesticides and in pest resistance to pesticides. Both of these problems may be ameliorated by controlled release formulations. For example, starch xanthide matrix encapsulation apparently overcame the enhanced degradation of thiocarbamate herbicides (in soils with a history of thiocarbamate treatment) by making the herbicide less available to the degrading microorganisms (Hall *et al*, 1989) at the same time as providing excellent weed control. A similar response was found in the lack of enhanced degradation of carbosulfan which releases the insecticide carbofuran in soil (Jukes and Suett, 1992). The related argument, that lower ambient concentrations reduce selection pressure, may apply to situations where resistance is a threat to continued pesticide use.

There has been a considerable growth in interest in controlled release since the 1970's (Riggle and Penner, 1990). Commercial formulations have been slow in their uptake but recently the pace of development has quickened. Microcapsule systems are being applied to water dispersable granules and in seed dressings (Seaman, 1990). These include mixtures of capsules (to give a range of release characteristics) and of active ingredients. These advances are due to the need for solvent-free concentrates to give either water capsule suspensions or dry dispersable concentrates (safer to the operator and environment). Protection from photodegradation is an additional consideration.

Registration requirements are also leading to totally biodegradable components. This has encouraged research into matrix systems which are more appropriate for natural polymers than microcapsules, exemplified by the work on starch and lignin described herein. Sadly, there has been few publications in this area, or on any aspect of field validation or characterization of controlled release from the commercial sector.

However, the current interest in controlled delivery demonstrates its role, and emphasizes the potentially greater role for the future. This is likely to be with microcapsule systems, granules containing degradable and natural polymers and on the newer developments of the pest control industry, especially biorational and microbial products (Wilkins, 1994).

References

Ali, M.A.; Wilkins, R.M. *Sci. Total Envir.* **1990**, *123/124*, 469-480.

Berisford, W.C.; Dalusky, M.J.; Bush, P.B.; Taylor, J.W.; Berisford, Y.C. *Phytoprotection* **1991**, *72*, 15-20.

Boehm, N.; Anderson, T.P. In *Controlled Delivery of Crop Protection Agents*; Wilkins, R.M., Ed.; Taylor and Francis: London, 1990; pp 125-148.

Bote, A.N.; Nadkarni, V.M.; Rajagopalan, N. *J. Contr. Rel.*, **1993**, *27*, 207-217.

Capri, E.; Walker, A. *Bull. Environ. Contam. Toxicol.*, **1993**, *50*, 506-513.

Carr, M.E.; Wing, R.E.; Doane, W.M. *Starch/starke* **1994**, *46*, 9-13.

Chen, J.L.; Flynn, A.; Jackson, W.R.; Lichti, G.; Park, D. *J. Contr. Rel.*, **1992**, *18*, 101-112.

Chen, J.L.; Horne, P.A.; Jackson, W.R.; Lichti, G.; Park, D. *J. Contr. Rel.* **1994**, *29*, 83-95.

Davis, R.W.; Kamble, S.T.; Tolley, M.P. *Bull. Environ. Contam. Toxicol.* **1993**, *50*, 458-465.

Duncan, R.; Seymour, L.W. *Controlled Release Technologies, a survey of research and commercial applications*; Elsevier: Oxford, 1989; pp. 122.

Fleming, G.F.; Wax, L.M.; Simmons, F.W.; Felsot, A.S. *Weed Sci.* **1992**, *40*, 606-613.

Francis, C.A.; Flora, C.B.; King, L.D., Eds. *Sustainable Agriculture in Temperate Zones;* Wiley-Interscience: N.Y., 1990; pp. 387.

Gish, T.G.; Shirmohammadi, A.; Wienhold, B.J. *J. Envir. Qual.* **1994**, *23*, 355-359.

Gorski, S.F. *Weed Technol.*, **1993**, *7*, 894-899.

Greene, L.C.; Meyers, P.A.; Springer, J.T.; Banks, P.A. *J. Agric. Food Chem.* **1992,** *40*, 2274-2278.

Greene, L.C.; Phan, L.X.; Schmitt, E.E.; Mohr, J.M. In *Polymeric Delivery Systems: properties and applications*, ElNokaly, M.A.; Piatt, D.M.; Charpentier, B.A., Eds.; ACS Symposium Series 520; Washington, D.C.; 1993, 244-256.

Hall, F.R. *Proceed. Inter. Symp. Control. Rel. Bioact. Mater.* **1993**, *20*, 202-203.

Hall, F.R., Reed, J.P., Trimnell, D. *Proc. Intern. Symp. Control. Rel. Biact. Mater.*, **1989**, *16*, 424-425.

Herbig, S.; Beyerinck, R.; Gillespie, J. *Proc. Intern. Symp. Control. Rel. Bioact. Mater.* **1994**, *21*,

Janssen, L.J.J.M.; Boersma, A.; TeNijenhuis, K. *J. Membr. Sci.* **1993**, *79*, 11-26.

Jukes, A.A.; Suett, D.L. *Proc. Brighton Crop Protection Conference*, **1992**, *3*, 1223-1228.

Latheef, M.A.; Dailey, O.D.; Franz, E. In *Pesticide Formulations and Application Systems*, Berger, P.D.; Davisetty, B.N.; Hall, F.R., Eds.; **ASTM 1183**; American Society for Testing and Materials; Philadelphia, 1993; pp 300-311.

Levy, R.; Nichols, M.A.; Miller, T.W. In *Pesticide Formulations and Application Systems*; Devisetty, B.N.; Chasin, D.G.; Berger, P.D. Eds.; ASTM-STP 1146; American Society for Testing and Materials: Philadelphia, 1993; pp 214-232.

Lohmann, D.; D'hondt, C. In *Pesticide Formulations: innovations and developments*; ACS Symposium Series 371, 1988, ACS, Washington, D.C.; pp. 208-220.

Lohmann, D.; Petrak, K. *CRC Critical Reviews in Therapeutic Drug Carrier Systems*, **1989**, *5*, 263-320.

Lohmann, D.G., *Proc. Intern. Symp. Control. Rel. Bioact. Mater.*, **1992**, *19*, 170-171.

Marrs, G.J.; Scher, H.B. In *Controlled Delivery of Crop Protection Agents*; Wilkins, R.M.,

Ed.; Taylor and Francis: London, 1990; pp 65-90.

Mills, M.S.; Thurman, E.M. *Environ. Sci. Technol.* **1994**, *28*, 73-79.

Negre, M.; Gennari, M.; Raimondo, E.; Celi, L.; Trevisan, M.; Capri, E. *J. Agric. Food Chem.* **1992**, *40*, 1071-1075.

Pepperman, A.B.; Kuan, J.C.W. *J. Contr. Rel.*, **1993**, *26*, 21-30.

Petersen, B.B.; Shea, P.J. *Weed Sci.*, **1989**, *37*, 719-723.

Prasad, M.P.; Kalyanasundaram, M *J. Contr. Rel.*, **1993**, *27*, 219-225.

Pussemier, L.; Debongnie, P.; Van Elsen, Y. *Med. Fac. Landbouww. Univ. Gent* **1992**, *57*/3b, 1165-1171.

Riggle, B.D., Penner, D. *Rev. Weed Sci.*, **1990**, *5*, 1-14.

Ripper, W.E.; Greenslade, R.M.; Heath, J.; Barker. K. *Nature*, **1948**, *161*, 484.

Saidi, J.A.; Ringo, D.F.P.; Owenya, F.S. *Trop. Pest Management* **1992**, *48*, 382-385.

Schoknecht, U.; Rudolph, D.; Hertel, H. *Pestic. Sci.* **1994**, *40*, 49-55.

Schreiber, M.M.; Hickman, M.V.; Vail, G.D. *J. Envir. Qual.* **1993**, *22*, 443-453.

Schreiber, M.M.; Hickman, M.V.; Vail, G.D. *Weed Tech.* **1994**, *8*, 105-113.

Seaman, D., *Pestic. Sci.* **1990**, *29*, 437-449.

Tocker, S. *Proceed. Inter. Symp. Control. Rel. Bioact. Mater.* **1993**, *20*, 200-201.

Tomelia, D.A.; Wilson, R. Canadian patent, CA 1 316 364, April 20, 1993, 79pp.

Tsuji, K. In *Controlled Delivery of Crop Protection Agents*; Wilkins, R.M., Ed.; Taylor and Francis, 1990, 99-124.

Vidal, G. *Phytoma* **1992**, *443*, 53-54.

Wienhold, B.J.; Gish, T.J. *Chemosphere* **1994**a, *28*, 1035-1046.

Wienhold, B.J.; Gish, T.J. *J. Envir. Qual.* **1994b**, *23*, 292-298.

Wilkins, R.M., Ed. *Controlled Delivery of Crop Protection Agents*; Taylor and Francis; London, 1990a; pp. 322.

Wilkins, R.M. *Proc. Brighton Crop Protection Conference*, **1990b**, *3*, 1043-1052.

Wilkins, R.M. *Proceed. Intern. Symp. Control. Rel. Bioact. Mater.*, **1991**, *18*, 78-79.

Wilkins, R.M. *Proceed. Intern. Symp. Control. Rel. Bioact. Mater.*, **1993**, *20*, 79.

Wilkins, R.M. In *Minutes 9th International Symposium on Microencapsulation*; Kas, H.S.; Hincal A.A., Eds.; Editions de Sante, Paris, 1994; pp 171-186.

Wilson, H.P.; Hines, T.E.; Hatzios, K.K.; Doub, J.P. *Weed Tech*, **1988**, *2*, 24-27.

Wing, R.E.; Carr, M.E.; Doane, W.M. In *Pesticide Formulations and Application Systems: 13th Volume, ASTM STP 1183*, American Society for Testing and Materials: Philadelphia, 1993; pp 293-299.

Biopesticide Formulations and Their Applications

T. R. Shieh, Sandoz Agro Inc., 975 California Avenue, Palo Alto, CA 94304

Many biologicals, including bacteria, baculoviruses, protozoa, fungi, and nematodes, have been under development or commercialization as biopesticides for control of insects, mites and aphids, weeds and plant pathogens for over thirty years. Currently over 90% of biological pesticides is based on *Bacillus thuringiensis* formulations which contain viable spore and delta-endotoxin crystal proteins that are toxic to the larvae of a number of lepidopterous species as well as dipterous and coleopterous species of economical importance. Difficulties and future potentials for wider commercial use of biopesticides are reviewed.

Chemical pesticides have been used successfully for many decades. They are likely to remain as a major contributing factor for a productive agriculture for many years to come. Along with the introduction of new crop varieties and mechanization, synthetic chemicals are largely responsible for the doubling of crop yields per acre since 1940 (National Research Council, 1989).

The benefit of agricultural chemicals, however, has come with some risks. Risks include, 1) contamination of ground water and running waters with some chemicals, 2) effects on nontarget organisms and risks posed to human health from acute and chronic exposure to pesticides, 3) elimination of natural enemies causing secondary pest outbreaks and 4) development of resistance to chemicals. These environmental concerns and public anxiety about the health effects of chemicals prompted the EPA to establish the amendments to the Federal Insecticide, Fungicide and Rodenticide Act requiring most pesticides already on the market to be re-registered and require more stringent toxicological testing for registration of new chemicals. In contrast, since the 1970's, the EPA has had a program allowing microbial pesticides to be registered relatively quickly and at lower costs.

During 18 years, between 1972 and 1990, the annual growth rate of total global agrochemical products averaged 4.1% level, which is below the average inflation rate during the same period. For 5 years from 1990 to 1995, the annual growth rate is projected to be only 2.3% level according to the Agricultural Service report of 1991 (McDougall, 1991).

The total B.t. products, which represent the majority of commercial biopesticides, grew at the rate of 37% during 1985 to 1990, and the rate of growth is estimated at 12.1% from 1990 to 1995 (Rigby, 1991)) (Table 1). These reports, however, also indicate that B.t. total market volume in 1990 represents only 1.4 % of total insecticides used, and occupy less then 0.5% of total agricultural chemical market value. Obviously, in spite of the faster annual growth rate, biopesticides are expected to occupy only a small portion of the total pesticide market value for a foreseeable future.

Currently, various developmental and commercial biopesticides used for control of insects, mites, aphids, weeds and plant pathogens include: bacteria, baculoviruses, protozoa, fungi and nematodes (Table 2).

Bacteria, viruses and protozoa species are microbials that must be ingested by target insects in order to be effective. Their infectivity is limited to a few families or species of insects, thus providing a relatively narrow insecticidal spectrum. Fungi and nematodes, on the other hand, can cause infection by penetrating the outer structure of a variety of insects thus providing a broader insecticidal spectrum.

While infection processes are a common mechanism for microbial pesticides, the principle mode of action of B.t. based pesticides is based on the toxic effects of delta-endotoxin protein which causes characteristic feeding inhibition and eventual toxemia to the midgut of susceptible larvae. Current status of biopesticides that have been registered for pest control against various targets during the past 30 years (Table 3) illustrate that most bacterial formulations for insect control are still in commercial use especially the B.t. preparations.

The *H. zea* NPV formulation for control of cotton boll worm was registered in the mid-70's, but was taken off the shelf only after one year of commercial distribution due to stiff competition from more effective synthetic pyrethroids which were registered in the same year.

Two baculovirus preparations were registered by the U.S. Forestry Service for control of forest pests in the late 70's. But only recently, renewed interest in baculoviruses by private industries, led to registration of *Spodoptera* NPV

1054–7487/95/0104$12.00/0

Table 1. GLOBAL PESTICIDE MARKET GROWTH ESTIMATES

	Market Values * 1972	Market Values * 1990	% Growth 1972 - 1990	Market Value 1995	% Growth 1990 - 1995
Total pesticides[1]	12,725	26,400	4.1	29,570	2.3
	1985	1990	1985 - 1990	1995	1990 - 1995
B.t. products[2]	21.8	105.0	37	189	12.5

* In millions

1 Adapted from Agro chemical science, June 1991. John McDougall
Update of agro chemical products, section parts 1 and 2

2 Adapted from Agro estimates - March 1991
Sarah Rigby in B.t. in Crop Protection, PJB publication Ltd.

Table 2. BIOPESTICIDE CLASSIFICATION AND MODE OF ACTION

Classification	Organisms	Uptake Route	Mode of Action	Spectrum
Insecticide	bacteria (B.t.)	ingestion	feed inhibition by toxin toxemia - toxin septicemia - spore	narrow
	baculoviruses	ingestion	infection, multiplication	narrow
	protozoa	ingestion	infection, multiplication	narrow
	fungi	ingestion penetration	infection, multiplication metabolic toxin	broad
	nematodes	ingestion penetration	invasion, multiplication symbiotic bacterium	broad
Acaricide	fungi	penetration	invasion of mycelium	broad
Mycoherbicide	fungi	penetration	invasion of mycelium	narrow/ broad
Microbial Antagonist	bacteria	antagonism	antibiotic, metabolites	narrow
	fungi	antagonism	antibiotic, metabolites	narrow

Table 3. REGISTRATION STATUS AND COMMERCIAL USE OF BIOPESTICIDES

Organisms	Target Insects	Commercial Status
Bacteria		
Bacillus popilliae	Japanese beetle	Commercial
Bacillus thuringiensis	Lepidopteran spp.	Commercial
Varieties and Toxins	Mosquitoes/Blackflies	Commercial
	Colorado potato beetles	Commercial
Srratia entomophilia	Grass grubs	Commercial
Viruses		
Heliothis zea NPV	*Heliothesis spp.*	Discontinued
Carpocapsa pamonella GV	Codling moth	Discontinued U.S.
Spodoptera exigua NPV	Beet armyworm	Registered 1994
Orgyia pseudotsugata NPV	Douglas fir tussak moth	Government
Llymantria dispar NPV	Gypsy moth	Government
Protozoa		
Nosema locustae	Grasshopper	Discontinued
Fungi		
Hirsutella thompsnii	Citrus rust mites	Discontinued
Phytophthora palmivara	Milkweed vine	Discontinued
Collectotrium gloeosporiodes	N. joint vetch	Discontinued
Metarhizium anisopliae	Cockroaches, Scarab	Commercial
Verticillum recanti	Aphids in greenhouse	Commercial
Gliocladium viren	Pythiums and rhizotonia	Registered
Nematodes		
Steinernema carpocapsae and *Heterorhabditis spp.*	Soil insects	Commercial

and codling moth GV for pest control in agriculture.

Nosema locustae spore preparation was registered by private industry with the assistance of USDA for suppression of grasshopper in rangeland. This formulation, however, was discontinued for commercial distribution due to production and product stability problems.

Several fungal formulations were registered in the 70's and 80's but were discontinued for commercial use in the last 2-3 years. However, interest in fungi persists and in recent years various types of formulations, based on conidia or mycelium have been registered. Their degree of commercial success, however, remains to be demonstrated in the market place.

In recent years, several entomopathogenic nematode formulations have become commercially available for control of various pests. The EPA regards entomogenous nematodes as naturally occurring parasites rather than an insecticide and allows registration of nematode products without toxicological data required for other microbials.

There are many other microorganisms that have been reported to have beneficial effects on crop production, but most have gained little acceptance in the market place.

Difficulties associated with the development and commercial use of biopesticides include:

1) Production difficulties where host insect rearing is required for multiplication of pathogens such as viruses and protozoas.
2) Formulation difficulties for where the preservation of viability of the living organisms must meet the commercial requirements.
3) Need to establish relatively expensive bioassay systems for quality control.
4) Inferior field efficacy when compared to that of synthetic chemicals.

Bacteria, such as *B. thuringiensis*, fungi and nematode juveniles are produced by conventional fermentation processes. *B. popillae*, baculoviruses and protozoa are produced in whole organisms that require the host insect rearing system for multiplication of pathogens.

Tissue culture technology similar to the submerged fermentation processes can be used for production of certain insect viruses; however, at the present time, it is limited to a few cell lines which produce relatively low virus yields with maximum of 10^8 polyhedral inclusion body per ML.

Information on the production yields and the application rates shown to achieve certain efficiency objectives for each microbial agent provides the strategic information for determining the economic and formulation structure of a proposed product.

The types of formulations applied to microbial agents are quite similar to that of chemical pesticides. The aim is to produce a suitable product that incorporates optimum effectiveness and economical use for a particular environment and circumstance. Microbial agents are produced by submerged or solid surface culture fermentation, or by multiplication of pathogens in a host insect population (Table 4). Such biological processes require multiple nutrient systems to support the complicated growth cycle of a pathogen as well as the host organism. The product yield and purity, measured in terms of potency or activity, varies widely depending on the strain of the organism and condition used for its production. Variations in growth conditions for a particular strain of organism effects the downstream recovery processes associated with concentration of active ingredient, and therefore effects the quality of technical material produced for final formulation.

Secondly, microbials are particulate matter. Unlike chemical molecules, insect pathogens are insoluble living entities with an organized cellular structure entrapped in a generally hydrophobic cellular membrane. The particle size can vary from less than half µm with granulosis virus, to over 1000 µm with juvenile nematode. The bacterial spore, crystal toxin and nuclear polyhedrosis virus are about 1 µm, whereas spore and conidia of protozoa and fungi are much larger. The hydrophobicity and particulate nature of microbials effects the wetability, dispersability and suspensibility of formulated products.

The most important factor to consider in development and commercialization of a biopesticide is the shelf stability characteristics. Since microbials are, in general, much more sensitive to temperature than most chemical

Table 4.

PRODUCTION METHODS, MICROBIAL SIZES, YIELD AND APPLICATION RATES OF TYPICAL BIOPESTICIDES

Organism Types	Production Methods	Organism Sizes	Yields	Rates
Bacteria				
B. thuringiensis	Fermentation			
Crystal toxin		0.1 x 1.5 µm	10-30% dry wt.	10-80 g/HA
Spore		1 x 4 µm	2-4 x 10^9/ml	0-10^{13}/HA
B. popilliae	Host insect			
Spore		0.5 x 1.8 µm	5 x 10^9/larvae	10^7/larvae
Baculoviruses	Host insect		5 x 10^9/larvae	
	(Tissue culture)		10^6 - 10^8/ml	
Nuclear polyhydrosis		0.5 x 1.5 µm		2.5-5 x 10^{11}/HA
Granulosis		0.2 x 0.5 µm		1.2 x 10^{11}-10^{13}/HA
Protozoa	Host insect		6 x 10^4/insect	2.5 x 10^9-2.5 x 10^{12}/HA
Spore		2.8 x 5.2 µm		
Fungi	Fermentation			
Condia		2 x 8 µm	2-4 x 10^9 ml	10^{11}-3 x 10^{13}/HA
Mycellium		300-800 µm	3-5^8/L	2KgHA
Nematode	Fermentation			
	(Host Insect)		1-2 x 10^5/ml	1.5-5.0 x 10^9/HA
Infective juveniles		500-1300 µmπ		

pesticides, the half-life of unformulated microbials are characteristically short (Table 5). *B. thuringiensis* and the nuclearpolyhedrous virus remain stable only under freezing temperatures. However, under refrigeration and at room temperature, the half-life of virus shortens significantly to about one year. At temperatures, of 40° to 50°C, the half-life reduces further depending on the materials tested. Conidia and mycelium fungi are much less stable even under freezing temperatures. At room temperature the half-life becomes less than two months with most fungi preparations. The half-life of nematode under aeration in water averages about 40 weeks, but at higher temperatures, less than 10% remain after a week.

The stability of the protozoan, *N. locustoe*, is extremely poor. Only 10% of initial spores are viable after 15 weeks in the freezer and quickly inactivate in the refrigerator. These data provide an explanation as to why B.t. is the most widely used biopesticide among the various microbials. Microbials are also sensitive to sunlight and atmospheric oxidation.

Laboratory studies performed by placing various microbials on glass plates and then exposing the plates to UV irradiation or to simulated sunlight, indicate a rapid loss of biological activities within a few hours.

The half-life of microbials sprayed on plant leaves and exposed to outdoor sunlight was also short. Viruses, protozoa and nematode preparations lose 50% of their activity within a few hours. In contrast, the half-life of *B. thuringiensis* and fungi preparations are in the range of one to four days (Table 6) depending on the intensity of sunlight, temperature level and other environmental factors such as pH of leaf surface.

It has been postulated that wavelengths of 300 to 380 nm (near UV) of the solar spectrum is largely responsible for the inactivation of biologicals. The lethal effects are attributed to photochemical transformations of pyrimidine bases (McLaren, Shugar, 1964; Patrik, Rahn, 1976) by producing linkages between successive pyrimidine bases to form dimers. Photochemical radiolysis also effects the breakdown of protein molecules. Under aerobic conditions, biologically active protein may be inactivated by cleavage of the peptide chain yielding protein fragments (Schuessler, Schillivg, 1984). It starts with the formation of •OH radicals

$$H_2O + O_2 \xrightarrow[\text{photon}]{} \cdot OH + HO_2$$

which attack the biomolecules to yield peroxy-

Table 5. SHELF STABILITY OF UNFORMULATED BIOPESTICIDES

	Half-Life in Weeks					
	Temperature C°					
	-5	8-10	22-27	30	38-45	50
B. thuringiensis						
powder	>300	>150	>100	-	15-60	10
liquid	>300	>150	>96	-	10-40	4-10
***H. zea* NPV**						
powder	>200	60	45	-	6-25	1-2
Trichoderma viride						
conidia	>20	-	7	5	-	-
mycelia	25	-	4	2	-	-
S. carpocapsae						
aerated in H_2O	-		40 <10% at 6 wk			
***N. Locustae* spore**						
in H_2O	10% at 150 wk		<1			
dry cadaver		1% at 52 wk				

Table 6. HALF-LIFE OF MICROBIAL ACTIVITIES EXPOSED TO UV OR SUNLIGHT

Microbials	Stimulated Conditions	Stimulated Half-life	On Plants Conditions	On Plants Half-life
B. thuringiensis	Glass plates under sunlight	30 min	Bean leaves under sunlight	1-4 days
NPV	Leaf tissue under UV	<1 hr	Leaves under sun 60-70o F	6 hr
	Glass plates under sunlight	<24 hr	93-98o F	4 hr
Fungi	Glass plates, simulated sunlight	2.4 hr	Cotton foliage under sunlight	>2 days
Protozoa	UV	2.1 hr	Bean leaves under sunlight	<24 hr
			Field tobacco	0.6-2.2 day
Nematode	Water suspension sunlamp or sunlight	16-32 min	Chrysanthemum under shade	3-4 hr

radicals and oxyradicals. These radicals finally decompose to fragments.

$$pH + \bullet OH \longrightarrow P\bullet + H_2O$$
$$P\bullet + O_2 \longrightarrow PO_2$$
$$2\,PO_2 \longrightarrow 2\,PO\bullet + O_2$$
$$PO\bullet \longrightarrow F_1 + F_2$$

Studies on the effect of sunlight inactivation of *B. thuringiensis* crystal toxin by Ramon spectroscopy and amino acid compositional analyses demonstrate the correlation between destruction of indole residues of tryptophan and oxygen mediated sunlight inactivation of crystal toxin activitie (Pusztai *et al.*, 1991).

In order to inhibit the free radical formation and oxidation process during the photo inactivation of protein toxin or living organisms, the following methods have been examined by various investigators with various degrees of success in laboratory evaluations.

a. Addition of massive amounts of sunlight reflecting particles such as active carbon or clay granules to shield sunlight from reaching biomolecules.
b. Addition of aromatic sun screening agents (Cohen *et al.*, 1991) and optical fluorescent brighteners (Shapiro, Robertson, 1992; Shapiro, 1992; Hamn, Shapiro, 1992) to absorb light with wave-lengths between 280 to 400 nm.
c. Addition of antioxidants to limit free radical formation (Pusztai *et al.*, 1991).
d. Reducing oxygen concentration by incorporating humectants or oil (Pusztai *et al.*, 1991).
e. Removal of exogenous or endogenous chromophores which create singlet oxygen species upon irradiation by light (Pusztai *et al.*, 1991).
f. Encapsulation with suitable materials to block sunlight or oxygen penetration (Speaker, Lesko, 1976; Speaker, 1988; Dunkle, Shasha, 1989; Fravel *et al.*, 1988; Liu *et al.*, 1993).

In spite of earlier successes with some of the above approaches, practical formulation of commercial products to achieve favorable cost/benefit ratios still remain to be demonstrated.

Another important factor that affects the efficacy of B.t. based pesticides is a characteristic anti-feeding effect. Ingestion of sublethal doses of B.t. toxin causes paralysis of the gut and cessation of feeding within minutes (Dulmage, Martibez, 1978; Gould *et al*,, 1991); however, depending on the dose ingested and the susceptibility of the individual larvae, cessation of feeding can be temporary and the larva can recover to continue normal development.

As demonstrated with spruce budworm larvae (Frankenhuyzen, Payae, 1993), the delivery of a given dose per unit foliage in the form of multiple droplets is less effective than when that same dose is contained in one single droplet (Table 7). It is apparent that the formulation and application strategy for B.t. based products should aim at delivering lethal dose droplets that will not allow the recovery of paralyzing larvae upon ingestion.

Currently, four subspecies of *B. thuringiensis* that contain various types of crystal toxin are produced commercially as biopesticides. Subspecies *kurstaki* and *aizawai* are active against Lepidoptera species and *israelensis* and *Tenebrionis* are active against dipteran and coleopteran species, respectively (Table 8). Typical dry B.t. formulation include: water dispersable granules (WG), wettable powder (WP), granules, and dust. Liquid formulations include suspension concentrations (SC), ultra-low volume (UL) and oil flowables (OF).

Dry formulations with moisture levels of less than 5% were observed to provide longer shelf-life and contain higher potency units per weight when compared to the aqueous SC formulations.

The oil formulations, however, are gaining more popularity for use in low volume aerial spraying on crops such as cotton and banana.

Both conventional ground and aerial spray application are used for control of Lepidoptera species in vegetables, fruit trees, cotton, soybean and alfalfa using *kurstaki* and *aizawai* strains. Subspecies *israelensis* has been used for control of blackflies and river and mosquito larvae in habitats such as ponds, sewers, and salt marshes.

In general, increasing the spray application volumes has often been found to increase the effectiveness of B.t. formulations presumably due to improved coverage.

For vegetables and fruit trees, up to 400

Table 7. EFFECT OF B.t. SPRAY DROPLET POTENCY CONTENT AND DOSE ACQUISITION ON THE MORTALITY OF SPRUCE BUDWORM LARVA*

No. Needles Consumed	No. Droplets	Potency Per Droplet	Total Potency Consumed	% Larvae With Full Dose	% Mortality
1	1	5.2	5.2	82	60
1	2	2.6	5.2	65	40
1	4	1.3	5.2	55	45
2	2	2.6	5.2	50	25
4	4	1.3	5.2	35	20

* Adapted from Dr. K. F. Frankenhuyzen, presentation at the XXVIth Annual SIP Meeting.

Table 8.
***BACILLUS THURINGIENSIS* STRAINS USED FOR COMMERCIAL INSECTICIDE**

Subspecies	Crystal Toxin Types	Target Pests
Kurstaki	1Aa, 1Ab, 1Ac, 11A, 11B	Lepidoptera
Aizawai	1Aa, 1Ab, 1B, 1C, 1D, 1F	Lepidoptera
Israelensis	IV	Diptera mosquitoes, blackflies
Tenebrionis	III	Coleoptera Colorado potato beetles

liters per hectre spray volume may be used with ground rig applications and up to 80 liters per hectre may be sprayed by air.

For cotton, soybean, and alfalfa, a maximum of up to 80 liters per hectre by ground and 10 liters per hectre by air is commonly used. Low volume aerial applications on these crops, however, tend to produce less efficacious results as the droplet coverage of the entire plant becomes more difficult.

For control of forestry insects, undiluted ultra-low volume liquid formulations containing high potency units are frequently used for large scale aerial spraying with fixed wing air craft or helicopter. In Canada, twin engine DC-3 airplanes are used (Table 9).

The UL formulation for forest application by air requires

a. a high potency concentration product,
b. sufficiently low viscosity characteristics to permit smooth mechanical pumping of fluid through spray nozzles to provide optimal particle size and swath coverage,
c. formulations containing sufficient anti-evaporant to prevent evaporation of droplet during the free fall to minimize drift, and
d. formulations which are not corrosive to aircraft and spray equipment.

For aerial treatment of rivers for blackfly larvae control, a concentrated formulation must spread quickly upon impact with the water surface allowing it to be carried downstream where blackfly larvae habitate on vegitation.

The World Health Organization has established the operationally acceptable B.t. formulation for blackfly larvae control to be at less than 1 mg/L/10 min level to produce 100% mortality (World Health Organization, 1985).

The majority of B.t. formulations today are being used for control of lepidopteran species in agriculture crops such as vegetables, alfalfa, cotton, and fruit trees. However, B.t. products are seldom used alone through the season for many reasons: high insect pressures which can not be controlled sufficiently with B.t., multiple pest species that are not controlled by B.t., and in some cases, the desire to rotate with different pesticides to minimize selective pressure for

Table 9. SPRAY APPLICATION OF B.t. FORMULATIONS

Host Plant or Habitat	Spray Methods	Volumes L/HA)
vegetables	air	20-80
	ground	80-400
fruit trees	air	80
	ground	200-400
cotton/soybean	air	10-40
	ground	50-80
alfalfa	air	10-40
	ground	50-80
corn	air	10 lb of granules
forestry	helicopter	2-8
	DC-3	1-2
river (blackfly)	air	spot drop to maintain 0.5 to 1.0 mg/L/10 min
mosquitoes	air	2-8
	ground	10-20

resistance. These problems lead to the use of B.t. in the so called IPM program (Bryant, 1994).

Selective use of B.t. early in the season, when pest pressure is mainly from lepidopterous species, offers an opportunity to exploit the selectivity advantages of B.t. and allow the population of naturally beneficial insects to increase to suppress other pests. This factor also prevents the build-up of resistance to conventional chemicals, such as pyrethroids, which are best held back until the later part of the growing season when the risk of damage to the harvestable product demands high control and broad spectrum activity.

Such an IPM oriented program has been used successfully for cole crop production and more recently in cotton production in the U.S., Mexico, Central America and Australia. B.t. is also used with conventional pesticides in over 60% of B.t. usage in field applications. Mixtures of B.t. with carefully chosen partners could mean that the chemical partner can be applied at a reduced dose, thereby reducing its environmental effects.

A wider commercial acceptance of biopesticides, not only B.t. related products, but viral, fungi and nematodes products will be determined by the degree of success the industry has in improving the efficacy and consistency of those products.

Technology developments required to achieve such goals include:

1. Virulence improvements and host range expansion.
2. Shelf-life and persistency extension.
3. Production process innovation.
4. Field application and delivery improvements.

The search for new strains from nature that provide higher virulence or different host specificity have shown a limited promise. Fifteen years ago, B.t. strains were active only against lepidopteran species. Since then, new strains have been discovered from nature that are used commercially for control of dipteran and coleopteran species. More recently, newer B.t. isolates have shown activities against other classes of target pests such as grubs, mites, nematodes, ants, and flatworms (Feitelson, 1993).

Much publicized new technology of the 80's to utilize the genetic engineering to create new organisms, so far has produced little advantage over natural strains in both virulency and host range specificity. However, the basic molecular biology knowledge has largely contributed to the understanding of the mode of action of toxins at the molecular level that could ultimately contribute to the improvement of virulence or expression of host range of a biological toxin.

Attempts to significantly improve the shelf-life of living organisms or biological toxins have proven to be difficult. Optimization of formulation characteristics, such as pH, ionic strength, selection of stabilizing agents, water content etc. have resulted in only limited success.n

As described previously, attempts to block the free radical formation by various means, including encapsulation processes have shown some possibility of extending microbial persistency. Formulation studies centered on improving the field persistency of microbial pesticides probably will continue.

Manufacturing processes include production, recovery, and formulation of a product. Improvements in yields will have a direct effect on product costs and allow production of a higher quality product. Experiences have shown that the feasibility of improving the production of microbials by fermentation or insect rearing processes.

To improve the field efficacy, microbials must be delivered to target pests with conventional spray practices. Improvement in application timing, and spray technology, such as the utilization of electrostatic spray may be advantageous for microbial deposition.

A significant improvement in the delivery system for B.t. toxin may be achieved by transfer of the toxin gene into plants colonizing *P. fluorescens* or endophytic bacteria, *Clavibacter xyli*. The idea is to let toxin carrying naturally occurring bacteria that inhabit the plant system to reproduce in plants and spread their insecticidal effects. Such alternative delivery system for B.t. toxin has been tested for control of the corn borer on corn plants.

A more practical and successful auto-delivery system for microbial based toxins has been demonstrated with the transgenic plants. Genes of B.t. toxin or viral protein coats have been transferred to several recipient plants. These plants then become resistent to plant feeding insects or viral diseases. As shown in Table 10, transgenic corn, cotton, tobacco, potato and tomato have been developed with permits issued for field testing since 1987. Transgenic plants that are resistant to various viral diseases are also under field evaluation by many companies (Biotechnology Permits, 1987-1993).

I believe the technology used for creation of various transgenic plants that are resistant to viral disease and insect pests, along with the creation of herbicide-resistant crops will be the noble delivey systems for crop protection of the 21st century.

Table 10.
PERMITS ISSUED FOR FIELD TESTING OF TRANSGENIC PLANTS FOR VIRAL DISEASE AND INSECT RESISTANT BY VARIOUS INDUSTRIES

Crops	Companies	Virus Resitance	Resistance Insect
Corn	Ciba Geigy		B.t. *Cry* gene
	Monsanto		B.t. *Cry* gene
	Pioneer	MCMV, MDMV-A MCDV, CNV, TMV	
	Northrup King	MDMV-B	B.t. *Cry* gene
Cotton	Calgene		B.t. *Cry* gene
	Monsanto		B.t. *Cry* gene
	Northrup King		B.t. *Cry* gene
Soybean	Pioneer	SMV	
Tobacco	Ciba Geigy		B.t. *Cry* gene
	Rohm Hass		B.t. *Cry* gene
	Sandoz		B.t. *Cry* gene
			B.t. *Cry* gene
Potato	Calgene	PLRV	B.t. *Cry* gene
	Frito-Lay	PVY	B.t. *Cry* gene
	Monsanto	PLRV, X, Y, CPB	B.t. *Cry* gene
Alfalfa	Pioneer	AMV, CMV, TMV	
Tomato	Monsanto	TOMV, TMV	B.t. *Cry* gene
	Sandoz		B.t. *Cry* gene
	Agrigenetics	AMV	B.t. *Cry* gene
	Upjohn	TMV, TOMV	
	Rogers NK	TOMV	B.t. *Cry* gene
Melon, Squash	Upjohn	CMV, PRV ZYMV	

REFERENCES

Biotechnology permits. Biotechnology, Biologicals and Environmental Protection USDA - Aphis, Hyattsville, MD 20781, 1987-1993.

Bryant, J. E. **1994**. Agriculture, Ecosystems and Environment

Cohen, E., Roger, H., Joseph, T., Braun, S., and Margulies, L. **1991**. *J. of Inv. Path.* 57, 343-351.

Dulmage, H. T. and Martibez, E. **1978**. *J. Inv. Path.* 32, 40-50.

Dunkle, R. L. and Shasha, B. S. **1989**. *Environ. Entomol.* 18, (6) 1035-1041.

Feitelson, J. S. **1993**. Advanced Engineered Pesticides, ed. L. Kim, p. 63-71.

Frankenhuyzen, K. V. and Payae, A. **1993**. Abstract, XXVIth SIP Meeting.

Fravel, D. R., Morois, J. J., Lumsden, R. D., and Connick, Jr., W. J. **1988**. *Phytopathology* 75, 774-777.

Gould, F., Anderson, A., Landis, D., and Van Mellart, H. **1991**. *Entomol. Exp. Appl.* 58, 199-210.

Hamn, J. J. and Shapiro, M. **1992**. *J. Econ. Entomol.* 85, (6) 2149-2152.

Liu, Y. T., Sui, M.-T., Ji, D. D., Wu, I. H., Chou, C. C., and Chen, C. C. **1993**. *J. Inv. Path.* 62, 131-136.

McDougall, J. **1991**. Agro Chemical Science. Update of the Agrochemical Products Section, Parts 1 and 2.

McLaren, A. D. and Shugar, D. **1964**. Phytochemicstry of proteins and nuclear acids. New York, MacMillan.

National Research Council. **1989**. Alternative Agriculture. National Academy press, Washington, D.C.

Patrik, M. H. and Rahn, R. O. **1976**. Photochemistry and Photobiology of Nuclear Acids, Vol. II, ed. S. Y. Wang, New York Academic Press, pps. 35-95.

Pusztai, M., Fast, P., Gringorten, L., Kaplan, H., Lessard, T., and Carey, P.R. **1991**. *Biochem. J.* 273, 43-47.

Rigby, S. **1991.** Agrow Estimates. B.t. in crop protection, PJB publication LTD.

Schuessler, H. and Schillivg, K. **1984**. Int. J. Radiat. Biol. 45, (3) 267-281.

Shapiro, M. **1992**. *J. Econ. Entomol.* 85, (5) 1682-1686.

Shapiro, M. and Robertson, J. L. **1992**. *J. Econ. Entomol.* 85, (4) 1120-1124.

Speaker, T. J. and Lesko, L. J. **1976**. U.S. Patent 3,959,457.

Speaker, T. J. and Speaker, T. J. **1988**. U.S. patent, 4,743,583.

World Health Organization **1995**. Review of the work of the onchocerciuis control program in the Volta River Basin area. 1974-1984. OCP/GVA/85.IR

ENVIRONMENTAL FATE AND BEHAVIOR

Movement of Pesticides to Non-Target Areas

David J S Arnold, AgrEvo UK Limited, Chesterford Park, Saffron Walden, Essex, UK, CB10 1XL

Losses of pesticides and movement to non-target areas raise natural concerns. Clearly there has to be a balance between the benefits of pesticide use and their environmental impact. Key routes of transport of compounds from agricultural uses include leaching processes to groundwater, run-off to surface waters, spray drift and volatilisation. Further redistribution of residues through long range transport may also be of significance. This paper describes routes and mechanisms for pesticide transport away from the target and describes ongoing research which is leading to the development of a better understanding of the interactions between pesticides and environmental compartments.

Crop protection, public and animal health have benefited to an incalculable extent from the development and use of pesticides over the past 30-40 years. The consequences of pesticide use to the environment are now, more than ever, key factors in balancing risks against benefits. The general lack of awareness of how pesticides with certain properties could be widely distributed in the environment is markedly reduced. In the modern industrialised nations of Europe and in North America pesticide regulation has developed into a rigorously controlled system with the aim of ensuring the safety of new chemicals to the operator, consumer and environment. This process requires a high degree of understanding of the environmental fate and behaviour of pesticides, and the potential impact of residues of the parent molecule and degradates/metabolites that move from their site of application or formation.

Much of the regulatory information on pesticides in the seventies and early 80's focused on persistence and efficacy. For example, there is a clear relationship between the need for an insecticide to persist long enough in soil to control a soil borne pest and the emergence of the undamaged crop. However, the longer the persistence the greater the potential for movement of a proportion of the residue away from its point of application. It is this factor that would receive much wider attention today.

That pesticides move to non-target areas is not disputed. There is a wealth of evidence in the literature documenting the presence of what are commonly referred to as persistent organohalide compounds in a variety of environmental compartments remote from their sites of application.

For example, information on the long range transport and environmental chemistry of toxaphene has recently been reported (Bidleman and Muir 1993). Toxaphene is a mixture of more than 180 chlorinated bornanes and bornenes. It was one of the most widely used pesticides in the seventies, controlling insect pests in cotton, smallgrain cereals, vegetables and soyabeans. In the USA, until its ban in 1982, total use of toxaphene amounted to around 3×10^5 metric tonnes. Derivatives were also, until recently, used on cotton in the former Soviet Union. Toxaphene is known to be persistent in the environment and whilst it demonstrates carcinogenic (Saleh 1991) and mutagenic properties (Hooper et al 1979). There is no evidence to link the presence of toxaphene in organisms with any increased risk either to man or the environment. Toxaphene has been detected in central Europe for example in Lake Baikal at concentrations of between 34 and 143 pg/l (McConnell et al 1993) and in fish and seals at up to 2.3 mg/kg (Kucklick et al 1993). The lighter, higher vapour pressure congeners in the water suggest that they arose from atmospheric deposition. Toxaphene has also been detected in N American lakes and in the air above the Arctic, so clearly there is evidence of widespread distribution of molecules of this type. We should, however, be cautious not to assume that this results in particular negative side-effects. There is still much to learn about the mechanisms of pesticide transport, the processes of transformation and whether the resultant residues pose any significant (ecological) threat.

Pesticides are in a continual state of flux. Once they are applied in agriculture there begins the process of dissipation. The main problem is that in many instances only a fraction of the applied material either hits the target, such as in the case of a post-emergence treatment, or is required to do the job, as in the case of a soil applied insecticide. Even that proportion reaching the crop may well be subject to losses to the atmosphere through physical and chemical processes as will be discussed later.

In order to examine the processes involved in the movement and redistribution of pesticides it can be helpful to visualise their contribution at different scales, ie. local (laboratory and field), regional and global. Each scenario will exert its own set of influences and consequences and, more importantly, significance. For example, the extent of microbial activity in the top soil of a cornfield may have a significant impact on the potential availability of a pesticide related to field leaching patterns but will be of little direct importance in relation to the movement of the same chemical between environmental compartments on a global scale.

It is important to recognise that from a pesticide regulatory view point most of the science is operating at the local scale. This is inevitable because it is possible to control, standardize and compare behaviour at this level. The further we move away from the site of application the greater the diversity and the lower the possibility of control. As variability increases because of factors such as different chemical use patterns, climate and geography, we need to focus more on our ability to predict pesticide movement in order to forecast the likelihood of undesirable behaviours and their frequency. This requires the development and use of appropriate models.

There is however another important factor. For new pesticide molecules there is a need to develop a data bank of environmental fate and effects information upon which to build increased levels of confidence in order to predict behaviour prior to registration. Any knowledge of where a compound may move to on a regional or global basis will be largely speculative based on prediction through modelling.

It is only when a compound has been used on a wide scale and generally at high use rates that the regional and global dissipation of the material becomes clear - if the compound is monitored. Only then will it be possible to determine if prediction of pesticide movement were judged correctly. Confidence in predicting behaviour improves as these data are more widely generated and used both in the calibration of models and validation of their predictions.

Environmental Concerns

Public awareness of pesticides has increased as a consequence of the widening debate on environmental issues. We would be foolish as scientists, either within or outside the agrochemical industry, to disregard such concerns. We have a responsibility to ensure that the consequences of introducing pesticides into the environment are not deleterious to the extent that they lead to short or long-term, non-reversible impacts. But this is different from the argument that the mere presence of a xenobiotic in any environmental compartment is undesirable.

It is a question of the level, however contentious this line of argument becomes. It is "how much" that triggers the extensive regulatory development programmes on the fate of new compounds destined for use in crop protection. The size of a residue must however be seen in terms of its impact and whether it gives cause for concern in classical concentration and exposure terms.

Probably the key concern in recent years has been the presence of pesticides in water used for abstraction for drinking water. In Europe the arguments over the setting of a maximum acceptable concentration (MAC) in groundwater for individual pesticides at 0.1 $\mu g l^{-1}$, the same as that for drinking water, (EEC 1980) has been extensively debated. Pesticides are still the only group of chemicals whose presence in drinking water is not linked to any assessment of toxicity.

Losses of pesticides to air and their possible sink in other environmental compartments in remote ecosystems is an increasingly relevant regulatory consideration of which one facet is the ecological quality of surface waters. Residues in food is not the subject of this paper but the chemical industry must maintain an awareness of the importance of public perceptions where pesticides in food are concerned because it frequently drives the debate on the acceptability of pesticides per se.

Factors Influencing Pesticide Movement to Non-target Areas

New molecules are synthesised to meet real agricultural needs and these dictate their structure and therefore their behaviourial properties. With an increased knowledge of the structural basis for chemical behaviour the intrinsic environmental properties of pesticides can be largely predicted (Arnold and Briggs 1990). When coupled with simple laboratory data, for example, on soil adsorption and kinetics, these provide early indications of persistence, leaching and volatilisation. However, synthesis of pesticides which balance efficacy and environmental acceptability is not simple. Transfer of the molecule to the field often leads to the expression of a different level of efficacy and environmental behaviour from that predicted in the glasshouse and laboratory.

Once a pesticide is applied to a crop its fate and behaviour are influenced by both abiotic and biotic processes largely under the control of the prevailing weather. These processes interact in a complex way which determine the availability of the pesticide and/or its degradates to move to non-target areas. Methods of application, formulation, crop type, season, and any other use pattern factors will also have a significant influence on the quantity of material available for re-distribution, both spatially and temporally.

Loss Processes and Re-distribution

The key processes involved in pesticide transport to non-target areas are leaching (to groundwater) run-off, erosion and drainage from soil surfaces into surface waters, and losses to air including spray drift during application and volatilisation from plant and soil surfaces.

Leaching to Groundwater

Groundwater is a vital resource. In many areas it is the sole source of drinking water and its 'quality' has to remain high. There is a wealth of information in the literature regarding the presence of certain pesticides in groundwater and a comprehensive report on the potential contamination of groundwater by pesticides was published by Aharonson (1987). The surprising thing is how low the actual concentrations are in the context of the mass of active ingredient applied (Clark and Gomme 1991).

Most of the ground water contamination does not come from the leaching of pesticides through their proper use in agriculture. A survey of farm wells by Frank et al (1990) investigated if normal use of pesticides led to contamination of well water. From data collected in 1986-1987 from 179 wells, in 1986 10 wells contained measurable levels of pesticides and in 1987, 4 wells.

In order to examine the potential for pesticides to move from their site of application to ground water, we need simply to look at the physical chemical properties such as water solubility, sorptive potential and stability. Generally the following criteria indicate potential to leach:

Water solubility (Sw) > 30 mg l^{-1}
Organic carbon adsorption coeff (Ko_C) < 500
Hydrolysis DT_{50}* > 4 days
photolysis DT_{50}* > 4 days
soil DT_{50}* > 21 days
Vapor pressure (Vp) < 10^{-4} Pa

*DT_{50} is the time for disappearance of half the initial residue.

However, as stated earlier, behaviour in soil is influenced by use patterns which may either ameliorate or accelerate leaching in a variety of different locations or situations.

Mechanisms

For many years pesticide leaching was viewed as a largely straightforward matter of downward chromatographic flow in soils of uniform structure. The application of soil physics particularly in terms of understanding structure and hydrology has confounded this view. The problem is not so simple. The soil can be considered as two hydrological zones, the upper (unsaturated) zone and the lower (saturated) zone. The level of the saturated zone changes as aquifers are recharged from precipitation and the water table rises and falls. This affects the distance a leachable pesticide needs to travel before it reaches groundwater. The fluctuations in matric potential of the unsaturated zone influence both the soil structure microbial activity and hydrological properties and therefore the behaviour of pesticides. Isolating processes to determine how they influence pesticide fate in the soil profile is not feasible other than through laboratory and possibly semi-field scale investigations but it is not difficult to see that adsorption and degradation will be key factors influencing the availability of apesticide or a degradation product for movement through the soil profile. If it is our aim to limit concentrations of pesticides in soil so that they do not leach then there is a potential conflict. Non-ionised pesticides that are strongly sorbed, eg with a K_{oc} value of > 3,000 will not readily leach. This in turn reduces bioavailability and tends towards a longer residence time in soil thus maintaining higher concentrations. Clearly ideal pesticides (from a leaching perspective) will have a K_{oc} of ca 500 but a DT_{50} of only a few days. However, this has to be balanced against the properties of soil degradation products. The active ingredient is often not the molecule of concern as with relatively non-mobile herbicide esters that hydrolyse to mobile biologically active acid moieties. In such cases it may be of benefit to have a slow rate of degradation of the non-mobile component and hence 'control' the release of the mobile moiety, allowing further degradative processes to reduce the amounts available for leaching. Adsorption and desorption can be readily measured in the laboratory but the processes that are in place in the field are more complex. In classical studies Walker (1987) modelled isoxaben leaching based on a simple adsorption constant but found that the degree of sorption increased with time. When an equation to describe this phenomenon was fed into a model, predicted and measured isoxaben concentrations were found to be similar.

The microbiology of soils is key to the removal of bioavailable pesticides. Surface soils have readily measurable microbial biomass and the degradation capacity can be demonstrated in the laboratory and field. We are beginning to learn more about subsoils in terms of their microbiology and degradative capacity (Siebert et al 1982) and the extent to which biodegradation takes place in the saturated zone. Biodegradation in subsurface soils can contribute to the attenuation of pesticides in their path through the soil profile. Populations of aerobic heterotrophs have been estimated in the saturated zone to be greater than 10^6 cells/gram and many pesticides have the potential to act as a carbon or nitrogen source. Pesticides are not, however, only moving in a downward direction. In summer months in the temperate climates there is a net upward movement of water through evapotranspiration and pesticides will be redistributed in the unsaturated zone either sorbed to particles or in solution and be subject to degradative processes in both upper and lower soil layers.

It therefore comes as no surprise that soil structure has a profound influence on pesticide movement. It is well known that in many soil types macropores may be formed which can act as rapid transport routes for water and hence (potentially) pesticides.

The extent to which macropore flow may contribute to pesticides entering groundwater by-passing potential soil adsorption sites is uncertain because heavy soils are frequently drained to avoid waterlogging during wet periods when cracks are closed, thus shedding water into drainage channels and streams. Macropores can exist even in sandy soils through channels left by decayed roots and via wormholes that interlace many soils, particularly those under reduced tillage systems and pasture. Recent research has shown enhanced retention of certain pesticides on to organic rich surfaces that line worm channels of *Lumbricus terrestris* (Stehouwer et al 1994).

Assessing and Predicting Leaching Potential

We have not moved very far in the past 5 years or so in terms of developing practical approaches to evaluate leaching although the reliance on the results of experimental work that are used for decision making by government regulatory authorities may be questionable.

All leaching studies are variations on the theme of applying chemical to the top and measuring it at the bottom. Regulatory triggers determine the consequences for the registrant. Are they, however, appropriate in the context of the need to protect groundwater?

According to German assessment criteria (BBA 1993), if 5% of a pesticide leaches in a laboratory leaching study or 2% in an aged leaching study, this triggers a lysimeter study. If, following an extensive 3 year lysimeter study in a sandy soil, the molecule of concern, or an active metabolite is detected at an average concentration of >0.1[1]µg/l in leachate at 1m depth over the study periodthis can be a "no go" decision point for the registrant. Thus limited information may be used to make judgements on pesticide behaviour in space and time on a regional scale. Detection of a residue in leachate from a sandy soil lysimeter is no indication of either the magnitude of the residue likely to be present in groundwater or the risk of it getting there. The problem for regulators is that they have few other means by which to compare the behaviour of one molecule with another other than through a standardised testing scheme. This approach must change and I believe it will through the development and use of models and improved soils/climatic information. If, however, we are to place greater reliance on the results of simulations industry has to be prepared to demonstrate that it is providing a high level of protection of groundwater. This can only be validated through an increase in monitoring - itself an expensive process.

Run-off, Drainage and Erosion

Run-off is a well documented mechanism for movement of pesticides from their site of application as a consequence of extensive crop irrigation or excessive rainfall. The process is less well defined, from a mechanistic point of view, than leaching and is not easily described by modelling. In many cases run-off leads to severe erosion of soil which may end up in surface water and give rise to undesired effects on aquatic biota that are of considerably greater impact than that of the pesticide itself. In the USA, sediments represent the largest volume pollutants in natural water (Cooper 1993).

In North America pesticide run-off is of concern from two main perspectives, contamination of wells and ecological impacts on aquatic organisms, particularly fish and amphibians. Run-off events from storms are often dramatic and on a large scale. In Europe run-off events are generally less dramatic but are increasingly viewed in a regulatory context as an important consideration in certain circumstances. The key word is circumstances. The transport of a pesticide in run-off is not as predictable in the same way that leaching can be predicted from the physico-chemical properties of the pesticide. So much is dependent upon use patterns, weather and geography. In a run-off event it may matter little whether a pesticide has a high or low K_{oc} value as it may be moved either adsorbed to soil particles or in the water phase. It is the impact it has in the variuos environmental compartments that is important.

However, run-off can occur on land with only slight inclines, for example, less than 2°. The pesticide may be transported to the lower end of a field where ponding occurs thus resulting in a potentially high loading over a smaller surface area. This would then become a leaching issue governed by the sorption and degradation rate of the compound. On steeper slopes overland flow can give rise to the dramatic consequences described earlier. The hydrology of the soil linked to soil structure and morphology clearly has a major impact on pesticide movement through run-off. In structured soils, particularly clays, macropores are principal routes for the lateral and vertical movement of water and generally weakly sorbed pesticides.

In soils where artificial drainage has been installed to prevent waterlogging, the residence time of a pesticide will be influenced by the soil water status. Rapid transport to drains and thus to surface water will occur if significant rainfall occurs shortly after pesticide application. It is thus a balance between the need to minimise waterlogged soils for optimum crop growth and the speed of transport of the pesticide into drainage water.

What is the significance of pesticides in run-off water? In an extensive review, Wauchope (1993) found that total losses of pesticides in run-off water were 0.5% or less except for highly persistent molecules. Pesticides with a water solubility of 10 ppm or higher were mainly dissolved in the water phase. Run-off events are discontinuous hence loadings to water courses will be intermittent. From an ecotoxicological point of view the concentration of the pesticide in water and/or sediment may be critical hence the impact of the extent and frequency of exceedence of the effect concentration needs to be known.

Evaluating the impact of a run-off event for regulatory purposes requires an assessment of risk. Simply requesting run-off studies to show that under the worst case scenario certain insecticides result in fish kills is hardly an effective use of resources and application of appropriate risk reduction techniques to reduce or avoid the potential impact is the way forward. The use of buffer zones or vegetative strips is being researched as a way of limiting the movement of sediments, nutrients and pesticides in to receiving waters (Muscutt et al 1993). The implementation of Best Management Practice in the USA is one good example of a collective approach to dealing with such issues. Here farmers, manufacturers and regulators devise programs and systems often based on a catchment or watershed to reduce or minimise pesticide contamination of water.

Losses to Air

Pesticide losses to air are difficult to assess, hence the subject area has been less well investigated from a regulatory viewpoint than, for example, losses to groundwater. However, on a regional and global scale the potential for long range transport of pesticides and consequential redistribution in environmental compartments is now receiving more attention.

Losses from sites of application can occur through two major processes:

1) Drift of sprayed droplets during application
2) Volatilisation and transport of pesticides from surfaces, water, soil, crops in the vapour phase.

Spray Drift

The losses from drift result from an inherent inefficiency in pesticide application. Initial losses are largely unrelated to the intrinsic physico-chemical properties of the pesticide, being dictated by droplet size, method of application and prevailing weather conditions. The fate of pesticides once airborne is then determined by physical and chemical interactions similar to those involved in pesticides volatilised from surfaces.

Drift has been studied extensively in different cropping systems (Davis et al 1993, Salyani & Cromwell 1992, Riley et al 1989) although there are widely differing estimates of losses in any given situation.

At the field scale, as a rule, drift is least from ground sprayers and highest from aerial sprayers. Buffer zones of varying distances have been cited as necessary in a regulatory context to prevent the movement of sprays to non-target areas (either hedgerows or water). These are precautionary measures but are based on a knowledge of agricultural practices and perceived vulnerability of ecosystems. For example, aerial spraying has been virtually eliminated from UK agriculture as the recommended buffer-zone between the limit of spraying and an adjacent body of water is 250m. With some fields only a few acres in area, this is an impossible restriction. In the vast wheat growing areas of North America where the sheer size of fields requires the use of aerial application, buffer zones of a few metres would be judged as meaningless.

It is difficult to assess the extent of drift and it is perhaps more apposite to ask what is the impact of drift from the target.

The environmental consequences of drift have been extensively studied in the UK under the auspices of English Nature (1993). Their report acknowledges the relationship between pesticide type, target, method of application and sensitivity of the off-target habitat, and sees buffer zones as an appropriate means of protecting vulnerable species. Exposure is the key; when products which have demonstrated high aquatic toxicity in the laboratory are applied in the field to bodies of water at simulated drift concentration, the toxicity is not expressed to anything like the same degree.

Mitigating factors such as adsorption, degradation, dilution contribute to a reduced level of effects.

Fine droplets from mist blowers in orchards will have the potential to travel further from the site of application compared with nozzle sprayers mounted a few inches from the crop canopy. Studies carried out in Germany demonstrate less than 1% drift from ground sprayers 5 metres from the source compared with 11-20% drift from orchard sprayers at the same distance. This concurs with data from Canada where drift losses of diclofop-methyl applied by ground sprayer to wheat were <0.1% of that applied (Smith et al 1986). This result was based on losses of the compounds both in droplets and in the vapour phase. Movement of volatile pesticides in the vapour phase can lead to unwanted side effects in non-target areas. It is well known that volatile herbicide esters can cause crop damage some kilometres from their site of application (Maber and Pinnell 1992). Drift of volatile components from spray applications is, however, a different phenomenon from pesticide losses from plant and soil surfaces in the vapour phase.

Volatilization From Plant/Soil Surfaces

The air/soil air/plant boundary layers provide an interface in which pesticides may be moved in the vapour phase. Soil processes will affect the movement and residence time of a pesticide once it reaches the soil. These processes are interactive and include sorption, desorption, advection, dispersion and degradation (biotic and/or abiotic). The heterogeneity of soils provides discontinuity in flow patterns for solute movement and, as previously described, can involve preferential flow paths in macropores. For volatile compounds, transport of the pesticides as a solute is confounded by vapour phase transport within the soil and volatilisation at the surface. Where conditions lead to water evaportation at the soil surface, volatile compounds can move at a faster rate in soil water than they can volatilise at the soil surface thus leading to a concentration gradient at the soil/air interface. This cycle of redistribution and loss will, as a consequence, reduce the amounts of pesticide available for leaching.

Recent European regulatory interest in volatilization of pesticides has been led by German registration criteria to demonstrate the rate and amount of pesticide loss from plant and soil surfaces. Whilst this is an area of knowledge that is under-researched, data for regulatory purposes are required from industry which are likely to be generated from widely differing and hence non-comparable experimental systems. Strangely, vapour pressure of the active ingredient is not a trigger value for evaluating potential loss from plant or soil.

Whilst it is possible to predict from chemical structure and physico chemical properties that losses will occur in the vapour phase, the duration and extent of losses will be widely variable at the field scale influenced by a host of factors including formulation type, mode of application, timing, target (crop, water, soil), topography, temperature, moisture, windspeed, sunlight and so forth. These influences and interactions cannot be reproduced in an experimental laboratory unit. What can be of value is to measure the behaviour of different chemicals and formulations within defined experimental conditions and to compare these data with what may be predicted from physico chemical properties.

Comparisons may also be made in the field where (in the simplest approach) residues remaining on soil or crops are taken as that portion which has not volatilised. However, such interpretations are probably flawed as they are likely to over-estimate the suggested loss.

Losses of pesticides from soil and plant surfaces may also be manipulated by changes in formulation type.

The saying "what goes up must come down" is certainly true in terms of the redistribution of pesticides but again we need to be cautious in ascribing the mere presence of a compound to negative environmental consequences. Once pesticides are airborne they may become associated with particulate matter or water droplets and be redeposited either as dry deposits or in precipitation. Airborne residues of commonly used low solubility herbicides, including atrazine, can be detected but generally at low nanogram per m^3 levels (Grover 1991). Concentrations are reduced by chemical and photochemical degradation processes in the atmosphere which impacts on their redistribution and hence availability for re-deposition.

Predicting Movement of Pesticides Through Modelling Fate and Behaviour

Modelling may hold the key to facilitating the extrapolation of experimental data at the local scale to regional or global patterns of behaviour.

Leaching

Leaching models are perhaps the most widely known tools for those involved in pesticide regulation. They are vital to the future assessment of risk as defined as the probability of a residue of a certain magnitude leaching to groundwater. Leaching models such as PRZM (Carsel 1984) have been modified to improve the quality of their simulations. For example, a variation of PRZM (PELMO) (Klein 1993) was developed to incorporate aspects of pesticide behaviour such as non-first order degradation patterns and soil depth related degradation rates. Such models are able to simulate pesticide behaviour in a way that links laboratory, semifield and field data and provides an estimate of leaching potential.

The term predictive model is misleading in the sense that all models have an inherent level of uncertainty and being precise about a compounds' behaviour is not straightforward. The calibration process is one that is frequently applied in the use of such models to reduce the level of uncertainty. The key is to carry out a sensitivity analysis to determine which factors most influence the leaching behaviour of the compound, e.g. DT_{50} in relation to soil type, temperature and K_{oc} measurements can be used as inputs to a model such as PRZM and combined with information from databases (soils and climate) to simulate leaching potential. These outputs, for example as maximum depths to which the pesticide can leach, can then be used to determine the probability of the pesticide reaching groundwater under a defined set of critiera. This process has been pioneered by DowElanco (Laskowski et al 1990) and offers far more realistic scenarios then the worst case assumptions beloved of some regulators.

Leaching models are also being adapted to account for the process of non-steady infiltration of solutes, for example by macropore flow (Jarvis and Leeds Harrison 1987, Jarvis1991).

Run-off

Run-off models are less familiar than leaching models and none have been developed for use in pesticide regulation.

Persistence is a key factor and residence time of pesticide residues in the topsoil naturally determines that available to be moved in a run-off event. Whilst soil degradation and field dissipation studies will provide data on persistence, simulating behaviour in a range of scenarios is possible using models such as PERSIST (Walker & Barnes 1981).

This model, which has been extensively used and validated, includes parameters which allow assessment of the influences of temperature and water content in degradation rates. The physical processes involved in run-off are, of course, largely independent of degradative processes, although transporting residues to 'new' sites could infer changed biotic and abiotic influences on the pesticide residues. The development of a fugacity model to predict concentrations of pesticides in surface water from run-off has recently been reported (Di Guardo et al 1994). The fate and behaviour of four herbicides following application and entry into surface waters in two hydraulically isolated basins was compared with data from simulations of a fugacity-derived model. The model was generally in agreement with measured values but did not account for certain processes such as initial losses of pesticides during application to soil or soil erosion. The usefulness of the model, however, was demonstrated.

Characterising the nature of the environment into which pesticides are introduced is an essential prerequisite to describing, simulating or predicting the potential losses of pesticides from their targets.

Modelling pesticide behaviour in relation to the soil and water environment (hydrology) has been limited by the lack of appropriate databases or systems that account for spatial and temporal variability. Insofar as they exist, such data bases need to be extensive and detailed. In the UK such a spatial environmental information system is under development, jointly funded by government and industry. The system known as SEISMIC (Hallet et al 1994) contains information on soils, hydrology of soil types, cropping patterns, weather data and climatic data. Information is based on a 5 km grid system. The information was designed for selection and transfer to models to simulate pesticide leaching or persistence and to assist in the definition of regions vulnerable to pesticide use.

In the USA computer modelling is being used in the development of agricultural management practices that protect water quality. For example, hydrologic/water quality models such as CREAMS (Knisel 1980) have been developed to predict the impacts of agriculture on surface water quality. Their use has been limited due to the lack of good spatial/temporal data but recent developments in the geographic information system (ARC/INFO) should provide the ability to account for the huge variability in environmental parameters.

Losses to Air

Short range spray drift and dry deposition of pesticides has been simulated using a simple mathematical model. (Miller and Hadfield 1989).. Not surprisingly drift was determined to be governed by spray nozzle type, yielding different droplet patterns, and by wind velocity. Deposition however decreased exponentially with increasing distance from the point of application. The model can be used to determine pesticide concentration in water related to distance from the emission. From a regulatory standpoint, modelling losses of pesticides to the atmosphere from soils and plant surfaces and their consequent movements is non-existent. However, the future probably lies in considering air as one compartment of a multimedia approach to modelling within the concept of fugacity.

The fugacity approach of Mackay (1979) describes an environmental model into which the characteristics of the substance can be placed to predict transport, rates of transformation, regions of accumulation and concentrations. It can be applied at various scales but it is argued that using a fugacity model could have defined the problems associated with persistent organochlorine compounds earlier.

Wania and Mackay (1993) modelled the distribution of toxaphene using a multimedia model based on four compartments : soil, sediment, air and water. The model describes interactions between compartments in any one climatic region. Each "building block" is linked by advection in the atmosphere thus providing a model for global distribution of the chemical.

Fugacity based equations assist the process of determining the direction and magnitude of fluxes between compartments. In the case of toxaphene the model was able to describe the directions of fluxes from air to water and water to air that were coincident with levels of toxaphene found in environmental compartments in different regions. If fugacity is able to determine the dominant environmental links for each pesticide use patterns could be designed to minimise the entry of the pesticide into that compartment.

Regulation of Pesticides - Research Needs

The development of new molecules acknowledges environmental impact potential, particularly in terms of persistence and mobility. The development of pesticides active at lower rates means reducing loadings at the point of application and therefore less material with loss potential. The pesticide registration process requires an extensive suite of studies to evaluate the fate, behaviour and impact of a chemical before its authorisation for use in agriculture. Many of these studies fail to address the particular relationship between use pattern and consequences of use. Improving the scope and accuracy of predictive models is one way forward. Modelling at different levels needs to be encouraged (for example volatility modelling at the field scale) in order to better understand the mechanisms of pesticide loss from surfaces. In the shorter term this might involve the generation of more rather than less laboratory and field data in order to both calibrate the models and validate model predictions.

To understand how pesticides move to non-target areas we need to encourage the establishment of regional soils crop and climatic data bases for incorporation into models. The agrochemical industry itself can assist in several ways to limit the movement of residues. Formulations need to be manufactured that stay where they are targeted. Applications systems need to focus on reducing losses of pesticides during placement. Research into reducing run-off through novel approaches to soil management should also provide use pattern related management practices to attenuate movement. The global nature of certain uses of pesticides requires better assessment of risk. Risk assessment leads to identifying or characterising the nature of the risk. The consequences of pesticide movement to non-target areas requires an improved characterisation of the systems likely to be affected so that better estimates of impacts can be made.

Pesticides by their very nature are designed to remove or reduce to manageable levels pests, weeds or pathogen infestations. The level of impact on non-target species is already assessed in the registration process using surrogates to represent trophic levels in the ecosystem. Calculation of predicted environmental concentrations provides estimates with which to derive potential exposure levels. These must then be balanced against the risk of the organism(s) being exposed (their sensitivity and resilience in terms of recovery). It is arguable whether off-target impacts from pesticides are at levels of concern as the process of risk reduction is an integral part of risk assessment. There is no such thing as an unchanged ecosystem. What we need to be more certain of is whether the perturbations that may be detected are firstly attributable to the side-effects of agrochemicals alone and secondly whether ultimately they are irreversibly damaging. The agrochemical industry faces a tough time as environmental pressures seek to reduce pesticide use. Regulation has to be based on good science and not gut feelings. Improvements in our understanding of the complex interactions between pesticide and environment enhance our ability to model the fate and behaviour of pesticides more effectively.

We need to encourage the improvement of existing environmental databases and establishment of such databases on a wider scale (e.g. Europe). Such databases provide the means to carry out a more realistic assessment of pesticide behaviour and will be used in pesticide regulation. The agrochemical industry is playing a leading rôle in this context and will continue to contribute scientifically and technically to improving our understanding of the fate and behaviour of pesticides and their environmental impact.

References

Aharonson, N. *Pure and Applied Chemistry*, **1987**, *59(10)*, 1419-1446.

Arnold, D.J.; Briggs, G.G. *Fate of Pesticides*; Eds; Hutson, D.; Roberts, T.R., Wiley & Sons Limited, 1990.

Arnold, J.G.; Williams, J.R. *Ag. Res. Serv.* Temple, 1985, SWRRB, Texas, USA, *Vol 1.*

BBA. *Criteria for Assessment of Plant Protection Products in the Registration Procedure*, Bibliothek der Biologische Bundesanstalt für Land- und Forstwirtschaft, **1993**, *Berlin 33 Germany*, ISBN 3-489-28500.

Bidleman, T.F.; Muir, D.C.G. *Chemosphere,* **1993**, *27(10)*, 1825-2094.

Carsel, R.F.; Smith, C.N.; Mulkey, L.A.; Dean, J.D.; Jovise, P. *Users Manual for the Pesticide Root Zone Model PRZM*, **1984,** EPA/600/3-84-109, USEPA Athens, Georgia.

Clark, L.; Gomme, J. *Pesticide Science*, **1991**, *32*, 15-33.

Cooper, C.M. *J. Env. Qual.*, **1993**, *22*, 402-408.

Davis, B.; Lahhani, K.; Yates, T.; Frost, A.J.; Plant, R.A. *Agriculture Ecosystems and Environment*, **1993**, *43(2)*, 93-108.

Di Guardo, A.; Calamari, D.; Zanin, G.; Consalter, A.; Mackay, D. *Chemosphere*, **1994**, *28(3)*, 511-531.

EEC. Directive (80/778/EEC). *Official Journal of the European Committees*, **1980**, *23*, L229/11.

English Nature, Peterborough, UK. *The*

Environmental Effects of Pesticide Drift, Ed. Cooke, A.S., **1993**, ISBN 1-85716-124-6.

Frank, R.; Braun, H.; Clegg, B.; Ripley, B.; Johnson, R. *Bulletin of Environmental Contamination and Toxicology*, **1990**, *44(3)*, 410-419.

Grover, R. *Environmental Chemistry of Herbicides*. Ed. Grover, R.; Cessna, A.J. CRC Press Inc., Florida, USA, **1991**, 2, 89-117.

Hallet, S.H.; Thanigasalam, P.; Hollis, J. in *Environmental Modelling - the Next 10 Years*. Eds. Stebbing, A.R.D.; Travis, K.; Matthiessen, P., **1994,** ISBN 0-9522575-0, SETAC-Europe (UK Branch).

Hooper, N.K.; Ames, B.N.; Saleh, M.A.; Lasida, J.E. *Science,* **1979,** *201,* 591-592 .

Jarvis, N.J. *Dept. Soil Sci.,* 1991, Reps & Diss, Swedish Univ. Agr. Sci., Uppsala, Sweden, 58.

Jarvis, N.J.; Leeds-Harrison, P.B. *J. Soil Sci.*, **1987**, *38(3)*, 487-498.

Knisel, W.G. *CREAMS : A field scale model for chemical run-off and erosion from chemical managements systems,* **1980,** Report 26, USDA Washington, DC.

Klein, M. *PELMO : Simulation Model Using PRZM for Scenarios in Germany*, **1993,** Fraunhofer Institut Schmallenberg, Germany.

Kucklick, J.R.; McConnell, L.L.; Bidleman, T.F.; Ivanov, G.P.; Walla, M.D. *Chemosphere*, **1993**, *27(10)*, 2017-2026.

Laskowski, D.A.; Tillotson, P.M.; Fontaine, D.D.; Martin, E.J. *Phil trans. Royal Society London*, **1990**, *329*, 383-389.

Maber, J.; Pinnel, G. *Weed Science Soc. of Victoria, Melbourne, Australia*, **1992**, *2*, 293-296.

Mackay, D. *Envir. Sci. Tech.*, **1979**, *13(10),* 1218-1223.

McConnell, L.L.; Kucklick, J.R.; Bidleman, T.F.; Walla, M.D. *Chemosphere*, **1993**, *27(10)*, 2027-2036.

Miller, P.C.H.; Hadfield, D.J. *Journal of Agricultural Engineering Research*, **1989**, *42(2)*, 135-147.

Muscutt, A.D.; Harris, G.L.; Bailey, S.W.; Davies, D.B. *Agriculture, Ecosystems and Environment*, **1993**, *45*, 59-77.

Riley, C.M.; Wiesner, C.J.; Ecobichon, D. *Bull. Env. Contamin. Toxicol.*, **1989**, *42(6)*, 896-898.

Saleh, M.A. *Rev. Environ. Contam. Toxicol.* **1991**, *118,* 1-85.

Salyani, M.; Cromwell, R.P. *Transaction of the America Society of Agricultural Engineers*, **1992**, *35(4)*, 1113-1120.

Siebert, U.; Führ, F.; Mittelstaedt, W. *Landwirtschaft Forschung*, **1982**, *35*, 5-13.

Smith, A.E.; Grover, R.; Cessna, A.J.; Shewchuk, S.R.; Hunter, J.H. *J. Envir. Qual.*, **1986**, *15(3)*, 234-238.

Stehouwer, R.C.; Dick, W.A.; Traina, S.J. *J. Env. Qual.*, **1994**, *23*, 286-292.

Walker, A. *Weed Research*, **1987**, *27*, 143-152.

Walker, A.; Barnes, A. *Pesticide Science*, **1981**, *12*, 123-132.

Wania, F.; Mackay, D. *Chemosphere*, **1993**, *27(10)*, 2079-2094.

Wauchope, R.D. *J. Env. Qual.*, **1978**, *7(4)*.

Fate of Pesticides in Subsurface Soils and Groundwater

John P. E. Anderson, Bayer AG, PF-E, Institut für Ecological Biology, 51368 Leverkusen, Germany

This paper briefly discusses sampling procedures, the microbiology, the climate, and the degradation of pesticides in subsoils and groundwater. In these habitats, pesticides are degraded by the same biological and hydrolytic processes as in top soils and surface waters. As is shown using data from experiments with vadose zone soils, the variables determining degradation rates are the same in subsoils and topsoils. When laboratory tests show that pesticide degradation rates in subsoils are slower than those in top soils, this is mainly due to the smaller populations of subsoil microorganisms, and not to a lack of capacity of these organisms to degrade these chemicals.

Broad surveys in the United States and Europe have shown that in areas of intensive agriculture, contamination of groundwater with trace quantities of pesticides is not uncommon. In a review of the literature covering non-point contaminations in the United States, Hallberg (1989) found that a total of 39 different pesticides had been detected in the groundwater of 34 states or provinces. Leistra and Boesten (1989), who similarly reviewed the literature covering Europe, reported that at least 14 different pesticides were found in the groundwater of 7 Western European countries. In the United States and Europe, the pesticides most often detected were triazine herbicides, phenoxy herbicides, volatile fumigants, and organophosphate or carbamate nematicides (GIFAP, 1992; Helweg, 1992; IVA, 1989).

What happens to pesticides when they enter subsurface soils and groundwater? Reasonable answers to this question can be obtained experimentally, through leaching and degradation studies, and theoretically, through the use of leaching and degradation models. To create models which reliably predict the fate of pesticides in subsurface environments, a basic understanding of the hydrogeological cycle and a more thorough understanding of the mechanisms that determine the rates of pesticide transport and degradation in soils and groundwater are necessary. The principles of hydrogeology and of transport of pesticides through the unsaturated zone to saturated zone and groundwater have been extensively reviewed elsewhere (Bouwer, 1984; Matthess and Isenbeck, 1987; McCarthy and Zachara, 1989; Nicholls, 1988; Smith, 1990; Yaron, 1989), and will not receive further attention at this time.

In this paper, discussion will concentrate on pesticide degradation in subsoils and groundwater. Since agricultural chemicals applied to crops or soils must pass through the unsaturated or ***vadose zone*** to enter groundwater, special attention is given to this zone. The first section of the paper discusses methods for "clean sampling" of subsoils and groundwater. The second section gives a general overview of the microbiology of subsoils and groundwater, and also takes a detailed look at the microbes detected down the profiles of 4 selected European soils. The third and fourth sections, which respectively discuss fluxes in climate and degradation of pesticides, concentrate on the *vadose* zone of the soil. The final section closes by summarizing information gaps and future needs in this relatively new area of pesticide research.

Due to space limitations, a comprehensive review of the literature has not been attempted. Rather, where possible, reference is made to review papers or particularly pertinent primary literature. Since systematic studies on pesticide degradation as related to vadose zone climate and microbiology are limited in number, the body of the paper consists of data from the author's own laboratory. These data were generated during the course of an ongoing joint project between the Bayer AG (D), INRA (F) and Zeneca Ltd. (UK). The project has been partially and generously supported under the STEP (Science and Technology for Environmental Protection) and ENVIRONMENT Programs of the European Council of Ministers.

1054–7487/95/0127$12.00/0

Sampling of Subsoils and Groundwater

McCarthy and Zachara (1989) very appropriately stated that our understanding of subsurface environments is limited by the techniques we have to characterize them. This becomes quite clear when we attempt to sample deeper subsoils or groundwater without contaminating them with surface material or soil from layers directly above.

In early work with subsoils or groundwater, samples from depths greater than 10 or 12 m were collected using equipment designed to drill for water or oil (Phelps et al., 1989). This type of equipment required lubricating fluids or drilling muds, which caused serious problems with contaminations from the surface or upper layers of soil. Recently, some advances have been made with such equipment using sterilized mineral lubricating muds, sterilized muds containing marker organisms, or drilling without muds or other lubricants. Other developments include corers or sampling devices that can be lowered into predrilled boreholes such as extruders, corers, or split spoon samplers (Phelps and Russel, 1993; Dobbins et al., 1992). Many of the newer techniques and devices show promise, but to date, there is no perfect way to collect uncontaminated, deep-soil or groundwater samples. Work on this problem continues.

A variety of techniques can be used for clean sampling of vadose zone soil profiles. The simplest of these is to dig a fresh pit, cut away exposed and contaminated surfaces with sterilized tools, and sample the profile with clean and sterilized tools. This approach was successfully used by Helweg, (1992), Lavy et al., (1973), Moorman and Harper, (1989) and Walker et al., (1989) in their investigations on subsoil degradation of MCPA, Atrazine, Metribuzine and Chlorsulfuron, respectively.

Quite often, pit digging is not practical or would unduly disrupt an agricultural or study site. In such cases, or where samples from relatively deep vadose zone layers (e.g. 2 to 15 m) are needed, soils have been collected using coring devices which are either pressed (Mueller et al., 1992), hammered, or drilled into the soil (Dictor, 1994; Lewis et al., 1992; Takagi et al., 1992). For our own work, in which we sampled vadose zone soils to 8 m depths, we used a motor driven "Humax" coring device (Max Hug, Dipl. Masch. Techn., CH-6003 Luzern 11, Postfach 47) developed for collection of subsoil samples for trace analyses of pesticides and other chemicals. With this device, which has a functional limit of 5 to 10 m (depending on soil texture, soil density and soil moisture), hollow steel tubes are drilled step by step into the soil, and cores (e.g. 25 cm to 1 m long, 5 to 25 cm in diameter) are removed in disposable plastic cylinders during continued drilling. For depths down to 10 m, new sections of tubing are added as needed.

Microorganisms in Subsoils and Groundwater

With bioremediation in mind, serious investigation of the microbiology of subsoils and groundwater came into vogue in the late 1970's. The pioneer work in this area dealt with procedures for detection, enumeration and isolation of subsoil microorganisms (Dobbins et al., 1992; Ghiorse and Wilson, 1988), and thereafter, with the taxonomy, morphology, physiological characteristics, biomass, and activity levels of the organisms that had been detected (Bone and Balkwill, 1988; Ghiorse and Wilson, 1988; Hirsch et al., 1992).

In saturated zone sediments or groundwater, microbial populations have been found to consist mainly of bacteria. In a summary of the literature, Hirsch (1992) reported that more than 24 genera of bacteria have been detected in subsoil and groundwater samples. In free aquifer water, Gram-negative bacteria appear to predominate, while in saturated zone sediments, Gram-positive bacteria are most common. The majority of these bacteria are found attached to soil surfaces. Both mobile and immobile bacteria are detected in groundwater, and both can be transported long distances by the lateral movement of water. Filamentous fungi are rare in the saturated zone and groundwater and appear to be limited to the upper 10 m of soil.

The numbers of genera of fungi and bacteria found in subsoils and groundwater are fewer than those found in topsoils or surface water. However, the physiological capacities of subsurface microorganisms, and their ability to degrade pesticides and other organic chemicals, seem to be the same as those found in microorganisms from surface habitats (Beloin et al., 1988; Dobbins et al., 1992; Hirsch et al., 1992).

In the vadose zone of soils, in comparison to topsoil, the numbers of bacteria and fungi present, their biomass, and their activity levels decrease rapidly with increasing depth (Bone and Balkwill, 1988; Beloin et al., 1988; Dictor, 1994; Dictor et al., 1992; Ghiorse and Wilson, 1988; McNabb and Dunlap, 1975; Mueller et al., 1992). In the soil profiles thus far examined, the greatest decreases in microbial life occurred at depths between 30 cm and 150 cm. Thereafter, microbial numbers and biomass can further decrease or increase; when increases occur, numbers never reach those found in topsoil layers (Lewis et al., 1992; Weyandt and Schweisfurth, 1992). The depths at which microbial densities fluctuate differ from soil to soil, and the reasons for density differences have not been systematically investigated. Weyandt and Schweisfurth (1992) speculate that increases in density of bacteria in deeper layers might be related to conditions established in old capillary fringe areas. From our own work on subsoil microbiology and nutrient fluxes, we speculate that the scarcity of filamentous fungi below 10 m is primarily due to a lack of readily available carbon. In comparison to bacteria, non-predatory filamentous fungi are relatively large microorganisms. Because of this, they require correspondingly large quantities of readily available carbon for continuous activity and growth.

In our own laboratory, we investigated the distribution of bacteria and fungi, and the quantities of metabolically active microbial biomass in the vadose zones of 4 middle European soils. For this purpose, soil samples were collected using the Humax device described above. Immediately after collection, soil cores were transported to the laboratory where they were re-cored under aseptic conditions. To investigate the distribution of microbial cells in the soils, samples from each core were bulked, mixed, diluted with sterile water, plated on a variety of nutrient media, and incubated under aerobic or anaerobic conditions. The quantities of metabolically active microbial biomass at different depths were estimated as described by Anderson and Domsch (1978).

The results of this work are briefly summarized in Tab. 1 and Fig. 1. In all 4 profiles, the greatest decrease in microbial numbers (Fig. 1) and microbial biomass (Tab.1) occurred within the first meter under the soil surface. Thereafter, in deeper layers in 3 of the 4 profiles, both microbial numbers and microbial biomass increased and then again decreased with depth. In regard to the fungi, it can be seen that living cells of these organisms occurred in all soil layers (Fig. 1). However, the active colonization of subsoil particles by the metabolically active fungal mycelia, measured by a modification of the method of Gams and Domsch (1967), stopped in the silty loam soil of the Mönchengladbach site at 4 m, and in the sandy soil of the Laacherhof site at 2 m. In the above discussion of the microbiology of the saturated zone, it was stated that fungi are more or less limited to the upper 10 m of soils. The data given here, and the work of Dictor (1994), which showed that active fungi were exclusively found in the upper 2 m of a silty clay soil, support this generalization and add to it by suggesting that metabolically active fungi, as opposed to living but inactive fungal spores, are limited to the upper 2 to 4 m of soils.

The lack of fungi in deeper soil layers or groundwater means that metabolism in these habitats will be dominated by bacteria. This is in contrast to topsoils where fungal populations make the greatest contributions to the microbial biomass (Anderson and Domsch, 1975a; 1980), and play a predominant role in the initial and/or final steps of biodegradation and mineralization of many compounds (Anderson and Domsch, 1974, 1975b), including pesticides (MacRae, 1989).

Will the lack of fungi in deeper subsoils and groundwater negatively influence pesticide degradation? From the data thus far published for pesticides (Tab. 2), and for other organic or compounds (Dobbins et al., 1992), it can be speculated that the routes of metabolism, and the metabolites produced, will not differ from those detected in topsoils, subsoils or groundwater. However, the lack of fungi in deeper soil layers could lead to slower mineralization of the metabolites of some compounds.

Vadose Zone Climate

Among the variables that strongly influence the rates of pesticide degradation in soils and water are their oxygen content and temperature. At a total of 4 field sites in Germany, France and England, we have measured fluxes in these variables in top and vadose zone soils for up to 5 years (Dictor, 1994; Takagi et al., 1992). To

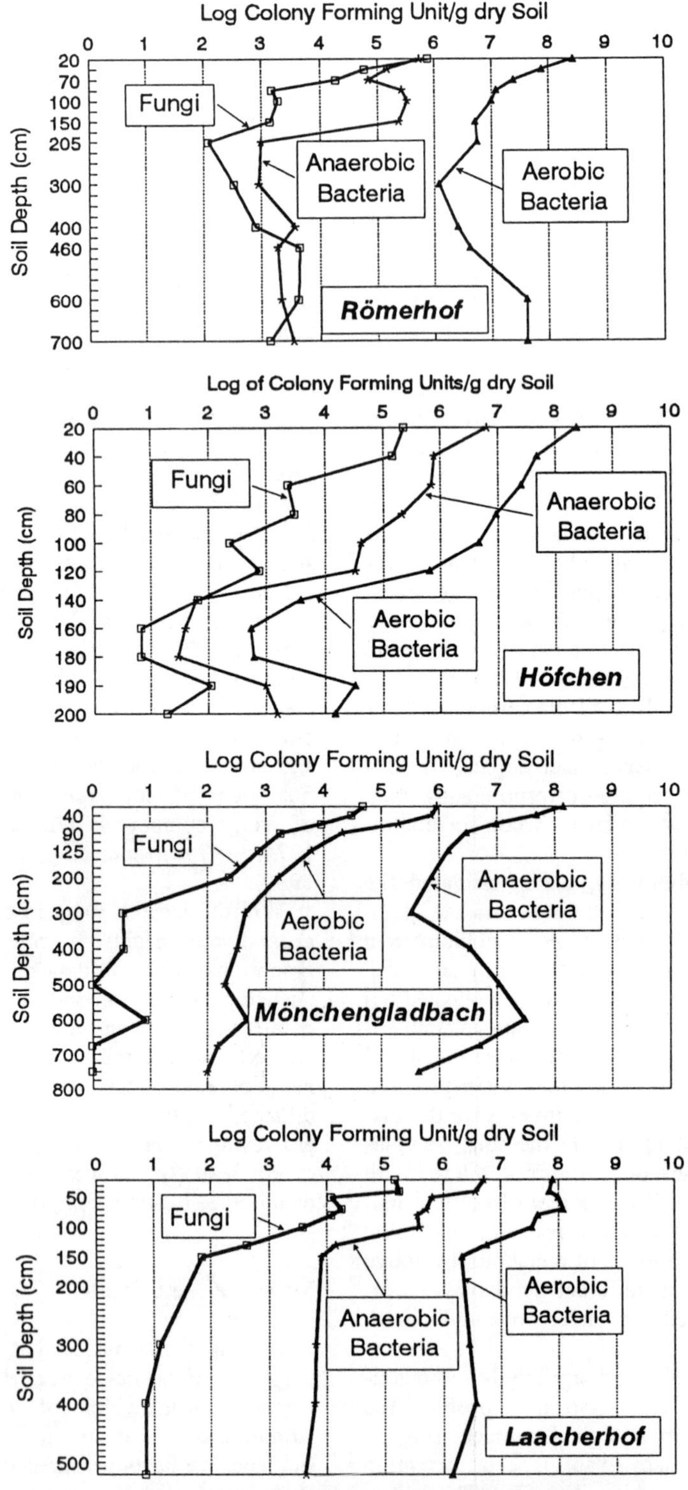

Fig. 1. Microbiological characteristics of the vadose zone of 4 German soils.

Table 1. Carbon content, pH values and microbial biomass in the profiles of 4 German soils.

Mönchengladbach (a)				Römerhof (b)			
Depth	% org. Carbon	pH (KCl)	Microbial Biomass (c)	Depth	% org. Carbon	pH (KCl)	Microbial Biomass (c)
0-25	0.88	5.5	462	0-20	0.87	6.9	660
25-50	0.46	5.3	144	20-45	0.39	6.7	174
50-75	0.20	5.4	48	45-70	0.22	6.5	110
75-100	0.24	5.4	35	70-90	0.20	6.5	88
100-145	0.09	5.3	15	90-100	<0.10	6.5	110
145-240	0.09	5.3	13	100-150	0.45	7.3	176
240-345	0.05	5.1	8	150-205	0.82	7.7	44
345-455	0.08	5.3	35	205-300	0.56	7.3	88
455-555	0.08	5.7	46	300-400	<0.10	6.7	44
555-635	0.06	6.1	41	400-460	<0.10	6.5	22
635-700	0.05	6.3	38	460-600	<0.10	6.3	10
700-800	0.03	6.5	7	600-700	<0.10	6.6	10
Laacherhof (a)				**Höfchen** (b)			
Depth	% org. Carbon	pH (KCl)	Microbial Biomass (c)	Depth	% org. Carbon	pH (KCl)	Microbial Biomass (c)
0-20	1.41	5.8	465	0-20	2.73	5.7	930
20-40	1.12	6.3	218	20-40	1.49	5.2	234
40-50	0.65	6.2	113	40-60	0.61	5.0	52
50-70	0.38	6.3	103	60-80	0.18	5.0	22
70-80	0.38	6.1	81	80-100	0.11	4.6	20
80-100	0.14	6.5	41	100-120	0.07	4.7	20
100-130	0.05	6.5	45	120-140	0.04	4.4	10
130-150	0.09	6.1	20	140-160	>0.01	4.2	< 10
150-300	0.03	6.2	44	160-180	0.02	4.1	<10
300-400	0.05	6.2	27	180-190	0.02	4.0	<10
400-520	0.02	6.2	36	190-200	0.02	4.1	<10

(a) adapted from Takagi et al. 1992
(b) Anderson and Andersch unpublished
(c) mg microbial C/kg dry wt soil

make these measurements, nests of tubular steel probes were drilled into the profiles and left with their bottom ends at depths between 1 and 8 m in the soil. The soil in the tubes was removed and the upper ends were capped and hermetically sealed. By means of valves in the caps, the air in the tubes could be evacuated, and fresh samples of soil atmosphere could be collected as required. The oxygen content of soil air samples was measured with an oxygen electrode and by gas chromatography. Subsoil temperatures were continuously monitored using thermometers inserted 6 cm into the soil at the base of the probes. A preliminary description of the probes and data from these investigations are given by Takagi et al. (1992).

The trends and general results found at all sites can be exemplified using data from a single year from a single soil profile, which is referred to in Tab. 1 and Fig. 1 as the Mönchengladbach

Table 2. Pesticides Subject to Abiotic or Biological Degradation in Subsoils and Groundwater

Common Name	Chemical Class	References (First Author and Date)	Common Name	Chemical Class	References (First Author and Date)
1,2-dibromo-3-chloro-propane	Halogenated propane	Leistra 1991	Fluometron	Substituted urea	Mueller 1992
1,2-dibromo-methane	Halogenated methane	Aelion 1987 Moore 1989 Pignatello 1987	HCH	Chlorinated hydrocarbon	Bachmann 1988 Ou 1985 Panda 1988
1,3-dichloro-propene	Halogenated propene	Boesten 1991 Leistra 1991	MCPA	Phenoxy carbonic acid	Helweg 1990 Helweg 1992
Alachlor	Acetamide	Moorman 1990 Pothuluri 1990	Mecoprop	Phenoxy carbonic acid	Agertved 1992
Aldicarb (Metabolites: Ald. sulfoxide, Ald. sulfone)	Oxime-carbamate	Miles 1985 Ou 1985 Smelt 1983 Vonk 1992	Methomyl	Carbamate	Smelt 1983
Atrazine	Triazine	Agertved 1992 Boesten 1993 Lavy 1973 Helweg 1992 McMahon 1992 Wood 1991 Wehtje 1983	Metribuzin	Triazinone	Locke 1991 Moorman 1989
Carbofuran	Carbamate	Lewis 1992 Ou 1985	Metsulfuron-methyl	Sulfonyl-urea	Walker 1989
Chlorsulfuron	Sulfonyl-urea	Walker 1989	Methyl iso-thiocyanate		Boesten 1991
2,4-D	Phenoxy carbonic acid	Dictor 1992 Dictor 1994 Helweg 1992 Lavy 1973	Oxamyl	Oxime carbamate	Smelt 1983
Ethylene-dibromide	Brominated ethylene	Aelion 1987	Parathion	Organo-phosphate	Adhya 1981 Adhya 1987
Fenthion	Organo-phosphate	Adhya 1981			

field site. The water table at this site fluctuates between 13 and 14 m, and the site is located at a latitude of 51° 08′ N and a longitude of 6° 21′ E, and is 75 m above sea level. According to USDA (1975) classification, this is a mesic soil; the texture down the profile ranges from sandy loam to sand. The quantities of oxygen found at different levels of the profile are shown in Fig. 2. The data show, that at all depths down to 8 m, aerobic conditions prevailed. The greatest fluxes in oxygen content in the vadose zone were at 1 m. Between April 1992 and April 1993 the quantities of oxygen found at a 1 m depth ranged from 89 to 67 % of that found 1 m above the soil surface. At 3, 5 and 8 m depths, the oxygen contents ranged from 77 to 83 %, 74 to 81 % and 79 to 83 %, respectively. The data given in Fig. 2 are in good agreement with those of Becker et al. (1989), Eichinger and Merkel (1985), and Torstensson et al. (1983), who

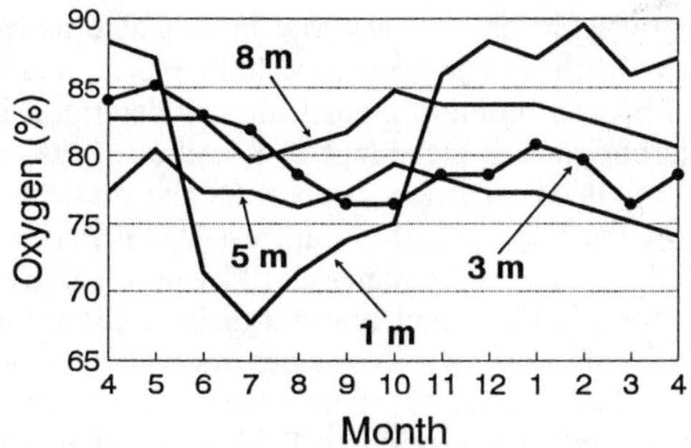

Fig. 2. O_2 content at different depths in the vadose zone of a field site in Mönchengladbach from April 1992 to April 1993 (values are in % of O_2 in the atmosphere 1 m above soil surface).

measured the oxygen in the soil atmosphere at different depths in a variety of light to heavy soils in Europe or the United States. It can be concluded from this information, that the macropores of vadose zone of soils contain enough oxygen to allow year-round oxidative activity of aerobic and facultatively aerobic microorganisms. However, the presence of anaerobic bacteria in all layers of the soils we and others have investigated (e.g. Fig. 1; Lewis et al. 1992), suggest the existence of large numbers of anaerobic microhabitats in top soils and in deeper soils layers. In well drained top soils, the relative importance of these habitats to pesticide degradation is believed to be small. In the vadose zone, where macropores and micropores are always moist and are often flooded with water, anaerobic degradation could be of importance. Further investigation of this subject is needed.

The lack of information on the importance of anaerobic microhabitats to pesticide degradation has been addressed by several investigators who incubated treated soils under anaerobic conditions. For example, Ou et al. (1985a; 1985b) followed the aerobic and anaerobic degradation of Aldicarb and Aldicarb sulfoxide in samples from the subsoil of orchards in Florida. Similarly, Smelt et al. (1983) followed the degradation of four carbamoyloxime pesticides in aerobic and anaerobic samples from fields in the Netherlands, and Walker et al. (1989) followed aerobic and anaerobic degradation of Chlorsulfuron and Metsulfuron in subsoils from 8 agricultural fields in the United Kingdom. In all cases, the pesticides tested were degraded under anaerobic conditions. With few exceptions, degradation was slower, and in some cases it was less extensive than under aerobic conditions.

The seasonal fluxes in subsoil temperatures at the Mönchengladbach site from April 1992 to April 1993, which exemplify the trends but not the actual temperatures at our other sites, are shown in Fig. 3. In the air 1 m above the soil, and in all horizons down to 8 m, temperature fluxes took the form of somewhat irregular sinus curves. As is found in the upper layers of all soils (USDA, 1975; Schachtschabel et al., 1989) the amplitudes of the curves decreased with increasing soil depth. At the Mönchengladbach site, between April 1992 and April 1993, the differences between the maximum and minimum surface air temperatures (weekly averages) were about 25 °C. One meter below the surface, the difference between maximum and minimum temperatures was about 14 °C, and at 3, 5 and 8 m, the differences were about 7, 2.5 and 1 °C, respectively. For modelling purposes, the phase shifts of the curves at different depths are of interest. At 3, 5 and 8 m, maximum and minimum temperatures were reached ca. 7, 15 and 27 weeks after the maximum was reached in the surface air.

Pesticide Degradation in Subsoils and Groundwater

In previous papers from our laboratory, we speculated that 4 major variables regulate the rates of degradation of pesticides in natural habitats such as soils, subsoils or groundwater (Anderson, 1990; Frehse and Anderson, 1983). These

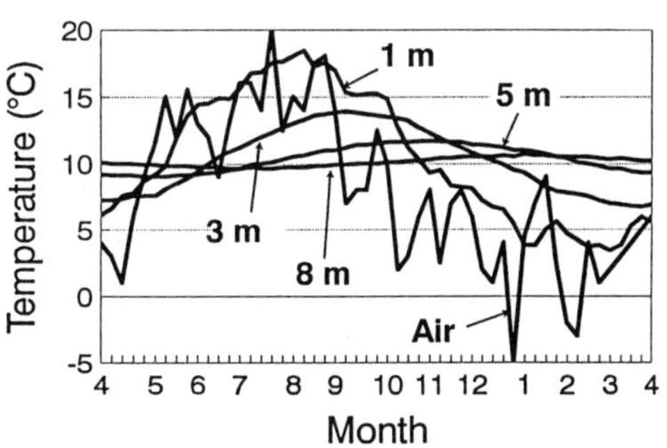

Fig. 3. Temperature at different depths in the vadose zone of the Mönchengladbach field site from April 1992 to April 1993.

are: (1) the structure of the pesticide; (2) the quantities and distribution of "reactive sites" in soil that can degrade it; (3) the availability of the pesticide to these sites; and, (4) and if the pesticide is being degraded by living cells, the activity level of these cells. Interacting with these variables to speed or slow degradation are the total amounts of pesticide in the habitat, available electron acceptors (e.g. oxygen), ambient temperature, available moisture and pH.

Structure of the Pesticide

In soils or groundwater, the structure of a pesticide will determine if it is degraded by abiotic reactions, biological reactions, or mixtures of these. The structural classes of pesticides thus far tested and found subject to degradation in subsoils and groundwater are listed in Tab. 2. Included in this list are acetamides, carbamates, chlorinated hydrocarbons, organophosphates, phenoxy acids, sulfonyl and substituted ureas, triazines and a variety of halogenated alkanes. In top soils, all of these chemicals are biodegradable: those thus far tested in subsoils and groundwater have also been found biodegradable. In addition, the carbamates, organophosphates and sulfonyl ureas are sensitive to pH induced hydrolytic degradation. It can thus be assumed, that chemicals which undergo biodegradation in topsoils, will also undergo biodegradation in subsoils and groundwater.

Pesticides that are subject to biodegradation in soils are occasionally toxic to the cells that can degrade them. If the intrinsic toxicity of a pesticide is high, cells can be overwhelmed if local concentrations are too high. This can slow or temporarily inhibit biodegradation. In subsoils or groundwater, where µg/kg or µg/l quantities of pesticides are occasionally detected, concentration related toxicity should not influence degradation rates. Rather, the paucity of pesticide molecules in these habitats might slow their degradation by failing to induce enzymes necessary for their degradation (Spain 1990).

The structure and elemental composition of a pesticide determines how much energy is needed to start and maintain its chemical or biochemical degradation, and if intact cells are involved in degradation, the energy and nutrients that can be won. Most deep subsoils and groundwater are oligotrophic or are very low in available nutrients. In these environments, both the structure and concentration of a pesticide will determine if microorganisms can adapt to it and use it as an energy and/or nutrient source. Of the pesticides shown in Tab. 2, Carbofuran, 2,4-D, MCPA and Parathion can all be mineralized by pure cultures or consortia of soil microorganisms (MacRae 1989). In fact, these pesticides, as well as Aldicarb, Fenamiphos, Isofenphos and Metribuzin, when applied repeatedly to field soils, have been shown to undergo accelerated microbial degradation (Adhya, 1987; Anderson, 1992; Moorman, 1990). Thus, if concentrations of such pesticides are high enough to induce degradation in subsurface environments, subsoil microorganisms should also be able to adapt to them and use them as primary or supplementary sources of nutrients (Moorman, 1990). Evidence for adaptation of subsoil or groundwater microorganisms to high or low concentrations of either pesticides or other organic chemicals, such as toluene, 1-methylnaphtalene, biphenyl, p-chlorophenol, p-nitrophenol, are given in the literature (Aamand et al., 1989; Aleion et al., 1987; 1989; Moorman, 1990).

Quantities and Distribution of "Reactive Sites"

In soils, subsoils or groundwater, pesticides can be degraded by a variety of "reactive sites", including dissolved ions, radicals, enzymes, intact microbial cells or consortia of microbial cells. Regardless of how these "reactive sites" are classified, their quantities and distribution in soils, subsoils or groundwater will directly influence the rates of degradation of the pesticides they attack.

Fig. 4 and 5 show the results of an experiment in which ^{14}C-carbonyl Carbofuran was added to samples from the profile of the Mönchengladbach site described in Tab. 1 and Fig. 1. In these studies, sterilized and microbiologically active samples of soil were used to obtain information about the relative contributions of chemically and biochemically "reactive sites" to degradation. ^{14}C-carbonyl Carbofuran was chosen as a model for these tests since it is sensitive to both chemical and enzymatic hydrolysis, and the first steps of degradation release $^{14}CO_2$. Formation of

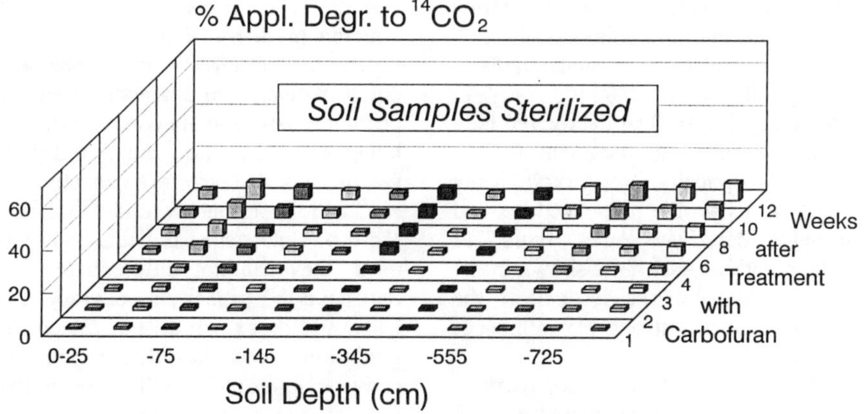

Fig. 4. Degradation of ^{14}C-carbonyl Carbofuran in sterilized samples of soil from different depths at the Mönchengladbach site. For exact depths refer to Table 1.

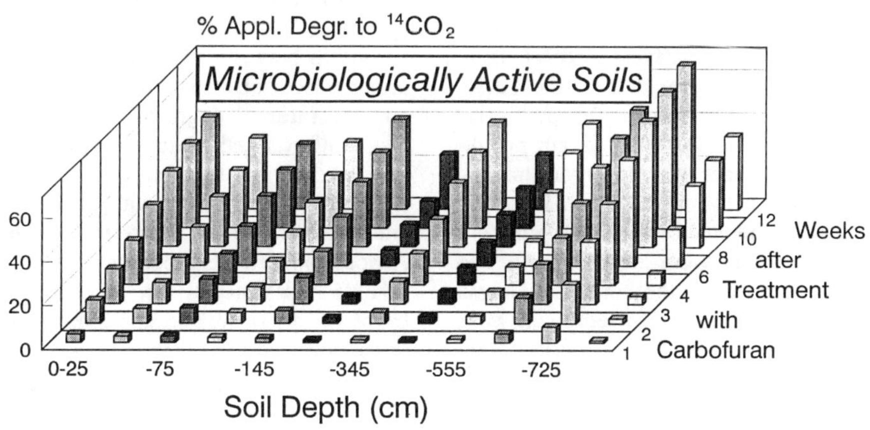

Fig. 5. Degradation of ^{14}C-carbonyl Carbofuran in microbiologically active samples of soil from different depths at the Mönchengladbach site. For exact depths refer to Table 1.

other radiolabelled metabolites or unextractable residues is minimal, so $^{14}CO_2$ values can be used to estimate degradation rates.

Comparison of Fig. 4 and 5 shows that in all layers tested, the microflora were primarily responsible for Carbofuran degradation. In the microbiologically active soils (Fig. 5), between 20 and 60 % of the applied radioactivity had been converted to $^{14}CO_2$ after 12 weeks. In the sterilized soils (Fig. 4), chemical hydrolysis with the release of $^{14}CO_2$ occurred, and was found positively correlated to the pH in the different layers (compare values in Tab.1 and Fig. 4). The quantities of Carbofuran degraded by chemical hydrolysis were relatively low, being between 2 and 8 % after 12 weeks. At the start of the experiment, the rates of Carbofuran degradation were positively but imperfectly correlated to the quantities of microbial biomass measured at the beginning of the experiment. As degradation progressed, correlations became weaker, and after 12 weeks, were no longer evident. Using a simple model to account for pH and biologically induced degradation, correlations improved, but could not fully account for the relatively low degradation rate in the microbiologically rich topsoil as compared to the quite similar rates in the microbiologically poor 100-145, 240-345 and 455-555 cm soil layers. The data given here suggest, but do not prove, that the quantities and distribution of chemically and biologically "reactive sites" in this profile influenced but did not exclusively determine the rates of Carbofuran degradation. As will be shown in the next section, *availability* of the pesticide to the "reactive sites" also strongly influences degradation rates.

Before addressing availability, other work in which "reactive sites" and degradation rates in subsoils have been examined should be briefly reviewed. Moorman and Harper (1989), working with 5-^{14}C-triazinone labelled Metribuzin in a 175 cm profile of a silty clay loam found that $^{14}CO_2$ was released during degradation in all layers. Mineralization was slow, with 15 to 21 % of the applied radioactivity being recovered as $^{14}CO_2$ from the 0 to 10 cm layers after 91 days, and 3 to 5 % being recovered from the 10-35 and 150-175 cm layers, respectively. Microbiological analyses of this 175 cm profile showed that the numbers of microorganisms decreased with increasing soil depth. However, direct correlation between microbial numbers and degradation rates were not evident. The authors concluded that the microorganisms necessary for mineralization of Metribuzin were present in all layers of the 175 cm profile. They suggested that the slower degradation rates at lower depths were due to limitation in the numbers of microorganisms and not to limitations in the species.

Moorman and Pothuluri (1990) and Pothuluri et al. (1990) studied Alachlor degradation in topsoils, vadose zone soils, and aquifer material collected from fields in Georgia and Oklahoma. They found that under aerobic conditions, the parent compound disappeared rapidly from surface zone soils, but was only slowly lost in vadose zone and aquifer samples. Microbial involvement in degradation was demonstrated with the addition of nutrients to the soil, which increased the rate of Alachlor loss. Based on dilution plate count estimates of the numbers of microorganisms, the authors concluded that the slower degradation of Alachlor in subsoils, as compared to top soils, was due to reduced microbial activity. The authors concluded that counts of microbial populations are not an effective predictor of degradation.

The lack of direct correlations between plate count numbers and degradation rates in the soils bears discussion. Plate count methods can detect living cells, but they can not differentiate between cells that are in an inactive state *in the soil*, and those that were metabolically active *in the soil*. Unless special media and techniques are used, they can not differentiate between cells that have the capacity to degrade a given pesticide, and those that do not. Since pesticide degradation in soils is a dynamic process, and inactive cells contribute little or nothing to this process, plate count numbers need not be correlated to degradation rates. This is true for topsoils, and will be particularly true in deeper subsoil layers or groundwater where nutrient quantities are usually limited, and many of the living cells will be in an inactive state.

Using a somewhat different microbiological approach, Walker et al. (1989) looked at the relationship between the microbial biomass in the upper 60 cm of eight different agricultural soils and the rates of degradation of the herbicides, Chlorsulfuron and Metsulfuron-methyl. In these tests, they found that degradation rates for both compounds decreased with increasing depths and were positively correlated with biomass. Adsorption and degradation rates for the compounds were negatively correlated with pH. However, there was an improvement in correlations when pH and biomass were included in multiple regressions. Similar to the data in our own subsoil studies with Carbofuran, these data support but do not prove the contention that the quantities and distribution of chemically and biologically "reactive sites" directly influence the rates of pesticide degradation in natural habitats.

Availability of Pesticide to "Reactive Sites"

In soils and groundwater, the availability of pesticides to "reactive sites" is usually a function of the quantity of chemical in solution at any given time. Stated differently, adsorption decreases availability and slows degradation. This has been demonstrated in soils for "chemically reactive sites" by Burkhard and Guth (1981) and "biologically reactive sites" by Anderson (1981). In both sets of studies, adsorption reduced the rates of degradation of the pesticides tested.

In our own studies with subsoils from the Mönchengladbach field site, the negative influence of adsorption on pesticide availability was partially overcome by loading the soils with an excess of Carbofuran. For this purpose, soils were treated with 30 mg Carbofuran a.i./kg dry wt soil. Preliminary experiments showed that this amount of the chemical did not have negative effects on its rate of degradation. As with other biodegradable pesticides we have tested (Anderson and Domsch 1980), increasing the concentration of Carbofuran increased the absolute quantities degraded per unit time by increasing the amounts in solution, and thereby the probability of contact between the initially fixed number of "reactive sites" and the pesticide molecules.

As a rough means of evaluating the availability of Carbofuran to "reactive sites" adsorption isotherms were also prepared using cores from a second sampling of the Mönchengladbach profile. The results of these tests are shown in Tab. 3. The data give a subtle explanation for the lack of direct correlation between the biomasses in each layer of subsoil and the rates of degradation. For example, degradation rates in the 0-25, 100-145 and 455-555 cm layers of this set of subsoil samples were quite similar (Fig. 5). For the 0-25 cm layer, the quantities of "reactive sites", given numerical values using biomass and pH measurements, were 462 and 5.5, respectively (Tab. 1). The availability of Carbofuran in this layer, given a numerical value using the adsorption constant, was 0.3026 (Tab. 3), which was the highest noted for all soil layers. For the 100-145 cm layer, the respective values were 15, 5.3 and 0.1058 and for the 455-555 cm layer, they were 46, 5.7 and 0.1355. In the 0-25 cm layer, the relatively high potential for degradation was countered by the relatively high degree of adsorption and therefore a relatively low degree of availability. In the 100-145 cm layer, the lower potential for degradation was balanced by the relatively high availability of the compound to the "reactive sites". In the 455-555 cm layer, biomass and pH values were much higher than in the 100-145 cm layer, but availability was reduced by the higher degree of adsorption.

Evaluation of the data for all layers of the Mönchengladbach site using a mathematical model that included biomass, pH and adsorption

Table 3. Freundlich adsorption constants for Carbofuran (22 °C) for soil samples from the Mönchengladbach field site.

Soil Depth (cm)	Adsorption Constant
0-25	0.3026
25-50	0.2484
50-75	0.1773
75-100	0.1024
100-145	0.1058
145-240	0.1134
240-345	0.0928
345-455	0.1068
455-555	0.1355
555-635	0.1135
635-700	0.0770
700-800	0.0199

values gave predicted degradation rates which were highly correlated to measured rates for the 0-25, 25-50, 50-75, 100-145, 145-240, 249-345, 455-555, 555-645 and 700-800 cm layers of the profile. However, in the 75-100, 345-455 and 635-700 cm layers, the predicted degradation rates were significantly *lower* than measured rates. The reasons for the direct correlations in 9 of the 12 layers, and the positive but poor correlations in remaining 3 layers are not yet known. Since our previous work (Anderson 1981; 1983; 1987) showed incubation of soils under warm (e.g. 22 °C) moist conditions in the laboratory causes drastic decreases in the soil microbial biomass with time, long term biomass studies are being conducted with new samples from the profile, and the model is being modified to compensate for biomass decreases (or increases) during the experiment. Also, since microbial biomass values and adsorption constants only provide crude estimates of the biodegradation potential of soils and the availability of pesticides for degradation, further work is being done to improve the quality of these measurements. With better understanding of how to quantitate "reactive sites" in soils, and how to measure the availability of pesticides to these sites, the predictive value of this and other pesticide fate models should improve.

Activity Level of the "Reactive Sites"

If a pesticide is degraded by living cells in top soil, subsoils or groundwater, the activity level of these cells will strongly influence the rate of degradation. This can be demonstrated using data from an experiment on the effect of temperature on the rates of biodegradation of Carbofuran in subsoil taken during a further sampling at the Mönchengladbach site. For this test, soil from the 60-100 cm layer of the profile was treated with 30 mg Carbofuran/kg soil, divided into 4 parts, and incubated at 22, 16, 10 and 4 °C.

The results of the experiment are shown in Tab. 4. The data show that by increasing or decreasing the amount of heat energy in the soil, the rates of degradation can be increased or decreased. During the 16 weeks of incubation in this test, the pH in the soils did not change, however the microbial biomass in all samples decreased slowly from 10 to 5 mg microbial C/kg dry wt soil. As expected, the adsorption of Carbofuran increased as temperature increased. This would tend to reduce the rates of degradation at higher temperatures. However, as shown in Tab. 4, the higher the temperature, the faster the rate of degradation. In our interpretation, the main effect of incubating the subsamples of soil at different temperatures was to change the energy levels in the system and thereby the activity levels of the chemically - and in particular - the biologically "reactive sites" that could degrade Carbofuran. Higher temperatures caused higher activity and speeded degradation, and lower temperatures decreased activity and slowed degradation. This principle has also been demonstrated with other pesticides (Frehse and Anderson, 1983).

Table 4. Influence of temperature on the rates of biodegradation of ^{14}C-carbonyl Carbofuran in soil from the vadose zone (60 to 100 cm) of the Mönchengladbach field site.

Weeks (a)	% of Applied Carbofuran Degraded to $^{14}CO_2$ at Different Temperatures			
	22 °C	16 °C	10 °C	4 °C
2	13.2	7.3	3.7	1.8
4	24.0	12.9	6.3	3.4
6	32.8	19.3	8.8	4.9
8	44.8	27.6	12.3	6.6
10	58.6	34.8	15.7	8.0
12	66.0	39.7	17.0	9.2
16	93.4	61.4	19.7	12.4

(a) Weeks after treatment with 30 mg a.i./kg soil

Conclusions

From the data and discussions presented in this paper, it can be concluded that in subsoils and groundwater, pesticides are subject to the same biological and hydrolytic degradation processes as in top soils and surface waters. The variables that determine rates of degradation in subsurface habitats are the same as those regulating the rates in top soils and surface waters. The assumption that pesticides entering subsoils are degraded at slower rates than in top soils is not always true. When tests were conducted under identical conditions in the laboratory, the pesticide used as a model in this work, and most other pesticides discussed in this paper, degraded in subsoils at rates that were slower, equal to or faster than those in top soils. If laboratory tests show that degradation rates in subsoils are slower than in top soils, this is mainly due to the smaller populations of microorganisms in these habitats, and not to a lack of capacity of the organisms to degrade the pesticides.

The data and discussions given here point to a number of information gaps and future needs in this relatively new area of pesticide research. To improve predictive models, the data base covering subsoil microbiology, fluxes in subsoil climate, and biological, physical, chemical, and structural variations in subsoil sites must be extensively expanded. For the models themselves, more detailed concepts and better definition of the variables to be measured and entered into the models are needed. For example, in soils, subsoils and groundwater, there are obvious relationships between pesticide degradation rates, the quantities and distribution of "reactive sites" in the soil, and the availability of pesticide molecules to these sites. How can we best quantify "reactive sites" and the availability of pesticide molecules to these sites? Are current measurements, including pH, microbial biomass and adsorption/desorption values sufficient for modelling, or must they be made more specific and improved? How can we use data from subsoil investigations and degradation tests to check the accuracy of models? What is the minimum degree of accuracy required for models used to predict the fate of pesticides in different soil layers? Direct and scientifically valid answers to these questions are important for all persons concerned with pesticides. False negative predictions can either stop development of valuable new chemicals, or trigger a cascade of additional and often pointless tests that can burden a chemical to the extent that it is dropped from further use or development. False positive predictions can promote development of chemicals that at some time or under some conditions exceed regluatory tolerance levels and must be withdrawn from use.

At this point in time, hard work and a great deal of creative effort are needed to determine how subsoil studies can best be conducted and how results can best be interpreted. What is not needed at this time are artificially "standardized" subsoil tests for the purpose of modelling and/or pesticide registration. The area is young and for the most part unexplored: with continued creative input and hard work, the basic information and the ideas needed to improve pesticide fate models should become available in the not too distant future.

References

Aarmand, J.; Jorgensen, C.; Arvin, E.; Jensen, B.K. *J. Contam. Hydrol.* **1989**, *4*, 299-312.

Adhya, T.K.; Sudhakar-Barik; Sethunathan, N. *Pestic. Biochem. Physiol.* **1981**, *16*, 14-20.

Adhya, T.K.; Wahid, P.A.; Sethunathan, N. *Biol. Fertil. Soils.* **1987**, *4*, 36-40.

Aelion, C.M.; Dobbins; D.C; Pfaender, F.K. *Environ. Toxicol. Chem.* **1989,** *8*, 75-86.

Aelion, C.M.; Swindoll, C.M.; Pfaender, F.K. *App. Environ. Microb.* **1987**, *53*, 2221-2217.

Agertved, J.; Rügge, K.; Barker, J.F. *Ground Water*. **1992**, 30, 500-506.

Anderson, J.P.E. *Soil Biol. Biochem.* **1981**, *13*, 155-161.

Anderson, J.P.E. *Soil. Biol. Biochem.* **1983**, *16*, 483-489.

Anderson, J.P.E. *Pest. Effects Soil Microfl.* **1987**, *Taylor & Francis, NY,* pp 45-60.

Anderson, J.P.E. *Adv. Appl. Biotech. Series, Biotech. Biodeg.* **1990**, *4*, 129-145.

Anderson, J.P.E.; Domsch, K.H. *Ann. Micr.* **1974**, *24*, 189-194.

Anderson, J.P.E.; Domsch, K.H. *Can. J. Microbiol.* **1975**a, *21*, 314-322.

Anderson, J.P.E.; Domsch, K.H. *Humus et Planta* **1975**b, *IV*, 211-214.

Anderson; J.P.E.; Domsch, K.H. *Soil. Biol. Biochem.* **1978**, *10*, 215-221.

Anderson, J.P.E.; Domsch, K.H. *Soil Sci.* **1980**, *130*, 211-216.

Anderson, J.P.E.; Domsch, K.H. *Arch. Environ. Contam. Toxicol.* **1980**, *9*, 259-268.

Anderson, J.P.E.; Lafuerza, A. *Proc. Internat. Symp. Environ. Aspects Pest. Microb.* **1992**, Sigtuna, Sweden, pp 184-192.

Bachmann, A.; Walet, P.; Wijnen, P.; de Bruin, W.; Huntjens J.L.M.; Roelofsen W.; Zehnder, A.J.B. *Appl. Env. Microb.* **1988**, *54*, 143-149.

Becker, K.-W.; Geries, H.; Meyer, B. *Mitteilgn. Dtsch. Bodenkundl. Gesellsch.* **1989**, *59*, 529-534.

Beloin, R.M.; Sinclair, J.M.; Ghiorse, W.C. *Microb. Ecol.* **1988**, *16*, 85-97.

Boesten, J.J.T.I.; van der Pas, L.J.T. *8th EWRS Symp. Braunschweig,* **1993**, pp 381-387.

Boesten, J.J.T.I.; van der Pas, L.J.T.; Smelt J.H.; Leistra, M. *Netherl. J. Agric. Sci.* **1991**, *39*, 179-190.

Bone, T.L.; Balkwill, D.L. *Microb. Ecol.* **1988**, *16*, 49-64.

Bouwer, H. **1984**, *Groundw. Poll. Microb.* 9-37.

Burkhard, N.; Guth, J.A. *Pestic. Sci.* **1981**. *12*, 45-52.

Dictor, M.C. **1994,** *These, Mini. lÁgric., Inst. Nat. Polytech., Lorraine Ensaia,* 131 pgs.

Dictor, M.C.; Soulas, G.; Takagi, K.; Anderson, J.P.E.; Lewis, K.; Lewis, F. *Proc. Internat. Symp. Environ. Aspects Pest. Microb.* **1992**, Sigtuna, Sweden, pp 284-287.

Dobbins, D.C.; Aelion, C.M.; Pfaender, F. *Crit. Rev. Environ. Control,* **1992**, *22*, 67-136.

Eichinger, L.; Merkel, B. *Symp. Wald Wasser, Grafenau, DWWK,* **1985**, 669-670.

Frehse, H.; Anderson, J.P.E. *IUPAC Pest. Chem. Human Welfare Environ.* **1983**, *4*, 23-32.

Gams, W.; Domsch, K.H. *Arch. Mikrobiol.* **1967**, *58*, 134-144.

GIFAP, *Fact Sheet, GIFAP-Bull.* **1992,** *18,* 1.

Ghiorse, W.C.; Wilson, J.T. *Adv. Appl. Microbiol.* **1988**, *33*, 107-172.

Hallberg ,G.R. *Agric. Ecosyst. Environ.* **1989**, *26*, 299-367.

Helweg, A. *Proc. Internat. Symp. Environ. Aspects Pest. Microb.* **1992**, *Sigtuna, Sweden,* 249-285.

Helweg, A. *7th Cong. Pest. Chem., IUPAC,* **1990**, *Hamburg* O7B-56.

Hirsch, P. **1992,** *Prog. Hydrogeochem.* 308-311.

Hirsch, P.; Rades-Rohkohl, E.; Köbel-Boelke, J.; Nehrkorn, A. **1992**, *Prog. Hydrogeochem.* 311-325.

IVA, **1989** *Pflanzenschutzwirkstoffe und Trinkwasser, Industvbnd Agrar e.V.* 36 pgs.

Lavy, T.L.; Roeth, F.W.; Fenster, C.R. *J. Environ. Qual.* **1973**, *2*, 132-137.

Leistra, M.; Boesten, J.J.T.I. *Agric. Ecosyst. Environ.* **1989**, *26*, 369-389.

Leistra, M.; Groen, A.E.; Crum, S.J.H.; van der Pas, L.J.T. *Pestic. Sci.* **1991**, *31*, 197-207.

Lewis, K.; Lewis, F.; Takagi, K.; Anderson, J.P.E.; Dictor, M.C.; Soulas, G. *Proc. Internat. Symp. Environ. Aspects Pest. Microb.* **1992**, *Sigtuna, Sweden,* 278-283.

Locke, M.A.; Harper, S.S. *Pestic. Sci.* **1991**, *31*, 221-237.

MacRae, I.C. *Rev. Environ. Contam. Toxicol.* **1989,** *109,*1-87.

Matthess, G.; Isenbeck, M. *Boreas* **1987**, *16*, 411-418.

McCarthy, J.F.; Zachara, J.M. *Environ. Sci. Technol.* **1989**, *23*, 496-502.

McMahon, P.B.; Chapelle, F.H.; Jagucki, M.L. *Environ. Sci. Tech.* **1992**, *26*, 1556-1559.

McNabb, J.F.; Dunlap, J. *Ground Water* **1975**, *13*, 33-44.

Miles, C.J.; Delfino, J.J. *J. Agric. Food Chem.* **1985**, *33*, 455-460.

Moore, A-T.; Vira, A.; Foget, S. *Environ. Sci. Technol.* **1989**, *23*, 403-406.

Moorman, T.B. *ACS Symp. Ser.* **1990**, *426*, 167-180.

Moorman, T.B.; Harper, S.S. *J. Environ. Qual.* **1989**, *18*, 302-306.

Moorman, T.B.; Pothuluri, J.V. *IUPAC & GDCh 7th Intern. Congress Pesticide. Chem. III:* **1990**, 07B-19, 63.

Mueller, T.C.; Moorman, T.B.; Snipes, C.E. *J. Agric. Food. Chem.* **1992**, *40*, 2517-2522.

Nicholls, P.H. *Pestic. Sci.* **1988**, *22*, 123-137.

Ou, L.T.; Edvardsson, K.S.; Rao, P.S.C. *J. Agric. Food Chem.* **1985**a, *33*, 72-78.

Ou, L.T.; Sture, K.; Edvardsson, V.; Thomas, J.E.; Rao, P.S.C. *J. Agric. Food Chem.* **1985**b, *33*, 545-548.

Panda, S.; Sharmila, M.; Ramanand, K.; Panda, D.; Sethunathan N. *Pestic. Sci.* **1988**, *23*, 199-207.

Phelps, T.J.; Fliermans, C.B.; Garland, T.R.; Pfiffner S.M.; White D.C. *J. Microbiol. Methods* **1989**, *9*, 267-279.

Phelps, T.J.; Russel, B.F. *Internat. Symp. Subsurf. Microbiol.* **1993**, A-01, Bath, UK.

Pignatello, J.J. *J. Environ. Qual.* **1987**, *16*, 307-312.

Pothuluri, J.V.; Moorman, T.B.; Obenhuber, D.C.; Wauchope R.D. *J. Environ. Qual.* **1990**, *19*, 525-530.

Shaffer, K.A.; Fritton, D.D.; Baker, D.E. *J. Environ. Qual.* **1979**, *8*, 241-246.

Schachtschabel, P.; Blume, H.-P.; Brümmer, G.; Hartge K.-H.; Schwertmann, U. *Lehrbuch der Bodenkunde*, **1989**.

Smelt, J.H.; Dekker, A.; Leistra, M.; Houx, N.W.H. *Pestic. Sci.* **1983**, *14*, 173.

Smith, C.J. *Environ. Fate Pestic.* **1990**, *John Wiley and Sons,* 47-99.

Spain, J.C. *ACS Symp. Ser.* **1990**, *426*, 181-190.

Takagi, K.; Anderson, J.P.E.; Lambertz, A.; Dictor M.C.; Soulas, G.; Lewis, K.; Lewis, F. *Proc. Internat. Symp. Environ. Aspects Pest. Microb.* **1992**, Sigtuna, Sweden, 270-277.

Torstensson, D.C.; Weeks, E.P.; Haas, H.; Fischer, D.W. *Radiocarb.* **1983**, *25*, 315-346.

USDA. *Soil Taxonomy Agric. Handbook* **1975**, *436*, 57-63, Washington.

Vonk, J.W.; Blom, A.J.M.; Smelt, J.H. *Proc. Internat. Symp. Environ. Aspects Pest. Microb.* **1992**, Sigtuna, Sweden, 306-311.

Walker, A.; Cotterill, E.G.; Welch, S.J. *Weed Res.* **1989**, *29*, 281-287.

Weyandt, R.; Schweisfurth, R. *Prog. Hydrogeochem.* **1992**, pp 365-377.

Wehtje, G.R.; Spalding, R.F.; Burnside, O.C.; Lowry, S.R.; Leavitt, J.C.R. *Weed Sci.* **1983**, *31*, 610-618.

Wood, M.; Harold, J.; Johnson, A.; Hance, R. *BCPC Monog.* **1991**, *47*, 175-183.

Yaron, B. *Agric. Ecosyst. Environ.* **1989**, *26*, 275-297.

Fate of Herbicides and Organochlorine Insecticides in Lake Waters

Derek C.G. Muir and Norbert P. Grift,
Freshwater Institute, Dept. of Fisheries and Oceans, Winnipeg MB
Canada R3T 2N6

Long residence times of several herbicides, especially atrazine, have been observed in the water column of large lakes. This paper reviews the recent information on pesticides in surface waters and examines the dissipation and air-water partitioning of eight pesticides found in precipitation and lake water over a three year period in a small (18.7 ha) oligotrophic lake in northwestern Ontario. Input of most herbicides (atrazine, alachlor, 2,4-D) to this lake occurred mainly via precipitation and gas absorption in May and June each year. Most herbicides had slow rates of disappearance (half-lives ranging from 25 d for trifluralin to 170 d for atrazine). Volatilization was the major route of removal of the semi-volatile insecticides such as endosulfan but low volatility, polar herbicides such as atrazine were lost mainly via degradation and water outflow.

The past few years have seen a large increase in the information available on levels of herbicides and insecticides in river and lake waters in the USA, Canada and western Europe. There are both regulatory and methodological reasons for this increase in data. The refinement of rapid methods of water extraction using solid phase cartridges coupled with GC-MS (Meyer et al. 1993; Pereira et al. 1990; Vitali et al. 1994) or with immunochemical screening for specific analytes (Thurman et al. 1990; Aga and Thurman 1993) has increased the numbers of samples that can be processed and permitted sampling with high spatial and temporal resolution. Use of high resolution capillary columns in gas chromatographic analysis along with mass spectrometry has increased the precision of analysis and lowered method detection limits in comparison with previous work performed in 1970's and early '80s. Concerns over deteriorating water quality, especially of sources of drinking water, have prompted sampling programs to determine if pesticide levels exceed US EPA maximum contaminant levels or the ECE drinking water directive (100 ng L^{-1}). The objective of this paper is to consider the ultimate fate of these herbicide and insecticide residues in lakes and ponds by examining published data on concentrations in river waters, precipitation and air, the residence times of selected compounds in receiving waters, and the importance of various removal processes such as volatilization, biotransformation, and sedimentation.

Herbicides in river waters

Atrazine, its degradation product deethyl atrazine (DEA), other triazine herbicides, and acetanilides (e.g. alachlor and metolachlor), are the pesticides most frequently reported in recent studies of surface waters. In the US midwest they are used on corn (maize) and soybeans and are flushed from cropland during the May and June application period following rainfall events (Thurman et al. 1992; Pereira and Hostettler 1993; Richards and Baker 1993). Studies in Canada (Bodo 1991; Frank et al. 1991; Maguire and Tkacz 1993), Sweden (Torstensson et al. 1990), Germany (Brauch 1993), France (Tronczynski et al. 1993) and other Mediterranean countries (Readman et al. 1993), as well as Switzerland (Buser 1990), have detected the same herbicides in rivers draining agricultural areas at similar concentrations. Less frequently reported are the phenoxyacid (2,4-D, MCPA), phenyl ureas (chlortoluron), benzoic acid herbicides (dicamba), and dinitroanilines (trifluralin) which, although widely used, are more rapidly degraded in soil than the triazines or acetanilides (Wauchope et al. 1992). Organophosphate, pyrethroid and carbamate insecticides are also detected infrequently due to rapid degradation and lower quantities used relative to herbicides (Frank et al.

1054–7487/95/0141$12.00/0

1991; Frank and Logan 1988; Richards and Baker 1993). Maximum herbicide concentrations occur during or immediately following the application period due to overland flow of water following heavy rains and coincide with maximum suspended solid concentrations and peak flow (Richards and Baker 1993; Schottler and Eisenreich 1994). Concentrations decline rapidly with dissipation half-times (DT50; time to reach 1/2 maximum concentration) of 5 to 30 days coinciding with the hydrograph. Tile drainage and ground water also contribute to observed triazine and acetanilide herbicide levels especially in periods of low flow and give rise to sustained concentrations near detection limits (<0.01 to <100 ng L^{-1}) throughout the year (Squillace et al. 1993; Bodo 1991). A study of watersheds in Ohio found no sustained trends in time-weighted mean concentrations of atrazine, alachlor and metolachlor over an eight year period in river water (Richards and Peters 1993). Bodo (1991) found declining concentrations of atrazine in southwestern Ontario watersheds between 1981 and 1990 coinciding with declining corn (maize) production in the region. Brauch (1993) reported declining loadings of atrazine in the river Rhine between 1989 and 1991 as a result of the ban on its use in Germany in 1990.

Pesticides in lake and estuarine waters

The pesticides in river waters are discharged into ponds, lakes, and estuaries but relatively few studies have examined the ultimate fate of the river-borne pesticides in these environments (Muir 1991). Pereira and Hostettler (1993) estimated that 205 t of atrazine (+ degradation products) were transported into the Gulf of Mexico in 1991. Tronczynski et al. (1994) estimated 3270 to 4400 kg yr^{-1} were transported by the Rhone River to the Mediterranean Sea. Munschy et al. (1993) found relatively low atrazine levels in offshore Atlantic Ocean waters of the Marennes-Oléron coast of France (5-28 ng L^{-1}) and much higher levels in estuarine areas (32 - 287 ng L^{-1}). Kucklick et al. (1993) found atrazine in the Winyah Bay estuary in South Carolina at concentrations ranging from 6 to 840 ng L^{-1} at an inland brackish water site and from 1 to 32 ng L^{-1} in saline water at a remote coastal site.

Of greatest interest are inland waters such as the Great Lakes, or shallow seas such as the Baltic or the Mediterranean, which receive discharges from large agricultural watersheds. Concentrations of atrazine, the most commonly reported herbicide in all studies to date, range from μg L^{-1} levels in lakes receiving agricultural runoff (Bacci et al. 1989; Spalding et al. 1994) to low ng L^{-1} levels in alpine lakes in Switzerland (Buser 1990) and in the Great Lakes (Schottler and Eisenreich 1994). Concentrations of herbicides in lake waters are generally much lower than in rivers but these residues persist much longer than in lotic environments (Table 1). Buser (1990) found that residence times of atrazine in Swiss lakes were linearly related to water residence times suggesting that dilution rather than degradation were the major removal processes. Schottler and Eisenreich (1994) found atrazine to be well mixed vertically in Lakes Ontario and Michigan indicating that residence times in the water column of these lakes must be in the order of years.

Atmospheric deposition of pesticides

The presence of atrazine in Swiss alpine lakes (Buser 1990) and in Siskiwit and Wallace lakes on Isle Royale in Lake Superior (Aga and Thurman 1993) indicates that atmospheric inputs are important for lakes and may be the only sources in some locations. Pesticide inputs from the atmosphere to lakes can occur via wet and dry deposition as well as via gas exchange from the atmosphere (Eisenreich and Strachan 1992). Herbicide concentrations observed in precipitation during the 1980-90s in the US midwest and Chesapeake Bay area, the Great Lakes region and western Europe are given in Table 2. Atrazine concentrations typically range from <0.1 to 5 μg L^{-1} (Capel 1991; Richards et al. 1987; Glotfelty et al. 1989) in precipitation and are strongly seasonal with highest levels during the application period. Richards et al. (1987) found concentrations of 13 herbicides and insecticides in rainwater from locations in Ohio were directly proportional to quantities used in the region.

Fluxes (μg m^{-2}) of herbicides in rainfall represent <1% of the rates of application (Capel 1991; Richards et al. 1987) but could be significant for non-target regions such as lakes. Eisenreich and Strachan (1992) estimated total

Table 1. Atrazine, alachlor and metolachlor concentrations in lakes and estuaries receiving discharges from large agricultural watersheds and approximate dissipation times

Location (Country/State)[a]		Year(s)	Chemical	Concentration (ng L^{-1})	DT50 days[b]	Reference
L. Baldegg	CH	1988-89	atrazine	200-300	180	Buser 1990
L. Sempach	CH		atrazine	240-260	~500	
L. Zurich	CH		atrazine	32-67	60	
L. Klöntal (848 m)[c]	CH		atrazine	<0.02-0.8	ND[d]	
L. Mutt (2446 m)	CH		atrazine	<0.02-0.05	ND	
Obsersee/Näfels (990 m)	CH		atrazine	<0.02-0.63	ND	
L. Chiusi	I	1986	atrazine	300-1,900	168	Bacci et al. 1989
Maskenthine L.	NL	1990	atrazine	500-14,000	193	Spalding et al. 1994
			alachlor	<100-7,000	21	
			metolachlor	<100-920	75	
Willow L.	NL	1990	atrazine	<100-9,600	124	
			metolachlor	<100-400	ND	
Ioannina L.	GR	1984-85	atrazine	2-223	30	Albanis et al. 1986
L. St. Clair	MI	1987	atrazine	460	ND	Frank et al. 1991
			metolachlor	200	ND	
L. Michigan		1991-92	atrazine	30-40	ND	Schottler & Eisenreich 1994
			alachlor	<1	ND	
			metolachlor	<1-5	ND	
L. Ontario		1991-92	atrazine	78-100	ND	
			alachlor	<1	ND	
			metolachlor	15	ND	
L. Huron		1991-92	atrazine	19-22	ND	
			alachlor	<1	ND	
			metolachlor	<1	ND	
L. Erie		1991-93	atrazine	25-115	ND	
			alachlor	<1-5	ND	
			metolachlor	10-28	ND	
L. Superior		1991-92	atrazine	3	ND	
			alachlor	<1	ND	
Wallace L.	MN	1992	atrazine	6.2-9.3	ND	Aga & Thurman 1993
Siskiwit L.	MN	1992	atrazine	14-16	ND	

[a] Country abbreviations (CH=Switzerland, NL = The Netherlands, I = Italy, GR = Greece).
[b] Dissipation half-life (time for 50% dissipation) reported by the authors or calculated from published data.
[c] Alpine lake remote from agricultural activity (elevation in parentheses).
[d] ND = not determined.

Table 2. Concentrations of atrazine, alachlor and metolachlor in precipitation

Location	Year(s)	Method[a]	Chemical	Concentration ($ng\ L^{-1}$)	Reference
Chesapeake Bay	Jun-Jul 1982	wet + dry	atrazine alachlor metolachlor	30-480 140-2,570 46-290	Glotfelty et al. 1990
Ohio	Apr-Aug 1985	wet + dry	atrazine alachlor metolachlor	<50-1,500 <100-4,900 <250-2,600	Richards et al. 1987
Iowa	Nov 1987-Sep. 1990	wet + dry	atrazine alachlor metolachlor	<100-40,000 <100-8,600 <100-2,700	Nations & Hallberg 1992
Minnesota	Mar 1989-Jun 1990	wet only	atrazine alachlor	<20-1,600 <20-22,000	Capel 1991
Midwest US	1990 - 91	wet only	atrazine alachlor	100-15,500 150-3,600	Pome et al. 1994
France (Marne R.)	1991	wet + dry	atrazine	10 - 350	Chrevreuil & Garmouma 1993
France (Paris)	1991	wet + dry	atrazine	<5 - 18	Chrevreuil & Garmouma 1993
Switzerland	1988-89	wet + dry	atrazine	<0.05-600	Buser 1990
Sweden (south)	1990-92	wet + dry	atrazine	<10-150	Kreuger & Staffas 1994
Germany (north)	1990-92	wet + dry	atrazine	10-150	Siebers et al. 1994
Italy (north)	1990-91	wet + dry	atrazine alachlor	<106-1100 <136-560	Trevison et al. 1993
S. Ontario	Apr 1991-Sep. 1992	wet only	atrazine metolachlor	<10-1000 <10-450	Hall et al. 1993

[a]Combined wet and dry deposition, i.e., uncovered sampler; wet only = automated sampler covered during dry periods.

atmospheric deposition to the Great Lakes (surface area) of 34 t for atrazine and 107 t of alachlor. Glotfelty et al. (1990) estimated that 0.64 to 1.2 t of atrazine entered Chesapeake Bay annually during 1981 - 1984 or about 10% of total inputs. Chevreuil and Garmouma (1993) concluded that atrazine supply to the River Marne basin in France via precipitation and gas exchange was equivalent to losses via runoff and percolation.

The importance of atmospheric deposition of currently used pesticides to surface waters can be seen in the case of Lake Erie where riverine inputs from the Canadian and US sides are relatively well documented (Table 3). The proportion due to wet/dryfall is estimated to range from 9.3% for

Table 3. Estimated quantities of major pesticides entering Lake Erie[a]

Compound	Source	Mean annual[b] concentration ($\mu g\ m^{-3}$)	Quantity ($kg\ yr^{-1}$)	% of total
atrazine	Detroit R.	22	3,750	9.3
	Tributaries	1,500	33,000	81
	Wetfall + dryfall	500 (May/Jun) 10 (other mo)	3,780	9.3
alachlor	Detroit R.	1	~200	0.8
	Tributaries	500	11,000	48
	Wetfall + dryfall	2,000 (May/Jun) 75 (other mo)	12,000	52
metolachlor	Detroit R.	1	~200	0.6
	Tributaries	1,000	22,000	65
	Wetfall + dryfall	2,000 (May/Jun) 75 (other mo)	12,000	35
endosulfan	Detroit R.	0.06[c]	12	8
	Tributaries	1	~110	70
	Wetfall + dryfall	1.5	34	22

[a] Calculations based on Schottler & Eisenreich (1994) and wetfall/dryfall data from Eisenreich & Strachan (1992)

[b] Concentration in inflowing water. Tributary data from Richards & Baker (1993); Frank et al. (1991) and Frank and Logan (1988); with tributary discharge of $2.2x10^{10}\ m^3yr^{-1}$; Detroit R. discharge of $1.707x10^{11}\ m^3yr^{-1}$

[c] Results for the St. Clair River (Chan 1993).

atrazine to 52% for alachlor based on mass input estimates by Schottler and Eisenreich (1994) and Eisenreich and Strachan (1992). Endosulfan, the major insecticide residue in tributary waters in southwestern Ontario in the mid-1980's (Frank and Logan 1988) has a similar proportion in wet/dryfall although quantities are much lower (Table 3).

These estimates do not include contributions from direct gas exchange between air and water. Glotfelty et al. (1990) noted that a disequilibrium existed between atrazine concentrations in the atmosphere above Chesapeake Bay and the water column such that the net flux would be to the water phase. The air-water distribution coefficients (K_{AW}) for atrazine is 1.2 x 10^{-7}, i.e., substantially favoring the water phase. For hydrophobic organics such as PCB congeners, which have K_{AW}'s ranging from 0.015 - 0.0017, net volatilization from water to air is usually observed. Jeremiason et al. (1994) estimated that the inputs of PCBs to Lake Superior waters via gas exchange were about 1/3 of the wet deposition inputs; however, the annual net flux of PCBs was from water to air.

This brief review of the recent data on current use pesticides in surface waters indicates that additional information is needed on the processes such as volatilization, deposition, degradation and partitioning, which control the residence times of these semi-volatile organic compounds in lake waters. The fate of these pesticides in these oligotrophic lakes is not well understood because most of the information on their degradation kinetics and transformation products is for soils or (eutrophic) pond waters. A three year study of a small (18.7 ha) oligotrophic lake in NW Ontario in which current use pesticides enter only via atmospheric deposition provided the opportunity to address this information gap.

Materials and methods

Lake characteristics: L375 is a 18.7 ha oligotrophic lake situated in the boreal forest on Precambrian shield granites (49°45'N, 93°47'W) at the Experimental Lakes Area in Northwestern Ontario (50 km east of Kenora ON; Fig. 1). This lake was selected as part of a study of the chemical limnology and bioaccumulation of organic contaminants in a series of lakes of increasing size (Fee and Hecky 1992; Muir et al. 1993). The lake has no direct road access and to our knowledge no pesticides have ever been used within the lake drainage area. L375 has a total volume of 2.17×10^6 m^3, mean depth of 11.6 m, maximum depth of 30 m, watershed area of 210 ha, and a water residence time of 3 to 5 years. Dissolved organic carbon and suspended carbon in the epilimnion of L375 ranged from 0.38-0.62 mg L^{-1} and from 0.35-0.55 mg L^{-1}, respectively, and pH from 7.0 to 7.6 during 1992.

Sample collection: Air, precipitation and lake water samples were collected over a 3 year period (1990-1992) to determine levels of current use pesticides entering the lake. Low concentrations of pesticides were anticipated because of the remote location of the lake (at least 100 km from the nearest agricultural area and 1000 km from the "corn belt" of the US midwest) so that the sampling program was designed to collect large volumes of air, water and precipitation.

Air sampling was conducted at a meterological station 15 km south of the lake using a high volume (Sierra Anderson PS-1) PUF sampler with Whatman GF/A glass fiber filter. Preparation and handling of the filters and PUF samples followed procedures described by Hoff et al. (1992). Approximately 300 m^3 of air was collected over a 24 h period every 6 days over the period July 1990 to June 1991. A set of large volume samples (1200 m^3) were collected for herbicide analysis from June to October 1992. PUFs and filters were placed in sealed glass jars and stored at 4°C until analysis. The glass fiber filters were not analysed for this study because the focus was on vapor phase concentrations of pesticides.

Rainfall was collected using two "Strachan" type covered collectors (Strachan and Huneault 1984) with collection area of 0.2025 m^2 located at the meterological site. Rain water was collected

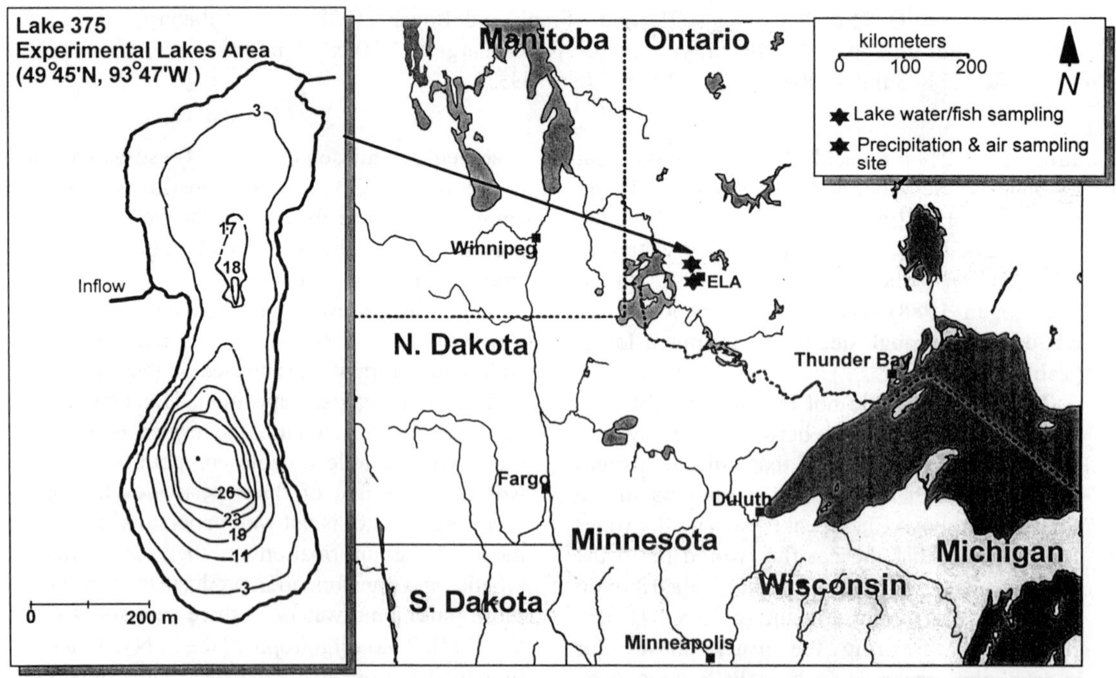

Fig. 1. Bathymetric map of Lake 375 (depth in meters) showing its location in Northwestern Ontario.

over 2 week (June) or monthly (July - October) periods in 19 L stainless steel softdrink cans. Volumes ranged from 5 to 19 L. Rain was filtered through a plug of fine glass wool placed at the bottom of the collection funnel to remove large particles.

Lake water was collected by pumping 80 to 100 L at 1 m depth with a submersible pump into 19 L stainless steel cans. Additional samples were collected at 10 m depth during 1992. Samples collected in 1990 and 1991 were extracted unfiltered while those collected in 1992 were filtered under nitrogen pressure through pre-cleaned Whatman GFC glass fiber (~1 μm).

Sample analysis: A broad spectrum approach was used for sample analysis in order to analyse neutral and acid herbicides (e.g. atrazine and 2,4-D) as well as hydrophobic organochlorine (OCs) compounds such as α-endosulfan and methoxychlor. PUFs collected in 1990 were Soxhlet extracted with hexane for 4 h to recover organochlorine compounds; selected samples from 1992 were extracted with dichloromethane (DCM) to recover herbicides as well as OCs. Internal recovery standards of aldrin, octachloronaphthalene and d5-atrazine were added at the extraction step. Extracts were evaporated to small volume, taken up in DCM and split 1:1 for separate analysis of OCs and herbicides. The OC extract was evaporated under vacuum and exchanged into hexane for chromatography on Florisil columns. Additional details on chromatographic cleanup and gas chromatographic analysis of OC compounds are given in Hoff et al. (1992). The herbicide extract was methylated with freshly prepared diazomethane and then cleaned up on Florisil columns. Neutral and methylated acid herbicides were eluted from Florisil with 10% ethyl acetate in hexane. By this procedure the herbicides 2,4-D acid and dicamba were analysed as their methyl ester derivatives.

Precipitation and lake water samples were extracted with DCM in their sampling containers. The water was adjusted to pH 2 to extract acid herbicides and phenolics and then to pH 10 to recover hydrophobic organics (Maguire and Tkacz 1989). Extracts were then split and treated as described for air samples.

OC pesticides were analysed by capillary GC-ECD using a 60 m x 0.25 μm DB-5 with H_2 carrier gas. Neutral and acid herbicides were quantified by capillary GC-mass spectrometry (Hewlett Packard 5971 MSD) using a 30 m x 0.25μm DB-5 column with He carrier gas. Detection criteria were correct ratios and retention times of two characteristic ions. On both instruments pesticides were quantified using external standard solutions.

Quality assurance: Compounds determined by GC-ECD (methoxychlor, α-endosulfan and trifluralin) were confirmed by GC-MS. Blank PUFs were analysed every 10 samples. Blanks for water consisted of 18 L of "Super'Q" distilled water and well water. All precipitation samples were collected in duplicate and analysed separately. A duplicate set of lake water samples were analysed once each year. Internal standard recoveries in PUF extraction were uniformly >90%. In precipitation and lake water internal standard recoveries ranged from 50-80% for d5-atrazine, and from 60-100% for aldrin and octachloronaphthalene. Results were not corrected for internal standard recoveries.

Results and Discussion

Concentrations in surface water and precipitation

Eight pesticides currently registered for use in the US and Canada were detected at low or sub-ng L^{-1} levels in the surface waters of L375 (Fig. 2). These compounds were also detectable in precipitation (Fig. 3) and in air. Concentration ranges and method detection limits for the eight compounds are given in Table 4. Concentrations of the pesticides in lake water generally followed an annual cycle with highest levels in May-July each year and lowest levels in August-September or in samples collected under ice (December, February). Atrazine and 2,4-D (acid) were present a highest concentrations (Fig. 2; Table 4) in lake surface waters with levels >1 ng L^{-1} in most samples. The observation of acid herbicides 2,4-D and dicamba in lake water was unexpected given the low volatility of these acids and the distance of the lake from agricultural sources. The OC insecticides, α-endosulfan and methoxychlor were

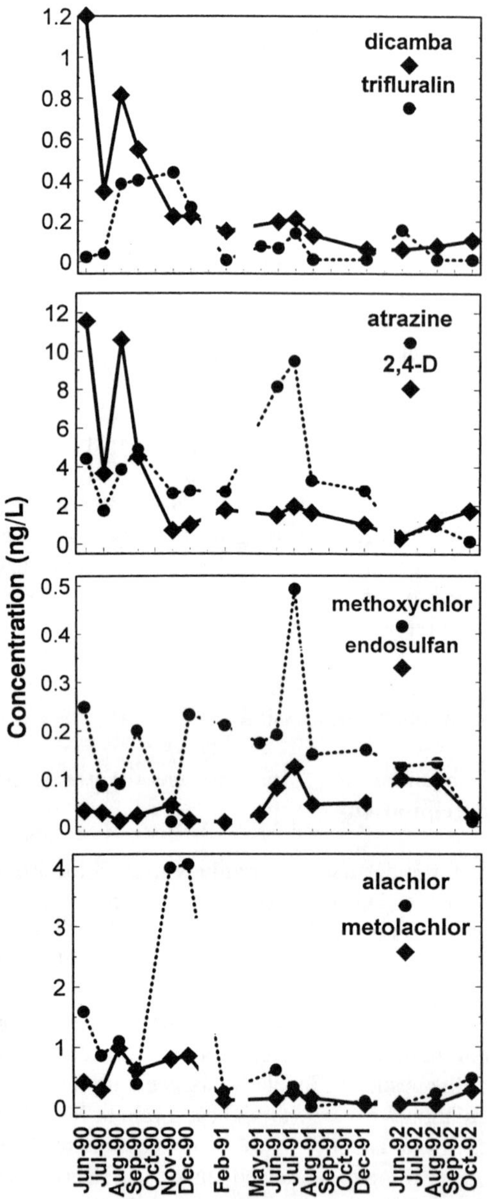

Fig. 2. Concentrations (ng L^{-1}) of eight current use pesticides in Lake 375 surface waters (1 m depth) over the period June 1990 to October 1992. Results are based on single monthly large volume samples.

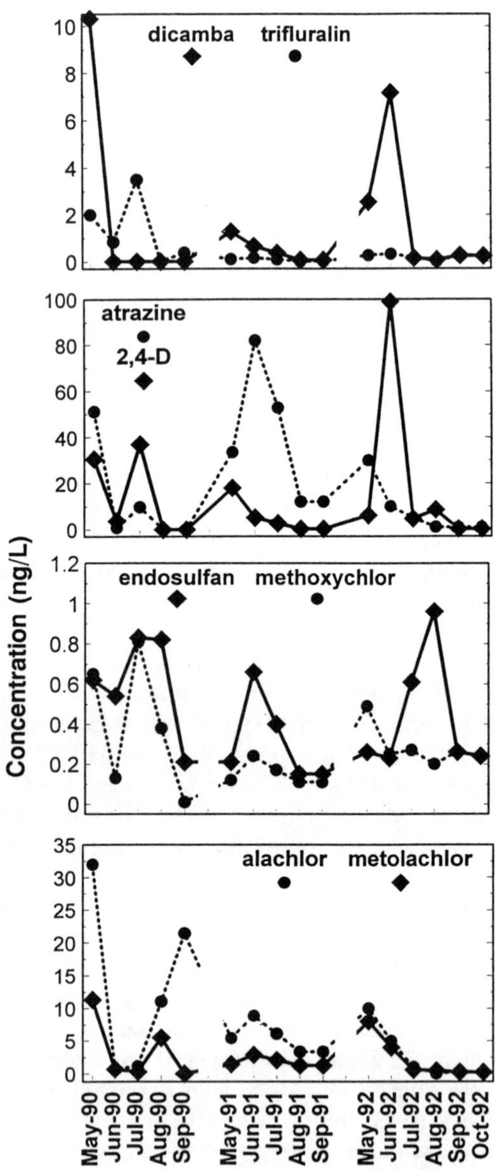

Fig. 3. Concentrations (ng L^{-1}) of eight current use pesticides in precipitation (wet only) collected at the Experimental Lakes Area meteorological site 15 km south of L375 over period May 1990 to October 1992. Results are average monthly concentrations calculated from 2 or 4 week periods of continuous collection.

Table 4. Range of concentrations and detection limits for eight current use pesticides detected in L375 water and precipitation collected during 1990-1992.

Compound	MDL[a] ng L^{-1}	water ng L^{-1}	precipitation ng L^{-1}	air[b] pg m^{-3}
alachlor	0.03	<0.03-1.6	<0.03-80	6.7-87
atrazine	0.03	0.16-9.5	<0.03-51	1.9-37
2,4-D	0.06	<0.06-11.6	<0.06-410	<1-3.3
dicamba	0.06	<0.06-1.64	<0.06-10.3	<1-1.3
endosulfan	0.02	<0.02-0.49	0.15-0.96	<1-107
methoxychlor	0.02	<0.02-0.12	<0.02-0.81	<1-22.5
metolachlor	0.03	<0.03-0.98	<0.03-11.3	32 - 81
trifluralin	0.02	<0.02-0.44	<0.02-3.5	<1-78

[a] Method detection limit for water based 3 x standard deviation in the blank with concentration calculated based on 1/500 injection of a 20 L water sample. Detection limits for air based on 300 m^3 samples for methoxychlor, endosulfan and trifluralin and 600 m^3 samples for other compounds.
[b] Results for June and October 1992 only using 600 m^3 sample.

present at lower concentrations than most of the herbicides and, with the exception of June-July 1991, showed less temporal variation.

The eight compounds were all present in the hypolimnion of L375 at three sampling times (June, August and October 1992). In June concentrations of most compounds were higher at 1 m depth than at 10 m. However by August concentrations were similar at both depths indicating full mixing of the compounds deposited earlier in the summer. All eight pesticides were undetectable in the particulate phase (>1 µm) of water samples collected in 1992. Therefore the concentrations reported in Fig. 2 and Table 4 were considered to be dissolved.

The concentrations observed for atrazine are similar to those reported for lakes on Isle Royale, which is about 400 km southeast of L375, and for Lake Superior itself (Table 2). Concentrations of atrazine are generally higher than observed by Buser (1990) in Swiss alpine lakes remote from agricultural activity. Much higher concentrations of atrazine, alachlor and metolachlor have been reported in lakes receiving runoff from agricultural areas (Table 2). The range of concentrations of 2,4-D and dicamba in L375 surface waters were similar to those reported by Waite et al. (1992) for ponds located within an agricultural watershed in Saskatchewan which received wet and dry deposition inputs as well as runoff.

The concentrations of the eight pesticides in precipitation generally showed the same temporal trends as L375 surface waters with highest levels in May/June. Atrazine, 2,4-D and bromoxynil were the major compounds observed with the latter two compounds achieving maximum concentrations in the 200-400 ng L^{-1} range (Table 4). Because rainfall amounts were also highest in May and June the fluxes (ng m^2) of pesticides entering the lake in this period dominated annual inputs.

Concentrations of herbicides such as atrazine, alachlor and metolachlor in precipitation at this remote location in May and August were at the low end of the range observed over the same months in the US midwest (Richards et al. 1987; Nations and Hallberg 1992). The range of concentrations of α-endosulfan and methoxychlor in precipitation was similar to those reported for

Lake Superior (Eisenreich and Strachan 1992; Chan and Perkins 1989) in the mid- and late 1980's. These compounds have also been detected in the 0.1-1 ng L^{-1} range in rain samples from the Canadian prairies and in southern Ontario (Eisenreich and Strachan 1992). While the presence of triazine and acetanilide herbicides in precipitation is well known (Table 3) there have been few other reports of acid herbicides. Waite et al. (1994) compared levels of bromoxynil, dicamba, 2,4-D and MCPA in dryfall, wetfall and air (PUF) in Regina (SK), within an agricultural area, and at Waskesiu (SK) in the boreal forest about 300 km north. All four polar herbicides were detectable at the remote location although wet deposition fluxes were about 10% to 25% of those at Regina. Atrazine, alachlor and metolachlor were not detected at either site.

The precipitation and air results from Saskatchewan and northwestern Ontario, when combined with those from the US midwest and the Great lakes area, indicate that there is a large gradient in levels of atrazine, alachlor and metolachlor during the summer months between the source regions in high use areas and nonagricultural areas to the north and east. The maximum loadings (mg ha^{-1}) of atrazine reaching the boreal forest at ELA in northwestern Ontario in the May-June period (1990-92) ranged from 0.01 to 0.08 g ha^{-1}. This is much lower than the 1.1 g ha^{-1} loading estimated for Lake Erie in May-June (Eisenreich and Strachan 1992) and the 500-1000 g ha^{-1} application rate on agricultural land in the US midwest. The source regions for the acid herbicides detected in rain and air at ELA are most likely in North Dakota, northwestern Minnesota, southern Manitoba and Saskatchewan (Fig. 1) where these herbicides are used for broad leaf weed control in cereal crops. Maximum wet deposition of 2,4-D of 1.1 g ha^{-1} at ELA in 1992 was about 10-fold higher than observed at Regina in 1993 (Waite et al. 1994).

Air-water exchange of pesticides in L375

The fugacity ratios (f_w/f_a) of the pesticides were calculated in order to evaluate the air-water equilibria in L375 and the direction of transfer through the air-water interface (Table 6). Calculations were based on water concentrations in June and October 1992 paired with average air concentrations for the same months at the sampling site 15 km south of L375. Fugacity ratios for methoxychlor, α-endosulfan and trifluralin were >1 in June indicating a net diffusive transfer from water to air. These ratios are similar to those for di- to pentachlorobiphenyls in Lake Superior (Baker and Eisenreich 1990). In October f_w/f_a value were <1 for trifluralin indicating net air to water transfer. The f_w/f_a values for the more polar herbicides were <1 both in warm conditions in June and at the colder temperatures prevailing in October 1992. The direction of transfer of air-borne polar herbicides is thus consistently from air to water.

The flux of pesticides into or out of L375 was estimated using the two-film model (Schwarzenbach et al. 1993):

$1/K_{ol} = 1/k_w + RT/k_aH$

where K_{ol} = the overall mass transfer coefficient (m h^{-1}), k_w and k_a are the mass transfer coefficients for the water film and the air film, respectively, H = Henry's Law constant, R = gas constant and T = temperature (K). Values of k_w and k_a were determined using:

$k_w = (D_w/D_w^{O_2})^{0.57} \cdot (4 \times 10^{-5} \cdot u_{10}^2 + 0.0004)$ and $k_a = (D_a/D_a^{H_2O})^{0.67} \cdot (2\,u_{10} + 0.3) \cdot H/RT$, where D_w and D_a= diffusivities of the pesticide in water and air ($cm^2\ s^{-1}$), respectively, and $D_w^{O_2}$ and $D_a^{H_2O}$ are diffusivities of oxygen and water. Mean daily temperatures and wind speeds (u_{10}) were 20°C and 3 m s^{-1} (at 10 m height) for June and 5°C and 2 m s^{-1} for October in 1992 (Beaty 1994).

K_{ol} values for the phenoxyacid and triazine herbicides were independent of k_w values and $\approx(k_a \cdot H/R \cdot T)$ because of low H for these compounds. The volatilization of these polar herbicides is thus controlled by the transfer through the air film as indicated by air-film resistances ($k_w/(k_a \cdot H/R \cdot T)$ of <0.1; Table 5). Methoxychlor, α-endosulfan and trifluralin are both air and water film controlled due to their intermediate resistances ranging from 0.16 to 1.2. In October all K_{ol}'s are about 10 times lower due to lower air and water temperatures which give rise to lower Henry's Law Constants. Values of H could be calculated for 10 of 12 compounds at 20°C using published solubility and vapor pressure data (Table 5). For H values at 5 °C the relationship described by Kucklick et al. (1991) for γHCH was used, which yields values about 1/3 those at 20 °C.

Table 5. Henry's Law constant, air-film resistance and fugacity ratios for eight pesticides detected in L375 surface waters

Compound	H (20°C) $Pa\ m^3\ mol^{-1}$	Film resistance[a] June	Film resistance[a] October	f_w/f_a June '92	f_w/f_a Oct '92	Reference
alachlor	$6.2x10^{-3}$	$1.9x10^{-3}$	$6.8x10^{-4}$	$1.8x10^{-3}$	$56.3x10^{-3}$	Glotfelty et al. 1990
atrazine	$2.9x10^{-4}$	$9.3x10^{-5}$	$3.2x10^{-5}$	$1.4x10^{-3}$	$3.1x10^{-3}$	Suntio et al. 1989
2,4-D	$1.4x10^{-5}$	$4.4x10^{-6}$	$1.5x10^{-6}$	$0.7x10^{-3}$	$1.4x10^{-3}$	Suntio et al. 1989
dicamba	$1.2x10^{-4}$	$3.8x10^{-5}$	$1.3x10^{-5}$	$2.4x10^{-3}$	$4.5x10^{-3}$	Worthing & Hance 1991
endosulfan	2.98	0.90	0.31	5.7	2.0	Suntio et al. 1989
methoxychlor	1.53	0.47	0.16	9.5	21.8	Suntio et al. 1989
metolachlor	$9.1x10^{-3}$	$2.8x10^{-3}$	$9.9x10^{-4}$	$3.3x10^{-3}$	$9.0x10^{-3}$	Howard 1992
trifluralin	4.02	1.24	0.43	15.6	0.4	Suntio et al. 1989

[a] Air/water film resistance = $k_w/(k_a \cdot H/R \cdot T)$.

Table 6. Observed dissipation rates and calculated volatilization rates of pesticides in L375

Compound	Overall dissipation[a] rate, k_{diss} (d^{-1})		Volatilization[b] rate, k_{vol} (d^{-1})		Volatilization or absorption flux[c], $ng\ m^{-2}\ d^{-1}$	
	1990	1991	June	October	June	October
alachlor	-0.034±0.007	-0.006±0.007	$8.2x10^{-4}$	$1.0x10^{-6}$	-24	-0.09
atrazine	-0.004±0.002	-0.007±0.006	$4.2x10^{-5}$	$5.2x10^{-8}$	-10	-0.03
2,4-D	-0.011±0.006	-0.005±0.001	$2.0x10^{-6}$	$2.5x10^{-9}$	-1.0	-0.03
dicamba	-0.009±0.001	-0.008±0.002	$1.7x10^{-5}$	$2.1x10^{-8}$	-0.4	-0.01
endosulfan	-0.008±0.006	-0.005±0.006	0.181	$3.3x10^{-4}$	15	0.04
methoxychlor	<0.001	-0.006±0.007	0.127	$2.0x10^{-4}$	14	0.02
metolachlor	-0.009±0.005	-0.011±0.001	$1.2x10^{-3}$	$1.5x10^{-6}$	-18	-0.50
trifluralin	-0.028±0.011	-0.014±0.015	0.221	$4.3x10^{-4}$	33	-0.14

[a] First order rate constant from the equation $\ln C = \ln C_o - k_{diss} t$ where C_t = water concentration at time t (days) and C_o = concentration in July or August each year (N = 4 or 5).
[b] Calculated using the two-film model as described by Schwarzenbach et al. (1993).
[c] Negative flux = transfer from air to water.

Net volatilization fluxes (F; $ng\ m^{-2}\ h^{-1}$) were calculated from:

$$F = K_{ol}(C_W - C_A \cdot RT/H)$$

where C_W = water concentration and C_A = air concentration ($ng\ m^{-3}$). Net volatilization fluxes of 14 to 33 $ng\ m^{-2}\ d^{-1}$ were obtained for methoxychlor, α-endosulfan and trifluralin in June (Table 6). Fluxes for July and August are also expected to be positive because of high water temperatures. In October F values were lower (negative for trifluralin) reflecting lower K_{ol}'s and lower air concentrations. All of the polar herbicides show net absorption from air to water with F values ranging from 0.4 $ng\ m^{-2}\ d^{-1}$ for dicamba to 24 $ng\ m^{-2}\ d^{-1}$ for alachlor in June and much lower fluxes in October (Table 6). The results for the acid herbicides are calculated on the assumption that they are behaving as neutral species in the atmosphere. The acid herbicides, 2,4-D and dicamba, with pK_a's of 2.6 and 1.9, respectively are anions at the prevailing pH of 7.0 - 7.6 in L375.

McConnell et al. (1993) found similar gas absorption and volatilization behavior of αHCH in the Great Lakes to that observed for trifluralin, endosulfan, and methoxychlor in this study. The flux direction of HCH at 6 to 11°C water temperatures was from air to water, while at 18-22°C the flux was from water to air. αHCH has a low H value (0.63 Pa m^3 mol^{-1} at 20°C) and according to the two-film model would be expected to undergo greater gas absorption than the three semivolatile compounds studied in L375.

Overall inputs and outputs

A first-order model (Mackay et al. 1992) was used to examine the importance of various removal and input processes for the pesticides in L375. The overall unsteady state mass balance of each pesticide in the water column can be written as a series of rate constants (d^{-1}):

$$dM_w/dt = I + M_s k_{sed} - M_w(k_{vol} + k_{out} + k_{trans} + k_{ws})$$

where I = inputs (inputs from precipitation, dryfall and gas absorption), M_s=mass in sediment, k_{sed} = water to sediment transfer, M_w= mass in the water column, k_{vol} = volatilization loss, k_{out} =outflow, k_{trans} = transformation (biodegradation, photolysis) and k_{ws}= sediment to water transfer. For the pesticides determined in this study sediment deposition and sediment to water transfer can be ignored. The three most hydrophobic compounds, α-endosulfan, methoxychlor, and trifluralin, were not detected (<1 ng/g dry wt) in surface sediments in L375 (Muir et al. 1994). If sedimentation and degradation of these compounds occurred, it was not measurable and therefore would be included in an overall "transformation" rate (k_{trans}). Outflow is also potentially significant for compounds with long residence times because of the 5 yr average water residence time (5.2 x 10^{-4} d^{-1}). Therefore the equation can be written as:

$$dM_w/dt = I - M_w(k_{vol} + k_{trans} + k_{out}) = I + k_{diss} M_w$$

where k_{diss} = overall first order dissipation or disappearance rate (d^{-1}) observed for each compound in lake waters. Values of k_{diss} were calculated for each compound using water concentrations from August to December (or February for 1990-91 data) (Table 6). With the exception of methoxychlor in 1990, the pesticides concentrations could be fitted by first order decay curves. Because of the limited number of sampling times the uncertainty of each rate constant is high (Table 6). Furthermore, although most inputs from precipitation and gas absorption occurred between May and July, small amounts also were deposited in October. Nevertheless, for most compounds there is an annual decline in concentrations following maxima in June or July. Volatilization during the summer months can account for losses of methoxychlor, α-endosulfan, and trifluralin, because k_{vol} values for these compounds ($k_{vol} = K_{ol} \div$ mean depth; Table 6) are greater than k_{diss}. But k_{vol} values for the more polar, less volatile herbicides are much smaller than k_{diss} Dissipation of these latter compounds must therefore be due to transformation processes such as biodegradation and photolysis.

Information on the biodegradation and photolysis of the eight pesticides is briefly summarized in Table 7. More detailed information is available in reviews of Wauchope et al.(1992), Howard (1991) and Muir (1991). In general all of the compounds have been found to be degraded more rapidly in soils and sediment-water systems than in sediment-free incubations, or in the water column of mesocosms or ponds, presumably due to higher biomass of degrading organisms associated with sediments. For example, 2,4-D acid degraded more rapidly (DT50 ~15-90 d) in river water with sediment (250 g L^{-1}) than in river water alone (DT50 ~15->180 d)(Watson 1977). Methoxychlor degradation in filtered pond and river waters was directly proportional to microbial biomass measured by standard plate counts (Paris and Rogers 1986). Temperature and substrate (pesticide) concentration are known to influence biotransformation rates. The prevailing low temperatures in L375 (except in the epilimnion during June to August) as well as the low concentrations of each compound could limit biotransformation(Alexander1985).Bio-degradation of 2,4-D acid and alachlor in lake water has been shown to be proportional to concentration (Boethling and Alexander 1979; Novik and Alexander 1986). Direct photolysis is an important pathway of degradation for trifluralin which has a UV absorption spectra that overlaps the sunlight spectrum (Zepp et al. 1977). Indirect photolysis due to photosensitized reactions involving humic materials could be an important removal pathway for all of the compounds in the upper water column although DOC levels were low (<1 mg/L). Hydrolysis is unlikely to be

Table 7. Dissipation half-lives of pesticides in soils and in sediment-water and sediment-free laboratory incubations.

Pesticide	Soil half-life (d)[a]	DT50 in sediment-water (d)[b]	DT50 in water column (d)[c]
alachlor	15 (14-49)	23 - <100	>42
atrazine	60 (18-119)	7 - ~30	>120[d]
2,4-D acid	10 (4-16)	15 - >180	≥20[e]
dicamba	14 (3-30)	60-100	>160
endosulfan	50 (10-200)	-[f]	<14[g]
methoxychlor	120 (60-210)	100-150[h]	<10[d]
metolachlor	90 (12-180)	-	-
trifluralin	60 (60-90)	3 - ~20	<1[i]

[a] Wauchope et al. 1992

[b] Sediment-water or soil-water incubations. Results from Muir (1991) unless otherwise indicated.

[c] In sediment-free natural waters under laboratory or field conditions. results from Muir (1991) unless otherwise noted.

[d] Yoo et al. (1981); lake mesocosms.

[e] Boylc (1980);Scott et al. (1988); large outdoor ponds.

[f] No published data available.

[g] Eichelberger and Lichtenberg (1971) - river water die-away test.

[h] Muir et al. (1984)

[i] Zepp et al. (1977) assuming photolytic degradation

important for the compounds determined in this study because they are neutrals with no ester groups (e.g. triazines and acetanilides) or acids (e.g. 2,4-D, dicamba). Information on the fate of these pesticides in lake waters under field conditions is very limited. Yoo et al. (1981) found atrazine to be very persistent in the water column of large enclosures (90-125 m^3) with a half-life of ≥120 days. Methoxychlor studied under the same conditions disappeared more rapidly than atrazine from the water column (DT50, 10-20 days) due to greater sorption to sediment. 2,4-D acid has been found to dissipate relatively rapidly in the water column of large eutrophic ponds (DT50 ≥ 14 - 35 d) (Boyle 1980; Scott et al. 1988). In summary, with the exception of atrazine and dicamba, the pesticides in this study have more rapid rates of dissipation in laboratory sediment-water systems and in controlled field studies, than was observed in L375. This may reflect the lower biomass of degrading organisms in oligotrophic waters as well as continued inputs via precipitation.

The relative importance of various input and removal pathways of endosulfan and atrazine in L375 is compared in Table 8. Inflowing and outflowing water were assumed to have concentrations equal to those in L375 surface waters. Transformation rates in the water column were assumed to be equal to k_{diss} (Table 6). Volatilization and absorption of atrazine and α-endosulfan were calculated for June and October by multiplying flux (Table 6) by lake surface area. Precipitation is the major input pathway for both endosulfan and atrazine accounting for 53% and 68%, of June inputs respectively. Gas absorption is also a major input route for atrazine (24%) while tributary input is significant for endosulfan (47%). No estimate was made for dryfall inputs; these are likely to be <1% at 20 °C for endosulfan and 35% of wet fall in the case of atrazine (Eisenreich and Strachan 1992). Subcooled liquid vapor pressures of alachlor, dicamba, endosulfan, and trifluralin (Suntio et al. 1988) are sufficiently high (0.005 to 0.027 Pa) that they are predicted to be in the vapor phase at 20°C according to the Junge equation (Eisenreich and Strachan 1992). Dryfall could be more significant for the less volatile herbicides (atrazine, 2,4-D, metolachlor) especially under colder conditions.

Outputs of endosulfan occur via volatilization

Table 8. Input and removal (mg per month) of atrazine and α-endosulfan from L375 in June and October 1992

Input/output	endosulfan $mg\ mo^{-1}$ (%)		atrazine $mg\ mo^{-1}$ (%)	
	June	Oct	June	October
Imports				
Precipitation[a]	7.2 (53%)	0.8 (76%)	170 (68%)	3.4 (57%)
Gas absorption[b]	0	0	57 (24%)	0.2 (3%)
Tributaries[c]	6.2 (47%)	0.3 (24%)	22 (8%)	2.4 (40%)
Outputs				
Volatilization[b]	100 (84%)	0.2 (23%)	0	0
Outflow[b]	18 (15%)	0.5 (54%)	62 (94%)	4.2 (71%)
Transformation[d]	1.1 (1%)	0.2 (23%)	3.7 (6%)	1.7 (29%)

[a] precipitation input (mg) = monthly rain volume (m^3) x concentration ($\mu g\ m^3$) ÷ 10^3.
[b] Gas absorption and volatilization (mg) = Flux ($ng\ m^2\ d^{-1}$) x days in the month x lake surface area ($1.87 \times 10^5\ m^2$) ÷ 10^6.
[c] Tributary input/output (mg) = monthy inflow/outflow volume (m^3) x concentration in L375 surface water ($\mu g\ m^3$) ÷ 10^3.
[d] Transformation (mg) = k_{diss} (d^{-1}) x M_w (mg) x days of the month.

(84% in June and 23% in October) and lake outflow (15% in June and 54% in October). Removal of atrazine is dominated by lake outflow in both months because transformation, the only other removal pathway, is assumed to be slow.

Conclusions

Flux and fugacity ratio calculations suggest that, under the climatic conditions prevailing in much of Europe and the US midwest, lakes are sinks for polar, relatively non-volatile triazine, acetanilide, phenoxyacid and benzoic acid herbicides. Lakes are both sinks and sources of semi-volatile compounds such as endosulfan, methoxychlor and trifluralin depending on water/air temperatures. The results from L375, which receives only atmospheric and atmospherically derived tributary inputs, indicate that residence times (i.e. $1/k_{diss}$) of atrazine, endosulfan, methoxychlor, and dicamba in oligotrophic lake waters are > 100 days while those of alachlor, metolachlor, 2,4-D, and trifluralin range from 25 to 100 days. The relatively long residence times for the polar herbicides are consistent with other from laboratory and field data. The persistence of the organochlorine insecticides, methoxychlor and endosulfan in lake waters is probably the result of continued inputs via precipitation which did not vary seasonally as much as the herbicides. Because microbial degradation and photolysis is limited in deep oligotrophic lake waters air-water exchange and outflow dominate the removal processes. There are large uncertainities in estimating the volatilization outputs and the gas absorption and dryfall inputs of current use pesticides. Little is known about the variation of Henry's Law Constant or vapor pressures with temperature for most of the current use compounds detected in surface waters in Europe and North America. The availability of this data would enable predictions of gas exchange to be refined. Further information is also needed on microbial degradation rates in oligotrophic waters because, along with outflow, this may be the only removal process for polar herbicides in large lakes.

Acknowledgments

We wish to thank M. Loewen and K. Koschansky (Freshwater Institite, Winnipeg) for technical help on the analysis of water, precipitation and air samples, K. Beaty and K. Ling (Freshwater Institute, Experimental Lakes Area) for air sample collection, R. Hoff (Atmospheric Environment Service, Environment Canada, Downsview ON) for providing air samplers and W.M.J. Strachan (National Water Research Institute, Environment Canada, Burlington ON) for providing precipitation samplers. We thank G.R.B. Webster (Dept. of Soil Science, University of Manitoba) and G.A. Stern (Freshwater Institite, Winnipeg) for helpful reviews of the manuscript.

References

Albanis, T.A.; Pomonis, P.J; Sdoukos, A.T. *Sci. Total. Environ.* **1986**, *58*, 243-253.

Alexander, M. *Environ. Sci. Technol.* **1985**, *19*, 106-111.

Bacci, E.; Renzoni, A.; Gaggi, C.; Calamari, D.; Franchi, A.; Vighi, M.; Severi, A. *Agric. Ecosys. Environ.* **1989**, *27*, 513-522.

Baker, J.E.; Eisenreich, S.J. *Environ. Sci. Technol.* **1990**, *24*, 342-352.

Beaty, K. Unpublished data on lake water temperatures, air temperatures and wind speeds at the Experimental Lakes Area. Freshwater Institute, Winnipeg MB.; **1994.**

Boethling, R.S.; Alexander, M. *Appl. Environ. Microbiol.* **1979**, *37*,1211-1216.

Boyle, T.P. *Environ. Pollut.* **1980**, *A21*, 35-49.

Brauch, H.-J. *Water Supply* **1993**, *11*, 31-38.

Buser, H-R. *Environ. Sci. Technol.*, **1990**, *24*, 1049-1058.

Capel, P.D. In: *Proceedings of the Technical Meeting, US Geological Survey Toxic Substances Hydrology Program.* **1991,** 334-337.

Chan, C.H. *Water Pollut. Res. J. Can.* **1993**, *28*, 451-471.

Chan, C.H.; Perkins, L.H. *J. Great Lakes Res.* **1989**, *15*, 465-475.

Chevreuil, M.;Garmouma, M. *Chemosphere* **1993**, *27*, 1605-1608.

Eisenreich, S.J.; Strachan, W.M.J. *Estimating Atmospheric Deposition of Toxic Substances to the Great Lakes. An update*. National Water Research Institute, Environment Canada, Burlington ON. **1992**

Eichelberger; J.W.; Lichtenberg, J.J. *Environ. Sci. Technol.* **1971**, *5*, 541-544.

Fee, E.J.; Hecky, R.E. *Can. J. Fish. Aquat. Sci.* **1992**, *49*, 2434-2444.

Frank, R.; Logan L.; Clegg, B.S. *Arch. Environ. Contam. Toxicol.* **1991**, *21*, 585-595.

Frank, R.; Logan, L. *Arch. Environ. Contam. Toxicol.* **1988,** *17*, 741-754.

Hall, J.C.; Van Deynze, T.D.; Struger, J.; C.H. Chan. *J. Environ. Sci. Hlth.* **1993,** *B28*, 577-598.

Hoff, R.M.; Muir D.C.G.;Grift, N.P *Environ. Sci. Technol.* **1992**, *26*, 266-275.

Howard, P.H. (Ed). *Handbook of Environmental Fate and Exposure Data for Organic Chemicals. Vol. III. Pesticides*. **1991**. Lewis Publishers

Kucklick, J.R.; Hinckley, D.A.; Bidleman, T.F. *Mar. Chem.* **1991**, *34*, 197-209.

Kucklick, J.R.; Bidleman, T.F 1993. *Mar. Environ. Res*. **1993**, *37*, 63-78.

Krueger, J.; Staffas, A. Presented at the 8th IUPAC Int'l Congress of Pesticide Chemistry, Washington DC., **1994**, Abstract No. 547.

Mackay, D. ; Song, S.;Diamond, M.,Vlahos, P.;Voldner, E.; Dolan, D. *Virtual Elimination of Toxic and Persistent Chemicals from Lake Superior: Role of Simple Mass Balance Models.* Draft report to the International Joint Commission, Windsor ON., October **1992**

Maguire, R.J.; Tkacz, R.J. *Arch. Environ. Contam. Toxicol.* **1993**, *25*, 220-236.

Maguire, R.J.; Tkacz, R.J. *Chemosphere,* **1989**, *19*, 1277-1287.

McConnell, L.L.; Cotham, W.E.; Bidleman, T.F. *Environ. Sci. Technol.* **1993**, 27, 1304-1311.

Meyer, M.T.; Mills, M.S.; Thurman, E.M. *J. Chromatog.* **1993**, *629*, 55-59.

Muir, D.C.G. In: Environmental Chemistry of Herbicides, Vol II., R. Grover and A.J. Cessna, Eds., CRC Press, Boca Raton FL, pp. 1-87. **1991**

Muir, D.C.G.; Grift, N.P.; Fee, E.; Hoff, RM.; Strachan, W.M.J. Presented at the annual meeting of the International Association of Great Lakes Research, Green Bay WI, June **1993.**

Muir, D.C.G.; Kenny, D.F.; Grift, N.P.; Robinson, R.D.; Titman, R.D.; Murkin, H.R. *Environ. Toxicol. Chem.* **1991**, *10*, 395-406.

Muir, D.C.G.;Yarechewski, A.L. *J. Environ. Sci. Hlth.*, **1984**, *B19*, 271-295.

Nations, B.K.; Hallberg, G.R. *J. Environ. Qual.* **1992**, *21*, 486-492.

Novik, N.J.; Alexander, M. *Appl. Environ. Microbiol.* **1985**, *49*, 737-743.

Paris, D.F.; Rogers, J.E. *Appl. Environ. Microbiol.* **1986**, *51*, 221-225.

Pereira, W.E.; Rostad, C.E.; Leiker, T.J. *Anal. Chim. Acta* **1990**, *228*, 69-75.

Pereira, W.E.; Hostettler, F.D. *Environ.Sci. Technol.* **1993**, *27*, 1542-1552.

Pomes, M.L.;Thurman, E.M.;Aga, D.S.;Goolsby, D.A. Presented at the 8th IUPAC Int'l Congress of Pesticide Chemistry, Washington DC., **1994**, Abstract No. 54.

Readman, J.W.; Albanis, T.A.; Barceló, D.; Galassi, S.; Tronczynski,J.; Gabrielides, G.P. *Mar. Pollut. Bull.* **1993**, *26*, 613-619.

Richards, R.P.; Kramer, J.W.; Baker, D.B.; Kreiger, K.A. *Nature*, **1987**, *327*, 129-131.

Richards, R.P.; Baker, D.B. *Environ.Toxicol. Chem.* **1993**, *12*, 13-26.

Schottler, S.P.; Eisenreich, S.J.; Capel, P.D. *Environ. Sci. Technol.* **1994**, *28*, In press.

Schwarzenbach, R.P.; Gschwend, P.M.; Imboden, D.M. *Environmental Organic Chemistry*. J. Wiley & Sons, Inc., New York, NY. **1993**.

Scott, B.F.; Painter, D.S.; Nagy, E.; Dutka, B.J. *National Water Research Institute Report*, Environ. Canada, Burlington ON **1988**.

Siebers, J.; Gottschild, D.; Nolting, H.-G. *Chemosphere,* **1994**,*28*, 1559-1570.

Spalding, R.F.; Snow, D.D.; Cassada, D.A.; Burbach, M.E. *J. Environ. Qual.* **1994**, *23*, 571-578.

Squillace, P.J.; Thurman, E.M. *Environ. Sci. Technol.* **1992**, *26*, 538-545.

Strachan, W.M.J.; Huneault, H. *Environ. Sci. Technol.* **1984**, *18*, 127-130.

Suntio, L.R.; Shiu, W.Y.;Mackay, D.;Seiber, J.N.; Glotfelty, D. *Rev. Environ. Contam. Toxicol.* **1988**, *103*, 1-59.

Thurman, E.M.; Meyer, M.; Pomes, M.; Perry, C.A.; Schwab, A.P. *Anal. Chem.*, **1990**, 62, 2043-2048.

Thurman, E.M.; Goolsby, D.A.; Meyer, M.T.; Mills, M.S.; Pomes, M. L.; Kolpin, D.W. *Environ. Sci. Technol.* **1992**, *26*, 2440-2447.

Torstensson, L. In: *Proceedings of the 8th International Symposium on Aquatic weeds*, European Weed Research Soc., Wageningen, The Netherlands. **1990**, 215-220.

Trevison, M.; Montepiani, C.; Ragozza, L.;Bartoletti, C.;Ioannilli, E.; Del Re, A.A.M. *Environ. Pollut.* **1993**, *80*, 31-39.

Tronczynski, J.; Munschy, C.; Duran, G.; Barceló, D. *Sci. Total Environ.,* **1993**, *132*, 327-337.

Tronczynski, J.; Munschy, C.; Moisan, K. Presented at the April 1994 meeting of the American Chemical Society, Agrochemicals Div., Abstract No. 125. 1994.

Vitali, P.; Venturini, E.; Bonora, C.; Caloir, R.; Raffaelli, R. *J. Chromatog.* **1994**, *660*, 219-222.

Worthing, C.R.; Hance, R.J. *The Pesticide Manual*. 9th Ed. British Crop Protect. Council, Farnham UK. **1991**.

Waite, D.T.; Grover, R.; Westcott, N.D.; Sommerstad, H.; Kerr, L. *Environ. Toxicol. Chem.* **1992**, *11*, 741-748.

Waite, D.T.; Cessna, A.J.; Kerr, L.A.;McIntosh, T.C.; Chau, D.F.;Gurprasad, N.P.;Banner, J.A. Presented at the April 1994 meeting of the American Chemical Society, Agrochemicals Div., Abstract No. 58. **1994**.

Watson, J.R. *Water Res.* **1977**, *11*, 153-163.

Wauchope, R.D; Buttler, T.M.; Hornsby, A.G.; Augustijn, M.; Burt, J.P. *Rev. Environ. Contam. Toxicol.* **1992**, *123*, 1-155.

Yoo, J.Y.; Solomon, K.R. *Can. Tech. Fish. Aquat. Sci.* **1981**, *1151*, 164 - 167.

Zepp, R.G.; Cline, D.M. *Environ. Sci. Technol.* **1977**, *11*, 359-366.

Origin and Fate of Pesticides in Air

James N. Seiber and James E. Woodrow
Center for Environmental Sciences and Engineering, University of Nevada, Reno, Reno, Nevada 89557 USA

Studies during the past few decades have firmly established that pesticides and their transformation products are present in air as a result of drift, post-application volatilization and erosion from treated fields, emissions from formulation and waste sites, home use, and other sources. It is now more urgent that we better understand the fate of these chemicals once they become airborne, particularly so because of the apparent tendency of stable chemicals to become distributed worldwide by such processes as long distance migration via winds and clouds, global fractionation, and cold condensation. Also, the identification of at least one pesticide (methyl bromide) as a potential stratospheric ozone depleter adds still another dimension to this subject. The sources and sinks for atmospheric residues will be reviewed, with particular attention to newer information on volatilization, ambient occurrence, and environmental significance. It is suggested that the occurrence and fate of pesticides in air should receive more intensive study in the future, because of new regulatory requirements and because the scarcity of information on the fate of atmospheric residues limits material accountability for pest control chemicals once released to the environment.

Concerns over chemical contamination of the air have historically focussed on the "criteria" air pollutants -- hydrocarbons, oxidants, SOx, NOx, carbon monoxide, particulate matter and lead. In recent years there has been growing interest regarding other, primarily organic chemical species, that may have direct toxicological or other long-term human and ecological health impacts when present in the air. These may include contaminants which are emitted directly to the air (primary pollutants) as well as those formed in the air from various precursors (secondary pollutants) (Pitts, Jr., 1993). The designation "Hazardous Air Pollutants" (HAPs) is applied to these chemicals, and 189 such materials are explicitly mentioned in the Clean Air Act Amendments (CAAA) of 1990. California's Toxic Air Contaminant (TAC) Act includes a definition and classification of TACs which are analogous to those of HAPs. Because of these two pieces of legislation, and ones like them in other states and nations, as well as other more general concerns over environmental quality, information will be increasingly required regarding emission sources, physical and chemical properties which influence transport, fate, and exposures, and toxicological impacts of actual and potential atmospheric contaminants (National Research Council, 1994). Pesticides are among the HAPs and TACs, and indeed, pesticides and pesticide transformation products are prominent on the CAAA-90 list of 189 HAPs (Table 1). California has collected monitoring data indicating exposure, and performed quantitative risk assessment, both according to outlined steps (California Air Resources Board, 1993) for one candidate pesticide (ethyl parathion) under its Toxic Air Contaminant Act. The banning of ethyl parathion in the U.S. by EPA removed this chemical from further interest as a TAC. Another pesticide, Telone or 1,3-D, was suspended in California following a determination that its use as a soil fumigant posed an imminent hazard from inhalation of vapors escaping from the soil -- a violation of another piece of California air toxics legislation known as the "Hot Spots" regulation (AB 2588; California Health and Safety Code 44300 et seq.). One chemical (methyl parathion) is in the final stages of evaluation as a TAC, and 21 others are in various stages of risk assessment and/or exposure assessment data collection (Table 2). The regulation of pesticides based upon their occurrence as air contaminants is a trend which is clearly increasing, to the point of presenting

Table 1. Pesticides and Pesticide Transformation Products Listed as Hazardous Air Pollutants (HAPs) under the U.S. Clean Air Act Amendments of 1990.

1,3-dichloropropene	ethylene thiourea
2,3,7,7-tetrachlorodibenzodioxin	heptachlor
2,4,5-triclorophenol	lindane
2,4-D salts and esters	methoxychlor
4,6-dinitro-o-cresol and salts	methyl bromide
acrolein	parathion
captan	pentachloronitrobenzene
carbaryl	pentachlorophenol
chloramben	phosgene
chlordane	phosphine
DDE	propoxur
dichlorvos	toxaphene
ethylene dibromide	trifluralin
ethylene oxide	

another criterion for use by society when considering first registering or continuing the registration of pesticidal chemicals.

Pesticides are frequently applied through the air, from which drift as particulate matter and/or vapors may occur, or they may enter the air by volatilization or dust erosion from surfaces after application is complete (Figure 1) (Seiber and Woodrow, 1983). A large data set is available which details the amounts which enter the air from drift (Lewis and Lee, 1976) and post-application flux or volatilization (Spencer et al., 1973; Taylor and Glotfelty, 1988; Taylor and Spencer, 1990). Some of this work has been driven by worker health and safety concerns, and some by non-target crop damage or illegal residue concerns (Breeze, 1993). But there is very little known, even today, concerning the eventual fate of the airborne residues as they migrate from the point of release until they are mineralized to natural constituents of the environment (Woodrow et al., 1983). This is not a trivial question. For example, for the heavily used rice herbicide, molinate, at least half of the applied material ends up as vapor, totalling a release of several hundred thousand kilograms annually in California's Sacramento Valley alone (Seiber et al., 1989); most of the diazinon used as a dormant spray in California's Central Valley (120,000 kg were reported for this use in 1990) volatilizes within a few weeks of application (Glotfelty et al., 1990a); and at least one-third, and perhaps more, of the one million-plus kilograms of methyl bromide applied as a soil fumigant in California volatilizes, even when it is injected into, and confined within the soil with a plastic tarp covering (Yagi et al., 1993; Majewski et al., 1994). For methyl bromide, there is the additional, more immediate concern that it is a depleter of stratospheric ozone, along the lines of the chlorofluorocarbons (Albritton and Watson, 1992). In this review we will deal primarily with more recent studies of pesticides in air, particularly those published since the review paper presented on this topic at the 1990 IUPAC Pesticide Congress (Majewski, 1991) and an earlier one presented at the 1982 Congress (Seiber and Woodrow, 1983). Certainly there have been many fine reviews of the subject in the intervening years, including roughly half of the chapters from an American Chemical Society Symposium in 1988 (Kurtz, 1990) and the review of de Voogt and Jansson (1993).

Post Application Volatilization of Pesticides

Influencing Factors.

Volatilization of pesticides after application to soil, water, and plant foliage is often a significant pathway for losses to the atmosphere, with subsequent movement to non-target areas. Losses can also occur as particulates through the action of

Table 2. Pesticides as Toxic Air Contaminants under California's AB 1807 Statutes (Status as of May 1994; California Air Resources Board)

Evaluation Completed
Ethyl Parathion
Evaluation Underway
Methyl Parathion
Data Gathering In Process
1,3-Dichloropropene
Azinphosmethyl
Benomyl
Captan
Carbofuran
Chloropicrin
Chlorothalonil
DEF
Garlon/2,4-D
Inorganic Arsenic
Mancozeb
Metam Sodium (MITC)
Methidathion
Methomyl
Methyl Bromide
Molinate
Monocrotophos
Naled
Oxydemeton-Methyl
Paraquat

wind erosion. However, losses of pesticides from treated surfaces typically occur in the form of vapors for those compounds that have appreciable vapor pressures. Factors that influence evaporative losses from treated surfaces include vapor pressure and environmental conditions that control movement away from treated surfaces (e.g., molecular diffusion and eddy dispersion) (Thornthwaite and Holzman, 1939; Parmele et al., 1972). A volatilizing pesticide will first diffuse through a stagnant air boundary layer immediately above the treated surface and then be carried away by turbulent flow of the air. As the wind speed increases, the depth of the stagnant air boundary layer will decrease, resulting in an increased rate of volatilization of the applied pesticide. The roughness of the surface will influence volatilization rate by leading to greater turbulence closer to the surface as the roughness increases, resulting in greater turbulent transfer of residue to the atmosphere.

Other factors that influence volatilization include sorption to soil and Henry's law constant (H, ratio of pesticide vapor pressure and water solubility). The effective vapor pressure of a pesticide can be reduced by soil sorption, which is primarily determined by soil organic matter content. However, moisture content of the soil can offset the effect of sorption on volatilization by enhancing pesticide volatility in moist soils, compared to these soils when dry, because water will compete for sorption sites. Often when pesticides are applied to soil surfaces, pesticide flux remains high until the surface dries. Flux will then be limited by soil sorption, and when surface residues become depleted, diffusion of pesticides from within the soil penetrated by the spray becomes more important. In this regard, depth of soil incorporation is a factor influencing volatilization; movement through soil to the surface, which is in part accomplished by diffusion through soil vapor spaces and through soil moisture, is slowed with greater incorporation depths. Diffusion coefficients for pesticides in water and vapor are important here, as well as soil porosity. Furthermore, the movement of water to the soil surface can carry pesticide residues to the surface (wick effect caused by capillary action). In this regard, the pesticide's H becomes important. This has been borne out by studies that have compared the volatilization of a number of incorporated pesticides for similar soils and found that volatilization depended primarily on the H (Jury et al., 1984b). When the water solubility was relatively high (low H), pesticide residues would accumulate at the surface faster than they would evaporate, resulting in increased flux over time. In contrast to this, pesticides with relatively low water solubilities (high H) would show declining flux over time regardless of the movement of water to the surface.

A pesticide's H is also important for moderately soluble residues in treated water, since distribution between the air-water interface is determined by this term. Depending on the conceptual model used to describe evaporation from water, flux can be directly proportional to H. However, complicating this may be situations where flux is

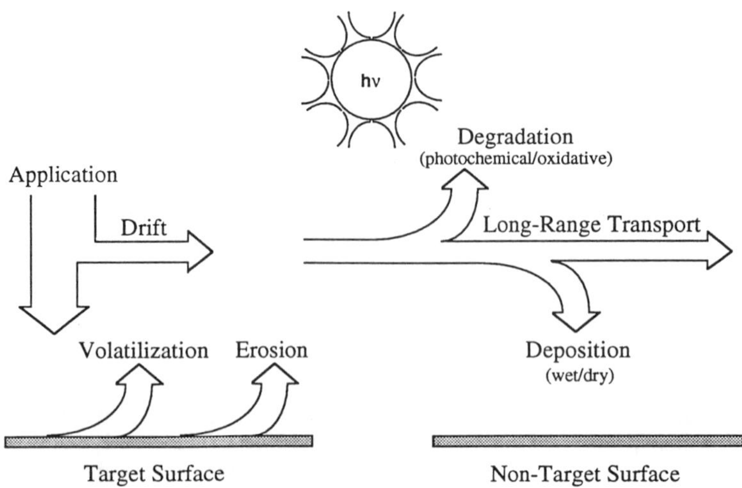

Fig. 1. Occurrence of pesticide residues in air.

also a function of concentration gradients in the water so that component diffusion to the water surface controls flux. The latter would be characteristic of deep water where mixing may not be efficient in contrast to turbulent water with efficient mixing and large surface-to-volume ratios where flux would be greater (e.g., flooded rice fields) (Tinsley, 1979). For water-miscible residues, volatilization of each component dissolved in water may be simply controlled by the component's partial pressure [derived from a component's saturation vapor pressure, solution mole fraction, and activity coefficient (Raoult's law)], assuming no concentration gradients and minimal interaction between the solution components.

Field Measurements. Studies of pesticide volatilization conducted under field conditions have primarily focused on residues applied to soil, and to a lesser extent those applied to foliage and water (Taylor et al., 1977; Willis et al., 1983; Seiber et al., 1986; Ross and Sava, 1986; Cooper et al., 1990; Glotfelty et al., 1990a). Techniques for measuring volatility losses vary from residue analysis of the treated surface (soil, water, foliage), from which volatility losses are assessed as the amount lost from the surface not accounted for by other dissipation mechanisms, and air enclosed in chambers placed just above defined surface areas (Sanders et al., 1985; Jenkins et al., 1991; Woodrow and Seiber, 1991; Batterman et al., 1992; Yagi et al., 1993) to micro-meteorological methods that relate residue levels in air above the treated surface to volatilization rates from the surface (Thornthwaite and Holzman, 1939; Parmele et al., 1972; Wilson et al., 1982; Glotfelty et al., 1984; Majewski et al., 1989, 1990, 1991, and 1993; Whang et al., 1993). In-depth reviews of various pesticide volatilization studies are available (Taylor and Glotfelty, 1988; Taylor and Spencer, 1990). Both the rates and amounts of volatilization losses vary greatly between chemicals, application variables, and meteorological conditions. Extremes include Beacon weed oil (a diesel fraction) for which volatilization $t^{1/2}$ when applied to soil is approximately 50 minutes (Woodrow et al., 1986), to toxaphene applied to cotton ($t^{1/2}$ = 200 hr, with 80+ % volatilized in 60 days) (Seiber et al., 1979), to only 3.4% of trifluralin volatilized in 90 days when incorporated to 7.5 cm in soil (Glotfelty et al., 1984).

Laboratory Work. Much laboratory work has been involved with simulating field conditions which affect volatilization of pesticides from soil, water, and foliage (Nash et al., 1977; Seiber et al., 1986; Spencer et al., 1988; Spencer and Cliath, 1990; Woodrow et al, 1990; Maguire, 1991). Techniques have included dynamic, flow-through chambers (Nash and Beall, 1977; Nash et al., 1977; Sanders and Seiber, 1983 and 1984) and static chambers, where vapor concentrations of pesticide residues are measured prior to the establishment of equilibrium (Rolston, 1986;

Woodrow and Seiber, 1991), and at equilibrium conditions (Sagebiel et al., 1992). These methods are useful for estimation purposes when no field data exist, and to rank or prioritize chemicals for more in-depth studies.

Modeling. In recent years, several investigators have attempted to correlate pesticide volatilization with the various influencing factors in descriptive equations that can be used to predict losses from treated surfaces. These equations, or models, range in complexity from the "effusion" type, which assume that volatilization is occurring from a non-adsorptive surface or from a well-mixed body of water (only pesticide vapor pressure is the determining factor) (Tinsley, 1979), to descriptive differential equations that include all possible factors that together will affect net volatilization (e.g., diffusion coefficients in water and vapor, mass-transfer coefficients, soil adsorption, H, vapor pressure, wind speed, etc.) (Jury et al., 1983, 1984a, 1984b, and 1984c; Lyman, 1990). Once again, these techniques are useful for ranking and prioritizing, but their ability to predict absolute volatilization rates in the environment has generally not been validated.

Ambient Levels of Occurrence

There is largely just sporadic data on the ambient occurrence of pesticides in air. Much of the data come from sampling of organochlorines in more remote atmospheres, in which they tend to be readily detectable. This reflects their stability which leads to long-term cycling in the biosphere including deposition in remote areas far removed from their source. Reviews of organochlorines in the atmosphere, and in atmospheric deposition samples, are in several of the chapters in Kurtz, 1990.

Kutz et al. (1976) carried out one of the more comprehensive monitoring programs for pesticides in air that could be related to use patterns in the U.S., involving 28 sites in 16 states, between 1970 and 1972. Organochlorine residues, from the DDT group, chlordane, and toxaphene dominated the data, partially reflecting the availability of sampling and analysis methods for these chemicals. A more recent study, by Goolsby et al. (1994), focussed on rain samples taken at 81 sites in 23 largely northcentral and northeastern U.S. states. Analyses were primarily for herbicides used in corn and soybean production, particularly the triazines and acetanilides. Residues were generally in the 100 to 1000 ng/l range during the Spring-Summer use periods. Details of studies of regional and national atmospheric pesticide levels, including vapor, aerosol/moisture, and particulate phases, will be found in an upcoming review (Majewski and Capel, personal communication).

There needs to be more data collected showing basin-wide levels of pesticides, and variations in observed levels caused by such factors as distance from use areas or source to sampling sites, weather conditions, quantity and timing of applications, and the like. Laws such as California's Toxic Air Contaminant Act and recent concerns over inadvertent residues caused by the movement of pesticides through the air (Turner et al, 1989) will likely stimulate interest in this subject. An example of the type of studies in this category was reported by Seiber et al. (1989) on the patterns of air contamination with chemicals used in rice paddies in the Sacramento Valley air basin of California. Table 3 summarizes ambient air concentrations of three of the chemicals--molinate, thiobencarb, and methyl parathion--during or just following the major use period, as measured at four small community locations within the general area of rice culture in this air basin (See Figure 2), and at a "control" location south of the rice growing area. The measured concentrations were consistent with predictions based upon the physicochemical properties of the pesticides and their usage rates and volumes. Molinate, the chemical with the highest H and, consequently, the highest rate of volatilization from water, and with the largest volume of use on rice in that area, gave the highest ambient air concentrations (up to 650 ng/m^3). Thiobencarb has about 1/10th the H, and is used in lesser quantities during the application season; its maximum air concentration (approximately 70 ng/m^3) reflects these conditions. Methyl parathion has an even lower H, and an only infrequent usage on rice; the highest average air concentration of this chemical was only about 6 ng/m^3. Methyl paraoxon was not a significant residue, either because of its very low H, or a low rate of thion-to-oxon conversion in the paddy waters, or both. This same sampling area of the Sacramento Valley was also home to field flux measurements for the three aforementioned chemicals, and for MCPA

Table 3. Summary of Ambient Air Concentrations for Three Rice Chemicals During the Period May 12 - June 12, 1986, Sacramento Valley, CA.

	Average (Highest) Concentration (ng/m^3)			
Location	Methyl Parathion	Methyl Paraoxon	Molinate	Thiobencarb
Maxwell	6.2 (25.7)	0.8 (3.1)	650 (1700)	29.0 (64.5)
Williams	3.1 (21.8)	0.6 (1.1)	200 (420)	12.9 (23.3)
Trowbridge	0.32 (1.1)	<0.5	150 (280)	67.8 (250)
Robbins	0.27 (0.72)	<0.5	60 (140)	14.2 (40.8)
Davis	<0.5	<0.5	<20	<2.0

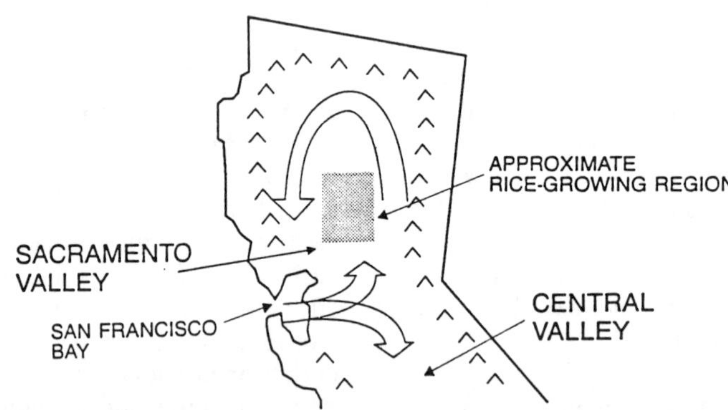

Fig. 2. Map of Sacramento Valley, California, showing typical summertime wind patterns (open arrows) (Woodrow et al., 1990).

and its breakdown product, 4-chloro-o-cresol (Seiber et al., 1986). The relative fluxes were in the order cresol > molinate > thiobencarb > methyl parathion > MCPA, again as expected based upon H constants. The fluxes provided a source term for modelling of typical air concentrations in the areas downwind from treated fields, to which a Gaussian plume dispersion model could be applied for predicting concentrations. The dispersion model predictions gave estimates of downwind air concentrations that were in reasonable agreement with measured values. This provides evidence that the transport-fate of these chemicals was largely due to physical processes, that is, volatilization followed by dispersion. Chemical reactions did not appear to dominate the transport-fate picture in this instance, with the exception of the MCPA-to-4-chloro-o-cresol conversion (Seiber et al., 1986). There are, of course, cases in which chemical conversion plays more important roles in the fate processes connected with the atmosphere. Of the rice chemicals described above, methyl parathion probably oxidizes and hydrolyzes to dimethylphosphoric or thiophosphoric acids plus p-nitrophenol in water. None of these products were likely to volatilize from water to air because of disadvantageous H. Vapor phase oxidation of the parent thion could produce methyl paraoxon; this product is relatively stable in the vapor phase but is of such a hydrophilic nature that it might be expected to re-enter the water phase, either in surface water or atmospheric cloud or raindrop water, whenever the opportunity arose. Some pesticide transformations may occur at such rates in the environment, either in/on the originating surface, or in the air itself, that toxic products are released which may be very important in some environmental situations (Wolfe and Seiber, 1993; Coats, 1991). Figure 3 lists some examples of

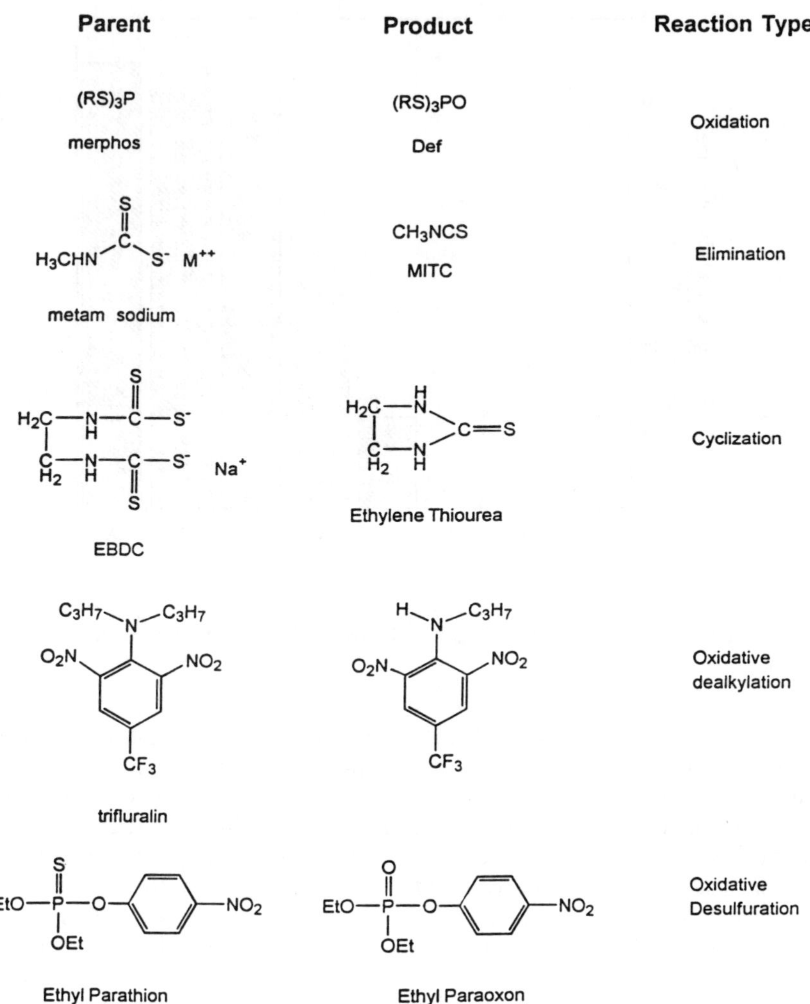

Fig. 3. Examples of pesticide transformations that may lead to release of toxic products to air under some conditions.

activations to volatile products, most of which have been confirmed by either chamber studies or by environmental sampling. Glotfelty et al. (1990b and references therein) provided several examples of correlating airborne pesticide levels with use patterns and wind movement. Detailed analysis of rainwater in Maryland showed microgram per liter concentrations of atrazine and alachlor during the spring time use period, primarily April- June for atrazine, and April-July for alachlor. Rainwater residues at other periods during the year were less than 0.05 micrograms per liter. Deposition of atrazine in rainwater in the Chesapeake Bay area showed very marked seasonality (Figure 4). Because atrazine first appeared in rainwater samples before the chemical was used in Maryland, the authors postulated that long range movement had occurred, from Atlantic Coast states further south where the Spring use period was a month or so ahead of the Maryland region. Toxaphene also appeared in rainwater samples in the March-July season, but residues were detectable at other times of the year as well, indicating the broader spectrum of use for this insecticide.

Several reports have appeared documenting fogwater as a good indicator of atmospheric contamination with pesticides (Glotfelty et al.,

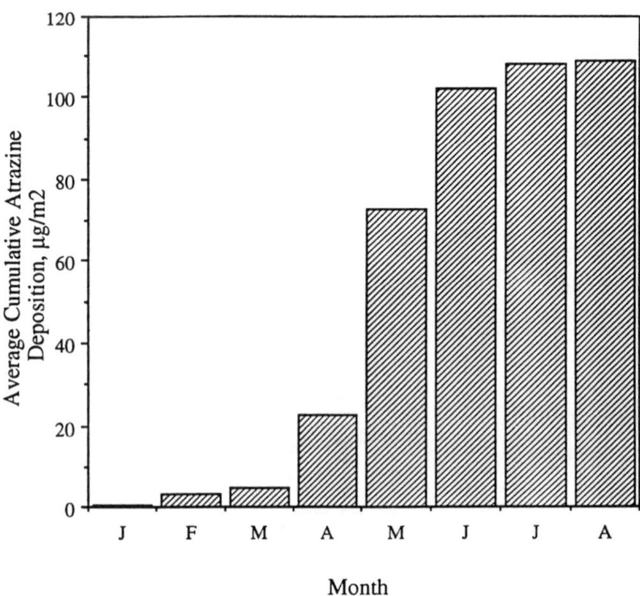

Fig. 4. Cumulative amount of atrazine ($\mu g/m^2$) deposited by rain in Wye River in 1982 (Glotfelty et al., 1990b).

1987; Glotfelty et al., 1990a; Seiber et al., 1993). Unlike cloudwater, fogwater samples atmospheric pollutants near ground level. Unlike rainwater, it apparently becomes enriched with residues with time, perhaps because it remains in suspended contact with the air mass, and perhaps because of impurities -- surface organic films, colloids, etc -- or because the high surface area of fogwater or, more properly, the high surface area-to-volume ratio allows the surface to act as a significant third phase (Hoff et al., 1993). Whatever the mechanism, it is clear that some pesticides, and other trace organics, may become enriched in fogwater well beyond levels predicted from simple vapor-to-water partitioning. Whereas early vintage fogwater collectors were quite exotic, there are now available readily fabricated fogwater collectors for general use (Collett, Jr., et al., 1990).

Most of the U.S. studies of pesticides in fogwater were done in California's Central Valley, because of the propensity for long periods of wintertime inversions, creating the phenomenon of "Tule" fog, and because of the heavy use of pesticides in this valley. One group of chemicals actively used during the wintertime fog season are the organophosphorus insecticide dormant sprays applied to almonds, peaches, nectarines and other orchard crops. Drift during application and post-application volatilization both contribute to airborne levels--primarily as vapors in the downwind ambient air (Schomburg et al., 1991)--which subsequently may become scrubbed by fogwater. An analysis of the dynamics of this system (Seiber et al, 1993) showed that air concentrations of ethyl parathion, chlorpyrifos, diazinon, and methidathion were often in the 100 ng/m^3 level, and that fogwater, sampled through a dynamic fogwater collector or simply by holding a funnel underneath fogwater-laden trees, ranged to 100 ppb. Oxons of the four OP insecticides were present as well, and were more abundant in day than in night samples, perhaps indicating the importance of photochemical oxidant (Faust, 1994).

Significance of airborne pesticides

There are a number of reasons to be concerned with airborne residues of pesticides. From a general viewpoint, once pesticides enter the air, we have lost control of them. It is difficult to defend against the charge that we are unable to account for the bulk of the applied material, when much or most of the pesticides wind up in the air. As virtually all studies to date have shown, the air has become both a receptacle and a transport vehicle for pesticide residues that simply must be better understood if we are to "close the loop" on accounting for what has been applied in agriculture, forestry, and home and commercial uses.

There exist several more specific reasons for concern over airborne pesticides: Atmospheric residues may pose a risk to people, and to plants and wildlife. An example of the former is provided by 1,3-dichloropropene which was suspended for use in California because airborne residues in the field environment were judged to be high enough to pose a significant health risk to workers and those living next to fumigated fields. An example of the latter was provided by the release of methyl isothiocyanate (MITC) during the hydrolysis of Metam sodium spilled in the upper Sacramento River in 1991, resulting in defoliation of vegetation and, apparently, some losses of birds, insects, and other animals in the zone of the plume. (State of California, 1991).

Deposition of airborne pesticides may transfer pesticides from the use area to a non-target area, with a variety of potential problems resulting. Deposition to non-target food crops may give an illegal residue, and prevent the marketing of the produce (Turner et al., 1989). Deposition of either pesticide vapors, or fogwater containing pesticide residues, to hawks and, potentially other birds and other forms of wildlife, may constitute an important route of exposure , at least in the vicinity of uses (Wilson et al., 1991).

Long range transport of pesticides in the air may transfer these materials to relatively pristine -- and fragile -- environments, with the potential for widespread ecological effects. The best studied examples continue to be the organochlorines, which contaminate a variety of sea life (whales, other fish, gulls, etc) including within both the Arctic and Antarctic circles (Kurtz, 1990). The propensity of the very stable pollutants with intermediate vapor pressures to move from the tropic and temperate zones of the earth to the polar regions is so marked that some authors have postulated that a global fractionation has occurred somewhat akin to what happens on a cold condenser surface in a sublimation experiment (Wania and MacKay, 1993).

There are other related concerns as well. A number of investigators (Taylor and Constable, 1994; Gaggi et al., 1993; Baldocchi et al., 1987) have shown that plant foliage, and particularly tree leaves and needles, may take up residues from the air. This is a potentially valuable sample set for determining patterns of contamination, but there may also be effects on trees particularly when the accumulated residue includes herbicides or other phytotoxic agents. The uptake and release of residue by foliage are relatively slow, so that equilibrium may take weeks to achieve, and depuration weeks or even months. We recently examined pine needles as atmospheric scrubbers. Separate analysis of the surface rinse, cuticular wax, and needle tissue can yield valuable information. Pine trees placed in and near a diazinon-treated orchard well after application had picked up diazinon residues in the high ppb-low ppm range. Early residues favored the surface layer of the needle, but with time, the cuticular wax and needle tissue were the more significant depots. Preliminary results from pine needles sampled at remote sites removed from agricultural pesticide use areas showed the presence of a number of organophosphate residues (Aston and Seiber, 1994), providing further support for using pine needles as environmental markers for airborne contamination (Frank et al, 1994).

Regional Transport and Deposition of Organophosphates. While most of the prior research on the long-range transport and deposition of pesticides, including the global fractionation and cold condensation aspects, have dealt with organochlorine pesticides (chapters in Kurtz, 1990; chapters in Calimari, 1993), other agrochemicals may display this phenomenon as well, if not globally at least regionally. Atrazine and alachlor are examples, cited above. Organophosphates represent another potential group, because of their large usages on the aerial portions of crops, which places them in position to volatilize, and because some are surprisingly stable. Although most OP atmospheric reactivities with hydroxyl radical are measured in days or hours (Winer and Atkinson, 1990), there may be sufficient stability, particularly in relatively unpolluted air, for movement-deposition to occur over intermediate distances. Also, the oxons of thionate insecticides are quite stable in air, in contrast to their behavior in soil and water, providing residence times sufficient for movement and deposition. We collected air and rainwater samples from two locations in the Sierra Nevada mountains, ranging up to 150 km from the major pesticide use areas in California's Central Valley (Zabik and Seiber, 1993). Samples were analyzed for the same OPs used in dormant spraying in the San Joaquin Valley. At the 533-m elevation, approximately 50 km from the nearest sources, measurable air concentrations, in the 1-to-100 pg/m^3 range, were recorded for chlorpyrifos, diazinon, diazinon oxon, parathion and paraoxon. These levels were approximately 1/10th to 1/100th those recorded in the Central Valley itself during the same Jan-March sampling period. At 1930 m elevation, approximately 150 km from the closest source orchards, only paraoxon (ca 10 pg/m^3) was found. In contrast, in rainwater from the same locations, there was little difference in concentrations, and all 5 OPs were readily measured in samples from both the 533 and 1930 m elevations. Rainwater concentrations from the two mountain locations were generally in the 10 to 100 ng/l (ppt) concen- tration range. We estimated that an input of 32 kg of OPs by wet deposition alone could be occurring in Sequoia National Park from usage in the Central Valley. This is still just a small fraction of the estimated 500,000 kg of these same OPs used in the Central Valley each year, but it does not include other inputs, from dry dust deposition in the summertime or from vapor-to-foliage exchange throughout the year, and it represents just one 160,000 ha area and just a few of the many pesticides used in the Central Valley. Whether there is an ecological risk posed by these rainwater residues is unknown. While most wildlife show effects only at concentrations much higher than those measured, chlorpyrifos has LC50s of 130 and 570 ng/l for the fathead minnow (Pimephales promelas) and stonefly (Pteronarcys californica), respectively, suggesting that ecological food chain effects could occur in puddles, vernal pools, and small lakes of evaporating rainwater. The possibility of larger watershed impacts, through release of material over broader areas by rainstorms and snowmelt, should be considered as well, and in fact the Sierra Nevada system may represent a good test for regional "cold condensation" effects, given the regional weather patterns and location of the central Sierras downwind to the large agricultural Central Valley of California. It is interesting to note that, in an earlier era, Cory et al. (1970) had found DDT/DDE in frogs in the Sierra Nevada range suggesting a potential link between chlorinated hydrocarbons and the rather baffling decline in frog populations in that area.

Methyl Bromide. Another aspect to the impact of airborne pesticide residues has emerged just in the past few years. This has to do with the growing concern over the role of methyl bromide in stratospheric ozone depletion. The concern is based upon the fact that bromine, one source of which is the photochemical conversion of methyl bromide at the shorter wavelengths of light in the stratosphere, is 30 to 60 times more efficient, on a per molecule basis, than chlorine as an ozone depleter (Albritton and Watson, 1992). The stability of methyl bromide in the atmosphere is such that it may survive atmospheric reactions and diffuse to the stratosphere where depletion can occur (Mellouki et al., 1992; MBGC, 1993). The situation is far from clear-cut. The natural and combustion sources of methyl bromide may far outweigh the agricultural input, and there is suggestion that the ocean may represent both a source and a sink, that is a buffer, for atmospheric methyl bromide (Butler, 1994). But methyl bromide has been categorized as a Class I ozone-depleting chemical by the U.S. Environmental Protection Agency and has been recommended for phaseout by the United Nations Environmental Programme (UNEP) under the Montreal Protocol.

All of this assumes that methyl bromide is released to the air in substantial amounts in connection with its use in agriculture, and that there are no other "sinks" which might substantially reduce methyl bromide levels in the atmosphere. In regards to the first point, we conducted a field test of the volatilization flux of methyl bromide from tarped and untarped agricultural fields (Majewski et al, 1994). The cumulative volatilization losses from the tarped field accounted for 23% of the nominal application within the first 5 days, and 36% of the nominal application within 9 days. In contrast, the untarped field lost 98% of the nominal application by volatilization in 5 days. Yagi et al (1993) estimated an even greater emission rate of methyl bromide from tarped fields, based upon the use of flux chambers positioned on top of the tarp. Whether the correct figure is 36% or 100% (environmental and management variables will undoubtedly influence the results for any given field), it is clear that a substantial portion of the

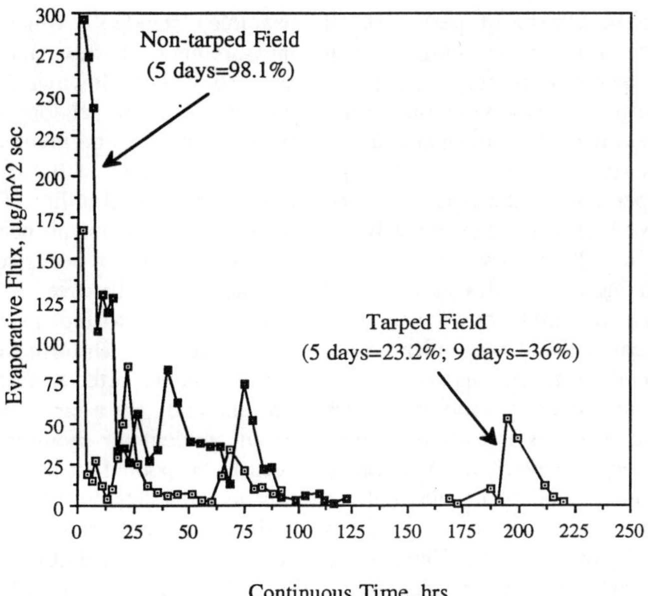

Fig. 5. Methyl bromide evaporative flux from nontarped and tarped fallow agricultural fields. Methyl bromide was injected to a depth of 25–30 cm. Flux was measured by the aerodynamic gradient method. The tarp was removed from the tarped field on the 8th day after application (Majewski et al., 1994).

50 million kg used in agriculture each year will enter the atmosphere under the present management regime.

What happens to the methyl bromide once it becomes airborne, and what fraction enters the stratosphere, is largely unknown. Potential sinks include soil (methyl bromide has a high affinity for dry soil, and degrades both chemically and biologically with a half-life of several days once in soil) and plant foliage, as well as water including ocean water. There is also the possibility of interception of some methyl bromide by fog and cloud droplets, and its reaction or deposition from these phases. It is important that we understand all aspects of the transport and fate of methyl bromide, not only to address the immediate issues which threaten the use of this material in agriculture (an industry which had already lost DBCP and ethylene dibromide because of groundwater contamination concerns, and 1,3-dichloropropene because of worker exposure concerns) but also because it represents an opportunity to better understand what happens to pesticides once they become airborne--again in the spirit of "closing the loop" in accounting for the materials used in agriculture.

Summary and Conclusions

Attention to the occurrence and distribution of pesticides in the air has been increasing. In part this is a result of growing societal concern over "toxics" in air reflected in several pieces of legislation in the 1980s and 1990s. But it is also because of infor- mation indicating that the air is a major receptacle of applied pesticides and the compartment about which we know the least regarding pesticide fate.

The area of post-application volatilization is one in which much new detailed information has been forthcoming recently. Better methods for determining flux, and better analytical methods for trapped residues, and the evolution of predictive models have all contributed. It is not unusual to see results showing that half or more of the applied material becomes airborne within just a few days of application. This is in addition to the often substantial release to the air as drift during application. There is quite a challenge ahead for manufacturers and pest management specialists to keep the applied material at the target. Purposeful design of low volatility chemicals, the further development of encapsulating delivery material, and the timing and placement of pesticides all represent possibilities. For the soil fumigants such as methyl bromide, the use of interceptive barriers to prevent release to the air may be worth further exploration. The downwind movement and fate of airborne pesticides is still not well enough understood, except perhaps for the organochlorine chemicals which have been studied extensively. If so much of the applied chemical winds up in the air, why are we finding only traces in air in the air basins of use and also areas downwind from those basins? We can assume that much of the airborne material degrades, although there are very little experimental data. And if it does degrade, what are the products and where do they wind up? It is unlikely that pesticides mineralize in the air compartment, certainly not without the intermediate formation of a variety of products some of which may be quite stable. The finding of pesticides in fogwater points out the largely overlooked potential of clouds in playing a role in transport and fate processes. Several measurements have been made of rainwater and some of snow--the end result of cloud accumulation -- but little is known of the clouds themselves as potential accumulators and/or reactors for airborne pesticides. Several investigators have suggested that tree foliage may play a prominent role in scrubbing pesticides from the air. A role for dust, and dust-associated deposition, may exist as well, certainly so for chemicals of lower vapor pressures and in the more arid and semiarid climates.

The environmental significance of airborne pesticides is even less clear, and represents a fruitful area for more research. Direct human and wildlife health considerations at or near the site of applications aside, where direct effects are most likely and indeed have been recorded in a few instances, what is the impact of airborne residues hundreds of kilometers from the site of release? If trees are receptacles of airborne pesticides, is there an effect on the forest ecosystem? Rain and snowfall deposition also pose threats, certainly so for the more remote, often more fragile environments. As with any risk associated with a chemical contaminant, it is the most vulnerable process in the ecosystem which must be discovered and understood, and become the focus of mitigation efforts.

Experiments need to be designed in such a way as to maximize the information payload. For example, studies which combine accurate measurement of flux should be carried out in synchrony with downwind monitoring including the use of sentinel organisms. Atmospheric scientists need to be involved, and environmental modelers must be included so that the results can have more general utility. Conversion products as well as parent chemicals need to be included in the analyses. The air needs to be more thoroughly sampled, not only over horizontal distances but also vertically, to the inversion layer and to the cloud cover. Unfortunately, most of the experiments reported to date are limited in scope -- a flux study, or a farm- worker exposure study, or an ambient air monitoring study. Resources need to be pooled so that the larger-scale studies can be implemented. Because of the rather dim outlook for public sector funding, it may be necessary to form industry-wide taskforces, such as exist for the alternatives to fluorocarbons (Alternative Fluorocarbons Environmental Acceptability Study -- AFEAS) and for methyl bromide (Methyl Bromide Global Coalition), to provide a funding mechanism. Expansion of the Spray Drift Task Force in the U.S. to the broader area of transport and fate might represent another mechanism.

Detailed knowledge of the behavior of pesticides in air, once of strictly academic concern, has become much more urgent, and the needs seem to be increasing at a faster rate than the level of scientific effort.

REFERENCES

Albritton, D.L. and R.J. Watson. 1992. Methyl bromide and the ozone layer: A summary of current understanding. United Nations Environment Programme on behalf of the contracting parties to the Montreal Protocol. pp 21.

Aston, L.S. and J.N. Seiber. 1994. Methodologies for the analysis of organophosphate pesticides in four compartments of needles of Pinus ponderosa. Eighth International Congress of Pesticide Chemistry. ACS/IUPAC, Session 8-C, Washington, D.C. July 4-9.

Batterman, S.A., B.C. McQuown, P.N. Murthy and A.R. McFarland. 1992. Design and evaluation of a long-term soil gas flux sampler. ***Environ. Sci. Technol.*** **26:**709-714.

Baldocchi, D.D., B.B. Hicks and P. Camara. 1987. A canopy stomatal resistance model for gaseous deposition to vegetated surfaces. ***Atmos. Environ.*** **21:**91-101.

Breeze, V.G. 1993. Phytotoxicity of herbicide vapor. ***Revs. Environ. Contamin. Toxicol.*** **132:**29-54.

Butler, J.H. 1994. The potential role of the ocean in regulating atmospheric CH_3Br. ***Geophys. Res. Lett.*** **21:**185-188.

California Air Resources Board, 1993. Program Update: ***Air Toxics Update #8***, Sacramento, California.

Calimari, D. (Ed) 1993. ***Chemical Exposure Predictions***, Lewis Publishers, Ann Arbor, MI.

Coats, J.R. 1991. Pesticide degradation mechanisms and environmental activation. In: ***Pesticide Transformation Products: Fate and Significance in the Environment*** (Somasandaram, L. and J.R. Coats (Ed), ACS Symposium Series 459, American Chemical Society, Washington, DC pp.10-31.

Collett, Jr., J.L., B.C. Daube, Jr., J.N. Munger and M.R. Hoffman. 1990. A comparison of two cloudwater/fogwater collectors: The rotating arm collector and the Caltech active strand cloudwater collector. ***Atmos. Environ.*** **24A:**1685-1692.

Cooper, R.J., J.J. Jenkins and A.S. Curtis. 1990. Pendimethalin volatility following application to turfgrass. ***J. Environ. Qual.*** **19:**508-513.

Cory, L., P. Fjeld and W. Serat. 1970. Distribution patterns of DDT residues in the Sierra Nevada Mountains. ***Pestic. Monit. J.*** **3:**204-211.

de Voogt, P. and B. Jansson. 1993. Vertical and long-range transport of persistent organics in the atmosphere. ***Revs. Environ. Contamin. Toxicol.*** **132:**127.

Faust, B.C. 1994. Photochemistry of clouds, fogs, and aerosols. ***Environ. Sci. Technol.*** **28:**217A-222A.

Frank, H., H. Scholl, D. Renschen, B. Rether, Laouedu and Y. Norocarpi. 1994. Haloacetic acids, phytotoxic secondary air pollutants. ***Environ. Sci and Pollut. Res*** **1:**4-14.

Gaggi, C., D. Calimari and E. Bacci. 1993. Bioconcentration of non-polar xenobiotics in terrestrial plant biomass. In: ***Chemical Exposure Predictions***, D. Calimari (Ed), Lewis Publishers, Ann Arbor, MI. pp 147-160.

Glotfelty, D.E., J.N. Seiber, and L.A. Liljedahl. 1987. Pesticides in fog. ***Nature*** **325:** 602-605.

Glotfelty, D.E., C.J. Schomburg, M.M. McChesney, J.C. Sagebiel and J.N. Seiber. 1990a. Studies of the distribution, drift, and volatilization of Diazinon resulting from spray application to a dormant peach orchard. ***Chemosphere*** **21:**1301-1314.

Glotfelty, D.E., A.W. Taylor, B.C. Turner and W.H. Zoller. 1984. Volatilization of surface-applied pesticides from fallow soil. ***J. Agric. Food Chem.*** **32:**638-643.

Glotfelty, D.E., G.H. Williams, H.P Freeman and M.M. Leech. 1990b. Regional atmospheric transport and deposition of pesticides in Maryland. In: ***Long Range Transport of Pesticides*** (Kurtz, D.A., Ed), Lewis Publishers, Chelsea, MI. pp 199-221.

Goolsby, D.A., E.M. Thurman, M.L. Pommes and W.A. Battaglin. 1994. Temporal and geographic distribution of herbicides in precipitation in the midwest and Northeast United States, 1990-91. In: ***Proceedings of the 4th National Pesticide Conference***, Richmond, VA, Nov 1-3, 1993. (To be published).

Hoff, J.T., D. Mackay, R. Gillham and W.Y. Shiu. 1993. Partitioning of organic chemicals at the air-water interface in environmental systems. ***Environ. Sci. Technol.*** **27:**2174-2180.

Jenkins, J.J., R.J. Cooper and A.S. Curtis. 1991. Field chamber technique for measuring pendimethalin airborne loss from turfgrass.

Bull. Environ. contamin. Toxicol. **47**:594-601.

Jury, W.A., W.F. Spencer and W.J. Farmer. 1984a. Behavior assessment model for trace organics in soil: II. Chemical classification and parameter sensitivity. *J. Environ. Qual.* **13**:567-572.

Jury, W.A., W.F. Spencer and W.J. Farmer. 1984b. Behavior assessment model for trace organics in soil: III. Application of screening model. *J. Environ. Qual.* **13**:573-579.

Jury, W.A., W.F. Spencer and W.J. Farmer. 1984c. Behavior assessment model for trace organics in soil: IV. Review of experimental evidence. *J. Environ. Qual.* **13**:580-586.

Jury, W.A., W.F. Spencer and W.J. Farmer. 1983. Behavior assessment model for trace organics in soil: I. Model description. *J. Environ. Qual.* **12**:558-564.

Kurtz, D.A. (Ed). 1990. ***Long Range Transport of Pesticides***, Lewis Publishers, Chelsea, MI.

Kutz, F.W., A.R. Yobs and H.S.C. Yang. 1976. National Surveillance Program for Pesticides in Air. In: ***Air Pollution from Pesticides and Agricultural Processes***, R.E. Lee, Jr. (Ed), CRC Press, Cleveland, OH. Chapter 4.

Lewis, R.G. and R.E. Lee, Jr. 1976. Air pollution from pesticides: Sources, occurrence and dispersion. In: ***Air Pollution from Pesticides and Agricultural Processes***. R.E. Lee, Jr. (Ed), CRC Press, Cleveland, OH. pp 5-50.

Lyman, W.J., W.F. Reehl and D.H. Rosenblatt. 1990. ***Handbook of Chemical Property Estimation Methods: Environmental Behavior of Organic Compounds***. American Chemical Society, Washington, D.C., Chapters 15 and 16.

Maguire, R.J. 1991. Kinetics of pesticide volatilization from the surface of water. ***J. Agric. Food Chem.*** **39**:1674-1678.

Majewski, M.S., D.E. Glotfelty and J.N. Seiber. 1989. A comparison of the aerodynamic and the theoretical profile-shape methods for measuring pesticide evaporation from soil. ***Atmos. Environ.*** **23**:929-938.

Majewski, M.S., M.M. McChesney and J.N. Seiber. 1991. A field comparison of two methods for measuring DCPA soil evaporation rates. ***Environ. Toxicol. Chem.*** **10**:301-311.

Majewski, M.S., M.M. McChesney, J. Prueger, J.E. Woodrow and J.N. Seiber. 1994. Methyl bromide volatilization from tarped and untarped fields. Submitted for publication.

Majewski, M., R. Desjardins, P. Rochette, E. Pattey, J. Seiber and D. Glotfelty. 1993. Field comparison of an eddy accumulation and an aerodynamic-gradient system for measuring pesticide volatiization fluxes. ***Environ. Sci. Technol.*** **27**:121-128.

Majewski, M.S., D.E. Glotfelty, K.T. Paw U and J.N. Seiber. 1990. A field comparison of several methods for measuring pesticide evaporation rates from soil. ***Environ. Sci. Technol.*** **24**:1490-1497.

MBGC. 1993. Methyl Bromide Global Coalition and National Aeronautics and Space Administration, State of the Science Workshop, October 28, Washington, D.C.

Mellouki, A., R.K. Talukdar, A.M. Schmoltner, T. Gierczak, M.J. Mills, S. Soloman and A.R. Ravishankara. 1992. Atmospheric lifetimes and ozone depletion potentials of methyl bromide (CH_3Br) and dibromomethane (CH_2Br_2). ***Geophys. Res. Letts*** **19**:2059-2062.

Nash, R.G., M.L. Beall, Jr. and W.G. Harris. 1977. Toxaphene and 1,1,1-trichloro-2,2-bis (p-chloro-phenyl) ethane (DDT) losses from cotton in an agroecosystem chamber. ***J. Agric. Food Chem.*** **25**:336-341.

Nash, R.G. and M.L. Beall, Jr. 1977. ***Terrestrial Microcosms and Environmental Chemistry***. National Science Foundation, NSF.RA 79-0026, pp.86-94.

National Research Council. 1994. ***Science and Judgment in Risk Assessment***. Committee on risk assessment of hazardous air pollutants. National Academy Press, Washington, D.C.

Parmele, L.H., E.R. Lemon and A.W. Taylor. 1972. Micrometeorological measurement of pesticide vapor flux from bare soil and corn under field conditions. ***Water Air Soil Pollut.*** **1**:433-451.

Pitts, Jr., J.N. 1993. Atmospheric formation and fates of toxic ambient air pollutants. ***Occup. Med. State Art Rev.*** **8**:621-622.

Rolston, D.E. 1986. Gas flux. In: ***Methods of Soil Analysis, Part I. Physical and Mineralogical Methods,*** American Society of Agronomy -- Soil Science Society of America, Madison, WI, Monograph #9.

Ross, L.J. and R.J. Sava. 1986. Fate of thiobencarb and molinate in rice fields. ***J. Environ. Qual.*** **15**:220-225.

Sagebiel, J.C., J.N. Seiber and J.E. Woodrow. 1992. Comparison of headspace and gas-stripping methods for determining the Henry's Law Constant (H) for organic compounds of low to intermediate H. ***Chemosphere*** **25**:1763-1768.

Sanders, P.F., M.M. McChesney and J.N. Seiber. 1985. Measuring pesticide volatilization from small surface areas in the field. ***Bull. Environ. Contamin. Toxicol.*** **35**:569-575.

Sanders, P.F. and J.N. Seiber. 1984. Organophosphorus pesticide volatilization: Model soil pits and evaporation ponds. In: ***Treatment and Disposal of Pesticide Wastes,*** R.F. Krueger and J.N. Seiber (Eds), ACS Symposium Series #259.

Sanders, P.F. and J.N. Seiber. 1983. A chamber for measuring volatilization of pesticides from model soil and water disposal systems. ***Chemosphere*** **12**:999-1012.

Schomburg, Charlotte J., Dwight E. Glotfelty and James N. Seiber. 1991. Pesticide occurrence and distribution in fog collected near Monterey, California. ***Environ. Sci. Technol.*** **25**:155-160.

Seiber, J.N., M.M. McChesney and J.E. Woodrow. 1989. Airborne residues resulting from use of methyl parathion, molinate and thiobencarb on rice in the Sacramento Valley, California. ***Environ. Toxicol. Chem.*** **8**:577-588.

Seiber, J.N., S.C. Madden, M.M. McChesney and W.L. Winterlin. 1979. Toxaphene dissipation from treated cotton field environments: Component residual behavior on leaves and in air, soil and sediments determined by capillary gas chromatography. ***J. Agric. Food Chem.*** **27**:284-291.

Seiber, J.N. and J.E. Woodrow. 1983. Methods for studying pesticide atmospheric dispersal and fate at treated areas. ***Residue Reviews*** **85**:217-229.

Seiber, J.N., B.W. Wilson and M.M. McChesney. 1993. Air and fog deposition residues of four organophosphate insecticides used on dormant orchards in the San Joaquin Valley, California. ***Environ. Sci. Technol.*27**:2236-2243.

Seiber, J.N., M.M. McChesney, P.F. Sanders and J.E. Woodrow. 1986. Models for assessing the volatilization of herbicides applied to flooded rice fields. ***Chemosphere*** **15**:127-138.

Spencer, W.F., W.T. Farmer and M.M. Cliath. 1973. Pesticide volatilization. ***Res. Revs.*49**:1.

Spencer, W.F. and M.M. Cliath. 1990. Movement of pesticides from soil to the atmosphere. In: ***Long Range Transport of Pesticides,*** D.A. Kurtz (Ed) Lewis Publishers, Inc., Chelsea, MI, pp.1-16.

Spencer, W.F., M.M. Cliath, W.A. Jury and L.Z. Zhang. 1988. Volatilization of organic chemicals from soil as related to their Henry's law constants. ***J. Environ. Qual.*** **17**:504-509.

State of California. 1991. ***Natural Resource Damage Assessment Plan. Sacramento River: Cantara Spill.*** The Resources Agency. California Department of Fish and Game. Sacramento, CA.

Taylor, A.W. and D.E. Glotfelty. 1988. Evaporation from soils and crops. In: ***Environmental Chemistry of Herbicides, Volume 1.*** R. Grover (Ed) CRC Press, Inc., pp.89-129.

Taylor, A.W., D.E. Glotfelty, B.C. Turner, R.E. Silver, H.P. Freeman and A. Weiss.

1977. Volatilization of dieldrin and heptachlor residues from field vegetation. ***J. Agric. Food Chem.* 25**:542-548.

Taylor, A.W. and W.F. Spencer. 1990. Volatilization and vapor transport processes. In: ***Pesticides in the Soil Environment: Processes, Impacts, and Modeling.*** SSSA Book Series #2, Soil Science Society of America, Madison, WI, pp.213-269.

Taylor, G.E., Jr. and J.V.H. Constable. 1994. Modelling pollutant deposition to vegetation: scaling down from the canopy to the biochemical level. In: R. Percy, N. Cape and R. Garrells (Eds). ***Air Pollution and Plant Cuticles***, Springer-Verlag, New York, New York (in press).

Thornthwaite, C.W. and B. Holzman. 1939. The determination of evaporation from land and water surfaces. ***Mon. Weather Rev.* 67**:4-11.

Tinsley, I.J. 1979. ***Chemical Concepts in Pollutant Behavior.*** John Wiley and Sons, New York, Chapter 4.

Turner, B., S. Powell, N. Miller and J. Melvin. 1989. ***A field study of fog and dry deposition as sources of inadvertant pesticide residues on row crops.*** Environmental Hazards Assessment Program. State of California, Department of Food and Agriculture. Sacramento, CA. EH-89-11, 42p + appendices.

Wania, F. and D. MacKay. 1993. Global fractionation and cold condensation of low volatility organochlorine compounds in polar regions. ***Ambio* 22**:10-18.

Whang, J.M., C.J. Schomburg, D.E. Glotfelty and A.W. Taylor. 1993. Volatilization of fonofos, chlorpyrifos, and atrazine from conventional and no-till surface soils in the field. ***J. Environ. Qual.* 22**:173-180.

Willis, G.H., L.L. McDowell, L.A. Harper, L.M. Southwick and S. Smith. 1983. Seasonal disappearance and volatilization of toxaphene and DDT from a cotton field. ***J. Environ. Qual.* 12**:80-85.

Wilson, B.W., M.J. Hooper, E. E. Littrell, P.J. Detrich, M.E. Hansen, C.P. Weisskopf, and J.N. Seiber. 1991. Orchard dormant sprays and exposure of red-tailed hawks to organophosphates. ***Bull. Environm. Contamin. Toxicol.* 47**:717-724.

Wilson, J.D., G.W. Thurtell, G.E. Kidd and E.G. Beauchamp. 1982. Estimation of the rate of gaseous mass transfer from a surface source plot to the atmosphere. ***Atmos. Environ.* 16**:1861-1867.

Winer, A.M. and R. Atkinson. 1990. Atmospheric reaction pathways and lifetimes for organophosphorus compounds. In: ***Long Range Transport of Pesticides.*** D.A. Kurtz (Ed) Lewis Publishers Inc., Chelsea, MI. pp.114-126.

Wolfe, M.F. and J.N. Seiber. 1993. Environmental Activation of Pesticides. In: ***Occupational Medicine: State of the Art Reviews.*** pp.561-573.

Woodrow, J.E. and J.N. Seiber. 1991. Two chamber methods for the determination of pesticide flux from contaminated soil and water. ***Chemosphere* 23**:291-304.

Woodrow, J.E., M.M. McChesney and J.N. Seiber. 1990. Modeling the volatilization of pesticides and their distribution in the atmosphere. In: ***Long Range Transport of Pesticides.*** D.A. Kurtz (Ed) Lewis Publishers, Inc., Chelsea, MI, pp.61-81.

Woodrow, J.E., D.G. Crosby and J.N. Seiber. 1983. Vapor-phase photochemistry of pesticides. ***Res. Revs.* 85**:111-125.

Woodrow, J.E., J.N. Seiber and Y.H. Kim. 1986. Measured and calculated evaporation losses of two petroleum hydrocarbon herbicide mixtures under laboratory and field conditions. ***Environ. Sci. Technol.* 20**:783-789.

Yagi, K., J. Williams, N.Y. Wang and R.J. Cicerone. 1993. Agricultural soil fumigation as a source of atmospheric methyl bromide. ***Proc. Natl. Acad. Sci. USA* 90**:8420-8423.

Zabik, J.M. and J.N. Seiber. 1993. Atmospheric transport of organophosphate pesticides from California's Central Valley to the Sierra Nevada mountains. ***J. Environ. Qual.* 22**:80-90.

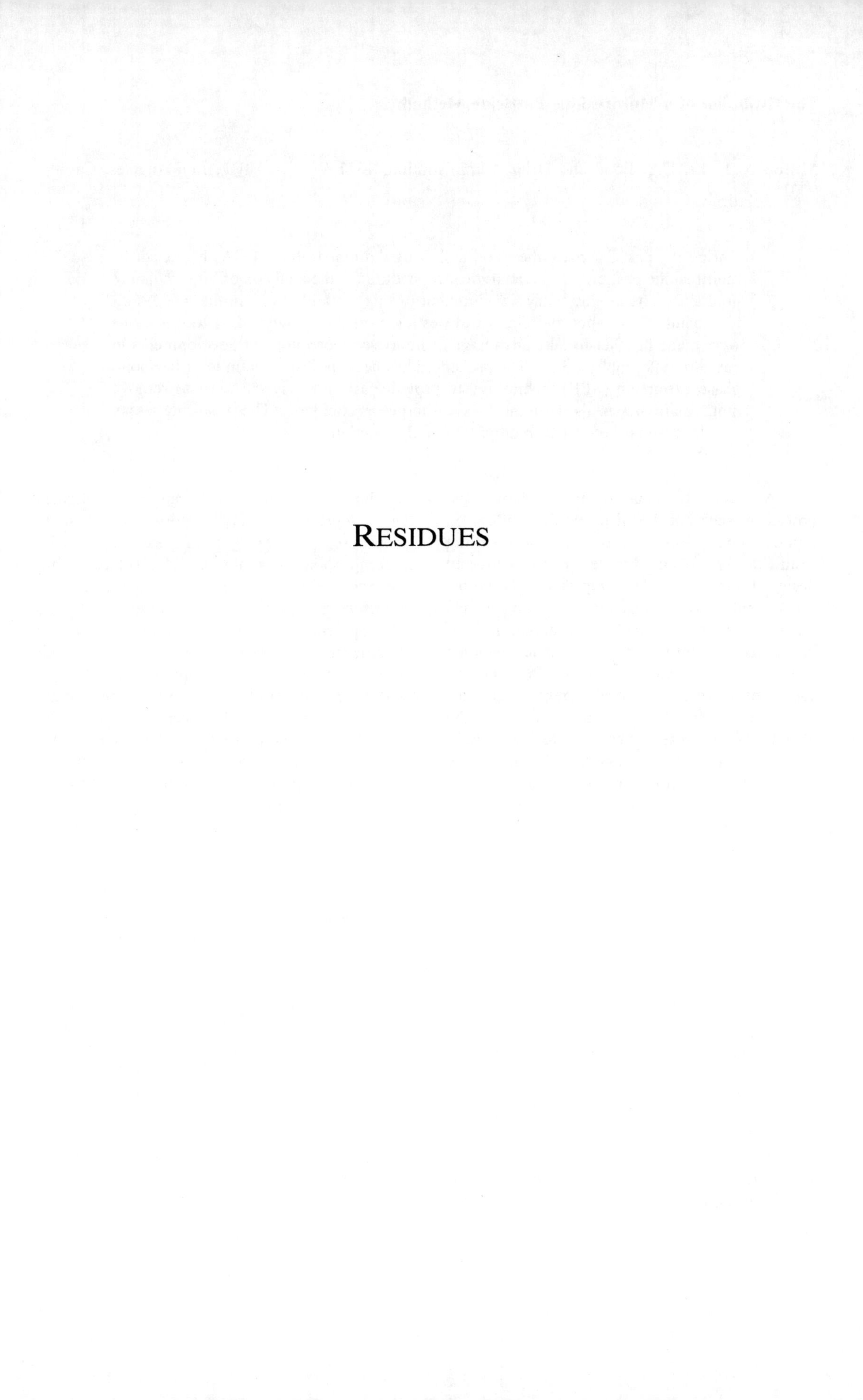

RESIDUES

The Evolution of a Multiresidue Pesticide Method

Milton A. Luke, U.S. Food and Drug Administration, 1521 W Pico Blvd., Los Angeles, Ca 90015

During the past 30 years the Food and Drug Administration (FDA) has used the multiresidue pesticide Luke analytical procedure for the analysis of high moisture products. The evolutionary development of this method began in the late 1960's to expand the number and classes of pesticide residues analyzed. Modifications were made in 1981 to take advantage of improved technological developments in gas chromatography (GC). The procedure has been modified again to utilize solid phase extraction (SPE) technology to provide fast and easy cleanups to remove more plant materials without loss of target pesticides. The resulting assay solutions allow better chromatography and detection.

FDA uses multiresidue methods as a practical process to seek out illegal pesticide residues on rapidly moving food shipments. The resulting residue data is also used to determine trends and design future pesticide compliance programs. As a matter of practicality, a multiresidue method is often used even when a sample is to be analyzed only for a specific residue. Multiresidue pesticide methods often have a variety of modifications for different products or pesticides. Modifications are a part of the evolutionary process that can eventually result in a new analytical procedure. Methodology changes and modifications are often made to utilize new technology to fill one or more of the following needs:

1. a more specific residue identification
2. a lower detection level
3. a lower cost of analysis
4. a shorter time of analysis
5. extend the scope of the analysis

The development of the multiresidue method by Mills et al. (1963), commonly referred to as the MOG procedure (Fig. 1a), was a major change from analytical procedures that utilized extractions in which a sample was tumbled with solvent in a container and quantification was accomplished by paper and thin layer chromatography. The method used Florisil, a synthetic adsorbent, to remove sample extractives that interfered in the GC analysis which used an electron capture detector. The method established a solid analytical approach for the analysis of a broad range of pesticides and replaced many individual residue procedures. The analysis of a sample could be accomplished for a number of residues within the relatively short time that is required in monitoring perishable produce. The development of the potassium chloride thermionic detector (KClTD) for phosphates and nitrogen extended the range and number of pesticides that could be determined by a single multiresidue method. However, the scope of the procedure did not include the more polar organonitrogen and organophosphorus pesticides that were being used. An attempt to analyze the MOG's acetonitrile extract solution for these polar compounds resulted in a procedure in which the assay solution was difficult to concentrate and in which the residual acetonitrile affected the response of the KClTD detector.

The substitution of acetone for acetonitrile in the sample extraction resulted in a method (Fig. 1b) with virtually the same assay solution. It eliminated a problem that occurred when high sugar samples were extracted with acetonitrile in which two liquid phases were obtained versus the normal one. However this variation did not result in a better analysis for the polar residues.

The next modification of the acetone method added a methylene chloride partition after two petroleum ether partitions (Fig. 2). The methylene chloride solution was concentrated and injected into a GC with a KClTD detector without cleanup. This modification provided a

1054–7487/95/0174$12.00/0

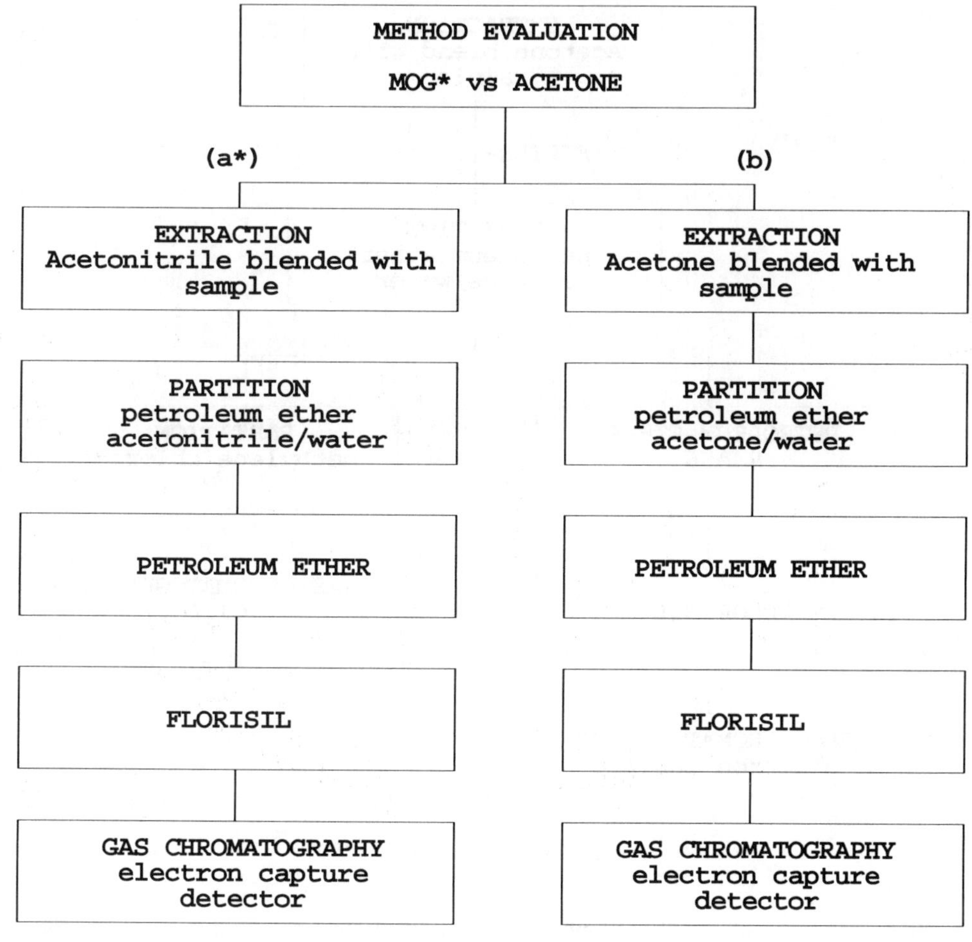

Fig. 1. Evolution Stage I: 1963 to 1969.

rapid method to determine more polar pesticides. and was used to detect and quantify mevinphos residues in domestic lettuce as well as other pesticide residues as they were encountered.

In 1972 a new pesticide, methamidophos (Monitor[R]), was introduced which had considerable usage in Southern California. The compound was too polar to be determined by the method illustrated in Fig. 2. A major revision of the method was made in which water was not added in the partitioning step so as to obtain a better recovery of the water soluble compound. The petroleum ether partitioning was also eliminated. The resulting method was used to take legal actions on lettuce samples containing high levels of methamidophos. The thought of analyzing the assay solution for other pesticide residues occurred since it was logical to assume that it could contain all residues that are less polar than methamidophos. The analytical scheme that was conceived is illustrated in Fig. 3. The acetone assay solution from the partition step could be injected directly on the GC system with the element selective KClTD detector to determine the phosphate and nitrogen residues. The extract could then be cleaned up on a Florisil column for the determination of chlorinated pesticides by electron capture detection. This modification to the acetone method with minimal cleanup was used to deal

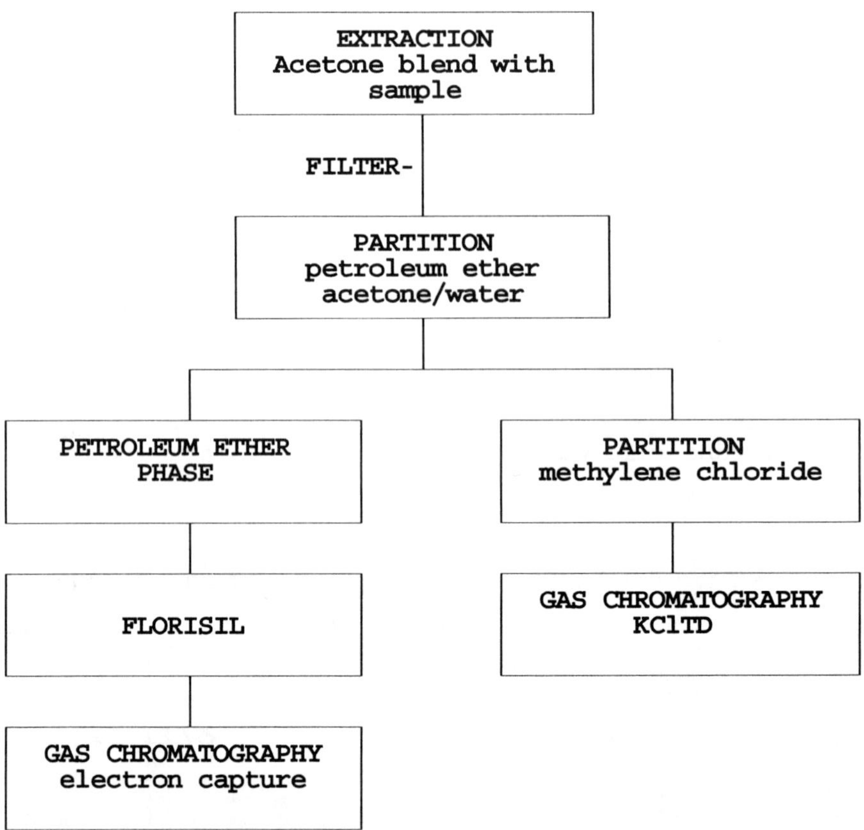

Fig. 2. Evolution Stage II: 1970.

with high residues of methamidophos in domestic lettuce and unregistered uses on Mexican peppers. These uses established the acetone extraction method as an extremely rapid and broad ranging multiresidue method. This was an evolutionary step using the current technology to the fullest. Application of the method to certain fruits yielded a white flocculent precipitate. Although this precipitate did not appear to affect the analysis, the acetone in the partitioning step was replaced with petroleum ether to reduce the amount of this precipitate in the final assay solution. This method was published in 1975 (Luke et al.) and is commonly referred to as the Luke Method (Fig. 4).

The next major technological advances were the development of the flame photometric detector (FPD) which could be used in a phosphorus or sulfur detection mode and the Hall electrolytic conductivity detector (ELCD) which could be used in a halogen or nitrogen detection mode. The Luke Method was modified (Fig. 5) to utilize this new technology (Luke et al.,1981). The use of the FPD provided a less complex chromatogram than that produced by the KClTD detector and did not require a cleanup of the assay solution. The use of the ELCD in the halogen mode eliminated the need for a Florisil cleanup to determine chlorinated pesticides. The modification to the method, besides the elimination of the Florisil cleanup, was a solvent reconcentration of the solution obtained from the partition step to reduce the methylene chloride level in the assay solution. Without this modification, the ELCD detector would be saturated by the methylene chloride from injections of the assay solution.

The Luke Method has a long history of regulatory use as well as publication in FDA's

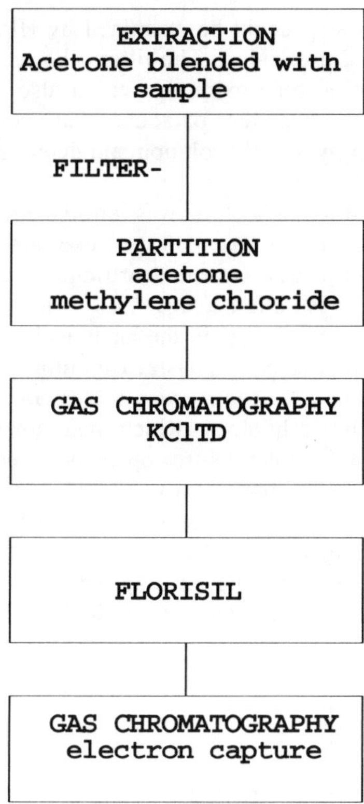

Fig. 3. Evolution Stage III: 1972.

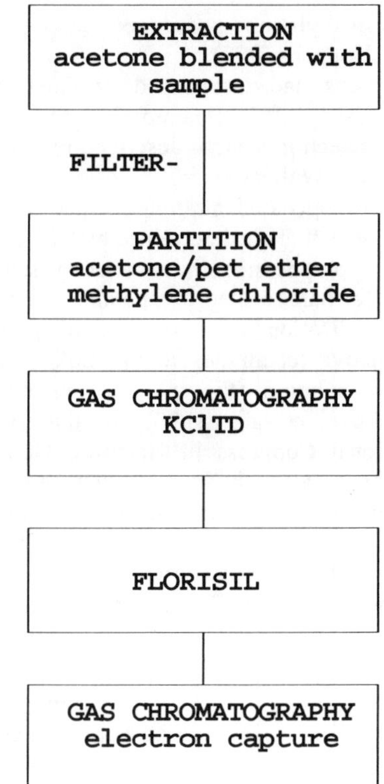

Fig. 4. Evolution Stage IV: 1975 (the Luke et al. method).

Pesticide Analytical Manual. The method was adopted as an official final action procedure in 1986 by the Association of Official Analytical Chemists. It has evolved by the application of other analytical techniques to the assay solution. The author's laboratory continues the analysis of the Luke assay solution by any of the following approaches:

1. Methyl carbamates are analyzed by a C-18 SPE cartridge cleanup followed by a high performance liquid chromatography (HPLC) determination with the detection system described by Moye (1977).
2. Phenylureas are determined by HPLC after a C-18 SPE cleanup using the detection system described by Luchtefeld (1987).
3. Chlorophenoxy acids are determined by GC with a ELCD detector after

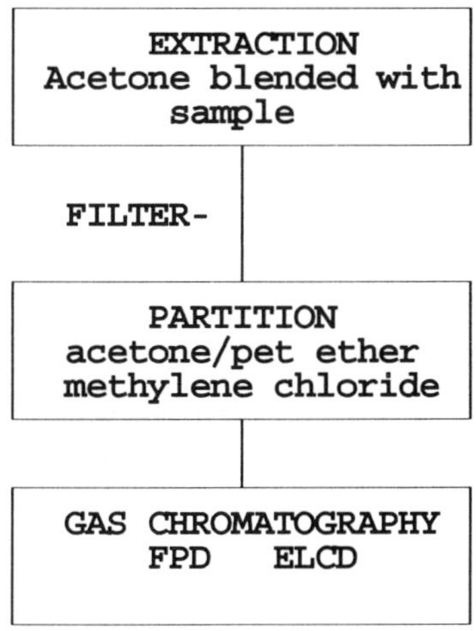

Fig. 5. Evolution Stage V: 1975 (the Luke et al. method).

methylation procedure described by Hopper (1992).

4. Benzimadazoles are determined after a C-18 SPE cleanup by the HPLC detection system described by Gilvydis and Walters (1990).
5. Halogenated pesticides that are heat sensitive are determined by HPLC using a photoconductivity detector sensitive for halogenated compounds.

The application of a number of pesticide determinative techniques to the Luke Method analytical solution (Fig. 6) was presented in 1990 (Luke et al., 1990) at the IUPAC International Congress in Hamburg Germany. The figure also includes other analytical determinations that could be made using the Luke procedure. Pesticides such as avermectin by TFA derivitization followed by HPLC determination. The water phase from the partition step could be examined by HPLC for additional pesticides. The filter cake remaining from the acetone extraction could also contain pesticides such as paraquat that could be extracted by an acid solution and determined by HPLC.

In 1993 a new method was introduced (Cairns et al.) which was based on the Luke Method but uses solid phase extraction cartridges to cleanup the assay solution (Fig. 7). This was a revolutionary change in the analytical approach which previously emphasized minimal cleanup procedures. The change was necessitated by the change in technology which had gone from packed GC columns to open bore columns which are more easily affected by the nonvolatile sample extractives that are deposited in their injection ports. In addition, a new detection system, the ion trap detector requires a low sample background to attain the sensitivity

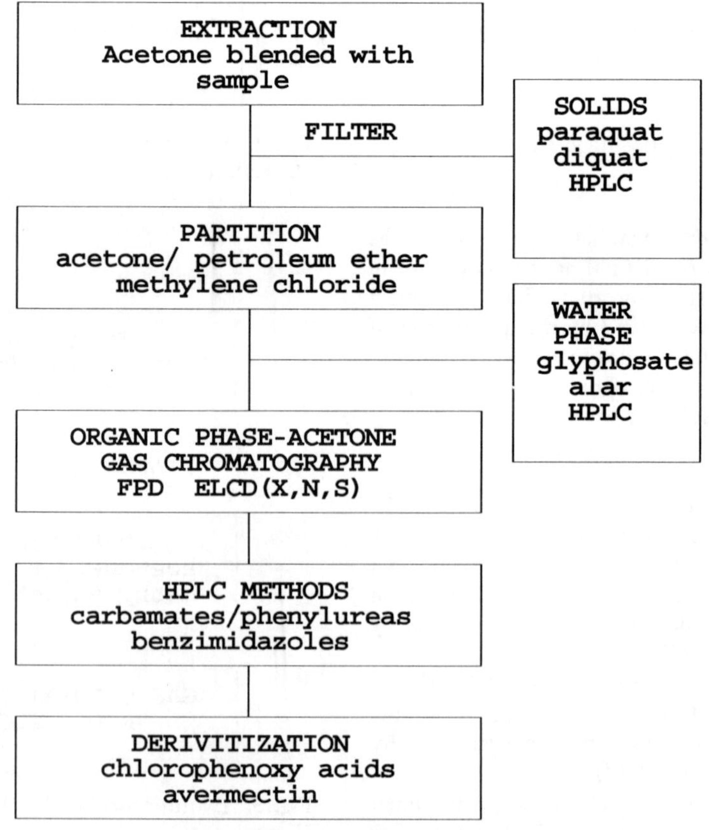

Fig. 6. Evolution Stage VI: 1990 (IUPAC presentation, Hamburg, Germany).

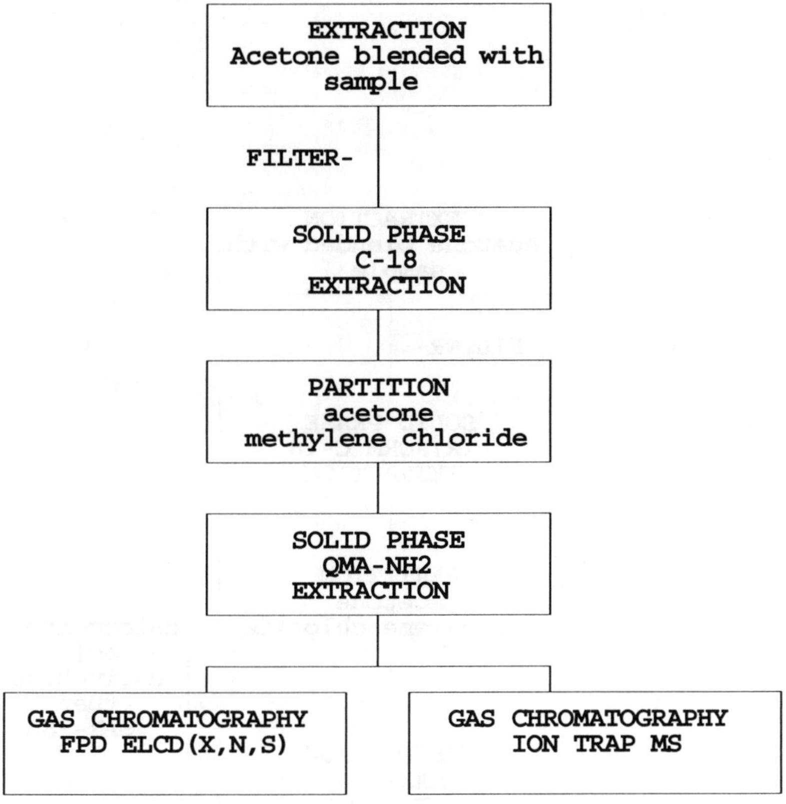

Fig. 7. Evolution Stage VII: 1993.

for residue monitoring. The new method uses SPE cartridges at two points in the assay. Each of the two cleanup steps allow the pesticides to elute through the SPE while holding back some of the plant extractives. The first uses a C-18 SPE to remove nonpolar sample compounds by a reverse phase cleanup. The second cleanup is done after the partitioning step. It uses a strong and a weak anionic extracting SPE cartridge to remove acidic phenols and saccharides. The partitioning step has also been changed in order to obtain a better recovery of the more polar residues.

The GC analysis is the same as before except the cleaner extract of the new method allows better chromatography. Injection port inserts as well as columns require less maintenance. The examination of the analytical extract using other multiresidue methods no longer requires additional cleanup. Methyl carbamates, phenylureas and benzimidazoles can be determined without the SPE cleanups that they previously required. All the pesticides described in the IUPAC 1990 article can be determined with the new method. The chlorophenoxy acids will need to be methylated prior to the final cleanup on the anionic extracting SPE cartridges. This new method was introduced at the state of Florida's 30th Annual Pesticide Residue Workshop.

There has now been one year of experience with the method on regulatory surveillance samples. Some crops have presented problems when the SPE cleanups did not remove enough of the sample extractives. The solid phase extraction cartridges were changed to provide a cleaner extract. The C-18 was replaced with a trifunctional C-18 which has a greater capacity for the same size and weight of cartridge material. The original weak and strong anionic extracting SPE cartridges (Waters' QMA and NH2 SPE cartridges) were replaced with SPE

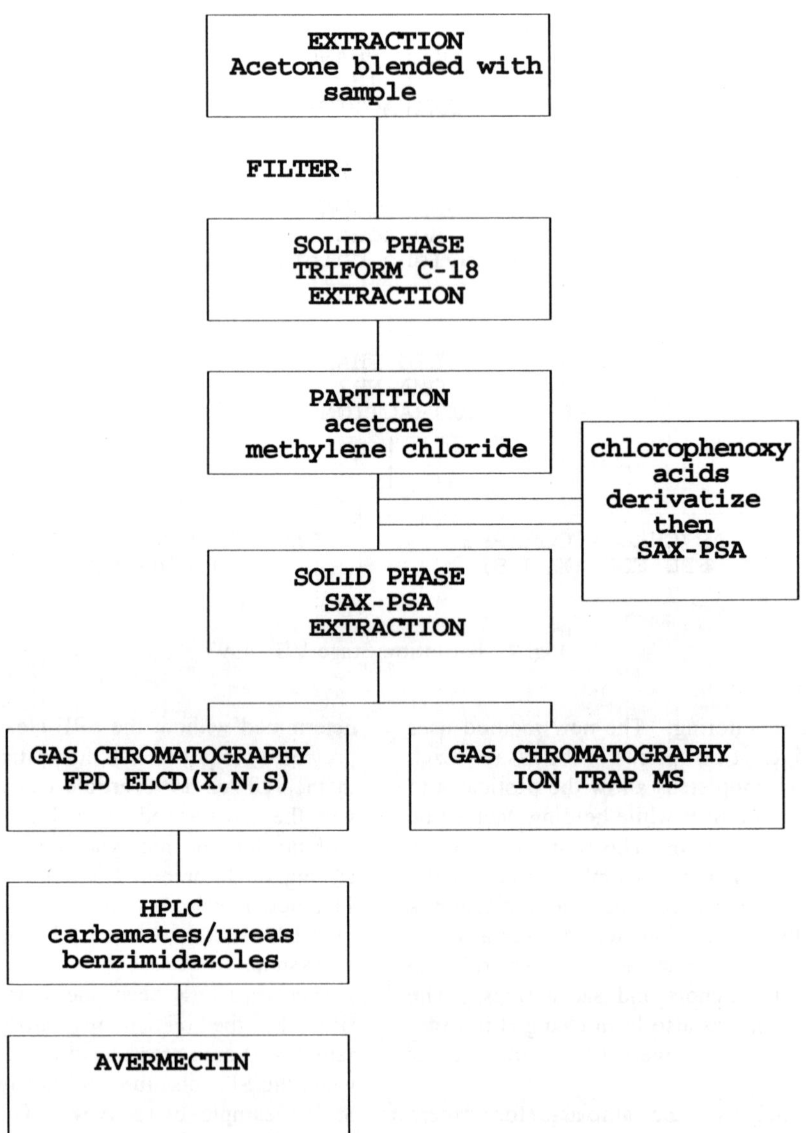

Fig. 8. Evolution Stage VIII: 1994.

Table 1. Intralaboratory recovery data for tomato spikes.

Pesticide	Percent Recovery		
	(0.1 ppm)	(0.3 ppm)	(0.5 ppm)
Acephate	81.98	109.89	89.48
DDE, p,p'-	89.52	98.87	83.92
Dimethoate	87.88	104.30	105.68
Chlorpyrifos	100.96	106.19	92.79
Endosulfan I	85.90	101.08	81.97
Endosulfan sulfate	104.50	106.43	90.81
Ethion	83.72	98.32	105.26
Guthion	99.05	105.12	107.75
Malathion	84.61	103.50	100.20
Permethrin, cis-	76.95	110.26	92.95

Table 2. Intralaboratory recovery data for lettuce spikes.

Pesticide	Percent Recovery		
	(0.1 ppm)	(0.3 ppm)	(0.5 ppm)
Acephate	101.98	99.65	115.46
DDE, p,p'-	81.38	108.30	94.71
Dimethoate	100.00	112.50	110.91
Chlorpyrifos	83.37	112.37	108.82
Endosulfan I	86.30	111.87	98.53
Endosulfan sulfate	87.75	119.29	106.57
Ethion	89.18	113.13	92.11
Guthion	106.67	124.23	102.98
Malathion	87.25	112.94	100.41
Permethrin, cis-	96.72	104.03	105.56

Table 3. Intralaboratory recovery data for blackberry spikes

Pesticide	Percent Recovery		
	(0.1 ppm)	(0.3 ppm)	(0.5 ppm)
Acephate	101.98	91.17	83.09
DDE, p,p'-	98.62	96.98	81.94
Dimethoate	109.17	103.52	95.91
Chloryprifos	106.73	101.72	83.97
Endosulfan I	93.25	96.40	92.87
Endosulfan sulfate	98.20	106.43	108.63
Ethion	90.90	94.28	99.80
Guthion	106.67	106.83	101.39
Malathion	98.04	98.60	102.24
Permethrin, cis-	91.09	112.45	90.17

cartridges produced by Varian (SAX and PSA). The resulting changes (Fig. 8) require slightly more solvent to elute pesticides from the SPE's.

The method has recovered the same pesticides recovered by the Luke Method. Intralaboratory studies have been made on three crops and nine pesticides fortified at 0.1 ppm, 0.3 ppm and 0.5 ppm to establish the reliability of the new method. The recoveries, in Tables 1-3, are within the acceptable limits of residue analyses. All of the determinations were made by GC with either the FPD detector or the ELCD detector in the halogen mode.

The Los Angeles Pesticide Residue Analytical Team performed all the development and recoveries. The Team currently consists of Ruth Dixon, Dorothy Wade, Herb Masumoto, David Kodama, Bill Langham, Kathleen Cortese, Irene Cassias, Sally Yee, Alondra Little, Walden Lee and Jerry Froberg.

Conclusion

The 1994 version of the new acetone method is part of a revolutionary process providing a cleaner analytical determinative extract to identify and quantify pesticide residues. The new acetone method extract can be examined by a variety of detectors for GC and HPLC without any further cleanup.

References

Cairns, T.; Luke, M.A.; Chiu, K.S.; Navarro, D.; and Siegmund, E.G. *Rap. Comm. Mass Spec.* **1993**, 7, 1070-1076.

Gilvydis, D.M. and Walters, S.M. *J. Assoc. Off. Anal. Chem.* **1990**, 73, 753-761.

Hopper, M.L. *J. Agric. Food Chem.* **1987**, 35, 265-269.

Luchtefeld, R.G. *J. Assoc. Off. Anal. Chem.* **1987**, 70, 740-745.

Luke, M.A.; Langham, W.S.; Kodama, D.; Masumoto, H.T.; Froberg, J.E.; and Doose, G.M. In *Pesticide Chemistry*; Frehse, H., Ed.; Advances inin International Research, Development and Legislation; VCH: New York, 1991; XIV, 373-382.

Luke, M.A.; Froberg, J. E.; and Masumoto, H.T. *J. Assoc. Off. Anal. Chem.* **1975**, 58,1020-1026.

Luke, M.A.; Froberg, J.E.; Doose, G.M. and Masumoto, H.T. *J. Assoc. Off. Anal. Chem.* **1981**, 64, 1187-1195.

Mills, P. A.; Onley, J.H.; and Gaither, R.A. *J. Assoc. Off. Anal. Chem.* **1963**, 46, 186-191.

Moye, H.A.; Scherer, S.J.; and St. John, P.A. *Anal Lett* ***1977***., 10, 1049-1073.

Isomer- and Enantioselective Residue Analysis of Chlorinated Pesticides using Chiral High Resolution Gas Chromatography and Various Mass Spectrometric Detection Techniques

Markus D. Müller and Hans-Rudolf Buser
Swiss Federal Research Station, CH-8820 Wädenswil, Switzerland

Whereas most chlorinated insectides have been banned in industrialized countries, production and use likely continues in other parts of the world. Residues of these compounds are detected at considerable levels in items such as fish and mussels intended for human consumption. In this paper we describe the achiral and chiral residue analysis of three groups of chlorinated insecticides (HCHs, chlordanes, toxaphenes) with a series of chiral components together with selected metabolites in fish samples and other biota collected at various places in the northern and southern hemisphere. The data presented illustrate the importance of enantioselective residue analysis as an essential tool for the assessment of origin of residues and possible toxicogical effects.

Chlorinated insecticides are still in use in many parts of the world for agricultural and vector control purposes (Bidleman *et al.* 1988). Due to the high amounts produced, the physicochemical properties such as high partition coefficients, volatility, slow degradation in soil and accumulation in the trophic chain, they are ubiquitous environmental contaminants. It is therefore not surprising to find some components and metabolites of chlorinated insecticides at considerable levels in biota of various trophic levels and in human tissue (Kucklick *et al.* 1994, Winter, 1992, Kutz *et al.* 1991).

In this paper, we present data on isomer and enantiomer composition of residues of hexachlorocyclohexanes (HCHs), chlordanes and toxaphenes, results which have been published elsewhere (see below). The technical materials of chlordane (a tricyclic chlorinated hydrocarbon with 7 to 10 chlorine atoms attached) and of toxaphene (mainly bornanes with 6 to 10 chlorine atoms) are complex mixtures of congeners and isomers. Toxaphene constitutes probably the most diverse pesticide mixture with hundreds of components. Most toxaphene components, several chlordane components and one HCH-isomer (α-HCH) are chiral and exist as two enantiomers. These enantiomers may show different biological properties (such as accumulation, toxicity) and behaviour in the environment.

In previous studies, we investigated the enantiomer composition of the various chiral isomers and congeners in environmental samples and biota and compared it to that in the technical mixtures. In many cases, marked differences in enantiomer composition in the residues indicate enantioselective processes in the environment.

Experimental.

High-resolution gas chromatography mass spectrometry. An achiral capillary column (a 25-m SE 54) and three types of chiral of chiral columns based on cyclodextrin derivatives were used: a 2,3,6-tri-O-methyl-β-cyclodextrin (PMCD) column, a 2,3,6-tri-O-ethyl-α-cyclodextrin (PECD) column and a 2,3,6-dimethylbutylsilyl-β-cyclodextrin (BSCD) column. In the cours of this work, a series of similar columns was used. Table 1 lists typical chiral columns with selected properties. The preparation of these columns has been described (Müller *et al.*, 1992, Buser and Müller, 1992a, Blum and Aichholz, 1990, Buser and Müller, 1993). Briefly, capillary columns (glass or fused silica) are statically coated with a mixture of an achiral polysiloxane (PS 086 or OV 1701 for glass and fused silica, respectively) and the desired cyclodextrin derivative. The resulting columns show good separation efficiency and inertness and tolerate temperatures up to 250 °C under temperature programmed conditions. A VG Tribrid double-focusing magnetic sector hybrid mass spectrometer (VG Analytical Ltd., Manchester, England) was used as a detector system. The ion

1054–7487/95/0183$12.00/0

Table 1. Chiral HRGC columns used.

Polysiloxane phase	Chiral selector (content)	Column material and dimensions	GC oven temp. program	separation of enantiomers
PS 086	PMCD 20 %	glass, 15 m 0.3 mm i.d.	60-140-250 °C (3°C/min)	hexachlorocyclohexanes
PS 086	PMCD 10 %	glass, 20 m 0.3 mm i.d.	100-120-250 °C (2°C/min)	octachlordanes
OV 1701	BSCD 30%	fused silica 16 m 0.25 mm i.d.	100-120-250 °C (2°C/min)	heptachlorepoxide, oxychlordane, toxaphenes
OV 1701	PECD 30 %	fused silica 16 m 0.25 mm i.d.	100-120-250 °C (3°C/min)	heptachlor, heptachlorepoxides, photoheptachlor

source was operated either in the electron ionization (EI, 70eV, 180° C) or negative ion chemical ionization mode (ECNI, 50 eV, 140 °C) mode using neat argon as buffer gas at a pressure of 0.5 - 1 x 10^{-4} mbar as measured by the ion gauge. Full scan spectra were recorded in the mass range of m/z 50-500 at a resolution of m/Δm of 500, and selected ion monitoring (SIM) of up to 12 ions simultanously (0.5 sec/scan) for improved sensitivity. The toxaphenes were analyzed using full-scan EI, ECNI and the MS/MS capabilities of the instrument. In „selected-reaction-monitoring“ (SRM), characteristic ion transitions were monitored by selecting definite mass differences between the magnet and the quadrupole analyzers. A collision energy of 18eV and argon as collision gas were used (Buser and Müller, 1994a).

All samples were comparatively analyzed using an achiral and selected chiral columns. None of the columns used showed sufficent chiral separation for all isomers and pairs of enantiomers under investigation.

Results and Discussion

Residues of HCHs.

The technical mixture of HCH is produced by chlorination of benzene under UV-light, leading to a mixture of stereomers, predominantly the α, β, γ and δ-isomer. Only the α isomer is chiral and exists as two enantiomers. The γ-isomer, lindane, is the only one to possess insecticidal activity. Both lindane and the technical mixture are used.

Residues in the environment and biota consist mainly of the α, β and γ-isomers. Lindane shows least persistence, whereas the β-isomer is the most stable and accumulating isomer (Kutz *et al.*1991) The α-isomer shows intermediate persistence. We investigated the isomer- and enantiomer-composition in technical HCH and environmental samples (air, rain, seal, whale) using a PMCD column with electron capture or mass spectrometric detection (Müller *et al.* 1992). The extraction and clean up of the samples has been described elsewhere. Figure 1 shows the analysis of four sample extracts. The chiral column separates the two α-HCH-enantiomers. The elution order on this column was later assigned as (+)/(-)-α-HCH (first- and second-eluting) using chiral HPLC with chiroptical detection (Müller and Buser, 1994)

Whereas the soil and rain samples show only small deviations from a 1:1 enantiomer ratio (ER), the seal extract shows ER values of 2.3 as determined by ECD and up to 2.0 as determined by HRGC-HRMS. Replicate determinations showed that enantiomeric ratios can be determined with good accuracy and precison, generally with less than 1 % standard deviation. As enantiomers have almost identical physicochemical properties except rotation of the plane of polarized light, they have identical response factors for an instrumental detector system, thus facilitating determination of enantiomer ratios. Table 2 gives concentrations and ER values of α-HCH for the samples analyzed. ER values >1.0 indicate an enantiomer excess of the firsteluted (+)-enantiomer in the environmental samples analyzed. This is not necessarily always the case, as in other work, ER values of 0.85-

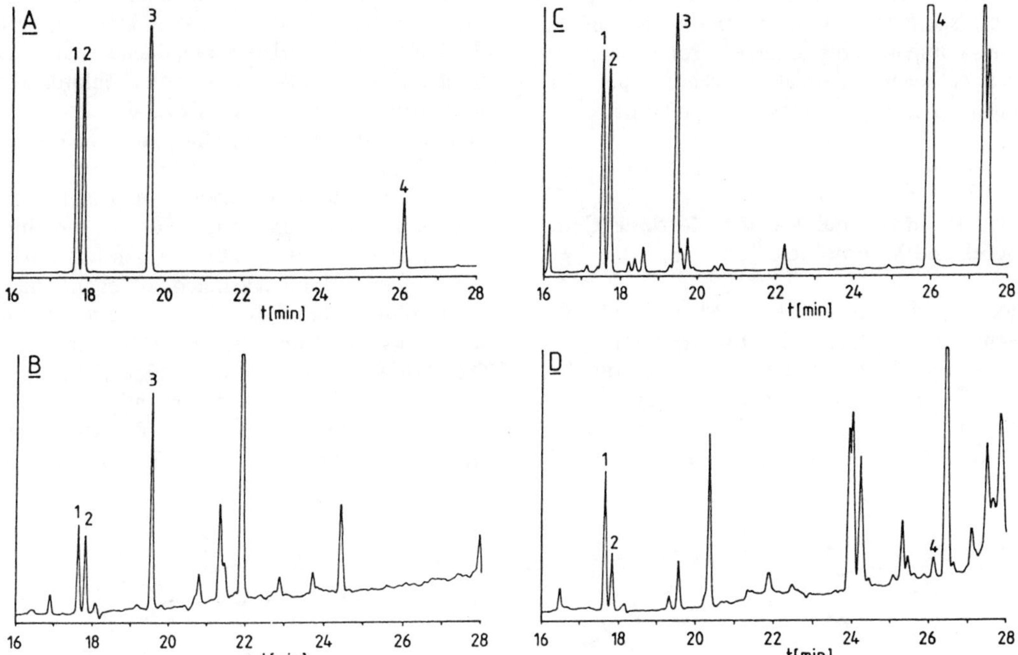

Fig. 1. HRGC-ECD chromatograms (15 m column with 20 % PMCD in PS086) of A: HCH standard (200 - 300 pg) ; B: rain; C: soil sample close to a former HCH factory; D: seal blubber. Peaks: 1 (+)α-HCH, 2 (-)α-HCH, 3 γ-HCH, 4 β-HCH (Reproduced from Müller *et al.* Copyright 1992 American Chemical Society).

Table 2. Concentrations of α- and γ-HCH and Enantiomer Ratios of α HCH in Samples Analyzed

Sample	ER (ECD)	α-HCH[a]	γ-HCH[b]
technical HCH	1.002		
rain	1.075	1.47	2.10
soil	1.099	214	232
ambient air	1.02	0.75	0.34
seal blubber	2.32	99	37
whale blubber	2.86	128	1130

[a], [b]: Concentrations are in ng/L for rain, ng/g for soil, seal and whale blubber and ng/m^3 for air.

0.87 and 1.3 - ∞ were determined for sea water and duck liver extracts, respectively (Pfaffenberger *et al.* 1992). The α-HCH- enantiomers seem to undergo different fates in the various compartments of the environment and in warm blooded animals, where the (-)-enantiomer seems to be less accumulating.

Chiral Residue Analysis of Chlordane Components and Metabolites.

Chlordane was a widely used insectide for agricultural and household use, and for wood preservation. Technical chlordane is a complex mixture of various isomers and congeners derived from hexachlorocyclopentadiene with up to 120 components identified up to now (Dearth and Hites, 1991). It is produced by the Diels-Alder cycloaddition reaction of hexachlorocyclopentadiene with cyclopentadiene and subsequent chlorination. Figure 2 shows the structures of nine main chlordane components and metabolites identified either in technical chlordane or in environmental samples. Several major chlordane components (heptachlor, cis- and trans-chlordane) and minor components (e.g. MC-5 and MC-6) and the metabolites heptachlor *exo*- and *endo*-epoxide are chiral. None of the chiral column showed sufficient resolution for isomers and congeners and separated all enantiomers. Each column (cf. Table 1) had ist merits and drawbacks (see later) (Buser and Müller, 1993). The PMCD column shows sufficient chiral selectivity to separate the major and minor octachlordanes (Buser *et al.*1992) The two ECNI SIM chromatograms in Figure 3 show the eluti-

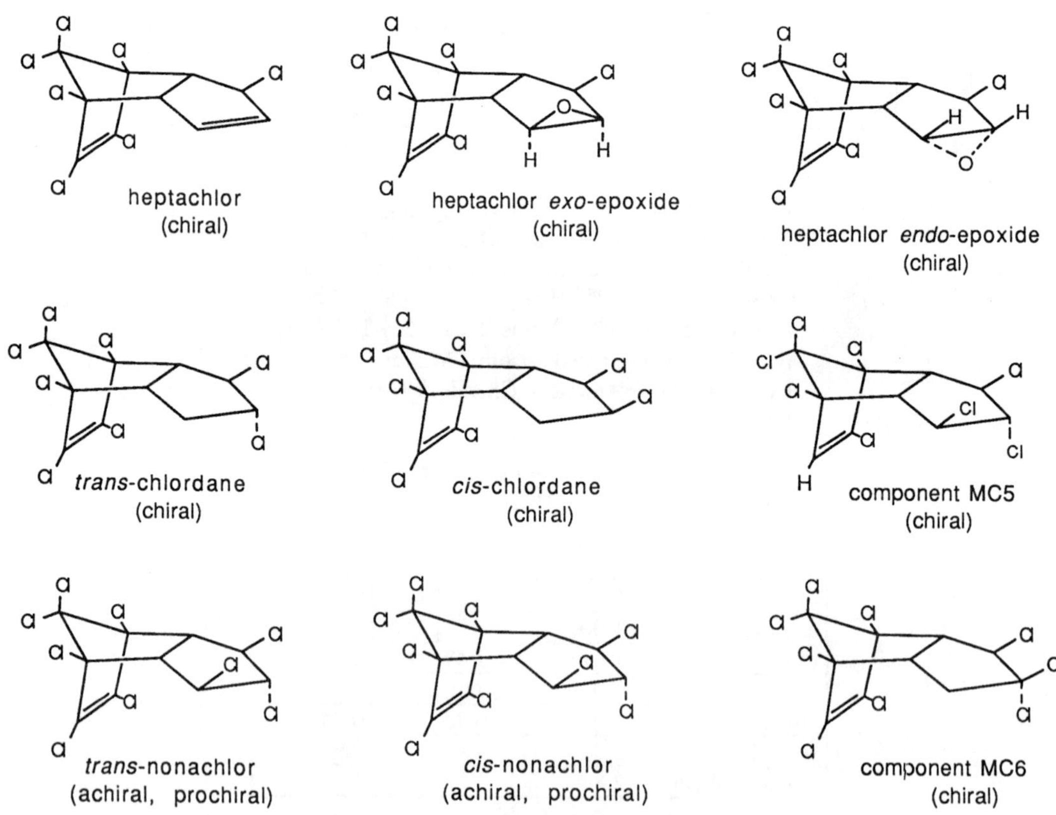

Fig. 2: Structures of some main chlordane components and metabolites. Only one enantiomer of chiral compounds is shown. (Reproduced from Buser and Müller. Copyright 1993 American Chemical Society).

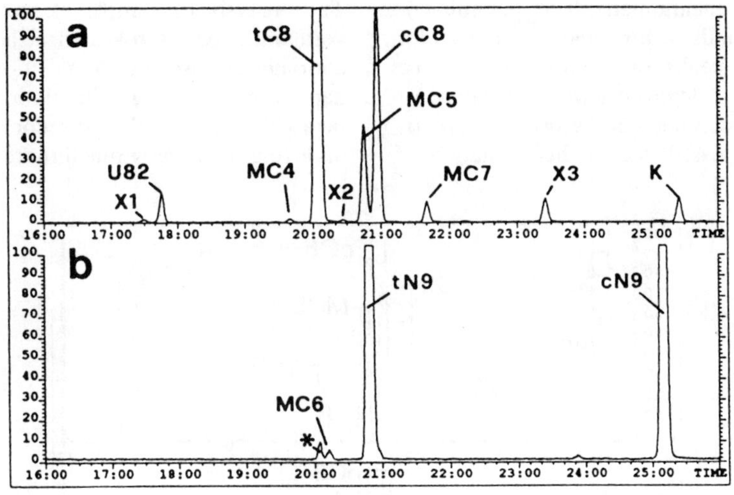

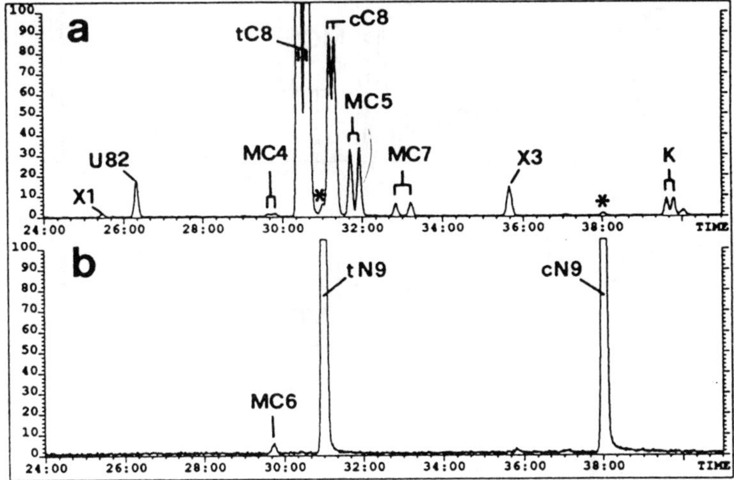

Fig. 3. ECNI–SIM chromatograms showing elution of (a) octachlordanes (m/z 410) and (b) nonachlordanes (m/z 444) in the technical chlordane mixture with components indicated (cf. Fig. 2). Top, achiral column; bottom, chiral column (10% PMCD in PS086). Signals for *trans*- and *cis*-nonachlor at m/z 410 are marked with an asterisk. For details see text. (Reproduced from Buser and Müller. Copyright 1992 American Chemical Society.)

on of octa- and nonachlordanes in the technical mixture on an achiral (left side) and chiral column (right side). Components MC-4, *trans*- and *cis*-chlordane, MC-5 and MC-7 are clearly separated into the enantiomers. The achiral nonachlor compounds *trans*- and *cis* nonachlor elute as single peaks on both columns. The asterisks in Fig. 3 indicate a weak response of the nonachlor-compounds *trans*- and *cis*-nonachlor at m/z 410 due to the formation of $(M-Cl)^-$ or $(M-Cl+H)^-$ -ions, despite the use of the buffer gas. Heptachlor, the main heptachloro compound, is not resolved on this column, but on a PECD column (Buser and Müller, 1993). The analysis of sample extracts from aquatic vertebrate species revealed clearly changed enantiomer and isomer composition for octachlordane. Figure 4 shows the elution of a series of octa chlordanes with some compounds compounds (e.g. U82) now much more prominent, and

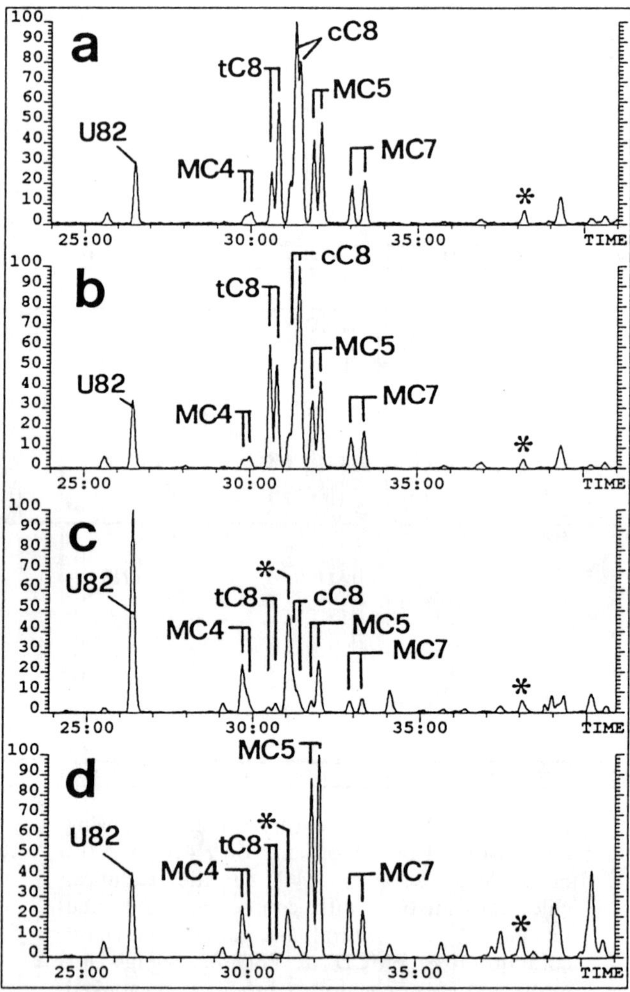

Fig. 4. ECNI-SIM chromatograms showing elution of octachlordanes (m/z 410) with components indicated. (a) Baltic herring, (b) Baltic salmon, (c) Baltic grey seal, (d) Antarctic penguin. Abbreviations and asterisks: see Fig. 3 and text. (Reproduced from Buser and Müller. Copyright 1992 American Chemical Society).

the ratio of trans- and cis- chlordanes are drastically changed. Similarily enantiomeric ratios such as that of cis- and trans-chlordane and of MC-4, MC-5 and MC-7 are also significantly changed.

Compound U82, an octachlordane with a 5+3-chlorine-substitution pattern and unknown structure and the metabolites heptachlor *endo* and *exo*-epoxide and oxychlordane are not resolved on the PMCD column, but on a BSCD column. Figure 5 illustrates the presence of non-racemic heptachlor *exo*-epoxide in four sample extracts. Detection was by ECNI MS. It is interesting to note that the first eluting enantiomer, now identified as the (+)-enantiomer (Müller and Buser, 1994), is clearly reduced in herring,

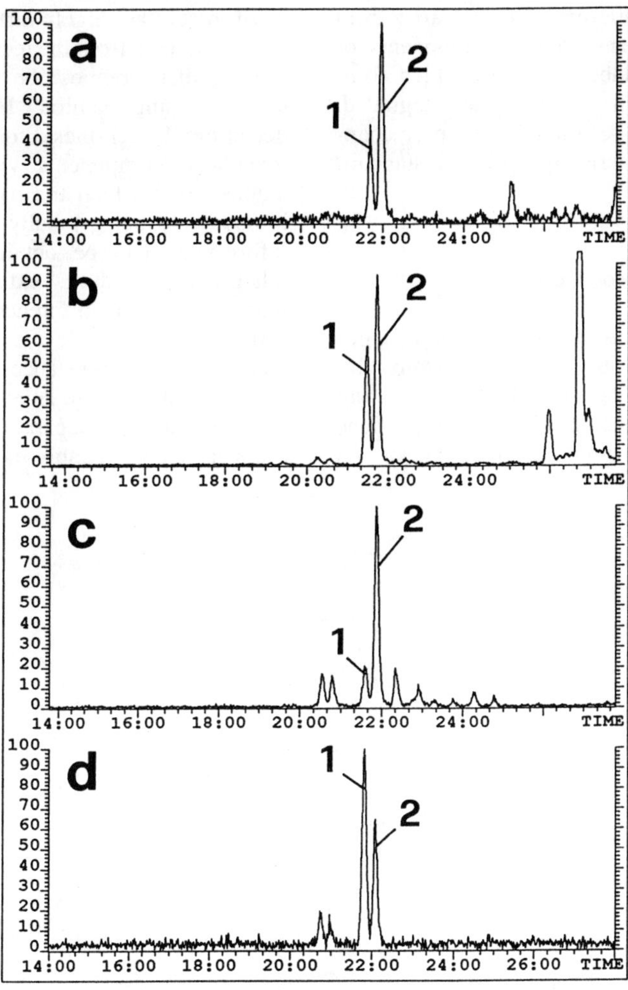

Fig. 5: EI-SIM chromatograms (m/z 353) showing the elution of the two enantiomers of heptachlor *exo*-epoxide on the PS086/BSCD column. (a) Baltic herring, (b) Baltic salmon, (c) Baltic grey seal, (d) human adipose tissue. Peak identification: 1 = (-)-heptachlor *exo*-epoxide, 2 = (+)-heptachlor *exo*-epoxide. (Reproduced from Buser and Müller. Copyright 1992 American Chemical Society).

salmon and seal whereas in human adipose tissue the second-eluting is reduced. The *endo*-isomer was not present in these samples, but would be clearly separated from the *exo*-isomer by this column.
The interpretation of the ER differences detected in the sample extracts is not easy without additional data. The ER values observed result from enantioselective processes in uptake, translocation and metabolism of these compounds. Therefore, changed enantiomer ratios can arise from uptake of enantiomer enriched components or from a different metabolic rate, by which compounds or metabolites are further degraded. These questions can be studied using pure enantiomers or enantiomerically enriched standard substances.

Analysis of Toxaphene Residues.

Toxaphene is produced by chlorination of camphene, leading to a very complex mixture with several hundred components. The isomeric and enantiomeric composition of several toxaphene samples has recently been described (Buser and Müller, 1994a) and the reaction mechanisms involved have been discussed. Briefly, the chlorination of camphene produces, by rearrangement reactions, polychlorobornanes, polychlorobornenes and polychloroisocamphanes (Fig. 6, Buser and Müller 1994b). Depending on structure, degree of chlorination and position of chlorine atoms, these isomers and congeners undergo metabolization in the environment to a highly variable extent. Therefore, the residues found in environment are highly different in composition of components from the technical material.
The isomer composition of toxaphenes was studied using achiral HRGC with ECNI-detection. Under these conditions, the various toxaphene components yield intense $(M\text{-}Cl)^-$ anions, from which the molecular composition is deducible but generally no further structural information can be obtained. EI-MS, on the other hand, produces extensive fragmentation with significant differences between isomers and congeners.
Under such circumstances, the application of chiral HRGC leads to an even more complicated situation, as the separation into enantiomers produces a considerably more complex chroma-

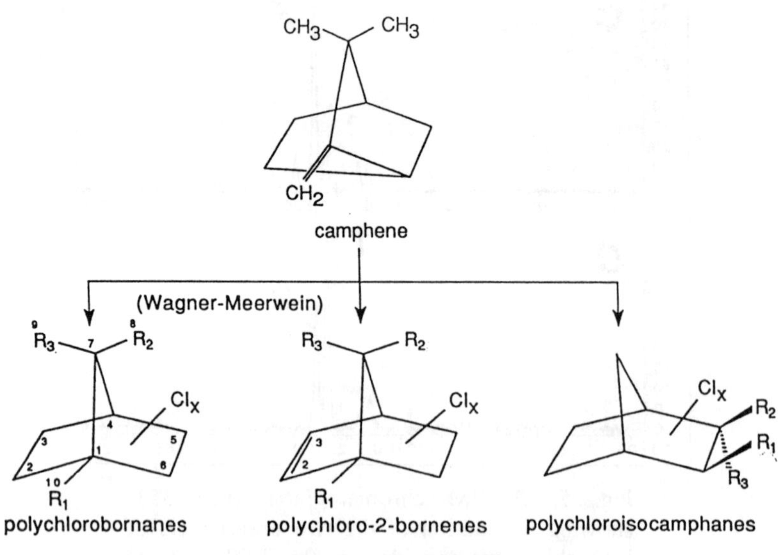

Fig. 6. Formation of polychlorobornanes, polychlorobornenes and polychloroisocamphanes from chlorination of camphene. (Reproduced from Buser and Müller. Copyright 1994 American Chemical Society).

togram. ECNI-MS provides high sensitivity but no isomer selectivity, which is necessary to distinguish between coeluting isomers and enantiomers. Isomer selectivity is achieved using EI-MS/MS with SRM. As the detection requires a specific parent-daughter-ion relationship, specific for a certain partial structure, an enhanced isomer selectivity is obtained in this detection mode. Particularily, eliminations of $C_2H_{4-x}Cl_x$ (x=1-3) from $(M-Cl)^+$ or $(M-HCl)^+$ were monitored. In this way, several important chiral toxaphene components such as TOX8 and TOX9 were resolved using a BSCD chiral column (Buser and Müller, 1994a). Fig. 7 shows the se-

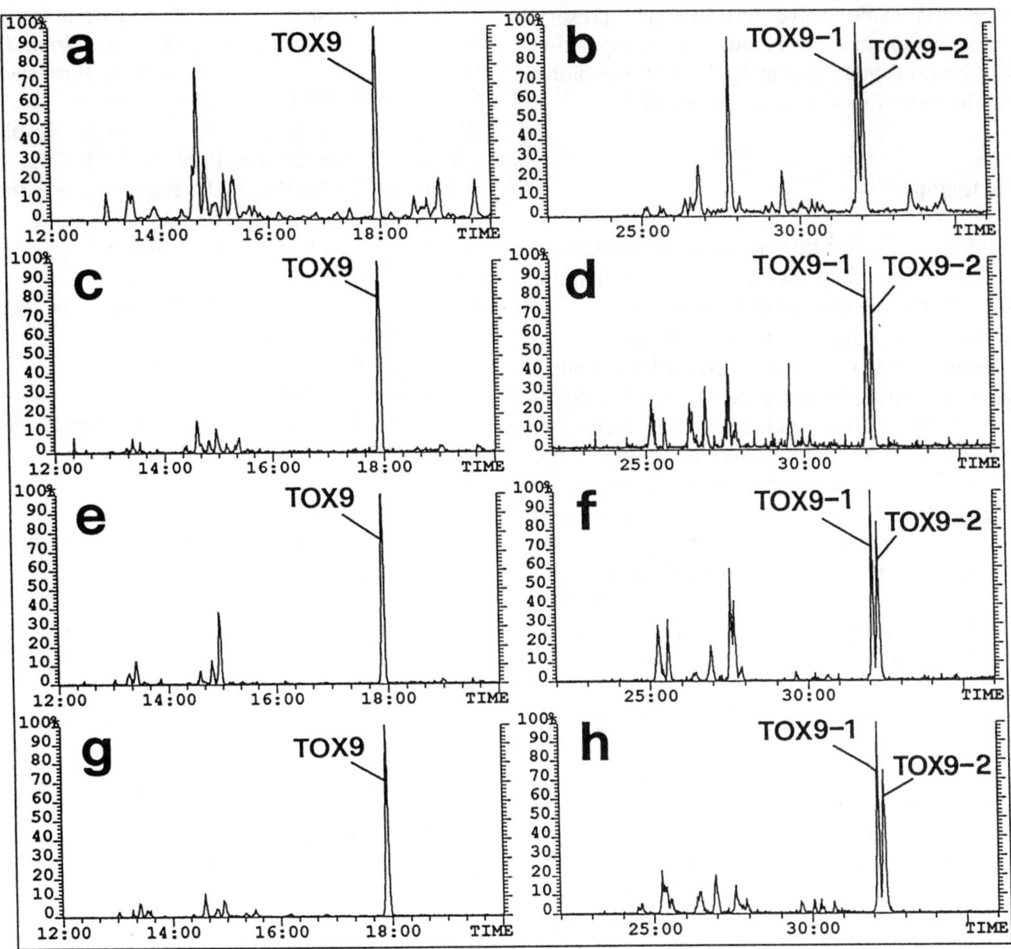

Fig. 7: SRM chromatograms $410^+ \rightarrow 314^+$ of nonachlorobornanes in technical toxaphene and in samples of aquatic species. Left panel: achiral HRGC (SE 54), right panel: chiral HRGC on OV1701-BSCD. (a) and (b) Technical toxaphene, (c) and (d) baltic herring, (e) and (f) Arctic seal, (g) and (h) Antarctic penguin. Note the separation of TOX-9 into enantiomers on the chiral column. (Reproduced from Buser and Müller. Copyright 1994 American Chemical Society).

paration of selected toxaphene components in technical toxaphene and in residues. The technical products from different producers showed remarkably similar isomeric and enantiomeric composition with some small but significant deviations from 1:1 enantiomeric ratios. Toxaphene residues analyzed in aquatic vertebrate samples showed extensive alterations from the original toxaphene mixtures with only a few but largely the same polychlorobornanes present. Only small changes in enantiomer composition could be observed, indicating little if any biological degradation of these compounds.

Conclusions

Enantioselective residue analysis of residues of chiral chlorinated pesticides has been demonstrated with three classes of insecticides.

The capillary columns used with α- and β-cyclodextrin derivatives as chiral selectors offer high separation efficiency and thermal stability required for residue analysis. When stereoisomers in complex mixures have to analyzed, not only sufficient enantiomer resolution has to be achieved, but also an adequate isomer resolution is important, especially where the detection system used cannot provide sufficient isomer selectivity. Mass spectrometry in the EI, ECNI- and EI-SRM mode were used for high sensitivity and selectivity. Enantiomer ratios could be determined with good precision and accuracy (a few percent standard deviation at low residue levels (e.g. of heptachlor in air at a level of a few pg/m^3). Enantioselective analysis of organochlorine residues were based on a thorough comparison of the technical materials of HCHs, chlordane and toxaphene and residues found in the environmental samples. The residues detected differ largely from the technical mixtures in isomer, congener and also partly in enantiomer composition. This findings clearly illustrate the fact that optical isomers of pesticides may undergo different fates in the environment.

References

Bidleman, T.F.; Zaranski, M.T.; Walla, M.D. Toxaphene, Usage, Aerial Transport aind Deposition. In *Toxic Contamination in Great Lakes,* Schmidtke, N:W. Ed. Lewis, Publishers: Chelsea, MI, **1988**.

Blum, W.; Aichholz, R. J. High Resolut. Chrom. Chrom. Comm. **1990**, *13*, 718-720.

Buser, H.-R.; Müller, M.D. Rappe, C. *Environ. Sci. Technol.* **1992**, *26*, 1533-1540.

Buser, H.-R.; Müller, M.D. *Anal. Chem.* **1992,** *64*, 3168-3175,.

Buser, H.-R.; Müller, M.D. *Environ. Sci. Technol.* **1993**, *27*, 1211-1220.

Buser, H.-R.; Müller, M.D. *Environ. Sci. Technol.* **1994a**, *28*, 119-128

Buser, H.-R.; Müller, M.D. J. Agric. Food Chem. **1994b**, *42,* 393-400.

Dearth, M.A.; Hites, R.A. *Environ. Sci. Technol.* **1991**, *25*, 245-254.

Kucklick, J.R.; Bidleman, T.F.; McConnell, L.L.; Walla, M.D.; Ivanov, G.P. *Environ. Sci. Technol.* **1994**, *28*, 31-37.

Kutz, F.W., Wood, P.H.; Bottimore, D.P. *Reviews of Environm. Contam. Toxicology*, **1991**, *120*, 1-82.

Müller M.D.; Schlabach, M.; Oehme M. *Environ. Sci. Technol.* **1992**, *26*, 566-569.

Müller, M.D.; Buser, H.-R. *Anal. Chem.* submitted for publication.**1994.**

Pfaffenberger, B.; Hühnerfuss, H.; Kallenborn, R.; Köhler-Gunther A.; König, W.A.; Krüner, G. *Chemosphere* **1992**, *25*, 719-725.

Winter, K.W., *Reviews of Environm. Contam. Toxicology*, **1992**, *127*, 23-67.

Opportunities for Pesticide Residue Analytical Method Development: The Potential for Aqueous Extractions of Pesticide Residues from Fruits and Vegetables

H. Anson Moye, Pesticide Research Laboratory, Food Science and Human Nutrition Department, Institute of Food and Agricultural Sciences, University of Florida, Gainesville, FL 32611-0720

Potential and demonstrated alternatives to the use of organic solvents for the extraction of pesticide residues from vegetables are discussed, including the use of solid phase extraction devices coupled to aqueous based extractions. An alternative to exhaustive extractions, equilibrium extractions, particularly solid phase microextractions (SPME) are considered as another way to reduce extraction time and eliminate organic solvents.

Pesticide analytical methods evolved from colorimetric to gas chromatographic determinative steps in the late 1960s (Wilderman, M., Shuman, H. 1968; Gutenman, W.H., Lisk, D.J., 1963; Klein, A.K., et al., 1959; Mills, P.A., et al., 1963; Hardin, L.J., Sarten, C.T., 1962; Klein, A.K., 1960). Since one of the first classes of pesticides to be dealt with was hydrophobic, the chlorinated hydrocarbons, solvents chosen for their extraction from a variety of commodities were typically lipophilic, such as petroleum ether, mixed hexanes, and benzene (Klein, A.K., et al., 1959; Hardin, L.J., Sarten, C.T., 1962; Klein, A.K., 1960). As more polar pesticides, such as the organophosphates, phenoxyacetic acids, and triazines came into use, somewhat more polar solvents, such as chloroform, acetone, acetonitrile, and methanol found favor (Luke, M.A., et al., 1975; Luke, M.A., et al., 1981; Lee, M.S., et al., 1990). Troublesome water, present in the commodity, typically was removed either by drying the extract with an anhydrous salt, such as sodium sulfate, or by solvent exchange into a water immiscible solvent. Such dry extracts could then be evaporated to small volumes by Danish-Kuderna devices over steam, or by rotory evaporation, giving enhanced limits of detection for analysis by gas chromatography. It was important to prevent water from reaching the gas chromatographic analytical column, since the liquid phase was an oil or a wax that was deposited upon the solid support by evaporating the solvent away from an appropriate solution of the liquid phase. The presence of water caused stripping of the liquid phase, with associated peak broadening and shortening, and detector fouling. One of the first detectors used for analysis of the chlorinated hydrocarbons, the electron capture detector, was also sensitive to the presence of water, generally producing a large and tailing solvent peak.

As pesticide molecular structures changed with the development of a wide array of classes of pesticides, generally accepted multiresidue methods for non-fatty foods, particularly fresh fruits and vegetables, utilized either acetone or acetonitrile as the extracting solvent, such as in the Luke method and the California Department of Food and Agriculture method, respectively (Luke, M.A., et al., 1981; Lee, M.S., et al., 1990). Either solvent is more adept at extracting polar metabolites, such as phenols, oxons, sulfoxides, and sulfones than those used previously. The gas chromatographically troublesome N-methylcarbamates are also easily extracted by these two solvents, and then separated by reversed phase high performance liquid chromatography, with detection by post-column fluorogenic labeling (Krause, R.T., 1980).

Organic solvents other than acetonitrile and acetone are also typically used in the Luke II and the California Department of Food and Agriculture methods, including n-hexane, methylene chloride, and petroleum ether. These solvents, and others used in methods developed by the United States Environmental Protection Agency (EPA), such as benzene, carbon tetrachloride, chloroform, toluene, methyl ethyl ketone, methyl isobutyl ketone, tetrachloroethylene, 1,1,1-trichloroethane, trichloroethylene, and xylene have been targeted to be reduced by 50% by the year 1995 in contract

1054–7487/95/0193$12.00/0

and government analytical chemistry laboratories. The EPA, and the Food Safety and Inspection Service of the United States Department of Agriculture, have recognized the potential hazard that these solvents present to users and the environment (Ellis, R., 1992). A summary of solvent used for the analysis of pesticide residues in food monitoring programs in the United States in 1987 is shown in Fig. 1 (Moye, H.A., et al., 1988).

There is a rapidly growing interest in the use of supercritical fluid extractions (SFE) for the extraction of a variety of analytes from various matrices, using gases in the supercritical region, such as CO_2, which is frequently modified with small percentages of methanol to increase its polarity. While there have been recent successes in the extraction of moderately polar pesticides from various substrates, including soils and sediments using modified CO_2, little has been done for the very polar ones (Robertson, A.M., Lester, J.N., 1994; Hopper, M.L., King, J.W., 1991).

This paper examines the potential for the use of innocuous water for the extraction of pesticides from high water content samples, such as fruits and vegetables, and will explore some emerging and conceptual options to the use of large amounts of organic solvents in the pesticide residue analytical chemistry laboratory.

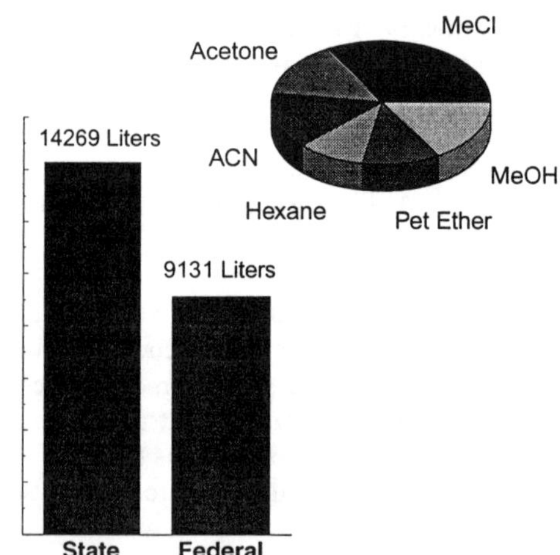

Fig. 1. Organic solvents used in the pesticide monitoring programs of state and federal laboratories in the United States in 1987. Only fresh foods are accounted for (Reproduced from Moye, H.A., et al., 1988).

Properties of Water

Water may be the most studied of all liquids, and is the subject of a five volume series edited by Franks (Franks, F., 1973). It was the subject of much controversy regarding "polywater" in the late 1960s (Franks, F., 1982), which was supposedly formed in freshly drawn quartz capillaries. This water was subsequently generally accepted as being no more than "polycontaminated" water. Many models of water, in order to explain its anomalous characteristics (latent heats of vaporization, etc.), consider it as existing in a "bulky" form, usually as free flowing liquid of individual molecules, and a "dense" form, being one or more polymers, in the form of pentamers, which cannot be separated from each other physically, because of a rapidly occurring equilibrium (Franks, F., 1973). A recent hypothesis, which has been successful in explaining the change in heat capacity of liquid water with temperature, treats water as being an equilibrium mixture of tertrameric and octameric water, but does not rule out the possibility of a pentamer-decamer structure (Benson, S.W., Siebert, E.D., 1992; Fig. 2). While providing a good model for the thermodynamic treatment of water, no convincing spectral evidence has been collected to support this thesis. However, there have been many strong arguments put forth which use some aspects of the "polymeric" or "ordered" nature of liquid water to explain many sorts of well documented solubilization phenomena. One example is "perturbing" the polymeric nature of water in order to achieve solubility characteristics for large macromolecules (Franks, F., 1973).

For its molecular weight, water has an anomously high boiling point, density, and latent heat of vaporization. Of all liquids at room temperature, it also has a very high dielectric constant (ϵ = 80.0), a recognized measure of the "polarity" of the molecule. Since the hydrogen-oxygen bond angle is 104.5°, the molecule also has a large dipole moment, 1.83 Debyes. Comprised only of hydrogen and oxygen atoms, water can form hydrogen bonds with acids, as a receptor, and bases, as a donor of protons. These and other physical and chemical characteristics are summarized in Table 1.

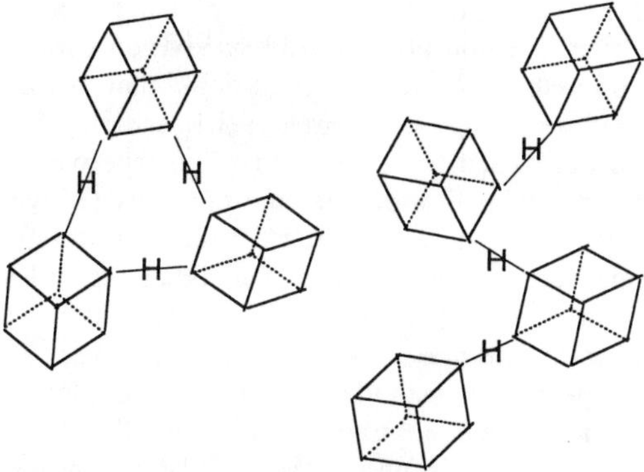

Fig. 2. A diagrammatic representation of the "polymeric" nature of water, showing the two types of octamers that can occur (Reproduced from Benson, S.W., Siebert, E.D., 1992, American Chemical Society).

Table 1. Properties of Water (25°C)

• Boiling Point	100°C
• Density	1.0
• Refractive Index	1.33
• Dielectric Constant (ε)	80.0
• Bond Angle	104.523°
• Bond Energy	459kJ / mol
• Dipole Moment	1.83 D
• Radius	1.4 Å

Modes of Extraction

When the pesticide residue analyst speaks of "extraction", he inevitably includes the concept of "exhaustive" extraction in his definition. Historically, great effort has been made to ensure that nearly complete extraction of any pesticide analyte, approaching 100%, is realized for foods and environmental samples, and that the extractions can be accomplished in a minimal amount of time. Such "exhaustive" extractions are typically accomplished by using organic solvents in which the analyte possesses a large distribution coefficient, K_d, a solvent to matrix ratio such that the rate of pesticide extraction is enhanced, and special solvent recycle techniques, such as the Soxhlet extractor, which repeatedly exposes the matrix to fresh solvent. Multiple extractions are frequently employed to further enhance the effects realized by a single extraction. In this mode of extraction, typically large volumes of solvent are used to further enhance the preferred solubility for the solvent by the pesticide.

Advantages of the "exhaustive" mode of extraction are the following: (1) nearly all the mass of the analyte is available for the determinative step, making the lowest limits of detection possible, and (2) a broad spectrum of compound classes can be simultaneously extracted. Disadvantages include: (1) large volumes of solvent are required, (2) interferences are frequently coextracted, (3) concentration steps are frequently required for sample introduction into the analytical instrument, typically a chromatograph, and (4) extractions are lengthy.

Another, less used mode of extraction is the "equilibrium" mode, whereby a solvent is selected which possesses an intermediate distribution coefficient, such that only a portion of the analyte is extracted from the matrix. Since the distribution constant is usually expressed in terms of concentrations of the solute distributed between two phases, with their respective volumes,

$$K_d = C_A/C_B$$

and the masses of analyte can be given as a function of the volume,

$$m_A = V_A C_A \quad \text{and} \quad m_B = V_B C_B$$

then, $m_A = K_d(V_A/V_B)\, m_B$ or $m_A = K_d\, V_A\, C_B$

It can be seen that for intermediate values of the distribution coefficient, $500 > K_d > 10$, the volume of the *A* phase can be reduced significantly, compared to the *B* phase, while still retaining a significant mass of analyte. Since, for any given temperature, the distribution coefficient is constant for any solvent pair, the amount of analyte extracted from *B* into *A* is then directly proportional to the concentration of the analyte in the *B* phase. If temperature and volumes can be well controlled, and the method for analyte detection is sufficiently sensitive, it should be possible to perform "equilibrium" extractions, if certain conditions are maintained to prevent depletion of the analyte in the *B* phase, such as proper adjustment of K_d, and restricting the volume of the *A* phase. How analytes are partitioned between two solvents for the two types of extractions is graphically depicted in Fig. 3.

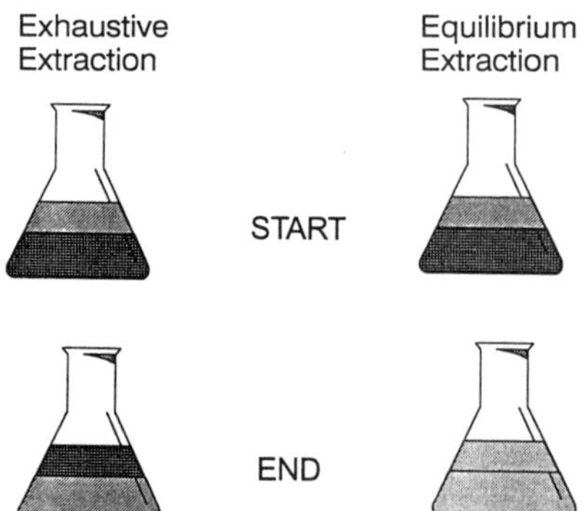

Fig. 3. A visual comparison of a solute distributing between two immiscible solvents for both exhaustive and equilibrium extractions.

The section below on *solid phase microextractions* describes how this technique can be successfully employed.

Aqueous Solvent Systems For The Extraction of Pesticides From Vegetables

Recent advances in solid phase extraction (SPE) technologies in the form of membrane disks have generated new interest in the extraction of an array of pesticide classes from water (McDonnell, T., Rosenfeld, J., 1993; Barcelo, D., et al., 1993; Hagen, D.F., et al., 1990). These devices have several advantages over the SPE cartridges, including high permeability, resulting in short extraction times, freedom from interferences due to plasticizers in the cartridge cylinder, and less propensity for plugging when filtering dirty samples (Hagen, D.F., et al., 1990).

Two recent and somewhat similar studies have shown the Empore® (3M) versions of these disks are capable of stabilizing many types of pesticides during storage. In one of these studies, (Sensemen, S.A., et. al., 1993) 12 pesticides representing several classes of both nitrogenous and organophosphorous types were extracted from laboratory water at a concentration of 20 µg/L and subjected to various storage regimens for periods up to 180 days. When the disks were held at 4°C, losses were less than from the water samples stored at that temperature. Storage at -20°C gave even better recoveries. The second study answered questions about whether a group of 36 pesticides in USEPA National Pesticide Survey Method no. 1 would survive on similar Empore® disks under conditions that would simulate storage in a remote water sampler that had been used to sample "dirty" surface waters under typical Florida heat and humidity (Moye, H.A., et al., 1994). After 30 days of storage under such conditions, it was found that pesticides on the disks at both 0.5 and 10 µg per disk were generally stable. When averaged over all treatments, 19 pesticides were recovered after 30 days at levels equal to or higher than 70%. Such ability to extract pesticides from "dirty" water, and to stabilize them, prompted further study on the disk's ability to extract pesticides from vegetables, using either water or water modified with surfactants, a cyclodextrin, or urea, a known water "perturber".

Ordered media in separations. Solutes can be partitioned from an aqueous environment into solubilized molecules which have available interior portions which are much less "polar" than water itself. One class of molecules that is growing in use for such purposes is the cyclodextrins, a cup-like arrangement of cyclic oligosaccharides, of which β-cyclodextrin is a member (cyclohepta-amylose). Partitioning between water and the interior of β-cyclodextrin has been demonstrated for a number of aromatic and aliphatic molecules, with formation constants, K_f, approaching 5×10^3 (Blyshak, L.A., et al., 1989), with a higher affinity for mononuclear aromatics than for aliphatics. Polynuclear aromatics have higher formation constants for the larger cupped γ-cyclodextrin (cycloocta-amylose, Mohseni, R.M., Hurtubise, R.J., 1990). Use of these species has grown for the chromatographic separations of chiral compounds by high performance liquid chromatography, a result of an equilibrium being established between the mobile phase and the stationary phase containing the cyclodextrin, illustrating the rapidity with which exchange into and out of the cyclodextrin occurs (Mohseni, R.M., Hurtubise, R.J., 1990).

Another form of ordered media, hydrophobic, or "normal" micelles, arises from the aggregation of molecules having both hydrophilic and hydrophobic portions, typically called surfactants (Arunyanart, M., Cline Love, L.J., 1984). Surfactants can be neutral, anionic, cationic, or zwitterionic in nature. Just as an aromatic or aliphatic solute can partition into the cup-like cavity of a cyclodextrin, so can it partition into the interior of a micelle, and line up with the "palisades" of the micelle. Micelles form when the "critical micelle concentration" of the surfactant in water is exceeded, and increase in number within the solution as more surfactant is added. Consequently, the amount of dissolved solute which the micelles can accommodate depends upon the mass of surfactant added to the solution beyond the critical micelle concentration. Destruction of the micelle within the water solution is accomplished by simple dilution so that the surfactant concentration drops below the critical micelle concentration, releasing the solute to the water.

Water content of fruits and vegetables; mixed solvent extractions. All fresh fruits and vegetables are composed predominantly of water (Wyatt, B.K., Menill, A.L., 1975). Some, such as celery, are nearly all water (94%), while apples, are less (84%). If 100 grams of such foods are extracted with 100 mL of acetone, as is done in the Luke method, the corresponding amount of water in the extract approaches 50% (Table 2). Yet, highly effective extractions for a very wide variation of pesticide polarities are achieved by this method for fruits and vegetables with low fat contents.

Relatively high percentages of water in mixed solvents have been shown to be efficient for the extraction of chlorinated hydrocarbons from hay and other dried samples using blending, with recoveries equivalent to those observed for Soxhlet extraction(1). Interestingly, improved recoveries for field incurred dieldrin from radishes have also been observed with increasing water content, with a maximum occurring at

Table 2. Water Content of Fruits, Vegetables and Luke I Extract Solution *a*

	% Water in	
Food	Fresh Crop	Extract Solution
Apples	84.1	46
Peas	82.7	45
Carrots	88.2	47
Potatoes	77.8	44
Radishes	93.6	48
Celery	93.7	48
Spinach	92.7	48

a 100mL of Acetone ; 100g Crop

Table 3. Extraction of Dieldrin from Acetonitrile - Water Mixtures

Percent Water	Percent Recovery
0	67
10	69
20	73
30	71
40	75
50	76

50:50 water:methanol or water:acetonitrile. How the polarity of the solvent mix was affected by various percentages of organic solvent and their associated recoveries is shown in Table 3. Unfortunately, this study did not pursue water:solvent mixtures beyond the 50% water level (Wheeler, W.B., et al., 1982).

Water based extractions of pesticides from vegetables. Recent studies in this laboratory have investigated the feasibility of employing water based systems for the extraction of the pesticides diphenamid, methomyl, thiofanox, thiram, and tolylfluanid (Fig. 4) from celery, lettuce, tomato, and white potato at 0.1, 1.0, and 10.0 µg/g (Szabo, N.J., Moye, H.A., 1994). Three surfactants were studied at concentrations twice the critical micelle concentration, one cyclodextrin (β-cyclodextrin; 11.3 g/L), one water "perturber" (urea at 5.0 M), and laboratory water. Table 4 summarizes the materials that were used. Crops were fortified and extracted by blending with the aqueous system. Extractions from the aqueous extracts were accomplished by passing them through a 4.7 cm C_{18} Empore disk with vacuum. Analysis of the disk eluates was performed by HPLC with post-column photolysis and fluorescence detection. Of the five pesticides studied, thiram and methomyl gave very low recoveries when extracted from water alone with the Empore disk and consequently were not studied with crop. Table 5 summarizes the average recoveries for extracting three of these pesticides from the water alone. Extractions were in triplicate at the 1.0 µg/g level.

Highest recoveries from the crops were observed for diphenamid, using the Brij 58 and Triton X-100 systems when recoveries were averaged for all four crops. When only lettuce was considered, water alone was just as efficient,

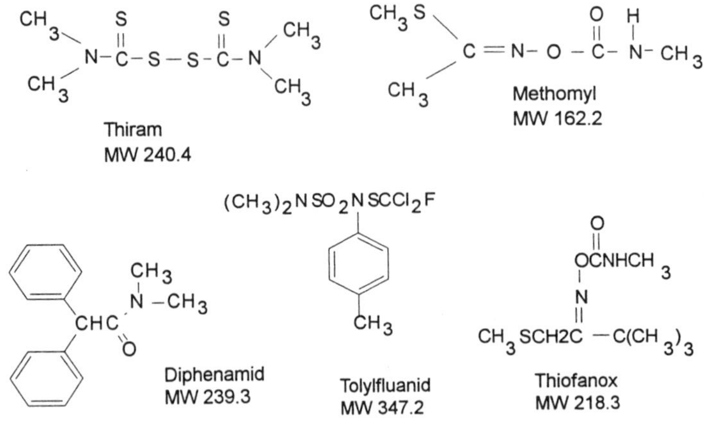

Thiram
MW 240.4

Methomyl
MW 162.2

$(CH_3)_2NSO_2NSCCl_2F$

Diphenamid
MW 239.3

Tolylfluanid
MW 347.2

Thiofanox
MW 218.3

Fig. 4. Chemical structures of the pesticides that were examined for aqueous based extractions from vegetables. (Data are from Szabo, J. J., Moye, H. A., 1994.)

Table 4. Characteristics of Water Systems

Modifier	Surfactant CMC (mM)	Extraction Conc. (g/L)
BCD (beta-cyclodextrin hydrate)	--	11.350
Urea	--	300.300
Brij 58 (polyoxyethylene 20 cetyl ether)	0.077	0.173
Triton X-100 (poly(phenol ether))	0.20	0.250
HDS (Sodium n-hexadecyl sulfate)	0.58	0.400
SDS (Dodecyl sodium sulfate)	8.27	4.772

Table 5. Average Extraction Efficiencies from Water with C18 Empore Disks

	Meth	Diph	Tolyl
Water	14.7	93.4	81.9
BCD	19.1	111.0	117.1
Brij 58	7.47	104.4	123.1
Triton	6.04	94.8	100.4
HDS	13.7	99.3	----
Urea	3.38	89.4	64.6

giving recoveries ranging from 77 to 107%. Disk plugging was observed only for tomato, which took several hours to complete a filtration. The presence of surfactants, cyclodextrin or urea did not increase the rate of filtration. When all loads and crops were averaged, there was no statistical difference for the extraction of diphenamid with HDS, urea, and water (100%). Water (alone) was by far the best solvent system for thiofanox (98%). And interestingly, β-cyclodextrin was best for tolylfluanid (60%). Consequently, if 70% is accepted as a minimum acceptable recovery, then water was the most satisfactory solvent for the extraction of both diphenamid and thiofanox. No overall advantages were seen for employing ordered media for the extraction of these two pesticides.

Water at Supercritical and Subcritical Temperatures and Pressures

If problems are experienced with water being too "polar" for certain "nonpolar" solutes during the extraction process, there is the potential for adjusting its dielectric constant by raising its temperature. If, for example, the optimum acetonitrile:water ratio for the extraction of dieldrin is 50:50, as has already been shown (Wheeler, W.B., et al., 1982), and the water content of radishes is taken into consideration (94%), then the water percentage for that radish extraction is 66%. The dielectric constant, ε, for an acetonitrile:water mixture of this composition is 58. This dielectric constant can easily be reached by heating 100% water to approximately 85°C, as can be seen in Fig. 5.

More dramatic reductions in the polarity of water can be used by heating and pressurizing it to its critical point (Fig. 6; Roger, C., 1991). Nearly complete removal of an array of alkyl and aromatic environmental pollutants from a highly contaminated soil has been achieved in 15 minutes of extraction by employing both

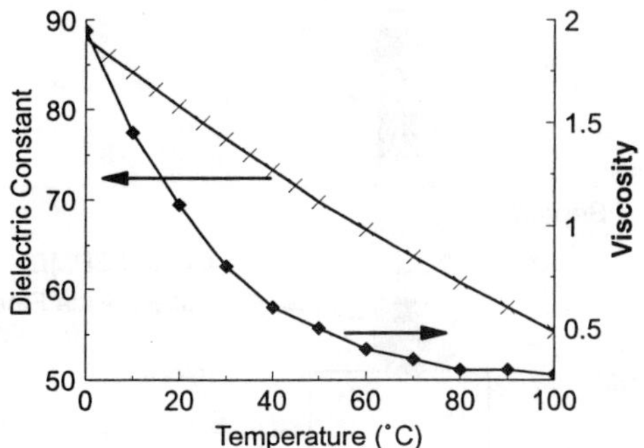

Fig. 5. The effect of temperature on the dielectric constant, ε, and viscosity, ή, of water.

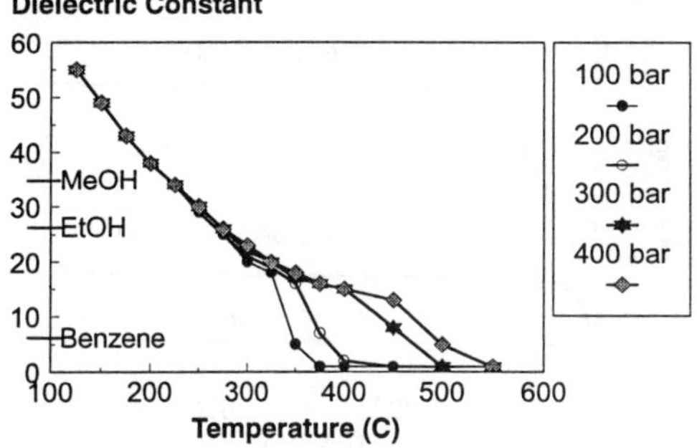

Fig. 6. The effect of temperature and pressure on the dielectric constant of water at subcritical and supercritical values (Data from Roger, C., 1991).

subcritical and supercritical water at 350 bar (Hawthorne, S.B., et al., 1994). Extraction efficiencies were more dependent on water temperature than on pressure, with 250° and 300°C being optimum for all of the aromatic species, from naphthalene to benzo[a]pyrene. Higher temperatures, 400°C, were required for the alkanes, out to tricontane (C30). A great degree of solvent selectivity was observed for the alkanes when they were extracted at 50 bar and 250°C, with efficiencies for octadecane (C18) being 90% and hexacosane (C26) being 6%. The one heterocyclic oxygen compound studied, dibenzofuran, was stable at temperatures up to 400°C, although best recoveries, 117% , were observed at 350 bar and 200°C (ε=36). For the larger molecular weight PAH compounds, such as benzo[a]pyrene, subcritical water (250°C, 50 bar) was more efficient than supercritical CO_2 (50°C, 659 bar), giving efficiencies for the contaminated soil of 176 and 98 % respectively. Higher extraction efficiencies than 100% were frequently reported for this study, when comparisons were made to residues claimed by the supplier of the standardized soil.

When soil extractions were performed at 350 bar at various temperatures, it was clear that the kinetics of extraction were much faster for the PAH compounds studied in the 200° to 300°C range than at 50°C. Viscosity differences may have been a factor for this particular study, as well as solubility ones.

Since supercritical water is somewhat corrosive, and has been used to degrade organic compounds by both hydrolysis and oxidation (Shaw, R.W., et al., 1991), the examination of water in the subcritical regions, even at very low pressures, such as 1 bar, offers some attractive opportunities. Since the viscosity of water diminishes significantly, along with the dielectric constant, as water at 1 bar is heated to 100°C, the kinetics of extraction ought to increase, since diffusion controls the extraction rate (Fig. 5). The relationship between the diffusion of an uncharged molecule in water to several parameters can be expressed by the Wilke-Chang equation (Lyman, W.J., et al., 1982),

$$D_{BW} = \frac{7.4 \times 10^{-8} (\phi_W M_W)^{1/2}\, T}{\acute{\eta}_W V_B^{0.6}}$$

where, D_{BW} = diffusion coefficient for solute, B, in water (W, cm^2/sec)
ϕ_W = solution association constant (2.6 for water)
M_W = molecular weight of water (18)
T = temperature (K)
$\acute{\eta}_W$ = viscosity of water (centipoises)
V_B = molar volume of solute (mL/mol)

Consequently, the rate at which a solute diffuses from a point source in two dimensions is directly proportional to the temperature (T), inversely proportional to the viscosity, and inversely proportional to the molar volume of the solute taken to the 0.6 power. Since the viscosity of water decreases somewhat with temperature between 25 ° and 100°C (Fig. 5), and the diffusion coefficient is directly proportional to temperature, the use of warm water for the extraction of pesticides from any matrix should be favorable both from a partitioning and also from a kinetics aspect. If there are significant differences in molecular size between the solute, B, and coextractants that are larger molecules, such as lipids, then the rate at which the smaller solute would be extracted is higher, giving an opportunity for selective extractions from a kinetics aspect alone.

Solid Phase Microextractions (SPME)

One special case of "equilibrium extractions" that completely eliminates the need for organic solvents in the extraction of an array of organic compounds from water is a technique that has come to be known as *solid phase microextraction* (SPME; Louch, D.L., et al., 1992). It relies upon the rapid extraction of a solute from water by a polymeric or adsorptive stationary phase that is a buffer coating upon an optical fiber (Fig. 7). Extraction is effected from a volume of water by immersing a short section of the buffer coating, typically 2 cm, into a stirred water sample, and continuing to stir until equilibrium is reached between the water and the buffer coating (Fig. 8). Typically, depending upon the composition of the buffer coating, its thickness, and the nature of the solute, equilibrium is reached in less than 15 minutes (Fig. 9; Louch, D.L., et al., 1992). Commercially available buffer coatings include polydimethylsiloxane, polyimide, polyacrylate, and a recently developed proprietary molecular sieve, Carboxen ®.

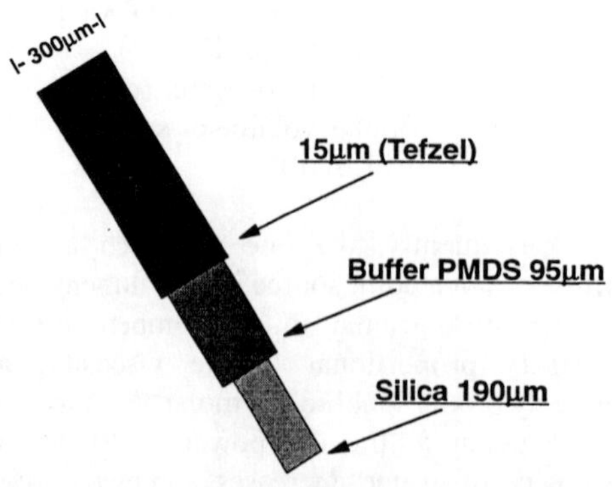

Fig. 7. A diagramatic representation of an optical fiber tip used in SPME.

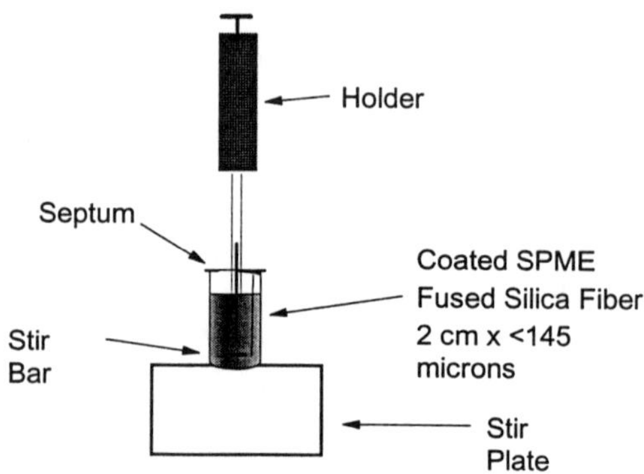

Fig. 8. Typical arrangement used for the SPME extraction of water.

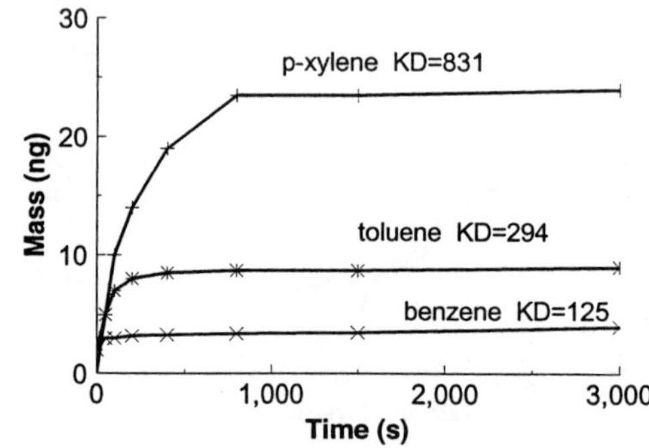

Fig. 9. The effect of the distribution coefficient, K_D, on the extraction kinetics of closely related aromatics from water (Reproduced from Louch, D.L, et al., 1992, American Chemical Society).

Since the optical fibers are of an extremely small diameter, nominally 300μm, they can be inserted directly into the injection port of a capillary column equipped gas chromatograph, where the solute is desorbed onto the front of a "cold" analytical column, which is then temperature programmed to develop the chromatogram. Since the exposure of the fiber to the high temperatures of the injection port provides a way to clean the fiber during the determinative step, the fiber can effectively be reused; as many as 300 uses have been reported (Potter, D.W., Pawliszyn, J., 1994). Critical to the whole operation is the ability to protect the fiber physically during its insertion into the gas chromatograph; this is accomplished with a fiber holder into which the fiber retracts while insertion through the septum is taking place. The fiber is then extended from the holder to allow volatilization of solute from the buffer coating.

While extremely low levels of detection have been reported for aromatic hydrocarbons extracted from water by polydimethylsiloxane coated fibers with analysis by gas chromatography with flame ionization or mass spectrometric detection (Potter, D.W., Pawliszyn, J., 1994), there are practical limitations as to how "hydrophobic" an analyte can be. Previous studies have shown that there is a fairly close match between a compound's octanol:water partition coefficient, K_{OW}, and the distribution coefficient between polydimethylsiloxane and water, K_D, (Table 6; Potter, D.W., Pawliszyn, J., 1994). It has also been shown that there is a negative linear relationship between the log of the distribution constant and the log of the limit of detection using polydimethylsiloxane coated fibers (Fig. 10). However, as the K_D exceeds approximately 10^3, several practical problems arise. Depending upon the film thickness, equilibration times become prohibitive, and since the mass of analyte partitioned onto the fiber is a function of the film thickness, using thinner films reduces the limits of detection. The commercially available 7μm films should be used with compounds having K_D values above 10^3. Also, when K_D values exceed this value, there is carryover from one sample to another, since even at the elevated temperatures of the injection port a small amount of analyte remains dissolved in the film; further heating eventually clears the fiber of such carryover, however, and the fiber can then be used for subsequent injections. Adsorption of the analyte to organic particulates suspended in water also becomes significant at such K_D values, although it can be argued that particulates in some cases should be treated separately than liquid water.

Table 6. Comparison Between Octanol-Water Partition Coeffcients (Kow) and Determined Distribution Constants (Kd) for PAH/PCB Target Analytes

Analyte	Log Kow	Log Kd (exp.)
Naphthalene	3.01-3.59	3.01
Anthracene	4.54	4.10
Benz[a]anthracene	5.61	4.96
Benzo[a]pyrene	6.44	4.86
PCB3	5.64	4.94
PCB5	6.85	4.89

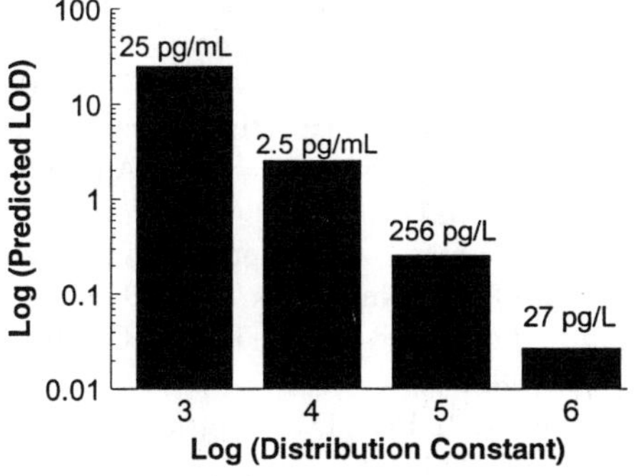

Fig. 10. The effect of analyte concentration on theoretical limits of detection by SPME with gas chromatographic determination. (Reproduced from Potter, S. W., Pawliszyn, J., 1994.)

Table 7. Solid Phase Micro Extraction of Pesticides from Water (50ng/L ; n=7)

Pesticide	Area Counts	%RSD
alpha-HCH	160671	14.2%
Heptachlor	223841	13.2%
Aldrin	278538	6.4%
Dieldrin	214452	14.4%
Endrin	175912	15.9%
4,4'-DDT	133421	23.9%
Methoxychlor	67878	25.8%

Preliminary results on an array of chlorinated hydrocarbon pesticides have been encouraging (Table 7), where extractions were performed for 15 minutes with a 100μm polydimethylsiloxane film and analyzed by gas chromatography; water was fortified at the 50 ng/L level (Shirey, R.E., 1994).

Conclusions

Even though water is considered a "polar" solvent, and hence unsuitable for the extraction of nonionic pesticides, this may not necessarily be true, even considering water at ambient temperatures and pressures. Recent data have shown that even certain pesticides with intermediate values of K_{OW} may be extracted from vegetables with acceptable recoveries, without the need for employing special techniques, such as the use of ordered media in the form of micelles or cyclodextrins, or the use of water "perturbers", such as urea. After such extractions, concentration of the pesticide may be effected by the use of solid phase extraction devices, such as the membrane extraction disks of the reverse phase type. The polarity of water, and its viscosity, can be significantly altered by heat and pressure. While supercritical water approaches the polarity of typical hydrocarbons, it also becomes somewhat corrosive and participates in hydrolytic and oxidative degradation processes of a number of oxygenated and halogenated compounds. Recently acquired data shows, however, that subcritical water has extremely powerful solvation properties for even very hydrophobic solutes, and is very efficient for the extraction of such compounds from naturally contaminated soil. As water is heated toward its boiling point at ambient pressure, both its polarity and viscosity are significantly reduced, offering potential improvements in not only solubilization of moderately polar analytes, but also in the kinetics of extraction.

Fruits and vegetables may be prime candidates for the use of water as an extraction solvent in the trace analysis of pesticide residues, since they are composed mostly of water, and since they typically contain only small concentrations of lipids. High percentages of water in conjunction with acetonitrile and methanol were more effective in extracting a chlorinated hydrocarbon insecticide, dieldrin, from radishes that had been grown in soil treated with the pesticide than were less aqueous solvent pairs. Further research with field incurred residues of other pesticide classes is needed to determine the extent to which water is effective.

Our experiences using water as an extraction solvent in conjunction with solid phase extraction disks have been encouraging. Problems with disk plugging due to coextractives have been minimal, and when water solubilities are moderate, those pesticides can be efficiently recovered from the water extracts. Further research is merited with heated water, water at subcritical temperatures, and with equilibrium extraction approaches, such as solid phase microextractions.

Use of "nonexhaustive" or "equilibrium" extraction techniques may be an alternative way of eliminating organic solvents, if water itself is the matrix under study, or if water can be used for the extraction of the matrix. Since only organics in relatively clean water have been extracted using this technique, there is much that can be explored in the way of coupling aqueous extractions of fruits and vegetables to the SPME approach. Effects of coextractives on extraction equilibria and kinetics will need to be determined. Undoubtedly, as the technique gains popularity, and it should because of its speed and simplicity, there will be additional optical fibers that will become commercially available, which should make it applicable to a wider range of compounds, including pesticides. Such a wider range of fiber buffer polarities will be critical for matching pesticide polarities, so that equilibriums can be reached within reasonable times.

Acknowledgements

The assistance of Keith Tolson in the preparation of this manuscript is much appreciated by the author.

Literature Cited

Arunyanart, M.; Cline Love, L.J. *Anal. Chem.* **1984**, *56*, 1557-1561.

Barcelo, D.; Durand, G.; Bouvot, V.; Nielen, M. *Environ. Sci. Technol.* **1993**, *27*, 271-277.

Benson, S.W.; Siebert, E.D. *J. Am Chem. Soc.* **1992**, *114*, 4269-4276.

Blyshak, L.A.; Dodson, K.Y.;Patonay, G.; Warner, I.M.; May, W.E. *Anal. Chem.* **1989**, *61*, 955-960.

Ellis, R. Symposium on New and Emerging Approaches to the Development of Methods in Trace-Level Analytical Chemistry. Berkeley Springs, WV, Aug. 17-21, **1992**, Food Safety and Inspection Service, United States Department of Agriculture, Washington, DC.

Franks, F. Water: A Comprehensive Treatise. Plenum Press, New York, NY, **1973**

Franks, F. Polywater. The MIT Press.Cambridge, MA, **1982**.

Gutenman, W.H.; Lisk, D.J. *J. Agric. Food Chem.* **1963**, *11*, 304-306.

Hagen, D.F.; Markell, C.G.; Schmitt, G.A. *Analyt. Chim. Acta.* **1990**, *236*, 157-164.

Hardin, L.J.; Sarten, C.T. *Journal of the A. O. A. C.* **1962**, *45*, 988-993.

Hawthorne, S.B.; Yang, Y.; Miller, D.J. *Anal. Chem.* **1994**, in press.

Hopper, M.L.; King, J.W. *Journal of the A. O. A. C.* **1991**, *74*, 661-666.

Klein, A.K.; Laug, E.P.; Sheehan, J.D. Jr. *Journal of the A.O.A. C.* **1959**, *42*, 539-544.

Klein, A.K. *Journal of the A. O. A. C.* **1960**, *43*, 703-706.

Krause, R.T. *Journal of the A. O. A. C.* **1980**, 63, 1114-1124.

Lee, M.S.; Papathakis, M.L.; Feng, H.-M.C.; Hunter, G.F.; Carr, J. Poster 8B-14, 7th International Congress of Pesticide Chemistry, Hamburg, Germany, Aug. 5-10, **1990**.

Louch, D.L.; Motlagh, S.; Pawliszyn, J. *Anal. Chem.* **1992**, *64*, 1187-1199.

Luke, M.A.; Froberg, J.E.; Masumoto, H.T. *Journal of the A. O. A. C.* **1975**, *58*, 1020-1026.

Luke, M.A.; Froberg, J.E.; Doose, G.M.; Masumoto, H.T. *Journal of the A. O. A. C.* **1981**, *64*, 1187-1195.

Lyman, W.J.; Reehl, W.F.; Rosenblatt, D.H. eds. Handbook of Chemical Property Estimation Methods; Environmental Behavior of Organic Compounds. McGraw-Hill Book Company, New York, NY, **1982**.

McDonnell, T.; Rosenfeld, J. *J. Chromatogr.* **1993**, *629*, 41-53.

Mills, P.A.; Onley, J.H.; Gaither, R.A. *Journal of the A. O. A. C.* **1963**, *46*, 186-191.

Mohseni, R.M.; Hurtubise, R.J. *J. Chromatogr.* **1990**, *499*, 395-410.

Moye, H.A.; Shen, S.; Ruby, A. Pesticide Residues in Foods: Technologies for Detection. US Congress, Office of Technology Assessment, **1988**.

Moye, H.A.; Ali, T.; Anderson, A.; Tolson, J.K. *J. Agric. Food Chem.* **1994**, submitted.

Potter, D.W.; Pawliszyn, J. *Environ. Sci. Technol.* **1994**, *28*, 298-305.

Robertson, A.M.; Lester, J.N. *Environ. Sci. Technol.* **1994**, *28*, 346-351.

Roger, C. Los Alamos National Laboratory Report LA-UR-91-554. **1991**.

Sensemen, S.A.; Lavy, T.L.; Mattice, J.D.; Myers, B.M.; Skulman, B.W. *Environ. Sci. Technol.* **1993**, *27*, 516-519.

Shaw, R.W.; Brill, T.B.; Clifford, A.A.; Eckert, C.A.; Franck, E.U. *Chemical and Engineering News,* **1991,** *Dec. 23*, 26-39.

Shirey, R.E.; Supelco, Inc., unpublished results, **1994**.

Szabo, N.J.; Moye, H.A.; *J. Agric. Food Chem.* **1994**, submitted

Wheeler, W.B., Thompson, N.P.; Edelstein, R.L. Littell, R.C.; Krause, R.T. *Journal of the A.O. A. C.* **1982**, *65*, 1112-1117.

Wilderman, M. Shuman, H. *Journal of the A. O. A. C.* **1968**, *51*, 892-895.

Wyatt, B.K.; Menill, A.L., Composition of Foods, USDA, **1975**

Residues: Biotechnology Based Methods

Bruce D. Hammock and Shirley J. Gee, Departments of Entomology and Environmental Toxicology, University of California, Davis, CA 95616

Biotechnologies are offering new methods for the analysis of both recombinant and classical chemicals. The most common of these are antibody based immunoassays. Since the last IUPAC meeting we have seen a large increase in the use of immunoassays, in the integration with other analytical procedures, in the availability of reagents in commercial kits, and in the regulatory acceptance of the technology. In this paper the advantages and limitations of the technology will be discussed as well as its current status and the research trends. The low cost of immunoassay coupled with the high sensitivity, accuracy and precision indicates that it will have an increasing impact on the field of environmental chemistry, particularly pesticide chemistry.

Over the next decade there will be a large increase in the use of biologically based systems in analytical chemistry. These technologies can be divided into biological assays, nucleic acid assays and binding protein assays (primarily immunoassays). The polymerase chain reaction allows one to detect very small quantities of nucleic acids often in complex matrices. A variety of cell based systems allow one to screen for biological activities based on a receptor interaction. Often the cell based system acts as a sensitive transducer. In many ways these systems combine the best aspects of biological assays and physical assays in being sensitive and reproducible yet monitoring a biological phenomenon rather than a chemical or chemical class. In contrast, assays with isolated enzymes or binding proteins are physical assays based on the law of mass action. These binding assays are not biological assays, and like any detectors have a defined selectivity that may be but is not necessarily associated with a biological activity. Even though the same receptor or enzyme system can be used in these assays as in the biological based assays, the transduction system is not a living system. With enzymes, transduction can be associated with binding of the substrate or inhibitor. Examples of this are the commercial glucose sensors for blood analysis and the assays based on the inhibition of acetyl cholinesterase as a detection system for organophosphates.

The Promise and the Reality of Immunoassay.

The most widely employed but by no means the only system will be that which is based upon antibodies. In 1980 the first status report on the use of immunochemical methods in environmental chemistry was published by the American Chemical Society (Hammock and Mumma, 1980). Even at this time, there were data showing excellent correlation between immunochemical methods and classical technologies. Examples demonstrating the power of immunochemical methods to provide analytical data of improved accuracy, precision, and sensitivity at a fraction of the cost of many classical methods were also given. Nevertheless, this article outlined the *potential* of immunochemical methods because none were used to generate real analytical results. By the time of the IUPAC meetings in Ottawa in 1986 (Hammock et al., 1987) and Hamburg in 1990 (Stanker et al., 1991) there were many examples of successful immunoassays for environmental chemicals. The manuscripts describing these assays still referred to the potential rather than the reality of the technology. A major change is illustrated by the extensive poster session and active work shop at the Washington, D.C. IUPAC meeting. No longer are analysts discussing the potential of the technology. Rather the technology is in use by many analytical chemists in the environmental field. In this chapter we will provide a comparison of the current state of the technology with the situation a few years ago (Table 1). Instead of a historical perspective we will use this comparison to predict future trends with the immunochemical technology and will address the problem of how new analytical technologies are accepted. We can expect that a variety of new analytical technologies including biosensors will impact the environmental field. These technologies not only will be applied to existing analytical needs, but to new needs as well such as the analysis of products of recombinant DNA research. As an international scientific community we need to have more rapid methods for evaluating new analytical technologies and integrating the valuable ones into environmental chemistry.

Fig. 1 shows the number of papers published per year on the use of immunochemical methods in environmental chemistry. Numbers alone are a poor indication of the impact of a technology, but they reflect the more important fact that 20 years ago all of the manuscripts came from several laboratories while today the manuscripts are from numerous private, public, and industrial laboratories. Many of the manuscripts are from well-respected analytical laboratories lacking prior experience in immunochemistry. As manuscripts from these laboratories describe the use of the technology to solve analytical problems, the field of environmental immunochemistry gains credibility. The diverse background of scientists developing as well as using these methods illustrates that in general immunochemical methods are user friendly.

Twenty years ago only a few industrial scientists saw the value of immunochemical technology. When assays were developed they were used in house because management often considered the technology proprietary and because no industry wanted to risk a delay in a new product registration while a regulatory agency considered the appropriateness of an analytical technology. Now it is a rare player in the agricultural chemicals field that does not have in-house expertise in immunoassay. Immunoassays increasingly are seen as a way to speed product registration, and federal agencies

Table 1. Immunochemistry in Environmental Analysis

THEN	NOW
• Promise of immunochemistry Few labs involved Limited industrial interest Few commercial assays Little regulatory interest No official approval	• Reality of immunochemistry Many labs involved Extensive industrial interest Numerous commercial assays Clear leadership from regulators Examples of official approval
• Adapting extraction and cleanup	• Development of extraction and cleanup
• Isolated technology	• Integrated technology
• Competing technology	• Complementary technology
• Technology transfer	• Technology development

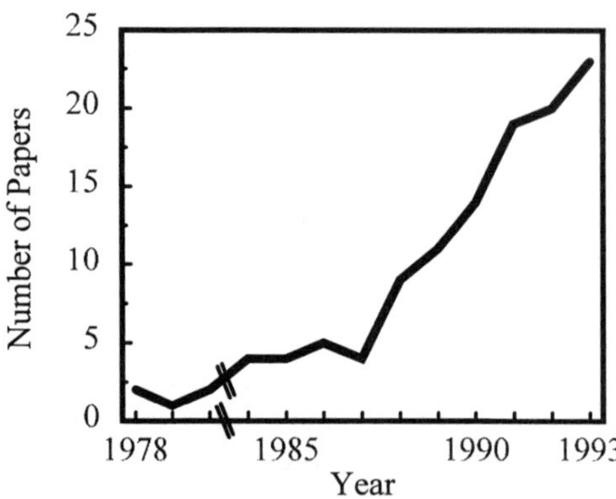

Fig. 1. A representative number of papers on immunoassay of hazardous compounds published over the last 20 years. Numbers are taken from a database compiled in the author's laboratory. Does not include every reference on the subject.

have even required that immunoassays be developed in selected situations to ask specific questions regarding new registrations. With several companies the development of immunoassays are seen as an important component of product stewardship and resistance management. Some companies now have immunoassays for many of their products. Many companies are using the assays as in-house research tools as well as methods to generate regulatory data.

Most of the academic and government laboratories that have developed assays for environmental chemicals have been very open with their reagents and provide them gratis to any qualified party. This practice has fostered the development of a field with extensive collaboration and healthy rather than bitter competition. It also has served to attract rather than repel new users. However open these investigators have been, immunochemical methods would not develop a wide user base without vendors providing convenient immunodiagnostic kits and excellent product support. The evolution of several companies capable of providing high quality reagents and support internationally has been a major boost to the technology. As immunodiagnostics is integrated into classical analytical chemistry we will see more analytical packages provided involving reagents and instruction for extraction and separation as well as for analysis. By having easy to use kits, increasing numbers of analytical chemists will be able to concentrate on innovative uses for the technology rather than on its development. A limitation as well as a strength of immunodiagnostics is that it is a compound directed method. Commercial vendors of assays are providing a great service by offering a collection of many class and compound selective assays allowing the analyst to address an icreasingly broad array of environmental chemicals.

Interestingly one of the first immunoassays for an environmental chemical was the development of a fluorescence polarization assay by Ron Baron of the U.S. Environmental Protection Agency (USEPA). With few exceptions, however there was no development of immunochemical technology, little understanding of the technology, and often an active resistance to it in regulatory agencies. This situation has changed dramatically with many local and national laboratories in several countries actively involved in both the development and the implementation of the technology. Such active involvement is important in the ability of the agencies to evaluate the appropriate use of the technology for residue analysis.

The process of validation of an analytical method is crucial to its acceptance by regulatory agencies. The validation and implementation of any analytical method will depend on its ultimate use. Method validation techniques for groundwater monitoring may differ from those used to provide registration data, or to determine regulatory compliance. Thus it is important to define, at the start of a project, the data quality objectives. Once this is accomplished, the analytical technique should be validated based on the performance characteristics of the method. That is the selectivity, limits of quantification and detection, accuracy, precision, linearity and recovery. It is also critical to consider the stability of the analyte. During the validation process the assay must be cross validated with another method that is as specific as possible, being careful to avoid method bias, i.e., assuming the "reference" method is absolutely accurate. Once performance charac-teristics are determined, these must be monitored regularly using quality control standards. Examples of such performance based studies have been published (Wittmann and Hock, 1993).

As a means to encourage acceptance of immunoassay technology the USEPA Environmental Monitoring Systems Laboratory at Las Vegas has organized an Immunochemistry Summit (Van Emon and Gerlach, 1994). This meeting brings together developers and users of the technology from many sectors and leads to a higher standard for the science. One of the action items determined from the first Immunochemistry Summit meeting was the need for a task force to address the various technical and procedural concerns of the chemical industry, kit manufacturers, and government scientists. Representatives of these groups met and formed the Analytical Environmental Immunochemical Consortium (AEIC). This group has the following objectives: to establish performance standards for test kits (Rittenburg and Dautlick, 1995), to provide a consensus voice for the environmental immunochemical industry, to provide scientific expertise to users of test kits and to regulatory agencies and to promote a continuous education program. The group has compiled a list of definitions of commonly used words in immunochemistry, a compendium of commercially available test kits and a document of general guidelines that manufacturers should use in the writing of test kit inserts (Mihaliak and Berberich, 1995). The degree of their success is measured by the fact that they have been invited to present several training programs in immunochemical methods to regulatory agencies and to the general analytical chemist population.

The California Environmental Protection Agency, Department of Toxic Substances has also taken a proactive stance toward the acceptance of new analytical technology. A certification Program has been initiated for the performance-based certification of hazardous waste treatment technologies and measurement technologies that are advancements to their respective fields. In this program, manufacturers of new technology submit data packages and validation reports. Expert scientists from both inside and outside the agency review the data and will offer to conduct, or participate in, validation studies if these are necessary. Other Cal/EPA departments and boards cooperate in the evaluation so that the recommendations for the application of certified technologies reflect the position of these departments as well. Certified technologies are held to specified quality standards for the period of certification and are subject to monitoring by the state. A goal of the program is to have the certificates recognized by federal as well as other state agencies to avoid the duplication of effort frequently encountered in the evaluation of new technology. Immunochemical test kits have been evaluated as some of the first new technologies (G.W. Fuhs, personal communication).

Concern for water quality has led to the elaboration of standards for water, waste water and sludge by the German Standards Institute. Because of the volume of samples that can be generated by monitoring water quality and the results of water treatment regimens, analytical methods were sought that could handle large sample loads. Immunoassay was considered as such a method. For such monitoring programs, comparability of results is absolutely necessary. Thus, under the standing committee on water analysis, an Immunoassay Study Group was organized by the German Society of Chemists to conduct interlaboratory studies to demonstrate the feasibility of immunoassays for water analysis. The first of the interlaboratory studies has been published (Hock et al., 1991). A second study on trinitrotoluene (TNT) and atrazine will be carried out in January of 1995.

Directly related to the increased governmental interest in the technology are the actual registrations of compounds and enforcement actions based on immunochemical technology. Actual numbers are hard to obtain since analytical technologies used for registration are sometimes seen as providing a competitive advantage. However, the opinion that immunoassay can only be a screening method for other technologies largely has been dispelled, and companies no longer need worry about being the first to submit analytical data based upon immunochemical techniques.. Several countries have registered chemicals based in part or exclusively on data generated with immunochemical tools.

The rapid acceptance of immunochemical technologies over the last few years has been propelled by many factors. Certainly a major driver has been the high cost and limited throughput of most methods now in common use such as GC, GC-MS and HPLC (Niessner, 1993). The ability to process numerous samples in parallel, at low cost not only makes immunoassay a valuable screening tool to prioritize samples prior to classical analysis, but also makes it a reasonable replacement technology in those cases when data on only a few analytes are needed. Another factor has been the analytical difficulties presented by many modern pesticides. These difficulties include chromatographic instability, very low biologically relevant concentrations, and structures that do not lend themselves to classical analysis. The latter is sure to become more important as peptides and proteins are used in agricultural pest control. Immunoassays can be applied to a wide variety of analytes, but fortunately immunoassay also can be seen as a complementary technique to GC since it works so well for many polar analytes such as peptides or pesticide metabolites. Another major driving force in the acceptance of the technology is the need for immediate data. The engineering community already is using biologically based assays for evaluating contaminated sites and to aid in clean up. We will see increased use of immunoassay by farmers with questions regarding potentially toxic residues effecting crop rotation, use by pest control operators to verify coverage, use to insure safe use of compounds in the field, use in integrated pest management programs and use to evaluate residues in crops. All of these applications require methods that can be run onsite with fast turn around time.

Immunoassay is Driving New Technologies for Sample Processing.

For many uses immunoassays can be developed of sufficient selectivity and specificity that a sample can be diluted to eliminate matrix effects and analyzed directly. However, to reach low levels of detection or to provide support for the predicted structure of the residue it sometimes is important to include extraction and clean up steps in the procedure. In the early days of immunoassay the tendency was to use procedures established for classical chromatographic methods and remove a sample for immunoassay at the earliest point in the procedure that eliminated matrix effects. Often this adaptation of classical clean up procedures to immunoassay involved a solvent exchange system from a volatile solvent to a water miscible solvent (Li et al., 1989). This procedure has the advantage facilitating direct comparisons of immunoassays with classical analytical methods following extraction and clean up. However, it fails to marshall the complete power of immunochemical methods. Currently we see innovative extraction and clean up methods developed with the advantages of immunochemistry in mind. The technology allows small samples to be used and very small volumes of solvents to be employed as well. In many cases the solvents used for extraction and solid phase clean up can be environmentally benign solvents or even water. Immunoassay is an ideal technology for use with supercritical fluid extraction since the extract can be concentrated into a small volume of water miscible solvent (Lopez-Avila et al., 1993). Multiple forces will drive analytical chemistry toward miniaturization of extraction and clean up, and these trends will emphasize the advantages of immunoassay. Antibodies will also become components of more sophisticated particulate and membrane solid phase systems facilitating both extraction and clean up. As regulatory agencies experience pressure to reduce their use of the common solvents used in residue analysis, immunological methods offer a viable alternative.

Immunoassays Complement other Analytical Technologies

Most of the laboratories involved in early work on immunodiagnostics also used other analytical technologies including GC, HPLC, and MS. Few of these workers viewed immunodiagnostics as a competing technology. Rather it was seen as a complementary technology to be applied judiciously to certain compounds and in key problem areas (Hammock, 1988). However, immunochemistry was viewed as a competing technology by others in the analytical community, and most papers presented their data as a comparison of immunochemical versus chromatographic technology to argue for acceptance. Fortunately the concept of competing technologies is dying out. Immunochemistry is being viewed as one more analytical tool to be used with the appropriate compound and problem. Integration of immunochemistry with other analytical methods and the development of hypenated methods is the trend (de Frutos and Regnier, 1993; Lucas et al., 1995).

An early example is the use of immunochemistry to screen samples for a group of compounds. Since most samples are negative and common immunoassay formats are designed to give false positives rather than false negatives with operator error and matrix problems, this is an attractive way to reduce sample load. These assays also allow the analyst to rank samples so that high levels of a compound do not contaminate other equipment. Even for immunoassays only used for screening, it is critical that the properties of the assay, including cross reactivity and sensitivity, have been carefully evaluated. In some cases, people not familiar with the technology view this appropriate use of immunoassay as an indication that it is only a screening tool rather than a highly quantitative as well as qualitative technology. The same reagents can be formatted and used both as a screening tool as well as a precise and accurate analytical method. Immunoassays are being used for large monitoring projects, pesticide registration, metabolism studies, affinity chromatography, food inspection, human and environmental exposure and numerous other applications (Table 2).

As immunoassays make routine appearance in more laboratories we will see increasing integration with other technologies. This could include immunoaffinity concentration and clean up procedures for a class of compounds followed by mass spectral or other analytical tools. Similarly immunoassays can have a confirmatory role. Using off line or even on line immunochemical detectors for standard and microbore HPLC (Krämer et al., 1994), capillary electrophoresis, or even thin layer chromatography not only can dramatically increase the sensitivity of an analytical procedure, but it can increase the confidence in the identification of a residue. For example it is unlikely that an unknown residue will have the same retention volume on a high resolution LC as a standard and an identical dose response curve on a resulting immunoassay with one or more antibodies. A demonstration that a peak on GC, LC or other technologies is removed by immunoprecipitation or immunoaffinity with a selective antibody also is evidence in favor of a correct structural assignment. In developing countries thin layer chromatography often is used as an analytical method. When coupled with immunoassay, TLC procedures become many orders of magnitude more sensitive. One or more immunoassay based detection systems can greatly increase the analysts' confidence in the tentative identification of a compound by immunoassay. Evaluation of the slope generated by an unknown or application of chemometrics to a multi-antibody analysis can provide still greater confidence in a structural assignment.

Technology Transfer vs. Technology Development.

The early workers in environmental diagnostics had all used immunochemical technology in other fields such as biochemistry or clinical chemistry. They noted that there were no fundamental differences in the problems or chemistries faced in environmental

Table 2. List of Relevant Reviews

AREA OF INTEREST	REFERENCE
Methods for developing immunoassays for small molecules	Tijssen, 1985; Masseyeff et al., 1993b; Masseyeff et al., 1993a
Advantages and disadvantages of immunoassay	Cheung et al., 1988
Review of applications	Niessner, 1993
Development of immunoassays for pesticides	Jung et al., 1989
Application to genetically engineered organisms	Hammock, 1988
Applications to residue analysis	Van Emon et al., 1985
Application to biomarkers of exposure	Vanderlaan et al., 1991
Application to food	Kaufman and Clower, 1991; Gazzaz et al., 1992
Application to natural toxins	Terplan, 1994
Application to plant viruses	Dewey et al., 1991
List of immunoassays developed for agrochemicals	Gee et al., 1994
List of companies providing test kits	Gee et al., 1994
Review of recombinant antibodies	Karu, 1993
Review of other new technologies	Nelson et al., 1995

and clinical chemistry where immunodiagnostics were well accepted. Thus most of the early papers involved surveying the medical and biochemical literature for new developments and applying them to environmental chemistry. It certainly is wise to be aware of the immunodiagnostic literature in general, but increasingly immunochemists working in the environmental area are making significant advances in the broader field of immunodiagnostics and biosensors. Many of these advances are illustrated in a new ACS book (Hammock and Gee, 1995; Nelson et al., 1995). There are several notable areas where the environmental field has moved ahead of the clinical field. The hesitancy of many classical analytical chemists to use radioactivity forced the environmental field toward the area of nonisotopic immunoassay long before the clinical field began to move in this direction. In order to obtain credibility, most applications of environmental immunoassay have involved careful sample preparation and handling and attention to sources of error influencing accuracy and precision of the resulting assay. With the wide acceptance of immunoassay in the clinical area many assays have tended to be only semiquantitative. Although the resulting data may be adequate for diagnosis, the field as a whole is not as quantitative as the environmental field, and the resulting data may not be as universally useful. Although the environmental field is still in its infancy, most workers in the field have set very high standards for themselves.

Environmental immunoassay in some

cases faces problems of greater complexity than clinical immunoassay. For instance environmental chemicals cover a wide range of polarity. Developing immunoassays for highly lipophilic metabolites is driving the development of novel hapten and handle chemistries. Increasingly hapten design is carried out with the aid of molecular modeling. Although there are many very sensitive clinical assays, most clinical assays are restricted to concentrations of analyte that are pharmacologically or physiologically relevant. In contrast the environmental field often is politically driven and analytical methods must detect compounds many orders of magnitude below levels that are of toxicological or environmental significance. Thus, new technologies leading to more sensitive assays or simple ways to integrate sample preparation are being actively explored in the environmental field.

One aspect of improved sensitivity involves development of novel assay formats or new ways to use existing reagents to yield an improved assay. The immunoassay is based on the Law of Mass Action and as such is a physical assay, rather than a bioassay, even though a biological molecule (the antibody) is used. There are a wide variety of formats in which the immunoassay can be found. A review of some of the types and their advantages and disadvantages are described by Van Emon et al. (1989). A common format is the enzyme linked immunosorbent assay (Fig. 2). In this format the antibody is immobilized on a solid phase (i.e., a microtiter plate, test tube or membrane). The analyte in the sample and the labeled analyte (i.e., an analyte mimic labeled with an enzyme) compete for binding sites on the immobilized antibody. After a fixed time, the unbound materials are removed by washing. Substrate for the enzyme is added. Any enzyme-labeled analyte that is bound then converts the substrate to a colored product. Since this is a competitive format, the more analyte that is present, the less of the labeled analyte can bind. The less labeled analyte bound, the less color formed. Thus the concentration of analyte is inversely proportional to the amount of color formed.

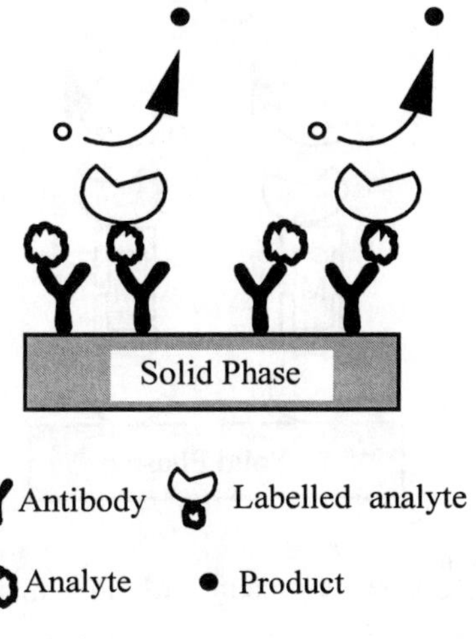

Fig. 2. Schematic representation of an ELISA format. The description is given in the text.

Antibodies can be used in many formats. A second commonly used format is called the double sandwich assay. This assay is typically used to measure proteins and would be suitable for monitoring products of biotechnology. In this assay an antibody is immobilized on the solid phase. The analyte is added and binds to the immobilized antibody. A second antibody is added that now binds to the analyte forming a sandwich. The second antibody is labeled (i.e., with an enzyme). Color development in this case is directly proportional to the amount of analyte present. This double sandwich strategy can be used with a variety of diverse binding molecules. For example, Fig. 3 shows a sandwich method for the detection of mercury. In this case, the binding molecules are not antibodies, but rather are chelating molecules. The assay is technically as simple as the assay that uses an antibody. Other methods for the detection of metals that do use antibodies are described by Wylie et al. (1991) and Chakrabarti et al. (1994).

ELISA technologies, however, dominate the environmental field and will continue to do so for years. In spite of its adaptability and many advantages, ELISA methods have limitations in the number of steps, speed, and sensitivity that can be reduced with promising

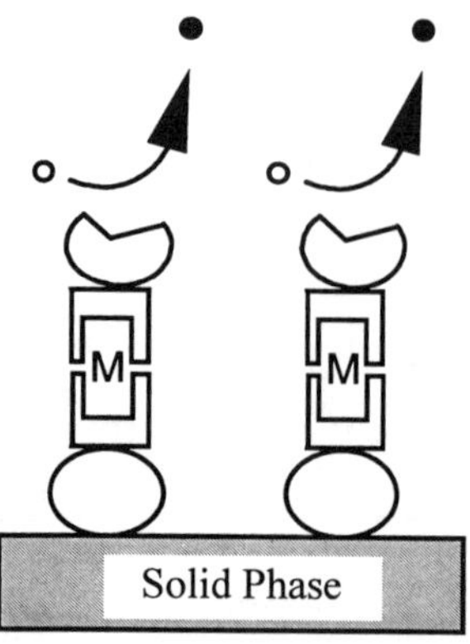

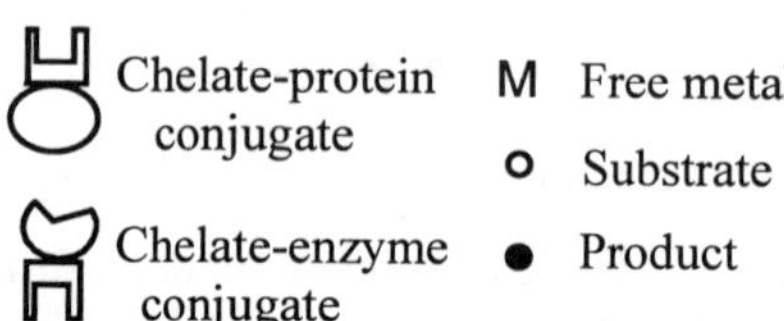

Fig 3. Schematic representation of a metal sandwich chelate assay. The description is given in the text.

new formats. Since often the same reagents can be used in multiple formats, we will see formats develop that will emphasize a particular advantage such as field portability, very high sample throughput, continuous data output, extreme sensitivity and other attributes. Miniaturization of the technology and more intimate association between the antibody and a physical transduction system leading to true biosensors is a trend. It is important to remember that the same reagents can be used in a classical ELISA format, a sophisticated biosensor system, and at the other extreme a disposable field dipstick detector.

Molecular biology not only will generate new analytes for immunoassay, but will lead to improvements in immunoassay itself. Currently most assays are based on antibodies generated in animals and are termed polyclonal assays. Usually such polyclonal based systems offer high sensitivity and selectivity at a low price. An increasing number of assays are based on monoclonal technology which offers the advantage of a theoretically unlimited supply of an antibody of defined properties. An exciting prospect is to replace monoclonal antibodies with recombinant antibodies or other binding proteins. There are many conceptual advantages including lower cost, reduction in animal usage, and of great importance to the analyst, the ability to design a binding site. This offers the possibility of antibodies of extraordinarily high affinity leading to still more sensitive assays. The ability to tailor a binding site should make it possible to develop more readily, class specific antibodies to address multianalyte residue problems. Recombinant technology is far from routine and certainly is not inexpensive at this time. However, the obvious advantages have led several laboratories to pursue this technology actively. The technology also can be applied to other proteins to generate binding molecules that are not antibodies, improve the utility of antibodies in biosensor applications, and develop enzymes as both better reporter systems in ELISA and as detector-transducer systems themselves. Of equal importance with the evolving recombinant technology and expression systems to generate, select and produce antibodies *in vitro* is the evolution of computer systems and software to predict the binding interactions of antibodies and antigens. This software will be critical for the design of superior binding molecules.

In classical analytical chemistry new methods for data handling and analysis have had a tremendous impact. Certainly one example is the quadrapole mass spectrophotometer where in its infancy the quality of its output was dramatically enhanced by signal averaging systems. The miniaturization of immunoassay technology and a variety of electronic transduction systems such as CCD cameras will lead to improved data quality in immunoassay as well. The large data sets that can be generated by immunoassay will drive innovative systems for tracking samples and for data analysis. Other data analysis systems will allow immunoassays to be used to address multianalyte questions. The first step in this process will involve the use of a small number of immunoassays to determine the amount of individual compounds in a mixed sample. In the longer term it is likely that an array of immunoassays coupled with chemometric approaches can be used as so called environmental tasters (Cheung et al., 1993).

Robotics systems will play an increasing role in analytical chemistry. These systems are seeing an early application in immunoassay due to the ability to process large numbers of samples in a parallel process with immunoassay, the consistently ordered array of immunoassays, and because a computer is usually needed to collect data from an ELISA reader. This computer is used only periodically to collect data so it is free to drive a robotics system, to perform more sophisticated data analysis and to aid the analyst in other ways. Increasingly in analytical chemistry we will see expert systems and artificial intelligence systems designed to aid the analyst with trouble shooting assays and even knowing when and how to apply an analytical technology. Training and troubleshooting systems for immunochemistry lend themselves to a computer based format and we likely will see immunoassay introduce this concept into the analytical field.

Other Sources of Information

A wealth of literature exists in this area, particularly since the technology has been used in the clinical arena for many years. There are a wide variety of textbooks on developing immunoassays for small molecules (Tijssen, 1985; Masseyeff et al., 1993b; Masseyeff et al., 1993a). Advances in immunochemical technology over the last 15 years have also been reported (Hammock and Mumma, 1980; Hammock et al., 1987; Stanker et al., 1991). In the primary literature there is a journal dedicated to *Food and Agricultural Immunology*, and immunochemical papers commonly appear in a variety of analytical journals (*Analytical Chemistry*) in areas such agrochemicals (*Journal of Agricultural and Food Chemistry)*. Several authors have compiled recent lists of companies manufacturing test kits. Table 2 provides some references to overview literature on some aspects of immunochemical technology. Of particular note, is a recent review by Niessner (Niessner, 1993) that provides a critical review of literature extending beyond agricultural chemicals to the broader field of environmental and human health chemistry.

Literature Cited

Chakrabarti, P.; Hatcher, F.M.; Blake, R.C.; Ladd, P.A. ; Blake, D.A. *Anal. Biochem.* **1994**, *217*, 70-75.

Cheung, P.Y.K.; Gee, S.J. ; Hammock, B.D. In *The Impact of Chemistry on Biotechnology: Multidisciplinary Discussions*; Phillips, M.P.; Shoemaker, S.P.; Middlekauff, R.D.; Ottenbrite, R.M., Eds.; American Chemical Society: Washington DC, 1988; 217-229.

Cheung, P.Y.K.; Kauvar, L.M.; Engqvist-Goldstein, A.E.; Ambler, S.M.; Karu, A.E. ; Ramos, L.S. *Anal. Chim. Acta* **1993**, *282*, 181.

de Frutos, M. ; Regnier, F.E. *Anal. Chem.* **1993**, *65*, 17A-25A.

Dewey, M.; Evans, D.; Coleman, J.; Priestley, R.; Hull, R.; Horsley, D. ; Hawes, C. *Acta Bot. Neerl.* **1991**, *40*, 1-27.

Gazzaz, S.S.; Rasco, B.A. ; Dong, F.M. *Crit. Rev. Food Sci. Nut.* **1992**, *32*, 197-229.

Gee, S.J.; Hammock, B.D. ; Skerritt, J.H. In *New Diagnostics in Crop Sciences*; Skerritt, J.H.; Appels, J., Eds.; CAB International: Wallingford U.K., 1994; in press.

Hammock, B.D. In *Biotechnology for Crop Protection*; Hedin, P.A.; Menn, J.J.; Hollingworth, R.M., Eds.; American Chemical Society: Washington, DC, 1988; 298-305.

Hammock, B.D. ; Gee, S.J. In *Advances in Immunological Methods for Pesticide Analysis*; Nelson, J.O.; Karu, A.E.; Wong, R.B., Eds.; American Chemical Society: Washington, D.C., 1995; in press.

Hammock, B.D.; Gee, S.J.; Cheung, P.Y.K.; Miyamoto, T.; Goodrow, M.H.; Emon, J.V. ; Seiber, J.N. In *Pesticide Science and Biotechnology*; Greenhalgh, R.; Roberts, T.R., Eds.; Blackwell Scientific Publications: 1987; 309-316.

Hammock, B.D. ; Mumma, R.O. In *Recent Advances in Pesticide Analytical*

Methodology; Harvey, J.r.J.; Zweig, G., Eds.; American Chemical Society: Washington, D. C., 1980; 321-352.

Hock, B.; Hansen, P.-D.; Kanne, R.; Krotzky, A.; Obst, U.; Oehmichen, U.; Schlett, C.; Schmid, R.; Weil, L. *Anal. Lett.* **1991**, *24*, 529-549.

Jung, F.; Gee, S.J.; Harrison, R.O.; Goodrow, M.H.; Karu, A.E.; Braun, A.L.; Li, Q.X. ; Hammock, B.D. *Pest. Sci.* **1989**, *26*, 303-317.

Karu, A.E. In *Hazard Assessment of Chemicals*; Saxena, J., Eds.; Taylor & Francis International Publishers: Washington D.C., 1993; 8, 205-321.

Kaufman, B.M. ; Clower, M., Jr. *J. Assoc. Off. Anal. Chem.* **1991**, *74*, 239-247.

Krämer, P.M.; Li, Q.X. ; Hammock, B.D. *J.A.O.A.C. Int.* **1994**, *77*, in press.

Li, Q.X.; Gee, S.J.; McChesney, M.M.; Hammock, B.D. ; Seiber, J.N. *Anal. Chem.* **1989**, *61*, 819-823.

Lopez-Avila, V.; Charan, C. ; Van Emon, J. *Environ. Lab.* **1993**, *5*, 30-34.

Lucas, A.D.; Gee, S.J.; Hammock, B.D. ; Seiber, J.N. *J. A.O.A.C. Int.* **1995**, *in press*,

Masseyeff, R.F.; Albert, W.H. ; Staines, N.A., Eds. *Methods of Immunological Analysis;* VCH Publishers: New York, 1993a.

Masseyeff, R.F.; Albert, W.H. ; Staines, N.A., Eds. *Methods of Immunological Analysis;* VCH Publishers: New York, 1993b.

Mihaliak, C.A. ; Berberich, S.A. In *Advances in Immunological Methods for Pesticide Analysis*; Nelson, J.O.; Karu, A.E.; Wong, R.B., Eds.; American Chemical Society: Washington, D.C., 1995; in press.

Nelson, J.O.; Karu, A.E. ; Wong, R., Eds. *Advances in Immunological Methods for Pesticide Analysis;* ACS Symposium Series; American Chemical Society: Washington, D.C., 1995.

Niessner, R. *Anal. Methods Instrumentation* **1993**, *1*, 134-144.

Rittenburg, J. ; Dautlick, J. In *Advances in Immunological Methods for Pesticide Analysis*; Karu, A.E.; Nelson, J.O.; Wong, R.B., Eds.; American Chemical Society: Washington, D.C., 1995; in press.

Stanker, L.H.; Watkins, B.E. ; Vanderlaan, M. In *Pesticide Chemistry: Advances in International Research, Development, and Legislation*; Frehse, H., Eds.; VCH: Weinheim, 1991; 397-407.

Terplan, G. In *Immunochemical Detection of Pesticides and Their Metabolites in the Water Cycle*; Hock, B.; Niessner, R., Eds.; VCH: Weinheim, 1994; in press.

Tijssen, P. *Practice and Theory of Enzyme Immunoassays*; Elsevier: Amsterdam, 1985.

Van Emon, J.; Seiber, J.N. ; Hammock, B.D. In *Bioregulators for Pest Control*; Hedin, P.A., Eds.; ACS Publishers: Washington D.C., 1985; 307-316.

Van Emon, J.; Seiber, J.N. ; Hammock, B.D. In *Analytical Methods for Pesticide and Plant Growth Regulators: Advanced Analytical Techniques.*; Sherma, J., Eds.; Academic Press: New York, 1989; 217-263.

Van Emon, J.M. ; Gerlach, C.L. *Environ. Lab.* **1994**, *6*, 24-28.

Vanderlaan, M.; Stanker, L.H.; Watkins, B.E. ; Roberts, D.W., Eds. *Immunoassays for Trace Chemical Analysis: Monitoring Toxic Chemicals in Humans, Food, and the Environment;* ACS Symposium Series; American Chemical Society: Washington, D.C., 1991.

Wittmann, C. ; Hock, B. *J. Agric. Food Chem.* **1993**, *41*, 1795-1799.

Wylie, D.E.; Carlson, L.D.; Carlson, R.; Wagner, F.W. ; Schuster, S.M. *Anal. Biochem.* **1991**, *194*, 381-387.

BIOTECHNOLOGY

Biotechnology in the Production of Disease and Pest Control Agents

Lene Lange, Industrial Biotechnology, Novo Nordisk, DK 2880 Bagsværd, Denmark

Fermentation is the basic technology for the production of both biocontrol agents and microbial metabolites. The production process has bearing for biological activity, reliability in field performance, toxicology and also for the commercial viability of the microbial products and is in this way central for the expansion of microbial solutions.

It is apparent that the globe is divided into many tribes with regard to trends in both research, development and practical use of microbial solutions. Highlights from China, Japan, Former Soviet Union, South America, Europe, USA and Australia will be given in brief. It is concluded that in the past the driving factors for using microbial plant protection have primarily been tradition for use of fermentation products in food and feed, shortage of foreign currency and non-availability of synthetic chemicals. It appears that a new development is taken place, favouring microbial solutions due to environmental concern and due to an accentuated wish to avoid pesticide residues in export commodities, in drinking water and in consumers goods.

It is undisputed that microbes represent a huge pool for future exploitation in plant protection, either directly as biocontrol agents, as the active ingredient, or as lead molecules for chemical synthesis. Biotechnological solutions opens up for new, interesting approaches.

Microbial plant protection is both an alternative and a supplement to synthetic chemicals. It has been presented recently in a number of excellent reviews, covering discovery, producing organisms, active compounds, activities etc (Omura 1989; Yamaguchi 1994; Pillmoor 1993). In the present paper focus will be on two different aspects, the technologies involved and a review of highlights of actual uses of microbials in plant protection seen from a global perspective. More biological solutions are brought into practice by farmers world wide than usually anticipated.

A third aspect, the characterization of secondary metabolites according to taxonomic position of the producing organism, the chemical structure, ecological role, and type of mode of action is dealt with in Lange and Lopez (1994).

Production

Fermentation technology has a central position for further development and expansion of the use of microbial products in plant protection. Fermentation is the basic technology for the production of both biocontrol agents (using the living organism directly) and microbial metabolites (using the products, being proteins, peptides or secondary metabolites, as the active compounds for field application (Lange 1992, 1993; Lange et al. 1993). The production process has bearing for biological activity, reliability in field performance, toxicology and also for the commercial viability of the microbial products.

The life cycles of most biocontrol agents have specific stages which are optimal for application in the field or in

1054–7487/95/0216$12.00/0

the green house. It is therefore important to control the fermentation process so that the wanted stage is produced in sufficient quantity (being e.g. conidia or chlamydospores in *Trichoderma,* sporangia or oospores in *Lagenidium*; or vegetative cells or spores for bacteria) . The optimization is here the result of a compromise between many parameters (e.g. active stage, ready to do its effect and at the same time with the sufficient shelf life).

For production of secondary metabolites the growth conditions are of utmost importance (substrate composition, aeration, temperature etc). Recent results further suggest that more subtle differences in growth conditions, resulting in differences in variations in microbial morphology also has bearing on production of secondary metabolites .

Most of the disappointments in the field of biocontrol are due to lack of reproducibility of the results in the field or green house, from lab to land or from one application to the other (e.g. use of *Pseudomonas* against soil-borne diseases in cereals). Much of the latter is caused by susceptibility to environmental differences from season to season, soil to soil, and year to year but also the variation in the product produced can be of importance. It is crucial to be able to distinguish between these reasons -best done by eliminating the batch to batch differences in the up-scaled production of the biocontrol agent.

The new registration requirements in Europe, potentially including testing of microbial plant protection agents also for possible non-target effects of bioactive metabolites further emphasize the need for fermentation know-how in the development process.

The up-scaled production process is also often the link between academia and industry, as most inventions made in university institutes need to be produced under industrialized conditions in order to be ready for full scale biological and commercial evaluation. In this way the professionalism of the up-scaled production may be crucial for the survival of a new product concept.

The Global Perspective

With modern communication one would presume more or less similar trends in the development in most fields. However, for microbial plant protection the globe has up to now been divided into many tribes.

The Far East (especially China and Japan) has a long tradition for using fermented products in both food, feed and medicine. It is therefore not surprising that it has been in these countries that microbial products also for plant protection purposes have had the most extended practical application.

Highlights from microbial plant protection in China:

**Bacillus thuringiensis* (Bt) is produced and used in many parts of China. The fermentation takes place on many local production plants, typically supervised by the regional university institute, and with further support from the central research effort at Academia Sinica.

*Entomopathogenic nematodes are used large scale for treatment of orchard and shadow trees (applying the nematodes in water into a drilled hole in the trunk) and for soil application (e.g. in orchards against peach borer). Cheap production conditions have been obtained by using freely available substrate (e.g. Beijing duck intestines).

*An efficient large scale production of *Beauveria bassiana* to control Asian corn borer and peach fruit moth has been developed (using a combination of liquid and solid fermentation).

*A number of Agromycins (viz. secondary metabolites from several species of Streptomyces) are used widely in China, especially for control of fungal diseases.

*The concept "Yield Improving Bacteria" is also used widely in large scale in China. Special formulations for seed treatment, root dipping and foliar application on a range of crop plants (e.g. wheat, rape, corn, cotton, beet, water melon etc) have been developed. The central institute for this work is Institute of Plant Ecological Engineering at Beijing Agricultural University. The concept has been further described by Chen et al. 1994.

The bacterial suspension seems to have both direct effect on the pathogen, as well as acting as inducer of SAR (systemically acquired resistance) and having growth stimulating effect. The extension of this concept to e.g. Europe has been questioned as to whether it will have the same effect under high yielding conditions.

Highlights from microbial plant protection in Japan:

Japan has more than 30 years experience with the use of microbial secondary metabolites in plant protection (Misato and Yamaguchi 1984). Among the products used commercially only a few have been commercialised outside Japan. The most prominent example is the Avermectins (Babu 1988) and to a much more limited scale Validamycin (from Takeda Chemical Industries).

The compounds commercialised in Japan are many. The following are all discovered, evaluated, developed, commercialised and used in large scale in Japan (Yamaguchi 1994; Lange and Lopez 1994).

Antifungal:

Blasticidin S (against rice blast); Kasugamycin (also against rice blast); Polyoxins (against rice sheath blight and fungal diseases of fruit trees); Validamycin A (rice sheath blight and fungal disease of vegetables); Mildiomycin (powdery mildew of rose and crape myrtle).

Insecticidal:

Polynactins (against mites in fruit trees and tea plants); Milbemectins (similar targets as Polynactins)

Herbicidal:

Bialaphos (weeds in orchard and mulberry fields)

All the mentioned products are found and produced in Actinomycete species, and are marketed as semipurified compounds, obtained from microbial fermentation of the original producing organism (not manipulated).

An interesting spin-off from the study of microbial metabolites for plant protection has been the use of the de-toxification gene found in microorganisms, protecting the producing organism from its own toxin. Such genes can be inserted in plants and hereby obtaining plants which are protected against the damage from a herbicide application or from the effect of the toxin in the pathogenesis of a bacterial or fungal pathogens (Yamaguchi et al. 1990).

Most of the discoveries of microbial metabolites with biological activities in Japan have been done in programmes also directed towards discoveries of relevance for development of human drugs. Recently the research seem so be preferably directed towards drug discovery as the present situation in Japan leaves a number of options for protecting the crops against major pests and diseases.

Also the commercial viability of the metabolites gives incentives for human drug development as end user prize for microbial fermentation products may in certain cases be a limiting factor for products applied in agriculture.

Highlights from use of microbial plant protection in Australia and New Zealand

*Btk for Lepidopteran larvae in cotton and vegetables.

*Enthomophagous nematodes (especially against black current borer)

*Verticillium against plant damaging insects

*Yield improving bacteria(see above)

*VAM (mychorriza, -stimulating plant growth)

*Red sterile fungus -inhabiting the outer cell layers of the root system in e.g. wheat, protecting against diseases (especially take all) and with a growth stimulating effect

**Serratia entomophila* (against grass grubs, Ocallaghan and Mahanty 1992).

The last decade or so has been very productive in this part of the world regarding innovative research and development within the field of microbial plant protection. Concepts (e.g. some of the nematode products) have been initiated here, and have later spread to the rest of the world; other concepts have been found and developed exclusively for this part of the world (as e.g. the *Serratia* product). The rather close proximity to China has been decisive for several trends, with the exchange of product ideas and experience both ways.

Highlights of microbial plant protection in Former Soviet Union and Eastern Europe:

As opposed to the traditional background for use of fermentation products in the Far East the reason for using large scale application of microbial products in the Former Soviet Union and in East Europe has primarily been the shortage of foreign currency for importation of chemical pesticides.

The most widely used products are

*Bt against Lepidopteran larvae.

**Trichoderma* and *Gliocladium* against soil-borne fungi

**Beauveria* and *Metarrhizium* against plant damaging insects

The products were typically produced by local university institutes and primarily used by the state farms.

The rather extensive experience in microbial plant protection (Khloptsera 1992) gained from these several years of large scale practice has not been much recognized in the scientific community. The explanation for this is the assumption that experience gathered from use on state farms, with rather low yielding agricultural conditions may not be transferable to high yielding conditions.

Highlights from the use of microbial plant protection in South America

The major part of the activities in biocontrol in South America has taken place in Brazil but the concept has recently been given higher priority also in e.g. Argentina and Uruguay as well as in Central America.

Most of the practical applications have been within the field of insect control but also disease control has received considerable research attention. The latest results are assembled in the EMBRAPA publication (Bettiol 1991). The practical application of microbial plant protection in this region may be influenced both by financial restrictions, favouring local solutions over import of synthetic pesticides but also (negatively) by the limited tradition for (industrial) fermentation.

*Baculovirus is used extensively to control lepidopteran larvae, e.g. soybean caterpillar. Last year baculovirus was e.g. applied to approx 1mill ha alone in Brazil, corresponding to 10% of the total soybean growing area. Most of the virus is produced by simply harvesting infected larvae, hanging nice and dry from the plants. A few factories rear the insects artificially and harvest from inoculated larvae but so far no production is build on in vitro culturing of insect tissue.

**Rhizobium* inoculation of leguminous crops (especially soybean, but also alfalfa, french bean, peas etc) is a widespread practice in South America. It is estimated that 80% of the leguminous cropping area in Brazil is treated with *Rhizobium* inoculum. The contribution to nodulation and savings on the fertilizer side are obvious and convincing. That *Rhizobium* inoculation also has a beneficial side effect, protecting against soilborne fungal pathogens (pre- and post-emergence damping-off) is also suggested.

Production of the *Rhizobium* inoculum takes place as a simple fermentation process and the culture is added in a standardized way to a sterilized peat product, ending up with a separately marketed package, ready for on-farm mixing with the seed.

*Bt products for control of lepidopteran larvae have had fierce competition from the very cheap Baculovirus products. The major use of Bt is consequently outside the target spectrum of the virus products, with a stronghold within the area of public health (mosquito vector control and nuisance control).

**Metarrhizium* and *Beauveria* have been widely tried in Brazil and Argentina, as well as in Central America. Several small scale factories started up in the 80´ties but only few are still running. However, the local use should not be underestimated.

**Trichoderma* especially against *Phytophthora* in apple has interesting but limited potentials (Valdebenito-Sanhueza 1991).

*The use of *Acremonium* against insect attack in coconut and oil palm plantations are typical on-farm operations (in larger plantations). It is a practice which saves considerable chemical spraying but survives apparently so well from season to season that a commercialisation of the product apparently has not been needed.

*A number of different Mycoherbicides have been tested both in Argentina and Brazil. However, so far no real (commercial) winner has turned up, even though using *Helminthosporium* against the *Euphorbia* weed in soy bean fields looks promising (Yorinori and Gazziero 1989).

Highlights from the use of Microbial Plant protection in Europe.

*Bt products (Btk both for forestry and agriculture; Btt for agriculture (against Colorado potato beetle) and Bti for public health) are the most widely commercialized microbial products in Europe. They are produced and marketed by Abbott, Novo Nordisk , Sandoz and Ciba Geigy. The total Bt market has grown considerably over the last years -even in a period where the over all ag-chem market is stagnating. The European sectors with the most promising growth of Bt use is forestry.

*Avermectin, with a Streptomycete secondary metabolite as active ingredient, has a wide range of applications in Europe.

**Trichoderma* for disease control has been marketed for a number of targets, especially green house cultures and forestry (Lynch 1990).

**Verticillium*, used against plant damaging insects in green house cultures (tomatoes, cucumber and pot plants)

*The *Metarrhizium* concept has recently been presented in a new form by Bayer, named Bio 10-20 (Stenzel et al. 1992).

Highlights from the use of microbial Plant protection in North America

*One of the most successful applications of microbials has been the use of BtK to control attack of lepidopteran larvae in the Canadian forest. Ultra low volume application, applied as aerial sprays (down to 1liter per ha) has further minimized the impact on nature, all in all resulting in maybe the most well targeted insect control programme of the entire world.

*Avermectins, many products used against numerous targets

**Trichoderma* and *Gliocladium* against soil-borne fungi (Lumsden and Lewis 1989).

**Coniothyrium* against sclerotial fungi

*Insect pathogenic fungi against plant damaging insects

*Nematodes against insects

*Insect viruses

**B.subtilis* as seed dresser (from Gustafson, Texas).

Environmental considerations for choosing microbials for plant protection?

As can be seen from this global overview, the most prominent of the incentives and drivers for using microbials in plant protection has not been a concern for adverse effects on nature and man by the use of synthetics. On the contrary, tradition, shortage of foreign currency, non-availability of new chemistry etc have been predominant.

It is to be expected that the recent experience with pesticide residues in crop, soil and drinking water will change this picture.

Registration requirements in US (and the EPA administration) has given a favourable position for true biologicals including metabolites, building on the assumption that biologicals over all are more environmentally friendly . as e.g. avermectin have had to go through a registration package much similar to the synthetic chemicals. The recent change in registration requirements in Europe for microbials is, intentionally or un-intentionally pulling in the other direction. It is presently presenting a threat to smaller products, as the commercial value of such products may not be matching the extra expenses needed for the new registration requirements. It is to be sincerely hoped that the farmer and the grower do not because of this loose good products.

The Technological Perspective

It is undisputed that microbes represent a huge pool for future exploitation in plant protection, either directly as biocontrol agents, or where microbial products (being either primary or secondary metabolites) act as the active ingredient, or as lead molecules for chemical synthesis of new compounds.

New advancements in the field of biotechnology opens up for a number of new approaches

*construction of transgenic plants, expressing e.g. insect deterrent molecules, or through changes in the plant surface, making it apparently incompatible to the intruding pathogen or expressing a higher level of systemically acquired resistance, fighting back the pathogens more efficiently by the host defence own weapons.

The genes of microbial origin may prove to be a very important source for the establishment of such new plant materials. The over all reaction in the public suggests that the broad use of transgenic crop plants will be accepted. However, full clarification on this as well as on the patent and breeders right status need still to be achieved.

*construction of gene spliced microorganisms for deliberate release, to act as insect control agents, antifungals or microbial herbicides are easily technologically within reach. Right now several potential product concepts are within the tube all over the world (e.g. the "yield improving bacteria" in which the Bt toxin is also produced -or *Rhizobium* inoculants carry genes for other biologically active compounds, or use of the insect viruses to introduce other biologically faster killing agents). However, it is still not crystallized whether the use of such manipulated microorganisms for deliberate release (especially adapted to compete under realistic field conditions) will be accepted by the public and by the registration authorities, in East and West, North and South.

*Construction of more efficient producing organisms for safe and commercially viable large scale production of microbial products has large potentials -and may be foreseen to be more readily acceptable by registration authorities and the public as such organisms are made and used in confined conditions (ie within the production plants) and are not adapted to compete in nature. The obvious achievements are in the area of higher yield,safer production organism, improved field performance of the end product, and better tools for resistance management (by combined use of different mode of action).

*The scope is still open for achieving similar progress within the field of secondary metabolites. Here the approach will be new biotechnological tools within the pathway engineering discipline.

*Another area lies in the microbial expression of non-microbial compounds, of plant or animal origin. The technologies are by now ready for this and future will show how much this will be implemented.

*So far simple advancement, as the dual culture of bacteria and nematodes for efficient large scale production of entomopathogenic nematodes has given significant contribution to the field of microbial plant protection.

*Similarly attempts have been made to grow virus in vitro in insect cell tissue culture. Here, however, the achievements have been very limited. It's as demonstrated definitely do-able but the low production yield and the high costs involved has slowed down the use of viruses more than it has stimulated. The best results have been obtained in the countries where they have gone directly to the use of the "low tech" approach, harvesting from diseased larvae (as e.g. in Brazil).

*An interesting new production approach has been taken in the Bayer Bio 10-20 product where the mycelial pellet provides the "surface" from which the fungi will sporulate (Stenzel et al. 1992). In this way both high efficacy in production, sufficient shelf life and further propagation on location under field conditions are achieved. A similar approach may be used towards other imperfect fungi, also for the ones which in nature only sporulate on fungal sclerotia (?)

*The most significant limiting factor so far for the widespread application of bacteria as disease control agents has been the too short shelf life of the non-spore forming bacteria (e.g. *Pseudomonas).* Improvements in the stabilization of this achieved by radically new approaches in the fermentation and formulation technology is still to come.

*A very important recent contribution from the microbes is that they have acted as source for new lead molecules for both the phenylpyrroles (from pyrrolnitrin) and the strobilurins, which have given rise to whole families of new product leads (Gehman et al. 1990; Anke et al. 1984, respectively). The necessary chemical alterations as compared to the original products have all resulted in increased persistence under field conditions (ie increased UV stability). Hopefully this has not lead to problems with leaving pesticide residue in the environment -as we would then be back to the same kind of problem as we have faced with the synthetics.

*Microbials as inducers of systemic resistance is an area which attracts a lot of attention in these years. Interesting results have been published regarding use of e.g. *Bacillus* preparations or yeast suspensions as inducers of SAR. This may open a new avenue for microbials to be used both as seed dressers and foliar sprays and maybe also as post harvest protection. The latter is much dependent on the public acceptance of microbes on the harvested end-product.

Conclusion

Different traditions both for agricultural practice, registration requirements and R&D philosophy have lead to very different pictures of microbial plant protection in the various regions of the world. Maybe the real breakthrough will come when the dominant incentive is to make plant protection as environmentally friendly and as safe as we ever can; combining the experience and creativity from all parts of the world; -and eventually bring it into reality as commercially viable, high performing products. To achieve this we need the best biotechnology added to the classical fermentation technology.

References

Anke, T.; Schramm, G.; Schwalge, B.; Steffan, B.; Steglich, W. 1984 Liebigs *Ann. Chem.*, 1616-1625.

Babu, J.R. In *Biologically active natural products potential produce in agriculture,* Cutler, H.; American chemical society, Washington DC,1988, 92-108

Bettiol, W. *Controle Biológico de Doencas de Plantas;* EMBRAPA Brazil 1991

Chen, Y.; Mei, R.; Lu, S.; Liu, L.; Kloepper, J.W. In *Management of soilborne Diseases*. Eds. Utkehde, R.; Gupta, V.K., New Delhi 1994 (in press).

Gehmann, K.; Nyfeler,R.; Leadbeater, A.J., Nevill, D.; Sozzi, D. In *Brighton Crop protect Conf. -Pests and Diseases* 1990, p399-406

Khloptsera, R.I.; *Biocontrol News and Information,* **1992**, *13,* 29N-32N

Lange, L. *Progress in Botany* **1992**, *53,* 252-270

Lange, L. *Developments in Industrial Microbiology Series, SIM,* **1993**, *33*

Lange, L.; Breinholt, J., Rasmussen,

F.W.; Nielsen, R.I. *Pestic.Sci.* **1993**, *39*, 155-160

Lange, L. ; Lopez, C.S. In: *Natural products as source of new agricultural chemicals,* Copping,L., Ed; Critical Reports on Applied Chemistry, SCI, 1994; (in press).

Lumsden, R.D.; Lewis, J.A.; In *Biotechnology of fungi for improving Plant growth,* eds. Whips, J.M. and Lumsden, R.D., 1989,177-190

Lynch, J.M., 1990 In *New Directions in biological control: alternatives for suppressing agricultural pests and diseases.* Liss, New York, pp 243-253

Misato, T.;Yamaguchi, I.; *Outlook on Agriculture* **1984** *13,* 136-139

Ocallaghan, M.; Mahanty, H.K. *Biocontrol Science & Technology* **1992** *2,* 297-305

Omura, S. *The search for bioactive-compounds from microorganisms* Bock /Springer series in Contemporary Bioscience, Springer Verlag, New York, 1989; 352pp.

Pillmoor, J.B., Wright, K., Terry, A.D.; *Pesticide Science*; **1993**, *39*, 131-140

Stenzel, K.; Hölters, J.; Andersch, W.; In *Brighton Crop Protection Conference Pest and Diseases,* **1992**,363-368

Valdebenito-Sanhueza, R.M. In *Control Biologico de Doencas de Plantas,* Bettiol, W. Ed.; 1991, pp303-305

Yamaguchi, I. In *Natural products as source of new agricultura*, Copping,L., Ed; Critical Reports on Applied Chemistry, SCI, 1994; (in press)

Yamaguchi, I; Kamakura, T.; Yoneyama, K. In: *Abstr 7th Cong Pesticide Chemistry*, Hamburg,1990, p 0305. pp388

Yorinori, J.T.; Gazziero, L.P.; In *Proc. VII Int. Symp. Biol. Contr. Weeds, Italy,* Delfosse, E.S. 1989 pp 571-576

Approaches to Identify Genes for Disease Resistance in Plants

Patrick Schweizer, Laurence Bindschedler, Philippe Meuwly, Wolfgang Mölders, Jean-Luc Coquoz, Anthony Buchala, Jean-Pierre Métraux, Institut de Biologie Végétale, Université de Fribourg, 1700 Fribourg, Switzerland

Plants can resist pathogens by inducing various mechanisms which interfere with the progression of the invading microorganisms. Induced resistance is expressed as a result of prior exposure to pathogens or to various pretreatments, ranging from chemicals to physical stress. In most of these cases, the expression of specific genes is associated with the defence response. Some of these genes are likely to be involved causally in the resistance response and represent interesting candidates to be used in programs for genetically engineered disease resistance. The isolation of genes involved in either induction or maintenance of resistance is based on the working hypothesis that they are expressed at low levels in non-induced control plants whereas their expression would be enhanced after induction. The goal of this paper is to present several experimental systems we are currently using in our laboratory to study and isolate genes conferring resistance to plants. The aspects which have a review character are not meant to be comprehensive but rather to provide the reader with an update in the selected area.

Systemic acquired resistance (SAR)

After infection by necrotizing pathogens, plants can develop resistance against subsequent infections by the same or other widely different pathogens (Madamanchi and Kuc, 1991; Métraux, 1994). Interestingly, resistance is expressed at the site of infection as well as in uninfected parts of the plant (systemic acquired resistance, SAR). The systemic nature of the response allows a clean dissection of resistance reactions taking place in the non-infected parts of the plants from reactions taking place at the infection site, where besides local acquired resistance (LAR), stress responses due to the infection are also taking place. Fig. 1 shows a typical defence reaction of cucumber inoculated on the lower leaf with the bacterium *Pseudomonas lachrymans* and challenged on the upper leaf with the anthracnose fungus *Colletotrichum lagenarium*. A time-lag is necessary between the infection with the inducing *P. lachrymans* and the challenge infection, the duration of which depends on the time required by the inducer organism and the plant to develop in the host, and to transport a signal to the upper leaf, respectively. In the experiment presented here, this time-lag is short. SAR can be expressed against a broad range of pathogens, independently of the microorganism used for the primary infection (Table 1). The host plant reacts actively to the initial infection as evidenced by studies showing changes in the accumulation of a wide array of proteins called pathogenesis-related or PR-proteins (Table 1; reviewed by Stintzi et al, 1993).

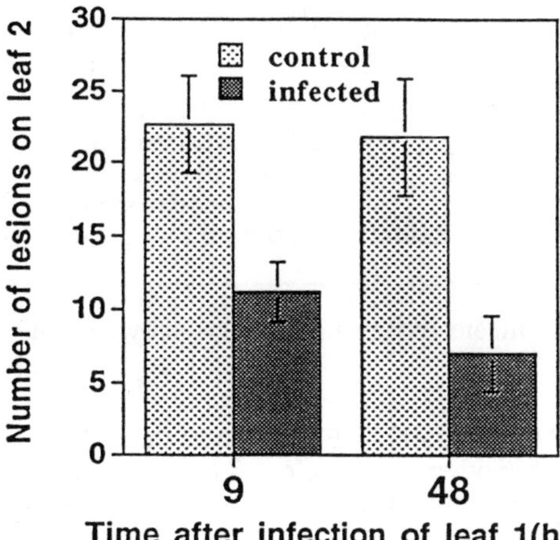

Fig.1: Systemic protection of cucumber against *Colletotrichum lagenarium* inoculated on the upper leaf (40 x 5 μl droplets containing 2 x 10^5 spores/ml) 9 to 48 hours after pretreatment of the lower leaf (leaf 1) with the bacterium *Pseudomonas lachrymans* (injection of the bacterial suspension containing 2 x 10^8 cfu/ml into the leaf at 10 panels each averaging 25 mm^2)(number of plants=3, ±SD).

SAR genes

In cucumber PR-proteins include chitinase and peroxidase, the former has antifungal properties and the latter seems to be associated with increased lignification of the cell wall (see review by (Madamanchi and Kuc, 1991;

1054–7487/95/0225$12.00/0

Table 1: Plants, inducing pathogens , afforded resistance and production of PR's (table compilated from Madamanchi and Kuc 1991; Weete 1992; Uknes et al, 1992, 1993; Métraux et al 1993).

Plant	Inducing pathogen	Protection against	PR's
Arabidopsis	Turnip crinckle virus	Pseudomonas syringae Turnip crinckle virus	+
Barley	Erysiphe graminis f.sp. hordei*	Erysiphe graminis f.sp. hordei	+
Bean*	Colletotrichum lindemuthianum* Colletotrichum lagenarium	Colletotrichum lindemuthianum* Colletotrichum lagenarium	+
Cucumber*	Colletotrichum lagenarium* Pseudoperonospora cubensis Pseudomonas lachrymans Tobacco necrosis virus	Colletotrichum lagenarium Cladosporium cucumerinum Fusarium oxysporum f.sp. cucum. Pseudomonas lachrymans Sphaerotheca fuliginea Tobacco necrosis virus	+
Muskmelon	Colletotrichum lagenarium	Colletotrichum lagenarium	+
Tobacco*	Tobacco mosaic virus Tobacco necrosis virus Thielaviopsis basicola Peronospora tabacina* Pseudomonas syringae	Thielaviopsis basicola Phytophthora parasitica Peronospora tabacina Pseudomonas syringae Phytophthora parasitica Pseudomonas tabaci Tobacco mosaic virus Tobacco necrosis virus	+
Sicklepod	Alternaria crassiae	Alternaria crassiae	?
Tomato	Phytophthora infestans	Phytophthora infestans	+
Rice	Pseudomonas syringae	Magnaporthe grisea	+
Watermelon	Fusarium oxysporum f.sp. cucum.	Colletotrichum lagenarium	+

* means those pathosystems showed protection under field conditions

Métraux, 1994). Fig. 2 shows the accumulation of another PR-protein, PR1, concomitant to the development of resistance in cucumber. In tobacco a number of genes, termed SAR genes, the response of which correlates in time with the onset of SAR have been identified and cloned (Ryals et al, 1992, Ward et al, 1991). Besides chitinases, β-1,3-glucanases, proteinase inhibitors, the identity of several others is not known. The evidence based on temporal and spatial correlations is insufficient to causally link all these proteins to the resistance of the plant. With genetic engineering becoming increasingly accessible, the relevance of SAR genes can be evaluated using plant transformation. The resistance of plants transformed with SAR genes placed under a constitutive promoter can be evaluated against various pathogens. Positive results are obtained if transformed plants show improved resistance with no deleterious effects on yield compared to the untransformed phenotype. The suppression of glucanases in tobacco using

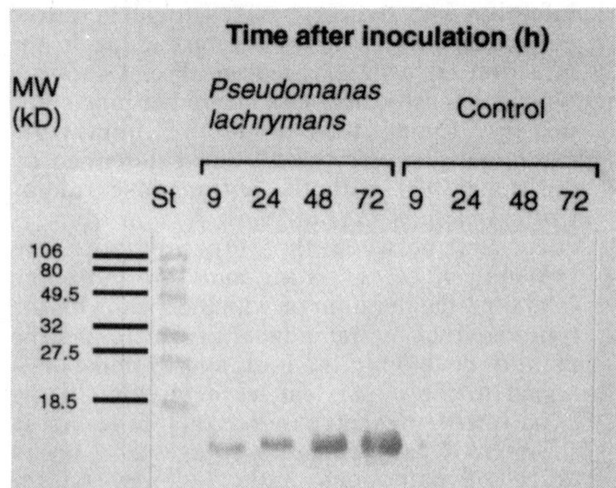

Fig.2: Time-course of the accumulation of PR1 in *Pseudomonas lachrymans*-infected cucumber leaves detected by immunoblotting using an antiserum against the major basic PR1 of tomato.

antisense technology did not increase the susceptibility of the plant (Neuhaus et al, 1992). Glucanase acts synergistically with chitinase against fungi *in vitro,* which might explain these results (Mauch et al, 1988). In other words, constitutive expression of both chitinase and glucanase might be required for improved disease control. Similarly, combinations of several hydrolytic enzymes have also been found to be very active against fungi (Sela-Burlage et al, 1993). Transformation of *Nicotiana sylvestris* to overexpress a basic chitinase from the closely related tobacco yielded no substantial resistance against attack by *Cercospora nicotianae* (Neuhaus et al, 1991). Overexpression of a bean chitinase in tobacco gave protection against the soil-borne fungus *Rhizoctonia solani* (Broglie et al, 1991). In this case, the pathogen might be confronted unexpectedly with an enzyme whose biochemical properties are different from the endogenous counterpart produced by the host plant. Tobacco overexpressing PR1a, another major PR-protein, shows resistance against oomycetes (Alexander et al, 1993). Overexpression of osmotin, a 24 kD PR-protein, delayed the expression of disease symptoms in potato infected with *Phytophthora infestans* (Liu et al, 1994). These data suggest that it is critical to test the expression of given genes or combinations thereoff in the appropriate host background and against appropriate pathogens. Taken together, these results indicate that the proteins expressed during SAR represent a potential source of antimicrobial plant components that can be used to improve substantially the defence of crop plants.

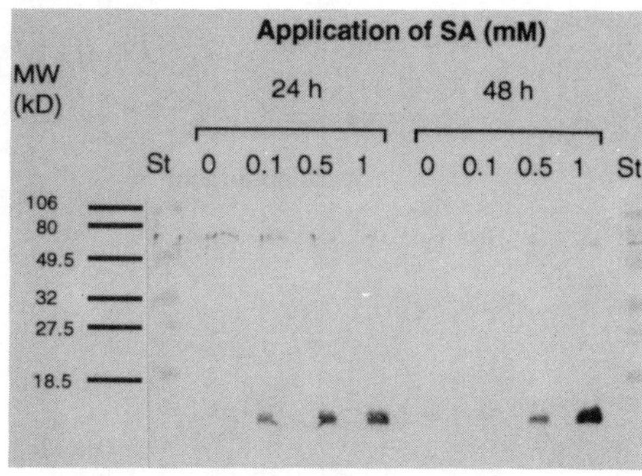

Fig.3: Time-course of the accumulation of PR1 in SA-treated cucumber leaves detected by immunoblotting using an antiserum against the major basic PR1 of tomato.

SAR signals

Systemic acquired resistance relies on the transmission of information between the infection sites and the uninfected parts of the plant. A putative endogenous signal was postulated to explain the systemic transmission of resistance during SAR. Salicylic acid (SA) has been proposed to be the signal for induced resistance based on the timing of its appearance, its endogenous levels and its effects on the plant (Malamy et al 1990, Métraux et al, 1990, Meuwly and Métraux, 1993, Rasmussen et al, 1991, and review by Raskin, 1992). SA acts as an inducer as demonstrated by its low antimicrobial potential and its marked ability to induce PR-proteins (Fig. 3). The signaling function of SA has been further investigated by transforming tobacco plants with a bacterial gene coding for salicylate hydroxylase, an enzyme catalysing the conversion of SA into the SAR-inactive molecule catechol. Such plants are unable to accumulate SA to a high level and cannot undergo SAR which indicates that SA is an indispensable ingredient in the transduction pathway (Gaffney et al, 1993). It remains to be seen whether SA is the primary systemic signal made within the first hours after infection or whether it is itself induced by another, yet unknown systemic signal (Lawton et al, 1993; Rasmussen et al, 1991). SA is a powerful substance to elucidate the signal transduction pathway of SAR. SA can also be used as a substituent for pathogen infection to search for genes involved in induced resistance. A number of other substances like methyl jasmonate, jasmonic acid, systemin and arachidonic acid have also been shown to reproduce the induction of defence reactions much like infection by pathogens or herbivores (Fig. 4, see below; Pearce et al, 1991). Such molecules will also add to the tools which can be used to search for genes causally linked to resistance.

Compounds such as SA represent interesting model compounds for the development of non-antibiotic crop protectants, which trigger the natural potential for resistance in various plants. By virtue of their mode of action and by analogy to pharmacopea used for that purpose in humans, such compounds might be termed immunostimulants. The synthetic inducer 2,6-dichloro-isonicotinic acid (INA, Fig 4) or the closely related N-phenylsulfonyl-2-chloro-

arachidonic acid

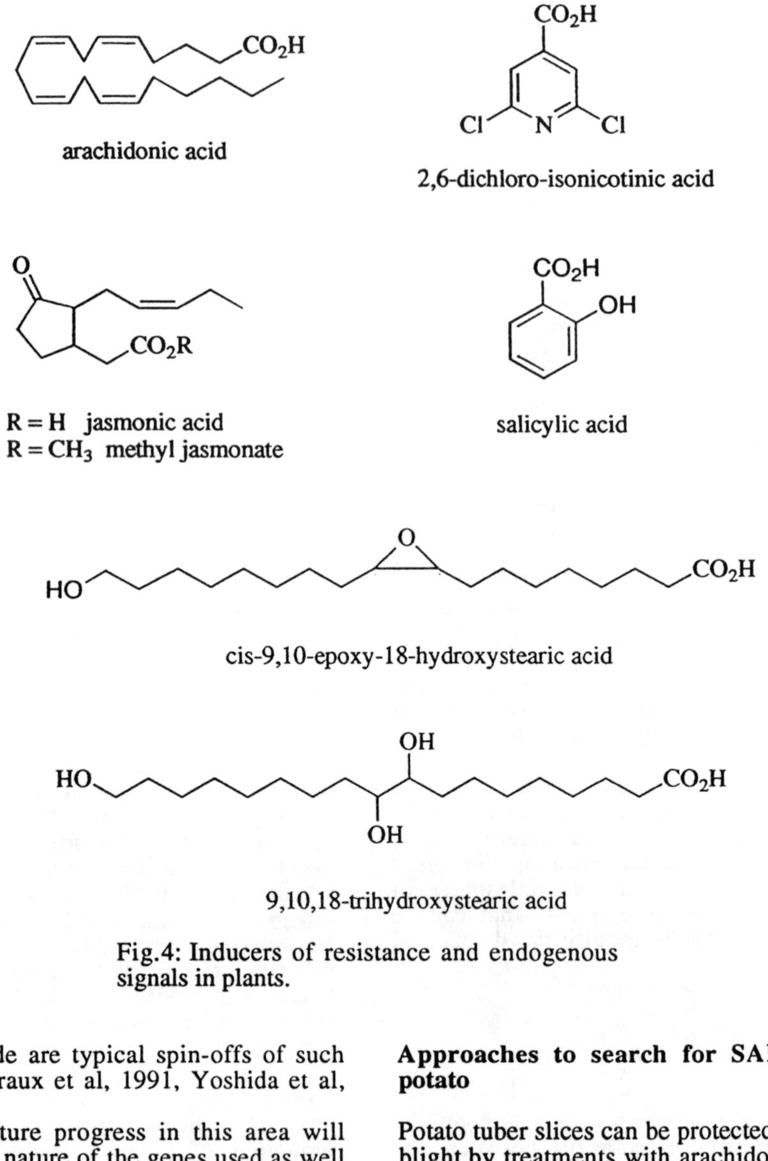

2,6-dichloro-isonicotinic acid

R = H jasmonic acid
R = CH_3 methyl jasmonate

salicylic acid

cis-9,10-epoxy-18-hydroxystearic acid

9,10,18-trihydroxystearic acid

Fig.4: Inducers of resistance and endogenous signals in plants.

isonicotinamide are typical spin-offs of such research (Métraux et al, 1991, Yoshida et al, 1990).

Clearly, future progress in this area will depend on the nature of the genes used as well as on their mode of expression in the plant (Lamb et al, 1992). In some cases constitutive expression might suffice whereas in others, organ specific expression might be needed. Another possibility is the use of promoters such as the prp1-1 promoter from potato (Martini et al, 1993), which are activated locally by a wide number of pathogens or immunostimulants. Resistance genes placed under the control of such promoters would only be expressed locally, at the site of attempted invasion. This is likely to be energetically less demanding for the plant than constitutive overexpression and might have advantageous consequences for crop yields.

Approaches to search for SAR genes in potato

Potato tuber slices can be protected against late blight by treatments with arachidonic acid and eicosapentaenoic acid. These fatty acids are components from the late blight fungus *Phytophthora infestans* and can be released by germinating sporangia during the early phases of infection (Ricker et al, 1992). Besides causing the accumulation of sesquiterpene phytoalexins at the site of application, these compounds were also shown to induce SAR against *P. infestans* in the foliage (Bostock et al, 1986; Cohen et al, 1991). Interestingly, arachidonic acid applied to the lower leaves of potato plants induces the accumulation of SA and resistance against *P. infestans* (Meuwly et al, 1993) as well as against the ascomycete *Alternaria solani* (Table 2). The same

Table 2: Protection of potato (*Solanum tuberosum cv Bintje*) against *Phytophthora infestans* or *Alternaria solani* by arachidonic acid

	Necrotic areas (mm^2/leaf) caused by	
Treatment	Phytophthora infestans	Alternaria solani
Control	282 (± 93)	61 (± 10)
Arachidonic acid^s (1500 ppm)	30 (± 9)	25 (± 6)
Arachidonic acid^i (10 ppm)	59 (± 21)	

General: Arachidonic acid was either sprayed on the leaves (^{s}) until runoff or applied by injection (^{i}) into the intercellular space of leaves using a cork-and-syringe device. Plants were all infected 4 days after treatment by spraying the leaves with a suspension of 5.10^4 sporangia/ml of *P. infestans*, or with a suspension of 5.10^4 spores/ml of *A. solani*, and resistance was measured 6 days later. The extent of the symptoms was determined by measuring the size of the necrotic areas (n = 5, ±SD).

treatment leads to the accumulation of PR1 protein, as evidenced by immunodetection using an antibody against the tomato P14a (data not shown), which has reportedly antioomycete properties (Niderman et al, 1993). Work is in progress in our laboratory to isolate genes induced by arachidonic acid and to determine their potential for disease resistance in transgenic plants.

Approaches to search for SAR genes in rice

The signaling events that lead to induction of defence genes and resistance in cereals, including rice, are poorly understood. We are interested in 2,6-dichloroisonicotinic acid (INA), cutin monomers and jasmonates as candidates for inducers of resistance. The approach is based on determining the resistance-inducing potential of INA, cutin monomers and jasmonates, and on following the pattern of induction of a set of selected markers for plant defence reactions. These markers are lipoxygenase (LOX, enzyme activity), pathogen-induced peroxidase (POX, mRNA levels), and extracellular PR1. The project is aimed at molecular cloning of rice genes, whose regulation are correlated with acquired resistance.

In monocotyledonous plants, comprising the worlds most important cereal crops, the mode of action of **INA** and the mechanism of SAR is less well understood than in dicotyledonous plants. However, INA induces resistance in rice to the rice blast fungus, *Magnaporthe grisea* (Métraux et al., unpublished), and we are presently characterizing changes in gene expression of INA-treated rice plants. Drench application of INA protected rice seedlings against infection by *M. grisea* (Table 3), which is in agreement with earlier results (Métraux et al., unpublished). The compound also induced accumulation of LOX, POX and PR1 (Table 3). The enhancement of LOX activity possibly leads to accumulation of antifungal, oxygenated C18 fatty acids (Kato et al, 1993), hexanal or hexenal (Croft et al, 1993). Another possibility of LOX function lies in participation in a jasmonate-mediated signaling pathway (Farmer and Ryan, 1992). Accumulation of the peroxidase mRNA could suggest cell wall modification (Reimmann et al, 1992). Finally, PR1 is a typical representative of the PR-response of plants. In general, the results demonstrate that INA induces SAR not only in dicotyledonous plants but also in rice.

In the last three years, increasing attention has been paid to the plant growth regulator **jasmonic acid (JA)** and its conjugates, especially after it was suggested that JA is involved in signal transduction in herbivore- (Farmer and Ryan, 1992) and pathogen- (Mueller et al, 1993) attacked cells. These recent findings make JA an excellent candidate as second messenger for different forms of plant stress. In pathogen-attacked rice, no direct evidence for such a function of JA is presently available, but several lines of indirect evidence make the hypothesis of JA-mediated signaling in rice attractive. First, *M. grisea* and INA enhance LOX activity (Table 3; Ohta et al,

Table 3: Host-gene expression and resistance induction in rice by different compounds.

Compound	Conc. (ppm)	Application	Induction of				
			LOX[a]	POX[b]	PR1[c]	% Resistance[d]	
INA	90	drench	++	+	+	78±6	(2)
MeJA	0.25[e]	gas (local)	n.t.	n.t.	++	-30±44	(2)
JA	1,000	spray local [f]	-	n.t.	++	n.t.	
JA	1,000	spray syst.[g]	-	n.t.	-	9±70	(10)
Palm.a.	10,000	spray local [h]	-	n.t.	-	15±50	(3)
HESA	250	spray local [h]	n.t.	-[i]	-[i]	55±5	(3)
CHA	1,000	spray local [h]	-	-	-	56±13	(3)
CHA	10,000	spray local [h]	-	-	-	66±28	(3)
M. grisea	-	spray local [k]	+	++	++	n.t.	

General: INA, 2,6-dichloro-isonicotinic acid; MeJA, methyl jasmonate; JA, jasmonic acid; Palm.a., palmitic acid; HESA, cis-9,10-epoxy-18-OH-stearic acid; CHA, cutin hydrolysate from apple fruit (alkaline hydrolysis); -, no induction; +, weak induction; ++ strong induction; n.t., not tested.

[a]Lipoxygenase enzyme activity 3 days after treatment/inoculation, using linoleic acid as substrate and measuring the increment of E_{234}.

[b]Peroxidase mRNA abundance on Northern blots 12 to 72 h after treatment/inoculation, using the rice cDNA "RIR3" as probe (Reimmann et al., 1992).

[c]Protein abundance on Western blots 3 days after treatment/inoculation, using an antiserum against the major basic PR1 ("P14a") from tomato as probe.

[d]Reduction of the number of acute lesions per leaf. Plants were challenge inoculated 3 days after treatment, except for MeJA (2 days after onset of gasing plants). The number of independent experiments is indicated in brackets. Mean values ± SE (2 Exp.) or ± SD (3 to 10 Exp.) are given.

[e]Plants were treated with 250 nl MeJA/l air, evaporating from a cotton ball in a closed plexiglas container.

[f]Whole plants were sprayed with JA in ethanol. Control plants were sprayed with solvent alone. Local protection by JA spray could not be tested due to direct antifungal activity of JA (Neto et al., 1991).

[g]The leaf-sheath zone at the base of the plants was sprayed, and the non-sprayed (covered) leaf blades were taken for analysis. This treatment was referred to as "spray systemic".

[h]Whole plants were sprayed with fatty acids in chloroform:methanol 2:1 (v/v), while control plants were sprayed with solvent alone.

[i]1000 ppm HESA was applied for these tests.

[k]Plants were inoculated with 5 x 10^5 spores/ml in (0.25% w/v gelatin, 0.05% w/v Tween 20). Control plants were mock inoculated.

1991) which is catalysing the first biosynthetic step from linolenic acid to JA (Farmer and Ryan, 1992). Second, precursors of JA accumulated in *M. grisea*-infected rice and exogenous application of these precursors led to accumulation of the phytoalexin momilactone A (Neto et al, 1991; Kodana et al 1991). Our laboratory is investigating the role of JA in local acquired resistance (LAR) and SAR in rice by a phytopathological and biochemical approach. Exogenous application of JA or methyl-jasmonate (MeJA) caused accumulation of PR1 (Table 3), and the accumulation of PR proteins was confined to the treated parts of the rice seedlings. However, when JA treated plants were challenge inoculated with *M. grisea*, no protection was obtained (Table 3). Both MeJA-treated leaves (LAR) and non-treated leaves of plants that were sprayed with JA at the leaf-sheath zone (SAR) were fully susceptible to *M.*

grisea. Therefore, JA itself is not sufficient to induce all host genes necessary for the establishment of LAR or SAR. Especially, a role of the JA-induced extracellular PR-proteins in defence against *M. grisea* is questionable. We are presently investigating whether JA is necessary for the onset of LAR or SAR by using inhibitors of JA biosynthesis and by measuring endogenous JA pools in pathogen- and INA-treated rice leaves. Nevertheless, knowing about the changes in host gene expression that are not correlated with induced resistance may help discriminating between relevant and irrelevant genes for later transgenic approaches aimed at higher resistance of rice to *M. grisea*.

The third type of inducers under investigation are cutin monomers. All aerial parts of terrestrial plants are covered by the polyester membrane cutin plus embedded waxes (reviewed by Kolattukudy, 1981). Pathogens, including *M. grisea* (Sweigard et al, 1992), possess cutinases, *i.e.* esterases capable of hydrolyzing cutin. In several cases rapid release of cutinase by fungal spores on leaf surfaces was demonstrated, which might facilitate spore adhesion or appressorial penetration of epidermal cell walls (reviewed by Mendgen and Deising, 1993). Cutin monomers as enzymatic breakdown products of cutin are inducers of cutinase gene expression in phytopathogenic fungi (see e.g. Bajar et al, 1991). However, the signaling potential of cutin monomers in plants is unknown. From an evolutionary point of view, and in analogy to the signaling potential of other breakdown products of plant cell walls (reviewed by Ryan, 1994), cutin monomers may well act as primary signals in plants, too, and we are currently testing this working hypothesis in the rice/*M. grisea* interaction. Exogenous application of synthetic cis-9,10-epoxy-18-hydroxystearic acid, the major component of cereal cutin (Espelie et al, 1979; Matzke and Riederer, 1990), protected rice against subsequent challenge infection by *M. grisea* (Table 3). The same result was obtained with a cutin hydrolysate from apple fruit that is rich in hydroxylated C18 fatty acids (Fig.5 and Table 3; Espelie et al, 1979). In contrast, no protection was obtained with palmitic acid (Table 3). None of the tested fatty acids had detectable antifungal activity against *M. grisea* *in vitro* at a concentration of 1 to 10 mg/plate (Fig.6). In the same *in vitro* test system, 100 µg Benomyl or 10 µg Prochloraz as fungicide standards inhibited conidial germination by 97% and 71%, respectively. Therefore, cis-9,10-epoxy-18-hydroxystearic acid and the cutin hydrolysate fulfilled the basal requirements for inducers of LAR or SAR, i.e. protection *in planta* but lack of antifungal activity *in vitro*. On the other hand, expression of the selected marker genes in rice was not influenced by these protective agents (Table 3), thereby obscuring their mode of action. It may

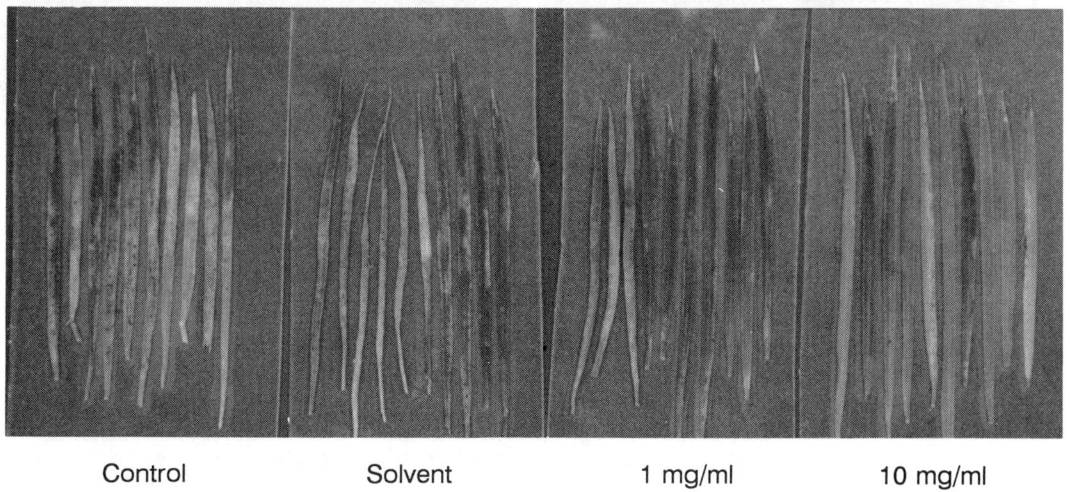

Fig.5: Protection of first leaves of young rice seedlings by spray application of apple cutin hydrolysate (CHA) 3 days prior to challenge inoculation with *Magnaporthe grisea* (2×10^5 spores/ml). Solvent = chloroform:methanol 2:1 (v/v). The picture was taken 5 days after inoculation. Some solvent-treated leaves completely collapsed due to confluent, acute lesions by *M. grisea*.

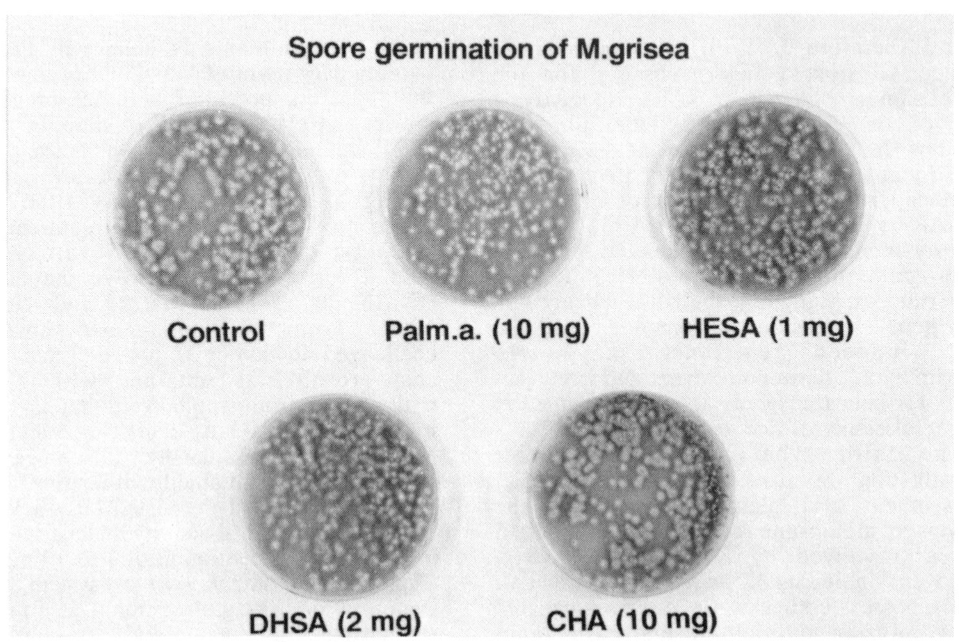

Fig.6: Lack of fungicidal activity of cutin monomers against *M. grisea in vitro*. Rice-agar plates were spray-coated with the indicated amounts of fatty acids. After evaporation of the solvent, the plates were inoculated with 500 conidia of *M. grisea*, and 4 days later the number of conidia-derived fungal colonies were counted. None of the tested fatty acids inhibited conidial germination and subsequent colony growth to any extent. The germination rate on the solvent-treated control plate was 60%. Palm.a., palmitic acid; HESA, cis-9,10-epoxy-18-OH-stearic acid; DHSA, erythro-9,10-dOH-stearic acid; CHA, apple fruit cutin hydrolysate.

be speculated that cutin monomers act on a different signaling pathway than e.g. INA or JA, and that they induce also a different set of, yet unknown, rice gene products. However, 2-dimensional analysis of *in vitro* translation products revealed no cutin-monomer induced changes in rice gene expression, suggesting a rather low abundance of putative cutin-monomer induced mRNAs.

In conclusion, based on their potential to induce resistance in rice to its most important fungal pathogen, *M. grisea*, INA and cutin monomers appear to be interesting molecules in the search for new defence genes in this world crop. The envisaged differential cloning strategy for inducer-activated mRNAs opens a way to finding new, cryptic defence genes that can be functionally tested in transgenic rice lines.

Approaches to search for SAR genes in barley

Plants under heat stress may become more susceptible (see e.g. Hazen and Bushnell, 1983; Ouchi et al, 1977) or more resistant (Abbatista et al, 1988; Stermer and Hammerschmidt, 1984 and 1987) to pathogen attack, and the reasons for this controversial effect are still unknown .

In barley, the second cereal crop under investigation in our laboratory, we are describing the mechanism of LAR after a pulse heat shock of 50°C for 30 to 60 seconds. This heat treatment induces a rapid, efficient and transient LAR in barley leaves to subsequent infection by the barley powdery mildew fungus *Erysiphe graminis* f.sp. *hordei* (Fig.7). Microscopic analysis revealed no inhibition of germination of *E. graminis* on heat-treated leaves, but arrest of fungal development after appressorium formation (Fig.8). The fact that *E. graminis* formed two to three appressorial lobes, each reflecting a penetration attempt, on heat-treated leaves indicates resistance of epidermal cell walls to penetration. We also followed regulation of two markers for LAR, peroxidase and PR1, in heat-treated and in challenge inoculated leaves (Table 4). Twentyfour hours after heat-treatment when resistance to penetration was manifest, neither peroxidase activity nor PR1 levels were enhanced, irrespective of whether the plants were challenge- inoculated or not. On the other

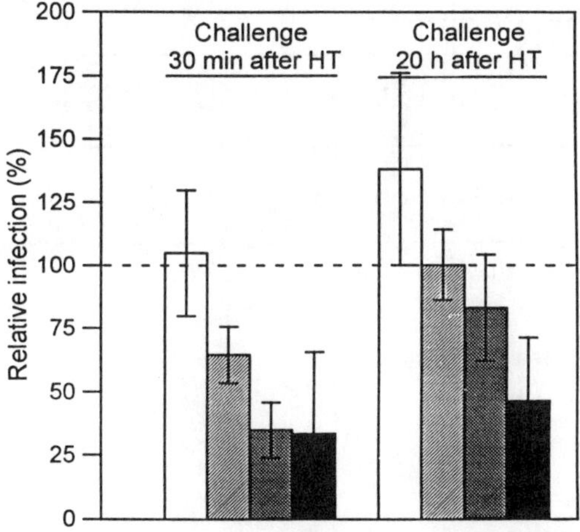

Fig.7: Heat-induced resistance in barley to *Erysiphe graminis* f.sp. *hordei*. First leaves were dipped in water of 50°C for 30 s (▨) 40 sec (▩) or 60 sec (■), prior to challenge inoculation. Control leaves were dipped in water of 20°C for 60 sec (□). Resistance is reflected by a low relative infection, compared to non-treated control plants (100%). HT = heat treatment. Mean values ± SD of 3 to 9 independent experiments are shown.

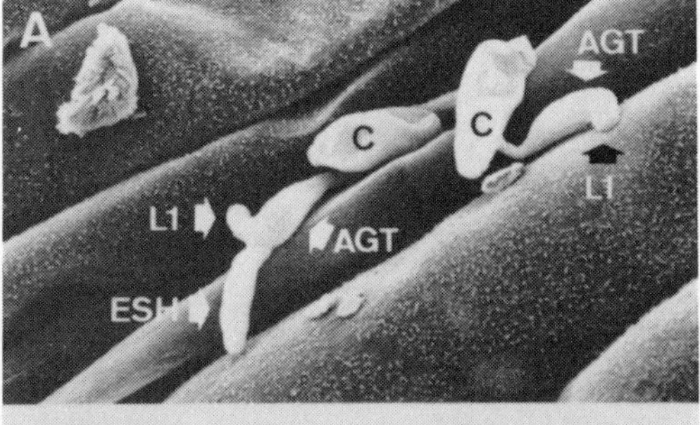

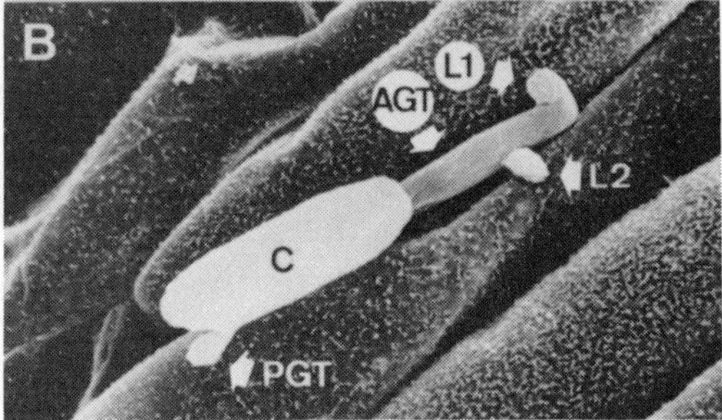

Fig.8: Normal fungal development of *Erysiphe graminis* f.sp. *hordei* on heat-treated barley leaves, but failure of haustorium formation (scanning electron microscopy of inoculated leaf surfaces). A: control leaves; B: heat-treated leaves (60 sec at 50°C). The failure of haustorium formation is reflected by the lack of secondary elongating hyphae (ESH). Repeated penetration attempts are reflected by the formation of two appressorial lobes (L1 and L2) on heat-treated leaves. C, conidium; PGT: primary germ tube; AGT: appressorial germ tube.

hand, 14 hours and 24 hours after heat treatment a number of new mRNAs are accumulating in the treated leaves, as reflected by the 2-dimensional pattern of *in vitro* translation products (Table 4). The pattern of heat-induced mRNAs differed largely from the pattern of mRNAs induced during the compatible interaction with *E. graminis*, with marginal overlap. The heat-induced pattern also differed from the pattern induced during a typical heat-shock, i.e. incubation of plants at 38°C for 4h, although both treatments induced HS18 family of heat-shock proteins. Remarkably, the typical heat shock did not induce resistance against *E. graminis* (reviewed by Nover et al, 1989). We do not know yet whether the pattern of heat-induced mRNAs includes "typical" heat-shock proteins of the HSP 18-, HSP70- or HSP90-families (reviewed by Nover et al, 1989).

In conclusion, the described heat-induced resistance seems to be based on a new defence mechanism that is correlated with activation of a set of host genes different from the well-known PR proteins or peroxidases, which are also induced during the compatible interaction with *E. graminis* and are therefore not indicative for resistance. Molecular cloning of heat-induced mRNAs is in progress and will provide putative transgenes for the creation of disease-tolerant plants.

Conclusions

Plant diseases have plagued farmers ever since the advent of agriculture. To manage pathogens various methods have been developed, ranging from simple phytosanitary measures to the use of chemicals and resistant varieties. Chemical pest control has gained a prominent place in modern agricultural practice. However, growing concerns about pesticide overuse leading to undesirable accumulation of residues in the environment led to the formulation of the Integrated Pest Management concept (as discussed by Schwinn 1988). This approach consists in a comprehensive use of all currently available methods including forecasting, sanitation methods, chemical control and use of

Table 4: Host-gene expression in heat-treated or infected barley leaves.

Treatment	Time (h)	Induction of		
		POX[a]	PR1[b]	mRNAs (number)[c]
Heat[d]	14	n.t.	n.t.	14
E. graminis[e]	14	n.t.	n.t.	14
Heat[d]	24	-	-	19
E. graminis	24	++	+	8
Heat + *E. gr.*[f]	24	-	-	n.t.

General: Primary leaves of 6-day old barley seedlings were heat-treated or inoculated, and the parameters were measured in the treated leaves (local effect).

[a]Peroxidase enzyme activity, using guaiacol as substrate.

[b]Protein abundance on Western blots, using an antiserum against the major basic PR1 ("P14a") from tomato as probe.

[c]Number of new or enhanced *in vitro*-translation products on 2-dimensional polyacrylamide gels.

[d]Primary leaves were heat treated by dipping in water of 50°C for 60 s.

[e]Primary leaves were inoculated by dusting conidia from heavily infected barley over the test plants. This inoculation resulted in a confluent lawn of mycelium on test plants 5 to 6 days after inoculation.

[f]Plants were heat treated and subsequently challenge inoculated as soon as the leaves were air dry.

resistant varieties. The use of crops engineered with genes for resistance can readily be included in this approach. The basic protection afforded by such plants is anticipated to decrease the need for intensive chemical treatments, with the exception, perhaps, of situations of extreme disease pressure.

Acknowledgements

We wish to acknowledge the financial support from the Swiss National Science Foundation (grant 31-34098), the Rockefeller Foundation as well as supports from CIBA and Sandoz-Agro Corporations. We thank Dr. L. Sticher for helpful comments and critical reading of the manuscript.

References

Abbattista Gentile, I.; Ferraris, L.; Matta, A. *J. Phytopathology* **1988**, *122*, 45-53.

Alexander, D.; Goodman, R.M.; Gutrella, M.; Glascock, C.; Weymann, K.; Friedrich, L.; Maddox, D.; Ahl-Goy, P.; Luntz, T.; Ward, E.; Ryals, J. *Proc Natl Acad Sci USA*. **1993**, *90*, 7327-7331.

Bajar, A.; Podila, G.K.; Kolattukudy, P.E. *Proc. Natl. Acad. Sci. USA* **1991**, *88*, 8208-8212.

Bostock, R.M..; Schaeffer, D.A..; Hammerschmidt, R. *Physiol. Molec. Plant Pathol.* **1986,** *29*, 349-360.

Broglie, K.; Chet, I.; Holliday, M.; Cressman, R.; Biddle, P.; Knowlton, S.; Mauvais, C.J.; Broglie, R. *Science*. **1991**, *254*, 1194-1197.

Cohen,Y.; Gisi, U.; Mösinger, E. *Physiol. Plant Pathol.* **1991**, *38*, 255-263.

Croft, K.P.C.; Jüttner, F.; Slusarenko, A.J. *Plant Physiol.* **1993**, *101*, 13-24.

Espelie, K.E.; Dean, B.B.; Kolattukudy, P.E. *Plant Physiol.* **1979**, *64*, 1089-1093.

Farmer, E.F.; Ryan, C.A. *Plant Cell* **1992**, *4*, 129-134.

Gaffney, T.; Friedrich, L.; Vernooij, B.; Negrotto, D.; Nye, G.; Uknes, S.; Ward, E.; Kessmann, H.; Ryals, J. *Science*. **1993**, *261*, 754-756.

Hazen, B.E.; Bushnell, W.R. *Physiol. Plant Pathol.* **1983**, *23*, 421-438.

Kato, T.; Yamaguchi, Y.; Namai, T.; Hirukawa, T. *Biosci. Biotech. Biochem.* **1993**, *57 (2)*, 283-287.

Kodama, O.; Akatsuka, T. *Agric. Biol. Chem.* **1991**, *55 (4)*, 1041-1047.

Kolattukudy, P.E. *Ann. Rev. Plant Physiol.* **1981**, *32*, 539-67.

Lamb, C.J.; Ryals, J.A.; Ward, E.R.; Dixon, R.A. *Biotechnology* **1992**, *10*, 1436-1445.
Lawton, K.; Vernooij, B.; Friedrich, L.; Gaffney, T.; Alexander, D.; D., N.; Métraux, J.P.; Kessmann, H.; Gut Rella, M.; Uknes, S.; Ward, E.; Ryals, J. *Plant signals in interactions with other organisms;* Am. Soc. Plant Physiol.: **1993**, pp. 126-133
Liu, D.; Raghothama, K.G.; Hasegawa, P.M.; Bressan, R.A. *Proc. Natl. Acad. Sci. USA* **1994,** *91*, 1888-1892.
Madamanchi, N.R.; Kuc, J. *The fungal spore and disease initiation in plants and animals;* Plenum Press: New York, **1991**, pp. 347-362.
Malamy, J.; Carr, J.P.; Klessig, D.F.; Raskin, I. *Science*. **1990**,*250*, 1002-1004.
Martini, N; Egen, M.; Rüntz, I.; Strittmatter, G. *Mol.Gen. Genet.* **1993**, *236*, 179-186.
Matzke, K.; Riederer, M. *Planta* **1990**, *182*, 461-466.
Mauch, F.; Mauch-Mani, B.; Boller, T. *Plant Physiol*. **1988**, *88*, 936-942.
Mendgen, K.; Deising, H. *New Phytol.* **1993**, *124*, 193-213.
Métraux, J.P.; Ahl-Goy, P.; Staub, T.; Speich, J.; Steinemann, A.; Ryals, J.; Ward, E. *Adv. Molec. Genetics Plant-Microbe Interactions;* Kluwer: Dordrecht, **1991**, pp. 432-439.
Métraux, J.P. *Immunology, a comparative approach;* Wiley: Chichester, **1994**, 1-28
Métraux, J.P.; Ryals, J.; Ward, E. *McGraw -Hill yearbook of Science and Technology;* McGraw-Hill: New York, **1993**, pp. 322-325.
Métraux, J.P.; Signer, H.; Ryals, J.; Ward, E.; Wyss-Benz, M.; Gaudin, J.; Raschdorf, K.; Schmid, E.; Blum, W.; Inverardi, B. *Science*. **1990**, *250*, 1004-1006.
Meuwly, P.; Métraux, J.P. *Anal Biochem.* **1993**,*214*, 500-505.
Meuwly, P.; Sticher, L.; Mölders, W.; Summermatter, K.; Coquoz, J.L.; Buchala, A.; Métraux, J.P. *Plant Signals in Interactions with other Organisms.;* Am. Soc. Plant Physiologists: **1993**, pp. 54-64
Müller, M.J.; Brodschelm, W.; Spannagl, E.; Zenk, M.H. *Proc. Natl. Acad. Sci USA* **1993**, *90*, 7490-7494.
Neto, G.C.; Kono, Y.; Hyakutake, H., Watanabe, M.; Suzuki, Y.; Sakurai, A. *Agric. Biol. Chem.* **1991**, *55*, 3097-3098.
Neuhaus, J.-M.; Flores, S.; Keefe, D.; Ahl-Goy, P.; Meins, J.F. *Plant Mol. Biol.* **1992**, *19*, 803-813.
Neuhaus, J.M.; Ahl-Goy, P.; Hinz, P.; Flores, S.; Meins, F. *Plant Mol. Biol.* **1991**, *16*, 141-151.
Niderman, T.; Bruyère, T.; Gügler, K.; Mösinger, E. In: Fritig, B.; Legrand, M., eds, *Mechanisms of plant defence responses* **1993**, Kluwer, p. 450.
Nover, L.; Neumann, D.; Scharf, K.D. In: Nover, L.; Neumann, D.; Scharf, K.D., eds, *Heat Shock and Other Stress Response Systems of Plants, Results and Problems in Cell Differentiation* **1989**, *Vol. 16*, Springer-Verlag, Heidelberg, FRG, chapter *B1.1*, pp. 5-13.
Ohta, H.; Shida, K.; Peng, Y.-L.; Furusawa, I.; Shishiyama, J.; Aibara, S.; Morita, Y. *Plant Physiol*. **1991**, *97*, 94-98.
Ouchi, S.; Nakabayashi, H.; Oku, H. *Ann. Phytopath. Soc. Japan* **1977**, *43*, 455-461.
Pearce, G.; Strydom, D.; Johnson, S.; Ryan, C.A. *Science* **1991**, *253*, 895-898.
Raskin, I. *Annu Rev Plant Physiol*. **1992**,*43*, 439-463.
Rasmussen, J.B.; Hammerschmidt, R.; Zook, M.N. *Plant Physiol*. **1991**, *97*, 1342-1347.
Reimmann, C.; Ringli, C.; Dudler, R. *Plant Physiol.* **1992**, *100*, 1611-1612.
Ricker, K.E.; Bostock, R.M. *Physiol. Molec. Plant Pathol.* **1992**, *41*, 61-72.
Ryals, J.; E. Ward, E.; Ahl-Goy, P.A.; Métraux, J.P. *The Biochemistry and Molecular Biology of Inducible Enzymes and Proteins in Higher Plants;* Cambridge Press: **1992,** pp. 205-229.
Ryan, C.A. *Proc. Natl. Acad. Sci. USA,* **1994**, *91*,1-2.
Schwinn, F.J. *Ecol. Bulletins*. **1988**, *39*, 82-88.
Sela-Burlage,M.B.; Ponstein,A.S.; Bres-Vloeman, S.A.; Melchers, L.S.; Van den Elzen, P.J.M.; Cornelissen, B.J.C. *Plant Physiol.* **1993**,*101*, 857-863.
Stermer, B.A.; Hammerschmidt, R. *Physiol. Mol. Plant Pathol.* **1987**, *31*, 453-461.
Stermer, B.A.; Hammerschmidt, R. *Physiol. Plant Pathol.* **1984**, *25*, 239-249.
Stintzi, A.; Heitz, T.; Prasad, V.; Wiedemann-Merdinoglu, S.; Kaufmann, S.; Geoffroy, P.; Legrand, M.; Fritig, B. *Biochimie* **1993**, *75*, in press
Sweigard, J.A.; Chumley, F.G.; Valent, B. *Mol. Gen. Genet.* **1992**, *232*, 174-182.
Uknes, S.; Mauch-Mani, B.; Moyer, M.; Potter, S.; Williams, S.; Dincher, S.; Chandler, D.; Slusarenko, A.; Ward, E.; Ryals, J. *Plant Cell* **1992**, *4*, 645-656.
Uknes, S.; Winter, A.M.; Delaney, T.; Vernooij, B.; Morse, A.; Friedrich, L.; Nye, G.; Potter, S.; Ward, E.; Ryals, J. *Plant Cell*. **1993**, *6*, 692-698.
Ward, E.R.; Uknes, S.J.; Williams, S.C.; Dincher, S.S.; Wiederhold, D.L.; Alexander, D.C.; Ahl-Goy, P.; Métraux, J.P.; Ryals, J.A. *Plant Cell*. **1991**, *3*, 1085-1094.
Weete, J. *Physiol. Mol. Plant Pathol.* **1992**, *40*, 437-445.
Yoshida, H.; Konishi, K.; Nakagawa, T.; Sekido, S.; Yamaguchi, I. *J. Pesticide Sci.* **1990**, *15*, 199-203.

Regulatory Aspects in the European Union of Foods Derived from Modern Biotechnology*

Dr. K. Mehta, Biotechnology and Agro-Processing Industry Division, European Commission, 200 Rue de la Loi, 1049 Brussels, Belgium

The present regulatory oversight in the European Union applicable to foods and food ingredients derived from Biotechnology is being reexamined in order to integrate food safety assessment and environmental assessment.
General principles of regulatory philosophy as they are currently in force are first briefly reviewed. The discussion of the main elements of the new legislation focuses on the scope of the legislation i.e. the categories of foods and food ingredients that would henceforth require a pre-market scientific assessment and the administrative procedure that applications to place new foods and food ingredients, falling under the scope of the foreseen legislation, would have to undergo. The close and complementary link with the regulatory oversight applied to transgenic seed varieties and the links with the present overall biotechnology framework is elucidated so as to clarify the important concept of a single notification and a single authorisation that underlies biotechnology regulatory oversight in the European Union.
More detailed presentation is outlined regarding the current and foreseen regulatory scope for foods and food ingredients derived from modern biotechnology. In particular it is pointed out that foods and food ingredients containing or consisting of viable genetically modified organisms are treated as a category distinct from foods and food ingredients that are produced from or isolated from genetically modified organisms. The essence of the approach in the Union is to ensure for products derived from biotechnology, a regulatory oversight that is identical to that applied to products destined for the same end-use and which in particular is based on the classical criteria of safety, efficacy and quality.
International regulatory practice is briefly reviewed by reference to the work in international fora such as OECD and the valuable role of internationally developed guidelines based on specific case studies of food assessment is highlighted.

The intensive scientific research over the last decade and the impressive rate of innovative activity in a number of product sectors where the application of modern biotechnology plays a crucial role, has brought biotechnology to the threshold of a period of major product commercialisations. This is particularly the case of products of agricultural biotechnology.

Issues of Regulatory Philosophy

A number of issues are raised for regulatory oversight. A basic question is whether a specific regulatory oversight is required.

Secondly, given the impact of modern biotechnology across several distinct sectors such as seed varieties, agricultural pesticides,

* The views expressed are those of the author and donot necessarily reflect that of the European Commission

1054–7487/95/0236$12.00/0

foods and food ingredients, there is clearly a need to ensure, wherever there are overlaps with existing legislation, that there are some essential linkages between the legislations.

A third issue is that the global nature of the market for products derived from biotechnology implies a need to take into account international practice and facilitate international harmonisation.

Current Situation in the European Union

The present regulatory situation in the Union is evolving. For the moment all foods and food ingredients containing or consisting of viable genetically modified organisms (gmos) fall under the scope of Directive 90/220 on the deliberate release of genetically modified organisms. The scope of this Directive covers deliberate release for R & D or field trial purposes. It also covers placing on the market of gmos unless there exists, for the product in question, specific product legislation pertaining to pre-market approval / authorisation. For example as there exists legislation for the placing on the market of pharmaceuticals, biopharmaceuticals would be subject to the same product legislation and fall out of the scope of the relevant part of the Directive 90/220. The essence of the approach in the Union is to ensure for products derived from biotechnology, a regulatory oversight that is identical to that applied to products destined for the same end-use and which in particular is based on the classical criteria of safety, efficacy and quality. At the same time Directive 90/220 serves as the relevant regulatory framework where product-specific legislation is not in place or not considered necessary.

Foods Derived from Modern Biotechnology

It is useful to distinguish between foods and food ingredients containing or consisting of viable gmos from those produced or isolated from genetically modified organisms. As the foregoing suggests the first category, foods and food ingredients containing or consisting of viable gmos, are already subject to regulatory oversight in the European Union because they fall within the scope of Directive 90/220. The major consideration for substituting an alternative and product-specific regulatory frame for such foods and food ingredients is so as to apply the same food safety assessment as would be applicable in general to equivalent foods and food ingredients. The second category, of foods and food ingredients derived from biotechnology which are produced from genetically modified organisms, is not currently subject to any regulatory oversight. The obvious reason for extending regulatory oversight to such foods and food ingredients is to satisfy the need for a food safety evaluation of foods and food ingredients that present significant differences as compared with conventional foods or food ingredients in terms of nutritional value, composition, metabolism, intended use or the level of undesirable substances therein.

The discussion above presents the árguments based on the grounds of food safety evaluation that justify regulatory oversight of foods and food ingredients. The issues raised are however not confined just to foods and food ingredients derived from modern biotechnology. There are other not hitherto consumed sources of foods and food ingredients other than those derived from biotechnology, that are being placed on the market. In addition, processing technologies not previously commercially used in food processing, are finding applications in food production. It is these considerations that underly the proposed legislation on Novel Foods and food ingredients. Its scope is wider than foods derived from gene technology : the legislation currently under discussion is product-specific rather than technology-specific. It is also worth noting that the proposed Novel food and food ingredients regulation introduces for whole foods and food ingredients, not hitherto consumed, a food safety assessment that todate has only been systematically applied to food additives, for foodstuff flavourings, for extraction solvents used in foodstuff manufacture etc. Such food safety assessment shall complement the existing requirement on food manufacturer to ensure that foods and food ingredients are wholesome.

Regulation on Novel Foods and Food Ingredients

The draft regulation currently under discussion addresses 3 key questions
- the scope defining the novel foods and food ingredients subject to pre-market notification and authorisation
- the procedure of notification and authorisation for the placing on the Union market
- provisions relating to labelling of novel foods and food ingredients.

Scope

The notion of novelty on one hand provides a neat way of ensuring that regulatory oversight is extended only to new foods and food ingredients, and where safety and public health considerations justify it; on the other hand it is a notion that is difficult to define, particularly given that technical progress is a continuing process. The approach to resolving these difficulties in the regulation is that of an implicit decision tree which is elaborated on the basis of two key criteria. The first criterion of novelty is a general one, what might be termed a sufficient condition : a food or food ingredient is novel and hence falls under the proposed regulatory oversight if it has not been consumed to a significant degree hitherto in the European union. If the food has been significantly consumed elsewhere then the relevant food assessment information would be appropriately taken into account. The second criterion, a necessary condition, is that of significant difference as compared to conventional foods or food ingredients. The significant difference is defined with respect to properties such as composition, nutritional value, metabolism, intended use or the level of undesirable substances therein. Such significant difference may be present in an unprocessed food or food ingredient or could result from the application of a method of processing not hitherto used commercially in food processing. In the latter case, the processed food or food ingredients that present significant differences in comparison with conventionally processed foods or food ingredients also fall under the scope of the regulation.

In addition to these general criteria defining novelty, foods or food ingredients must fall in a certain number of categories laid down in the text of the draft regulation.

The categories of foods and food ingredients currently in discussion are :

(a) Foods and food ingredients containing or composed of <u>viable</u> genetically modified organisms and which have not hitherto been placed on the market of the European union

(b) Foods and food ingredients produced from genetically modified organisms except for those, which by comparison with conventional foods or food ingredients, are not significantly different in a number of properties that are listed.

(c) Foods and food ingredients with a new or intentional modified molecular structure

(d) Single-cell proteins

(e) Foods and food ingredients that have undergone a process not currently used in commercial food production and where such process gives rise to significant change in properties in comparison with conventionally processed foods and food ingredients.

It is useful to make 3 important clarifying observations. The Novel Foods and food ingredients regulation, as the title implies, covers food and food ingredients. Additives, flavourings and extraction solvents, and processing aids whatever their method of manufacture, fall under other specific and existing legislation irrespective of the technology of production. In the second place, the regulation doesnot just cover biotechnology foods and food ingredients and hence is not technology-specific. Thirdly concepts such as significant difference have to be interpreted in line with internationally established scientific guidelines and practices.

From the viewpoint which is the focus of this conference, it can be concluded from the foregoing that biotechnology foods or food ingredients that contain or are composed of viable genetically modified organisms would fall in the scope of the new regulation and would be subjected to scientific evaluation and authorization unless previously placed on the market in the European Union. This would be the case of e.g. a transgenic tomato. As regards foods and food ingredients produced from but not containing or composed of genetically modified organisms, they would be required to be pre-notified and subjected to a scientific evaluation unless, on the basis of available scientific evidence, they do not significantly differ from conventional foods or food ingredients.

Furthermore, the proposed Novel Foods and food ingredients regulation, in conformity with the current regulatory policy of single notification and single authorisation in the European Union regarding biotechnology foresees a close link with the proposed amendments to the directives on seed varieties as well as the basic gmo directives. A submission for approval of a transgenic seed variety under the seed directive would imply that the modified plant and the foods derived would in parallel be scrutinised as a novel food and the decision on the seed variety submission would state that it is safe for the consumer and from the point of view of the environmental assessment required under Directive 90/220. In this way duplicate notifications and authorisations can be avoided.

The attached table graphically illustrates the integration between the existing and proposed legislation in order to implement the single notification and single authorisation policy.

Procedure for Scientific Evaluation, Notification and Authorization

The preceeding presentation of the scope of the Novel Foods and food ingredient regulation had the aim of defining the categories of foods and food ingredients which need to undergo a pre-market scientific evaluation.

The procedure that is currently proposed for this scientific evaluation has a number of key elements. In the first place, it has a preponderant decentralised element since the initial assessment is undertaken by national competent food assessment bodies nominated by Member States authorities. Secondly because the decision to place on the market is valid throughout the Union, all other Member States' authorities and the European Commission are submitted from the outset a summary of the application. Finally a formal decision is foreseen for those applications that raise issues of public health.

LINK BETWEEN THE PROPOSED NOVEL FOODS REGULATION, THE PROPOSED SEEDS DIRECTIVE AND DIRECTIVE 90/220/EEC.

	PROPOSED SEEDS DIRECTIVE	PROPOSED NOVEL FOODS REGULATION	DIRECTIVE 90/220
Transgenic seed intended for food use	**Applies,** including: -a food safety assessment as in NF, - a risk assessment similar to 90/220.	**Integrated**	**Integrated**
Transgenic plant (not seed) intended for food use.	**Does not apply**	**Applies,** including: -a risk assessment similar to 90/220.	**Integrated**

Explanatory note:

Colour-shaded areas show pieces of legislation which do not formally apply, since their risk/safety assessments have been integrated into the piece of legislation which actually applies.

For the more straight forward applications the procedure involved is that of pre-market notification and if the initial assessment report is favourable and if no objections are raised within a fixed period then the applicant is advised that marketing can commence.

Bearing in mind these general key principles, the procedure in detail particularly for transgenic foods or food ingredients would follow the main steps indicated below.

the applicant would submit a complete file, including the relevant tests, scientific data to the European Commission. For transgenic foods this would include a written consent of the competent authority for Directive 90/220 or the results of any field trials for research and development purposes and which constitute the required environmental risk assessment. The applicant is also required to submit a proposal for labelling in conformity with the provisions of the Novel Foods and food ingredient regulation.

an initial scientific assessment by a competent national food assessment body would be carried out within two months of the submission of the request. This body would in particular evaluate that the food is safe, that it is similar to foods or food ingredients it is intended to replace, and that the consumer is not misled and finally also give an opinion whether a full scientific evaluation is needed or not.

the initial scientific assessment report would then be circulated by the European Commission to all Member States for observations within a fixed period. If either the initial scientific assessment report or any subsequent observations raise issues of effects on public health then the application is referred to the Scientific Committee for Food. This is an advisory committee at the Community level for food safety questions, and has the function of undertaking full scientific evaluation required under food legislation. The resulting opinion would be the basis for any regulatory decision.

for a food or food ingredient containing or composed of a viable genetically modified organism the procedure foresees also the necessary consultations with the bodies set up for the environmental assessment.

The final decision on an application would therefore cover the food safety as well as the environmental aspects by consulting appropriately the respective bodies that exist for these aspects. The decision would also indicate where necessary the conditions of use and the labelling requirements.

Labelling Questions

The provisions on labelling cover not just foods derived from modern biotechnology but all Novel Foods and food ingredients. As a general rule the provisions on the labelling of foodstuffs contained in the Directive 79/112 have to be respected. In addition a case by case approach is foreseen whereby those Novel foods or food ingredients that present significant differences in characteristics when compared with equivalent conventional foods or food ingredients that specific additional labelling would be required. The applicant would include a justified proposal which would be examined and the decision on the application would indicate the additional labelling that is required. The current proposal for the Novel Food and food ingredient does not foresee a technology-specific label for foods but a functional label.

Conclusions

The regulatory aspects that are raised for foods derived from modern biotechnology are indeed those that have been discussed in international fora such as OECD. Certain broad principles such as the concept of substantial equivalence that emerged from those discussions in the past suggest the regulatory approach to follow. It is clear that since many of the products that result from the application of modern biotechnology fall in categories of products that are already under regulatory oversight, it is necessary that an integrated regulatory framework is put into place by amending and by complementing the existing legislation.

This is the approach that currently underlies the legislation that is presently under discussion in the European Union and elsewhere. It is also possible to identify similar elements in the regulatory oversight : categories of foods and food ingredients that are covered vary but always include transgenic foods and seeds; a pre-market scientific evaluation is postulated and finally a case by case approach is retained for labelling. There exists further scope for international harmonisation and which could probably be more fully exploited when there are at international level more detailed guidelines and case studies on the scientific evaluation of foods from modern biotechnology.

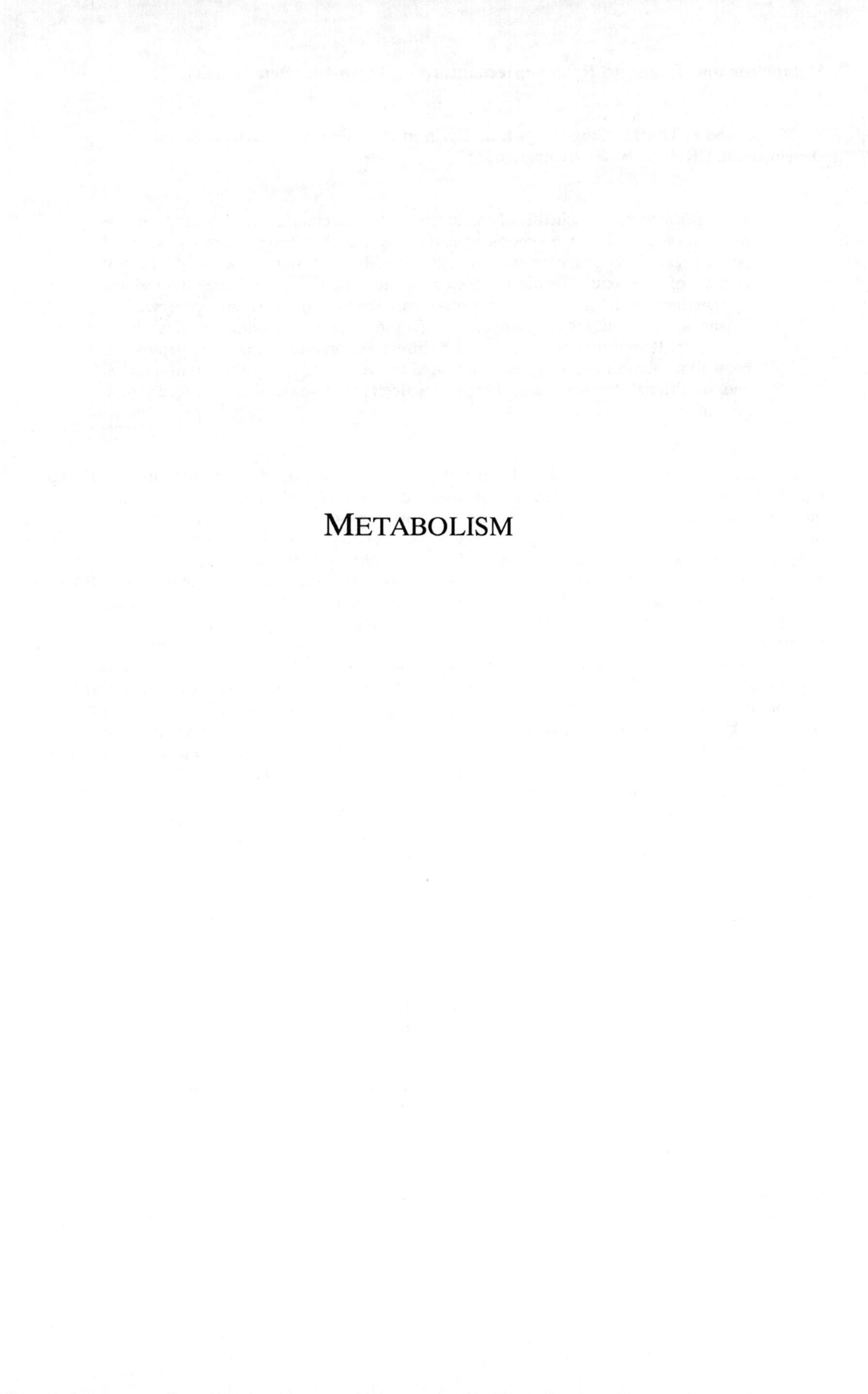

METABOLISM

Metabolism and Transport Related Mechanisms in Organ-Directed Toxicity

W. Mücke and **P. Thanei,** Ciba-Geigy Ltd., Division Plant Protection, Research and Development, CH-4002 Basle, Switzerland

The manifestation of chemically induced organ-directed toxicity depends on the inherent properties of the xenobiotic and on certain characteristics of the potential target organ. The link between exposure of a living system to a xenobiotic and arrival of a toxic chemical species in the target organ are interrelated biotransformation and transport processes. These processes are governed by balances of various competing factors (e.g. absorption/elimination, activating/detoxifying enzymes). Slight differences in any of the basic parameters controlling these processes, as determined by sex, species, strain, environmental and nutritional factors may change drastically the toxicological profile of a xenobiotic.

The extent of a chemical-induced toxicological response is generally determinated by the concentration of the toxicant in the target organ and the duration of its exposure.

Prior to arrival at the target organ the toxicant or its precursor has to pass - passively or actively - various barriers. Somewhere during this transport the ultimate toxicant may be generated.

The survival of the ultimate toxicant will depend on its chemical stability in its actual environment and on the balance of the activating and deactivating enzyme activities present. The ultimate toxicant must arrive at an organ or tissue where biochemical processes are operative with which it is able to interact. In the tissue, defense mechanisms may be available to prevent damages. If interactions of the toxicant with essential biochemical mechanisms in the tissue occur, repair mechanisms may eliminate lesions.

In consequence, only when the ultimate toxicant arrives in a susceptible tissue, will toxicological events be observed. These will only occur if the toxicant is able to maintain a toxic concentration for a defined period of time and is able to overcome the defense and repair mechanisms operative on the cellular and molecular level.

Factors such as dose, route of administration, species, strain, sex, age and developmental state, health status, nutritional and environmental factors have been shown to influence whether or not a toxicological response occurs and, if so, its severity.

Considering the complexity of the subject, it is not surprising that the mechanisms of organ-directed toxicity only for some xenobiotics - mostly drugs or model compounds - are understood in detail . To broaden the data base for this presentation, results reported for all kinds of xenobiotics are being used. However information obtained from studies with pesticides are emphasized.

This paper will concentrate on relevant biotransformation and transport mechanisms. However, in short sections, two other important aspects, i.e. the types of toxic mechanisms and the role of organ physiology and biochemistry will be stressed. Defense and repair mechanisms will not be discussed.

Absorption and Biotransformation

Unless the chemical displays its toxicity directly on the body surface leading to local damages or lesions, the xenobiotic has to be absorbed, i.e. to pass through body membranes, to arrive in the systemic circulation for the manifestation of toxicity. The principle routes of entry are the gastro-intestinal tract, the skin, and the respiratory tract. These organs, despite being a more or less effective physical barrier against xenobiotics, also contain a wide array of xenobiotic-metabolizing enzymes, which allow them to toxify or detoxify chemicals.

In the literature, the enzyme activities in the intestinal tract, skin and the respiratory tract are generally considered low - as determined in-vitro - compared to those present in the liver. However these results may not reflect their

1054–7487/95/0244$12.00/0

relative contribution in an actual in-vivo situation, as additional factors (e.g. availability of cofactors/cosubstrates, position of the organ within the body) need to be considered.

Route of Entry: The Gastro-Intestinal Tract

In the gastro-intestinal tract, the xenobiotic is transported from the intestinal lumen across the epithelial cell and finally collected in the nearby blood or lymph capillaries. The absorption of xenobiotics across the epithelial cell is principally passive. However, there is evidence, that a few xenobiotics are "hitchhiking" on active transport mechanisms designed to assist essential components for endogenous synthesis. One of these examples appears to be paraquat which is absorbed from the intestinal tract by an energy-dependent, saturable mechanism associated with the carrier-mediated transport system for choline (Heylings, 1991; Nagao et al., 1993).

In contrast, active transport mechanisms from the vascular side into the lumen, representing thus an excretory function, have been demonstrated for some xenobiotics including cardiac glycosides, amines, quaternary ammonium compounds and organic acids (Lauterbach et al., 1989). A similar system which transports glutathione yet accepts certain glutathione conjugates has been also described (Vincenzini et al., 1991).

Evidence for non-biliary excretion into the intestinal lumen has been found in toxicokinetic studies with the persistent organochlorine insecticides, mirex (Pittman et al., 1976), dieldrin (Heath and Vedecar, 1964) and chlordecone (Boylan et al., 1979). This excretory mechanism has been recently confirmed for water-soluble metabolites of propachlor and was assumed to provide a route by which metabolites of xenobiotics may reach the intestinal lumen in animals which are poor biliary excretors (Aschbacher and Struble, 1987; Struble, 1991).

The metabolic capacity of the gastro-intestinal epithelial cells has been recently reviewed (Laitinen and Watkins, 1986; Combes, 1989; Schwenk, 1989; Siegers, 1989; Chadwick et al., 1992). Xenobiotic metabolizing enzymes have been detected in all sections, i.e. from the esophagus through the large intestine; however their activities vary to a high extent. The presence of activating enzyme systems in connection with the low activities of deactivating enzymes, i.e. sulfo-transferase, UDP-glucuronosyltransferase and glutathione-S-transferase in the stomach and large intestine has been considered responsible for the high susceptibility of these organs to carcinogenicity (Chadwick et al, 1992).

The most important region of the gastro-intestinal tract for absorption and consequently for metabolism of xenobiotics in the mucosa is the small intestine (jejunum). Generally the epithelial cells of this region also exhibit the highest concentration of xenobiotic-metabolizing enzymes within the alimentary tract. In contrast, metabolism in the gut lumen proceeds mainly in the lower gut region (colon), whose resident organisms show the highest metabolic activity. Hence it appears that, for most compounds, pre-systemic biotransformation is mostly due to metabolism in the epithelial cell and the contribution of the gut microorganisms is less important (Ilett et al., 1990). However, the contrary situation may be prevalent for poorly absorbed xenobiotics which may proceed to the lower part of the intestine. Besides pre-systemic biotransformation in the intestinal lumen, the intestinal enzymes may act upon the metabolites excreted into the intestinal lumen by biliary and non-biliary mechanisms.

The intestinal mucosa contains a wide variety of enzymes including cytochrome P450 isoenzymes and conjugating enzymes (Combes, 1989; Thies and Siegers, 1989). The metabolic capacity of the intestinal microflora is not limited to, but dominated by hydrolytic (e.g. ß-glucuronidase, arylsulfatase) and reductive (e.g. nitro- and azo-reduction, reductive deamination) reactions (Scheline, 1973; Goldman, 1978; Illing, 1981; Haenel and Grutte, 1984; Pelkonen and Hanninen, 1986). Cleavage of sulfuric and glucuronic acid conjugates, formed in the liver or other tissues and biliary excreted into the gut, may lead to reabsorption of the exocons and enterohepatic circulation and ultimately to a prolonged exposure of a potential target organ.

Methemoglobinemia as a consequence of intestinal microbial reduction of nitro derivatives is a well-established mechanism (Reddy et al., 1976). Also in the teratogenesis of nitrazepam, microbial nitro-reduction represents an essential step (Takeno and Sakai, 1991).

Cysteine conjugate ß-lyase which catalyzes the cleavage of the thioether linkage of S-alkyl-

and S-aryl-linked cysteine conjugates has been detected in a variety of gastrointestinal bacteria (Larsen, 1985). Cysteine conjugates which are biliary eliminated or formed from glutathione conjugates by γ-glutamyltranspeptidase and dipeptidase present in the luminar membrane of the bile duct and in the intestinal mucosa (Pinkus et al., 1977; Grafstrom et al., 1979; Tate, 1980) are cleaved to yield reactive thiol derivatives. A significant portion of these thiols react with the fibrous material present in the intestinal contents, presumably via sulfinic acids or thiosulfinates to yield non-extractable faecal residues (Bakke et al., 1981; Gustafsson et al., 1981). The major portion is reabsorbed into the intestinal mucosa and transformed to S-glucuronides, methylthio-, methylsulfinyl- and methylsulfonyl derivatives.

Many compounds have been demonstrated to follow the glutathionate pathway including significant participation of the microbial cysteine conjugate ß-lyase (Rafter et al., 1983), e.g. naphthalene (Bakke and Davison, 1988), 2,6-dichlorobenzonitrile, 2,6-dichlorobenzamide, 2,6-dichlorothiobenzamide (Bakke et al., 1988a/b), 2,4',5-trichlorobiphenyl (Bakke et al., 1982). The most prominent group, among the agricultural chemicals following this fate, are the chloroacetanilide herbicides, i.e. metolachlor, dimethachlor, propachlor, alachlor, acetochlor and butachlor. In principle, this pathway which leads from a detoxified product (glutathione conjugate) to a reactive species (thiol derivative) represents a re-toxifying process mediated by the enzymes of the intestinal micro flora.

Route of Entry: The Skin

The major function of the skin is to maintain the integrity of and to protect the living system from the environment. However, the skin is not a hermetic barrier as it needs exchange with the external environment to fulfill its physiological functions (e.g. water balance, thermo-regulation). This permeability of the skin allows certain xenobiotics to penetrate this organ and reach the systemic circulation. This leads potentially to toxic effects in the target organ.

The skin consists in principle of three major layers, each of which contain various cell types, i.e. epidermis (includes stratum corneum and the viable epidermis), dermis and the hypodermis.

It is generally accepted that the stratum corneum is the rate-limiting barrier to the penetration of xenobiotics through the skin. As this skin layer consists mainly of keratinized cells surrounded by lipids, the penetration process is assumed to be passive and governed by diffusion. Although percutaneous penetration is depending on biological factors (e.g. species, anatomical site, age, temperature, extent of hydration) the dominating determinant is given by the physico-chemical properties of the xenobiotic (Bronaugh and Maibach, 1985; Mukhtar, 1992).

However, skin does not only present a protective physical barrier against the external environment, it also contains a wide array of enzymes for the degradation and detoxification of xenobiotics.

The presence of numerous phase I and phase II metabolic pathways, including hydrolytic, reductive and oxidative mechanisms, as well as conjugation with glucuronic acid, sulfuric acid, amino acids and glutathione has been established. Multiple isoenzymes of glutathione S-transferase and cytochrome P-450 have been detected. It appears that most of the xenobiotic-metabolizing enzymes are preferentially localized in the epidermis (Noonan and Wester, 1985; Kao and Carver, 1990; Mukhtar et al., 1992; Raza al., 1992; Mukhtar, 1992).

As the activities of the enzymes are generally low, it is difficult to assess qualitatively and quantitatively the contribution of cutaneous metabolism in-vivo, as the overall pattern of metabolites is likely to be dominated by extra-cutaneous biotransformation. Hence, most of the results were obtained from in-vitro studies where for the majority of the compounds, similar metabolite patterns - as those obtained after oral administration - were observed.

In most cases, due to the limited permeability of the skin, the cutaneous penetration rate will be the rate-limiting step in the sequence of events leading ultimately to toxicity. Consequently, the exposure level of the potential target tissue may be generally lower as compared to levels arising from entrance via the gastro-intestinal tract. However, if the skin is the target organ, as for some chemical classes including polycyclic aromatic hydrocarbons, nitrosamines, nitrosamides and aromatic amines, the opposite situation is obvious (Agarwal and Mukhtar, 1992).

Route of Entry: The Respiratory Tract

The respiratory tract, consisting principally of the nasal cavity, the nasopharynx, the trachea and the lungs, is exposed to a wide variety of gases , vapors and particles.

External exposure of the respiratory tract to pesticides is not significant except for a few volatile compounds, e.g. fumigants. In contrast, systemic exposure is of significance, not only in the lungs, which receive 100% of the cardiac output, but also in the nasal mucosa. This is demonstrated by the fact, that out of seven selected chemicals investigated, all caused olfactory tissue lesions upon systemic exposure, but only two upon inhalation (Dahl and Hadley, 1991).

Much interest has been focused on the metabolic capacity present in the tissues of the nasal cavity and the lungs (Baron et al., 1988; Cohen, 1990; Smith and Brian, 1991; Dahl and Lewis, 1993; Reed, 1993).

An impressive list of xenobiotic metabolizing enzymes present in nasal cavity tissues of various species has been recently compiled and their significance discussed (Dahl and Hadley, 1991). The list contains many phase I and phase II enzymes including various cytochrome P450 isoenzymes. Interestingly, the cytochrome P450 isoenzymes were found to be resistant to induction by the classical agents active on the enzymes in the liver. Highest activity of the xenobiotic metabolizing enzymes are generally found in the olfactory epithelium.

One prominent example among pesticides, which exert toxicity on nasal cavity tissue, is alachlor. Alachlor induces the formation of nasal tumors in rats, but not in mice. This species-specificity and the economic importance of this herbicide stimulated studies of the mechanism underlying this species-specific, target toxicity.

In mammals, alachlor is extensively metabolized by various metabolic pathways including O- and N-dealkylation, p-hydroxylation of the phenyl moiety, conjugation of the phase I metabolites with glucuronic and sulfuric acid, as well as conjugation with glutathione followed by the corresponding catabolic reaction. The combination of these pathways lead to a rather complex metabolite pattern (Sharp, 1988; Feng and Patanella, 1988 and 1989). As the starting point in the metabolic activation of alachlor which leads ultimately to the formation of nasal tumors, the generation of 2,6-diethylaniline from its progening secondary arylamides is important. The 2,6-diethylaniline is oxidized to 4-amino-3,5-diethylphenol and subsequently to the strong electrophilic 3,5-diethylbenzo-quinone-4-imine, which binds covalently to sulfhydryl groups of proteins in the nasal turbinate tissue and thus becomes the ultimate toxicant.

Detailed comparative investigations with respect to the kinetics of this sequence of events revealed significant differences between rats and mice. In nasal turbinate tissue of rats, the arylamidase activity was 14 to 20 and the aniline hydroxylase 2 times higher than those determined in mice. It was suggested that these metabolic differences are responsible for the species-specific formation of nasal tumors in rats (Malik and Wilson, 1987; Feng et al., 1990; Wilson et al., 1992; Ward et al., 1992).

Among the many metabolites of alachlor, only two, i.e. 2-chloro-2',6'-diethyl acetanilide and 2-thiomethyl-2',6'-diethyl acetanilide were identified as suitable substrates for arylamidase. The latter one was found to be the preferred substrate (Feng et al., 1990). As the precursor of this metabolite is very likely generated during enterohepatic circulation by the bacterial cysteine conjugate ß-lyase, this example may represent a metabolic activation mediated by the gastro-intestinal content.

The mammalian lung which is composed of more than 40 different cell types, is an organ with extreme cellular complexity. Although significant species differences in the cellular distribution of the enzymes were observed, it appears, that the non-ciliated bronchiolar epithelial cells (*Clara Cells*) are the dominant site for the metabolism of xenobiotics (Devereux, 1993; Baron and Voigt, 1993). The presence of various cytochrome P450 isoenzymes in many mammalian species, including man, has been reported (Dahl and Lewis, 1993).

The metabolic capacity for hydrolytic, oxidative, and reductive processes as well as for phase II reactions including methylation, acetylation and conjugation with sulfuric and glucuronic acid and glutathione has been confirmed (Cohen, 1990).

Classical examples of chemical-induced pulmonary toxicity include furan derivatives including 4-ipomeanol (Boyd, 1980; Smith and Nemery, 1986; Gram, 1993), pyrrolizidine alkaloids (Smith and Nemery, 1986, Huxtable, 1993), butylated hydroxytoluene (Smith and Nemery 1986; Smith 1993; Bolton et al., 1993) and nitrosoureas (Smith and Nemery, 1986; Smith, 1993). All these compounds require metabolic activation prior to induction of the tissue damage. The nitrosoureas, used as chemotherapeutic agents in cancer treatment, very likely represent an exception, as they degrade spontaneously giving rise to strong electrophilic, i.e. alkylating products.

In contrast, the herbicide paraquat does not require any biotransformation to induce pulmonary lesions, i.e. destruction of alveolar Type I and Type II epithelial cells followed by extensive fibrosis (WHO, 1984; Smith and Nemery, 1986; Smith, 1987; Smith, 1988). Until now, it was assumed that the herbicide is not metabolized at all in mammalian systems (Daniel and Gage, 1966). However, one metabolite, i.e. the corresponding pyridone which was generated by rat liver homogenate, was recently identified (Fuke et al., 1993).

From the lower part of the intestinal tract, paraquat is partly absorbed by an energy-dependent, saturable mechanism associated with the transport system for choline (Daniel and Gage, 1966; Smith et al., 1974; Heylings, 1991; Nagao et al., 1993). It appears that the competition between this saturable transport system and the transport of the intestinal content through gastro-intestinal tract limits the extent of absorption. Consequently poor absorption, typically below 20% of the dose, has been found (Daniel and Gage, 1966; Smith et al., 1974). After absorption, paraquat is widely distributed between the various tissues and rapidly renally and biliary eliminated (Daniel and Gage, 1966; Hughes et al. 1973; Ameno et al., 1994).

The compound is selectively extracted from the systemic circulation by the lung. This leads to accumulation. This selective accumulation process has been characterized as energy-dependent and competitive to the uptake of some endogenous polyamines, e.g. spermine, spermidine, putrescine, cadaverine (Sharp et al., 1972; Rose et al. 1974; Waddell and Marlowe, 1980).

The ultimate biochemical mechanism of paraquat, which leads to destruction of alveolar epithelial cells, remains to be discovered. However, it is generally believed, that the primary step in the sequence of events is the generation of electrons by a redox cycle. This leads to continuous production of the superoxide radical anion ($O_2^{\cdot-}$). This redox cycle is maintained by a single electron reduction of the cation generating the free radical, which in the presence of oxygen will reconstitute the cation along with production of $O_2^{\cdot-}$ (Smith, 1987; Aust et al., 1993).

Another group of chemicals, which are related to pesticides and which induce pulmonary toxicity, are the trialkyl phosphorothioates. These are found as impurities in organophosphorus insecticides (Nemery, 1987; Gandy et al., 1993). As a structural requirement to induce a delayed lung toxicity being independent of acetylcholine esterase inhibition (Mallipudi et al. 1979; Verschoyle et al. 1980; Verschoyle and Cabral, 1982; Durham et al. 1988), the presence of at least one thiolo-alkyl group and the absence of a thiono-sulfur has been postulated (Umetsu et al. 1981; Aldridge et al. 1985).

When O, S, S-trimethyl phosphorodithioate (O, S, S-TMP) or O, O, S-trimethyl phosphorothioate (O, O, S-TMP) is orally applied to rats, the compounds are absorbed rapidly, distributed within the body and then rapidly eliminated. There is no accumulation of the compounds and/or their metabolites in the lung (Aldridge et al. 1984; Gray and Fukuto, 1984).

In the rat, the biotransformation of O, O, S-TMP proceeds preferentially by removal of the thiomethyl group. The chemical is ultimately degraded to CO_2 (Gray and Fukuto, 1984). The nature of the ultimate toxicant as well as the site of metabolic activation, i.e. pulmonary versus extra-pulmonary tissue, leading to pulmonary toxicity remains unknown (Verschoyle and Cabral 1982; Nemery, 1987; Gandy et al., 1993). However, the presence of a reactive sulfoxide of the parent compound has been evidenced and this is proposed as the ultimate toxicant (Segall and Casida, 1982; Gray and Fukuto, 1984)

Distribution

Once a xenobiotic or its metabolite has entered the body and arrived in the systemic circulation, it needs to reach its potential target organ. The principal vehicle for the distribution within the body is the blood which reaches all living cells within the system. If the blood components do not represent the target for irreversible covalent interaction, the xenobiotic may be transported in reversible association with plasma proteins (Tillement et al., 1984; Fichtl et al., 1991; Souich, 1993).

The concentration of the xenobiotic at the target site is directly related to the concentration of the free xenobiotic in the plasma. Consequently, protein binding, in principle, lowers the concentration of the xenobiotic in the target tissue and thus the potential for the expression of toxicity.

Species differences in plasma protein binding are frequently observed and need serious consideration in the extrapolation of toxic effects between species. In contrast, protein binding lowers the renal clearance rate, which in turn will lead to a prolonged exposure of the target tissue.

For the specific protection of some very sensitive tissues, mammals possess specialized anatomical barriers. The especially tight endothelium of the cerebral microvessels (blood-brain barrier) inhibits or limits the uptake of preferably polar xenobiotics into the cerebrospinal fluid system (Bradbury, 1979; Dermietzel and Krause, 1991; Audus et al., 1992; Brightman, 1992). The placental membranes protect, to some extent, the developing fetus against the adverse effects of xenobiotics. However, as the barrier function of the placenta appears to be restricted to polar compounds, it is assumed, that the protection of the fetus by xenobiotic-metabolizing enzymes is of more significance (Pelkonen, 1984; Lien, 1985; Brown et al., 1986; Juchau, 1990; Juchau et al., 1992).

The distribution process throughout the body is - with some exceptions - passive in nature. It is driven by diffusion and is thus dependent on the physico-chemical properties of the xenobiotic and/or its metabolites.

Lipophilic and persistent compounds, e.g. organochlorines and difluorobenzoylureas will be transferred to the fat tissue and may accumulate to a high level without displaying toxicity to the adipose tissue itself. Upon termination of exposure, these sequestered compounds may be slowly metabolized and eliminated without reaching an effect level in a potential target organ.

However, changes in the living conditions (e.g. health status, environment, nutrition, physical exercises), which triggers a mobilization of the fat depots, may lead to an accelerated release of the stored xenobiotics and thus to an increase in the concentration above an effect level. Numerous experiments in this context, mainly using DDT, have been published.

Besides deposition of persistent xenobiotics in the fat by partitioning, the formation of lipophilic conjugates, e.g. hybrid di- and triacylglycerols, in which one of the fatty acids is replaced by a xenobiotic carboxy acid, fatty acid esters of xenobiotic hydroxy derivatives followed by storage in the adipose tissue has been found for various compounds, including pyrethroids, and phenoxy propanoic acid derivatives (Caldwell and Marsh, 1983, Quistad and Hutson, 1986; Chang et al., 1986; Myamoto et al., 1986).

After lipolysis of the hybrid acylglycerols, the xenobiotic acid may be released and eliminated or reutilized by the liver for the synthesis of acylglycerols. The recycling mechanism will lead to an extended exposure of the target organ. This may be important as some of the corresponding xenobiotic carboxy acids are assumed to induce hepatomegaly and peroxisomal proliferation in rodents (Bentley et al., 1993).

Most of the compounds entering the systemic circulation will be biotransformed leading to metabolites, including reactive intermediates. The chemical and the biochemical stability of the potential ultimate toxicant formed is one of the most important factors determining the fate of this metabolite and thus the ultimate toxicological consequences.

Short-lived metabolites may immediately react with the enzyme, which leads to enzyme inactivation (suicide inhibition), or with other constituents within the cell (cell injury).

Long-lived metabolites may be able to leave the cell and also the tissue. They may reenter the systemic circulation and may ultimately arrive at their target organ. After arrival, they have to migrate via the interstitial fluid into the cell and enter specific cell organelles to exert toxicity. Ultra-long-lived metabolites may be stable enough to be renally or biliary excreted and may manifest their toxicity in the urinary or intestinal tract.

With exception of the suicide inhibition, which is solely due to chemical reactivity, the half-live of the toxic metabolite is determined by its inherent chemical stability and the presence and activity of detoxifying enzymes.

Elimination

The mammalian body is provided with several very effective routes for the elimination of xenobiotics. Any disturbance of these excretory mechanisms by chemicals or other factors may influence the toxicokinetic parameters of a xenobiotic and consequently the manifestation of toxicity. Even under physiological conditions, the elimination process can induce toxicological effects preferably in the organs involved.

The physiological processes, involved in the formation and elimination of the urine and its ingredients, make the urinary tract vulnerable to toxic effects. The concentration process may elevate the amount of a metabolite above its solubility. This leads to precipitation, which in turn - by several mechanisms - may result in the manifestation of severe nephrotoxicity.

A well established example is ß-naphthylamine, which is renally eliminated as the O-glucuronic acid conjugate of its hydroxylamine derivative. In the bladder, the conjugate is cleaved and the tumorigenic species released. This induces bladder cancer in primates including man and dogs.

The biliary excretion of a xenobiotic is mainly governed by its molecular weight, the charge, the polarity and some structural characteristics (Smith, 1973; Siegers and Watkins, 1991). A highly significant feature of biliary excretion is the possible enterohepatic circulation, which leads to a plethora of possibilities for chemicals to reach potential target organs (again) with or without additional biotransformation. Micro-organisms in the intestinal tract may also have a role in this process (Watkins, 1991; Siegers, 1991). In consequence, transport via the enterohepatic circulation leads to many possibilities by which chemicals can exert toxic effects.

The elimination of xenobiotics via the milk deserves special consideration, since the potential target is the entire nursing offspring with its limited ability to metabolize xenobiotics. If at all, it is usually an extremely small portion of the maternal burden of a chemical which is eliminated by this route. Although many polar xenobiotics have been detected in the milk, its high fat content obviously favors the secretion of lipophilic compounds, as the transfer from the plasma is preferably by diffusion. Hence plasma protein binding of the xenobiotic reduces its transfer into the milk (Levine, 1986; Gallenberg and Vodicnik, 1989).

Toxic Mechanisms

Basically a toxic effect in a living system is defined by alteration of a certain function or by structural damages. The chemically induced toxic responses may be initiated by numerous mechanisms ranging from local irritation to covalent binding to macromolecules. This section will be limited to metabolism related mechanisms, i.e., to toxic mechanisms mediated by reactive metabolites.

In principle, the reactive chemical species will covalently bind to cellular constituents (e.g. DNA, RNA, proteins, lipids, glutathione) or may generate reactive oxygen species by direct reaction with oxygen or participation in redox cycling mechanisms (oxidative stress). It is assumed that, in most cases, the toxic effects of reactive metabolites are a consequence of simultaneous and/or sequential alterations in several cellular processes.

Reactive intermediates are generally electrophilic species like epoxides, radicals, acetyl hydroxyl amines, quinones and acyl-glucuronides. They originate mostly from oxidation of precursors by cytochrome P450 isoenzymes.

Reduction of various compounds, including halogenated hydrocarbons and nitro compounds, to free radicals under anerobic conditions, is known.

Quinone compounds can undergo redox-cycling generating reactive oxygen species.

Glutathione and cysteine conjugates are precursors for various reactive intermediates, e.g. reactive thiol compounds which are formed by cleavage of the thioether linkage by the enzyme β-lyase. The glutathione conjugate of dihaloalkanes can give rise to reactive episulfonium ions. These result in the preferential covalent binding to RNA and DNA.

Radicals, reducing agents, redox cycling chemicals and other processes can generate various species of reactive oxygen, e.g. superoxide, peroxide and hydroxyl radical.

Due to the vital function of oxygen being the ultimate electron acceptor of respiration, the cell is equipped with very efficient systems to detoxify reactive oxygen species. These systems include superoxide dismutase, catalase and glutathione peroxidase which efficiently degrade superoxide and peroxide to innocuous water. Besides these enzymatic defense mechanisms, ascorbic acid in cytosol and carotene and tocopherol in membranes, respectively, protect the cell from oxidative stress.

In cases where the defense capacity is exceeded, oxidative stress occurs, i.e. reaction of reactive oxygen species with cellular components takes place, e.g. thiol oxidation in proteins, lipid peroxidation or direct oxidation of DNA.

Proteins containing thiol groups are prone to covalent binding and oxidation. Also, other functional groups from aminoacid side chains are susceptible for reaction with electrophiles. Due to their nucleophilic character, RNA and DNA bases can be modified by electrophiles or directly oxidized by reactive oxygen species.

The most prominent defense mechanism is effected by glutathione which acts as a scavenger for electrophiles, and as an antioxidant for reactive oxygen species or other oxidants. Glutathione reacts with electrophilic compounds by substitution or addition (epoxides, quinones), with oxidants by dimerization and with radicals by H-abstraction. The cellular glutathione can be depleted by a reactive species which is then available for reaction with cellular components.

The lipid components of membranes can react with reactive oxygen species and other reactive intermediates like radicals. As a consequence, the membrane integrity is damaged and its function impaired.

Mitochondrial proteins are prone to modification by oxidation and covalent binding of reactive metabolites, since they are rich in thiol groups and are exposed to high concentrations of reactive species due to their location at the site of generation.

Modification of proteins will result in various functional changes, e.g. enzyme inhibition, impaired function of transport proteins and impaired homeostasis. Therefore, covalent binding to proteins can have various deleterious effects. These are dependent on the site and extent of modification, and whether the protein is essential for cell viability.

Inhibition of enzymes, which participate in ATP production, damages the cellular energy supplying system.

Calcium homeostasis plays a crucial role in some types of hepatocellular injury since increase in intracellular calcium triggers a series of enzymatic reactions. These ultimately result in protein and lipid breakdown and DNA fragmentation.

Modified proteins can trigger the production of antibodies with the adduct, as hapten like antigen, leading ultimately to adverse immune reactions.

After metabolic activation, various compounds can bind to DNA. Reactive oxygen species may directly generate 8-hydroxy-deoxyguanosine. Modified nucleotides can lead to mutations and carcinogenicity if there is no enzymatic repair to replace the modified nucleotides.

Target Organ: Physiology and Biochemistry

The susceptibility of an organ or tissue in a living system to express toxicological effects upon exposure to a chemical depends on the physico-chemical properties of the chemical (i.e. the disposition of the chemical in this particular system), and the individual physiology and biochemistry of the organ (target organ).

Statistically, most frequently the liver and the kidneys, followed by other organs, e.g. blood forming system, lymphatic system, endocrine system, nervous system, and lungs have been identified as target organs. Some major characteristics of organs or systems being potentially vulnerable to manifestations of chemical-induced toxicity will be discussed.

The outstanding role of the liver is determined by its position within the body and by its plethora of xenobiotic-metabolizing enzyme systems. Compounds entering the body via the gastro-intestinal tract are generally absorbed into the hepatic portal vein and directly carried to the liver. Hence the liver, a route of entry for xenobiotics to the systemic circulation and consequently to other organs, is exposed to a higher xenobiotic concentration than the other organs, which may be exposed after dilution of the chemical within the systemic circulation.

The sinusoidal architecture of the liver provides a system designed for fast and efficient uptake by the liver parenchymal cells of xenobiotics and other constituents present in the portal blood. Many xenobiotics are very efficiently absorbed during the first pass through the liver (first pass effect). This leads to a high concentration of the chemical in the organ.

Besides extraction of nutrients from the portal blood followed by the synthesis of a wide variety of endogenous compounds, catabolism of endogenous compounds, uptake of xenobiotics followed by biotransformation, the secretion of bile is a major function of the liver.

Xenobiotics may be biliary eliminated to a high extent. This may lead to a high concentration in the bile fluid, to damage and lesions in the bile duct and to adverse interactions with endogenous compounds transported with the bile fluid, e.g. bilirubin.

Besides exposure of the liver to compounds entering the body via the intestinal tract, exposure from the systemic circulation is significant, as the mammalian liver receives - although accounting for only 2 - 5 % of the body weight - about 25% of the cardiac output.

The major role of the kidney, besides the control of composition and volume of body fluids, is the elimination of waste products from the body. This includes catabolic products of endogenous and exogenous compounds.

A significant portion of the blood delivered to the kidneys is processed by glomerular filtration in *Bowman's Capsule*. The large volume of the ultra-filtrate formed contains most of the xenobiotics. Additionally, xenobiotic acids and bases (e.g. glucuronic and sulfuric acid conjugates) may be secreted into the proximal tubule by active transport mechanisms. Re-absorption of nonpolar xenobiotics from the distal tubule into the bloodstream may also occur. Large amounts of water (as well as salts and endogenous nutrients) are reabsorbed. The concentrated urine which contains the polar xenobiotic compounds, is then ready for elimination from the body.

However, the processing of the blood plasma and its ingredients in the kidney (i.e. the active and passive concentration processes involved), may lead to elevated concentration of a potential toxicant in the tubular fluid or in the urine. This may eventually result in the manifestation of toxicity.

It is obvious, that organs, which are routes of entry to the body, may be exposed to a rather high concentration of a xenobiotic. This makes them potential targets for toxicity. Additionally, in all of these systems, a wide variety of xenobiotic metabolizing enzymes are present. These enzymes may produce potentially reactive metabolites.

However the lung is not only exposed to xenobiotics which enter directly. It is also exposed via the systemic circulation, as it receives 100% of the cardiac output.

A similar situation needs to be considered for the intestinal tract, as the intestinal mucosa is additionally exposed to the metabolites excreted with the bile fluid. The duration of exposure may be enhanced by entero-hepatic circulation. Furthermore, biotransformation of the xeno-biotic and its biliary metabolites by the intestinal flora may significantly influence the composition of metabolites to which the mucosa is exposed.

Special attention and considerations have to be given to organs which are potentially vulnerable to toxicity due to their high rate of proliferation and cell differentiation, e.g. blood forming tissues, fetal tissues.

In mammals, there are some biochemical processes which are exclusive to one particular organ which may become the target for an ultimate toxicant resulting in target organ toxicity, e.g. tissue-specific synthesis of hormones, tissue-specific action of hormones, organ-specific active transport systems, synthesis/release/deactivation of neurotrans-mitters.

However, many of the biochemical processes assumed to be involved in the mechanism of organ-specific toxicity are ubiquitous, i.e., they are present in many organs, e.g. nutrient supply, energy supply, synthesis of endogenous compounds, defense mechanisms, repair mechanisms. In these cases, the ultimate interaction of the toxicant with an individual step in the sequence of biochemical processes will trigger the organ toxicity. This step must be critical to a major function of that particular organ.

The toxicant affects an individual step, which is already or becomes - due to this interaction - a rate-limiting step in a vital biochemical process in this organ. This limits the capacity of that process and subsequently leads to an impaired function of the tissue. Ultimately organ toxicity may be manifest. This concept has been introduced by Aust (Aust, 1986) and named "Rate-Limiting Concept".

The above discussion shows that there are numerous mechanisms and complex processes by which chemicals entering the body may exert toxic effects. As full understanding of these processes as possible is essential to help to explain mechanisms of toxicology. In addition some of these processes may be specific to certain species. Again a full understanding may help to explain species specific toxicity.

References

Agarwal, R.; Mukhtar, H. in Mukhtar, H. (ed.), *Pharmacology of the Skin,* CRC Press Inc.: Boca Raton, Florida, **1992,** 371-387.

Aldridge, W.N.; Dinsdale, D.; Nemery, B.; Verschoyle, R.D. *Fundam. Appl. Toxicol.* **1985,** *5,* 47-60.

Aldridge, W.N.; Verschoyle, R.D.; Peal, J. A. *Pestic. Biochem. Physiol.* **1984,** *21,* 265-274.

Ameno, K.; Fuke, C.; Shirakawa, Y.; Ogura, S.; Ameno, S.; Kiriu, T.; Kinoshita, H.; Ijiri, I. *Arch. Toxicol.* **1994,** *68,* 134-137.

Aschbacher, P.W.; Struble, C.B. *Xenobiotica,* **1987,** *17,* 1047-1055.

Audus, K.L.; Chikhale, P.J.; Miller, D.W.; Thompson, S.E.; Borchardt, R.T. in Testa, B. (ed.) *Advances in Drug Research,* Academic Press Inc.: London, **1992,** *23,* 3-64.

Aust, S.D.; Chignell, C.F.; Bray, T.M.; Kalyanaraman, B.; Mason, R.P. *Toxicol. and Appl. Pharmacol.* **1993,** *120,* 168-178.

Aust, S.D. in Cohen, G.M. (ed.) *Target Organ Toxicity,* CRC Press Inc.: Boca Raton, Florida, **1986,** *1,* 121-127.

Bakke, J.E.; Rafter, J.J.; Larsen, G.L.; Gustafsson, B.E. *Drug Metab. Dispos.* **1981,** *9,* 525-528.

Bakke, J.E.; Davison, K.L. *Xenobiotica,* **1988,** *18,* 1057-1062.

Bakke, J.E.; Larsen, G.L.; Feil, V.J.; Brittebo, E.B.; Brandt, I. *Xenobiotica,* **1988a,** *18,* 817-829.

Bakke, J.E.; Bergman, A.L.; Larsen, G.L. *Science,* **1982,** *217,* 645-647.

Bakke, J.E.; Larsen, G.L.; Struble, C.; Feil, V.J.; Brandt, I.; Brittebo, E.B. *Xenobiotica,* **1988b,** *18,* 1063-1075.

Bakke, J.E.; Gustafsson, J.A.; Gustafsson, B.E. *Science,* **1980,** *210,* 433-435.

Baron, J.; Burke, J.P.; Guengerich, F.P.; Jakoby, W.B.; Voigt, J.M. *Toxicol. and Appl. Pharmacol.* **1988,** *93,* 493-505.

Baron, J.; Voigt, J.M. in Gram, T.E. (ed.) *Metabolic Activation and Toxicity of Chemical Agents to Lung Tissue and Cells,* Pergamon Press: Oxford, **1993,** 41-75.

Bentley, P.; Calder, I.; Elcombe, C.; Grasso, P.; Stringer, D.; Wiegand, H.J. *Fd. Chem. Toxic.* **1993,** *31,* 857-907.

Bolton, J.L.; Thompson, J.A.; Allentoff, A.J.; Miley, F.B.; Malkinson, A.M. *Toxicol. and Appl. Pharmacol.* **1993,** *123,* 43-49.

Boyd, M.R. in Hodgson, E.; Bend, J.R.; Philpot, R.M. (eds.) *Reviews in Biochemical Toxicology,* Elsevier North-Holland Inc.: New York, **1980,** *2,* 71-101.

Boylan, J.J.; Cohn, W.J.; Egle, J.L.; Blanke, R.V.; Guzelian, P.S. *Clin. Pharmacol. Ther.* **1979,** *25,* 579-585.

Bradbury, M. *The Concept of a Blood-Brain Barrier,* John Wiley & Sons: Chichester, **1979.**

Brightman, M.W. in Bradbury, M. (ed.) *Physiology and Pharmacology of the Blood-Brain Barrier, Handbook of Exper. Pharmacol.,* Springer-Verlag: Berlin, **1992,** *103,* 1-22.

Bronaugh, R.L.; Maibach, H.I. (eds.) *Percutaneous Absorption,* Marcel Dekker Inc.: New York and Basel, **1985.**

Brown, L.P.; Flint, O.P.; Orton, T.C.; Gibson, G.G. *Drug Metab. Rev.* **1986,** *17,* 221-260.

Burwen, S.J.; Schmucker, D.L.; Jones, A.L. in Jeon, K.W.; Friedlander, M. (eds.) *International Review of Cytology, A Survey of Cell Biology,* Academic Press Inc.: San Diego, **1992,** *135,* 269-313.

Caldwell, J.; Marsh, M.V. *Biochem. Pharmacol.* **1983,** *32,* 1667-1672.

Chadwick, R.W.; George, S.E.; Claxton, L.D. *Drug Metabol. Rev.* **1992,** *24,* 425-492.

Chang, M.J.W.; Leighty, E.G.; Haggerty, G.C.; Fentiman Jr., A.F. in Paulson, G.D.; Caldwell, J.; Hutson, D.H.; Menn, J.J. (eds.) *Xenobiotic Conjugation Chemistry,* ACS Symposium Series, Amer. Chem. Soc.: Washington DC, **1986,** *299,* 214-220.

Cohen, G.M. *Environ.Health Perspect.* **1990,** *85,* 31-41.

Combes, R.D. in Koster, A.S.; Richter, E.; Lauterbach, F.; Hartmann, F. (eds.) *Intestinal Metabolism of Xenobiotics, Progr. Pharmacol. Clin. Pharmacol.,* Gustav Fischer Verlag: Stuttgart, **1989,** *7/2,* 119-145.

Dahl, A.R.; Hadley, W.M. *CRC Crit. Rev. Toxicol.* **1991,** *21,* 345-372.

Dahl, A.R.; Lewis, J.L. *Annu. Rev. Pharmacol. Toxicol.* **1993,** *32,* 383-407.

Daniel, J.W.; Gage, J.C. *Brit. J. Ind. Med.* **1966,** *23,* 133-136.

Dermietzel, R.; Krause, D. in Jeon, K.W.; Friedlander, M. (eds.) *International Review of Cytology, A Survey of Cell Biology,* Academic Press Inc.: San Diego, **1991,** *127,* 57-63.

Devereux, T.R.; Domin, B.A.; Philpot, R.M. in Gram, T.E. (ed.) *Metabolic Activation and Toxicity of Chemical Agents to Lung Tissue and Cells,* Pergamon Press: Oxford, **1993,** 25-40.

Durham, S.K.; Gandy, J.; Imamura, T. *Toxicol. Pathol.* **1988,** *16,* 392-395.

Feng, P.C.C.; Patanella, J.E. *Pestic. Biochem. Physiol.* **1988,** *31,* 84-90.

Feng, P.C.C.; Patanella, J.E. *Pestic. Biochem. Physiol.* **1989,** *33,* 16-25.

Feng, P.C.C.; Wilson, A.G.E.; McClanahan, R.H.; Patanella, J.E.; Wratten, S.J. *Drug Metab. and Dispos.* **1990,** *18,* 373-377.

Fichtl, B.; Nieciecki v., A.; Walter, K. in Testa, B. (ed.) *Advances in Drug Research,* Academic Press: London, **1991,** *20,* 118-166.

Fuke, C.; Ameno K.; Ameno, S.; Kiriu, T.; Shinohara, T.; Ijiri, I. *Jpn. J. Legal Med.* **1993,** *47,* 33-45.

Gallenberg, L.A.; Vodicnik, M.J. *Drug Metab. Rev.* **1989,** *21,* 277-317.

Gandy, J.; Durham, S.K.; Imamura, T. in Gram, T.E. (ed.) *Metabolic Activation and Toxicity of Chemical Agents to Lung Tissue and Cells,* Pergamon Press: Oxford, **1993,** 153-163.

Goldman, P. *Annu. Rev. Pharmacol. Toxicol..* **1978,** *18,* 523-539.

Grafstrom, R.K.; Ormstad, P.M.; Orrenius, S. *Biochem. Pharmacol.* **1979,** *28,* 3573-3579.

Gram, T.E. in Gram, T.E. (ed.) *Metabolic Activation and Toxicity of Chemical Agents to Lung Tissue and Cells,* Pergamon Press: Oxford, **1993,** 145-152.

Gray, A.J.; Fukuto, T.R. *Pest. Biochem.* **1984,** *22,* 295-311.

Gustafsson, J.A.; Rafter, J.J.; Bakke, J.E.; Gustafsson, B.E. *Nutr. Cancer,* **1981,** *2,* 224-231.

Haenel, H.; Grutte, F.K. *Nahrung,* **1984,** *28,* 647-657.

Heath, D.; Vedecar, M. *Br. J. Med.* **1964**, *21,* 269-279.

Heylings, J.R. *Toxicol. Appl. Pharmacol.* **1991,** *107,* 482-493.

Hughes, R.D.; Millburn, P.; Williams, R.T. *Biochem. J.* **1973,** *136,* 979-984.

Huxtable, R.J. in Gram, T.E. (ed.) *Metabolic Activation and Toxicity of Chemical Agents to Lung Tissue and Cells,* Pergamon Press: Oxford, **1993,** 213-237.

Ilett, K.F.; Tee, L.B.G.; Reeves, P.T.; Minchin, R.F. *Pharmac. Ther.* **1990,** *46,* 67-93.

Illing, H.P.A. *Xenobiotica,* **1981,** *11,* 815-830.

Juchau, M.R.; Lee, Q.P.; Fantel, A.G. *Drug Metab. Rev.* **1992,** *24,* 195-238.

Juchau, M.R. in Kacew, S. (ed.) *Drug Toxicity and Metabolism in Pediatrics,* CRC Press Inc.: Boca Raton, Florida, **1990,** 15-35.

Kao, J.; Carver, M.P. *Drug Metab. Rev.* **1990,** *22,* 363-410.

Laitinen, M.; Watkins, J.B. in Rozman, K.; Hanninen, O. (eds.) *Gastro-Intestinal Toxicology,* Elsevier Science Publisher B.V.: Amsterdam, **1986,** 412-434.

Larsen, G.L. *Xenobiotica,* **1985,** *15,* 199-209.

Lauterbach, F.; Schorn, M.; Sprakties, G.; Sund, R.B. in Koster, A.S.; Richter, E.; Lauterbach, F.; Hartmann, F. (eds.) *Intestinal Metabolism of Xenobiotics, Progr. Pharmacol. Clin. Pharmacol.*, Gustav Fischer Verlag: Stuttgart, **1989,** *7/2,* 231-242.

Levine, W.G. in Cohen, G.M. (ed.) *Target Organ Toxicity,* CRC Press Inc.: Boca Raton, Florida, **1986,** *1,* 55-88.

Lien, E.J. in Jucker, E. (ed.) *Progress in Drug Research,* Birkhäuser Verlag: Basel, **1985,** *29,* 68-92.

Malik, J.M.; Wilson, A.G.E. Poster presented at ISSX/SOT North American Symposium on *Endogeneous Factors in the Toxicity of Xenobiotics,* Clearwater: Florida, USA, November 8-13, **1987.**

Mallipudi, N.M.; Umetsu, N.; Toia, R.F.; Talcott, R.E.; Fukuto, T.R. *J. Agric. Food Chem.* **1979,** *27,* 463-466.

Mukhtar, H. (ed.) *Pharmacology of the Skin,* CRC Press Inc.: Boca Raton, Florida, **1992.**

Mukhtar, H.; Agarwal, R.; Bickers, D.R. in Mukhtar, H. (ed.), *Pharmacology of the Skin,* CRC Press Inc.: Boca Raton, Florida, **1992,** 89-109.

Mukhtar, H. in Mukhtar, H. (ed.), *Pharmacology of the Skin,* CRC Press Inc.: Boca Raton, Florida, **1992,** 139-147.

Myamoto, J.; Kaneko, H.; Okuno, Y. in Paulson, G.D.; Caldwell, J.; Hutson, D.H.; Menn, J.J. (eds.) *Xenobiotic Conjugation Chemistry,* ACS Symposium Series, Amer. Chem. Soc.: Washington DC, **1986,** *299,* 221-241.

Nagao, M.; Saitoh, H.; Zhang, W.D.; Iseki, K.; Yamada, Y.; Takadori, T.; Miyazaki, K. *Arch. Toxicol.* **1993,** *67,* 262-267.

Nemery, B. in Costa, L.G.; Galli, C.L.; Murphy, S.D. (eds.) *Toxicology of Pesticides: Experimental, Clinical and Regulatory Perspectives,* Springer Verlag: Berlin, Heidelberg, **1987,** 297-303.

Noonan, P.K.; Wester, R.C. in Bronaugh, R.L.; Maibach, H.I. (eds.), *Percutaneous Absorption,* Marcel Dekker Inc.: New York, Basel, **1985,** 65-85.

Pelkonen, K.; Hanninen, O. in Rozman, K.; Hanninen, P. (eds.) *Gastro-Intestinal Toxicology,* Elsevier Science Publisher B.V.: Amsterdam, **1986,** 193-212.

Pelkonen, O. *Dev. Pharmacol. Therap.* **1984,** *7 (Suppl. 1),* 11-16.

Pinkus, L.M.; Ketley, J.N.; Jakoby, W.B. *Biochem. Pharmacol.* **1977,** *26,* 2359-2363.

Pittman K.; Weiner, M.; Treble, D.H. *Drug Metab. Rev.* **1976,** *4,* 288-295.

Quistad, G.B.; Hutson, D.H. in Paulson, G.D.; Caldwell, J.; Hutson, D.H.; Menn, J.J. (eds.) *Xenobiotic Conjugation Chemistry,* ACS Symposium Series, Amer. Chem. Soc.: Washington DC, **1986,** *299,* 204-213.

Rafter, J.J.; Bakke, J.; Larsen, G.; Gustafsson, B.; Gustafsson, J. in Hodgson, E.; Bend, J.R.; Philpot, R.M. (eds.), *Reviews in Biochemical Toxicology,* **1983,** *5,* 387-408.

Raza, H.; Agarwal, R.; Mukhtar, H. in Mukhtar, H. (ed.), *Pharmacology of the Skin,* CRC Press Inc.: Boca Raton, Florida, **1992,** 131-137.

Reddy, B.S.; Pohl, L.R.; Krishna, G. *Biochem. Pharmacol.* **1976,** *25,* 1119-1122.

Reed, C.J. *Drug Metab. Rev.* **1993,** *25,* 173-205.

Rose, M.S.; Smith, L.L.; Wyatt, I. *Nature,* **1974,** *252,* 314-315.

Scheline, R.R. *Pharmacol. Rev.* **1973,** *25,* 451-523.

Schwenk, M. in Koster, A.S.; Richter, E.; Lauterbach, F.; Hartmann, F. (eds.) *Intestinal Metabolism of Xenobiotics, Progr. Pharmacol. Clin. Pharmacol.,* Gustav Fischer Verlag: Stuttgart, **1989,** *7/2,* 155-169.

Segall, Y.; Casida, J.E. *Tetrahedron Lett.* **1982,** *23,* 139-142.

Sharp, D.B. in Kearney, P.C.; Kaufman, D.D. (eds.) *Herbicides: Chemistry, Degradation, and Mode of Action,* Marcel Dekker: New York, **1988,** 301-333.

Sharp, C.W.M.; Ottolenghi, A.; Posner, H.S. *Toxicol. Appl. Pharmacol.* **1972,** *22,* 241-251.

Siegers, C.P. in Koster, A.S.; Richter, E.; Lauterbach, F.; Hartmann, F. (eds.) *Intestinal Metabolism of Xenobioticcs, Progr. in Pharmacol. Clin. Pharmacol.,* Gustav Fischer Verlag: Stuttgart, **1989,** *7/2,* 171-180.

Siegers, C.P. in Siegers, C.P.; Watkins III, J.B. (eds.) *Biliary Excretion of Drugs and Other Chemicals, Progr. Pharmacol. Clin. Pharmacol.,* Gustav Fischer Verlag: Stuttgart, **1991,** *8/4,* 389-397.

Siegers, C.P.; Watkins III, J.B. (eds.) *Biliary Excretion of Drugs and Other Chemicals, Progr. Pharmacol. Clin. Pharmacol., Vol. 8/4,* Gustav Fischer Verlag: Stuttgart, **1991.**

Smith, B.R.; Brian, W.R. *Toxicol. Pathology,* **1991,** *19,* 470-481.

Smith, L.L.; Nemery, B. in Cohen, G.M. (ed.) *Target Organ Toxicity,* CRC Press Inc.: Boca Raton, Florida, **1986,** *2,* 45-80.

Smith, A.C. in Gram, T.E. (ed.) *Metabolic Activation and Toxicity of Chemical Agents to Lung Tissue and Cells,* Pergamon Press: Oxford, **1993,** 285-303.

Smith, L.L. in Hodgson, E.; Bend, J.R.; Philpot, R.M. (eds.) *Reviews in Biochemical Toxicology,* Elsevier Science Publishing Co. Inc.: New York, **1987,** *2,* 37-71.

Smith, L.L.; Wright, A.F.; Wyatt, I.; Rose, M.S. *Br. Med. J.* **1974,** *4,* 569-71.

Smith, R.L. *The Excretory Function of Bile,* Chapman and Hall Ltd.: London, **1973.**

Smith L.L. *Adv. Drug React. Ac. Pois. Rev.* **1988,** *7,* 1-17.

Souich du, P.; Verges, J.; Erill, S. *Clin. Pharmacokin.* **1993,** *24,* 435-440.

Struble, C.B. *Xenobiotica,* **1991,** *21,* 85-95.

Takeno, S.; Sakai, T. *Teratology,* **1991,** *44,* 209-214.

Tate, S.S. in Jacoby, W.B. (ed.), *Enzymatic Basis of Detoxification,* Academic Press: New York, **1980,** *2,* 95-120.

Thies, E.; Siegers, C.P. in Koster, A.S.; Richter, E.; Lauterbach, F.; Hartmann, F. (eds.) *Intestinal Metabolism of Xenobiotics, Progr. Pharmacol. Clin. Pharmacol.,* Gustav Fischer Verlag: Stuttgart, **1989,** *7/2,* 199-214.

Tillement, J.P.; Houin, G.; Zini, R.; Urien, S.; Albengres, E.; Barré, J.; Lecomte, M.; D'Athis, P.; Sebille, B. in Testa, B. (ed.) *Advances in Drug Research,* Academic Press: London, **1984,** *13,* 60-95.

Umetsu, N.; Mallipudi, N.M.; Toia, R.F.; March, R.B.; Fukutu, T.R. *J. Toxicol. Environ. Health,* **1981,** *7,* 481-497.

Verschoyle, R.D.; Cabral, J.R.P. *Arch. Toxicol.* **1982,** *51,* 221-231.

Verschoyle, R.D.; Aldridge, W.N.; Cabral, J.R.P. in Holmstedt, B.; Lauwerys, R.; Mercier, M.; Roberfroid, M. (eds.) *Mechanisms of Toxicity and Hazard Evaluation,* Elsevier North Holland Biomedical Press: Amsterdam, **1980,** 631-634.

Vincenzini, M.T.; Favilli, F.; Stio, M.; Iantomasi, T. *Biochim. Biophys. Acta,* **1991,** *1073,* 571-579.

Waddell, W.J.; Marlowe, C. *Toxicol. Appl. Pharmacol.* **1980,** *56,* 127-140.

Ward, D.P.; Kier, L.D.; Brewster, D.W.; Wilson, A.G. *Toxicol. Lett.* **1992,** *Suppl.,* 315.

Watkins III, J.B. in Siegers, C.P.; Watkins III, J.B. (eds.) *Biliary Excretion of Drugs and Other Chemicals, Progr. Pharmacol. Clin. Pharmacol.,* Gustav Fischer Verlag: Stuttgart, **1991,** *8/4,* 357-388.

WHO, *Paraquat and Diquat, Environmental Health Criteria 39,* Geneva, **1984**.

Wilson, A.G.E.; Feng, P.C.C.; Hopkins, W.E.; Lee, A.A. *Toxicol. Lett.* **1992,** *Suppl.*, 315-316.

Plant Metabolism and the Design of New Selective Herbicides

W. J. Owen, **G. deBoer**, DowElanco Discovery Research, 9330 Zionsville Road, Indianapolis, IN 46268-1054, U.S.A.

The selective action of a number of important classes of herbicides can be attributed to differential metabolism between crop and target weed species. A number of examples exist where tolerance is conferred by differences in rate of metabolism to non- or less phytotoxic products. In other cases different routes of metabolism are employed by the crop and weed species, and the relative toxicities of the corresponding products contribute to crop selectivity. Important routes of metabolic attack in plants include alkyl and aryl hydroxylations, *O*- and *N*-dealkylations, hydrolysis and conjugation with sugar moieties or with glutathione. Examples of what we have learned about the relative abilites of various plant species to carry out some of these transformations will be discussed, taking examples from the literature as well as from DowElanco's triazolopyrimidine sulfonanilide chemistry. With the development of greater interest in the enzymology of plant glutathione transferases and microsomal mixed function oxidases in particular, the basis is now being laid for a better understanding of the substrate specificities of herbicide detoxification enzymes from various sources in the absence of interference from factors such as rate of uptake, translocation and subcellular partitioning.

A broad range of factors can contribute to herbicide selectivity, including soil placement, rates of absorption and subsequent translocation, localisation within the plant and at the subcellular level, differences in active site sensitivity between weeds and crops or the deliberate creation of herbicide-resistant crops through mutational breeding or gene-transfer techniques (Ashton & Crafts, 1981). However, the most important crop selectivity mechanism is differential metabolism between crops and weeds. This mechanism accounts for crop selectivity to numerous classes of herbicides including triazines, acetanilides, phenoxy, aryloxyphenoxypropionates, diphenyl ethers, substituted phenylureas, imidazolinones, sulfonylureas and the triazolopyrimidine sulfonanilides (Owen, 1991). Of the various modes of metabolic attack on xenobiotics encountered in plants, alkyl and aryl hydroxylations, *O*- and *N*- dealkylations, hydrolysis and conjugation with the tripeptide glutathione (homo-glutathione in legumes) or with sugar moieties are frequently encountered. The first four of these mechanisms, considered for the most part to be mediated by microsomal mixed function oxidase enzymes, are particularly common in the context of metabolic selectivity (Cole, 1983) and are a focus of attention in the present paper. The susceptibility of various functional groups to oxidative metabolism will be illustrated by examples from the literature and from studies carried out within DowElanco on the ALS-inhibiting triazolopyrimidine sulfonanilides. Commercial triazolopyrimidine products or advanced candidates exist with selectivities to wheat, barley, rice, corn, soybeans and cotton. Our approach has been to try to exploit metabolic trends suggested by our existing knowledge of the detoxification capabilities of crop species by incorporating structural features considered to enhance the likelihood of selectivity being achieved. A detailed structure/selectivity study for particular modes of metabolic attack was not attempted since this would have required the synthesis of a large number of analogues that such a rational approach would aim to avoid. The extent to which this approach was successful will be

1054–7487/95/0257$12.00/0

discussed along with reference to some interesting new transformations to emerge from this work and from the literature.

The ultimate goal would be to be able rationally to design metabolism-based selective herbicides. One of the key obstacles to achieving this has been the paucity of knowledge on the activities and specificities of plant enzymes and isoenzymes involved in xenobiotic metabolism. However, some progress is being made in this respect, particularly for *mfos* and glutathione *S*-transferases, and reference will be made to these developments and the implications for rational design of the growing ability to evaluate the comparative metabolic capabilities of crop plants in the absence of other factors which contribute to selectivity.

Oxidative metabolism and crop selectivity

The difficulties associated with attempts to rationalise the design of selective agrochemicals based on simple knowledge of the susceptibilities of functional groups to specific modes of metabolic attack are well exemplified by the fungicide griseofulvin. This compound contains three ring methoxy substituents and in the course of its metabolism by various fungi *O*-demethylation occurs specifically at a different ring position. This is difficult to rationalise on simple chemical reactivity grounds and serves as a good illustration of the importance of substrate specificities of enzymes responsible for xenobiotic metabolism. This is further exemplified by the herbicide bentazon. Selectivity in this case has been attributed to rapid ring hydroxylation in tolerant crop species to yield less phytotoxic products. However, whereas hydroxylation is confined to the 6-position in species such as rice and wheat (Mine *et al*, 1975; Retzlaff & Hamm, 1976), in soybean both the 6- and 8-hydroxy derivatives are formed (Otto *et al,* 1979). Direct hydroxylation of aromatic rings is a prominent mechanism in cereals, notably wheat, and particularly well documented additional examples include the aryloxyphenoxy-propionate diclofop-methyl (Tanaka *et al,* 1990), the sulfonylurea chlorsulfuron (Sweetser *et al,* 1982), and the phenoxy-alkanoate 2,4-D (Guroff *et al,* 1976).

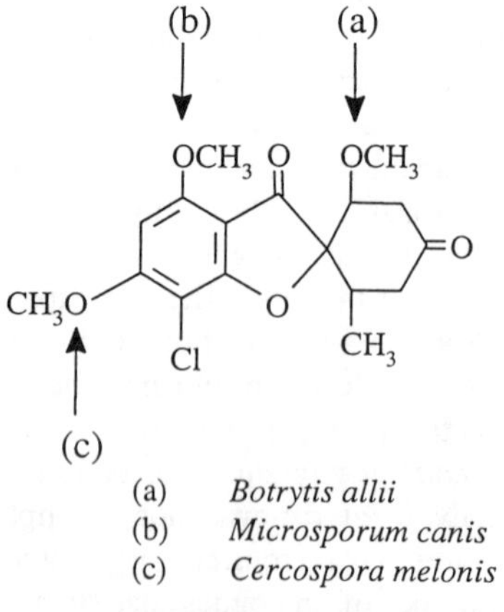

(b) (a)

OCH_3 O OCH_3

=O

CH_3O O

Cl CH_3

(c)

(a) *Botrytis allii*
(b) *Microsporum canis*
(c) *Cercospora melonis*

Fig. 1. *O*-demethylation of griseofulvin by various fungi

An interesting recent example is the metabolism of nicosulfuron in corn which, somewhat unusually, detoxifies this sulfonylurea by hydroxylation of the pyrimidine ring in the 5-position (Brown *et al,* 1991a) in preference to *O*-dealkylation of two adjacent methoxy substituents.

In contrast, tolerance of rice to the structurally similar bensulfuron-methyl results from selective *O*-demethylation of the 4-methoxy pyrimidine substituent rather than from ring hydroxylation (Brown *et al*, 1991a). Bensulfuronmethyl has a methylene group in the sulfonylurea bridge which proved essential for metabolism-based selectivity in rice. This clearly illustrated how oxidative attack at a particular site may be significantly influenced, if not dependent on, substituents remote from the site of metabolism. In some cases, however, it may be unclear as to whether substantial changes in selectivity amongst close analogues are a consequence of an impact of substituents on metabolism at a susceptible site or on overall physical/chemical characteristics with resulting effects on sub-cellular partitioning.

A particularly interesting feature of sulfonylureas is that seemingly subtle modifications to the parent structure can result in targeting of analogues for selective use in

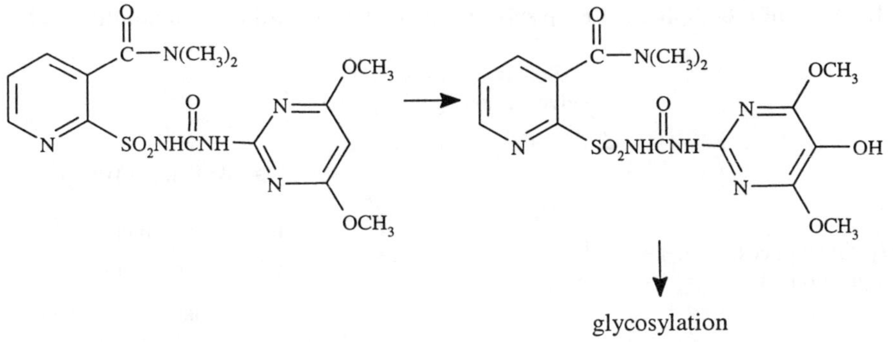

glycosylation

Fig. 2. Metabolism of nicosulfuron in corn

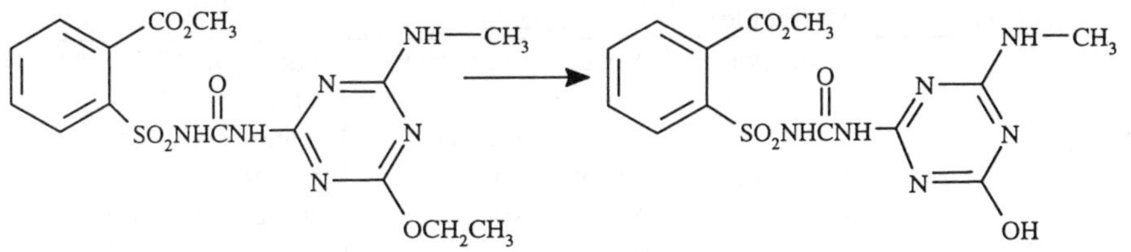

Fig. 3. *O*-dealkylation of ethametasulfuron in canola

different crops. Thus ethametsulfuron-methyl controls broad-leaved weeds in oilseed rape (canola) and constitutes another recent example where selectivity is based on *O*-dealkylation, in this case of the triazine 4-ethoxy substituent (Brown *et al,* 1991a). Interestingly, though the same ring system contained a 6-methylamino substituent *N*-dealkylation did not feature in the detoxification of ethametsulfuron in canola. Recent examples of the role of alkyl hydroxylation in metabolic selectivity are to be found amongst experimental soybean/wheat selective sulfonylureas such as DuPont's DPX-A5969 and DPX-L8747 (Brown *et al,* 1991b). Earlier well documented examples of the role of alkyl hydroxylation and of *O*-and *N*-dealkylation in conferring crop selectivity in particular to wheat are to be found within the substituted phenylurea and imidazolinone herbicide classes (Owen, 1989).

Metabolism based selectivity of triazolopyrimidines

The triazolopyrimidine sulfonanilides, like the sulfonylureas, represent a large class of herbicides within which comparatively minor changes in substitution patterns can result in selectivity being achieved in a number of crop/weed situations. Metabolism is a common factor accounting for their selectivity and the established sites of a variety of modes of oxidative attack in different plant species are summarised in Table 1. For convenience, the structures of specific triazolopyrimidine sulfonanilides referred to throughout the text are given in Fig. 4. Triazolopyrimidine I is an experimental analogue which displays broad selectivity between monocots and dicots (GR_{50} values generally 1-2 orders of magnitude higher vs monocots), with the notable exception of sorghum which is extremely susceptible to I. The striking relationship between GR_{50} and rate of metabolism based on disappearance of parent (Table 2) strongly implicated metabolism as a key element in selectivity.

Examination of metabolite profiles in all nine species over a 145h time course suggested that differences in metabolic pathways were unlikely to be a factor, with no striking differences between the profiles of monocots and dicots, and a number of major metabolites

Table 1. Sites of metabolism of crop selective triazolopyrimidine sulfonanilide herbicides

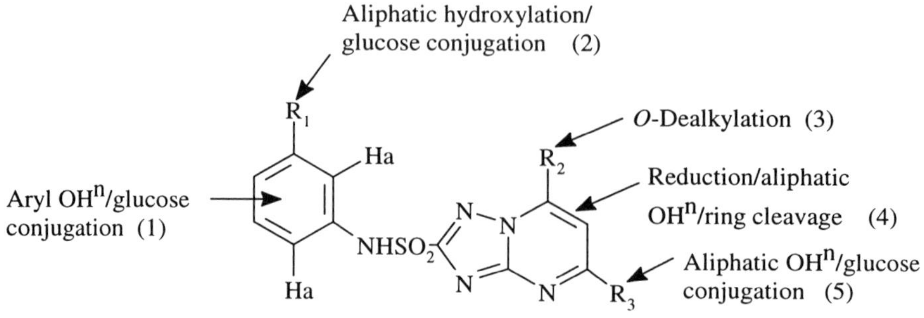

Aliphatic hydroxylation/ glucose conjugation (2)

R_1

O-Dealkylation (3)

Ha

R_2

Reduction/aliphatic OH^n/ring cleavage (4)

Aryl OH^n/glucose conjugation (1)

N N

$NHSO_2$

Aliphatic OH^n/glucose conjugation (5)

Ha

N N R_3

Metabolic Reaction	Selective Herbicide	Plant Species	Reference
(1)	Flumetsulam	Wheat, corn, barley	Frear, *et al*, 1993
		lambsquarter	Chang, *et al*, 1992
	I	Wheat, corn, rice	Hodges, *et al*, 1990
(2)	Metosulam	Wheat, rice) -	Swisher, *et al*, 1991
	IV	Wheat, rice)	Gerwick, *et al*, 1994
(3)	III	Soybeans	Gerwick, *et al*, 1994
	VIII	Cotton	DowElanco unpublished
(4)	Flumetsulam	Soybeans	Chang, *et al*, 1992
	II	Soybeans	Gerwick, *et al*, 1994
(5)	I	Rice, wheat, corn	Hodges, *et al*, 1990
	Flumetsulam	Wheat, corn, barley	Frear, *et al*, 1993

common to all species examined (Cole, 1983). The key difference was in rate of metabolism which was 3-5 fold faster in monocot species. The major route of metabolism involved alkyl hydroxylation at the 5-position, but not the 7-position, of the triazolopyrimidine ring followed by conjugation to the glycoside.

Not unexpectedly, hydroxylation of the aniline ring, in an as yet unspecified position, constituted a minor route. Interestingly, in extracts examined in more detail, only the polar glucoside of the phenolic aglycone was detected suggesting a very rapid rate of conjugation in this case. Although a similar pathway occurs in the susceptible dicots, evidently metabolism rates in the order of 0.1 to 0.2 μmol/h/gdw were insufficient to protect these species. The I_{50} value for the 5-hydroxy metabolite of I in the ALS assay was 0.4 μM compared to 0.1 μM for the parent compound. This implied that this metabolite had the potential to be herbicidal, since other triazolopyrimidines with I_{50}s in the same range or greater have demonstrated good whole plant activity. This points to the subsequent conjugation step as being the key detoxification step since the glycoside was inactive *vs* ALS even at 200 μM. It may be that the susceptibility of sorghum to I may be due to a diminished ability to conjugate the hydroxylated product, but this possibility has not been fully explored.

In contrast, flumetsulam, one of the active ingredients in Broadstrike[1], demonstrates a high degree of soybean selectivity in addition to multi-crop selectivity *vs* corn, wheat and barley and a broad spectrum of activity *vs* broadleaf weeds. With the notable exception of lambsquarter, comparison of whole plant GR_{80}s with the half life of the parent herbicide in a number of species revealed a clear relationship (Table 3). Despite this being the case, however, crop selectivity of flumetsulam turned out to owe as much to differences in the

[1] Trade mark of DowElanco

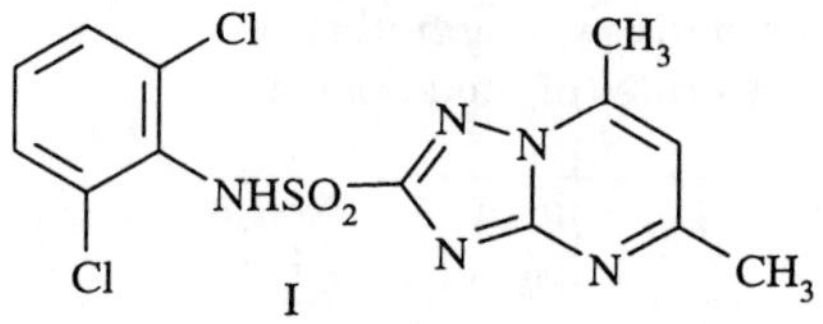

I

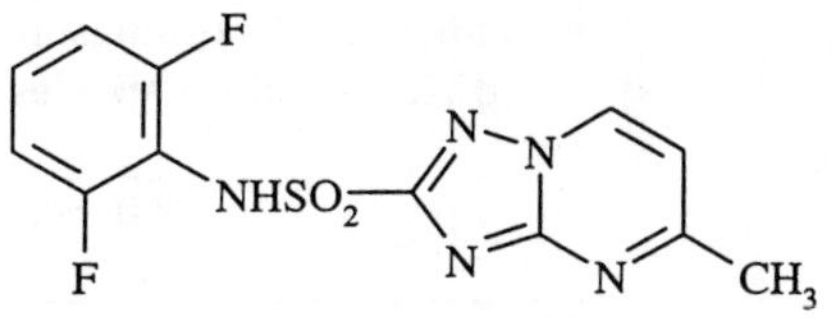

Flumetsulam

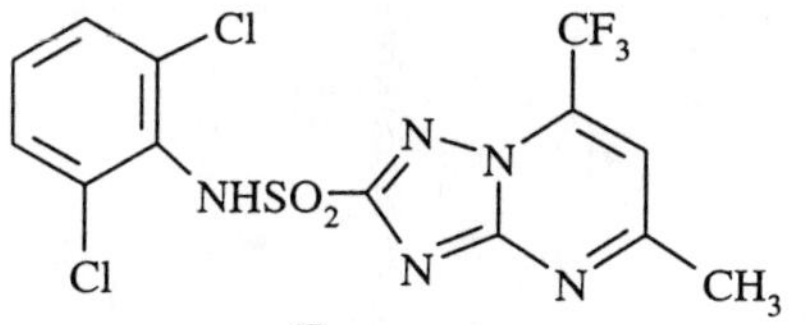

II

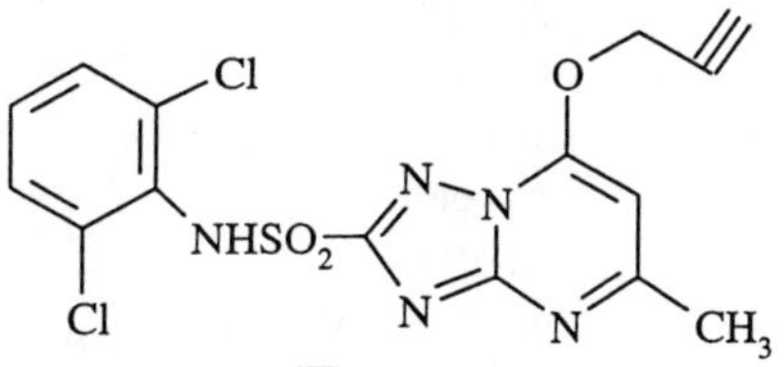

III

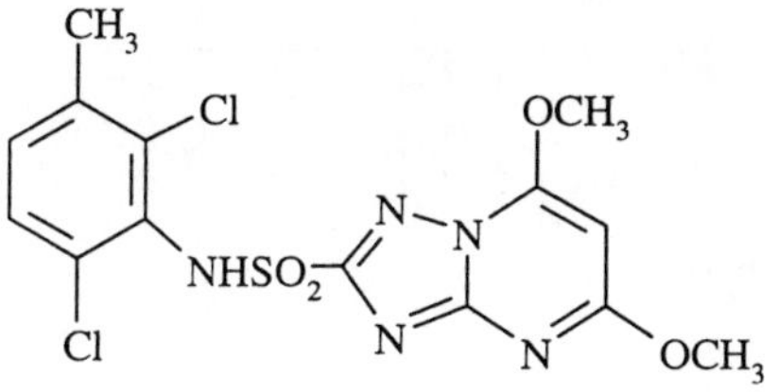

Metosulam

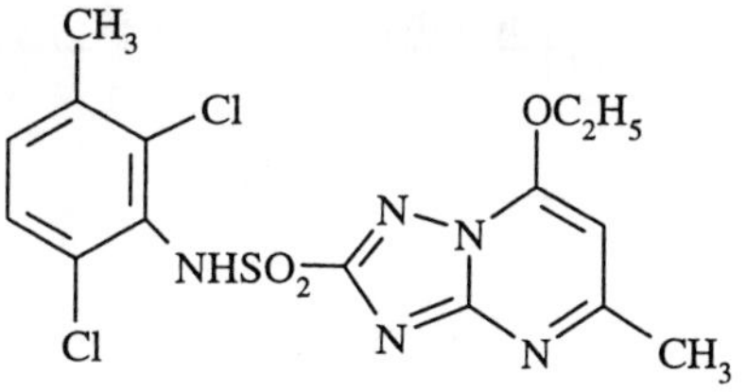

IV

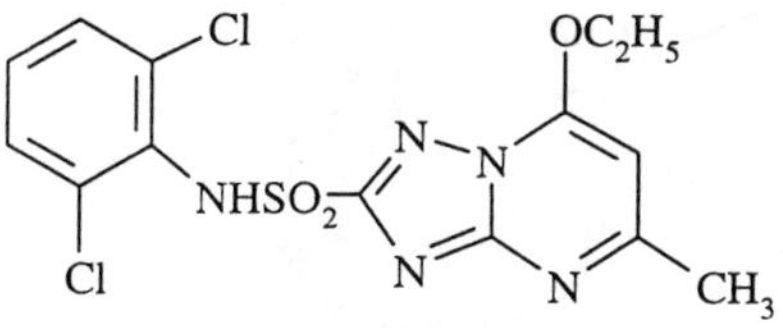

V

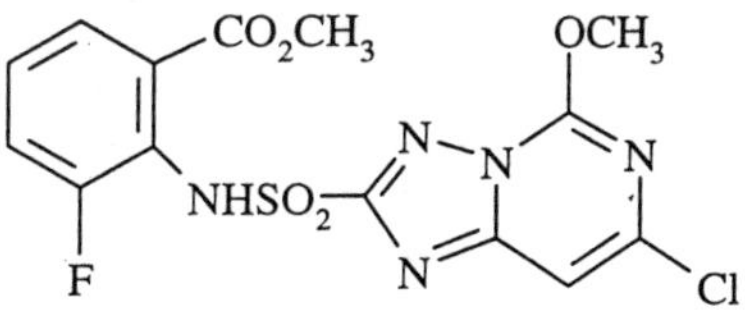

VI

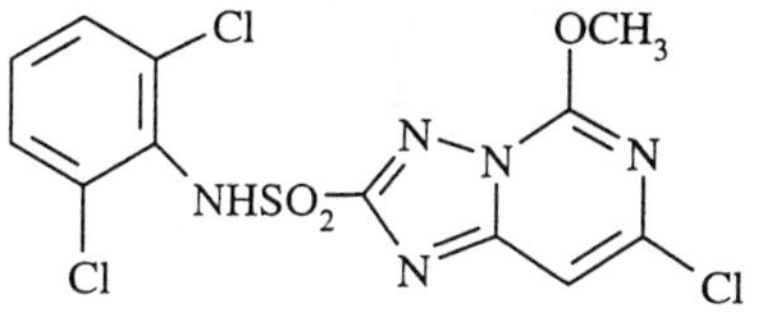

VII

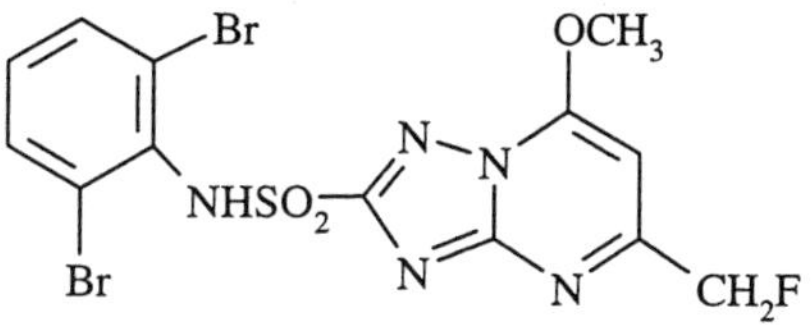

VIII

Fig. 4. Structures of Triazolopyrimidine sulfonanilides referred to in text

route of metabolism as to differences in rate (Chang *et al*, 1992). The dominant metabolic route in most species, except for soybean and lambsquarter, involved the hydroxylation of the 5-methyl position of the triazolopyrimidine ring, in a manner analogous to I, and resulting in a similar degree of detoxification.

Though oxidative metabolism of flumetsulam was also a dominant mechanism in lambsquarter, a susceptible species, this involved aryl hydroxylation at the 4-position of the aniline ring followed by glucosylation, which may also represent an additional mechanism in wheat, barley and corn (Frear *et al*, 1992). On the other hand, in tolerant soybean a somewhat unexpected transformation accounted for selectivity. Rather than utilising the sites susceptible to oxidative attack as in other species, hydroxylation occurred primarily at either position 6 or 7 of

Table 2. Comparison of the relative susceptibility of different plant species to I with rates of metabolism

Species	GR_{50}(mg litre $^{-1}$)	Rate of Metabolism (nmol.$g^{-1}h^{-1}$)
Rice	415	0.94
Wheat	2000	0.83
Sorghum	10.2	0.72
Corn	550	0.60
Pigweed	4.2	0.24
Velvetleaf	15.0	0.20
Jimsonweed	14.0	0.18
Soybean	2.5	0.17
Morning-glory	10.8	0.16

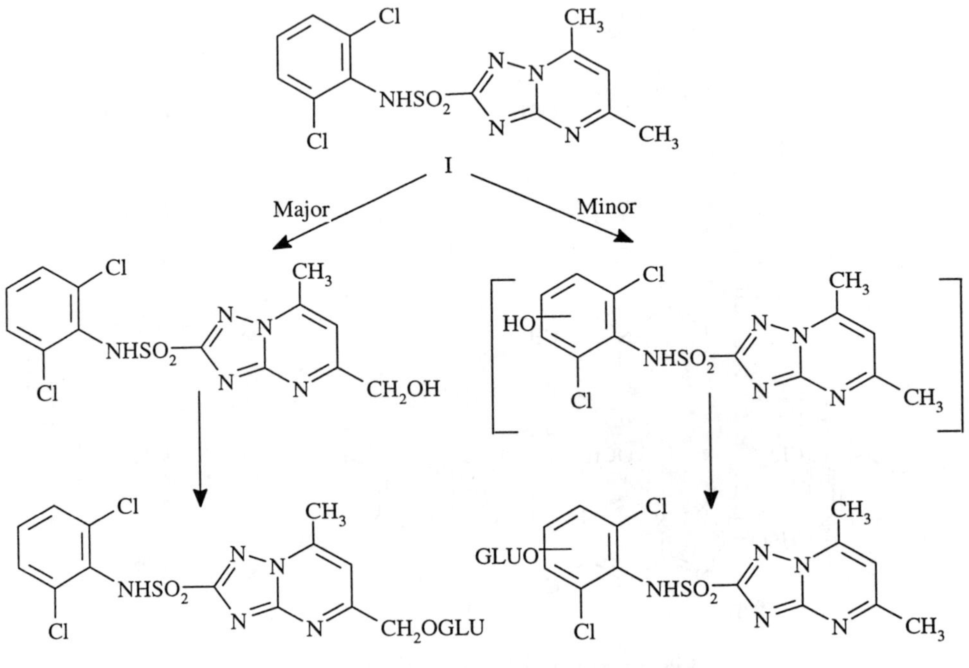

Fig. 5. Plant Metabolism of Compound I

the triazolopyrimidine ring, after initial pyrimidine reduction, to give hydroxytetrahydroflumetsulam. This was subsequently followed by facile ring opening to give hydroxybutyl-DFATSA. Ring opening offered an opportunity for further degradation by *N*-dealkylation. Soybean utilised this route to generate the aminotriazole, DFATSA, in competition with glycosylation of its hydroxybutyl precursor.

The primary lambsquarter metabolite is 8-fold less active than flumetsulam *vs.* ALS, whereas the initial product of soybean metabolism, hydroxytetrahydroflumetsulam, is 70-fold less active and DFATSA is totally inactive. The half-lives of flumetsulam in soybean and lambsquarter are relatively close, suggesting that selectivity between these two species is more likely to be a consequence of the different metabclic pathways employed.

Detoxification of flumetsulam by cleavage of the heterocycle constituted a novel mechanism

Table 3. Comparison of phytotoxicity and metabolism of Flumetsulam

Species	GR_{80}(g/ha)	T½ (h)
Corn	>280	2
Soybean	> 280	15
Velvetleaf	22	No metabolism detected
Lambsquarter	20	23
Pigweed	25	104
Morning-glory (ivyleaf)	85	115
Morning-glory (pitted)	199	25

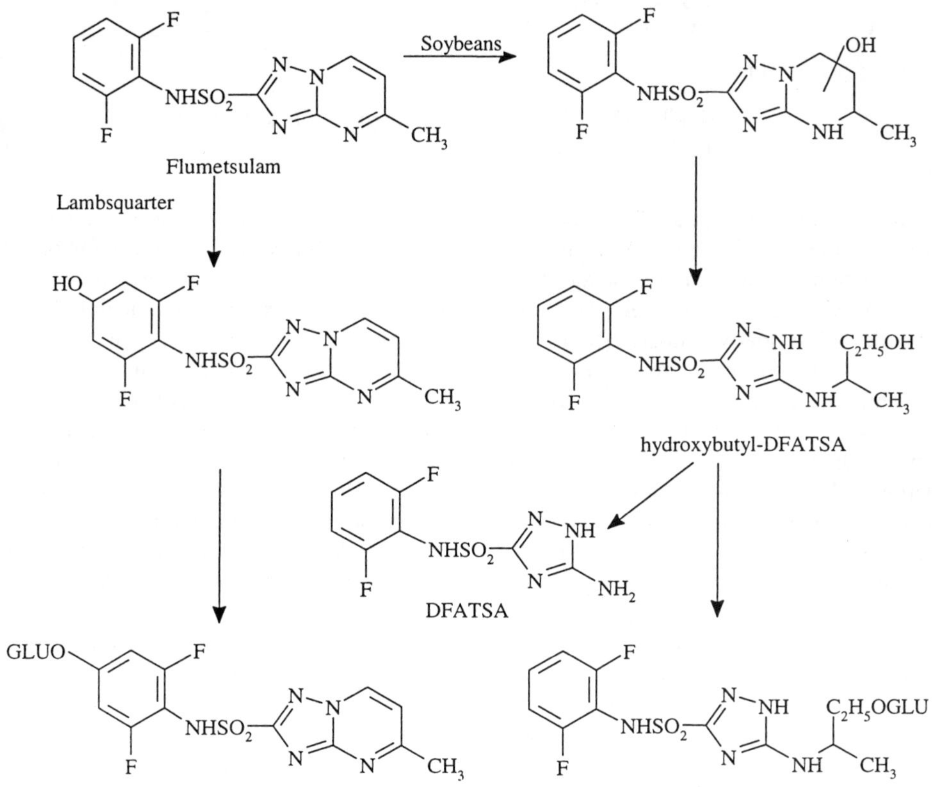

Fig. 6. Pathways of Flumetsulam metabolism in lambsquarter and soybean

apparently unique to soybean amongst the species studied. This mechanism was also utilised in the case of the soybean selective experimental analogue, II, from which the primary metabolite generated was the amino-triazole derivative (Fig.7) which was 500-fold less active *vs* ALS than the parent compound. Comparing the metabolism of I and II in soybean it is somewhat surprising that in the case of 5,7 dimethyl substitution on the triazolopyrimidine ring metabolism proceeds via 5-methyl hydroxylation, whereas replacement of the 7-methyl with a trifluoromethyl substituent dramatically modified the detoxification mechanism to ring reduction, hydroxylation and subsequent cleavage. However, soybean did not use this novel mechanism in the case of another very closely related and soybean selective analogue III (Fig.7). It is unclear whether the presence

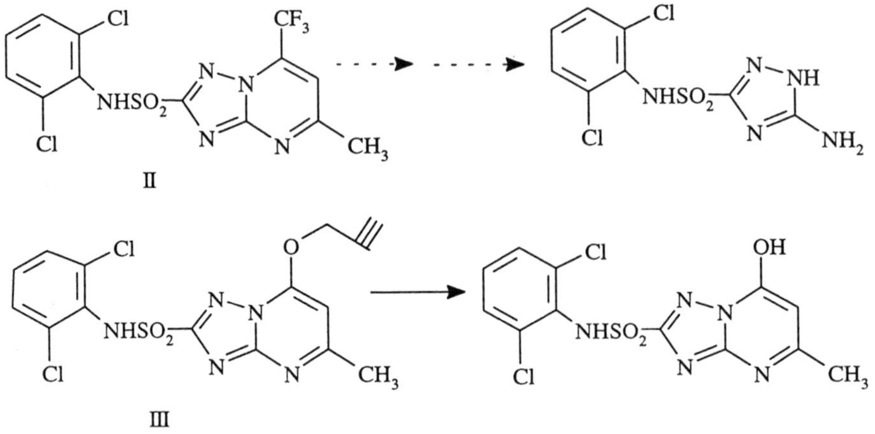

II

III

Fig. 7. Metabolism of Triazolopyrimidines II and III in soybeans

of the alkoxy substituent at the 7-position of the triazolopyrimidine ring rendered reduction and hydroxylation-mediated ring cleavage less favourable or whether it presented an alternative mode of oxidative degradation not available in other soybean selective analogues. Thus the major metabolite of III in soybeans was the 5-methyl, 7-hydroxy-triazolopyrimidine product of 7-alkoxy *O*-dealkylation, which is 15-fold less active *vs* ALS than the parent herbicide. To our knowledge, this is the first report of this mode of oxidative metabolism constituting a primary route of herbicide detoxification in soybeans. From an electronic standpoint it was somewhat unexpected that, in contrast to 7-trifluoromethyl, the alkoxy substitution at position 7 failed to promote the established ring cleavage route but this might well have been due to its facile *O*-dealkylation. Though *O*-dealkylation was recently shown to operate in the metabolism of the soybean selective sulfonylurea, thifensulfuron, this was confined to further modification of a late stage metabolite rather than being a primary detoxification mechanism (Brown *et al*, 1993).

Previous reference was made to the hydroxylation of heterocycle substituents as a means of detoxification of I in wheat. In further attempts to produce additional wheat selective triazolopyrimidine analogues, use was made of the known ability of this species to hydroxylate alkyl substituents on phenyl rings (Owen, 1989). Two such compounds are metosulam, under commercial development for broad leaf weed control in small grain cereals and rice, and the related experimental analogue IV. The 3-methyl substitution on the aniline ring confers broad grass selectivity without adversely affecting broadleaf weed activity. Both these compounds were metabolised by alkyl hydroxylation of the 3-methyl substituent followed by glucose conjugation. Although in each case the hydroxylated products were some 10-fold less inhibitory *vs* ALS than the corresponding parent herbicides, their I_{50} values were none-the-less of the order of 10-15 μM. This suggested that conversion of the 3-hydroxy metabolites to their glucose conjugates was crucial for effective detoxification. A similar metabolic route was apparent for metosulam and IV in several broadleaf weeds, and selectivity resulted from differences in rates of detoxification. It is interesting to note that for these particular analogues, there was little evidence of either *O*-dealkylation of the heterocycle alkoxy groups, a mechanism also prevalent in metabolic selectivities in wheat, or hydroxylation of the heterocycle methyl substituent of IV (c.f. Compound I and flumetsulam). However, chemical substituents which confer selectivity in bio-assays need not necessarily be modified during the detoxification process. Instead they may either alter the physical/chemical properties of the molecule so

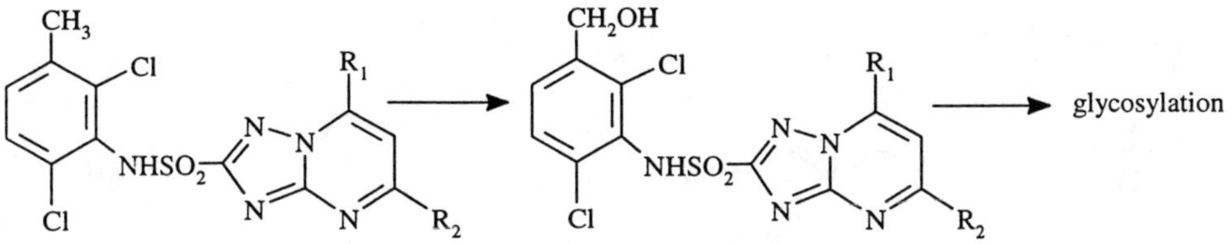

$R_1 = R_2 = -OCH_3$ = Metosulam; $R_1 = -OC_2H_5$, $R_2 = -CH_3$ = IV

Fig. 8. Fate of metosulam and Compound IV in wheat and rice

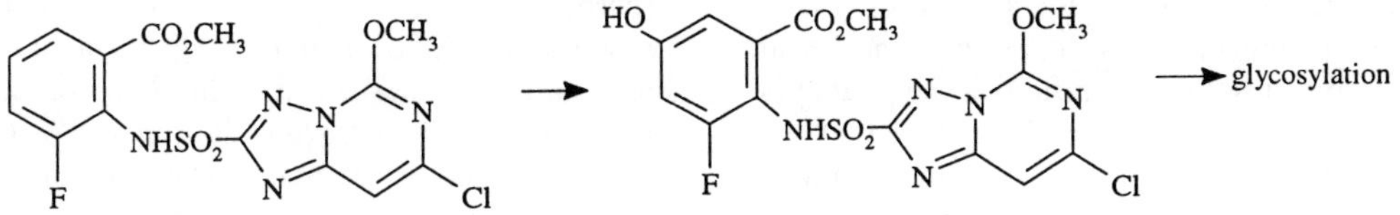

Fig. 9. Pathway of Triazolopyrimidine VI metabolism in corn

as to allow better delivery to the site of metabolism, or modify the reactivity at another site in the molecule, rendering it more susceptible to degradation.

The important role of the 3-methyl substitution in conferring grass selectivity to these triazolopyrimidines is also evident from a comparison with the 2,6-dichlorophenyl analogue V with the same substitution pattern on the heterocycle as IV. Corn metabolised IV very rapidly in relation to the 2,6-dichloro analogue. Though wheat metabolised both compounds comparatively rapidly, this difference between the two analogues was still retained. However, whether wheat metabolism of V proceeds by hydroxylation of the aniline ring, or by oxidative attack on either of the triazolopyrimidine substituents in the absence of the 3-methylphenyl substituent remains to be established. Certainly, aryl hydroxylation products appeared in the metabolite profiles of I and flumetsulam in corn (see above) and this mechanism was established as the primary detoxification route contributing to the selectivity of the experimental corn selective analogue VI with the 1,5c triazolopyrimidine configuration. The 2-carboxymethyl, 6-fluoro substitution pattern on the aniline ring of this analogue would appear to render the 4-position more susceptible to aryl hydroxylation. This receives support from comparison with the 2,6-dichloro aniline analogue VII, which is non-selective in corn, probably due to its 12-fold slower rate of metabolism. Ring hydroxylation was previously shown to feature significantly in the detoxification of the corn-selective Ciba-Geigy sulfonylurea primisulfuron, which also possesses a carboxymethyl substituent at the 2-position of the targeted aniline ring (Fonné-Pfister *et al*, 1990).

Differential metabolism has also been implicated in the selectivity of triazolopyrimidines in cotton. Thus cotton metabolised the dibromo analogue VIII by *O*-dealkylation of the 7-methoxy group in a manner analogous to the metabolism of III in soybean. Clearly the fluoromethyl substituent in the 5-position plays an important role in selectivity, as the corresponding 5-methyl compound exhibits greatly reduced cotton tolerance. However, in this case the introduction of this substitution turned out to activate the 7-methoxy to oxidative attack rather than to serve as a direct target for metabolic transformation. Again, the effectiveness of this mechanism in bringing about detoxification is evident from the 137-fold difference in potency of the parent herbicide and the 7-hydroxy metabolite *vs* ALS (I_{50} values = 0.86 μM and 0.11 μM for VIII and its metabolite, respectively).

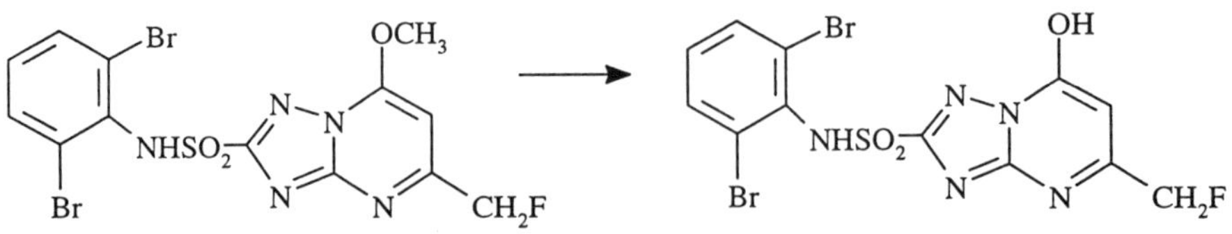

Fig. 10. Metabolism of Triazolopyrimidine VIII in cotton

Enzymological aspects

In many of the cases of differential metabolism cited above and in the literature at large the involvement of specific enzymes with different substrate specificities or catalytic activities is inferred from the knowledge that minor changes in structure can have dramatic effects on selectivity. In others, however, it is not possible to dissect out the extent to which factors such as differences in uptake, translocation and subcellular partitioning between species have influenced metabolism rates or metabolite profiles. These questions can only be resolved by resorting to the use of isolated enzyme systems involved in the metabolism of xenobiotics in plants in order to obtain unequivocal information regarding their specificities for herbicide substrates. However, in the case of cytochrome P450-mediated detoxifications, progress towards this has until comparatively recently been severely hampered by an inability to obtain plant microsomal preparations active in oxidative metabolism of herbicides.

Improvements in extraction procedures such as the inclusion of glycerol, thiol reagents, PVP/XAD, protease inhibitors and, occasionally, lithium carbonate to grinding media and/or the addition of a sephadex G25 filtration step, which presumably removes endogenous inhibitors, have yielded extracts active in the metabolism of several herbicides from a number of crop sources (see 18). An additional problem has been that, though plant P450-containing monoxygenases involved in oxidative detoxification of xenobiotics are generally constitutive, their activities are barely detectable and need to be stimulated (induced) to facilitate *in vitro* studies. Thus the use of pre-treatments with agents known to elevate P450 levels, such as phenobarbital, 2,4-D or certain herbicide safeners, was crucial to these developments. As a consequence, wheat microsomal preparations have been obtained which are capable of oxidatively metabolising chlorsulfuron, triasulfuron, diclofop, chlortoluron, diuron, isoproturon, linuron and 2,4-D, via alkyl and aryl hydroxylations and *N*-dealkylations. In the case of diclofop, it has recently been demonstrated that wheat microsomes generated all three ring-hydroxy isomers. This finding was in excellent accord with data generated in whole plants (Zimmerlin & Durst, 1992).

In addition, microsomal preparations from sorghum have recently been shown to catalyse the *O*-demethylation of the methoxypropyl side chain of metolachlor (Moreland & Corbin, 1991). Studies with safeners, known to enhance monoxygenase activity, have shown some interesting differential effects between herbicide substrates. Thus whilst a range of safeners enhanced metabolism of bentazon and the endogenous substrate lauric acid by sorghum microsomes, *O*-dealkylation of metolachlor was depressed and hydroxylation of cinnamic acid unaffected (Moreland *et al*, 1993a). These differential responses suggest the involvement of different P450-linked monoxygenase isoforms in herbicide metabolism. This has received support from studies with such P450 inhibitors as tetcyclacis, piperonyl butoxide, ABT, SKF-525A and tridiphane in a number of laboratories.

Interesting specificities are also beginning to emerge. Whereas sorghum microsomes were unable to metabolise triasulfuron and primisulfuron, similar preparations from corn oxidised both these substrates and nicosulfuron, but not imazethapyr (Barrett & Maxson, 1991), in addition to those mentioned above (Moreland *et al*, 1993b). The extent to which these herbicide monoxygenase activities are mediated by microsomal enzymes with defined physiological roles, but broad substrate specificities, remains unclear. Based on differential effects of P450 inhibitors, it would appear that cinnamic acid hydroxylase is not involved in herbicide hydroxylations, though there is emerging evidence that microsomal oxidation of diclofop may be mediated by lauric acid hydroxylase (Zimmerlin & Durst, 1992).

Apart from these developments, there has been encouraging progress made on the purification immunocharacterisation and molecular biology of cytochrome P450 - linked monoxygenases from a number of sources. These include cinnamic acid hydroxylase from Jerusalem artichoke, a 2,4-D monoxygenase from tulip and monoxygenases involved in sulfonylurea oxidation in *Streptomyces griseolus*. These studies are now paving the way for obtaining more detailed information on the degree of multiplicity of plant P450s and possibly for the expression of enhanced amounts of herbicide metabolising isoforms for enzymological studies.

Multiplicity of glutathione-S-transferases (GSTs) involved in herbicide metabolism, particularly in corn, has been established for some time. This is based on differential substrate specificities and safener induction of isozymes, together with the identification of several GST genes. This mode of metabolism has been especially prominent in corn and soybean selectivity but has featured little in species such as wheat. A possible exception is fenoxaprop-ethyl which is converted to a glutathione conjugate in wheat and barley, but, in this case it has not been unequivocally established whether this reaction is enzyme-mediated (Romano *et al,* 1993). This is of particular interest since in a recent study with eight phyllogenetically diverse plant species, which included cotton and corn, wheat demonstrated by far the highest GST activity with respect to the artificial substrate 1-chloro-2,4-dinitrobenzene (Anderson *et al*, 1993). Therefore, in the context of the rational design of selective herbicides, there is certainly a need to understand more about the substrate specificities of the GSTs of crop species, especially for those in wheat.

Conclusions

Detailed elucidation of the structure/selectivity relationships for a detoxification reaction in a specific crop will inevitably be dependent on the synthesis of numerous analogues, an exercise which the 'rational' approach aims to avoid. In the absence of such detailed investigations, one is left with the more simplistic approach of incorporating into new synthetic analogues structural features which have been identified as conferring metabolism-based crop, selectivity to existing herbicides. Despite the obvious lack of enzyme specificity information, not only between species but also between different herbicide chemistries within the same species, this paper has demonstrated that this approach can be successful.

However, because of a very distinct substrate specificity associated with herbicide metabolising enzymes, this extrapolated knowledge of a metabolic capability of a crop can only be utilised as a guide and cannot influence exact synthetic direction. It is not uncommon, therefore, that selectivity is established for a crop other than that being targeted, or that the metabolic transformation conferring selectivity may be somewhat unexpected due to lack of information on the influence of substituents distant from the incorporated 'metabolic handle', as in the case of flumetsulam. However, elucidation of the metabolic basis for selectivity early in a synthetic effort can help focus analoguing programmes towards the best candidate. Significantly more knowledge will be required before true rational design of metabolism-based selective herbicides becomes a common process. Progress being made with the enzymology is encouraging, particularly if this might eventually lead to the possibility of over-expression of Cyt P-450s. Not only would this provide access to larger quantities of these enzymes for assays and further characterisation, but may also allow herbicide metabolism studies to be undertaken without the requirement for radiolabelled substrates, which are frequently unavailable.

References

Anderson, M. P.; Beach, J.; Tucker, A., *Plant Physiol.* **1993**, *102*, Suppl., p.57.

Ashton, F. M. ; Crafts, A. S., *Mode of Action of Herbicides*, 2nd ed.; Wiley Interscience: New York, **1981**.

Barrett, M.; Maxson, J. M., *Z. Naturforsch.* **1991**, *46c*, 897-900.

Brown, H. M.; Brattsten, L. B.; Lilly, D. E.; Hanna, P. J., *J Agric. Food Chem.* **1993**, *41*, 1724-1730.

Brown, H. M.; Dietrich, R. F.; Kenyon, W. H.; Lichtner, F. T.; *Proc. Br. Crop Prot. Conf. - Weeds* **1991**a, Vol.2, 847-856.

Brown, H. M.; Fuesler, T. P.; Ray, T. M.; Strachan, S. D., In *Pesticide Chemistry - Advances in International Research, Development and Legislation*; Proc. of the 7th International Congress of Pesticide Chemistry (IUPAC); H. Frehse, Ed.; VCH: Weinheim, **1991**b, 257-266.

Chang, M.; Brown, S. M.; Swisher, B. A.; DeBoer, G. J.; Zakett, D.; McKendry, L. H.; Roth, G. A.; Stanga, M. A., *Abstr. Agro 30*, 204th ACS National Meeting **1992**.

Cole, D. J., In *Progress in Pesticide Biochemistry and Physiology*, Vol.3; D. H. Hutson and T. R. Roberts, Eds.; Wiley: Chichester, **1983**, 199-254.

Fonné-Pfister, R.; Gaudin, J.; Kreuz, K.; Ramsteiner, K.; Ebert, E., *Pest. Biochem. Physiol.* **1990**, *37*, 165-173.

Frear, D. S.; Swanson, H. R.; Tanaka, F. S., *Pest Biochem Physiol.* **1993**, *45*, 178-192.

Guroff, G.; Daly, J. W.; Jerina, D. M.; Ranson, J.; Witkop, B.; Udenfriend, S., *Science* **1967**, *157*, 1524.

Hodges, C. C.; DeBoer, G. J.; Avalos, J. *Pestic. Sci* **1990**, *29*, 365.

Mine, A.; Miyakado, M.; Matsunaka, S., *Pest. Biochem. Physiol.* **1975**, *5*, 566.

Moreland, D. E.; Corbin, F. T., *Z. Naturforsch.* **1991,** *46c*, 906-914. Moreland, D. E.; Corbin, F. T.; McFarland, J. E., *Pest. Biochem. Physiol.* **1993**a, *45*, 43-53.

Moreland, D. E.; Corbin, F. T.; McFarland, J. E., *Pest. Biochem. Physiol.* **1993**b, *47,* 206-214.

Otto, S.; Beutel, P.; Drescher, N.; Huber, R., In *IUPAC: Advances in Pesticide Science*; H. Greissbuhler, Ed.; Vol.3; Pergamon Press: Oxford, **1979**, 551.

Owen, W. J.; In *Herbicides and Plant Metabolism;* A.D. Dodge, Ed.; Cambridge University Press: Cambridge, **1989**, 171-198.

Owen, W. J.; In *Resistance 91 : Achievements and developments in combating pesticide resistance*; I. Denholm; A. L. Devonshire; D. W. Hollomon, Eds., Elsevier Applied Science: London, **1992**, 340-353.

Owen, W. J., In *Target Sites for Herbicide Action*; R. C. Kirwood, Ed.; Plenum Press: New York/London, **1991**, 285-314.

Retzlaff, G.; Hamm, R, *Weed Res.* **1976,** *16*, 263.

Romano, M. L.; Stephenson, G. R.; Tal, A.; Hall, J. C., *Pest Biochem. Physiol.* **1993**, *46*, 181-189.

Sweetser, P. B.; Schow, G. S.; Hutchison, J. M. *Pest. Biochem. Physiol.* **1982,** *17*, 18.

Tanaka, F. S.; Hoffer, B. L.; Shimabukuro, R. H.; Wien, R. G.; Walsh, W. C., *J. Agric. Food Chem.* **1990**, *38*, 559-565.

Zimmerlin, A.; Durst, F., *Plant Physiol.* **1992**, *100*, 874-881.

Advances In Metabolite Identification

Bruce C. Onisko, Zeneca Ag Products, Western Research Center, Richmond, CA 94804

Several mass spectrometric techniques are illustrated with experiments used in the identification of plant metabolites of the herbicide cycloate and the insecticide fonofos. (1) Data from a gas chromatograph coupled to a radioactive flow detector was used to help locate metabolite peaks in GC-MS data. (2) GC-MS-MS analysis using product ion scans was used to confirm metabolite structure in samples with coeluting impurities. (3) Packed capillary HPLC columns were used with the on-column focusing technique to analyze limited-quantity samples. (4) Parent ion scans were used to locate all compounds in a mixture that contained a specific substructural fragment. (5) LC-MS-MS analysis of a ^{13}C-labeled metabolite was useful for elucidation of a complex structure.

The use of mass spectrometry in metabolite identification has been reviewed four (Voyksner, 1990) and eight years ago (Lamoureux, 1986; Rosen, 1986). Identification of metabolites of unknown structure is still one of the most challenging problems of analytical chemistry. Metabolites of pesticides in plants, animals and soil are found in very low concentration. When metabolism studies are conducted using amounts that are biologically relevant, the quantity of metabolites that can be isolated is very limited. Mass spectrometry is the most useful spectroscopic technique to obtain structural data from sub-microgram samples.

Since it is almost impossible to obtain a metabolite sample that is chemically pure, most mass spectrometric analyses are coupled to a separation method. Gas chromatography (GC) and liquid chromatography (LC) are the most common methods employed. GC-MS analyses have the advantage of higher chromatographic resolution and sensitivity. However, metabolites (or their derivatives) need to be volatile and thermally stable for successful GC-MS analysis.

Many pesticide metabolites, and most pesticide conjugates, are polar and thermally unstable and therefore require LC-MS analysis. New sources for mass spectrometers such as fast atom bombardment (FAB) and electrospray (ESI) are able to ionize these fragile molecules without the need for derivatization (Caprioli, 1990; Fenn, 1989). Ironically, these soft ionization methods are so gentle that few structurally-useful fragment ions are generated. This problem has been solved by the development and use of tandem mass spectrometers which induce the fragmentation of ions by collisions with an inert gas.

This paper will illustrate the use of mass spectrometry for identification of pesticide metabolites. Experiments used in the identification of plant metabolites of the herbicide cycloate and the insecticide fonofos will be used as examples.

I. GC Techniques

A. Use of GC-RAM Data to Locate Metabolites by GC-MS.

Ironically, the hardest part of metabolite identification by GC-MS is often conceptually the simplest - locating the GC peak of interest in a chromatogram. The problem can be compared to "finding a needle in a haystack". The use of isotopically labeled pesticides can help dramatically. Even with an isotopically labeled precursor, however, locating the peak of interest is still difficult since there are often hundreds of GC peaks as candidates, and there is a good chance that many of these peaks are mixtures.

We have taken a straight-forward approach to simplify the problem of finding metabolite spectra in GC-MS data. Radiolabeled metabolite samples are first analyzed using a gas chromatograph coupled to a radioactivity monitor (GC-RAM). Commercially available GC-RAM detectors work by catalytic oxidation of organic compounds to carbon dioxide and water. The water is removed and then the radioactivity of the remaining carbon dioxide is measured. Samples containing as little as 1,000 dpm of carbon-14 give acceptable results.

The results of the GC-RAM analysis of a sample from a cycloate spinach metabolism study are shown in Fig. 1. Two radioactive peaks were resolved. The peak widths were broad because the RAM detector volume is large to ensure sensitive radiochemical detection. Chromatographic resolution, as shown by the response of the FID detector, was quite good. The result of GC-MS analysis of the same sample using a Finnigan-MAT model TSQ 70 triple quadrupole mass spectrometer is also shown in Fig. 1. By analyzing a mixture of standards on both instruments it was possible to predict when radioactive peaks should elute in the GC-MS data. For the spinach sample shown, the two peaks were predicted to elute in scans 696 and 746. The unknown cycloate metabolites were located in scans 693 and 752 (see Fig. 1).

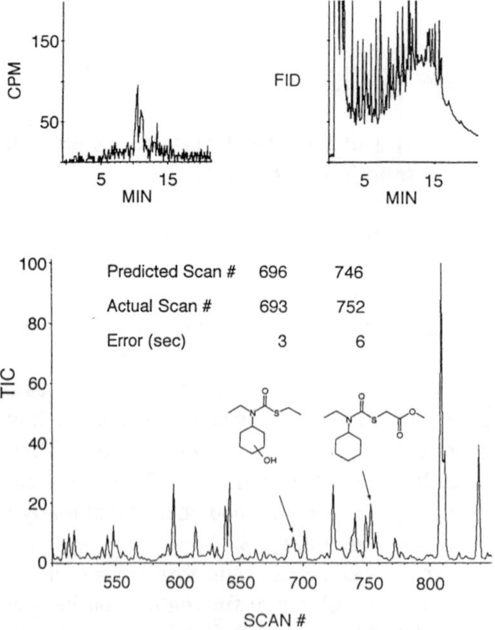

Fig. 1. GC-RAM trace (top left), GC-FID trace (top right) and GC-MS TIC trace (bottom) from analysis of a spinach extract containing two carbon-14 labeled cycloate metabolites.

The approach using GC-RAM data and external standards has saved a considerable amount of analysis time. With this method, the problem of locating the metabolite was reduced to the examination of the mass spectra of about ten GC peaks. The technique was successful despite the wide radiochemical peaks. This is because the retention times of the radioactive peak maxima can be measured with reasonable accuracy (ca. 0.1 minutes) and reproducibility.

Other problems in identification of unknowns by GC-MS may be encountered. Some molecular ions are so unstable that only fragment ions are detected. Production of more stable protonated molecules by use of chemical ionization may be successful in some instances. Often the metabolite peak is obscured by coeluting impurities. The analyst can try (1) improving the gas chromatographic separation (lower oven programming rates, smaller injection amounts, etc.), (2) further purification of the sample, or (3) tandem mass spectrometric approaches.

B. Confirmation of Metabolite Identity by GC-MS-MS.

A derivatized cycloate metabolite fraction from a wheat crop rotation study was analyzed by GC-MS using a Finnigan-MAT Model TSQ 70 triple quadrupole mass spectrometer. The results obtained are shown in Fig. 2. Because of coeluting impurities, the mass spectrum obtained shows no resemblance to the mass spectrum of the standard, even after data averaging and background subtraction. The result of the GC-MS-MS analysis of the same metabolite sample is shown in Fig. 3. To perform this experiment the first mass analyzer was set to transmit ions with a mass to charge ratio equal to that of the molecular ion of interest. The selected precursor ions were accelerated prior to collision-induced dissociation (CID) by applying a 20 V potential difference between the first quadrupole and the collision cell. (An inert gas such as Ar is usually used to generate fragments; none was needed in this case because of the instability of the radical cations selected for CID). The fragment ions were then separated according to their mass to charge ratio by the second mass analyzer. The resulting CID spectrum (see Fig. 3)

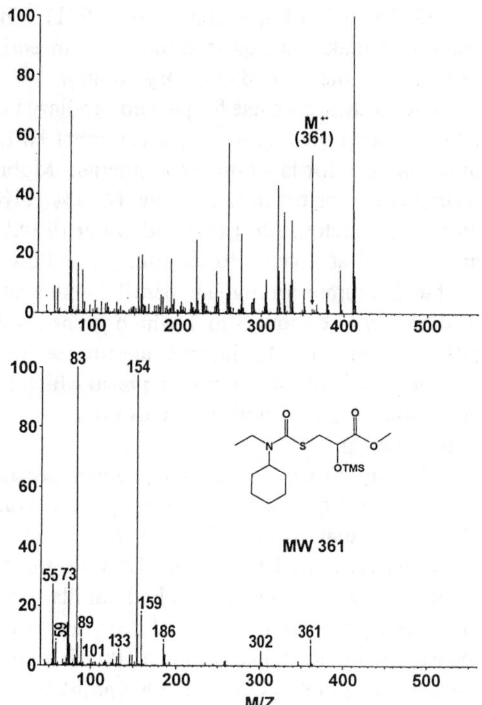

Fig. 2. Electron impact mass spectra from GC-MS analysis of a derivatized cycloate wheat metabolite fraction (top) and synthetic reference standard (bottom).

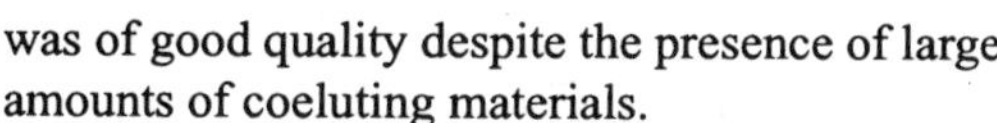

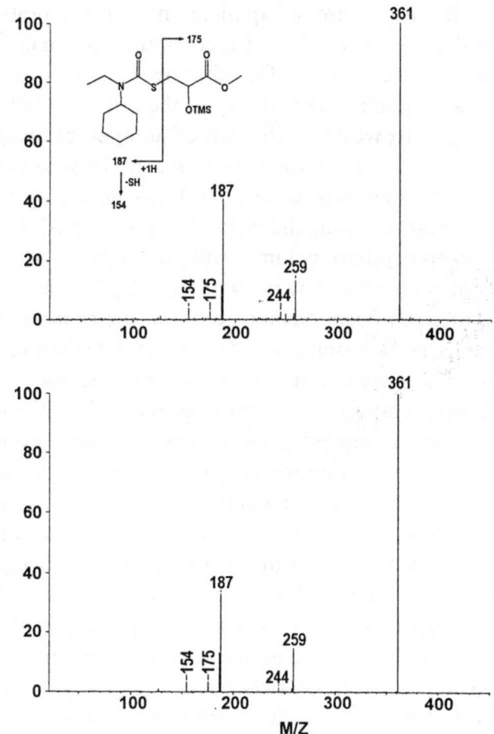

Fig. 3. CID mass spectra from GC-MS-MS analysis of a derivatized cycloate wheat metabolite fraction (top) and synthetic reference standard (bottom).

was of good quality despite the presence of large amounts of coeluting materials.

This technique has saved considerable time and effort. In a fonofos peanut metabolism study we successfully confirmed the identity of four metabolites in a crude extract by GC-MS-MS. The individual components were detected by sequentially selecting the desired molecular ions during the GC analysis. Without GC-MS-MS capability, solving this type of problem would have required further purification and analysis to obtain chemically pure GC peaks.

The GC-MS-MS technique may fail with triple quadrupole mass spectrometers if the impurities that coelute have ions of the same nominal mass to charge to ratio. High resolution precursor mass selection using tandem four sector or hybrid mass spectrometers may solve such problems (Mathews, 1987).

I. LC Techniques

A. Packed Capillary HPLC.

LC-MS analysis is required for metabolites that are not sufficiently volatile or thermally stable for analysis by GC-MS. Fast atom bombardment (FAB) and electrospray (ESI) are two ionization methods that generate ions of very low internal energy from molecules in solution (Caprioli, 1990; Fenn, 1989). Neither requires an external source of heat, and so both are well suited as LC-MS interfaces for metabolite identification. The low flow rate requirements of many of the commercially available interfaces, however, presents a problem. Flow FAB sources can tolerate a flow rate of no more than about 10 μL/min. Until recently (Chowdhury, 1990), most ESI sources had similar flow rate requirements.

In sample-limited applications, it is not practical to use conventional LC columns (4.6 mm i.d.) with an LC-MS interface that operates at a low flow rate since the majority of the sample would be split to waste. Selection of an appropriately scaled LC column solves the problem. To decrease the flow rate by a factor of 100, it is necessary to decrease column diameter by a factor of 10. Packed capillary columns of 0.322 mm i.d. are a good match for flow rates of 5 to 10 µL/min.

A significant advantage of the use of packed capillary LC columns is that the analytes eluting from the column are quite concentrated, and so detector response is greatly improved. To achieve this effect, however, it is necessary to apply the sample to the column as a very concentrated solution. In the conventional use of packed capillary LC, this is done by injection of a small volume (60 to 100 nL) of a sample that is highly concentrated. However, this can only be accomplished by injecting a small percentage of the total sample.

An injection technique called on-column focusing has been described recently (Ling, 1991; Brunmark, 1991). This method uses large injection volumes of samples in highly aqueous solvent to concentrate the sample on the front of the packed capillary reverse phase column. Using this technique, injection volumes up to 20 µL have been used. Peak widths at half height were only about 20% larger than peak widths that were observed after a 60 nL injection (Ling, 1991). This technique makes it possible to inject an entire sample onto the packed capillary column.

The hardware we use for packed capillary LC-FAB is shown in Fig. 4. A conventional HPLC pump is used for isocratic experiments. Mobile phase containing trifluoroacetic acid (0.1%), glycerol (5%), acetonitrile (x%) and water (95-x%) are pumped at about 130 µL/min. The flow is split to 2 columns arranged in parallel with a ratio of diameters of about 3 to 1 which gives a split ratio of about 10 to 1. Samples are injected onto the analytical column in mobile phase which has been diluted 2-fold with water to effect on-column focusing.

This experimental setup is convenient because most of the required hardware is already available in labs that do LC-MS. The HPLC pump is set to deliver a total flow about 10 times the desired flow rate. After the packed capillary column has equilibrated, the actual flow rate can be measured with a stopwatch and a 100 µL syringe used as a volumetric cylinder. One problem with the setup is that the capillary column flow will change if the resistance to flow of the column changes. This can occur if samples are injected that contain particulates or matrix material that precipitates in the desired mobile phase. Filtration of samples dissolved in mobile phase has minimized this problem.

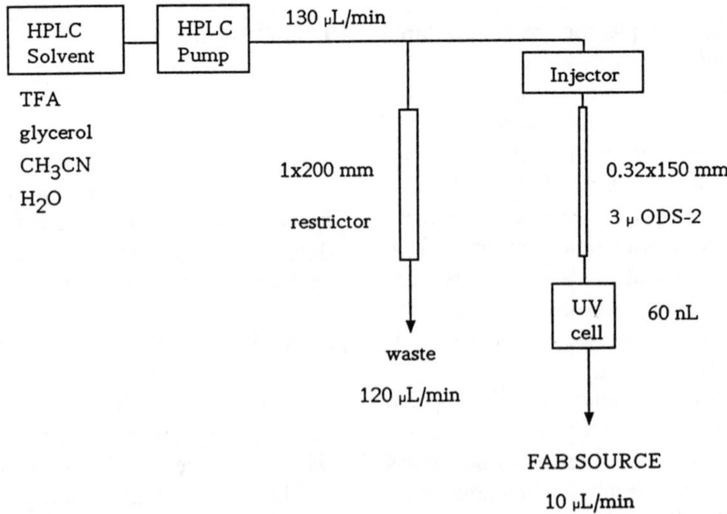

Fig. 4. Hardware schematic for packed capillary HPLC-MS experiments.

B. Use of Parent Scans in LC-MS-MS.

In ideal cases, LC-MS gives a set of chemically pure chromatographic peaks. The task is to determine which peak is the metabolite of interest. (As discussed above for GC-MS, a recognizable isotope pattern is most useful.) Once a candidate protonated molecule is found, a CID spectrum of that ion can be obtained, and the fragmentation pattern can be interpreted to suggest a structure for the unknown metabolite.

In the real world, overlapping LC peaks make the job of finding the desired peak very challenging. One technique that can help to solve such problems is LC-MS-MS using parent scans. To do this experiment, the first mass analyzer scans in the usual fashion to sequentially transmit ions of increasing mass to charge. These ions are fragmented by CID, and only fragments of a specific m/z ratio are transmitted by the second mass analyzer to the detector. This scan mode allows the precursors of a specific fragment to be located.

Choice of the correct fragment ion mass is critical. Luck and intuition are involved. Possible structures for the unknown metabolite can be suggested from a knowledge of (1) extraction properties, (2) chemical reactivity, and (3) chromatographic behavior as compared to the elution properties of related metabolites. Knowledge of the CID fragmentation of related compounds is helpful for predicting masses of significant fragment ions in the CID spectra of the candidate structures.

To illustrate this approach, the results of the LC-MS-MS analysis of radish leaf cycloate metabolites from a crop rotation study is shown in Fig. 5. Experimental details have been published elsewhere (Onisko, 1994). The sample had been obtained by preparative gradient LC of a radish leaf extract. The fraction used in this experiment was a minor component that contained about 10% of the radioactivity of the entire extract. HPLC analysis showed that it contained several closely related metabolites. LC-UV-MS analysis showed the elution of a broad unresolved UV peak that contained a large number of protonated molecules, none of which was obviously related to cycloate. For the LC-MS-MS experiment shown in Fig. 5, a sample containing approximately 15 ng of each metabolite was injected onto the capillary column

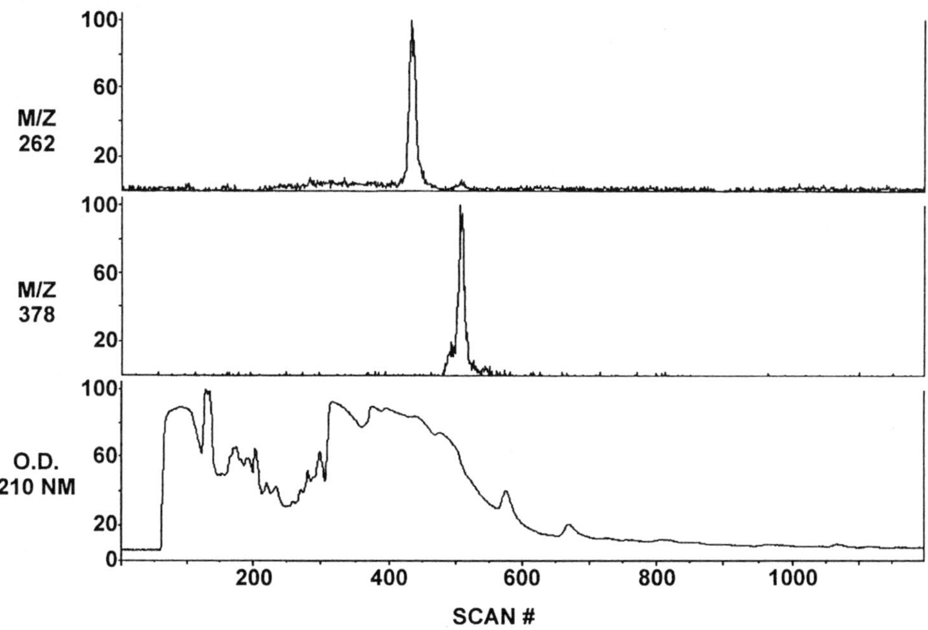

Fig. 5. LC-MS-MS analysis of cycloate metabolites in a radish leaf fraction showing UV trace (bottom), and selected ion intensities for ions of m/z 262 (top), and 378 (middle). Adapted from Onisko, 1994.

and the instrument was scanned to detect precursors of an ion of m/z 170 using 0.5 mTorr Ar as collision gas. The m/z 170 ion was chosen because it had been found to be a fragment of significant intensity in the CID spectra of ring-hydroxycycloate metabolites. Peaks in the TIC (total ion current) trace had spectra that suggested protonated molecules of m/z 262 and 378. Subsequent daughter scan analyses gave the CID spectra shown in Fig. 6. The postulated structures are glutathione conjugate degradates. Thioacetic and O-malonyl-thiolactic acid conjugates are well known plant metabolites of many xenobiotics (Lamoureux, 1989). The concentration of each of these metabolites in radish leaf was estimated to be only 20 ppb.

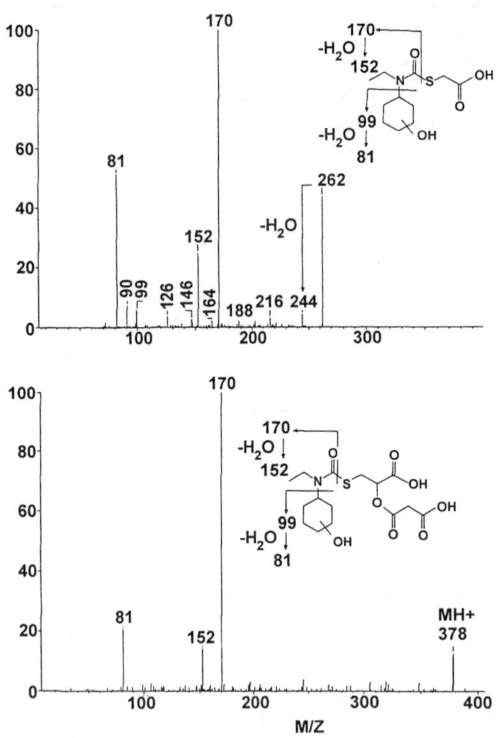

Fig. 6. CID mass spectra for ions of m/z 262 (top) and 378 (bottom) from the LC-MS-MS analysis of cycloate metabolites in a radish leaf fraction. Reproduced with permission from Onisko. Copyright 1994 John Wiley & Sons, Ltd.

C. Use of Stable Isotopes with LC-MS and LC-MS-MS.

The task of locating unknown metabolites in both GC-MS and LC-MS data is greatly simplified if the pesticide has an isotopic pattern that is easily distinguished from that of natural products. Some metabolites contain elements that rarely occur in natural occurring organic compounds. If this element has isotopes of significant natural abundance, such as chlorine, then metabolites of this pesticide are easy to recognize. In the absence of such elements, it is useful to synthesize the pesticide in isotopically labeled form. Most metabolism studies are conducted with material radiolabeled with carbon-14 for quantitation and chromatographic detection. If a single site label of at least 15 Ci/mole is used, the label can be useful for mass spectrometric purposes. It is often the case, however, that radiochemical detection needs can be met with material of considerably lower specific activity. Furthermore, uniformly-labeled material may be easier or cheaper to synthesize. For these cases, carbon-14 isotope peaks may not be of sufficient intensity to be useful for mass spectrometric detection. Synthesis of material labeled with carbon-13 can solve this problem.

In addition to assisting in the location of metabolites in GC-MS and LC-MS data, the use of isotopically labeled pesticides in tandem mass spectrometric experiments can be particularly useful for structural elucidation of metabolites with complex fragmentation patterns. This use can be illustrated by the identification of a metabolite of fonofos from a crop rotation study. The fonofos used in this study was uniformly labeled with carbon-14 in the phenyl ring. Another sample of fonofos was synthesized with all six carbons of the phenyl ring replaced with carbon thirteen. The two lots of fonofos were mixed such that the final sample had a ^{12}C to ^{13}C ratio of about 1.5/1.

A beet leaf extract was analyzed by reversed phase gradient HPLC and found to contain fifteen ^{14}C-labeled unknown metabolites of fonofos. The most non-polar of these unknowns was isolated by preparative LC, methylated with diazomethane, purified again by LC and named metabolite "A15M". The derivatized sample was

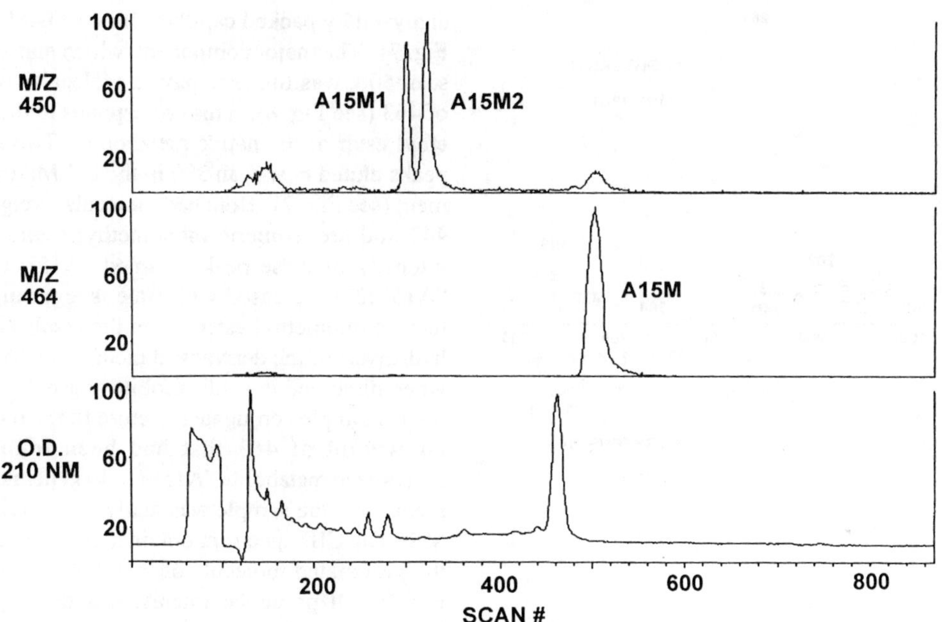

Fig. 7. LC-MS analysis of a fonofos beet leaf fraction showing UV trace (bottom), and selected ion intensities of ions of m/z 450 (top) and 464 (middle).

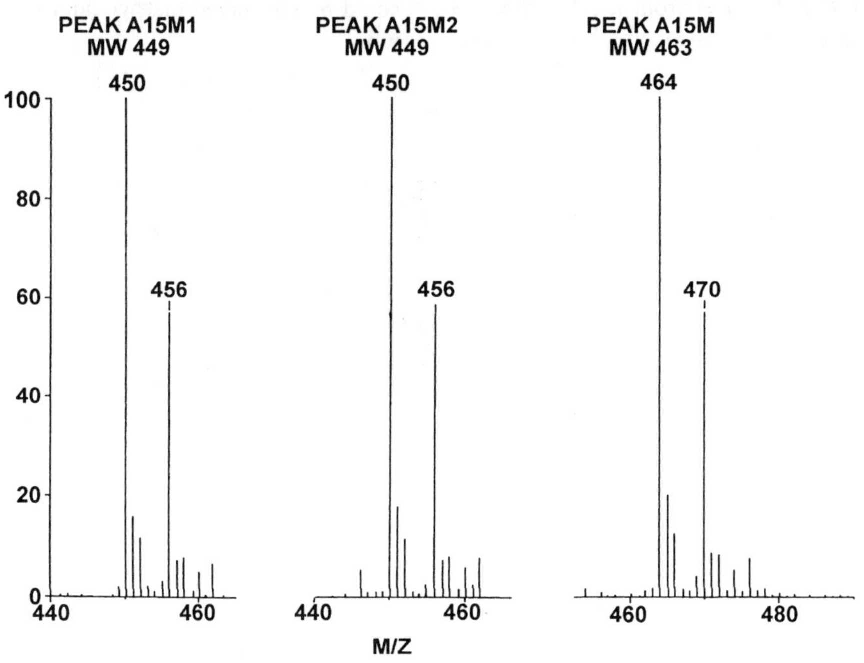

Fig. 8. Mass spectra of derivatized metabolite peaks A15M1 (left), A15M2 (center) and A15M (right) from the LC-MS analysis of a fonofos beet leaf fraction.

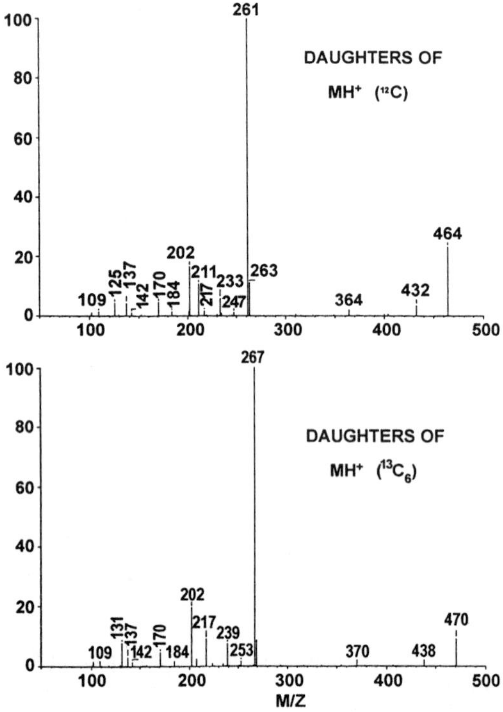

Fig. 9. CID mass spectra for ions of m/z 464 (top) and 470 (bottom) from the LC-MS-MS analysis of a derivatized fonofos beet leaf fraction.

analyzed by packed capillary flow FAB-MS (see Fig. 7). The major component, which eluted near scan 500, was found to have a molecular weight of 463 (see Fig. 8). This corresponds to the dimethyl ester of the native metabolite. Two minor peaks eluted near scan 300 in the LC-MS experiment (see Fig. 7). Both had molecular weights of 449 and are isomeric monomethyl esters. The intensity of these peaks (named "A15M1" and "A15M2") increased with time, suggesting that these monomethyl esters were the result of slow hydrolysis of the derivatized metabolite "A15M" when dissolved in acidic mobile phase.

No simple conjugate structure fit the molecular weight of 463 that had been found for derivatized metabolite "A15M". To generate fragment ions, the sample was analyzed by LC-MS-MS. The CID spectrum obtained by fragmenting the protonated molecule of m/z 464 is shown in Fig. 9. Of particular interest was the fragment ion of m/z 109. An ion of this mass to charge ratio had been observed in the CID spectra of many of the other fonofos metabolites and had been assigned the structure of the C_6H_5S ion. However no reasonable structures of molecular weight 463 could be conceived that would yield the C_6H_5S ion.

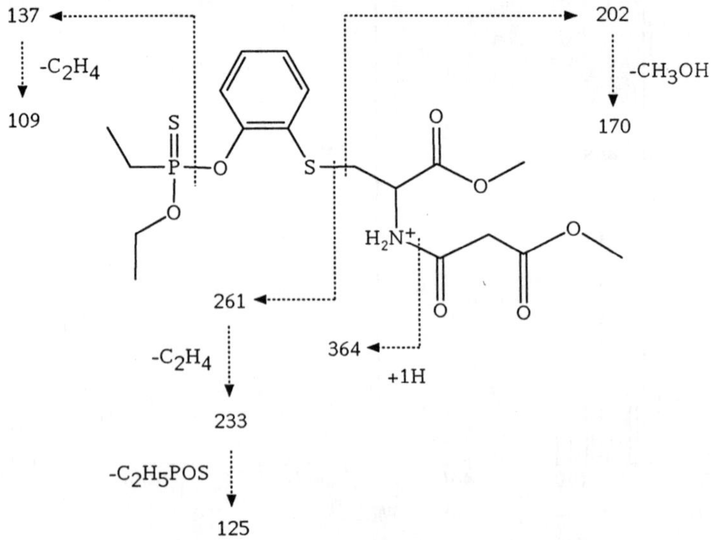

Fig. 10. Fragmentation scheme showing the origin of the major CID fragments of derivatized metabolite A15M.

The sample was again analyzed by LC-MS-MS, but this time a CID spectrum was obtained by fragmenting the protonated molecule of m/z 470 that contains six carbon-13 atoms in the phenyl ring. The result is shown in Fig. 9. Several fragment ions (see for example peaks of m/z 125, 211, 233, and 261) were found to contain the phenyl ring. Surprisingly the ion of interest of m/z 109 did not. The ion of m/z 109 was observed in the CID spectra derived from ions of both m/z 464 and 470.

Alternate structures for the fragment ion of m/z 109 were proposed, one of which corresponded to C_2H_6POS. This substructure ultimately led to the proposed structure shown in Fig. 10, which can explain the origin of the major CID fragments of the derivatized metabolite "A15M". This structure was subsequently confirmed by synthesis. The ortho isomer cochromatographed with the metabolite and gave an identical CID spectrum.

Future Trends

In the last decade, techniques used in the mass spectrometric identification of metabolites have progressed remarkably. There is no indication that this trend will not continue. The dramatically-improved chromatographic resolution shown by capillary zone electrophoresis (CZE) shows great promise. Mass analyzers, such as ion traps and Fourier transform mass analyzers, that can perform sequential MS-MS experiments without the need to physically link multiple mass spectrometers will be used to advantage. Mass spectrometric identification can be assisted by artificial intelligence techniques. Conceivably, high resolution CID libraries will be created. Their use with instruments that perform sequential MS-MS experiments will greatly simplify structural elucidation. The future approach might involve stepwise fragmentation of ions until an acceptable library fit to the CID spectrum of a reference ion is achieved. Reassembling the pieces would give the structure of the unknown.

Acknowledgements

The author gratefully acknowledges J. Kalbfeld, A. Tolentino and G. Cooper for radiosynthesis, F. Walker, N. Kerlinger and M. Prisbylla for metabolite synthesis, R. Staub, J. Barnes, W. Goldsby and H. Kan for thiocarbamate work, and C. Spillner, D. Tambling and D. Diaz for fonofos work.

References

Brunmark, P.; Dalene, M.; Sango, C.; Skarping, G.; Erlandsson, P.; Dewaele, C. *J. Microcol. Sep.* **1991**, 3, 371-375.

Caprioli, R.M. *Continuous-Flow Fast Atom Bombardment Mass Spectrometry*, John Wiley and Sons, New York, **1990**.

Chowdhury, S.K.; Katta, V.; Chait, B.T. *Rapid Commun. Mass Spectrom.* **1990**, 4, 81-87.

Fenn, J.B.; Mann, M.; Meng, C.K.; Wong, S.K.; Whitehouse, C.M. *Science* **1989**, 246, 64-71.

Lamoureux, G.L.; Frear, D.S. *Proc. 6th International Congress of Pesticide Chemistry* **1986**, 455.

Lamoureux, G.L.; Rusness, D.G. in *Glutathione. Chemical, Biochemical, and Medical Aspects*; Dolphin, D.; Avramovic, O.; Poulson, R.; Eds.; John Wiley and Sons, New York, **1989**; pp 153-196.

Ling, B.L.; Baeyens, W.; Dewaele, C. *J. Microcol. Sep.* **1991** 4, 17-22.

Mathews, W.R.; Johnson, R.S.; Cornwell, K.L.; Johnson, T.C.; Buchanan, B.B.; Biemann, K. *J. Biol. Chem.* **1987** 262, 7537-7545.

Onisko, B.C.; Barnes, J.P.; Staub, R.E.; Walker, F.H.; Kerlinger, N. *Biol. Mass Spectrom.* **1994** in press.

Rosen, J.D. *Proc. 6th International Congress of Pesticide Chemistry* **1986**, 301.

Voyksner, R.D. *Proc. 7th International Congress of Pesticide Chemistry* **1990**, 383.

Biotransformation in Aquatic Animals : Pesticides Revisited

Kevin M. Kleinow, Department of Pharmacology and Toxicology, School of Veterinary Medicine , Louisiana State University , Baton Rouge , LA 70803

Aquatic species , much like their mammalian counterparts are capable of a wide variety of biotransformation reactions. Inherent enzymatic activities , substrate specificities and the relative contributions of the various biochemical pathways act in concert with unique physiological features to fashion the basis of selective toxicities in aquatic species. The scope of biotransformation reactions , comparative relationships of biotransformation to toxicity as well as metabolic cascades of selected pesticides in aquatic species are discussed.

Environmental issues pertaining to pesticide use,application and safety have taken central stage in many arenas. With annual pesticide use estimated at approximately 1.5 billion pounds world wide (Wilkinson ,1987) , considerable potential exists for the inadvertent exposure of non-target species. This is especially true for the aquatic environment where agricultural runoff , groundwater contamination and errors in application expose non-target aquatic species to pesticides.

Much like for their mammalian counterparts, a variety of factors determine toxicity in aquatic animals. On an environmental basis availability of pesticides to fish and invertebrates is dependent upon water solubility , water quality, sorption to organics and inorganics , interaction with light and distribution. Organismic factors such as species , age , sex , trophic level , toxicant load and physiology are also major influencing considerations. Physiological factors important for xenobiotic disposition such as uptake, excretion and biotransformation are often unique and varied in aquatic species , in part , due to unique and varied adaptations to the aquatic environment. The purpose of this paper is to outline biotransformation in aquatic species and provide examples of pesticide biotransformation.

Cytochrome P450 Monooxygenases

Cytochrome P450 mediated monooxygenase (MO) reactions have been identified in a large number of fish and invertebrate species. As in mammals , the diversity of P450 mediated reactions in fish is extensive.They include hydroxylation , dealkylation , S-oxidation , N-oxidation and epoxidation (Stegeman,1989). While the scope of reactions are qualitatively similar to mammals,preferential suites of metabolites appear to be formed in aquatic species. For example, fish liver microsomes ,while capable of forming most of the mammalian benzo(a)pyrene metabolites,show a relative prediliction to the oxidation of the 7,8- and 9 , 10 positions (Stegeman,1981). Similar differences are evident in the metabolism of 2-methylnaphthalene where *in vivo* methyl group oxidation,an important route of biotransformation in the rat ,was relatively unimportant compared to dihydrodiol formation in the rainbow trout.(Melancon et al,1982; Melancon and Lech ,1984).Such examples can be extended to the biotransformation of a variety of other xenobiotic and endogenous substrates (Hansson et al ,1979;Gustafsson,1978). Differences in bio-transformation are even more pronounced in invertebrates. Mollusks ,for example,while ex-hibiting low monooxygenase activity *invitro* form predominantly quinone derivatives of benzo(a) pyrene (BaP) (Livingstone and Farrar,1984 ; Stegeman,1985).

Characterization of the P450 system in fish has delineated both similarities and differences with mammalian systems . Among these are lower P450 dependent specific activities (Bend and James,1978 ; Stegeman,1981), a selective response to inducers (Kleinow et al,1987 ; Gooch et al ,1989), lower temperature optima (Bend and James,1978; Gurumurthy and Mannering,1985) and ideal temperature com-pensation (Ankley et al,1985; Koivusarri and Andersson ,1984).

Microsomal cytochrome P450 levels in the liver of uninduced fish vary from less than 0.1 nmol/mg to nearly 2.0 nmol/mg protein. Generally, P450 levels in fish are lower than those in mammals , clustering between 0.2 and 0.5 nmol/mg protein (Stegeman,1981). Consistent with the lower total P450 levels are lower activities towards various prototypic substrates. In general monooxygenase activity is highest in the liver followed by the kidney for

1054–7487/95/0278$12.00/0

most fish species.Activity has been detected in a variety of other tissues including gill, brain, lens, ovary, testis, red muscle, spleen, gastrointestinal tract, blood and pancreas.(Stegeman ,1981)

Cytochrome P450 content has been reported for a variety of aquatic invertebrates including the lobster, spiny lobster, shore crab, blue crab, fiddler crab, spider crab, stone crab, spiny crab, barnacle ,and the crayfish (reviewed by James, 1989). Cytochrome P450 values for hepato-pancreas in these species range from 0.043 in the american lobster to as high as 0.91 nmol/mg protein in the spiny lobster. Cytochrome P450 content is often lower in extrahepatopancreatic tissues such as the green gland , stomach , testes, antennal gland, gill and heart .Many of these same tissues ,however,have greater specific activities than found in the hepato-pancreas (Burns ,1976; James et al ,1979; Khan et al ,1972).Monooxygenase activity is highly variable in crustacea (James ,1989) and quite low for species such as the mussel (*Mytilis edulis*), oyster (*Crassostrea virginica*),copepod (*Calanus helgolandicus*), starfish, sea urchin and gumboot chiton (Landrum et al,1981; Livingstone et al ,1989 ;Anderson, 1978; Walters et al ,1979).

Multiple P450 isozymes have been characterized via physical, chemical and/ or catalytic properties in the scup (*Stenotomus chrysops*) , cod (*Gadus morhua*) , trout (*Oncorhychus mykiss*) and the perch (*Perca fluviatilis*) (Zhang et al , 1991). PAH (ie BNF, 3-MC) inducible forms of P450 are well represented in fish (Stegeman and Kloepper-Sams,1987). P450E from scup , P450c from cod , and P450LM$_{4b}$ from rainbow trout appear to be isozymes orthologous to the mammalian representatives of the PAH inducible P4501A1. It is if these isoforms are functionally similar among the species under consideration

Differences appear to exist between mammals and aquatic species in regards to the temporal relationships associated with the process of induction. As compared to mammals, fish appear to have a greater time lag between the appearance of mRNA and the synthesis of the P450 1A1 protein. Similarly, once formed in fish the protein levels are maintained for extended periods of time (Kloepper-Sams and Stegeman, 1989).

A further distinction between fish and mammalian monooxygenase systems is the general lack of response of fish to classical phenobarbital- type inducers (DDT , kepone , mirex , lindane) (reviewed by Kleinow et al, 1987). Phenobarbital type induction is not apparent at the transcriptional, translational and catalytic levels in the trout , however ,the hybridization of rat cDNA pSP450- oligo (a PB inducible P450 isozyme in the rat) with genomic DNA from trout indicates a degree of sequence similarity(Kleinow et al , 1990). The results of this study as well as those of others (Miranda et al, 1990) lend support to the existence of phenobarbital like noninducible isozymes in the trout.

Equivocal results on the inducibility of invertebrate P450s are evident in the literature (James,1989). Phenobarbital generally shows no inductive effect in invertebrates. Currently it appears that polycyclic aromatic hydrocarbons exhibit either low or undetectable levels of induction in these species (Batel et al ,1988; James and Little ,1984; Lindstrom-Seppa and Hanninen ,1986).

The liver microsomal P450 dependent system in fish has a lower temperature optimum than mammalian systems (Gurumurthy and Mannering , 1985). In addition, fish hepatic MO activity responds to acclimation temperature in a compensatory manner (Ankley et al , 1985 ; Koivusarri and Andersson , 1984 ; Stegeman , 1979). That is to say, MO activities are nearly identical when measured *in vitro* at the respective acclimation temperatures (Koivusarri et al ,1981). In contrast , when MO activities are measured at a similar temperature, those fish acclimated to colder temperatures exhibit greater activity than those acclimated to warmer temperatures. In part, this temperature compensation can be explained by an increase in the reductase/ cytochrome P 450 ratio (Blanck et al ,1989). Temperature also appears to influence the rate of MO induction . Lower temperatures increase the time to reach maximal enzyme activity (Egaas and Varanasi , 1982 ; Andersson and Koivusarri , 1985). In a similar fashion, cold temperature has also been shown to have a pronounced effect on residue retention (Collier et al , 1978 ; Kleinow et al , 1994). Xenobiotic residues are maintained at cold temperatures for longer periods of time as compared to warm temperatures.

Flavin-Containing Monooxygenase

Flavin containing monooxygenase enzymes (FMO) in mammals carry out N or S oxidations for a variety of substrates including secondary and tertiary amines, thiocarbamates, N-hydroxylamines, thioamides, primary aryl amines, hydrazines, thiols and sulfides (Ziegler, 1988). FMO activity has also been presumptively identified in a variety of aquatic species including both fish and invertebrates.

Much of the early work with FMO activities in fish have centered around the oxidation of trimethylamine (TMA), an endogenous osmolite involved in osmoregulation in marine species (Goldstein and Dewitt- Harley ,1973; Strom, 1980). Recent studies have demonstrated significant FMO activity toward dimethyl-aniline(DMA) and methimazole in the liver , gill , kidney and intestine of the freshwater rainbow trout (Schlenk, 1989; Schlenk and Buhler, 1991b).Antibodies against mammalian FMO IA1 and IB1 isozymes also recognize proteins of similar molecular weights from the trout liver (Schlenk et al, 1993a). In contrast to mammals , FMO activity in the liver of the trout does not appear to differ with sex (Schlenk et al, 1993a) .However, FMO isozyme content and DMA oxidase activity does appear to increase with age (Schlenk et al, 1993a). Induction of FMO activity appears to vary with the fish species. Schlenk and Buhler (1993a) were unable to demonstrate any change in trout hepatic FMO isozyme content or enzyme activity with exposure to TMA. In contrast, salinity and TMA were shown to induce FMO activity in the guppy(*Poecilia reticulata*) and eel (*Anguilla japonica*) (Daikoku et al ,1988).

The carbamate pesticide, aldicarb,appears to be made more toxic in rainbow trout by S-oxygenation to the more potent cholinesterase inhibitor aldicarb sulfoxide (Schlenk and Buhler, 1991b). Likewise , bioactivation of the thioether herbicide thiobencarb appears to be catalyzed by a form of FMO in the stripped bass (*Morone saxatilis*) (Cashman et al , 1990). Other evidence suggests that the channel catfish does not have an active FMO form(Schlenk et al ,1993b). It has been suggested that this may be the reason for the relative insensitivity of channel catfish to thioether pesticides which require metabolic activation (Schlenk et al ,1993b).

In the species thus far examined the FMO activity in fish is considerably greater than that observed in invertebrates.The first indication of FMO activity in invertebrates was evident with studies examining TMA in whole organism homogenates of the copepod ,*Callanus* (Strom,1980). Subsequent studies have either directly or indirectly identified FMO activity in the mussels *Mytilus edulis*(Kurelec ,1985), and *Mytilus galloprovincialis* (Britivic and Kurelec, 1986) , the chiton , *Cryptochiton stelleri* (Schlenk and Buhler,1990), the Pacific oyster, *Crassostrea gigas* (Knezovich and Crosby , 1985) and a variety of marine sponges (Kurelec et al,1986). Studies including those with the marine clam, *Mercenaria mercenaria* (Anderson and Doos, 1983) , the mussels *Mytilus edulis* (Kurelec,1985;Marsh,1992) and *M. gallo-provincials* (Britivic and Kurelec, 1986) have implicated FMO activity in the bioactivation of aromatic amines to mutagenic metabolites.

Hydrolytic Reactions

Hydrolytic reactions are known to occur in plasma , liver and kidney of many aquatic species. While little information is known regarding the character of the enzymes responsible , hydrolytic reactions are well represented in the biotransformation of pesticides in aquatic animals. One of the few published comparative *in vivo* studies between fish and mammals deals with the pyrethoid permethrin (Glickman and Lech ,1981). In these studies the rate and capacity of trout to hydrolyze permethrin was much lower than in mammals. Recent studies in coho salmon with triclopyr BEE ,an auxin type herbicide, has demonstrated that it too is hydrolyzed (Barron et al, 1990). The ester form of triclopyr appears to facilitate absorption from water. However, accumulation was limited by rapid deesterification to triclopyr acid as the principal metabolite . Other hydrolysis reactions have been identified with malathion in pinfish (Cook and Moore ,1976), methyl parathion in crayfish and prawns (Foster and Crosby ,1987), fentrothion in blue crab(Johnston and Corbett, 1986)and 2,4 D esters in a variety of fish species(Rodgers and Stalling, 1972)

Glycosylation

Fish are capable of conjugating glucuronic acid to a wide variety of endogenous and xenobiotic substrates (Clarke et al,1991) . As with mammals, fish utilize the high energy nucleotide , UDP- glucuronic acid (Tsuyuki and Idler ,1960) as the donor of glucuronic acid to nucleophilic groups of the acceptor substrate. To date glucuronidation in fish has been primarily associated with O and N glucuronidation (Statham et al ,1975 ; Suzuki et al ,1977) ,with the former accounting for the majority of the known reactions. While O, N, S and C glucuronides have been identified in mammals, little investigational effort has focused on S and C glucuronidation to confirm their existence or absence in fish.

As in mammals, UDPGT activity in fish is microsomal (Castren and Oikari ,1983). The maximal expression of UDPGT activity *in vitro* occurs at 37 °C , in the presence of Mg^{++} at a neutral pH (Clarke et al,1992a). Fish exhibit latency in *in vitro* UDPGT activity in the liver and kidney and little if any latency in the intestine [(plaice)(Clarke ,1990)]. Latencies in fish liver and kidney UDPGT activities of 30 to 40 % compare to corresponding mammalian values of nearly 95 %.

Studies in the rainbow trout and the plaice, (using substrates conjugated by different

UDPGT isoforms in mammals) have demonstrated a broad substrate specificity for glucuronidation(Clarke et al, 1988 ; Gregus and Klaassen, 1988; Clarke, 1990). These findings as well as cross reactivity of fish UDPGT proteins with antibodies against mammalian isoforms suggest that multiple isoforms do exist in fish (Andersson et al ,1985 ; Clarke et al, 1992a). Recent work has isolated several UDPGT forms to homogenity. Differences in UDPGT activities not only exist between fish and mammals, but also between fish species and with the substrate. In plaice , for example, 1- naphthol UDPGT activity was 3 fold lower than that in Wistar rats. In contrast, 4-nitrophenol activity was 2 fold greater in plaice than rats when both were measured at 37 °C (Clarke ,1990, Clarke et al,1991). Additionally, phenol UDPGT activity of rainbow trout at 25 ° C was 5% of Sprague Dawley rats at 37°C (Gregus et al, 1983) . Comparisons of 4 - nitrophenol UDPGT activity at 25 ° C indicates a 5 fold greater activity in plaice than in rainbow trout (Gregus et al ,1983, George and Young, 1986).

The liver in most fish species is the most important site of glucuronidation. UDPGT (4-nitrophenol as the substrate) activity has also been detected in the gills , kidney , intestine and heart (Koivuaari, 1980; Lindstrom- Seppa et al, 1981). The tissue distribution of UDPGT activity appears to vary between species. The plaice (Clarke et al 1992b) and rainbow trout (Lindstrom- Seppa et al,1981) exhibit the the highest specific activity in the liver , with the vendace and roach demonstrating peak activities in the gills. Intestinal activity in the plaice appears to be relatively low as compared to that in the liver and kidney (Clarke et al,1992b).Hepatic UDP glucuronosyl-transferase has been shown to be inducible by β - naphthoflavone (Andersson et al ,1985).

Glucuronides of a variety of pesticides have been identified in aquatic species. Among those converted to glucuronide conjugates are pyrethroid insecticides (Glickman et al ,1981) , the lampreycide 3- trifluoromethyl-4- nitrophenol (Lech ,1974; Lech and Statham, 1975) , the organophosphorus pesticide, fenitrothion(Takimoto et al, 1987) and the carbamate insecticide, 1-naphthyl-N-methylcarbamate carbaryl (Sevin) (Statham et al , 1975).

A variety of studies with aquatic species have indicated that glycosylation may occur with conjugates other than glucuronic acid. In large measure, these reports have centered on invertebrate species. However , several studies have suggested that glucosides may be minor metabolites in teleost fish (Roubal et al ,1977 ; Varanasi et al,1979). Glycosylation in invertebrates has been shown to occur along a number of pathways including glucuronidation ,and glucoside and galactoside conjugation (Foster and Crosby ,1986; 1987). The relative distribution along these pathways for the compounds thus far examined appears to be strongly dependent on the invertebrate species. P- nitrophenyl β-D - glucoside formation from p-nitrophenol in the metabolic cascade of methylparathion, for example, shows such species variability with glucoside conjugation accounting for 2.0, 59.9 and 30.7 % of the total metabolites for the prawns *Sicyonia ingentis* , *Macrobrachium rosenbergii* and the crayfish *Procambarus clarkii* , respectively. Both glucuronidation and galactoside conjugation represented less significant pathways with contributions for each ,when present, of 2.3 % or less. *M. rosenbergii* and *Procambarus clarkii* consistently demonstrated the highest levels of glucoside conjugation with a variety of prototypic substrates.

Sulfate Conjugation

Sulfate conjugation has been identified in a variety of aquatic species including both vertebrates and invertebrates. Large differences in the extent of sulfate conjugation appear to exist with individual substrates , fish species and the type of sample.Prototypic substrates for sulfate conjugation in mammals such as 7-ethoxy-coumarin and acetaminophen exhibit little sulfation in the trout (Andersson et al ,1983; Gregus et al, 1983 ;Parker et al ,1981). Similarly , a variety of flatfish species fail to sulfate polyaromatic hydrocarbons to any significant degree (Varanasi and Gmur ,1981; Morrison et al ,1985). Other compounds such as phenol , m-cresol , 1-naphthol and O-chlorophenol are conjugated with sulfate in species such as the bream, guppy, roach, rudd and tench ,while only small amounts of phenol sulfate conjugates are formed in the perch (Layiwola and Linnecar ,1981; Layiwola et al, 1983).

Sulfate conjugate excretory patterns in a variety of fish species appear to be similar to those observed with mammals. The percentage of ^{14}C recovered as sulfates from the water after a 48 hour exposure ranged from 55-64% for m-cresol, 55-77% for 1-naphthol and 40-59% for O-chlorophenol (Layiwola et al, 1983). Corresponding % of recovered ^{14}C values for bile samples ranged from 8-20% for m-cresol, 4-12 % for 1-naphthol and 2-16% for O-chlorophenol. Similar information is also available for penta-chlorophenol. Sulfate conjugates of penta-chlorophenol appear as relatively minor components in the bile, but are significant contributors to the water fraction (presumably a collection of urinary and branchial elimination) (Glickman et al ,1977; Kobayashi et al ,1976; Stehly and Hayton ,1989). These results suggest that the percieved importance of sulfate conjugation maybe dependent in part upon the substrate, how the values were reported and the preferred excretory pathway of the conjugate. Sulfate conjugation has been shown to facilitate excretion of benzo(a)pyrene and BaP metabolites via organic anion transport mechanisms in the flounder kidney (Pritchard and Bend ,1991).

Substrates used among a number of fish species have delineated large species differences. Sulfate conjugation of phenol ranges from 5 to 45 % of the dose (Layiwola and Linncar, 1981) while the *in vitro* sulfation of carbaryl ranges between 1.2 to 5.1 %(Chin et al ,1979)and naphthalene between 1 and 19 % of the dose (Krahn et al ,1980 ; Varanasi et al, 1979) . Evidence of a dose dependent percentage of phenol conjugation in goldfish (Nagel et al ,1983), saturation of 1- naphthol sulfation in plaice hepatocytes (Morrison et al , 1985) as well as low Km and Vmax values for 3,7, and 9 OH BaP conjugation in catfish (Kleinow et al ,1994) (relative to glucuronidation) suggest that sulfation in fish as in mammals may be a low capacity high affinity system. Comparison of sulfation in the trout and the rat has demonstrated that for the substrate 2-naphthol trout had a 2-7 fold lower activity as compared to the rat (Gregus et al , 1983) .

Sulfate conjugation has been identified in a variety of invertebrate species including the crayfish (Foster and Crosby,1986) , lobster (*Homerus americanus*)(Elmamlock and Gessner,1978),spot shrimp (*Pandalus platyceros*)(Sanborn and Malins , 1980), gumboot chiton (Landrum and Crosby,1981) , ridgeback prawn (*Sicyonia ingentis*), malaysian prawn (*Macrobrachium rosenbergii*)(Foster and Crosby, 1987) and sea urchins. Investigations with phenolic substrates in the american and florida lobster have indicated several interesting features. The first of these is that sulfotransferase activity is evident in the antennal gland but not the hepatopancreas (James et al ,1987 ; Schell and James ,1989.) Hepatopancreas cytosol and microsomes in fact appear to inhibit sulfotransferase activity (Schell and James ,1989). Other studies with *Homerus americanus* have demonstrated that sulfate conjugation appears to be capacity limited (James et al ,1991) .The net effect is a dose dependent contribution of sulfate conjugation. Biotransformation of phenol to sulfate conjugates present a larger contribution at lower doses than at higher doses where excretion of the parent phenol via the gills predominate. An interesting consequence of sulfate conjugation of phenol in the lobster was a longer rather than shorter half-life as compared to the parent compound (James et al, 1991).Sanborn and Malins (1977) demonstrated similar results with naphthalene and naphthalene metabolites in larval spot shrimp. Adult and larval spot shrimp were shown to produce sulfate conjugates of naphthol as 7% and 39 % of the total , respectively (Sanborn and Malins ,1980). Other species such as the starfish (*Pisaster ochraceus*) and the gumboot chiton (*Cryptochiton stelleri*) have also demonstrated sulfate conjugation producing p-nitrophenyl sulfate following exposure to p-nitroanisole (Landrum and Crosby ,1981).

Glutathione Conjugation

Glutathione-S-transferase (GST) constitutes a significant portion of the soluble protein in the liver of fish. While a range of electrophilic substrates are conjugated with glutathione (GSH) by the action of GST, the diversity of substrates appears to be somewhat less for fish than for rodents(Nimmo,1987). Among the prototypic substrates GSH conjugated in fish are styrene 7,8 oxide, benzo(a) pyrene-4,5 oxide, octene-1,2 oxide, 1-chloro-2,4-dinitrobenzene, 1,2-epoxy-3-(p-nitrophenoxy) propane, ethacrynic acid and 1,2- dichloro-4-nitrobenzene (Bend and James ,1978; James and Bend ,1980; Stegeman ,1981 ; Bend et al 1977; James, et al ,1988; Forlin et al ,1986; Gregus et al ,1983). GST activities with these substrates in fish appear to extend from below to within the range of mammals. Perhaps the most universal activity in fish is with the prototypic substrate 1-chlorodinitrobenzene (CNDB) which is GSH conjugated in all of the thirty plus fish species examined. The specific activities with this substrate vary substantially between species with values ranging from 0.3 nmol/mg cytosolic protein/ min for the little skate (Bend and James, 1978; James and Bend ,1980) to values of 4000 nmol/mg cytosolic protein/ min for the sheepshead minnow (James et al ,1988). A considerable range in GST activities have also been reported for individual species such as the rainbow trout (87- 2500 nmol/mg cytosolic protein/ min) when CNDB is used as the substrate (George et al, 1989; Forlin et al, 1986 ; Gregus, et al ,1983 ; Andersson et al ,1985). Biologic variation or differences in assay conditions between studies maybe responsible for such results.

GST activity in fish as in mammals is expressed in the liver and other organs. For example, as measured with benzo(a)pyrene 4,5 oxide in the little skate (*Raja erinacea*) GST activities of 16.5, 31.6, 0.7 and 10.6

nmol/min/mg protein were evident for liver, kidney, gill and testes ,respectively.The relative activity in tissues appear to vary with the fish species and the particular substrate (Bauermeister et al ,1983; Braddon et al, 1985; Leaver et al ,1992b ; Lauren et al ,1989; Bend et al ,1977).

Much as with their mammalian counterparts, GST activity in fish is primarily cytosolic. However, a small amount of conjugating activity appears with microsomal fractions from the liver of rainbow trout (Nimmo et al ,1981 ; Lauren et al ,1989), plaice (George et al ,1990) and pike (Morgenstern et al ,1984). GST activity has also been reported in the microsomal fraction of the channel catfish gill(Gallagher et al ,1991).

Induction of GST activity has been reported for phenol (Chattergee and Bhattacharya, 1984), Aroclor 1254 (Ankley et al, 1986), benzo(a) pyrene (Fair ,1986), trans-stibene oxide (Scott et al ,1992; Leaver et al ,1992a), 3-methylcholanthrene (Leaver et al ,1992a), B - naphthoflavone (Andersson et al, 1985) and Clophen A-50 (Andersson et al, 1985) in a variety of fish species. Usually induction is 2 fold or less with these agents.

One pesticide which has received some attention in fish in regards to glutathione conjugation is the broad spectrum fungicide, chlorothalonil (Gallagher,1991).This compound contains four electrophilic chlorine atoms which readily undergo nucleophilic substitution with glutathione. The toxicity of chlorothalonil to channel catfish has been shown to be increased 3 fold after depletion of hepatic and gill GSH. These results support the conclusion that GSH provides a protective role against chlorothalonil toxicity.

Acetylation

Acetylation of xenobiotics has been demonstrated in a variety of aquatic vertebrate and invertebrate species. For fish species, acetylation of the fish anesthetic ethyl-m-aminobenzoic acid (MS-222) is one of the most well known. The dogfish shark while eliminating the bulk of this anesthetic across the gill unchanged has been shown to eliminate a small fraction as the N-acetyl deriviative (Stenger and Maren,1974). In the rainbow trout approximately 77-96 % of the MS-222 eliminated in the urine was acetylated (Hunn et al ,1968). Sulfonamide drugs such as sulfadimethoxine and sulfadimidine have been shown to be acetylated in rainbow trout and catfish (Squibb et al ,1988; Kleinow et al, 1992; Bergsjo and Bergsjo ,1978). Metabolite analysis in trout indicates that N- acetyl sulfadimethoxine comprises 86 % of the total radioactivity in the bile and 55 % in the liver, 20 hours after oral dosing. N_4 Acetylated sulfadimethoxine accounted for 52 % of the urinary elimination products(Kleinow et al, 1992). The elimination and metabolic profile for sulfadimethoxine in catfish appears to be similar to the trout (Squibb et al ,1988).

Acetylation of aryl or alkylamines is also evident for some invertebrate species. The biotransformation of p-nitrotoluene , p-nitro-anisole and p- nitrophenol by the sea urchin involves a reduction step followed by acetylation (Landrum and Crosby, 1981). Similarly, in gumboot chitons approximately 37-54 % of p-nitroanisole is reduced to p-anisidine and subsequently acetylated (Landrum and Crosby,1981). Studies with sulfadimethoxine in the lobster indicate little if any acetylation (James and Barron ,1988). This is in direct contrast to vertebrates where N-acetyl sulfadimethoxine is a major metabolite.

Amino Acid Conjugation

A biotransformational feature which has yet to be adequately explored is amino acid conjugation. While subject to change with more information, it appears that taurine conjugation of carboxylic groups is more important in aquatic species than conjugation with other amino acids such as glycine and glutamine (James et al , 1986). Evidence now exists for taurine conjugation in a variety of marine teleosts (including winter flounder, sheepshead, pinfish, mullet and drum) (Pritchard and James, 1979; Pritchard et al, 1978 ; James ,1978) ,marine elasmobranchs (dogfish shark, stingray,skate) (Guarino et al ,1977; Pritchard et al, 1978), a couple of freshwater fish species (catfish , trout) (Plakas and James ,1990; Burke et al ,1987), and a marine crustacean (spiny lobster) (James ,1982). The bulk of these studies have been performed with substrates such as phenylacetic acid (and its derivatives) and benzoic acid . Recent work with the freshwater red swamp crayfish has extended this list to include a taurine conjugate of the herbicide triclopyr (Barron et al, 1991).

Methyl Parathion

The metabolism of parathion and methyl parathion has been examined in a variety of invertebrate species including the rice field crayfish (*Procambarus clarkii*), the ridgeback prawn (*Sicyonia ingentis*) the Malaysian prawn (*Macrobrachium rosenbergii*)(Foster and Crosby, 1987) and lobster (*Homerus americanus*)(parathion)(Carlson, 1973). A number of significant differences between mammals and these organisms as well as differences between these representative decapods are evident.

Generally methyl parathion is very toxic to these animals with 96h LC_{50} values approximating 3 µg/ l (Cheah et al,1980). Lobster exposed to parathion demonstrated a LD_{50} value of approximately 0.3 mg /kg while values such as 7 mg/ kg have been reported for rats (Dubois et al,1949).

Methyl parathion is extensively metabolized by crayfish and prawn species(Fig.1). In the crayfish the most significant route of metabolism appears to be dearylation to p- nitrophenol (44%) with subsequent conjugation to the PNP -β - D - glucoside (31 %).This dearylation defined as a cleavage of the P-O (Ar) bond may proceed through oxidative or hydrolytic mechanisms. Desmethylation to form O- methyl - O (p-nitrophenyl) phosphorothioic acid was a minor pathway (5 %). No evidence for nitroreduction or ring hydroxylation was given under the experimental conditions. PNP sulfate (5.8 %) and the β - D- glucuronide (2.3 %) were evident, in small quantities.

The major metabolites of parathion in prawns *Macrobrachium rosenbergii* and *Sicyonia ingentis* included p- nitrophenol and polar conjugates. *Sicyonia* produced a greater variety of conjugates with a p-nitrophenyl unknown conjugate (52.1 %) and p-nitrophenyl sulfate (17.9 %) representing the major conjugates. The p-nitrophenyl β – D - glucoside (2.0 %) and galactoside (1.1 %) were also excreted by *S. ingentis* . P-nitrophenyl β -D- glucoside was the most prominent conjugate (59.9 %) from *M. rosenbergii* followed by lesser amounts of the sulfate (2.1 %). The p-nitrophenyl unknown conjugate and p- nitrophenyl β- D - galactoside conjugates were not evident in *M. rosenbergii* . O- demethylation to form desmethyl methyl parathion was a minor route in both prawn species with 2.3 and 4.0 % evident for *S. ingentis* and *M.rosenbergii* ,respectively. Desmethyl formation was a minor pathway of methyl parathion metabolism in the crayfish and both prawn species.

Methyl paraoxon was either not detected or only present in very small amounts for each of these species. However , only oxon product and not total degradation were measured. Methyl parathion was highly toxic to all three decapods indicating the presence of activation through oxidation. Either methyl paraoxon was not the active form of methyl parathion in these decapods or, once formed, it reacted rapidly with its target such that the oxon product was not

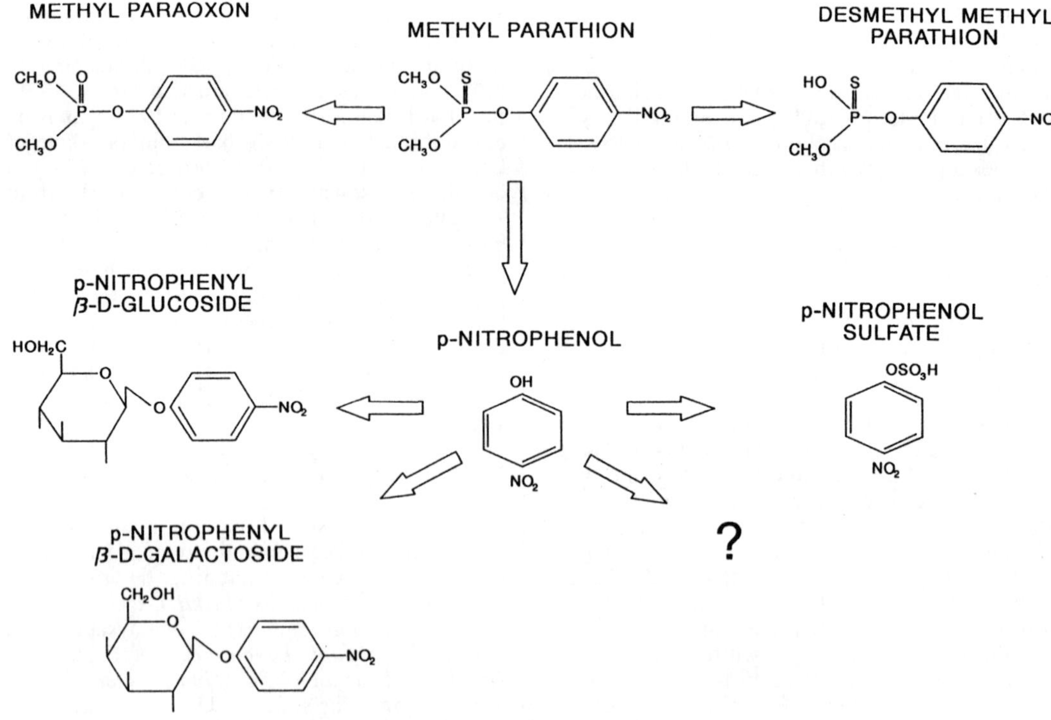

Fig.1. Generalized metabolic cascade for methylparathion in aquatic species.(Adapted from Foster and Crosby,1987)

observed.Similar results have been reported in the lobster where no paraoxon formation could be detected (Elmamlouk and Gessner, 1976; Carlson, 1973).

Permethrin

In general, pyrethroids exhibit a much higher toxicity to fish as compared to their mammalian counterparts. Toxicity studies with technical permethrin demonstrate intraperitoneal 24h LD_{50} values of 14 and 514 mg/kg for trout and mice ,respectively (Glickman et al ,1981). Trout exhibit 24 and 96 h LC_{50} values of 18 (Glickman et al,1981) and 5μg/ l (Marking and Bill, 1979). Comparative studies by Glickman et al (1981) demonstrated that the [1 R,S- cis] and [1 R,S trans isomer] of permethrin were 5x and 110x more potent to rainbow trout than mice ,respectively The difference in species specificity appears to be due to both a high intrinsic sensitivity of the trout CNS to pyrethroid insecticides (Eells et al ,1993) as well as differences in pyrethroid metabolism (Glickman et al , 1981a)(Fig. 2). For example following the inhibition of trans-permethrin metabolism in trout a 60 fold difference in toxicity as compared to mammals still exists (Glickman and Lech ,1982), supporting a multifactoral etiology. Mechanistic studies have demonstrated that trout synaptosomes exhibit significantly greater maximal membrane depolarization (3 fold greater) than those of rat brain synaptosomes upon permethrin exposure (Eells et al,1993).It has been suggested that this species difference may be due to binding domains in the voltage sensitive sodium channels.

In vitro metabolism studies with cis and trans permethrin have indicated that trout and carp microsomes exhibit no qualitative differences from those of mammals (Glickman et al ,1979). Hydrolysis of trans permethrin appears to be the predominant route of metabolism by carp and trout hepatic microsomes as it is for mammals. Oxidation, which also occurs with trans-permethrin metabolism in microsomes from fish, is the major route of metabolism of the cis-isomer. The 4' position on the aromatic ring appears as the predominant site of hydroxylation for both isomers. The methyl groups cis and trans to the carboxy moiety are also hydroxylated ,but to a lesser extent. Trans hydroxylation is the most evident. Carp microsomes demonstrate the greatest specificity between these two sites especially with cis-permethrin.

In vivo, considerable differences exist bet-

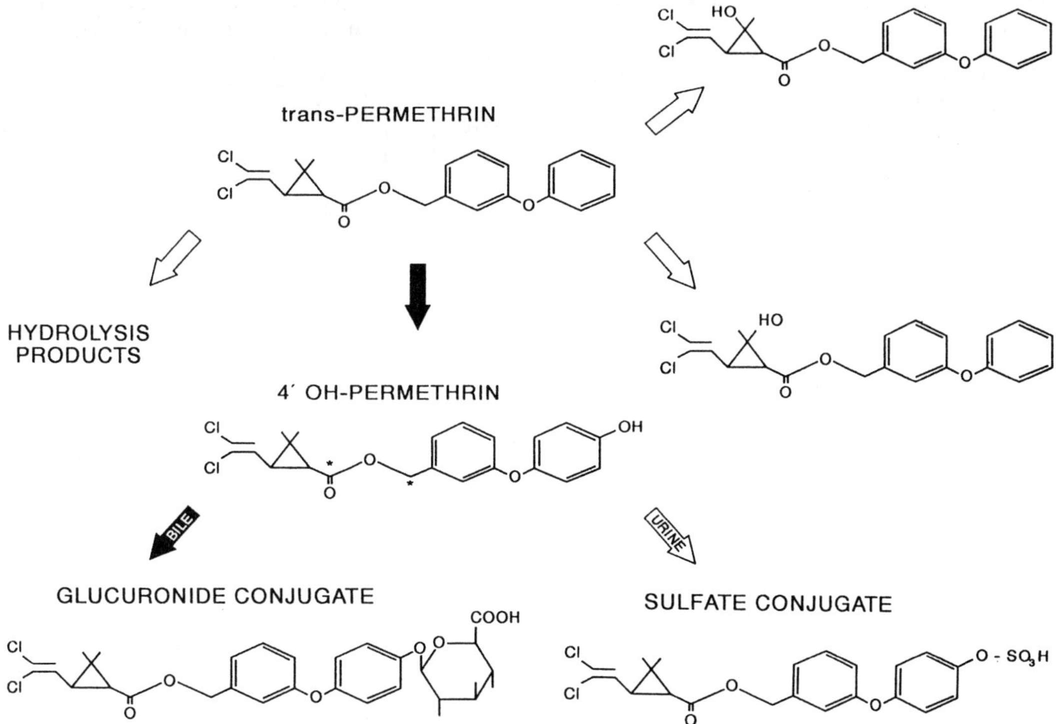

Fig. 2. Metabolic cascade for trans- permethrin in the trout.(Adapted from Glickman et al,1979)

ween fish and mammals. The rate and capacity of trout to hydrolyze cis and trans permethrin is much lower than their mammalian counter-parts (Glickman and Lech ,1981). Bile , the major route of elimination of cis and trans -permethrin in the trout contained predominantly glucuronide conjugates of the intact esters. 4'OH permethrin was the major conjugated component. Urine was a minor contributor to the excretion of both the cis and trans isomers .Only 3.3 and 7.1 % of the total cis and trans permethrin ip dose was eliminated in the urine by 48h. Urinary metabolites were highly polar and approximately 33(cis) and 45 % (trans) were hydrolyzed by aryl sulfatase.In general it appears that trout exhibit low hydrolytic activity towards permethrin relative to other metabolic pathways (oxidation and conjugation) as well as relative to mammals.

Aldicarb

The disposition of the carbamate pesticide aldicarb has recently been examined in the rainbow trout (Schlenk et al ,1992). Aldicarb is rapidly absorbed after oral administration (99 % in 3 h) ,and the highest tissue concentrations were found in the stomach , caeca , liver , heart and bile. In contrast to mammals, where urine (Andrawes et al ,1967) is the major route of excretion , the gills accounted for 96.2 % of the elimination in the trout.

In both *in vitro* and *in vivo* experiments aldicarb sulphoxide was the major metabolite. Evidence by Schlenk and Buhler (1991a,b) suggests that flavin containing monooxygenases are responsible for this S-oxidation step. *In vivo,* this metabolite accounted for approximately 7.6 % of the total dose or more than one half of the total metabolites recovered in the water and urine (14.3 %). Other metabolites such as the hydrolytic products sulfoxide oxime (0.88 %), aldicarb oxime (5.4 %) and aldicarb nitrile (0.56 %) were also evident *in vivo* (Schlenk et al,1992)(Fig.3). Only trace amounts of these same metabolites were evident with *in vitro* microsomal incubations. These results were interpreted to suggest that hydrolysis was due to the action of plasma esterases. While the formation of the oximes and nitriles are apparent detoxification reactions ,as in mammals ,aldicarb sulfoxide and sulphone formation potentiates acetylcholinesterase inhibition.

Conclusions

Aquatic animals represent a vast array of species with a multitude of physiological strategies to deal with the environment in which they live and xenobiotics which they may encounter. It is clear that in context of biotransformation and disposition of xenobiotics that aquatic species are not wet mammals. Diversity in specific activities, response to inducers, response to thermal considerations as well as isozyme characteristics and expression contribute to the metabolic profiles observed Although qualitatively similar to mammals in regards to the scope of available bio-transformation reactions aquatic animals often present unique suites of metabolites in response to the relative importance of a particular pathway. Unusual pathways of biotransformation also exist especially in invertebrate species. As more data becomes available it is evident that dispositional axioms generally held for mammalian systems may not be operative for all aquatic species (e.g. sulfation facilitates

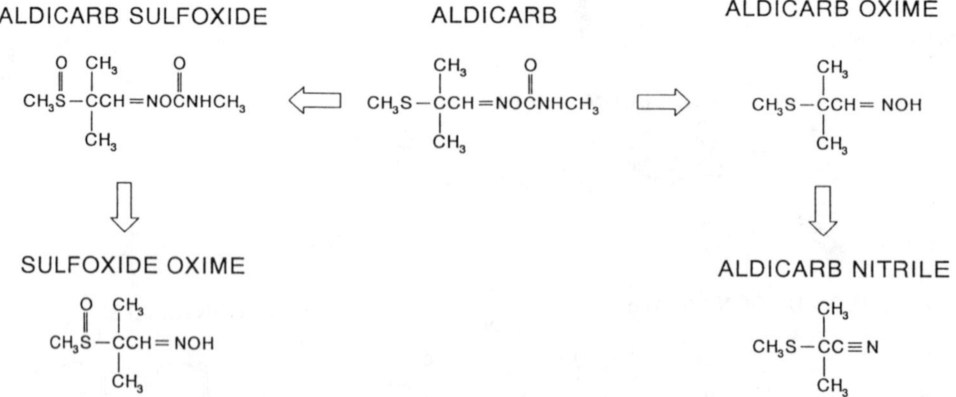

Fig. 3. Metabolic cascade for aldicarb in the trout.(Adapted from Schlenk et al ,1992)

elimination). Intrinsic sensitivity, the aqueous nature of the environment as well as differences in metabolic and dispositional capabilities form the basis of selective toxicities among aquatic species and with their mammalian counterparts.

References

Anderson, R.S. *E.P.A. Ecol. Res. Ser. monograph* **1978,** (EPA-600/3-78-009), 18p.

Anderson, R.S.; Doos, J.E. *Mutat. Res.* **1983,** *116,* 247-256.

Andersson, T.; Forlin, L; Hansson, T. *Drug Metabol. Disp.* **1983**, *11*, 494-498.

Andersson, T.; Koivusarri, U. *Toxicol. Appl. Pharmacol.* **1985**, *80,* 43-50.

Andersson, T.; Pesonen, M.; Johansson, C. *Biochem. Phamacol.* **1985,** *34,* 3309-3314.

Andrawes, N.R.; Dorough, H.W.; Lindquist, D.A. *J. Econ. Entomol.* **1967,** *60,* 979-987.

Ankley, G.T.; Blazer, V.S.; Reinert, R.E.; Agosin, M. *Aquatic. Toxicol.* **1986,** *9,* 91-104.

Ankley, G.T.; Reinert, R.E.; Wade, A.E.; White, R.A. *Comp. Biochem. Physiol.* **1985,** *81C,* 125-129.

Barron, M.G.; Hansen, S.C.; Ball, T. *Drug Metab. Disp.* **1991,** *19,* 163-167.

Barron, M.G.; Mayes, M.A.; Murphy, P.G.; Nolan, R.J. *Aquatic Toxicol.* **1990,** *16,* 19-32.

Batel, R.; Bihari, N.; Zahn, R.K. *Comp. Biochem. Physiol.* **1988,** *90C,* 435-438.

Bauermeister, A.; Lewendon, A.; Ramage, P.; Nimmo, I.A. *Comp. Biochem. Physiol.* **1983,** *74C,* 89-94.

Bend, J.R.; James, M.O. In *Biochemical Biophysical Perspectives in Marine Biology;* Malins, D.C.; Sargeant, J.A., Eds.; Academic Press: New York, **1978**; Vol. 4, pp 125-188.

Bend, J.R.; James, M.O.; Dansette, P.M. *Ann. N.Y. Acad. Sci.* **1977,** *298,* 505-521.

Bergsjo, T.; Bergsjo, T.H. *Acta Vet. Scand.* **1978,** *19,* 102-109.

Blanck, J.; Lindstrom, P.; Agren, J.J.; Hanninen, O.; Rein, H.; Ruckpaul, K. *Comp. Biochem. Physiol.* **1989,** *93C,* 55-60.

Braddon, S.A.; McIlavaine, C.M.; Balthrop, J.A. *Comp. Biochem. Physiol.* **1985,** *80B,* 213-216.

Britvic, S.; Kurelec, B. *Comp. Biochem. Physiol.* **1986,** *85C,* 111-114.

Burke, A.B.; Millburn, P.; Huckle, K.R.; Hutson, D.H. *Drug Metab. Disp.* **1987**,*15*, 581-582.

Burns, K.A. *Mar. Biol.* **1976,** *36,* 5-11.

Caldwell,J. In: *Metabolic Basis of Detoxification*; Jakoby, W.B.; Bend, J.R.; Caldwell, J., Eds.; Academic Press: New York, **1982**, 271-290.

Carlson, G.P. *Bull. Environ. Contam. Toxicol.* **1973,** *9,* 296-300.

Cashman, J.R.; Olsen, L.D.; Nishioka, R.S.; Gray, E.S.; Bern, H.A. *Chem. Res. Toxicol.* **1990,** *3,* 433-440.

Castren, M.; Oikari, A. *Comp. Biochem. Physiol.* **1983,** *76,* 365-369.

Chattergee, S.; Bhattacharya, S. *Toxicol. Letters* **1984,** *22,* 187-198.

Cheah, M.L.; Avault, J.W.; Graves, J.B. *La Agric.* **1980,** *23,* 8-11.

Chin, B.H.; Sullivan, L.J.; Eldridge, J.E. *J. Agric. Food Chem.* **1979,** *27,* 1395-1398.

Clarke, D.J. Ph.D. Thesis, University of Dundee, Scotland, **1990**.

Clarke, D.J.; Burchell, B.; George, S.G. *Comp. Biochem. Physiol.* **1992,** *102,* 425-432.

Clarke , D.J.; Burchell, B.; George, S.G. *Mar.Env. Res.* **1988,** *24,* 105-109.

Clarke, D.J.; Burchell, B.; George, S.G. *Toxicol. Appl. Pharmacol.* **1992,** *115,* 130-136.

Clarke, D.J.; George, S.G.; Burchell, B. *Aquatic Toxicol.* **1991,** *20,* 35-56.

Collier, T.K.; Thomas, L.C.; Malins, D.C. *Comp. Biochem. Physiol.* **1978**, *61C,* 23-28.

Cook, G.H.; Moore, J.C. *Agr. Food Chem.* **1976,** *24,* 631-634.

Daikoku, T.; Murata, T.M.; Sakaguchi, M. *Comp. Biochem. Physiol.* **1988,** *89A,* 261-264.

Dubois, K.P.; Doull, J.; Salerno, P.R.; Coon, J.M. *J. Pharmocol. exp. Therap.* **1949,** *97,* 79-91.

Eells, J.T.; Rasmussen, J.L.; Bandettini, P.A.; Propp, J.M. *Toxicol. Appl. Pharmacol.* **1993,** *123,* 107-119.

Egaas, E.; Varanasi, U. *Biochem. Pharmacol.* **1982**, *31,* 561-566.

Elmamlouk, T.H.; Gessner, T. *Comp. Biochem. Physiol.* **1976,** *53C*,19-24.

Elmamlouk, T.H.; Gessner, T. *Comp. Biochem. Physiol.* **1978,** *61C,* 363-367.

Fair, P.H. *Arch. Environ. Contam. Toxicol.* **1986,** *15,* 257-264.

Forlin, L.; Haux, C.; Karlsson-Norrgren, L.; Runn, P.; Larsson, A. *Aquatic Toxicol.* **1986,** *8,* 51-64.

Foster, G.D.; Crosby, D.G. *Environ. Toxicol. Chem.* **1986,** *5,* 1059-1070.

Foster, G.D.; Crosby, D.G. *Xenobiotica* **1987,** *17,* 1393-1404.
Gallagher, E.P.; Canada, A.T.; DiGiulio, R.T. *Aquatic Toxicol.* **1992,** *23,* 155-168.
Gallagher, E.P.; Kedderis, G.L.; DiGiulio, R.T. *Biochem. Pharmacol.* **1991,** *42,* 139-145.
George, S.G; Buchanan, G.; Nimmo, I.; Hayes, J.D. *Mar. Environ. Res.* **1989,** *28,* 41-46.
George, S.G.; Young, P.; Leaver, M.; Clarke, D. *Comp. Biochem. Physiol.* **1990,** *96C,* 185-192.
George, S.G.; Young, P. *Mar. Environ. Res.* **1986,** *24,* 93-96.
Glickman, A.H.; Hamid, A.R.; Richert, D.E.; Lech, J.J. *Toxicol. Appl. Pharmacol.* **1981,** *57,* 88-98.
Glickman, A.H.; Lech, J.J. *Toxicol Appl. Pharmacol.* **1981,** *60,* 186-192.
Glickman, A.H.; Lech, J.J. *Toxicol. Appl. Pharmacol.* **1982,** *66,* 162-171.
Glickman, A.H.; Shono, T.; Casida, J.E.; Lech, J.J. *J. Agric. Food Chem.* **1979,** *27,* 1038-1041.
Glickman, A.H.; Statham, C.N.; Wu, A.; Lech, J.J. *Toxicol. Appl. Pharmacol.* **1977,** *41,* 649-658.
Goldstein, L.; Dewitt-Harley, S. *Comp. Biochem. Physiol.* **1973,** *45B,* 895-903.
Gooch, J.W.; Elskus, A.A.; Kloepper-Sams, P.J.; Hahn, M.E.; Stegeman, J.J. *Toxicol. Appl. Pharmacol.* **1989,** *98,* 422-433.
Gregus,Z.; Klaassen,C.D. *J.Pharm. Pharmacol.* **1988,** *40,* 237-242.
Gregus, Z.; Watkins, J.B.; Thompson, T.N.; Harvey, M.J; Rozman, K.; Klaassen, C.D. *Toxicol Appl. Pharmacol.* **1983,** *67,* 430-441.
Guarino, A.M.; James, M.O.; Bend, J.R. *Xenobiotica* **1977**, *7,* 623-631.
Gurumurthy, P.; Mannering, G.J. *Biochem Biophys. Res. Commun.* **1985,** *127,* 571-577.
Gustafsson, J.A. *Steroid hydroxylations catalyzed by cytochrome P-450, In Methods in Enzymology,* Vol. 52, 377-388, Academic Press, New York, **1978.**
Hansson, T.; Rafter, J.; Gustafsson, J.A. *Gen. Comp. Endocrinol.* **1979,** *37,* 240-245.
Hunn, J.B.; Schoettger, R.A.; Willford, W.A. *J. Fish Res. Bd. Can.* **1968,** *25,* 25-31
James, M.O. In *Conjugation Reactions in Drug Biotransformation;* Aitio, A., Ed.; Elsevier: North Holland, Amsterdam, **1978**; pp 121-129.
James, M.O. In *Xenobiotic Conjugation Chemistry;* Menn, J.J.; Caldwell, J.; Hutson, D.; Paulsen, G., Eds.; A.C.S. Symposium Series No 299; American Chemical Society: Washington, DC, **1986**; pp 29-47.
James, M.O. *Drug Metab. Disp.* **1982,** *10,* 516-522.
James, M.O. *Xenobiotica* **1989,** *19,* 1063-1076.
James, M.O.; Barron, M.G. *Vet. Human. Toxicol.* **1988,** *30 (suppl. 1),* 36-40.
James, M.O.; Barron, M.G; Schell, J.D. *Bull. Mt. Desert Isl. Biol. Lab.* **1987,** *27,* 9-11.
James, M.O.; Bend, J.R. *Toxicol. Appl. Pharmacol.* **1980,** *54,* 117-133.
James, M.O.; Heard, C.S.; Hawkins, W.E. *Aquatic Toxicol.* **1988,** *12,* 1-16.
James, M.O.; Khan, M.A.Q.; Bend, J.R. *Comp. Biochem. Physiol.* **1979,** *62,* 155-164.
James, M.O.; Little, P.J. *Comp. Biochem. Physiol.* **1984,** *78C,* 241-245.
James, M.O.; Pritchard, J.B. *Drug Metab. Disp.* **1987**, *15,* 665-670.
James, M.O.; Schell, J.D.; Barron, M.G.; Li, C-L.J. *Drug Metabol. Disp.* **1991,** *19,* 536-541.
Johnston, J.J.; Corbett, M.D. *Toxicol. Appl. Pharmacol.* **1986,** *85,* 181-188.
Khan, M.A.Q.; Coello, W.; Khan, A.A.; Pinto, H. *Life Sci.* **1972,** *11,* 405-415.
Kleinow, K.M.; Beilfuss, W.L.; Jarboe, H.H.; Droy, B.F.; Lech, J.J. *Can. J. Fish. Aquatic. Sci.* **1992,** *49,* 1070-1077.
Kleinow, K.M.; Haasch, M.L.; Williams, D.E.; Lech, J.J. *Comp. Biochem. Physiol.,* **1990,** *96C,* 259-270.
Kleinow, K.M.; Jarboe, H.H.; Shoemaker, K.E.; Greenlees, K.J. *Can. J. Fish. Aquatic Sci.* (In Press).
Kleinow, K.M.; Melancon, M.; Lech, J.J. *Environ. Hlth. Perspect.* **1987,** *71,* 105-119.
Kleinow, K.M.; Smith, A.; Altman, A.; Holmes, E.; Venugopalan, C. *Toxicologist* **1994,** *14,* 431.
Kloepper-Sams, P.J.; Stegeman, J.J. *Arch. Biochem. Biophysics.* **1989**, *268,* 525-535.
Knezovich, J.P.; Crosby, D.G. *Environ. Toxicol. Chem.* **1985,** *4,* 435-446.
Kobayashi, K.; Kimura, S.; Akitake, H. *Bull. Jpn. Soc. Sci. Fish.* **1976,** *42,* 171-177.
Koivusaari, U.; Andersson, T. *Comp. Biochem. Physiol.* **1984,** *78B,* 223-226.
Koivusarri, U.; Harri, H.; Hanninen, O. *Comp. Biochem. Physiol.* **1981**, *70C,* 149-157.
Koivusaari, U.; Lindstrom, S.P.; Hanninen, O.

Adv. Physiol. Sci. **1980,** *29,* 433-440.
Krahn, M.M.; Brown, D.W.; Collier, T.K.; Friedman, A.J.; Jenkins, R.G.; Malins, D.C. *J. Biochem. Biophys. Methods* **1980,** *2,* 223-246.
Kurelec, B. *Biochem. Biophys. Res. Commun.* **1985,** *126,* 773-778.
Kurelec, B.; Britivic, S.; Krca, S.; Muller, W.E.G.; Zahn, R.K. *Comp. Biochem. Physiol.* **1986,** *86C,* 17-22.
Landrum, P.F.; Crosby, D.G. *Xenobiotica,* **1981,** *11,* 351-366.
Lauren, D.J.; Halarnkar, P.P.; Hammock, B.D.; Hinton, D.E. *Biochem. Pharmacol.* **1989,** *38,* 881-888.
Layiwola, P.J.; Linnecar, D.F.C. *Xenobiotica* **1981,** *11,* 167-171.
Layiwola, P.J.; Linnecar, D.F.C.; Knights, B. *Xenobiotica* **1983,** *13,* 107-113.
Leaver, M.J.; Clarke, D.J.; George, S.G. *Aquatic Toxicol.* **1992a,** *32,* 265-278.
Leaver, M.J.; Scott, K.; George, S.G *Mar. Environ. Res.* **1992b,** *34,* 237-241.
Lech, J.J. *Biochem. Pharmacol.* **1974,** *23,* 2403-2410.
Lech, J.J.; Statham, C.N. *Toxicol. Appl. Pharmacol.* **1975,** *31,* 150-158.
Lindstrom-Seppa, P.; Hanninen, O. *Archives of Toxicol.* **1986,** *Suppl. 9,* 374-377.
Lindstrom-Seppa, P.; Koivusaari, U.; Hanninen, O. *Comp. Biochem. Physiol.* **1981,** *69C,* 259-263.
Livingstone, D.R.; Farrar, S.V. *Sci. Total Environ.* **1984,** *39,* 209-235.
Livingstone, D.R.; Kirchin, M.A.; Wiseman, A. *Xenobiotica* **1989,** *19,* 1041-1062.
Marking, L.; Bills, T.D. *Quarterly Report;* U.S. Dept. of Interior, Fish and Wildlife Service, National Fishery Research Laboratory, **1979**.
Marsh, J.W.; Chipman, J.K.; Livingstone, D.R. *Aquatic Toxicol.* **1992,** *22,* 115-128.
Melancon, M.J.; Lech, J.J. *Comp. Biochem. Physiol.* **1984,** *79C,* 331-336.
Melancon, M.J.; Rickert, D.E.; Lech, J.J. *Drug Metab. Dispos.* **1982,** *10,* 128-133.
Miranda, C.L.; Wang, J.-L.; Henderson, M.C.; Buhler, D.R. *Biochim. Biophys. Acta.* **1990,** *1037,* 155-160.
Morgenstern, R.; Lundqvist, G.; Balk, L.; DePierre, J.W. *Biochem. Pharmacol.* **1984,** *33,* 3609-3614.
Morrison, H.; Young, P.; George, S.G. *Biochem. Pharmacol.* **1985,** *34,* 3933-3938.
Nagel, R. *Xenobiotica* **1983,** *13,* 101-106.
Nimmo, I.A.; Coghill, D.R.; Hayes, J.D.; Strange, R.C. *Comp. Biochem. Physiol.* **1981,** *68B,* 579-584.
Nimmo, I.A. *Fish Physiol. Biochem.* **1987,** *3,* 163-172.
Parker, R.S.; Morrissey, M.T.; Moldeus, P.; Selivonchick, D.P. *Comp. Biochem. Physiol.* **1981,** *70B,* 631-633.
Plakas, S.M.; James, M.O. *Drug Metab. Disp.* **1990,** *18,* 552-556.
Pritchard, J.B.; Bend, J.R. *Environ. Hlth. Persp.* **1991,** *90,* 85-92.
Pritchard, J.B.; Cotton, C.U.; James, M.O.; Giguere, D.; Koschier, F.J. *Bull. Mt. Desert Isl. Biol. Lab* **1978,** *18,* 58-60.
Pritchard, J.B.; James, M.O. *J. Pharmacol. Exptl. Therap.* **1979,** *208,* 280-286.
Rodgers, C.A.; Stalling, D.L. *Weed Sci.* **1972,** *20,* 101-105.
Roubal, W.T.; Collier, T.K.; Malins, D.C. *Arch. Environ. Contam. Toxicol.* **1977,** *5,* 513-529.
Sanborn, H.R.; Malins, D.C. *Proc. Soc. Exp. Biol. Med.* **1977,** *154,* 151-155.
Sanborn, H.R.; Malins, D.C. *Xenobiotica* **1980,** *10,* 193-200.
Schell, J.D.; James, M.O. *J. Biochem. Toxicol.* **1989,** *4,* 133-138.
Schlenk, D. *Aquatic Toxicol.* **1993,** *26,* 157-162.
Schlenk, D. Ph.D. Thesis, Oregon State University, Corvallis, **1989**.
Schlenk, D.; Buhler, D.R. *Marine Biology* **1990,** *104,* 47-50.
Schlenk, D.; Buhler, D.R. *Aquatic Toxicol.* **1991a,** *20,* 13-24.
Schlenk, D.; Buhler, D.R. *Xenobiotica* **1991b,** *21,* 1583-1589.
Schlenk, D.; Erickson, D.A.; Lech, J.J.; Buhler, D.R. *Fund. Appl. Toxicol.* **1992,** *18,* 131-136.
Schlenk, D.; Miranda, C.L.; Buhler, D.R. *Biochimica et. Biophysica Acta* **1993a,** *1156,* 103-106.
Schlenk, D.; Ronis, M.J.J.; Miranda, C.L.; Buhler, D.R. *Biochem. Pharmacol* **1993b,** *45,* 217-221.
Scott, K.; Leaver, M.J.; George, S.G. *Mar. Environ. Res.* **1992,** *34,* 233-236.
Squibb, K.S.; Michel, C.M.F.; Zelikoff, J.T.; O'Connor, J.M. *Vet. Human Toxicol.* **1988,** *30 (suppl. 1),* 31-35.
Statham, C.N.; Pepple, S.K.; Lech, J.J. *Drug Metab. Dispos.* **1975,** *3,* 400-406.
Stegeman, J.J. *J. Fish. Res. Bd. Can.* **1979,** *36,* 1400-1405.
Stegeman, J.J. In *Polycyclic Hydrocarbons and Cancer;* Gelboin, H.G.; Ts'o, P.Q.P., Eds.; Academic Press: New York, **1981**; Vol. 3, pp 1-60.

Stegeman, J.J. *Mar. Biol.* **1985,** *89,* 21-30.
Stegeman, J.J. *Xenobiotica* **1989,** *19,* 1093-1110.
Stegeman, J.J.; Kloepper-Sams, P.J. *Environ. Hlth. Persp.,* **1987**, *17,* 87-95.
Stehly, G.R.; Hayton, W.L. *Xenobiotica* **1989,** *19,* 75-81.
Stenger, V.G.; Maren, T.H. *Comp. Gen. Pharmacol.* **1974,** *5,* 23-25.
Strom, A.R. *Comp. Biochem. Physiol.* **1980,** *65B,* 243-249.
Suzuki, A.; Shimura, M.; Kikuchi, T.; Sekizawa, Y. *Bull. Jpn. Soc. Sci. Fish.* **1977,** *43,* 837-847.
Takimoto, T.; Oshima, M.; Miyamoto, J. *Ecotoxicol. Environ. Saf.* **1987,** *13,* 104-117.
Tsuyuki, H.; Idler, D.R. *Can. J. Biochem. Physiol.* **1960,** *38,* 1177-1183.
Varanasi, U.; Gmur, D.J *Aquatic Toxicol.* **1981**, *1,* 49-67.
Varanasi, U.; Gmur, D.J.; Treseler, P.A. *Arch. Environ. Contam. Toxicol.* **1979,** *8,* 673-692.
Walters, J.M.; Cain, R.B.; Higgins, I.J.; Corner, E.D.S. *J. Mar. Biol. Assoc.* **1979,** *59,* 553-564.
Wilkinson, C.F. In *Silent Spring Revisited;* Marco, G.J.; Hollingworth, R.M.; Durhams, W., Eds.; American Chemical Society: Washington, DC, **1987**; pp 25-48.
Zhang, Y. S.; Goksoyr, A.; Andersson,T. ;Forlin, L. *Comp. Biochem. Physiol.* **1991,** *98B* ,97-103.
Ziegler, D.M. *Drug Metab. Rev.* **1988,** *19,* 1-32.

MODE OF ACTION

The Influences of Molecular Mechanisms of Action on Herbicide Design

John B. Pillmoor, Stephen D. Lindell, Geoffrey G. Briggs and Kenneth Wright
AgrEvo UK Limited, Chesterford Park, Saffron Walden, Essex, CB10 1XL, UK.

Biochemistry has the potential to help the herbicide discovery process in a number of ways. The inputs can be grouped into four categories of lead inspiration, lead generation, lead elaboration, and lead optimisation. These are reviewed and then exemplified further by considering two enzymes, acetolactate synthase (EC 4.1.3.18) and imidazole glycerol phosphate dehydratase (EC 4.2.1.19) in detail. Finally, the prospects for the future, in particular the likely impact of molecular biology, are considered.

The objective of this paper is to review how biochemical knowledge has helped in the herbicide discovery process and to highlight how we expect it to continue to help in the future. To set the scene, a list of the herbicide sites of action of commercial importance is given in Table 1 with an estimate of the number of compounds that are cited in the most recent edition of the Pesticide Manual (Worthing and Hance, 1991) that act at each site. Some fifteen defined sites of action are evident from this analysis and at least one site (p-hydroxy-phenyl pyruvate dioxygenase, Schulz *et al*, 1993 and Prisbylla *et al*, 1993) could be added from recent work.

The table also reflects a good level of biochemical knowledge with only 14% of the compounds having a mode of action which is at present totally unknown. In all cases where the mode of action is known, this was elucidated after the discovery of the first compounds with biological activity. Nevertheless, once knowledge about the mode of action of a compound becomes available, this can then be used to influence further chemical synthesis in a number of ways. In this review we first consider the ways that biochemistry can potentially help in the herbicide discovery process and then exemplify some of these by considering two enzymes, acetolactate synthase and imidazole glycerol phosphate dehydratase, in more detail.

How Might Biochemistry Help?

Biochemistry can help in a number of ways, depending on the actual project and the level of biochemical knowledge that is available. The key inputs can be grouped into the following four categories:

Lead Inspiration

The ultimate application of biochemistry will be in the design of new herbicides from first principles by selecting an appropriate enzyme and designing novel inhibitors with suitable physicochemical properties to allow them to gain access to the target enzyme *in vivo*. Although it is true that no commercial herbicide has yet been discovered solely by this approach, degrees of success have been achieved in that a target enzyme has been rationally chosen and novel inhibitors, with at least some biological activity, have been discovered (see Pillmoor *et al*, 1991 and Abell *et al*, 1993 for reviews). Some key examples, are also noted below:

- pyruvate dehydrogenase (Baillie *et al*, 1988);
- ketol acid reductoisomerase (Aulabaugh and Schloss, 1990);
- glutamine synthase (Wright *et al*, 1991);
- dehydroquinate synthase (Frost *et al*, 1992);
- isopropylmalate isomerase (Hawkes *et al*, 1993a).

This mechanistic design approach to the discovery of new herbicides is exemplified further below in the section on the enzyme imidazole glycerol phosphate dehydratase.

Lead Generation

Random screening of compounds and natural products against target enzymes is well

Table 1

The Biochemical Sites of Action of the Main Commercial Herbicides

Site of Action		Number of Compounds[a]
Amino Acid Biosynthesis		
- acetolactate synthase	- sulfonylureas	15
	- imidazolinones	4
- enol pyruvoyl shikimate phosphate synthase		1
- glutamine synthetase		2
Lipid Biosynthesis		
- acetyl CoA carboxylase	- aryloxyphenoxy compounds	5
	- cyclohexanediones	7
- thiocarbamates		15
- others		2
Tetrapyrrole Biosynthesis		
- protoporphyrinogen oxidase		13
Carotenoid Biosynthesis		9
Cellulose Biosynthesis		1
Folate Biosynthesis		1
Photosystem II		
- triazines		19
- ureas		22
- others		15
Photosystem I		2
Oxidative Phosphorylation		3
Microtubule Assembly and Function		12
Auxin Mimics		20
Non-Specific chloracetanilides		10
Unknown		28
	Total	206

[a]: Based on the herbicides included in the Ninth Edition of the Pesticide Manual (Worthing and Hance, 1991)

established in the pharmaceutical industry as a means of identifying novel lead compounds (Schindler, 1992 and Hodgson, 1993). To date, however, there are few published examples of the application of this approach in the agrochemical industry. Nevertheless, there appears to be a growing interest in this area and we certainly believe that high throughput biochemical screening could have a role to play in the discovery of new agrochemical leads, providing that a sufficiently large and diverse number of compounds can be rapidly screened. Indeed, the recent report of the discovery of a novel inhibitor of acetolactate synthase from a microbial fermentation broth clearly indicates the potential of this approach (Haraguchi *et al*, 1992).

Lead Elaboration

Once the mode of action of a group of biologically active compounds is established, the knowledge about the target enzyme can be used to make large jumps in the synthesis direction. For example, hybrid inhibitors that combine the binding interactions of two different groups of compounds that bind to the enzyme can be considered. This approach has been adopted with the enzyme acetyl CoA carboxylase where more potent inhibitors were obtained by combining elements of the aryloxyphenoxy-propionic acid herbicides with either the cyclohexanedione herbicides or the acetyl CoA substrate (Rendina *et al*, 1994).

In addition, biochemical testing can identify compounds that have good biochemical activity but have little or no biological activity. In the absence of the biochemical data such compounds would not be considered further and an example of this is given later when considering acetolactate synthase. In cases where the biochemical activity is sufficiently good (for example when the K_i/K_m ratio for a competitive inhibitor is less than 10^{-3}), the reasons for the lack of biological activity may warrant investigation. There are, however, few published examples where such work has been undertaken, the general assumption probably being that the compound does not reach the enzyme due to poor uptake and transport or rapid detoxification. An exception to this is the work on ketol acid reductoisomerase (Wittenbach *et al*, 1991). Following the design of extremely potent inhibitors of the enzyme which showed only relatively weak herbicidal activity, a detailed investigation was undertaken. This revealed that the compound did successfully reach and inhibit the enzyme *in vivo*, leading to the conclusion that the enzyme itself is a poor herbicide target.

Lead Optimisation

Biochemistry can help the optimisation of a new series both by providing results rapidly and by contributing to structure-activity correlations.

Many biochemical assays can cope with large numbers of compounds and provide data within a few hours in contrast to most biological screens where the throughput is lower and, more crucially, the reporting time can be of the order of weeks. Consequently, the biochemical test can provide rapid feedback to the chemist so that effort can be directed to the active areas and away from inactive lines of synthesis. This can help to minimise both the time and the money wasted on synthesising and screening inactive compounds. In addition, the biochemical data can help to identify those compounds which justify entering a 'fast track' in the biological screens.

With respect to structure-activity correlations, the biochemical tests allow the absolute potency of a compound to be quantified without the problems of uptake, transport and metabolism that can hinder the interpretation of biological data. This type of data is essential if molecular modelling work or quantitative structure activity correlations are to be undertaken successfully. Such inputs can allow a clear picture of what is required to obtain activity at the enzyme level to be obtained. Once this is achieved, then the modification of a series to take into account other factors such as uptake, metabolism and environmental properties can be approached more logically. In many cases, of course, this will require compromises to be made between the structural requirements for biochemical potency and those for efficient delivery.

Acetolactate Synthase

Inhibitors of this enzyme have tended to dominate the herbicide field since the first sulfonylurea patents were filed in 1976 (Levitt, 1976 a,b). As is often the case, it was some time after the discovery of the first biologically active compounds before their mode of action was established and this turned out to be the inhibition of acetolactate synthase, an enzyme involved in the biosynthesis of the branched chain amino acids (Ray, 1984). A number of structurally different groups of herbicides, all acting through inhibition of the same enzyme, has now been discovered and a wide range of compounds has been commercialised (see Stidham, 1991 for general review). Some of the inputs that biochemistry can make are exemplified below with work undertaken with this enzyme both at Chesterford Park and elsewhere as reported in the literature.

Lead Elaboration

The initial phase of our work with acetolactate synthase involved a search for alternative structures to the sulfonylureas. Little herbicidal activity was obtained with the first compounds. However, the biochemical test showed that some inhibition of the plant enzyme was being achieved and this provided the incentive to drive the project forward. For example, Compound 1 (see Fig. 1 for structures) was initially synthesised from an intermediate in another project. Despite being inactive as a herbicide, it showed 30% inhibition of the plant enzyme at 100 μM. This led to a range of anilides being investigated of which a 2-chloro-6-methyl anilide substituent (Compound 2) improved the enzyme inhibition by a factor of ten. The corresponding 1-phenyl-5-propyl compound (Compound 3) showed similar inhibition of the enzyme but this time with a hint of herbicidal activity and classical symptomology for inhibitors of acetolactate synthase.

Further synthesis, including the development of a suitable route to allow the introduction of heterocyclic groups at position 1 of the triazole, subsequently achieved compounds that were both potent inhibitors of the enzyme and had sufficient levels of herbicidal activity to warrant field testing. An example is Compound 4 which showed 70% inhibition of the enzyme at 1 μM and good herbicidal activity at 8 g ai/ha. Unfortunately the soil persistence of this compound proved to be unacceptably long and it was not progressed further. Additional details of this work have been published elsewhere (Percival, 1991).

Lead Optimisation

During all our work with acetolactate synthase, the speed of the assay certainly helped us to concentrate on active lines of synthesis and avoid inactive ones, thereby saving us both time and money.

However, possibly the biggest help provided by the biochemical data was to enable good structure-activity correlations to be made. For example, it quickly became clear that a wide range of heterocycles at position 1 of the triazole, small substituents at position 5 of the triazole, and anilides with at least one electron withdrawing group in the *ortho*-position would give biochemically active compounds. From our own work and by analogy with the published sulfonylureas and sulfonanilides, many thousands of potentially active compounds were possible. Synthesis of all the possible combinations was clearly neither feasible nor desirable. Fortunately, the biochemical data showed clear trends which were not at all obvious on the basis of the biological data alone. Indeed, as our experience grew, it was possible to estimate the enzyme inhibition data by assigning a numerical factor to the key substituents at the three positions, some examples of which are given in Table 2. By then multiplying the factors together, a reliable estimate of the potency of a proposed compound could then be made, as shown in Table 3.

Lead Inspiration and Generation

Despite all our knowledge about the structure and mechanism of acetolactate synthase, it has not been possible to employ successfully a mechanistic approach to the design of completely novel inhibitors with biological utility. The knowledge that acetolactate synthase employs thiamine pyrophosphate as an essential cofactor did lead to the effects of thiamine thiazolone pyrophosphate being investigated. This compound (Compound 5 in Fig. 1), originally synthesised as an inhibitor of another thiamine pyrophosphate dependent enzyme, pyruvate dehydrogenase (Gutowski and Lienhard, 1976), was also shown to be a potent inhibitor of acetolactate synthase (Ciskanik and Schloss, 1986). However, the compound was biologically inactive, presumably due to lack of uptake and/or metabolic instability. Along similar lines, we wondered if our inhibitor of pyruvate dehydrogenase (Compound 6, Fig. 1), which was known to interact with the thiamine pyrophosphate of this enzyme (Baillie *et al*, 1988) would also be an effective inhibitor of acetolactate synthase. Unfortunately, for reasons that are not clear, this was not the case.

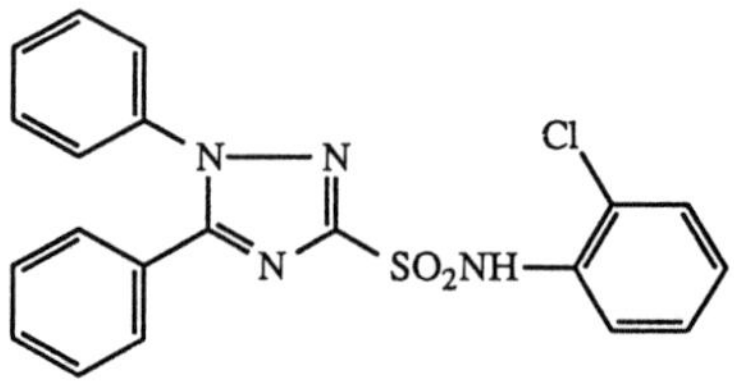

(1)

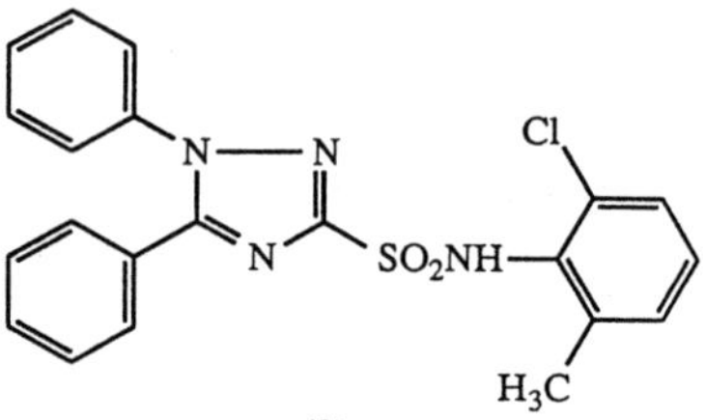

(2)

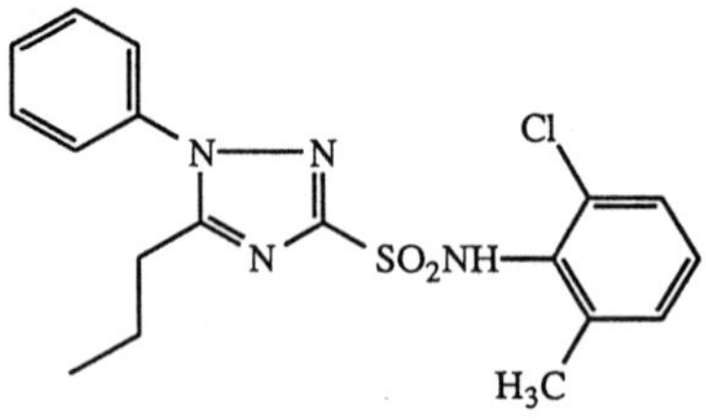

(3)

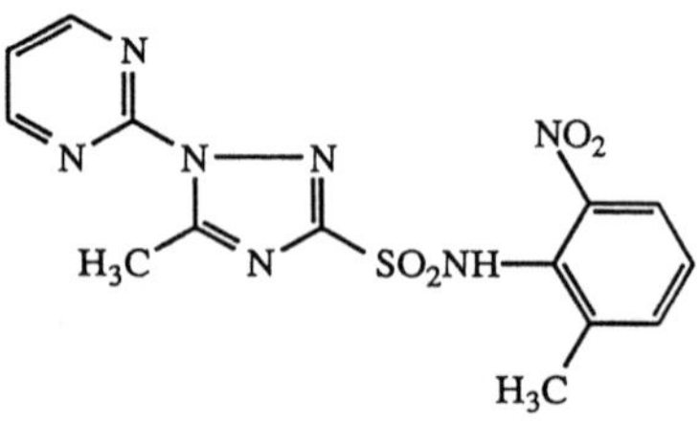

(4)

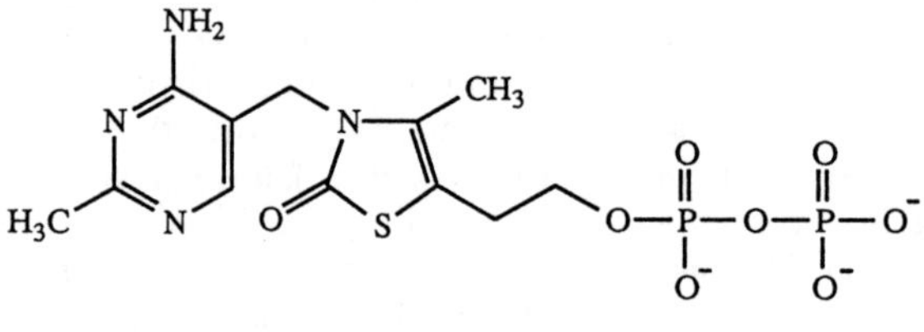

(5) Thiamin thiazolone pyrophosphate

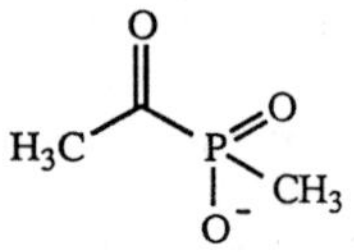

(6) Acetylmethyl-phosphinate

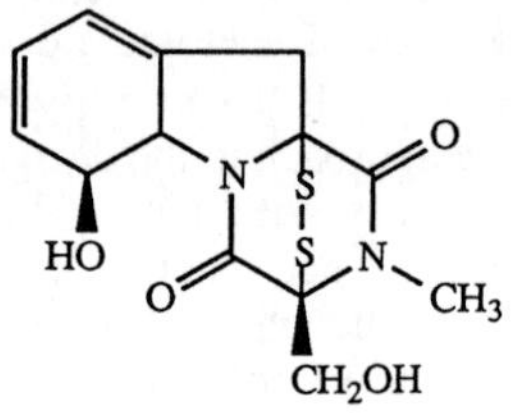

(7)

Fig. 1 **Acetolactate synthase inhibitors**

All the known herbicides that inhibit acetolactate synthase appear to bind at what has been termed an "extraneous binding site", which lies outside of the active site (Schloss and Aulabaugh, 1990). Despite our good understanding of the structural requirements for binding at this site, our knowledge about the three-dimensional nature of the binding site is insufficiently advanced to allow truly novel inhibitors to be designed from first principles. This could, of course, change in the future should a crystal structure of the enzyme become available.

Finally, as mentioned earlier, a novel inhibitor of the enzyme has been reported following a programme to screen natural products against the enzyme (Haraguchi *et al*, 1992 - Compound 7, Fig. 1). This clearly gives encouragement to the biochemical screening approach as a means of identifying novel inhibitors and more examples of this type might be expected in the future.

Table 2

Factors for a Range of Substituents at the Three Variable Positions of the Acetolactate Synthase Inhibitors

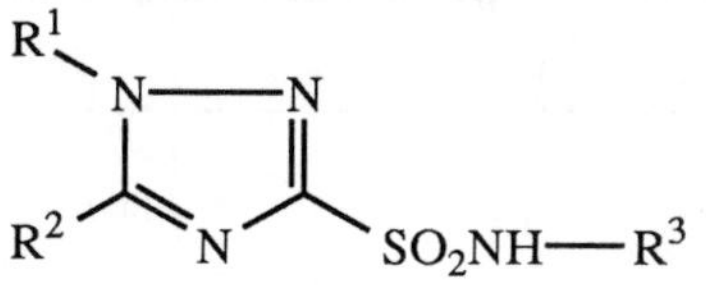

R^1 N N R^2 N SO_2NH—R^3

R^1		R^2		R^3	
Group	**Factor**	**Group**	**Factor**	**Group**	**Factor**
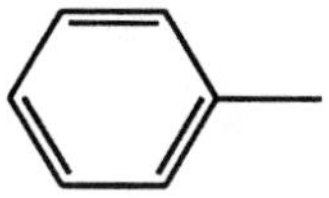	2.0	C_2H_5–	1.5	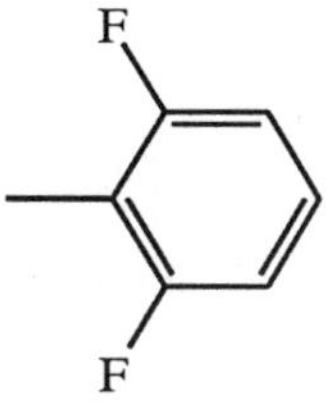 F, F	1.0
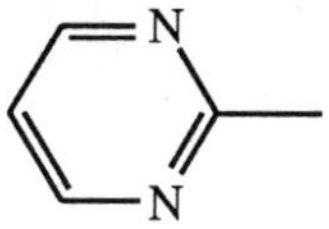 N, N	1.0	CH_3–	1.0	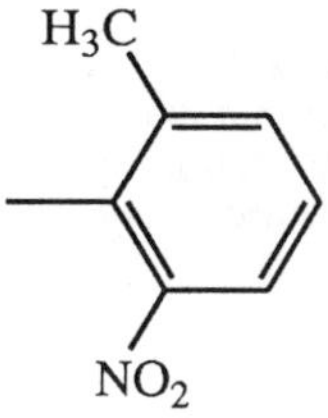 H_3C, NO_2	0.2
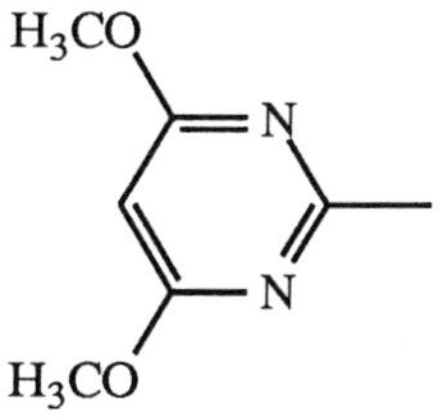 H_3CO, N, N, H_3CO	0.05			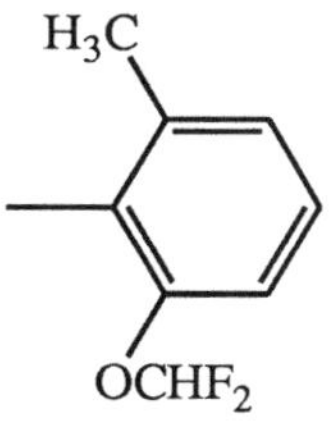 H_3C, $OCHF_2$	0.1

Imidazole Glycerol Phosphate Dehydratase (IGPD)

Our work on IGPD will be used to exemplify some aspects of the mechanistic design approach to discovering new potential herbicides. Our interest in IGPD resulted from an evaluation of all amino acid biosynthesis from which histidine biosynthesis was identified as an attractive pathway for study.

IGPD, which is the sixth enzyme in the pathway and catalyses the conversion of imidazole glycerol phosphate to imidazole acetol phosphate (Fig. 2) was chosen as a promising enzyme to work on with respect to the possible design of inhibitors. Although the commercial

Table 3

Comparison of the Predicted and Measured Biochemical Potency for Selected Combinations of Substituents for the Acetolactate Synthase Inhibitors

Substituent			Predicted I_{50} (μM)	Measured I_{50}(μM)
R^1	R^2	R^3		
Phenyl	Methyl	2,6-difluoro	2.0	2
"	"	2-methyl, 6-nitro	0.4	0.7
"	"	2-methyl, 6-difluoromethoxy	0.2	0.2
"	Ethyl	2,6-difluoro	3.0	10
"	"	2-methyl, 6-difluoromethoxy	0.3	0.1
Pyrimidine	Methyl	2,6-difluoro	1.0	1
"	"	2-methyl, 6-nitro	0.2	0.5
"	"	2-methyl, 6-difluoromethoxy	0.1	0.2
"	Ethyl	2,6-difluoro	1.5	8
4,6-dimethoxy pyrimidine	Methyl	2,6-difluoro	0.05	0.2
"	"	2-methyl, 6-difluoromethoxy	0.005	0.005

a: Activity measured in a crude enzyme preparation from etiolated maize shoots. Inhibitor and enzyme were incubated together for 10 minutes prior to addition of substrate (40 mM). This was followed by incubation for 30 minutes after which the reaction was halted and the amount of product measured.

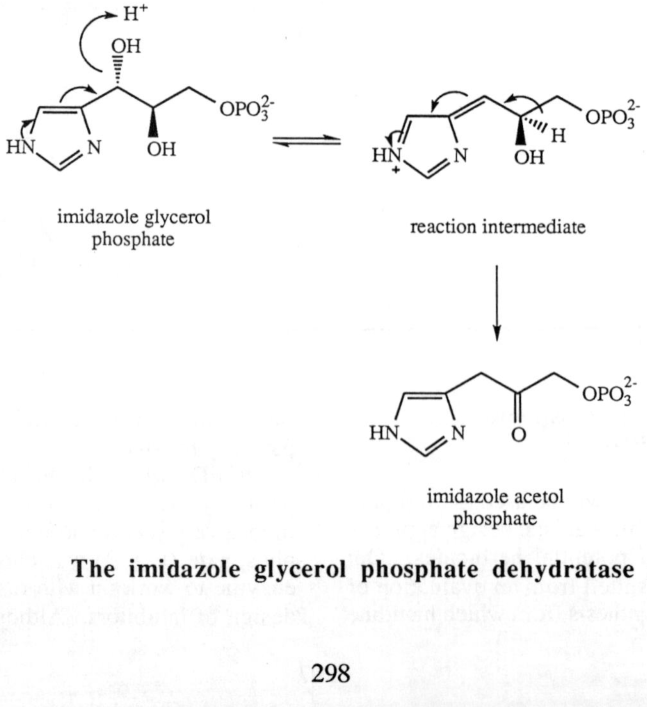

imidazole glycerol phosphate

reaction intermediate

imidazole acetol phosphate

Fig. 2 **The imidazole glycerol phosphate dehydratase reaction**

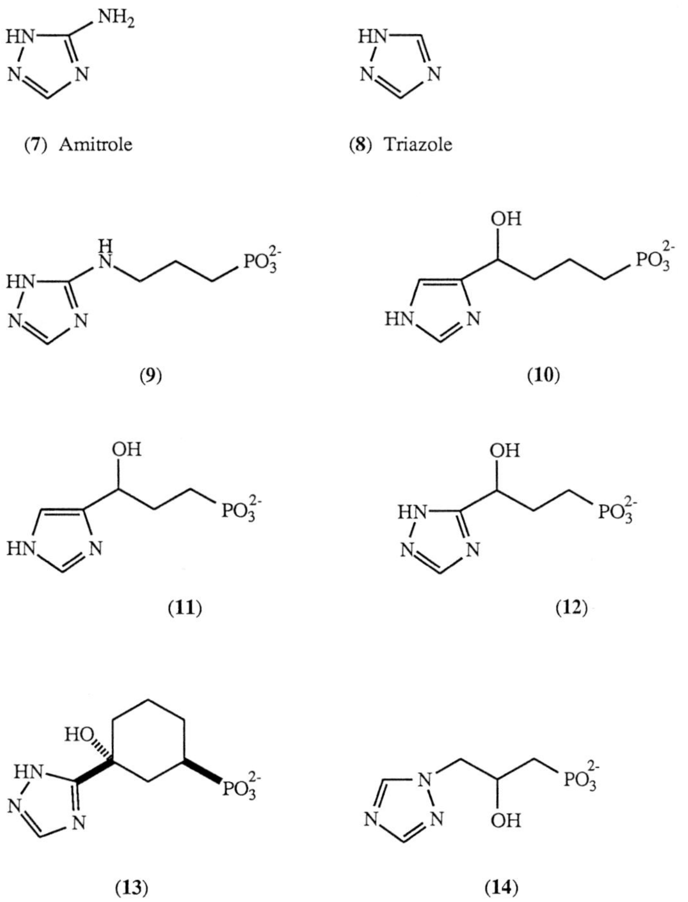

(7) Amitrole

(8) Triazole

(9)

(10)

(11)

(12)

(13)

(14)

Fig. 3 Imidazole glycerol phosphate dehydratase inhibitors

herbicide amitrole is reported to inhibit plant IGPD (Wiater *et al*, 1971), this is unlikely to be its primary mode of action as the compound is also known to have a number of other effects (Heim and Larrinua, 1989). Nevertheless, both amitrole and triazole (Compounds 7 and 8 in Fig. 3) are reasonably good competitive inhibitors (Wiater *et al*, 1971) and provided a starting point for our work. We speculated that they might be transition state-type inhibitors, mimicking the heterocyclic portion of the reaction intermediate shown in Fig. 2. We were encouraged in this thinking by the close similarity of the calculated isopotential contour maps for the diazafulvene moiety from the reaction intermediate with those for amitrole. These maps (Figs. 4 and 5) show the distribution of positive and, where appropriate, negative potential for the lower energy tautomers of the different heterocyclic systems. The comparison holds between the charged methylene imidazolium ion and the aminotriazolium ion (Fig. 4) and between the two neutral species (Fig. 5). While we are not certain which is the correct comparison, we favour that shown in Fig. 5 because there seems no obvious reason why amitrole (pK_a 4) should be protonated *in vivo*. The above, plus the known potentiation of the inhibition of Compounds 7 and 8 by phosphate (Wiater *et al*, 1971), inspired the synthesis of Compound 9 (Fig. 3). At the same time we also synthesised a simple substrate mimic, Compound 10, as a reference point against which to compare inhibition by Compound 9.

These compounds were modest inhibitors, and only the triazole (Compound 9) exhibited any biological activity, and then with symptomology more reminiscent of amitrole. Nevertheless, the biochemical data offered us encouragement. In particular, Compound 9,

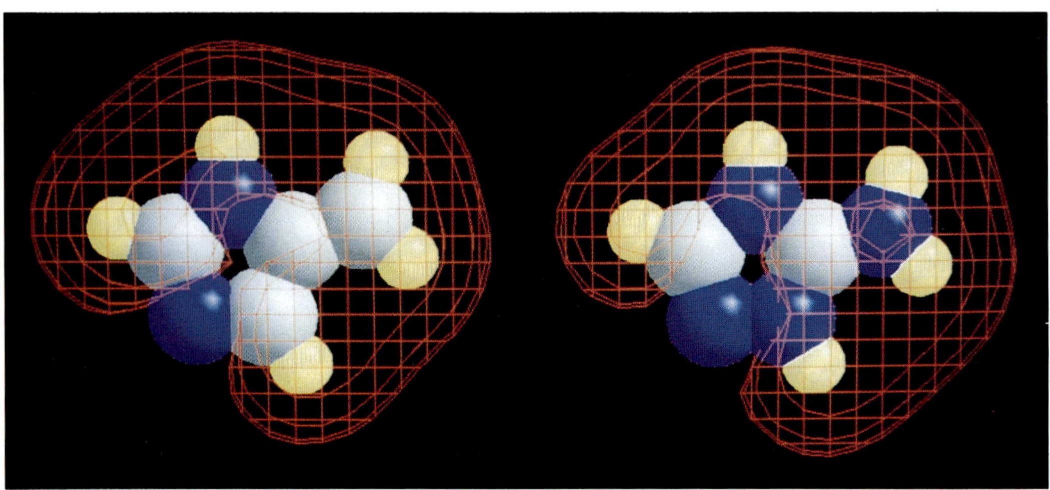

Fig. 4 100 kcal electrostatic isopotential contour maps (in the plane of the rings) for the methylene imidazolium (left) and aminotriazolium (right) ions

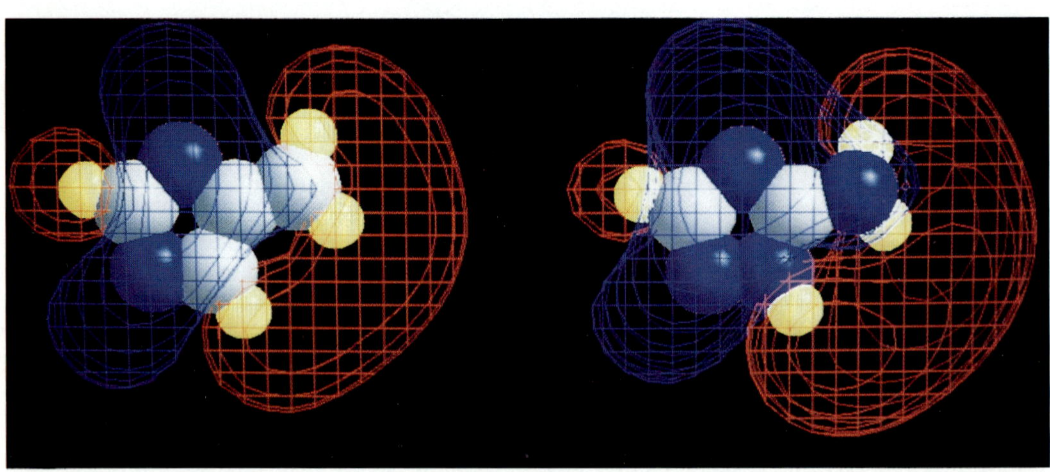

Fig. 5 10 kcal (red) and -10 kcal (blue) electrostatic isopotential contour maps (in the plane of the rings) for methylene imidazole (left) and amitrole (right)

which showed 20% inhibition of the *Escherichia coli* enzyme at 100 μM, was some 10 times more potent an inhibitor than Compound 10. This provided some support for our theory that the triazole might be mimicking a high energy intermediate in the reaction mechanism. From this and other work we were starting to build up a picture of what was required for binding to the enzyme. For example, we made the shortened chain imidazole (Compound 11) and were surprised to find that it was as potent an inhibitor as Compound 9. This led us to speculate that the shortened chain might be mimicking a folded conformation of the reaction intermediate in which the phosphate acts as an internal base and abstracts the C2-hydrogen to give the product. A similar mechanism and inhibitor have been reported previously for dehydroquinate synthase (Widlanski *et al*, 1989).

During this work it became clear that other companies were also interested in this enzyme. Ciba Geigy filed patent applications (Mori *et al*, 1992; Hayakawa *et al*, 1993) containing Compounds 12 and 13 (Fig. 3) which we immediately recognised as potential IGPD inhibitors. Synthesis of the above compounds and subsequent testing indicated that they were good inhibitors of the enzyme, showing 20% and 60% inhibition at 1 μM respectively. Moreover, they were also good herbicides with symptomology similar to other amino acid biosynthesis inhibitors. By observing their effects on germinating rice seedlings in agar, we showed that their biological activity could be reversed specifically with histidine, thereby providing further evidence that their herbicidal activity is due to the inhibition of IGPD. More recently, the knowledge about this enzyme as a herbicide target has been further extended following publication of work from Zeneca (Hawkes *et al*, 1993b). These authors have demonstrated that IGPD is the target for the herbicidally active Compound 14 (Fig. 3) and report on a number of other inhibitors (including Compounds 11 and 12). A detailed mechanistic analysis of the enzyme led these authors to also conclude that the phosphate dianion of the substrate acts as an internal base and to propose that the triazole inhibitors mimic the diazafulvene reaction intermediate.

From the work summarised here it is clear that IGPD has attracted considerable input as a potential herbicide target and it will be interesting to see if any company can bring their ideas to commercial fruition.

Prospects for the future

Biochemistry will undoubtedly continue to provide the types of input already mentioned above. Indeed, the inputs are likely to increase as the potential help that biochemistry can provide gains wider recognition and it becomes expected that there should be a biochemical input to the majority of projects.

A key development that is likely in the future is in the field of molecular biology. The rapid increase in our ability to clone and manipulate plant genes provides very exciting possibilities in both identifying those enzymes that offer the best potential as herbicide targets and then providing copious quantities of enzyme to aid the biochemical studies (see Pillmoor and Foster, 1994 for review).

In the field of validating new targets, the antisense approach is now gaining acceptance as a means of experimentally determining the physiological consequences of inhibiting a particular enzyme (Foster and Höfgen, 1993). Cloning and overexpression of target enzymes has already helped us to investigate biochemical targets that were difficult or even impossible to analyse by standard biochemical approaches. For example, in one of our projects it used to take a biochemist seven days and five kilograms of plant material to isolate 0.8 units of enzyme activity, only enough enzyme to test ten new chemicals. Following the cloning and overexpression of the plant enzyme in *Escherichia coli*, only six grams of cells and one day yields 170 units of enzyme, enough to test 2,000 new compounds. Such an approach can also improve the prospects of crystallisation and three dimensional analysis. To date this has only been reported for three proteins that are established as herbicide modes of action: photosystem II (Deisenhofer and Michel, 1989), the glyphosate target enol pyruvate shikimate phosphate synthase (Stallings *et al*, 1991) and the glufosinate target glutamine synthetase (Yamashita *et al*, 1989). Although these crystal structures have not yet, apparently, led to the design of novel inhibitors, the potential for this approach is clear from the successful work with structure-based drug design in the pharmaceutical industry (see Bugg *et al*, 1993 for review).

Acknowledgements

The authors gratefully acknowledge the inputs made by many colleagues at Chesterford Park in the work reported here. In particular, C. G. Earnshaw who suggested the use of isopotential contour maps in the IGPD work and kindly provided Figs. 4 and 5 and A. Percival and his co-workers who undertook the synthesis of the acetolactate synthase inhibitors. E. A. Saville-Stones, D. S. Carver, R. M. Turner, J. E. Dancer, R. G. Hughes and T. P. Monk were the key people involved in the IGPD work.

References

Abell, L. M.; Schloss, J. V.; Rendina, A. R. In *Pest Control with Enhanced Environmental Safety*; Duke, S. O.; Menn, J. J.; Plimmer, J. R. Eds.; ACS Symposium Series 524; American Chemical Society: Washington, DC, 1993; pp 16-37.

Aulabaugh, A.; Schloss, J. V. *Biochemistry* **1990**, 29, 2824-2830.

Baillie, A. C.; Wright, K.; Wright, B. J.; Earnshaw, C. G. *Pestic. Biochem. Physiol.* **1988**, 30, 103-112.

Bugg, C. E.; Carson, W. M.; Montgomery, J. A. *Scientific American* **1993**, 269, 60-66.

Ciskanik, L. M.; Schloss, J. V. *Federation Proc.* **1986**, 45, 1607.

Deisenhofer, J.; Michel, H. *Science* **1989**, 245, 1463-1473.

Foster, S. G.; Höfgen, R. In *Opportunities for Molecular Biology in Crop Production;* Beadle, D. J.; Bishop, D. H. L.; Copping, L. G.; Dixon, G. K.; Hollomon, D. W., Eds.; BCPC Monograph No.55; British Crop Protection Council: Farnham UK, 1993; pp 75-81.

Frost, J. W.; Piehler, L. T.; Montchamp, J. L. In *Biosynthesis and Molecular Regulation of Amino Acids in Plants*; Singh, B. K.; Flores, H. E.; Shannon, J. C., Eds.; American Society of Plant Physiologists: Maryland, USA, 1992; pp 163-173.

Gutowski, J. A.; Lienhard, G. E. *J. Biol. Chem.* **1976**, 251, 2863-2866.

Haraguchi, H.; Hamatani, Y.; Shibata, K.; Hashimoto, K. *Biosci. Biotech. Biochem.* **1992**, 56, 2085-2086.

Hawkes, T. R.; Cox, J. M.; Fraser, T. E. M.; Lewis, T. *Z. Naturforsch.* **1993a**, 48c, 364-368.

Hawkes, T. R.; Cox, J. M.; Barnes, N. J.; Beautement, K.; Edwards, L. S.; Kipps, M. R.; Langford, M. P.; Lewis, T.; Ridley, S. M.; Thomas, P. G. *Proc. British Crop Protect. Conf. - Weeds - 1993* **1993b**, 739-744.

Hayakawa, K.; Mori, I.; Iwasaki, G.; Matsunaga, S. I. European Patent 0,528,760,A1, 1993.

Heim, D. R.; Larrinua, I. M. *Plant Physiol.* **1989**, 91, 1226-1231.

Hodgson, J. *BioTechnology*, **1993**, 11, 683-688.

Levitt, G., U.S. Patent 4,127,405, 1976a.

Levitt, G., U.S. Patent 4,169,719, 1976b.

Mori, I.; Iwasaki, G.; Scheidegger, A.; Koizumi, S.; Hayakawa, K.; Mano, J. International Patent WO 92/19629, 1992.

Percival, A. *Pestic. Sci.* **1991**, 31, 569-580.

Pillmoor, J. B.; Foster, S. G. In *Molecular Biology in Crop Protection*; Marshall, G.; Walters, D. R., Eds.; Chapman and Hall: London, 1994 in press.

Pillmoor, J. B.; Wright, K.; Lindell, S. D. *Proc. British Crop Protect. Conf. - Weeds - 1991* **1991**, 857-866.

Prisbylla, M. P.; Onisko, B. C.; Shribbs, J. M.; Adams, D. O.; Liu, Y.; Ellis, M. K.; Hawkes, T. R.; Mutter, L. C. *Proc. British Crop Protect. Conf. - Weeds - 1993* **1993**, 731-738.

Ray, T. B. *Plant Physiol.* **1984**, 75, 827-831.

Rendina, A. R.; Abell, L. M. In *Natural and Engineered Pest Management Agents*; Hedin, P. A.; Menn, J. J.; Hollingworth, R. M. Eds.; ACS Symposium Series 551; Americal Chemical Society: Washington, DC, 1994; pp 407-24.

Schindler, P. In *Drugs from Natural Sources*; Coombes, J. D., Ed.; IBC Technical Services: London, 1992; pp 20-35.

Schloss, J. V.; Aulabaugh, A. *Z. Naturforsch.* **1990**, 45c, 544-551.

Schulz, A.; Ort, O.; Beyer, P.; Kleinig, H. *FEBS Letts.* **1993**, 318, 162-166.

Stallings, W. C.; Abdel-Meguid, S. S.; Lim, L. W.; Shieh, H. S.; Dayringer, H. E.; Leimgruber, N. K.; Stegeman, R. A.; Anderson, K. S.; Sikorski, J. A.; Padgette, S. R.; Kishore, G. M. *Proc. Natl. Acad. Sci. USA* **1991**, 88, 5046-5050.

Stidham, M. A. In *Topics in Photosynthesis*; Baker, N. R.; Percival, M. P. Eds.; Elsevier: Amsterdam, 1991; Vol. 10; pp 247-266.

Wiater, A.; Klopotowski, T.; Bagdasarian, G. *Acta. Biochimica Polonica* **1971**, 18, 309-314.

Widlanski, T.; Bender, S. L.; Knowles, J. R. *J. Am. Chem. Soc.* **1989**, 111, 2299-2300.

Wittenbach, V. A.; Aulabaugh, A.; Schloss, J. V. In *Pesticide Chemistry*; Frehse, H., Ed.; VCH: Weinheim, 1991; pp 151-160.

Worthing, C. R.; Hance, R. J. *The Pesticide Manual*; British Crop Protection Council, Farnham, UK, 1991; 9th Edition.

Wright, K.; Pillmoor, J. B.; Briggs, G. G. In *Topics in Photosynthesis*; Baker, N. R.; Percival, M. P., Eds.; Elsevier: Amsterdam, 1991; Vol. 10; pp 337-368.

Yamashita, M. M.; Almassy, R. J.; Janson, C. A.; Cascio, D.; Eisenberg, D. *J. Biol. Chem.* **1989**, 264, 17681-17690.

Influence of Molecular Mechanism of Action on Herbicide Use Rate

Mark A. Stidham, Herbicide Discovery, Agricultural Research Division, American Cyanamid Company, Princeton, New Jersey 08543-0400 USA

One of many factors determining the sensitivity of a plant to any given herbicide is the herbicide site of action. Paradoxically, it is difficult to correlate herbicide use rates with the molecular mechanism of action. Herbicide absorption, translocation, and metabolism can dominate any attempt to describe the lower limit for any given class of herbicide. To make any conclusions about this topic we must consider how to isolate the role of a given herbicide target in plant metabolism. Factors such as the concentration of target in the plant, the distribution of the target, and the consequences of herbicide action at the target all contribute to the influence of molecular mechanisms of action on use rate.

As commercial herbicides have become progressively more potent, the search for new herbicides has an added emphasis on low use rate. The pressure for low use rate herbicides comes from a variety of concerns. No matter the motivation, the industry has turned to biochemistry to help predict the areas of chemistry with the highest promise for low use rate herbicides. The question is usually posed in one of two contexts.

In one context, the question addresses a choices of herbicide targets as new sites of action for herbicide design or *in vitro* screening. Here we could speculate about the value of any particular enzyme as a site of herbicide action. Could an inhibitor of enzyme x be a low use rate herbicide? An example of this sort would be considering inhibitors of the second enzyme in branched amino acid biosynthesis (Schloss, 1990). Design or discovery of *in vitro* inhibitors raised the question, could potent inhibitors that are weak herbicides ever become like inhibitors of acetohydroxyacid synthase (AHAS) or protoporphyrinogen oxidase (PROTOX)?

In the second context a new class of herbicide chemistry with a known mode of action is queried as to its potential for improvement. For example, we could consider a new photosystem II inhibitor with activity at about 1 kg/ha. Could analog synthesis ever result in a herbicide of use rate any lower than the current most potent PSII inhibitors?

The present article begins with the QSAR approach used in chemistry. Whereas QSAR seeks to isolate physicochemical parameters that describe biological activity, the modification I describe divides the components of herbicidal activity that contribute to the use rate. This approach will let me consider how to integrate the molecular mechanism of action with the other critical factors involved in herbicide action. In the following parts, I will list some of the considerations to make while addressing the relation of use rate and mode of action, and this is followed by some observations from experience. The conclusion of the article reflects on what we have at present and what we can anticipate in the future.

The Equation

The equation that contains a term for the ideal rate of a herbicide of a specified mode of action will have the form:

$$\text{activity}_{\text{observed}} = \text{activity}_{\text{ideal}} \times [1 + \boldsymbol{a}(\textit{transport}) + \boldsymbol{b}(\textit{metabolism}) + \boldsymbol{c}(\textit{site interaction})]$$

The coefficients a, b, and c are related to a particular compound with a specific mode of action, while the terms *transport, metabolism,* and *site interaction* describe functions for the particular site of action. The coefficients are such that as they approach zero, the activity of the compound improves. Thus, a compound with ideal transport properties would have a small value ***a*** while a compound with poor transport properties would have a large value ***a***. Perhaps more easily understood is a compound with a very low inhibition constant ***c***, where the relative contribution of the site interaction term would approach zero.

From this kind of treatment, the importance of the various aspects of herbicide action can be specified for each given molecular mechanism of herbicide action. For example, PSII inhibitors would need translocation terms which describe the delivery of inhibitors to chloroplasts. Transport terms for AHAS inhibitors would describe delivery to the meristem. The metabolism term is also customized so that half-life dependence reflects the relative duration of action required. Finally, the term for site interaction needs to be weighted, as one could imagine a range of target sites from those requiring virtually complete inhibition to those requiring only partial inhibition.

At this point, consider a theoretical best case. A particular herbicide is ideally absorbed and translocated in the plant to the site of molecular action. There is no significant metabolism, and the interaction at the molecular target site occurs with high affinity and is irreversible. At this ideal can we begin to consider factors pertaining only to the molecular site of action. After considering this ideal case, I will try to illustrate from known examples the limits of our ability to define use rates for a given site of action.

Concentration of target in the plant

At the extreme, a single herbicide molecule is absorbed by the plant and translocated to the molecular site of action. Certainly, a single molecule is not sufficient to cause a herbicidal effect. For example, a meristem inhibitor might have a herbicidal action once a critical number of cells had been inhibited. Assuming that a critical process has some finite level of enzymes involved, the herbicide would have to produce some fractional inhibition in the cell and enough herbicide would need be present to inhibit a critical number of meristematic cells. Perhaps it is the concentration of targets on a per plant basis rather than a per cell basis that is most important.

In medicinal chemistry, the DNA-cleaving molecules of the enediyne class are among the most potent antibiotics (Lee, 1991; Zein, 1988). These molecules act at the level of picograms per ml, perhaps causing only a small number of double strand breaks in the DNA in an actively dividing cell. A specific DNA sequence may represent the lowest concentration of target sites in a plant.

At the other end of the spectrum are targets present at relatively high concentration. In Table 1 are the calculated concentrations of enzymes assuming a protein molecular mass of 50 kDa and a total soluble protein level of 10 mg per gram fresh weight (1%). For reference, glutamine synthetase, a target for phosphinothricin, constitutes about 1% of the total soluble protein of a mature spinach leaf.

Target Distribution

Herbicide targets can be coarsely divided into two groups: targets required for cell division in the meristem and targets present in mature tissue. The meristem of a plant constitutes only a small fraction of the total cellular mass and is accessed through contact or through phloem elements. Because of its small mass, the opportunity to access through contact may be limited to seed germination and root tips. Mature tissue is accessed either through surface contact or through xylem elements.

These two groups constitute different problems for herbicide action. The accessibility through contact is high for targets localized in mature tissue, and the xylem is a fairly passive distribution system that can feed all of the mature tissue. The challenge for herbicides acting at these targets is one of coverage. Imagine an ideal PS II inhibitor. A single molecule might block a single site in a single chloroplast, but this single molecule could not be imagined to have much of a herbicidal effect. Clearly, for "critical area" is needed for the equation. This term would need to be such that a smaller plant would require a smaller effected area. Targets that are localized into specific tissues may be accessed more readily than those widely distributed.

For targets localized in meristem tissue, the opportunity for contact is low. With the exception of root tip inhibitors, the main mechanism of herbicide delivery to these targets is through the phloem elements. This drawback of low contact area is an asset when considering the critical area required for effect; a smaller number of cells should require a smaller amount of herbicide. The added advantage of meristem localized targets is that phloem mobile materials can concentrate in the meristem. Thus, while direct contact may be low, the meristem can be thought of as the end of a funnel whereby

Table 1. Calculated Cellular Concentration for Proteins of Various Relative Abundance. Calculations are based on 10 mg protein per gram fresh weight of tissue and a molecular mass of 50 kDa for any given protein.

% of extractable protein	Fold enrichment to purity	Cellular Concentration (μM)
0.01	10,000	0.02
0.04	2,500	0.1
1.0	100	2

herbicide absorbed by mature tissue over a large area is removed and concentrated into the much smaller meristem.

Target Enzyme Turnover

A related consideration to target concentration is target half-life. An enzyme that has a long half life may be a better target than one that is rapidly turned over. In antiparasitic chemotherapy, ornithine decarboxylase is a differential site of action of difluoromethlyornithine due to the relative turnover rates of the enzyme between host and parasite (Wang, 1991)

Relative Role of Target in Metabolic Pathway

Target position in a metabolic pathway may be an important consideration in determining the use rate of a given herbicide. However, this consideration is difficult to discuss, as the chain of secondary events resulting from inhibition are difficult to anticipate or even interpret. The discussion of PROTOX below illustrates this point.

Amplification of a Physiological Signal

One could also imagine a single molecule activating some process in the plant that would be amplified by the plant into a herbicidal effect. The systemin peptide (Pearce, 1991) is an example of a such a signal amplification and translocation event. Femptomolar levels of systemin produced in response to wounding or pathogen invasion results in significant cellular responses in cells distant from the wound site.

Other Considerations

The necessary dosage for herbicidal effect in these cases depend on other factors that I have not counted. In the case of the light-dependent herbicides, light intensity, oxygen tension, and the metabolism of reduced oxygen all would influence the efficacy of the herbicides. In most other cases, temperature and growth rate are significant mitigating factors. Inclusion of these factors would greatly complicate the equation. Suffice it to say that these complications can also play a dominant role in determining herbicide activity. As such, it is even more difficult to isolate the molecular mechanism of action as a factor in determining herbicide use rate.

Also in this ideal treatment, I have allowed for no variation in plant species or plant growth stages. There must be variation among plant species as to the pathways that are limiting and the relative importance of a pathway at different growth stages. However, in the practical world, the role of target site variation must be considered small in comparison to the other factors in the equation.

Survey of known target sites

There are more than a dozen molecular targets for commercial herbicides, and there are many herbicides for which the molecular mechanism of action has yet to be discovered. The number of targets for which there are low use rate herbicides is much smaller. AHAS and PROTOX are targets for the most potent herbicides. The two targets could not be more different in their attributes (Table 2). One attribute of note that is common between the two sites is the number of classes of herbicide chemistry acting on those sites. With the notable exception of the photosystem II inhibitors, these targets of low use rate herbicides accept the widest range of chemistry as inhibitors.

To list "range of chemistry" as an attribute is somewhat arbitrary; the range of chemistry is related to the amount of synthesis effort spent in attempts to extend the known chemistry. Hovever, this brings up the main conclusion of this paper. That is, the lower limit use rate for herbicides acting at any given molecular target appears to be inversely proportional to the number of classes of herbicide chemistry for that mode of action.

This conclusion avoids answering the questions posed in the introduction and places the onus for the discovery of new low use rate herbicides back on the chemists. The conclusion directs attention to enzymes that are inhibited by a wide range of chemistry. The test to perform is to screen a large chemical library on a range of enzymes and advance those targets that are inhibited by the widest range of chemistry. It is unreasonable to suggest that all targets are potential sites of low use rate herbicides. However, it is quite arguable that there are few if any cases where the non-ideal terms of the equation have been minimized. This means that the most potent herbicides have yet to be discovered. However, biochemistry is quite limited in its ability to detail which targets are most likely to yield low use rate herbicides.

Table 2. Attributes of targets of low use rate herbicides.

Attribute	AHAS	PROTOX
Localization	plastid stroma	chloroplast envelope
metabolite accumulated upon inhibition	substrate	product
Order in pathway	early	late
Metabolic regulation	end-product feedback inhibited	no metabolite regulation
Tissue localization for most profound herbicidal action	meristematic tissue	mature leaf
Time to plant death	weeks	hours
First affected tissue	meristems	leaves
Enzyme concentration	low	low
Number of classes of herbicide chemistry	large	large
Light-dependence	None	Complete

The Instructive Case of PROTOX as a Herbicide Target Site

To expand on the above assertions, consider the case of PROTOX. Nitrofen was developed over thirty years ago, and its use rate was in the kilogram per hectare range. The use rate of the diphenylether class of herbicides dropped over time with the introduction of significant commercial herbicides. Today, herbicides related to nitrofen are effective at doses 0.1% those of nitrofen. Most of the progress came with no significant understanding of the molecular mechanism of action. However, the chemistry was easily manipulated. In addition, other herbicide chemistry was discovered that mimicked diphenylether mode of action.

In the late eighties when the mode of action of the diphenylethers was discovered, it became apparent that unrelated chemistry also had PROTOX as the molecular site of action. The cascade of events explaining the mode of action of PROTOX inhibitors recapitulates the themes of this paper (for review see Nandihalli, 1993 and Duke, 1994). Inhibition of steps early in the pathway by gabaculine or 4,6-dioxoheptanoic acid have little if any herbicidal consequences. In contrast, inhibition of PROTOX, an enzyme late in the pathway, results in many physiological effects. First, is the depletion of heme and protochlorophyllide. The result of depletion of these two end products is the increase of carbon flow through the pathway. Second, PROTOX inhibition results in a temporary build up of the substrate protoporphyrinogen IX (protogen). The activation of the inhibited pathway and the buildup of the substrate for PROTOX are not surprising events, but they are the preludes to the main herbicidal event. That is, the accumulated protogen is enzymatically oxidized at a distant site, and the resulting free protogen now acts as a light-dependent catalytic converter of triplet oxygen to singlet oxygen. In sum, inhibition of PROTOX leads to accumulation of its *product* and subsequently generates singlet oxygen that destroys the tissue! A more unexpected chain of events is difficult to imagine.

Conclusions: Can Use Rates Drop Below Today's Levels?

A theoretical exercise can help project expectations about use rates. Consider a herbicide of molecular mass 250 g/mole. A 1 kg/ha application of such a herbicide converts to a rate of 4 mole/ha. Table 3 gives some ranges of herbicide concentrations on a per plant and an internal concentration level assuming ideal coverage and uniform complete absorption. As rates decrease from 1000 to 10 g/ha, the amount of herbicide received per plant ranges from 40 μ mole to 0.02 μmole per plant. This calculation assumes the plants are completely intercepting all herbicide sprayed so that each plant in a plant density of 200 per m^2 intercepts a proportionately less amount of herbicide than each plant in a density of 10 plants per m^2 . At the low extreme, interception of 0.02 μmole per plant of fresh weight 100 g each calculates to 0.2 μM if distributed equally throughout the plant.

These calculations ignore real world constraints that make the amount of herbicide per plant much lower than these numbers. For example, if 10% of the applied herbicide actually made contact with plants, and then 10% of that amount were absorbed, these numbers would be off by a factor of 100. At the low extreme again,

Table 3. Calculations of herbicide concentration in plants at various application dosages, plant densities, and plant sizes.

Herbicide dose (g/ha)	1,000	100	10
Plant density (plants/m^2)		µmole/plant	
200	2	0.2	0.02
50	8	0.8	0.08
10	40	4	0.4

Dose (µmole/plant)	40	0.8	0.02
Fresh weight (g)		[herbicide] (µM)	
100	400	8	0.2
10	4,000	80	2
1	40,000	800	20

a 100 g plant treated at 10 g/ha could have an internal herbicide concentration of 2 nM instead of 0.2 µM On the other side of error, especially for meristem inhibitors, translocation efficiencies of 90% of the absorbed material are common. Such translocation results in a concentration in meristem cells of many orders of magnitude relative to the whole plant.

These numbers indicate that even at 10 g/ha and high plant density, herbicide concentrations in a plant can be in the micromolar range. With appropriate stability and enzyme target affinity, this concentration is well above that necessary to produce phytotoxic effects for many known modes of action.

Summary

Herbicide use rates have dropped over time due to synthesis efforts to explore known chemistry. The factors that limit herbicide use rate are well known, and molecular mechanism of action is one of them. However, given present understanding of the intricacies of cellular metabolism, and the vast potential of synthetic space that is unexplored, there is no compelling reason to conclude that use rates cannot continue to decrease and that the number of targets for low use rate herbicides will increase.

References

Duke, S.O., Nandihalli, U.B., Lee, H.J., Duke, M.V., *Am. Chem. Soc. Symp. Ser.* **1994**, *559*, 191-204.

Lee, M.D.; Ellestad, G.A.; Borders, D.B., *Acc. Chem. Res.*, **1991**, 24, 235-243.

Nandihalli, U.B., Duke, S.O., *Am. Chem. Soc. Symp. Ser.* **1993**, *524*, 62-78.

Pearce, G.; Strydom, D.; Johnson, S.; Ryan, C.A.; *Science* **1991**, 253, 895-897

Schloss, J.V.; Aulabaugh, A., *Z. Naturforsch.* **1990**, 45c, 544-551.

Wang, C.C., *J. Cell. Biochem.* **1991,** 45, 49-53.

Zein, N.; Sinha, A.M.; McGahern, W.J.; Ellestad, G.A., *Science* **1988**, 240, 1198-1201.

Molecular Neurobiology and Insecticide Discovery

David M. Soderlund, Department of Entomology, New York State Agricultural Experiment Station, Cornell University, Geneva, NY 14456

The rapid growth of information in the area of insect molecular neurobiology provides the opportunity to develop and implement new tools and strategies for the discovery of neuroactive insecticides. This chapter reviews recent advances in insect molecular neurobiology that are relevant to known and postulated sites of insecticide action and considers their potential application to insecticide discovery efforts.

The insect nervous system is the target organ of most of the variety of synthetic insecticides that have been discovered and developed for commercial use during the past five decades. This unintentional convergence of discovery efforts reflects the repeated empirical observation that the most effective chemicals for insect control are those that produce rapid and profound disruption of the neuronal integration of insect locomotion and feeding. These considerations also suggest that the insect nervous system will remain an important target of future insecticide discovery efforts.

The continued exploitation of the insect nervous system as a target for insecticide action is limited by three important constraints. First, most insecticides discovered and used extensively during the past five decades have been found to act at a small number of target sites (Soderlund *et al.*, 1989; Benson, 1992) (Table 1). Of these, only three sites (acetylcholinesterases, voltage-sensitive sodium channels, and γ-aminobutyric acid (GABA) receptors) account for the action of all of the major classes of insecticides of both historic and current significance in practical insect control. This observation suggests that the number of readily exploitable targets in the insect nervous system is limited. Second, the pharmacology of neuronal target sites is highly conserved between insects and non-target vertebrates, so that the discovery of intrinsically selective neuroactive insecticides is particularly challenging. Finally, the small size of insect nervous systems limits the use of subcellular preparations of nervous tissue in biochemical screening programs directed at identifying new compounds with neurotoxic activity.

Recent advances in the molecular neurobiology of *Drosophila melanogaster* have provided a number of new tools that can be employed in the search for novel neuroactive insecticides. Use of this experimental system has led to the molecular cloning and characterization of genes that encode elements of the principal target sites for the major insecticide classes. Progress in this area also extends to the cloning of additional *D. melanogaster* genes that code for other neuronal macromolecules that might be exploitable In insecticide discovery. In reviewing some of these recent developments I will place particular emphasis on the molecular biology of insect voltage-sensitive sodium channels and γ-aminobutyric acid (GABA) receptors, two targets of proven practical significance for which the greatest amount of information is currently available. I will also discuss some of the ways in which progress in molecular neurobiology can enhance insecticide discovery.

Voltage-Sensitive Sodium Channels

Pharmacology. Voltage-sensitive sodium channels carry a transient sodium current across nerve membranes that is responsible for the rising phase of the nerve action potential (Hille, 1992). The fundamental role of the sodium channel in nerve excitation makes it a highly sensitive target for toxicants that disrupt normal nerve function. In this context, it is not surprising that the sodium channel is probably the most widely-exploited target site for naturally-occuring neurotoxins that are employed by plant and animal species in their defensive or predatory strategies.

Many of these naturally-occuring neurotoxins have been employed as probes to define further the pharmacology of sodium channels and elucidate the number and properties of neurtoxin recognition sites associated with this target (Catterall, 1988; Catterall, 1992). Results of these studies identify five distinct recognition sites (Table 2, Sites 1-5), each of which is labeled specifically by a neurotoxin-derived radioligand, plus several additional sites that are inferred on the basis of the known actions of other drugs and neurotoxins that modify sodium channel function. Thus, the sodium channel should not be viewed as a single pharmacological

1054–7487/95/0309$12.00/0

Table 1. Sites Of Insecticide Action In The Nervous System

Target macromolecule	Active compounds
Acetylcholinesterase	Organophosporus esters Methylcarbamates
$GABA_A$ receptor	Chlorinated cyclodienes Lindane Bicycloorthobenzoates Phenylpyrazoles (Avermectins?)
Voltage-sensitive Na^+ channel	DDT and analogs Pyrethroids *N*-alkylamides Dihydropyrazoles
Nicotinic acetylcholine receptor	Nereistoxin/cartap Nitromethylene heterocycles
Octopamine receptor	Formamidines
GABA-insensitive Cl^- channels	Avermectins

SOURCE: Adapted from Soderlund *et al.* (1989) and Benson (1992).

target site but rather as a array of functionally related but pharmacologically distinct target sites, each of which is formed by a unique domain of the channel macromolecule.

The sodium channel is recognized as an important insecticide target site principally because DDT, DDT analogs and pyrethroids produce their toxic effects by modifying sodium channel function in the insect nervous system (Soderlund and Bloomquist, 1989; Narahashi, 1992; Bloomquist, 1993a). Detailed pharmacological studies of pyrethroids imply the existence of a unique binding domain (Lombet *et al.*, 1988) (Table 2, Site 6) on the sodium channel that recognizes these compounds. Two additional insecticide classes, the *N*-alkylamides and dihydropyrazoles, have also been found to affect sodium channel function by binding to sites other than Site 6 (Table 2) (Soderlund and Knipple, 1994). These results confirm the significance of the sodium channel as a target site and exemplify the opportunity to exploit binding domains other than Site 6 to achieve useful insecticidal activity.

Molecular Biology. The molecular structure of vertebrate voltage-sensitive sodium channels was deduced initially by means of protein purification and reconstitution experiments and subsequently by the molecular cloning of sodium channel genes (Catterall, 1988; Catterall, 1992; Kallen *et al.*, 1993). Purification studies identified a large α subunit as a common feature of all sodium channels. In addition, smaller β1 and β2 subunits have been identified as components of some vertebrate sodium channels. Molecular cloning has revealed a family of sodium channel α subunit genes in vertebrate brain, skeletal muscle and cardiac muscle. On the basis of these data, models of sodium channel α subunit structure have been proposed that predict the spatial association of four repeated homology domains, each of which contains a series of six membrane-traversing helices, to form a pseudotetrameric channel.

Information on the structure and diversity of sodium channel genes in insects is less complete. Two genes possessing structural homology to vertebrate sodium channels have been isolated from the *D. melanogaster* genome and characterized. One of these, *DSC1*, was isolated independently by several groups by low stringency hybridization of electric eel sodium channel cDNA probes (Salkoff *et al.*, 1987; Okamoto *et al.*, 1989; Ramaswami and Tanouye, 1989; Soderlund *et al.*, 1989). The other, which is encoded by the *para* locus, was isolated by molecular genetic methods (Loughney *et al.*, 1989). Genetic and physiological criteria implicate the *para*$^+$ gene product as the

Table 2. Neurotoxin and Insecticide Binding Sites on the Voltage-Sensitive Sodium Channel

Site	Active Neurotoxins	Physiological effect
1	Tetrodotoxin Saxitoxin	Inhibit ion transport
2	Veratridine Batrachotoxin Aconitine *N*-Alkylamides	Cause persistent activation
3	α Scorpion toxins Sea anemone toxins	Prolong inactivation
4	β Scorpion toxins	Enhance activation
5	Brevetoxins Ciguatoxin	Enhance activation
6	DDT and Analogs Pyrethroids	Prolong inactivation
7	Gonioporatoxin	Prolong inactivation
8	Pumiliotoxin-B	Causes persistent activation
9	Local anesthetics Anticonvulsants Dihydropyrazoles	Inhibit ion transport

SOURCE: Adapted from Soderlund and Knipple (1994).

physiologically predominant sodium channel of the *D. melanogaster* nervous system (Stern *et al.*, 1990). In contrast, *DSC1* maps to a cytogenetic locus that has not been identified by either genetic or physiological criteria as being involved in sodium channel function (Salkoff *et al.*, 1987). Genetic criteria also implicate the *para*$^+$ gene product as the site of action of pyrethroids in *D. melanogaster* (Hall and Kasbekar, 1989), and linkage analyses of pyrethroid resistance traits involving reduced neuronal sensitivity provide evidence that *para*-homologous genes are sites of action of DDT and pyrethroids in the house fly (Williamson *et al.*, 1993; Knipple *et al.*, 1994), the tobacco budworm (Taylor *et al.*, 1993), and the German cockroach (Dong and Scott, 1994). The molecular cloning and sequence analysis of the *para* genes of susceptible and pyrethroid-resistant house fly strains is currently being pursued by two research groups (Soderlund and Knipple, 1994; Williamson *et al.*, 1994).

Numerous studies have documented the ability of mRNA isolated from vertebrate tissues or synthesized *in vitro* from cloned sodium channel cDNAs to direct the synthesis of functional sodium channels in oocytes of the clawed frog, *Xenopus laevis* (reviewed in: Lester, 1988; Catterall, 1992). These studies show that only the large α subunit is required to form functional channels and that the pharmacological properties of channels obtained by this method are comparable to those of sodium channels in native membranes. However, coexpression of an α subunit from mammalian brain with the smaller β1 subunit is required to obtain expressed channels with normal fast inactivation kinetics in this system (reviewed in: Catterall, 1992). Recent studies have employed site-directed mutagenesis coupled with heterologous expression assays to identify structural domains of vertebrate sodium channels that are involved in ion pore formation (Pusch *et al.*, 1991; Heinemann *et al.*, 1992), channel inactivation (Moorman *et al.*, 1990; Patton *et al.*, 1992; West *et al.*, 1992), and the binding of tetrodotoxin and saxitoxin (Noda *et al.*, 1989; Terlau *et al.*, 1991;

Satin *et al.*, 1992; Kontis and Goldin, 1993). To date there are no published reports of the expression of insect sodium channels in *X. laevis* oocytes although there does not appear to be any intrinsic barrier to the efficient expression of insect voltage-sensitive ion channels in this system (Schwarz and Yool, 1992).

Implications for Insecticide Discovery. Several considerations suggest that the voltage-sensitive sodium channel is a particularly appropriate system in which to develop tools derived from molecular biology for insecticide discovery. First, the sodium channel is not only a target of proven value in insect control but also a target macromolecule with several pharmacologically distinct binding domains that remain to be exploited as sites of insecticide action. Second, the *para* locus in *D. melanogaster* has been unambiguously identified as a gene encoding a physiologically and toxicologically relevant sodium channel. Moreover, the molecular identification of this toxicologically relevant target has been extended, by means of genetic linkage studies of target site-mediated resistance traits, beyond *D. melanogaster* to include *para*-homologous sodium channel genes of other economically important insect species. Third, the various neurotoxin recognition sites of the sodium channel appear to represent separate binding domains associated with a single gene product, the large α subunit. Thus, the sodium channel is a much simpler system to exploit with confidence than those in which the physiological target is comprised of two or more discrete subunits, making the subunit composition of native receptors difficult to determine. Finally, polymerase chain reaction (PCR)-based homology probing strategies permit the isolation of *para*-homologous sodium channel gene sequences from virtually any arthropod species of interest (Doyle and Knipple, 1991). Therefore, discovery strategies based on target site molecular biology need not be limited to model insect species but instead may be tailored to focus directly on insect species that are the targets of insecticide development.

The most direct application of sodium channel molecular biology to insecticide discovery would be to create cell lines that express sodium channel genes from species of interest for use in high-throughput screens. At present, this approach is limited not only by the lack of sodium channel-expressing cell lines but also by the lack of suitable assays. The modification of sodium channel function is most reliably detected using electrophysiological techniques, but such assays are not able to accomodate the level of throughput that would make target site-based screening a productive strategy. Radioligand binding assays are more attractive from a procedural point of view, but they have two significant shortcomings as means of detecting new agents that act at the sodium channel. First, radioligand displacement assays report on binding site occupancy but do not necessarily provide insight into the functional consequences of binding. Thus, it is possible to obtain compounds from radioligand binding screens that are potent antagonists of ligand binding but possess no intrinsic pharmacological activity. Second, a complete screen for agents affecting sodium channels would require a large battery of radioligands to detect interactions at all of the binding domains that might be of interest.

GABA Receptors

Pharmacology. GABA receptors mediate synaptic inhibition in both vertebrate and insect nervous systems (Sattelle, 1990; Macdonald and Olsen, 1994). The binding of GABA activates a chloride channel that is intrinsic to the receptor complex, and the resulting influx of chloride hyperpolarizes the postsynaptic nerve terminal, making it less susceptible to depolarization by excitatory inputs. In addition to a binding site for GABA, the GABA receptor also has several other sites that bind channel-blocking convulsants, benzodiazepine anxiolytics, depressant barbiturates, and certain neuroactive steroids (Sieghart, 1992; Macdonald and Olsen, 1994) (Table 3).

Research in the mid-1980's provided strong evidence that the chlorinated cyclodiene insecticides and lindane exert their primary insecticidal and neurotoxic effects by binding to the GABA receptor domain labeled by picrotoxinin and TBPS, thereby blocking GABA-dependent chloride transport (Bloomquist, 1993b; Casida, 1993) (Table 3). Knowledge of the mode of action of cyclodienes stimulated renewed interest in the GABA receptor as an insecticide target, yielding several novel chemical classes of insecticides that act as GABA receptor blockers (Bloomquist, 1993b; Casida, 1993). The first commercial product to emerge from this effort is fipronil, a phenylpyrazole (Colliot *et al.*, 1992).

Early studies of the pharmacology of avermectins (principally avermectin B_1 and ivermectin) provided evidence that these compounds also affect GABA receptor function. Depending on the organism and preparation examined, avermectins either activate GABA receptors, enhance GABA-dependent activation, or inhibit GABA-dependent activation by binding to a domain that is distinct from those that bind

Table 3. Drug and Neurotoxin Binding Sites on the GABA Receptor

Site	Ligands
Neurotransmitter	GABA Muscimol Bicuculline (antagonist)
Chloride channel	Picrotoxinin TBPS Chlorinated cyclodienes Lindane Bicycloorthocarboxylates Phenylpyrazoles
Benzodiazepine	Diazepam Flunitrazepam Flumazenil (antagonist) β-carbolines (inverse agonists)
Barbiturate	Pentobarbital
Steroid	Alfaxalone
Avermectin	Avermectin

SOURCE: Adapted from Sieghart (1992) and Macdonald and Olsen (1994).

other known classes of GABA receptor ligands (Wann, 1987; Turner and Schaeffer, 1989) (Table 3). Although an action on GABA receptors was initially widely accepted as the mode of action of avermectins, it is now clear that these compounds are also capable of activating other classes of chloride channels that are not GABA receptors (Bloomquist, 1993b). The relative significance of GABA-gated and GABA-insensitive chloride channels in the insecticidal actions of avermectins remains to be elucidated.

Molecular Biology. Photoaffinity labeling and the purification of affinity-labeled proteins from vertebrate brain demonstrated that GABA receptors exist as heteromultimeric assemblies in the cell membrane (Macdonald and Olsen, 1994). Native receptors are postulated to be pentameric by analogy to nicotinic acetylcholine receptors (Macdonald and Angelotti, 1993). A large effort directed at the molecular cloning of vertebrate GABA receptor subunit genes has yielded more than 20 genes coding for five structurally-related but distinct subunit classes (α, β, δ, γ, and ρ), and multiple isoforms (e.g., α_1, α_2, etc.) have been identified in most of these classes (Burt and Kamatchi, 1991; Macdonald and Angelotti, 1993; Macdonald and Olsen, 1994).

Although GABA receptors and other ligand-gated chloride channels are also important neurotransmitter receptors in insects, only two insect genes thought to encode GABA receptor subunits have been characterized to date. The use of molecular genetic methods to map and clone the *Rdl* (resistance to dieldrin) locus of *D. melanogaster* revealed that the product of this locus exhibited significant inferred amino acid sequence similarity to known vertebrate GABA receptor subunit genes (ffrench-Constant *et al.*, 1991; ffrench-Constant and Roush, 1991). The identity of this gene with the locus that confers cyclodiene resistance provides strong evidence that the *Rdl* gene product participates in the formation of a toxicologically relevant insect GABA receptor. A second *D. melanogaster* gene, designated LCCH3, was isolated by homology probing techniques (Henderson *et al.*, 1994) and shown to exhibit a high degree of inferred amino acid sequence identity to vertebrate GABA receptor β subunit genes (Henderson *et al.*, 1993). LCCH3 was mapped to cytogenetic interval 13F, which is also known to contain the the *slrp* (*slow receptor potential*) locus (Henderson *et al.*, 1994). The altered electroretinogram and behavior phenotypes of *slrp* mutants are consistent with defects in neurotransmission (Homyk and Pye, 1989). Thus, the possible identity of LCCH3 and *slrp* suggests a physiological role for the LCCH3 gene product. However, there is no evidence as to whether the LCCH3 protein contributes to the formation of a toxicologically relevant insect GABA receptor.

The heterologous expression of functional GABA receptors from cloned vertebrate subunits has provided considerable insight into the diversity of subunits required to confer the functional and pharmacological properties observed for native GABA receptors (Macdonald and Angelotti, 1993; Macdonald and Olsen, 1994). With the notable exception of the ρ subunit, expression of single vertebrate GABA receptor subunits yields homomultimeric receptors that respond weakly to GABA and do not exhibit some of the characteristic pharmacological responses of GABA receptors. Expression experiments with combinations of subunits show that a heteromultimer composed of α and β subunits will respond to GABA and channel blocking agents. However, coexpression of α, β, and γ subunits is required to produce receptors that reflect the full pharmacological profile of native brain receptors. Among αβγ heteromultimers, the replacement of one subunit isoform with another can alter the pharmacological properties of the expressed receptor. It is thought that the pharmacological diversity of native GABA receptor subtypes in vertebrate brain is due to cell- or region-specific differences in the the expression of GABA receptor subunit isoforms.

Heterologous expression studies with insect GABA receptor-like genes have met with varying degrees of success. Unlike most vertebrate GABA receptor subunits, the *Rdl* gene product produces homomultimeric GABA receptors in *X. laevis* oocytes (ffrench-Constant *et al.*, 1993) or in the baculovirus/insect cell expression system (Lee *et al.*, 1993) that give robust responses to GABA and are blocked by an array of chloride channel blockers. These results not only provided functional confirmation that the *Rdl* gene product is a GABA receptor subunit but also permitted the identification of the specific mutation that confers the cyclodiene insecticide resistance associated with the *Rdl* locus. In constrast to the results obtained with *Rdl*, efforts to express the LCCH3 gene product in *X. laevis* oocytes either alone or in combination with vertebrate GABA receptor subunits or the *Rdl* gene product have failed to date to produce functional receptors that respond to GABA or any other candidate neurotransmitter (Knipple *et al.*, 1994). These findings implicate the existence of at least one additional, as-yet-uncharacterized subunit that must combine with LCCH3 to form a functional receptor. Although the failure of LCCH3/*Rdl* coexpression experiments may be interpreted to imply that these subunits do not coassemble *in vivo*, this interpretation is clouded by the ability of *Rdl* to form functional receptors alone only when expressed at high levels (ffrench-Constant *et al.*, 1993). Therefore, the native subunit composition of the receptor(s) incorporating the *Rdl* and LCCH3 gene products remains an open question.

Implications for Insecticide Discovery. Some aspects of insect GABA receptor molecular biology are now known in sufficient detail to make immediate contributions to insecticide discovery and evaluation. In particular, the functional expression of *Rdl* homomultimers provides means of developing cell lines to screen for compounds acting at the chloride channel site. Moreover, knowledge of the specific mutations that confer cyclodiene resistance permits the development of screens to identify compounds with high intrinsic activity against cyclodiene-resistant insects, an important consideration given the ubiquity of cyclodiene resistance in pest species (Georghiou, 1986). As with the sodium channel, homology probing techniques (Thompson *et al.*, 1993) should also permit the facile isolation of *Rdl* homologs from species of economic and public health importance to facilitate the development of screens on GABA receptors of target pest species rather than on model species. Finally, the chloride channel site of insect GABA receptors in is effectively labeled with [^{3}H]*n*-propyl-ethynylbicycloorthobenzoate (EBOB) (Deng *et al.*, 1991). This ligand is therefore a useful biochemical probe for high-throughput screening to identify new compounds acting at the chloride channel site.

Whereas the expression of *Rdl* alone appears to provide a tool for the evaluation of compounds acting at the the chloride channel site, the use of expressed insect GABA receptors to evaluate potential insecticides at other possible target domains of the GABA receptor is currently more problematic. Although *Rdl* forms functional homomultimeric receptors in expression systems, both the precedent of heteromultimeric vertebrate GABA receptors and circumstantial evidence derived from both the primary structure and functional expression of *Rdl* suggest that the toxicologically relevant native insect GABA receptor is likely to be a heteromultimer formed from *Rdl* and at least one other as-yet-unidentified subunit (Knipple *et al.*, 1994). Until the subunit composition of insect GABA receptors is firmly established, reliance on *Rdl* homomultimers as surrogates for the complete pharmacology of native insect GABA receptors may provide misleading results.

Other Target Sites

Although information on target site molecular biology is most advanced for sodium channels and GABA receptors, the body of knowledge required to apply techniques derived from molecular biology in the search for insecticides acting at other targets continues to grow. The following sections summarize the present status of information for other known insecticide target sites and for some other neuronal macromolecules that have been postulated to be suitable insecticide targets.

Acetylcholinesterase. The structure of the *Ace* gene, which encodes the neuronal acetylcholinesterase of *D. melanogaster*, has been known for several years (Hall and Speirer, 1986). More recently, a specific mutation at the *Ace* locus has been correlated with reduced sensitivity to cholinesterase inhibitors (Fournier *et al.*, 1992). Because the catalytic activity of acetylcholinesterase permits the development of simple and highly sensitive assays employing native tissue preparations (Devonshire and Moores, 1984), cloned acetylcholinesterase genes expressed in cell lines appear to offer little advantage over native tissue preparations in screens for anticholinesterases.

Nicotinic Acetylcholine Receptors. Centuries of use of nicotine as a botanical insecticide have validated the practical utility of the nicotinic acetylcholine receptor as an insecticide target (Eldefrawi, 1985), and the recent successful development of imidacloprid as a commercial product (Elbert *et al.*, 1990) provides further evidence for the relevance of this target site. Four genes encoding putative acetylcholine receptor subunits have been cloned from *D. melanogaster* (Hermans-Borgmeyer *et al.*, 1986; Bossy *et al.*, 1988; Jonas *et al.*, 1990; Sawruk *et al.*, 1990a; 1990b), and an additional acetylcholine receptor gene has been cloned from locust nervous tissue and has been shown to form functional homomultimers when expressed in *X. laevis* oocytes (Marshall *et al.*, 1990). PCR-based homology probing has demonstrated the existence of sequences homologous to both the ALS and SAD genes of *D. melanogaster* in several other insect species (Sgard *et al.*, 1993). In this system, as with GABA receptors, the subunit composition of native receptors is not known. Therefore, expressed homomultimeric receptors may not reliably mimic that pharmacological profile of native receptors.

Octopamine Receptors. Octopamine-like actions of formamidine insecticide/acaricides were first implicated by the ability of of chlordimeform (CDM) and its *N*-demethyl metabolite (DCDM) to stimulate light emission in fireflies (Hollingworth and Murdock, 1980). Subsequent studies have confirmed the action of DCDM, other insecticidal formamidines, and related compounds as agonists of octopamine receptor-coupled adenylate cyclase activity in preparations from insect nervous tissue (reviewed in: Evans, 1986). Two independent research efforts have characterized the molecular structure and pharmacological properties of a *D. melanogaster* receptor that recognizes both octopamine and tyramine as a ligand (Arakawa *et al.*, 1990; Sadou *et al.*, 1990). Although initially characterization as an octopamine receptor (Arakawa *et al.*, 1990), this receptor's much higher affinity for tyramine has led to the proposal that this amine, rather than octopamine, is the native ligand (Sadou *et al.*, 1990). Assays of adenylate cyclase activity and calcium fluxes in transfected or transformed cells show no stimulation of cAMP synthesis by either octopamine or tyramine but significant inhibition by both amines of forskolin-stimulated cAMP synthesis concurrent with an elevation of intracellular calcium levels (Arakawa *et al.*, 1990; Sadou *et al.*, 1990; Robb *et al.*, 1994). The pharmacological profile and the nature of the G protein-mediated coupling exhibited by this receptor suggest that it is not likely to be an important site of action of formamidine insecticide/acaricides. Nevertheless, this gene and genes encoding other *D. melanogaster* G protein-coupled receptors (Onai *et al.*, 1989; Witz *et al.*, 1990; Sadou *et al.*, 1992) provide potential points of entry in the search for genes encoding toxicologically relevant octopamine receptors.

GABA-Insensitive Chloride Channels. Although effects of avermectins on both vertebrate and invertebrate GABA receptors are well-documented, it is now evident that avermectin also activates other chloride channels that are not associated with GABA receptors (Bloomquist, 1993b). Recent heterologous expression studies of *Caenorhabditis elegans* mRNA show that avermectins activate chloride currents that are carried by GABA-insensitive, glutamate-gated chloride channels (Arena *et al.*, 1991; Arena *et al.*, 1992). Expression cloning efforts based on these studies, together with concurrent photoaffinity labeling efforts (Rohrer *et al.*, 1992), hold the promise of yielding a novel avermectin target site gene in the near future.

Excitatory Glutamate Receptors. Knowledge that glutamate acts as the principal excitatory neurotransmitter at insect neuromuscular junctions (Usherwood, 1994) has stimulated interest in the glutamate receptor as a potential insecticide target site. Further, polyamine toxins from wasp and spider venoms that block excitatory glutamate receptors have been evaluated as potential leads for the design of insecticides (Blagbrough *et al.*, 1992; Eldefrawi *et al.*, 1993), so far without commercial success. Ionotropic glutamate receptors in vertebrates are probably multimeric, and the genes that code for glutamate receptor subunits exhibit the structural landmarks of the ligand-gated ion channel gene superfamily (Hollman and Heinemann, 1994). Three *D. melanogaster* genes that encode putative glutamate receptor subunits are expressed preferentially in either nerve (DGluR-I and DNMDAR-I) or muscle (DGluR-II) (Schuster *et al.*, 1991; Ultsch *et al.*, 1992; Ultsch *et al.*, 1993). Expression of either DGluR-I or DGluR-II in *X. laevis* oocytes produced homomultimeric receptors that responded to a variety of agents known to act at vertebrate glutamate receptors but were virtually insensitive to glutamate (Schuster *et al.*, 1991; Ultsch *et al.*, 1992). These results suggest that additional complementary glutamate receptor subunits are required to produce fully functional receptors.

Voltage-Sensitive Calcium Channels. Voltage-sensitive calcium channels mediate the transient influx of calcium into nerve cells in response to a change in membrane potential (Bertolino and Llinas, 1992). Polypeptide toxins isolated from spider venoms that block insect nerve nerve terminal calcium channels, thereby interfering with neurotransmitter release, identify these channels as targets of potential toxicological relevance for the chemical disruption of normal nerve function (Adams *et al.*, 1992). Protein purification and molecular cloning studies have shown that vertebrate neuronal calcium channels are heteromultimeric (Hofman *et al.*, 1994). The large α_1 subunit, which is structurally related to the α subunits of voltage-sensitive sodium channels, is sufficient to form functional channels in expression experiments, but coexpression with the four other subunits (α_2, β, γ, and δ) yields channels with the full complement of biophysical and pharmacological properties evident in native channels. A recently-described putative calcium channel gene from *D. melanogaster* (Zheng *et al.*, 1994) exhibits significant sequence homology to vertebrate calcium channel α_1 subunits. The functional role of this channel in the insect nervous system and its pharmacological significance as a potential insecticide target site remain to be established.

Neurotransmitter Transporters. A family of specific, high affinity uptake transporters limit the duration of chemical signals between neurons and their target cells by removing amino acid (e.g., GABA, glutamate) and biogenic amine (e.g., dopamine, serotonin) neurotransmitters from synapses following their stimulus-dependent release (Amara and Kuhar, 1993). In vertebrates, many of these transporters are implicated as target sites for therapeutic agents and drugs of abuse. The observation that the insect antifeedant properties of cocaine and other inhibitors of neurotransmitter uptake are correlated with the inhibition of octopamine reuptake (Nathanson *et al.*, 1993) suggests that octopamine uptake transporters may provide an alternative target for the disruption of octopaminergic neurotransmission. The cloning of the first insect gene from this family, a cocaine-sensitive serotonin transporter from *D. melanogaster*, by PCR-based homology probing (Corey *et al.*, 1994) provides a point of entry for the further molecular characterization of these potential targets in insects.

Conclusions

The four years since the last International Congress of Pesticide Chemisty have witnessed explosive growth in the field of insect molecular neurobiology. As a result, genes coding for elements of almost all known insecticide target sites have been cloned from *D. melanogaster*, and straightforward methods exist to obtain homologous genes from insect species that are the specific targets of insecticide development efforts. The challenge for the agrochemical research community is to determine how to both catalyze and capitalize on the further advance of knowledge in these areas to enhance and streamline the process of insecticide discovery.

To exploit the heterologous expression of target site genes as a surrogate for insect neuronal preparations, it is essential to establish first that the expressed target site faithfully reflects the pharmacology of the target in its native environment. This consideration suggests that targets comprised of a single, fully functional gene product can be exploited more expediently than those comprised of a heteromultimeric assembly of several gene products. Biogenic amine receptors, neurotransmitter transporters, and possibly sodium and calcium channels (depending on the pharmacological fidelity of their expressed α subunits) are examples of single gene products, whereas members of the ligand-gated ion channel gene superfamily (GABA, glutamate, and nicotinic acetylcholine receptors) are examples of multimeric targets comprised of the products of two or more genes. Establishing the native subunit composition of ligand-gated ion channels is an important and unresolved issue not only for insect receptors but also for their counterparts in vertebrate nerve and muscle. In the absence of such information, the use of expressed homomultimeric receptors as models for native receptors must be approached with caution.

The successful exploitation of expressed target site genes in a screening program also depends on the availability of appropriate screening assays. Ideally, functional assays are preferable to radioligand binding assays because the former provide information on the functional modification of the target in addition to binding site occupancy and also because they provide the opportunity to detect compounds that act at completely novel domains on the target under investigation. In practice, the established functional assays for most target sites of practical interest are not readily adaptable to high-volume screening efforts. The use of radioligand binding assays is limited by the availability of suitable ligands either among compounds optimized for high affinity binding to insect targets (e.g., $[^3H]$EBOB for insect GABA receptors) or among radiolabeled insecticides that exhibit high affinity binding to toxicologically relevant sites (e.g., $[^3H]$avermectin B_{1a}). The development of appropriate screening assays may prove to be a challenge equal to or greater than the cloning and expression of target site genes for use in such assays.

The high degree of pharmacological conservation observed for many neuronal targets of insecticide action poses a significant challenge for the development of selective neurotoxicants. However, tools derived from molecular neurobiology may provide means of searching for intrinsically selective agents acting at targets of interest. For example, the establishment of cell lines that express either mammalian or insect target macromolecules and the use of these cells to screen for differential pharmacological activity represents an approach for identifying intrinsically selective agents or of pursuing intrinsic selectivity as a criterion for structure optimization within an active series of compounds.

So far I have discussed only the ways in which applications of molecular neurobiology might enhance conventional approaches to insecticide discovery and optimization because these applications have the potential to make a significant impact on the way in which discovery research is pursued in the near term. In the longer term, the elucidation of the molecular structures of insecticide targets should also provide a point of entry for the design of specific ligands based on knowledge of the topology of binding domains. The ultimate success of this approach will depend not only on gene cloning and characterization but also on the development of paradigms for the translation of linear sequences of transmembrane proteins into reliable three-dimensional structural models (Osguthorpe *et al.*, 1992).

References

Adams, M. E.; Bindokas, V. P.; Venema, V. J. In *Neurotox '91: Molecular Basis of Drug and Pesticide Action*; Duce, I. R., Ed.; Elsevier: New York, NY, 1992; pp 33-44.

Amara, S. G.; Kuhar, M. J. *Annu. Rev. Neurosci.* **1993**, *16*, 73-93.

Arakawa, S.; Gocayne, J. D.; McCombie, W. R.; Urquhart, D. A.; Hall, L. M.; Fraser, C. M.; Venter, J. C. *Neuron* **1990**, *2*, 343-354.

Arena, J. P.; Liu, K. K.; Paress, P. S.; Cully, D. F. *Mol. Pharmacol.* **1991**, *40*, 368-374.

Arena, J. P.; Liu, K. K.; Paress, P. S.; Schaeffer, J. M.; Cully, D. F. *Mol. Brain Res.* **1992**, *15*, 339-348.

Benson, J. A. In *Neurotox '91: Molecular Basis of Drug and Pesticide Action*; Duce, I. R., Ed.; Elsevier: New York, NY, 1992.

Bertolino, M.; Llinas, R. R. *Annu. Rev. Pharmacol. Toxicol.* **1992**, *32*, 399-421.

Blagbrough, I. S.; Brackley, P. T. H.; Bruce, M.; Bycroft, B. W.; Mather, A. J.; Millington, S., Sudan, H. L.; Usherwood, P. N. R. *Toxicon* **1992**, *30*, 303-322.

Bloomquist, J. R. In *Reviews in Pesticide Toxicology*; Roe, M. and Kuhr, R. J., Ed.; Toxicology Communications: Raleigh, NC, 1993a; Vol. 2; pp 181-226.

Bloomquist, J. R. *Comp. Biochem. Physiol.* **1993b**, *106C*, 301-314.

Bossy, B.; Ballivet, M.; Speirer, P. *EMBO J.* **1988**, *7*, 611-618.

Burt, D. R.; Kamatchi, G. L. *FASEB J.* **1991**, *5*, 2916-2923.

Casida, J. E. *Arch. Insect Biochem. Physiol.* **1993**, *22*, 13-23.

Catterall, W. A. *Science* **1988**, *242*, 50-61.

Catterall, W. A. *Physiol. Rev.* **1992**, *72*, S15-S48.

Colliot, F.; Kukorowski, K. A.; Hawkins, D. W.; Roberts, D. A. In *Brighton Crop Protection Conference Pests and Diseases*; 1992; pp 29-34.

Corey, J. L.; Quick, M. W.; Davidson, N.; Lester, H. A.; Guastella, J. *Proc. Natl. Acad. Sci. USA* **1994**, *91*, 1188-1192.

Deng, Y.; Palmer, D. J.; Casida, J. E. *Pestic. Biochem. Physiol.* **1991**, *41*, 60-65.

Devonshire, A. L.; Moores, G. D. *Pestic. Biochem. Physiol.* **1984**, *21*, 341-348.

Dong, K.; Scott, J. G. *Insect Biochem. Mol. Biol.* **1994**, *24*, 647-654.

Doyle, K. E.; Knipple, D. C. *Insect Biochem.* **1991**, *21*, 689-696.

Elbert, A.; Overbeck, H.; Iwaya, K.; Tsuboi, S. In *Brighton Crop Protection Conference Pests and Diseases*; 1990; pp 21-28.

Eldefrawi, M. E. In *Comprehensive Insect*

Physiology Biochemistry and Pharmacology; Kerkut, G. A. and Gilbert, L. I., Ed.; Pergamon: Oxford, 1985; Vol. 12; pp 263-272.

Eldefrawi, M. E.; Anis, N. A.; Eldefrawi, A. T. *Arch. Insect Biochem. Physiol.* **1993**, *22*, 25-39.

Evans, P. D. In *Neuropharmacology and Pesticide Action*; Ford, M. G., Lunt, G. G., Reay, R. C. and Usherwood, P. N. R., Ed.; Ellis Horwood, Ltd.: Leicester, 1986; pp 315-341.

ffrench-Constant, R. H.; Mortlock, D. P.; Shaffer, C. D.; MacIntyre, R. J.; Roush, R. T. *Proc. Natl. Acad. Sci. USA* **1991**, *88*, 7209-7213.

ffrench-Constant, R. H.; Rocheleau, T. A.; Steichen, J. C.; Chalmers, A. E. *Nature* **1993**, *363*, 449-451.

ffrench-Constant, R. H.; Roush, R. T. *Genet. Res.* **1991**, *57*, 17-21.

Fournier, D.; Bride, J.-M.; Hoffmann, F.; Karch, F. *J. Biol. Chem.* **1992**, *267*, 14270-14274.

Georghiou, G. P. In *Pesticide Resistance: Strategies and Tactics for Management*; National Academy Press: Washington, DC, 1986; pp 14-43.

Hall, L. M.; Kasbekar, D. P. In *Insecticide Action: From Molecule to Organism*; Narahashi, T. and Chambers, J. E., Ed.; Plenum: New York, 1989; pp 99-114.

Hall, L. M. C.; Speirer, P. *EMBO J.* **1986**, *5*, 2949-2954.

Heinemann, S. H.; Terlau, H.; Stühmer, W.; Imoto, K.; Numa, S. *Nature* **1992**, *356*, 441-443.

Henderson, J. E.; Knipple, D. C.; Soderlund, D. M. *Insect Biochem. Mol. Biol.* **1994**, *24*, 363-371.

Henderson, J. E.; Soderlund, D. M.; Knipple, D. C. *Biochem. Biophys. Res. Commun.* **1993**, *193*, 474-482.

Hermans-Borgmeyer, I.; Zopf, D.; Ryseck, R.-P.; Hovemann, B.; Betz, H.; Gundelfinger, E. D. *EMBO J.* **1986**, *5*, 1503-1508.

Hille, B. *Ionic Channels of Excitable Membranes;* second ed.; Sinauer: Sunderland, MA, 1992, pp 23-82.

Hofman, F.; Biel, M.; Flockerzi, V. *Annu. Rev. Neurosci.* **1994**, *17*, 399-418.

Hollingworth, R. M.; Murdock, L. L. *Science* **1980**, *208*, 74-76.

Hollman, M.; Heinemann, S. *Annu. Rev. Neurosci.* **1994**, *17*, 31-108.

Homyk, T. J.; Pye, Q. *J. Neurogenet.* **1989**, *5*, 37-48.

Jonas, P.; Baumann, A.; Merz, B.; Gundelfinger, E. D. *FEBS Lett.* **1990**, *269*, 264-268.

Kallen, R. G.; Cohen, S. A.; Barchi, R. L. *Mol. Neurobiol.* **1993**, *7*, 383-428.

Knipple, D. C.; Doyle, K. E.; Marsella-Herrick, P. A.; Soderlund, D. M. *Proc. Natl. Acad. Sci. USA* **1994**, *91*, 2483-2487.

Knipple, D. C.; Henderson, J. E.; Soderlund, D. M. In *Molecular Action and Pharmacology of Insecticides on Ion Channels*; Clark, J. M., Ed.; American Chemical Society: Washington, DC, 1994; in press.

Kontis, K. J.; Goldin, A. L. *Mol. Pharmacol.* **1993**, *43*, 635-644.

Lee, H.-J.; Rocheleau, T.; Zhang, H.-G.; Jackson, M. B.; ffrench-Constant, R. H. *FEBS Lett.* **1993**, *335*, 315-318.

Lester, H. A. *Science* **1988**, *241*, 1057-1063.

Lombet, A.; Mourre, C.; Lazdunski, M. *Brain Res.* **1988**, *459*, 44-53.

Loughney, K.; Kreber, R.; Ganetzky, B. *Cell* **1989**, *58*, 1143-1154.

Macdonald, R. L.; Angelotti, T. P. *Cell. Physiol. Biochem.* **1993**, *3*, 352-373.

Macdonald, R. L.; Olsen, R. W. *Annu. Rev. Neurosci.* **1994**, *17*, 569-602.

Marshall, J.; Buckingham, S. D.; Shingai, R.; Lunt, G. G.; Goosey, M. W.; Darlison, M. G.; Sattelle, D. B.; Barnard, E. A. *EMBO J.* **1990**, *9*, 4391-4398.

Moorman, J. R.; Kirsch, G. E.; Brown, A. M.; Joho, R. H. *Science* **1990**, *250*, 688-691.

Narahashi, T. *Trends Pharmacol. Sci.* **1992**, *13*, 236-241.

Nathanson, J. A.; Hunnicutt, E. J.; Kantham, L.; Scavone, C. *Proc. Natl. Acad. Sci. USA* **1993**, *90*, 9645-9648.

Noda, M.; Suzuki, H.; Numa, S.; Stühmer, W. *FEBS Lett.* **1989**, *259*, 213-216.

Okamoto, H.; Sakai, K.; Goto, S.; Takasu-Ishakawa, E.; Hotta, Y. *Proc. Japan Acad. Ser. B* **1989**, *63*, 284-288.

Onai, T.; FitzGerald, M. G.; Arakawa, S.; Gocayne, J. D.; Urquhart, D. A.; Hall, L. M.; Fraser, C. M.; McCombie, W. R.; Venter, J. C. *FEBS Lett.* **1989**, *255*, 219-225.

Osguthorpe, D. J.; Lunt, G. G.; Cockroft, V. B. In *Neurotox '91: Molecular Basis of Drug and Pesticide Action*; Duce, I. R., Ed.; Elsevier: New York, NY, 1992; pp 241-253.

Patton, D. E.; West, J. W.; Catterall, W. A.; Goldin, A. L. *Proc. Natl. Acad. Sci. USA* **1992**, *89*, 10905-10909.

Pusch, M.; Noda, M.; Stühmer, W.; Numa, S.; Conti, F. *Eur. Biophys. J.* **1991**, *20*, 127-133.

Ramaswami, M.; Tanouye, M. A. *Proc. Natl. Acad. Sci. USA* **1989**, *86*, 2079-2082.

Robb, S.; Cheek, T. R.; Hannan, F. L.; Hall, L. M.; Midgley, J. M.; Evans, P. D. *EMBO J.* **1994**, *13*, 1325-1330.

Rohrer, S. P.; Meinke, P. T.; Hayes, E. C.;

Mrozik, H.; Schaeffer, J. M. *Proc. Natl. Acad. Sci. USA* **1992**, *89*, 4168-4172.
Sadou, F.; Amlaiky, N.; Plassat, J.-L.; Borrelli, E.; Hen, R. *EMBO J.* **1990**, *9*, 3611-3617.
Sadou, F.; Boschert, U.; Amlaiky, N.; Plassat, J.-L.; Hen, R. *EMBO J.* **1992**, *11*, 7-17.
Salkoff, L.; Butler, A.; Wei, A.; Scavarda, N.; Giffen, N.; Ifune, K.; Goodman, R.; Mandel, G. *Science* **1987**, *237*, 744-749.
Satin, J.; Kyle, J. W.; Chem, M.; Bell, P.; Cribbs, L. L.; Fozzard, H. A.; Rogart, R. B. *Science* **1992**, *256*, 1202-1205.
Sattelle, D. B. *Adv. Insect Physiol.* **1990**, *22*, 1-113.
Sawruk, E.; Schloss, P.; Betz, H.; Schmitt, B. *EMBO J.* **1990a**, *9*, 2671-2677.
Sawruk, E.; Udri, C.; Betz, H.; Gundelfinger, E. D. *FEBS Lett.* **1990b**, *273*, 177-181.
Schuster, C. M.; Ultsch, A.; Schloss, P.; Cox, J. A.; Schmitt, B.; Betz, H. *Science* **1991**, *254*, 112-114.
Schwarz, T. L.; Yool, A. J. In *Neurotox '91: The Molecular Basis of Drug and Pesticide Action*; Duce, I. R., Ed.; Elsevier: New York, 1992; pp 165-177.
Sgard, F.; Obosi, L. A.; King, L. A.; Windass, J. D. *Insect Mol Biol.* **1993**, *2*, 215-223.
Sieghart, W. *Trends Pharmacol. Sci.* **1992**, *13*, 446-450.
Soderlund, D. M.; Bloomquist, J. R. *Annu. Rev. Entomol.* **1989**, *34*, 77-96.
Soderlund, D. M.; Bloomquist, J. R.; Wong, F.; Payne, L. L.; Knipple, D. C. *Pestic. Sci.* **1989**, *26*, 359-374.
Soderlund, D. M.; Knipple, D. C. In *Molecular Action and Pharmacology of Insecticides on Ion Channels*; Clark, J. M., Ed.; American Chemical Society: Washington, DC, 1994; in press.
Stern, M.; Kreber, R.; Ganetzky, B. *Genetics* **1990**, *124*, 133-143.
Taylor, M. F. J.; Heckel, D. G.; Brown, T. M.; Kreitman, M. E.; Black, B. *Insect Biochem. Mol. Biol.* **1993**, *23*, 763-775.
Terlau, H.; Heinemann, S. H.; Stühmer, W.; Pusch, M.; Conti, F.; Imoto, K.; Numa, S. *FEBS Lett.* **1991**, *293*, 93-96.
Thompson, M.; Steichen, J. C.; ffrench-Constant, R. H. *Insect Mol Biol.* **1993**, *2*, 149-154.
Turner, M. J.; Schaeffer, J. M. In *Ivermectin and Abamectin*; Campbell, W. C., Ed.; Springer-Verlag: Berlin, 1989; pp 73-88.
Ultsch, A.; Schuster, C. M.; Laube, B.; Betz, H.; Schmitt, B. *FEBS Lett.* **1993**, *324*, 171-177.
Ultsch, A.; Schuster, C. M.; Laube, B.; Schloss, P.; Schmitt, B.; Betz, H. *Proc. Natl. Acad. Sci. USA* **1992**, *89*, 10484-10488.
Usherwood, P. N. R. *Adv. Insect Physiol.* **1994**, *24*, 309-341.
Wann, K. T. *Phytother. Res.* **1987**, *1*, 143-150.
West, J. W.; Patton, D. E.; Scheuer, T.; Wang, Y.; Goldin, A. L.; Catterall, W. A. *Proc. Natl. Acad. Sci. USA* **1992**, *89*, 10910-10914.
Williamson, M. S.; Bell, C. A.; Denholm, I.; Devonshire, A. L. In *Molecular Action and Pharmacology of Insecticides on Ion Channels*; Clark, J. M., Ed.; American Chemical Society: Washington, DC, 1994; in press.
Williamson, M. S.; Denholm, I.; Bell, C. A.; Devonshire, A. L. *Mol. Gen. Genet.* **1993**, *240*, 17-22.
Witz, P.; Amlaiky, N.; Plassat, J.-L.; Maroteaux, L.; Borrelli, E.; Hen, R. *Proc. Natl. Acad. Sci. USA* **1990**, *87*, 8940-8944.
Zheng, W.; Feng, G.; Ren, D.; Eberl, D. F.; Hannan, F.; Dubald, M.; Hall, L. M. *J. Neurosci.* **1994**, in press.

Influences of Molecular Mechanisms of Action on Fungicide Design

Franz Dorn, Division Plant Protection, R&D Disease Control, Ciba-Geigy Limited, CH-4002 Basel, Switzerland

Biochemistry has become an important function in fungicide innovation following the discovery of the first sterol biosynthesis inhibitors. Knowledge of molecular mechanisms of action of sterol biosynthesis inhibition has greatly stimulated design and synthesis work in that area. The high expectations that rational design would develop into an important complement of the screening approach have not been met so far. Several new target sites for fungicides are proposed for further investigation for which inhibitors with excellent *in vivo* activity are known. Contributions of biochemistry to the phenylpyrroles, the pyrimidinamines and the strobilurins are reviewed and the position of biochemistry in industrial fungicide research is reassessed.

In the discovery of fungicides the search for compounds with a novel mode of action has become ever more important in view of the increasing number of products on the market and the growing problems of resistance for some fungicide classes. A novel mode of action generally leads to a competitive advantage for a new product because it implies a characteristic novel spectrum of activity and a specific resistance behaviour. The role of biochemistry in discovery and development of fungicides up to 1988 has been reviewed (Kuhn, 1989). In the following the position of biochemistry in the innovation process is reassessed focussing on its influence on design and synthesis of novel compounds, and a review of biochemical knowledge of newer fungicide products and developments is presented.

Biochemistry and Sterol Biosynthesis Inhibition

For the early contact fungicides as well as for the first systemic fungicides e.g. carboxin (White and Georgopoulos, 1992), benomyl (Davidse, 1986) or metalaxyl (Davidse, 1987) information about their mode of action was published years after their successful market introduction. Discovery and development of the first sterol biosynthesis inhibitors of the azole, morpholine and allylamine class took place without biochemical support. The rapid progress in elucidation of the different modes of action of these three fungicide classes has established biochemistry as a discipline in industrial fungicide research and led to the formation of biochemistry teams in many companies. Knowledge of the modes of action for these three sterol biosynthesis inhibitor classes at a molecular level has been a tremendous stimulus for design and synthesis work.

For sterol 14 α-demethylation inhibition models were postulated for the interaction of RR-diclobutrazol (Marchington, 1983) and propiconazole (Nyfeler and Huxley, 1986) with the active site of cytochrome P-450 which allowed binding and matching studies and other structure activity investigations. Such studies at ICI suggested tertiary alcohols e.g. of structure **1** to contain all the correct requirements for fungicidal activity and eventually led to the synthesis of flutriafol and hexaconazole shown in Fig. 1 (Worthington, 1987). Numerous accounts of structure activity studies with azoles including computer assisted molecular modelling work have subsequently been published (Köller, 1992a).

For the morpholine class of inhibitors the concept of high energy intermediates (HEI) promoted by Benveniste (Rahier et al, 1985) has stimulated intensive synthesis programs in the search for improved inhibitors of sterol Δ^8 - Δ^7-isomerase and Δ^{14}-reductase. Fig. 2 shows a

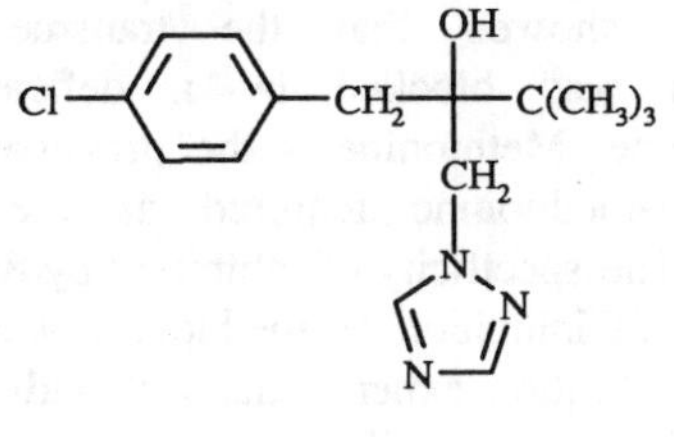

1

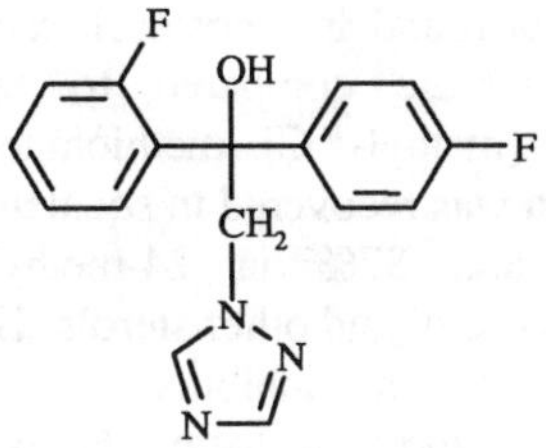

Flutriafol

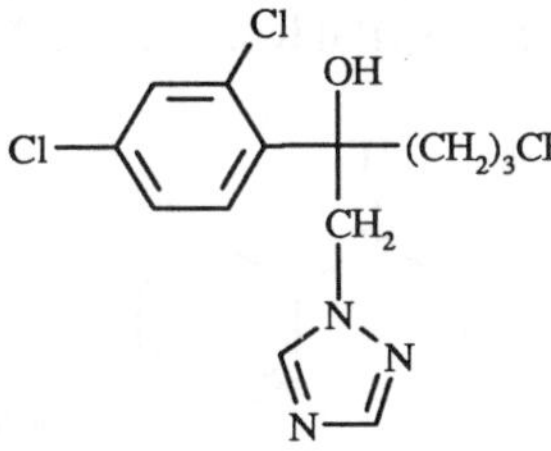

Hexaconazole

Fig. 1 Discovery of Flutriafol and Hexaconazole

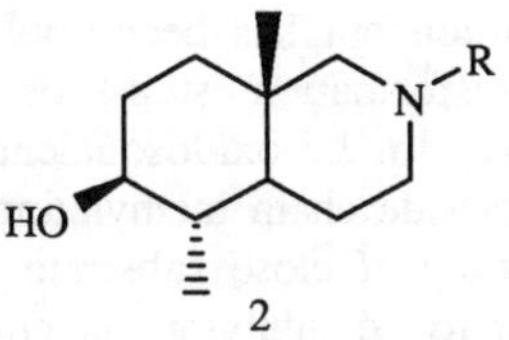

2

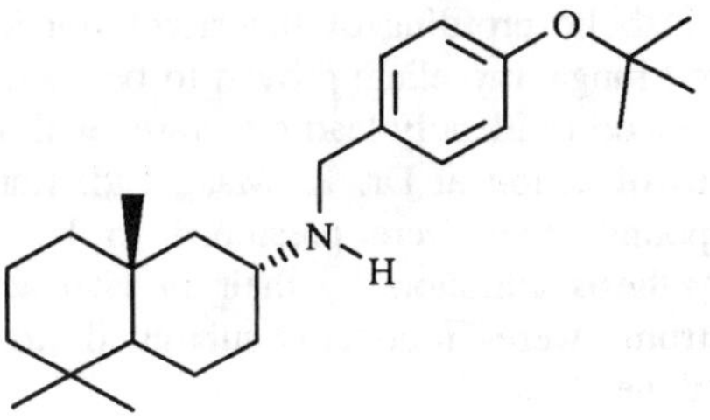

3

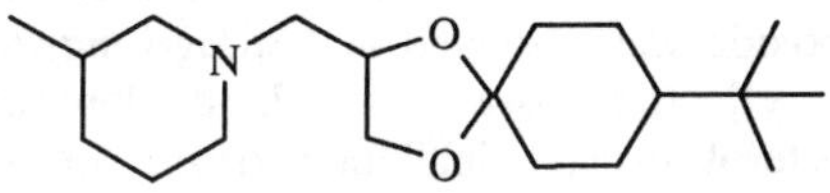

4

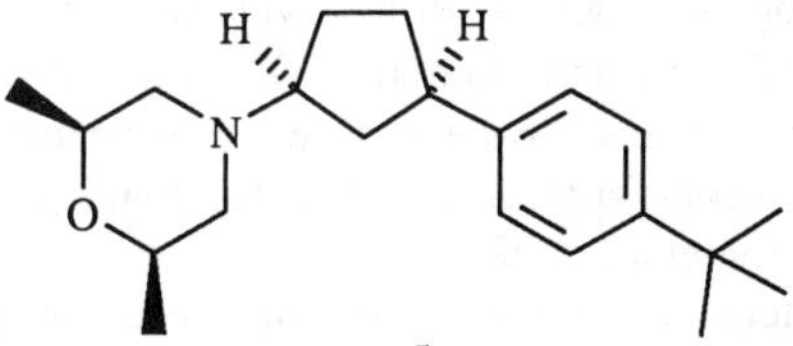

5

Fig. 2 Biorationally Designed Inhibitors

selection of typical structures. The original work was centered around structures of type **2**, in which a 3-β-hydroxy-8-azadecalin system simulates the AB-ring structure of sterols (Taton et al, 1987). A research team at BASF prepared a wide range of amine compounds comprising structural features of sterols. Δ^{14}-Reductase inhibition was correlated with fungicidal activity in greenhouse screening. Compound **3** e.g. showed a 25 times lower IC_{50} value than fenpropimorph in the enzyme assay (Akers et al, 1991). Ketal **4** is one of the most active compounds as sterol Δ^{14}-reductase inhibitor in an optimization program of Bayer covering oxa- and dioxa-spiro-decanes (Berg et al, 1993). The HEI-hypothesis and computer modelling studies related to the natural substrate undergoing the Δ^8-Δ^7-isomerisation reaction led Ciba scientists to investigate cyclopentane analogs of fenpropimorph (Huxley-Tencer et al, 1992). Compound **5** indeed proved to be more active in the Δ^8-Δ^7-isomerase assay by a factor of 20 and showed stronger activity *in vivo* against *Erysiphe graminis* than fenpropimorph. The same cyclopentanes form part of design work performed at ICI (Urch, 1991) and Du Pont (Basarab et al, 1992). None of these biorationally designed compounds, however, reached the activity of fenpropimorph in the field. The exact reasons for the lack of correlation between lab and field activity are not known.

The allylamines have been found to inhibit the epoxidation reaction of squalene. The antimycotic activity was optimised in a large synthesis program (Nussbaumer et al, 1991). Work to make allylamines available as plant protection fungicides has not produced sufficiently active compounds (Stanetty and Wallner, 1993).

The HEI-concept has been applied to other cationic transformation steps of sterol biosynthesis, e.g. to 2,3-oxidosqualene cyclisation and to sterol side chain methylation at position 24. The variety of close substrate analogs that were shown to be inhibitory in enzyme assays has been reviewed (Köller, 1992a).

Metabolic profiling of the sterol fraction of treated fungal mycelium proved to be a successful method to identify lead structures with novel modes of action at Dr. R. Maag Ltd. Random compounds that were presumed to be sterol biosynthesis inhibitors by their *in vivo* activity spectrum were routinely submitted to this technique.

For aminoacetophenone **6** (Fig. 3) accumulation of large quantities of squalene-2,3-epoxide and smaller amounts of squalene-2,3-22,23-diepoxide was observed in *Ustilago maydis*. In the cell free assay for 2,3-oxidosqualene-lanosterol cyclase inhibition (from *Saccharomyces cerevisiae*) an IC_{50} value of about 100µM was determined. Optimisation work led to analogs such as **7** with IC_{50} values as low as 0.1 µM (Guerry and Jolidon, 1992). *In vivo* activity improvements were more pronounced for antimycotic applications than for fungicidal uses (Jolidon et al, 1992).

Sterol analysis of *U. maydis* treated with anilide **8** showed accumulation of lanosterol indicating inhibition of lanosterol transmethylation at C-24. Prevention of the accumulation of lanosterol and reversal of growth inhibition by the addition of methionine to treated fungal material showed that the transmethylation reaction was blocked by a deficiency of methionine. Methionine is the precursor of S-adenosyl-methionine required as the methyl donor. The specificity of inhibition by **8** and the reversal of inhibition by methionine was shown by incorporation experiments with radiolabelled materials. In the inhibition experiment 80% of radioactivity from [1-^{14}C] acetate in the sterol fraction was found in lanosterol, representing a fourfold increase compared to the control. Feeding L-[methyl-^{14}C] methionine, 0.5% of radiocarbon was recovered in squalene, 12.5% in lanosterol and 87% in 24-methylene-24,25-dihydrolanosterol and other sterols. Hence **8** was identified as an inhibitor of methionine biosynthesis. Studies to locate the exact site and the mechanism of inhibitory action of **8** in the biosynthesis of methionine are the subject of two recent publications (Baldwin et al, 1994).

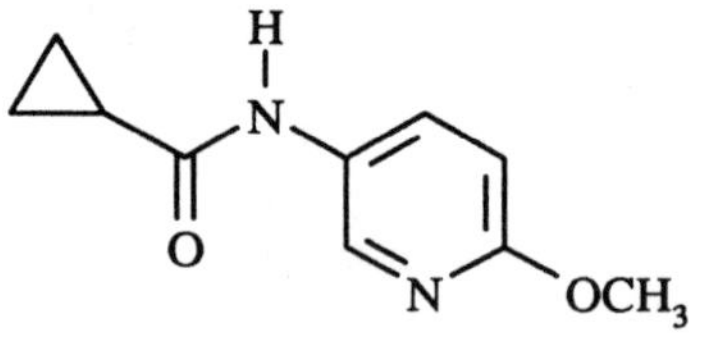

H N O N OCH_3

8

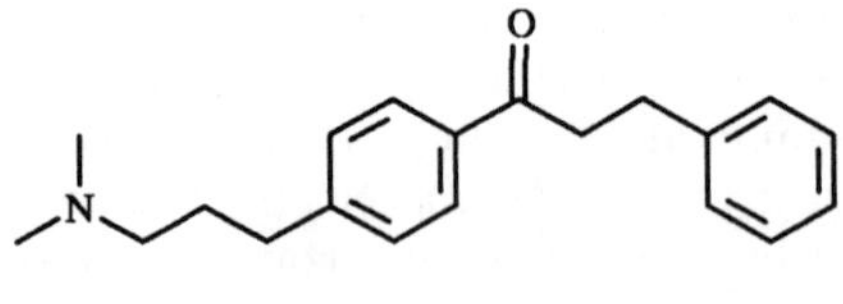

6

I N O

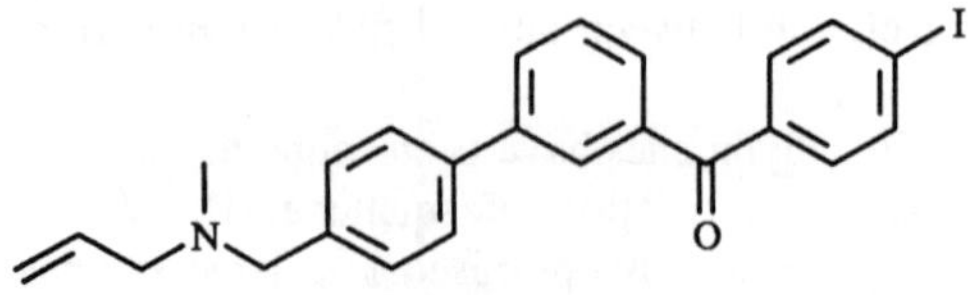

7

Fig. 3 Examples of 2,3-Oxidosqualene-Lanosterol Cyclase Inhibitors

Biorational Design

The success story of biochemistry in the area of sterol biosynthesis inhibition has raised high expectations that biorational design based on knowledge of structure and function of target enzymes and receptors would develop into an important complement of the classical screening approach. Progress in QSAR methods and in computer assisted molecular modelling techniques as well as the growing expertise in structure elucidation of proteins have further nourished these expectations. Schwinn and Geissbühler (1986) appraised biorational design as the most important instrument of the future to increase the efficiency of the innovation process for fungicides.

Unfortunately there are no reliable indications about the nature and the number of biorational design projects which have been undertaken in the past years in the fungicide area, either by companies in house or in collaboration with research groups from academia. Reports on breakthroughs are missing. Even in areas that have long been identified as ideal target sites for fungicides like e.g. chitin or polyamine biosynthesis no spectacular new findings on protein structures or fungicide leads have been published (Köller, 1992b).

Three new starting points which might be attractive for rational design projects have evolved from recent research work. Each case is documented by a fungicide with novel mode of action and excellent field performance which, however, will not be introduced to the plant protection market: Soraphen A (Fig. 4) from *Sorangium cellulosum* owes its broad spectrum of activity at low rates to its inhibitory effect on the enzyme acetyl-CoA-carboxylase. This mode of action was investigated in cell free systems from *U. maydis* (Pridzun, 1991). Dimefluazole (Fig. 4) acts on Oomycetes such as *Phytophtora* spp. and *Plasmopara viticola* (Gouot and Souche, 1987). In biochemical studies with *Pythium aphanidermatum,* the site of action of dimefluazole could be located in the respiratory pathway between cytochromes b and c (Pillonel, 1994). The third fungicide is the methionine biosynthesis inhibitor of structure **8** discussed earlier. It shows good field performance against e.g. *Botrytis cinerea* on grape vines and *E. graminis* on cereals.

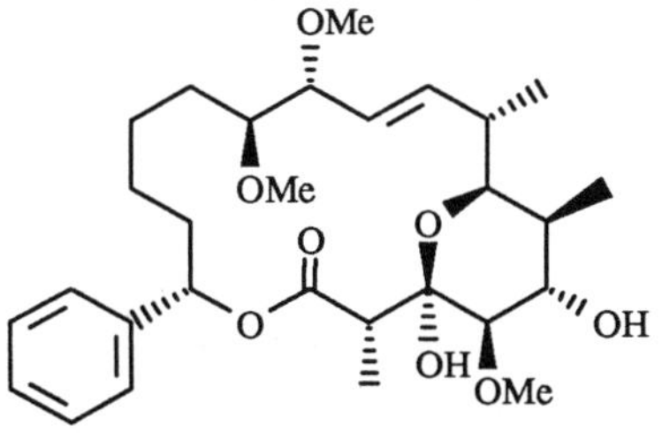

Soraphen A

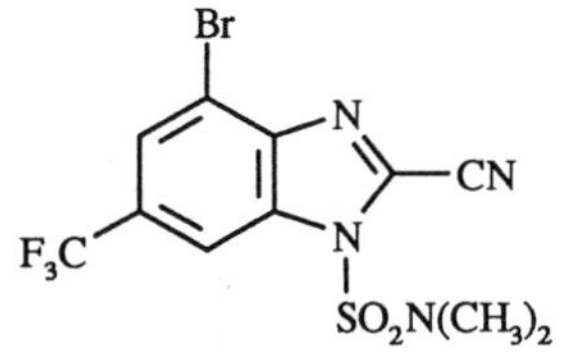

Dimefluazole

Fig. 4 Fungicides with Novel Mode of Action

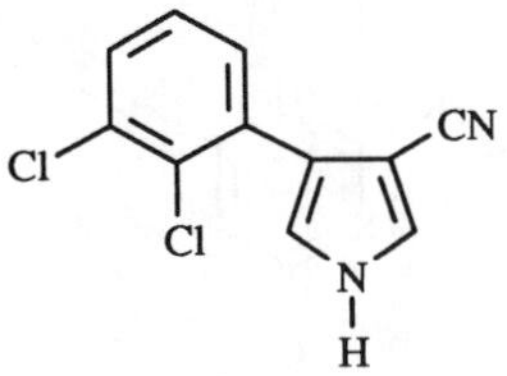

Fenpiclonil

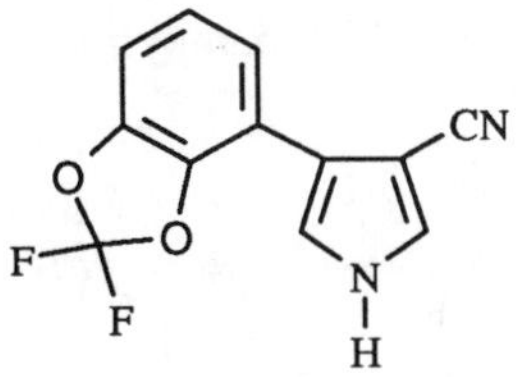

Fludioxonil

Fig. 5 The Phenylpyrroles

Mode of Action of New Fungicides

Biochemistry has played a minor role in discovery and development of three fungicide groups which recently have been or shortly will be introduced to the market.

The phenylpyrroles fenpiclonil and fludioxonil (Fig. 5) are derived from the antifungal secondary metabolite pyrrolnitrin (Nyfeler and Ackermann, 1992). The effect of fenpiclonil on membrane dependent transport processes has been described (Jespers et al, 1993). The primary mode of action of the phenylpyrroles may be based on inhibition of transport associated phosphorylation of glucose as shown for *Fusarium sulphureum* (Jespers, 1994).

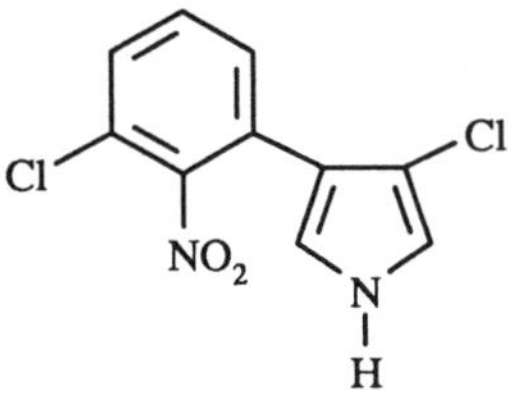

Pyrrolnitrin

The pyrimidinamine class comprises the fungicides mepanipyrim, pyrimethanil and cyprodinil (Fig. 6). Different hypotheses have been put forward for their mode of action. Miura et al (1993) and Milling et al (1993) reported the

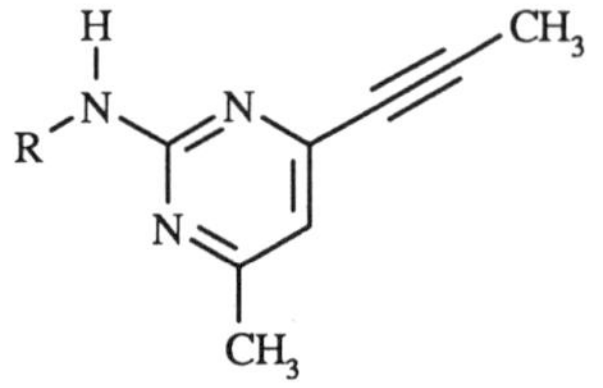

R = phenyl

Mepanipyrim

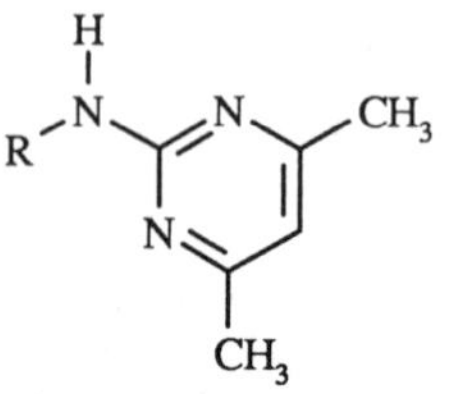

Pyrimethanil

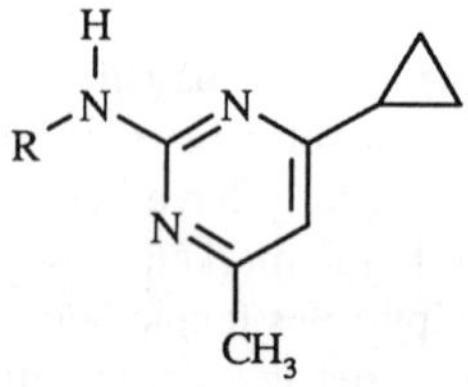

Cyprodinil

Fig. 6 The Pyrimidinamines

inhibitory effect of mepanipyrim and pyrimethanil respectively on hydrolytic enzyme secretion by *Botrytis* spp. Masner et al (1994) found that mycelial growth inhibition of *B. cinerea* by cyprodinil and the other pyrimidinamines can be reversed if methionine is added to culture media lacking amino acids. Further antagonism studies suggest that the pyrimidinamine fungicides act on methionine biosynthesis in a different way as the anilide **8**.

The effect of the antifungal antibiotic mucidin on mitochondrial respiration of various organisms was shown soon after its isolation from *Oudemansiella mucida* (Subik et al, 1974). In the optimisation work which started with the discovery of strobilurin A (Schramm, 1980) and led to the development products ICI A5504 (Godwin et al, 1992) and BAS 490F (Ammermann et al, 1992), *in vitro* enzyme assays do not seem to have played a major role (Köhle et al, 1993). First results on the protein sequence of cytochrome b in the binding region for strobilurins have been presented by Wiggins (1993). A model for the central part of cytochrome b has been generated which should allow interaction studies between the mitochondrial bc_1 complex and strobilurins (Link, 1993).

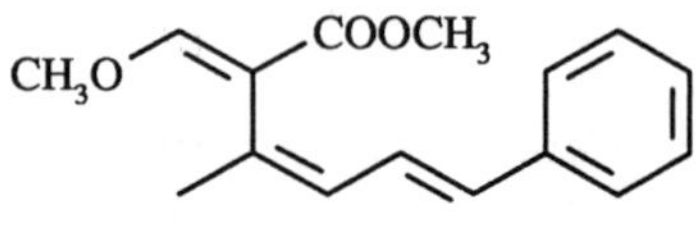

Strobilurin A

Host-Pathogen Interactions

Increasingly attention has been directed to the investigation of mechanisms that influence host-pathogen interactions.

Starting from the structures of known melanin biosynthesis inhibitors like pyroquilon and tricyclazole the design of substituted phtalazines as novel inhibitors has been reported (Omata et al, 1989). Simpson et al (1991) propose the investigation of alternative steps of melanin biosynthesis as target sites for inhibitors based on synthetic and mechanistic studies.

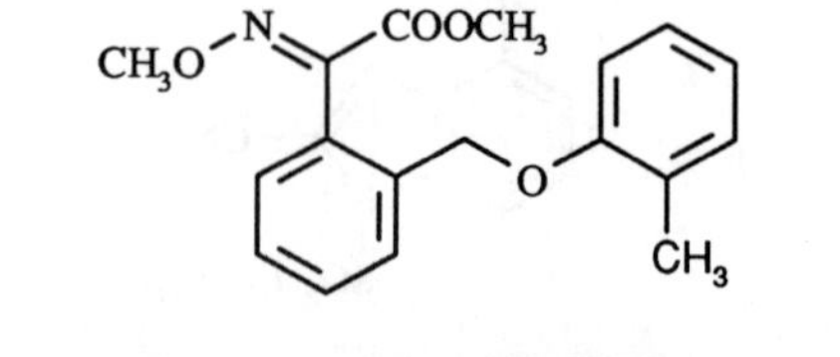

ICI A5504

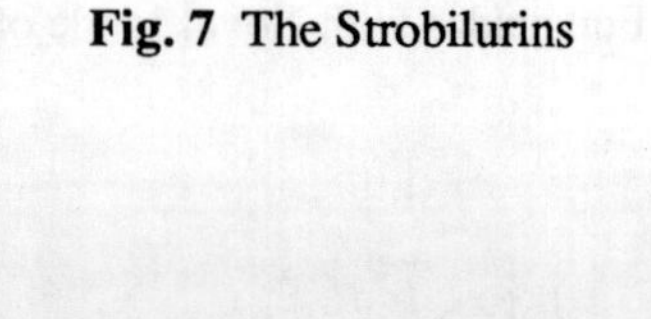

BAS 490F

Fig. 7 The Strobilurins

Literature abounds with research work on cutinases (Köller, 1992b). Interesting inhibitor structures with proven effect under *in vivo* conditions, however, have not been found.

The tremendous effort in pharmaceutical research on proteinases, in particular on aspartic proteinase secreted from *Candida* spp., could well form the basis for innovative research in the fungicide field (Tuite, 1992).

The phenomemon of systemic acquired resistance of plants to pathogen infection has continued to receive wide attention. Investigations of the inducing mechanisms have revealed that endogenous salicylic acid plays a crucial role in the induction process. Its function as systemically transmitted signal, however, remains questionable (Gaffney et al, 1993). Two classes of exogenous chemical inducers of systemic acquired resistance which fit the restrictive definition for inducing agents formulated by Kessmann et al (1993) have been identified in random screening systems: Derivatives of substituted isonicotinic acids, especially of 2,6-dichloroisonicotinic acid **9** (Staub et al, 1992), and derivatives of benzothiadiazole-7-carboxylic acids **10** (Schurter et al, 1989). Other chemical inducing agents claimed in recent patent literature comprise pyridinium betaines **11** (Lunkenheimer et al, 1992a), pyridinium salts **12** (Lunkenheimer et al, 1992b), diketopiperazines **13** (Kurohashi, 1993) and phosphate salts (Kuc, 1991).

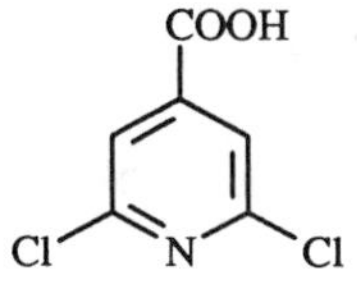

9 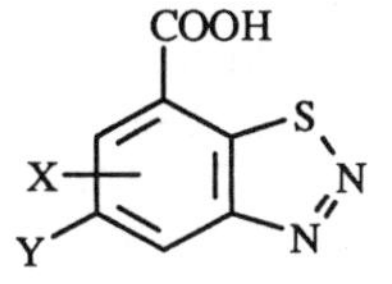 **10**

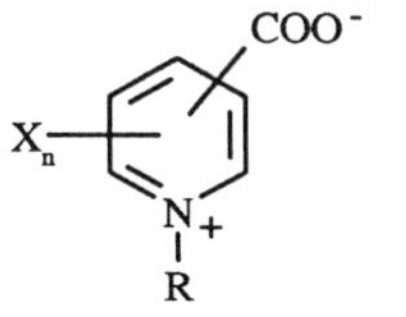

11 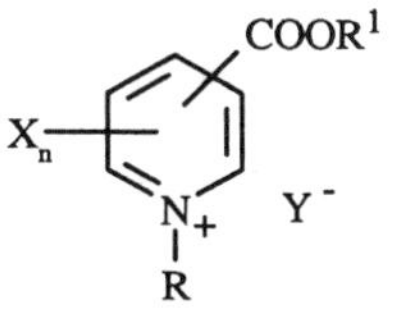 **12**

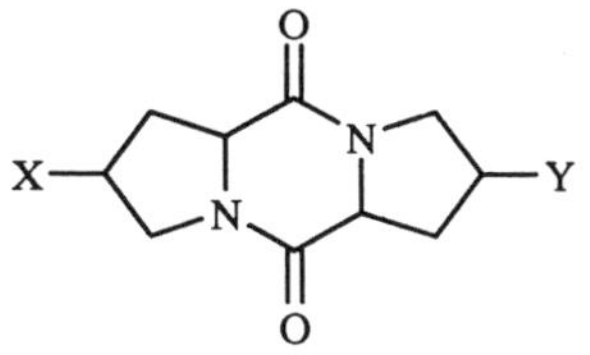

13

Fig. 8 Inducers of Systemic Acquired Resistance

Role of Biochemistry

The foregoing overview of projects and developments of the past years shows examples where knowledge of mechanisms of action has stimulated and advanced fungicide innovation. At the same time novel fungicide classes have been discovered and optimised in the traditional teamwork mainly between synthesis chemists and screening biologists.

Nevertheless, biochemistry has become an integral part of industrial fungicide research and development. Its main tasks may be seen as follows:

1. Characterisation of the mode of action of novel lead compounds discovered in random screening: The originality of the mode of action is often an important argument in the selection of fungicide leads for optimisation work. A set of bioassays used routinely for this purpose is shown in the Table.
2. Biochemical support of research projects: Biochemistry investigates at an early stage of an optimisation program, whether *in vitro* data can be provided for meaningful structure activity studies.
3. In depth investigation of the mode of action of developmental fungicides: It should be the goal of every company to present mode of action data when introducing a new product.

Biochemistry also takes the lead in biorational design projects. Tools and methods have been further developed and refined in the last years. Techniques of molecular biology allow novel approaches to find answers to some of the question asked in mode of action research. Bridging the gap between *in vitro* activity and field performance for biorationally designed compounds, however, remains the big unre-

Table List of Bioassays for Mode of Action Characterisation

Bioassays	Typical Inhibitors
In vitro growth test (18 fungi)	
Amino acid synthesis (antagonism test)	Pyrimidinamines
Cell-free respiration	Strobilurins, Carboxamides
Cell-free chitin synthesis	
Fatty acid synthesis (ACC cell-free test)	Soraphen A
Ergosterol synthesis: a) with intact cells b) cell-free test Squalene epoxydase C-14 demethylase Δ^{14}-reductase Δ^8-Δ^7-isomerase	 Allylamines Azoles Morpholines Morpholines
Lipid peroxidation	Dicarboximides
Uptake of radioactive glucose	Phenylpyrroles
Uptake of radioactive amino acids	Phenylpyrroles
Incorporation of radioactive uracil	Phenylamides

solved issue. It seems that this highly challenging approach to fungicide innovation requires further time until the first success story can be presented.

References

Akers, A.; Ammermann, E.; Buschmann, E.;Götz, N.; Himmele, W.; Lorenz, G.; Pommer, E.; Rentzea, C.; Röhl, F.; Siegel, H.; Zipperer, B.; Sauter, H.; Zipplies, M. *Pestic. Sci.* **1991**, *31,* 521-538.

Ammermann, E.; Lorenz, G; Schelberger, K.; Wenderoth, B.; Sauter, H.; Rentzea, C. *Proc. Br. Crop Prot. Conf. Pests and Dis.* **1992**, *1*, 403-410.

Baldwin, B.C., Whittington-Smith, R.; Grindle, M.; Kelly, S.L. Poster 422 *Eighth International Congress of Pesticide Chemistry,* Washington, **1994**.

Baldwin, B.C.; Brownell K.H.; Rees, S.B.; Wiggins, T.E.; Furter, R.; Masner, P.; Zobrist, P. Poster 423 *Eighth International Congress of Pesticide Chemistry,* Washington, **1994**.

Basarab, G.S.; Livingston, R.S.; Vollmer, S.J.; Johnson, C.B.; Kranis, K.T. In *Synthesis and Chemistry of Agrochemicals III;* Baker, D.R.; Fenyes, J.G.; Steffens, J.J., Eds.; ACS Symposium Series No. 504; American Chemical Society: Washington, DC, 1992; pp 414-427.

Berg, D.; Holmwood, G.; Krämer, W.; Pontzen, R.; Weismüller, J.; Tiemann, R. *J. prakt. Chem.* **1993,** *335*, 569-584.

Davidse, L.C. *Ann. Rev. Phytopathol.* **1986**, *24,* 43-65.

Davidse, L.C. In *Modern Selective Fungicides;* Lyr, H., Ed.; Fischer: Jena, 1987, 275-282.

Gaffney, T.; Friedrich, L.; Vernooij, B.; Negrotto, D.; Nye, G.; Uknes, S.; Ward, E.; Kessmann, H.; Ryals, J. *Science*, **1993**, *261*, 754-756.

Godwin, J.R.; Anthony, V.M.; Clough, J.M.; Godfrey, C.R.A. *Proc. Br. Crop Prot. Conf., Pests and Dis.* **1992**, *1*, 435-442.

Gouot, J.-M.; Souche, J.-L., European Patent 239,508, 1987.

Guerry, P.; Jolidon, S., European Patent 464,465, 1992.

Huxley-Tencer, A.; Francotte, E.; Bladocha-Moreau, M. *Pestic. Sci.* **1992**, *34,* 65-74.

Jespers, B.K.; Davidse, L.C.; De Waard, M.A. *Pesticide Biochemistry and Physiology* **1993**, *45*, 116-129.

Jespers, B.K. Doctorate Thesis, Agricultural University, Wageningen, 1994.

Jolidon, S.; Polak-Wyss, A.; Hartmann, P.G.; Guerry, P. *Proc. 3rd Int. Symp. Mol. Aspects Chemother.* **1991**, 143-152.

Kessmann, H.; Staub, T.; Hofmann, C.; Ahl Goy, P.; Ward, E.; Uknes, S.; Ryals, J. *10th Symposium of Research Comittee for the Bioactivity of Pesticides,* Nagano, 1993, pp 29-37.

Köhle, H.; Gold, R.E.; Ammermann, E. Meeting on *Design of Mitochondrial Electron Transport Inhibitors as Agrochemicals,* Bath, 1993.

Köller, W. In *Target Sites of Fungicide Action;* Köller, W., Ed.; CRC Press: Boca Raton, FL, 1992a; pp 119-206.

Köller, W. In *Target Sites of Fungicide Action;* Köller, W., Ed.; CRC Press: Boca Raton, FL, 1992b; pp 255-310.

Kuc, J.A., European Patent 445,867, 1991.

Kuhn, P.J. *Pestic. Sci.* **1989,** *25,* 123-135.

Kurohashi, M., Japanese Patent 43,408, 1993.

Link, T.A. Meeting on *Design of Mitochondrial Electron Transport Inhibitors as Agrochemicals,* Bath, 1993.

Lunkenheimer, W.; Wollweber, D.; Dehne, H.-W.; Dutzmann, S.; Hänssler, G.: Mauler-Machnik, A.: Schulz, U., European Patent 493,670, 1992a.

Lunkenheimer, W.; Wollweber, D.; Dehne, H.-W.; Dutzmann, S.; Hänssler, G.: Mauler-Machnik, A.: Schulz, U., European Patent 493,671, 1992b.

Marchington, A.F. *10th Int. Congr. Pl. Prot.* **1983,** *1,* 201-208.

Masner, P.; Muster, P.; Schmid, J. *Pestic. Sci.* **1994,** in press.

Milling, R.J.; Richardson, C.J.; Daniels, A. *6th International Congress of Plant Pathology,* Montreal, **1993,** abstract 3. 7. 15.

Miura, J.; Kamakura, T.; Maeno, S.; Hayashi, S.; Yamagushi, J. *6th International Congress of Plant Pathology,* Montreal, **1993,** abstract 3. 7. 5.

Nussbaumer, P.; Ryder, N.S.; Stütz, A. *Pestic. Sci.* **1991,** *31,* 437-455.

Nyfeler, R.; Huxley, P. *Proc. Br. Crop Prot. Conf. Pests and Dis.* **1986,** *1,* 207-215.

Nyfeler, R.; Ackermann, P. In *Synthesis and Chemistry of Agrochemicals III;* Baker, D.R.; Fenyes, J.G.; Steffens, J.J., Ed.; ACS Symposium Series No. 504; American Chemical Society: Washington, DC, 1992; pp 395-404.

Omata, K.; Tomita, H.; Nakajina, T.; Natsume, B. *Pestic. Sci.* **1989,** *26,* 271-281.

Pillonel, C. *Pestic. Sci.* **1994,** submitted.

Pridzun, L. Doctorate Thesis, Technical University, Braunschweig, 1991.

Rahier, A.; Taton, M.; Schmitt, P.; Benveniste, P.; Place, P.; Anding, C. *Phytochemistry* **1985,** *24,* 1223-1232.

Schramm, G. Doctorate Thesis, Friedrich-Wilhelms University, Bonn, 1980.

Schurter, R.; Kunz, W.; Nyfeler, R. European Patent 313,512, 1989.

Schwinn, F.; Geissbühler, H. *Crop Protection* **1986,** *5,* 33-40.

Simpson, T.J.; Dillon, P.M.; Donovan, T.M. *Pestic. Sci.* **1991,** *31,* 539-554.

Stanetty, P.; Wallner, H. *Arch. Pharm* (Weinheim) **1993,** *326,* 341-350.

Staub, T.; Ahl Goy, P.; Kessmann, H. In *Proceedings of the 10th International Symposium on Systemic Fungicides and Antifungal Compounds;* Lyr, H.; Polter, C., Eds.; Ulmer: Stuttgart, 1992; pp 239-249.

Subik, J.; Miroslav, Behun, M.; Musilek, V. *Biochem. Biophys. Res. Commun.* **1974,** *57,* 17-22.

Taton, M.; Benveniste, P.; Rahier, A. *Phytochemistry* **1987,** *26,* 385-392.

Tuite, M.F. *Trends Biotechnol.* **1992,** *10,* 235-239.

Urch, C.J.*Synthesis and Chemistry of Agrochemicals II;* Baker, D.L.; Fenyes, J.G.; Moberg, W.K., Eds.; ACS Symposium Series 443; American Chemical Society: Washington, 1991; pp 515-526.

White, G.A.; Georgopoulos, S.G. In *Target Sites of Fungicide Action;* Köller, W., Ed.; CRC Press: Boca Raton, FL, 1992; pp 1-29.

Wiggins, T.E. Meeting on *Design of Mitochondrial Electron Transport Inhibitors as Agrochemicals,* Bath, 1993.

Worthington, P.A. In *Synthesis and Chemistry of Agrochemicals;* Baker, D.R.; Fenyes, J.G.; Moberg, W.K.; Cross, B., Eds.; ACS Symposium Series No. 355; American Chemical Society: Washington, DC, 1987; pp 302-317.

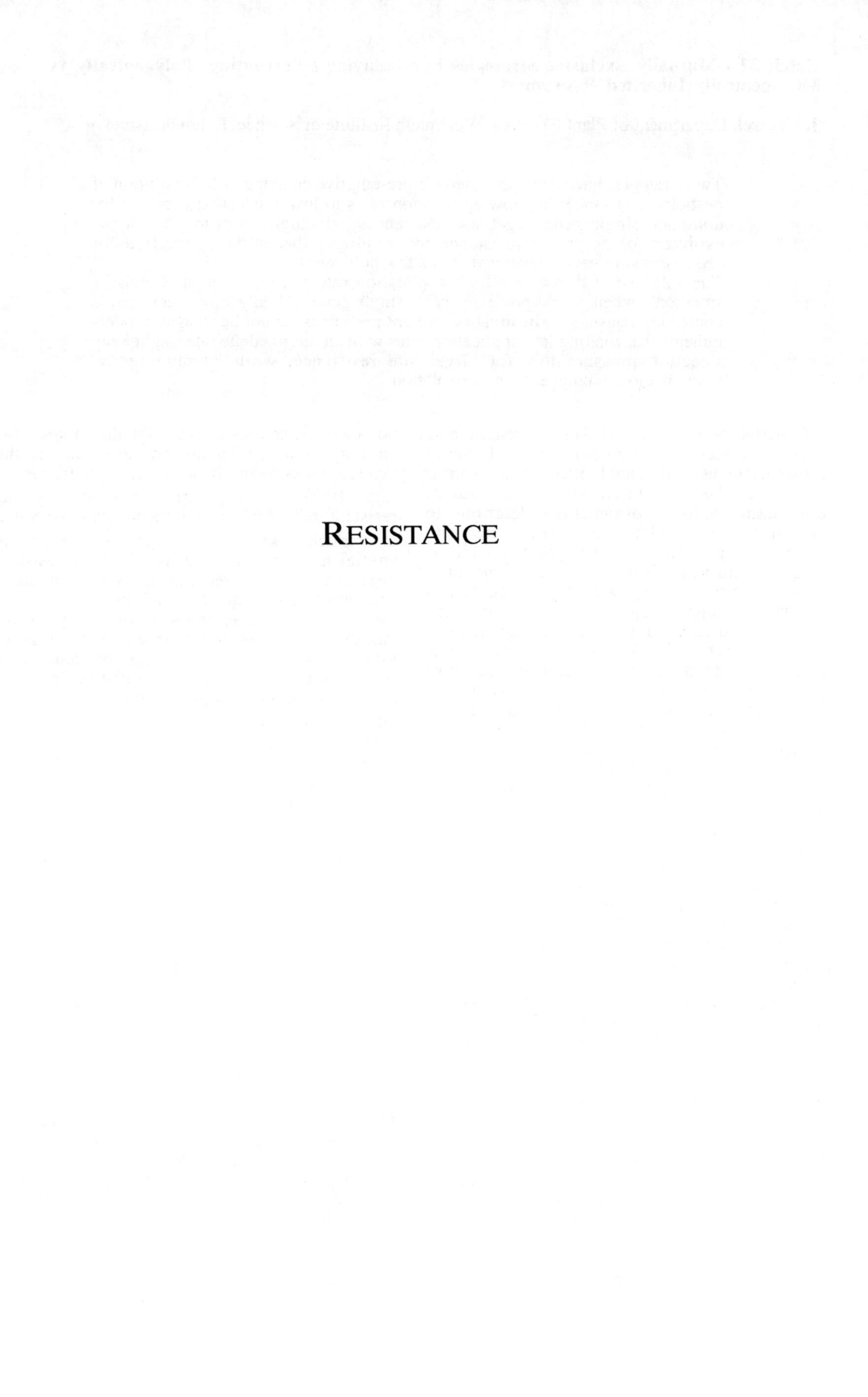

RESISTANCE

Catch 22 - Mutually Exclusive Strategies For Delaying / Preventing Polygenically vs. Monogenically Inherited Resistances

J. Gressel, Department of Plant Genetics, Weizmann Institute of Science, Rehovot, Israel

Two strategies have been proposed for pre-emptive delaying of the evolution of pesticide resistance: (a) low application rates to lower selection pressure for dominant single gene target site resistances; (b) high rates to prevent the evolution of polygenic resistance by requiring the unlikely selection for organisms simultaneously mutated in a few polygenes.

The sole use of a low or a high application rate can result in a "Catch 22 situation" when a pest possesses both single gene and polygenic mechanisms conferring resistance. In situations where pesticides cannot be rotated, models indicate that rotating low application rates with an intermediate rate might keep selection pressure low for target site resistance, while precluding the accumulation of polygenes in a population.

There has been much discussion by practitioners about the effects of dose rates on the evolution of resistance to pesticides, antibiotics and anti-cancer drugs; all chemicals used for the purpose of eliminating cells or organisms deleterious to human pursuits and welfare. Simultaneously, theoretical geneticists have tried to deal with the first data emerging from use of these biocides, which showed in many cases that resistance was inherited on single major genes. The same often holds for resistance to the metals in mine tailings (cf. Macnair et al. 1993). This is contrary to some evolutionary theoreticians who presume that most evolutionary change in nature is polygenic. Evolutionists do like to debate the importance of major genes vs. polygenes (Dove, 1993). This conundrum was 'solved' by Lande (1983) with the following explanation: "Empirical data on natural and domesticated populations, and analysis of the models, suggest that strong selection sustained over several generations is usually required for adaptive evolution by a major gene mutation, in order to overcome deleterious pleiotropic effects generally associated with major mutations. This helps to explain why adaptive evolution by major mutations occurs much more frequently in domesticated and artifically disturbed populations than in natural ones."

The early concept of vast-overkill with antibiotics was quickly abandoned as it enhanced the selection of rare resistant individuals by removing all susceptibles (i.e. the antibiotics exerted extremely high selection pressure). Conversely, the use of very high rates of insecticides has been proposed for situations where resistance is semi-dominantly inherited on a major gene coding for a resistance, based on the type of dose responses seen in Fig. 1. A somewhat lower rate would not control heterozygotes, which can then inter-breed producing offspring, 25% and of which will be homozygous and resistant to high dose-rates. This strategy is based on the concept that if the heterozygotes naturally occur at a low frequency (e.g. 10^{-6}) in a population, the homozygous resistant individuals would theoretically be found at a much lower frequency (e.g.$<10^{-12}$). Many species are at too low a density to have resistant organisms within breeding distance of a mate. The numbers of individuals required to test this assumption are huge. In one effort using a lower eukaryote, it was found that the recessively inherited homozygous resistant mutants were found at a ten fold lower frequency than the heterozygotes, i.e. 10^{-7} and not the expected 10^{-12} (Williams, 1976).

Thus, it is not clear that there will be much benefit to using ultra high rates to control heterozygotes. Lower doses were advised to lower selection pressure, allowing the often more

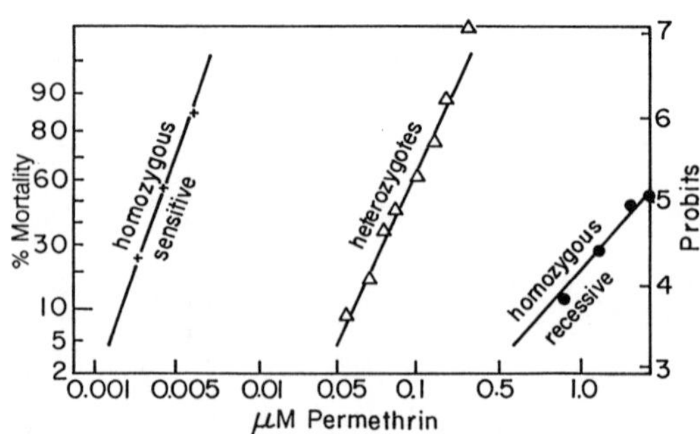

Fig.1. Dependence of resistance on dominance and dose. Dosage-response lines for larvae of *Culex quinquefasciatus* homozygous susceptible, heterozygous, homozygous-recessive resistant tested with permethrin. Redrawn from Georghiou and Taylor (1986).

1054–7487/95/0330$12.00/0

fit susceptibles to decrease the rate of evolution resistance in populations (Gressel and Segel, 1978). Indeed, major gene resistance to triazines evolved first and mainly where high rates of the more persistent pesticides in this class were used, and rarely were low rates and/or less persistent triazines were used.

Still, there are field data that justify the approach of not using low doses. Resistance quickly evolved when low dose-rates were used and quickly later evolved to stepped increases in dose rates. Resistance has not yet evolved in parallel situations where high doses were uniformly used. A recent cogent example can be seen in India, where the weed *Phalaris minor* evolved resistance to the phenylurea herbicide isoproturon in Green Revolution wheat growing areas. In the large "pockets" (ca. 30 km diameter) where this has occurred, farmers consistently underdosed the herbicide but not yet in nearby areas where full dose rates were continually used. For example, a typical farmer initially used only half the recommended rate of isoproturon. This successfully controlled *Phalaris* for the first three years, but with inadequate control in the fourth. He then used 0.75 the recommended rate successfully for two years, and unsuccessfully in the third. The full dose rate was then successful for one year but inadequate the next. Fifty percent above the recommended rate worked for a year, but not this past year. This strategy of continually increasing dose rates might be feasible for some insecticides used in high value crops, but is less feasible for fungicides and herbicides where there is far less margin between a utilizable rate and phytotoxicity to the crops. For less valuable crops, economics can also play a role in limiting the rates used for any pesticide. The evolution of single gene resistance is typified by what is seen in the field as a sudden jump from a seemingly totally sensitive population to one with where a high proportion of individuals is resistant to a high dose of the pesticide as seen in Fig. 2. When the enrichment of evolution is plotted on a logarithmic scale (based on a simple population dynamics model), the increase is linear (Fig. 2A). An example of such field data are shown in Fig. 2B. The farmer cannot tell when one pest in a million, or one in a hundred is resistant, only when a goodly proportion show resistance.

Where polygenic inheritance is involved, it has been shown time and again that the initial use of low dose rates facilitates rapid evolution of resistance. The nature of polygenic inheritance is such that there are small increments of increase in resistance in such a population (Fig. 3). Perhaps the appearance of a measurable proportion of individuals with the first increment of resistance is delayed (as in Fig. 2B) until the first gene dose for resistance has been sufficiently enriched in the population. After the first increment of resistance appears, some individuals can withstand the evolutionary pressure of higher pesticide doses, enriching for more gene doses. Initial models on evolution of quantitative resistance to pesticides were described, but not fully developed (Via, 1986). It has recently been stated that "the impact of quantitative trait locus studies on evolution has yet to be felt" (Cheverud and Routman, 1993). Presumably they mean that while there is considerable circumstantial and epidemiological evidence for polygenic controls, the genetic proofs are minimal. In this respect they are clearly on base. For example, there are yet no data to support that the resistances shown in Fig. 3A, B are polygenic, although there are data for polygenicity of the mild resistance shown in Fig. 3C. There is also some genetic evidence for polygenically inherited incremental increases in some fungicide resistances (cf. Hollomon et al., 1990; Brent et al., DeWaard, 1992), insecticides (cf. Mouches et al., 1986; Field et al., 1988), herbicides (Fig. 3C; Wang et al., 1991) and anti-cancer drugs (cf. Schimke, 1984). This incremental resistance can be due to either multiple genes or due to selection for sequential amplification of a single gene. For the sake of this discussion gene amplified traits are considered as polygenic. Thus, these two causes have similar implications for the management of resistance.

High initial doses have been proposed as an initial strategy in cancer chemotherapy (Harnevo and Agur 1991; 1992; Agur and Harnevo, 1993), because low and then increasing doses have been shown to select for amplification (Schimke, 1984). In the case of anticancer drugs, this modelling has suggested that after the first high initial doses are used, the dose can actually be dropped due to an interplay between remaining cancer cells with the inherent immunological resistance of the patient. This could be extrapolated to agriculture, where the crop has some mechanisms to fend off small infestations arthropods and pathogens and can compete successfully with late germinating weeds. This strategy has the potential problem in medicine of high doses being lethal to a certain portion of patients.

While high dose rates are useful to delay the evolution of populations with polygenically inherited traits, the strategy often suggested to delay target site resistance is to lower the dose rate (Gressel and Segel 1978; 1982). This decreases selection pressure, as a greater proportion of susceptible individuals remains after treatment, diluting and competing with the infinitesmal proportion of any selected resistant individuals in the population. There are other ways to lower the selection pressure of a

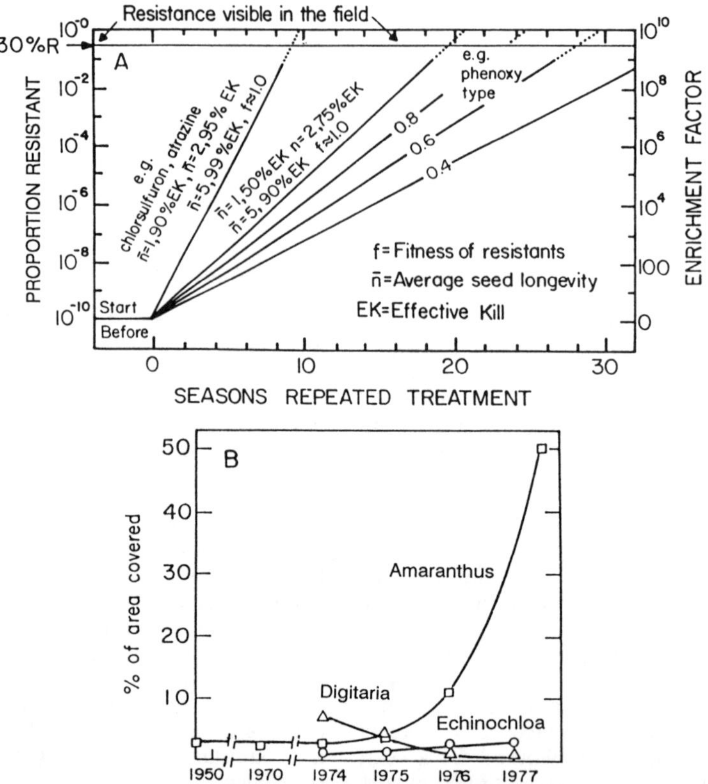

Fig. 2. Evolution of major gene traits following repeated treatments of a pesticide
A. Modelling: A scenario of repeat treatments with highly persistent herbicides that controls throughout a cropping season (acute line) vs. a short lived herbicide that allows the expression of the fitness difference between resistant and sensitive individuals after the herbicide has been dissipated (obtuse lines). Note logarithmic scale for increase. Based on equations in Gressel and Segel (1978).
B. Actual field data on resistance showing changes in weed populations in a monoculture maize treated annually with atrazine. *Amaranthus retroflexus, Echinochloa crus-galli,* and *Digitaria sanguinalis,* the foremost weeds, were counted. The field in the Agricultural Combine of Babolna, Hungary, was in corn treated with atrazine from 1970. (Data are plotted from Table 1 in Nosticzius et al. (1979).

pesticide where a single gene target site resistance is expected, depending on the compound and the pest situation. These include using related chemistries with less persistence, or fewer treatments with the same compound. This would allow later germinating susceptible members of the same weed species, or later influxes of the same species of fungi or insects to dilute the proportion of resistant individuals in a population.

This author has counted more than 50 models dealing with the evolution and management of resistance in pests; and most modellers seem to believe that the pest group they work with is biologically different from all others; ignoring the rest. Most models for the evolution of resistance and its management deal only with major gene effects (e.g., Taylor, 1983; Gressel and Segel 1978, 1982; Roush and Daly, 1990) and only a few deal with polygenic resistance (e.g. Shaw, 1989) and gene amplification (Tabashnik, 1990). None deal with the simultaneous existence of both genetic mechanisms in the same organism.

A "Catch-22" situation can transpire where the sole use of high rates or the sole use of low rates can each have the opposite of the desired effect. This occurs when the same pest species possesses genes for both polygenically inherited (or amplified) incremental resistance along with major gene(s) for resistance. The Catch 22 is that the use of high rates to delay polygenic resistance selects for single gene resistance, the use of lower rates to delay single gene resistance selects for polygenic (or amplified) resistance.

This is what seems to have happened in the

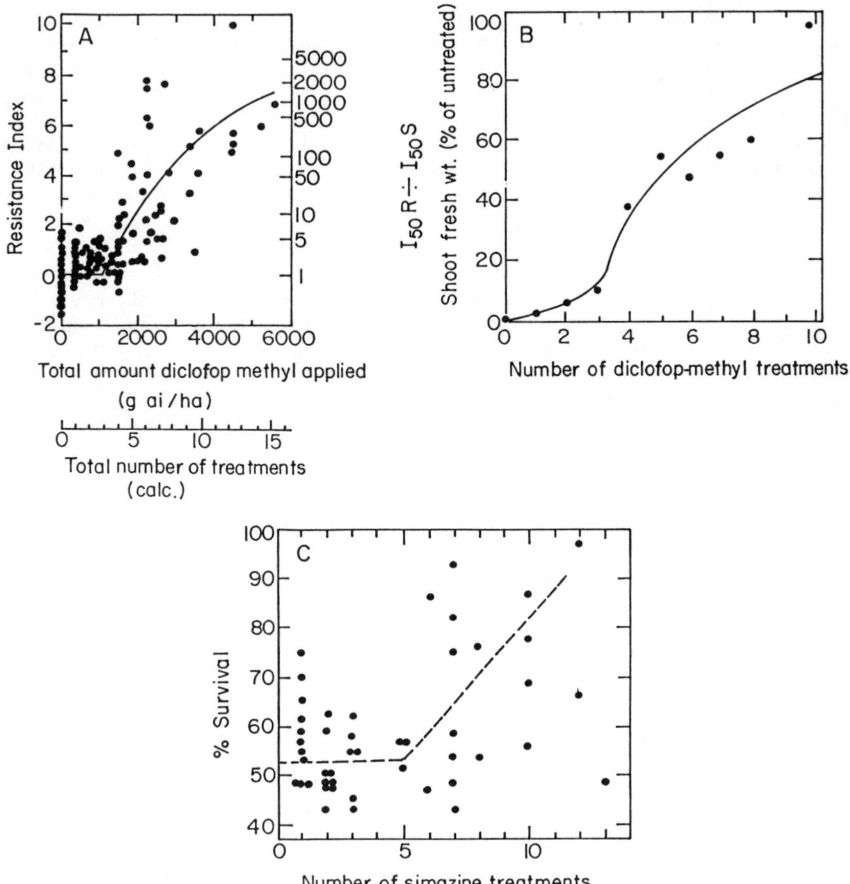

Fig. 3. Slow incremental increase in the dose level of resistance in repeatedly treated populations. The lines to show the way dose required for control may increase was drawn by this author for demonstration purposes only.
A. *Lolium rigidum* treated with diclofop-methyl in South Australia. The dose level need to control resistance in populations is shown as a function of the total amount of diclofop-methyl applied over the years. The typical rate used was 375g ha^{-1} diclofop-methyl. The resistance index is the difference in log between the I_{50} of resistant and susceptible populations. A population with a resistance index of 2 is resistant to field rates of diclofop-methyl. This index loses accuracy above a 500-fold increase in resistance. The populations of seeds were collected and tested by Ian Heap at the Waite Institute, Adelaide, Australia. This figure was modified and redrawn from Heap (1988).
B. *Lolium rigidum* treated with aryloxyphenoxy propionic acid herbicides in western Australia. 123 populations were treated with a fixed rate of herbicide. Redrawn from Gill and Diggle (1993). C. Increased resistance of *Senecio vulgaris* to simazine as a result of repeated treatments. *Senecio* seed was collected at 46 locations in England where the previous treatment history was known. The last treatment of each population was with 0.7 kg/ha simazine under controlled conditions, yielding the results in this figure. This figure was drawn from data in Figs. 3 and 4 in Holliday and Putwain (1980).

genus *Lolium* (ryegrass) over vast areas. The phenomenon of this genus will be used as the prime example throughout this discussion and for the modelling of alternate control strategies. In the plain states /provinces of the U.S.A / Canada diclofop-methyl was used at high use rates to control *Lolium,* and resistance evolved concurrently at many different locations. Resistance was traced to a modified acetyl-CoA-carboxylase, the target enzyme of the pesticide The only cross resistance found was to similar chemicals attacking the same target (Gronwald et al., 1992).

The same pesticide, but at a third the use rate was used to control ryegrass in Australian wheat and thousands of cases of resistance were found (Powles et al., 1992). Initially, none of the Australian populations had target site resistance and different populations had different qualitative and quantitative cross resistances to other herbicides having different chemistries and targets (Powles, 1993). Early on it was posited that the resistance in Australia was due to elevated levels of enzymes capable of degrading herbicides, probably one or many NADPH dependent cytochrome P450 monooxygenases (cyt P450s), similar to those present in wheat (Gressel, 1988). Wheat is known only to use monooxygenases to naturally degrade the selective herbicides used in this crop. Indeed, indirect evidence has shown that the Australian *Lolium* uses cyt P450s as well as other possible mechanisms (Powles, 1993).

Two studies have shown that the *Lolium* gradually became resistant to higher and higher doses of diclofop-methyl (Fig. 3A, B). This matches previous epidemiological evidence with simazine in *Senecio* (Fig. 3C). Researchers measured the level of resistance in populations from fields with different treatment histories of treatments with low levels of the diclofop-methyl (Fig. 3A, B). Surprisingly, farmers kept treating with the herbicide even after resistance was rampant, and thus the level of resistance continued to increase until it seemingly reached a plateau at a high level. Whether the level of resistance stabilized at a high level or could continue can be conjectured. These incremental data are quite different from those seen in Fig. 2, where resistance to a single high dose suddenly appeared. The data in Fig. 3 are thus not compatible with a single gene controlling resistance. They are compatible with selection of naturally mutated polygenes each conferring incremental levels of resistance or of the amplification of alleles, each conferring increments of resistance. Additional mutations and genetic recombinations further increase the levels of resistance.

The structural genes for cyt P450s are known to exist in large gene families containing tens to hundreds of genes. They have been described in insects (Bradfield et al., 1991; Gandhi et al., 1992; Feyereisen et al., 1994) and plants (O'Keefe et al., 1987) although they are best described in mammalian systems. Single gene mutations in the structural of cyt P450 are known to vastly modify substrate specificities. Additionally, genes that control the expression levels (synthesis and turnover) of groups of cyt P450s are known. Other pesticide detoxification mechanisms such as glutathione transferases can also be under polygenic controls. The vast arrays of different cross resistances to other herbicides in *Lolium* (Heap 1988; Powles et al., 1992; Powles, 1993) support polygenicity. Similar large differences in fungicide cross resistances have been shown to be controlled, in some cases, by many independent genes (Peever and Milgroom, 1993).

As soon as one resistant allele is selected for, the dense stands of *Lolium,* the pest can easily cross with nearby neighbors with the same or different alleles in this obligate outcrosser. Weed kill at the low rate of 375 g/ha used in Australia is far from complete. Some of the pests do receive the ca. 200g/ha threshold rate needed for kill, due to unequal dispersal, as depicted simplistically (Fig. 4, low dose). Some pests (not shown) will be in "refuges" and not receive any dose. Others (shaded area) receive an amount that is normally lethal. This concentration will not be lethal to a small number of pests that have a single (mutant) gene dose for resistance. The pest with the first mutant allele can also cross with unkilled, totally susceptible pests, receiving under 200g/ha receiving under 200 g/ha enriching them with a resistant allele. This first allele must be in sufficient quantity to meet with a neighbor with obligate outcrossers such as *Lolium.* Most grass species allow at least some self pollination so the mutant can find itself as mate. In the following generation a fourth of the outcrossed population contain one gene dose, a quarter of offspring from selfed individuals will contain two gene doses and half will have one.

The data in typical dose-response curves (eg. Fig. 1) depict only putatively dead/alive individuals at a fixed time after treatment under controlled conditions and thus "lose" data on sick pests that recover. After a few years of treatment of pristine populations with diclofop-methyl at low rates in Australia, there were often *Lolium* plants that appear very sick or even 'dead'. Many such plants recovered to produce some seed (Ian Heap, pers. comm.). These may well be the plants with the first gene dose .yet are not classified as "resistant".

It was suggested that the Catch 22 possibility described above is applicable to what has happened in *Lolium* (Gressel, 1993). In North-

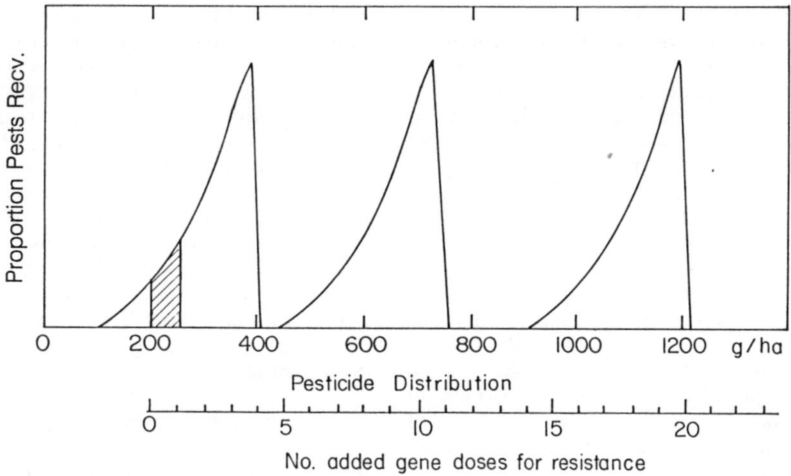

Fig. 4. Presumed distribution of a pesticide on pests following spraying in the field at 400, 750, 1200 g/ha, illustrating the proportion of pests receiving each dose. Double spraying is ignored, as are untouched escaped organisms (in "refuge"). An added scale shows how many additive independent mutant gene dosages would be required to withstand each dose. Assumptions: each mutant gene dose provides protection for 50g/ha beyond the threshold of 200g/ha. The cross hatched area shows the sensitive population from which one gene dose will be selected.

America only the manufacturer's recommended application rate of ca 1200 g/ha diclofop-methyl was used. In Australia such a rate was not economical in wheat cultivation and the rate of 375 g/ha was chosen because it gives adequate but hardly perfect control of *Lolium.* Thus, only a small increment of resistance was needed to change populations from susceptible to those with a modicum of resistance. The Australian conditions were conducive to rapid evolution with increases in levels of a polygenic type resistance for the following reasons: (a) The "pest" was often at very high initial population densities; it was often used as a pasture species prior to planting wheat, leaving behind fields sown with orders of magnitude more *Lolium* than wheat. Still resistance evolved first in monoculture wheat and not in wheat/pasture rotations; (2) *Lolium* is self-incompatible, thus mutants with different polygenes for resistance are likely to combine; (3) *Lolium* produces copious amounts of pollen (exacerbating allergies, and) facilitating the easy transfer of genes by wind pollination.

At the field dose rate used in North-America, the single major gene for resistance coding for acetyl CoA carboxylase is inherited as a semi-dominant trait (Betts et al., 1992). A higher rate that would require homozygous resistance would also kill the wheat crop.

The evolution of the single gene resistance follows the models for evolution of resistance previously proposed (Gressel and Segel 1978, Maxwell et al., 1990; Mortimer et al., 1992). Resistant populations appear "suddenly" seemingly with little advance warning (Fig. 2B). The plants are then resistant to massive doses of the herbicide that do not change with further treatments. This is because of the exponential enrichment of the resistant allele from its low natural frequency for a number generations where the increase in frequency of resistance is experimentally imperceptible (Fig. 2A). As there is always a small proportion of susceptible individuals missed by the herbicide treatment, the resistant individuals are not noticed until they become about 30% of the whole population. and then cannot be ignored. The gradual incremental resistance to increasing dose rates of herbicide seen in Australia (Fig. 3A, B) does not meet the single gene model, although the appearance of the first sickly, slightly resistant individuals a few generations after the beginning of use of the herbicide formally fits such a model.

Overcoming Catch 22: An Hypothesis

As: (1) Polygenic resistance can be delayed by treatments of moderate or high levels of pesticide, if done before too many mutated polygenes have accumulated following treatments with low levels of pesticide; and (2) the rate of evolution of target-site resistance is a function of selection pressure, and low and intermediate use rates of pesticide have low selection pressures for major gene resistance; It has been hypothesized (Gressel, 1993) that: a rotation of a number of treatments with low rates with a treatment at an intermediate rate will suppress the rate of evolution of both polygenic resistances resulting from low use rates and major gene resistance resulting from high use rates.

The intermediate treatment is expected to control individuals that have accumulated a few mutated polygenes, setting the situation back to the initial state (if there is a refuge of untreated sensitive individuals). The occasional use of moderate pesticide rates does add to the cost of production. Still, the alternate pesticides needed when resistant populations become predominant may cost far more if they exist. The possibility of broad cross resistances with polygenic mechanisms argues that losses will be greater if higher dose rates are not occasionally used.

This model allows for the use of far lower pesticide rates, i.e. the types of rates that do not select only for single gene resistance, and brings with it advantages in resistance management, economy, and lowers the environmental impact by lowering chemical input.

The model is not to be construed as a call for stoppage of crop, cultural practice, and pesticide rotations, which this author and many others feel provide the best possible pre-emptive resistance management. The model could best be used in situations where alternative crops and pesticides are impractical. This would be the case in many wheat growing areas, where because of land, season and/or rainfall, only wheat can be economically cultivated, and where evolved polygenically inherited resistance results in cross resistance to all wheat selective herbicides.

This model has implications beyond, preemptive pest resistance management in crops - it could well be considered in management schemes for antibiotic and anti-cancer drug therapies in medicine where resistance, including multi-drug resistance problems are rampant.

Assumptions for the model

In keeping with the example of *Lolium,* we will use actual data and assumptions relating to that pest. This will allow us to compare the model with what is known / has happened with *Lolium.*

Genes and Their Frequencies

Major Gene: In the case of *Lolium* and diclofop-methyl, various genetic analyses indicate that there is only one copy of the target-site gene, acetyl-Co-A-carboxylase (Betts et al., 1992). It is a very large gene, but probably can only be mutated at a few sites under the selection pressure of inhibitors of this enzyme. One site of mutation gives rise to plants with cross-resistance to other aryloxyphenoxy propionates (-fops), and another type mutation to both -fops and cyclohexanediones (-dims) that affect the same enzyme. Under selection pressure of -dims, mutants were found that were resistant to -dims alone and others to fops and -dims. It is not known if each type has more than one allowable mutation site - i.e. more than one allele. We will assume that 10^{-6} plants in a pristine population have target site resistance due to one mutation per gamete per generation and a slight unfitness of the gene that prevents its accumulation. This intuitive guess is based on the time it took and number of treatments, etc., to get resistance compared to other herbicides.

Polygenic. Because of the ubiquity of cyt P450 genes in families, the possibility of other mechanisms contributing to resistance(Powles 1993), and the variety of cross-resistance spectra (Heap, 1988; Powles et al., 1992), we can guess that there are at least 20 different polygenic genes and up to hundreds of genes that can each contribute to resistance. It is assumed that each mutated gene can independently contribute increments of resistance. It is also assumed for the case of this analysis that each allele can contribute an average increment of resistance to ca. 50 g/ha. Clearly it is important to have real data for this from areas where resistance has evolved, to allow deliniating strategies to delay evolution elsewhere. The possible interdependence among some alleles, and chromosomal linkage groups are ignored in this exercise. It is assumed in this discussion that the frequency of mutation at each of the polygenes is also 10^{-6}, per gamete per generation, but we will also assume that there are 100 genes that can be mutated. Thus, at any time $10^{-6}X10^{2} = 10^{-4}$ of the plants will bear a single mutant gene. The likelihood of any plant having initially having two such mutations is $10^{-4}X10^{-4}=10^{-8}$ with 3 mutations 10^{-12}, etc.

Selection at a low dose rate could also select for target site resistant alleles - yet the frequency differences between 10^{-6} and 10^{-4} suggest that such an event would be relatively rare. Indeed, following years of low dose selections in Australia, resulting in the accumulation of plants containing mixtures of many polygenic, low contribution alleles, only recently were populations found that also contained target site resistances (Tardiff et al., 1993).

Dose response curves

A pesticide dose response curve generated in the laboratory under ideal conditions typically is linear when plotted using probit techniques. (Fig. 1). This is not quite the case in the field where it is more skewed; at higher doses fewer than expected organisms are killed in the field. In the laboratory weed seed is often pretreated to obtain highly uniform germination, it is evenly seeded at a constant depth in special soil mixtures of uniform and ideal tilthe and watered to obtain good uniformity. The seedlings are then sprayed with a highly uniform spray giving uniform distribution of pesticide. In the field, the seeds germinate at less uniform times and two leaf and four leaf seedlings of the same species often have very different dose response curves. Some seedlings are shielded or shaded from spray by other seedlings or by clods of soil or rocks. A sprayer bouncing across a field cannot provide the same uniform pesticide distribution pattern as a laboratory sprayer. Similarly, insect larvae in the same instar are used for dose-response curves in the laboratory (e.g. Fig. 1). There are often large variances in susceptibility among different instars, with more advanced instars being less sensitive. In the field, there is often not the synchrony achieved in the lab, and various instars are treated simultaneously. Again a skewed dose-response probit curve will be obtained. Fungi at different stages of development, germination, penetration and establishment are differently affected by fungicides. This would also cause skewing of dose response curves.

Thus, if *Lolium* is 99% controlled by 250-300 g/ha diclofop-methyl in the laboratory, it is easy to see that it takes 375 g/ha to get 90-95% control in the field and 1200 g/ha to get the 97-99% achieved in North America. In both cases there are some total escapes due to refuges in the field, as well as late germination after the herbicide is dissipated. Presumed doses reaching different plants are depicted in Fig. 4. At 375 g/ha, the typical rate used in Australia for *Lolium,* 5-10% of the plants receive no effective amount of herbicide, and their offspring will be controlled by 375g/ha the following season if they intercross only with each other. Another 10-20% of the population is subjected to selection for a single gene dose (shaded area), because they receive 250-300 g/ha herbicide. Only a small proportion of the individuals receiving 200-250g/ha, (ca. 10^{-4}) have a mutation to allow survival, i.e., those resistant to this dose due to one mutant gene survive. Those that survive may be severely injured but they recover. If they could self pollinate (in *Lolium* they cannot) - then 25% of their offspring would have two gene doses and 50% one gene dose. If they are sufficiently close to another plant with the same mutated allele or one linked on the same chromosome (rare) 25% of their joint offspring will have two gene doses. If they cross with plants with other unlinked mutations, all joint offspring will have two resistant alleles, which will segregate in the following generation in different manners, depending upon their "mates".

The most likely crosses by the rare individuals that survive the 250-300 g/ha treatment are with the far more ubiquitous healthy plants in the below 250 g/ha class that did not receive an effectual dose. Half the offspring from such crosses will now have one gene dose. They will vastly increase the proportion of the population with one gene dose the following year, and many more plants receiving 250-300 g/ha will have a modicum of resistance, spreading much more pollen, increasing the chances of crosses resulting in two gene doses. It is the task of modellers to show how the gene doses become enriched in the population.

When the high dose rate of 1,200 g/ha is used, it is clear that >97% of the plants are killed (Fig.4). Most of the survivors were in refuges and received no pesticide at all. An infinitely small proportion of plants received 250-300g range, so that the selection for a single mutant minor gene dose would be minimal. Assuming one mutant minor gene dose is required for every 50g of pesticide above 200g/ha, then 20 gene doses would be needed to survive 1,200 g/ha. There would theoretically be one plant with 20 gene doses at a frequency of $10^{-4x20} = 10^{-80}$ in a pristine population. Thus, the only likely resistant survivors could be those with a major gene mutant. In the case of *Lolium,* only a target site mutation seems to be a major gene mutant. If the field is treated with a moderate dose (e.g. 700 g/ha in this case), 3-5 % of the plants receive less than a lethal dose, because they are escapees in refuges. Virtually all other plants, receive a dose that would still require many simultaneous mutations for there to be survival. Probably, for safety sake, an intermediate dose should be chosen to require 4-8 simultaneous mutations for a plant to be resistant.

Modelling

M. Mangel (U. California - Davis), along with colleagues and in collaboration with the author modelled the effect of repeatedly using an intermediate dose after a few low doses. One of the scenarios is plotted, using one of his models, in Fig. 5. In this scenario, it can be seen that if a high dose were used (Fig. 5, acute slope) resistance would be apparent after 12 generations. If a median dose alone were used, resistance would take far longer (lower straight line). A low dose (not shown) would facilitate evolution as shown in Fig. 3. Application of three low doses, allows for a considerable proportion of the population (but still less than 5%) to become resistant, but the intermediate dose, represses the level back down.

The model illustrated does not consider that the

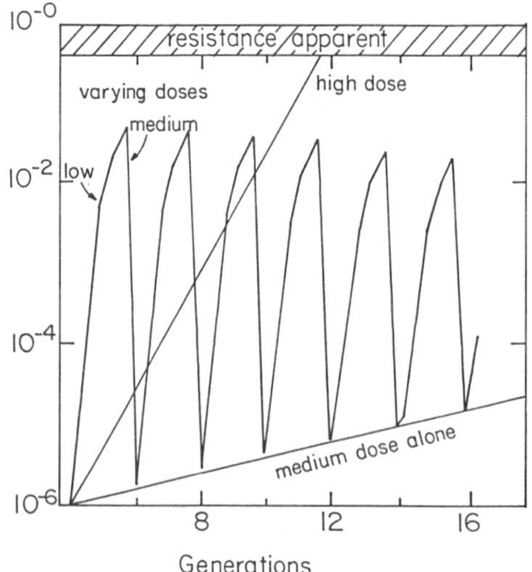

Fig. 5. Effects of varying low with medium doses on the enrichment of resistance. The effect of high dose, with high selection pressure and medium dose with lower selection pressure (straight lines) are plotted from equations similar to those used for Fig. 2A. The lines for 3 low doses interspersed with one medium dose were modified from those generated by the courtesy of Marc Mangel, U. California, Davis.

intermediate dose leaves about 5% of the population as escapees, which can recombine genetically if plant distances are sufficient for crossing. Thus, evolution of resistant populations might be faster than shown, but still much slower than the low dose alone, or high dose alone regimes. In addition to modelling, it is imperative to obtain genetic data to further understand what is happening in the field.

The advantage of such models is that there are easy experimental designs to test them. Experimental verification of such models can be facilitated by mixing pristine wild type material with pests known to contain different gene doses for resistance. This can ascertain whether the concept of using intermediate doses after a few low doses is more than a theoretical management tool. If it is a valid tool, it must be practiced over wide areas in concert, with all but the most immobile of pests, to prevent mixing of populations allowing for enrichment of genes for resistance. In weeds, there is good evidence in some instances that each case of resistance is due to evolution within a given field, and not due to gene flow (Darmency and Gasquez, 1981). This is not the case in insects, where flying and human transport allow for easy gene flow (Raymond et al., 1991).

Polygenic resistance seems more dangerous to pest management than monogenic resistance, whether due to a variety of genes coding for overlapping metabolic mechanisms or to amplifications resulting in multi-drug resistance. This is because resistance can be to a large spectrum of chemicals with different modes of action. Thus, we must weigh the risks of each alternative Catch 22 situation. Many more cases of polygenic resistance should be expected as more farmers cut doses to adhere to strictures to lower pesticide levels. It may well be wise to consider using no pesticide when infestations are low than to use a low dose, unless the low doses are interspersed with high doses.

Acknowledgements

The author appreciates the use of the model data kindly supplied by Marc Mangel, University of California, Davis. Many colleagues have supplied input and ideas but they are not responsible for the conclusions drawn herein. J. G. holds the Gilbert de Botton Chair in Plant Sciences.

References

Agur, Z.; Harnevo, L. in *Biomedical Modelling. and Simulation,* Eisenfeld, Witten and Levin *(in press.)*

Betts, K.J.; Ehlke, N.J.; Wyse, D.L.; Gronwald, J.W.; Sommers, D.A. *Weed Sci.* **1992,** *40,* 184-189.

Bradfield, H.Y.; Lee, Y.H.; Keeley, L.L. *Proc. Natl. Acad. Sci. USA* **1991**, *88*, 4558-4562.

Brent, K.J.; Hollomon, D.W.; Shaw, M.W. *Managing Resistance to Agrochemicals,* (Green, M.B., LeBaron, H.M., Moberg, W.K. Eds. ACS Symp. Ser. Washington, **1990** pp. 303-319

Darmency, H.; Gasquez, J., *New Phytol.* **1981,** *89,* 487-496.

Dove, W.F. *Genetics* **1993,** *134*, 999-1002.

De Waard, M.A. in *Resistance '91-Achievements and Developments in Combating Pesticide Resistance* (Denholm, I., Devonshire, A.L., Hollomon, D.W. eds.,) Elsevier: London **1992**. pp. 48-60.

Cheverud, J.M.; Routman, E. *J. Evol. Biol.* **1993**, *6*, 463-480.

Feyereisen, R.; Andersen, J.F.; Cariflo, F.A.; Cohen, M.B.; Koener, J.F.; Repecko, S.; Scott, J.A.; Snyder, M.J. (in press)

Field, L.M.; Devonshire, A.L.; Forde, B.G. *Biochem. J.* **1988**, *251*, 309-312.

Gandhi, R., Varak, E.; Goldberg, M.L. *Cell Biol.* **1992**, *11*, 397-404.

Georghiou, G.P.; Taylor, C.E. in *Pesticide Resistance: Strategies and Tactics for Management,* National Acad. Press: Washington, **1986,** pp. 157-169.

Gill, G.; Diggle, A.J. in *Integrated Weed Management for Sustainable Agriculture,* (Malik, R.K., ed). Ind. Soc. Weed Sci.: Hisar, **1993**, pp. 203-208.

Gressel, J.; *International Crop Science I.* (Buxton, D.R.; Shibles, R..; Forsberg, B.L.; Blad, K.H..; Asay, K.H.; Paulsen, G.M.; Wilson, R.F. eds.) Crop Sci. Soc. America: Madison WI. **1993**, pp 121-127.

Gressel, J; *Oxford Surv. Plant Mol. Cell Biol.* **1988**, *5,* 195-203.

Gressel, J.; Segel, L.A. *J. Theor. Biol.* **1979**, *75,* 349-371.

Gressel, J.; Segel, L.A. in *Herbicide Resistance in Plants,* (LeBaron, H.M.; Gressel, J., eds.) Wiley: New York 1972 pp.325-347.

Gronwald, J.W.; Eberlein, C.V.; Betts, K.J.; Baerg, R.J., Ehlke, N.J.; Wyse, D.L. *Pestic. Biochem. Physiol.* **1992**, *44*, 126-139.

Harnevo, L.E.; Agur, Z. *Math. Biosci..* **1991**, *103*, 115-138.

Harnevo, L.E.; Agur, Z. *Canc. Chemoth. Pharma.* **1992**, *30,* 469-476.

Heap, I.M. *Ph.D. Dissertation*, Waite Agric. Inst., Univ. of Adelaide, **1988.**

Holliday, R.J.; Putwain, P.D. *J. Appl. Ecol.* **1980,** *17,* 799-808.

Hollomon, D.W.; Butters, J.A.; Hargreaves, J.A. in *Managing Resistance to Agrochemicals,* (Green, M.B., LeBaron, H.M., Moberg, W.K., eds.) ACS Symp. Ser.; Washington, **1990,** pp.199-214.

Lande, R. *Heredity* **1983,** *50*, 47-65.

Macnair, M.R.; Smith, S.E.; Cumbes, Q.J. *Hered.* **1993,** *71*, 445-455.

Malik, R.K.; Singh, S. in *Integrated Weed Management for Sustainable Agriculture* (Malik, R.K. , ed). Ind. Soc. Weed Sci: Hisar, **1993**, pp. 225-238.

Maxwell, B.P.; Roush, M.L.; Radosevich, S.R.; *Weed Technol.* **1990.** *4,* 2-13.

Mortimer, A.M.; Ulf-Hansen, P.F.; Putwain, P.D. in *Achievements and Developments in Combating Pesticide Resistance.* (Denholm, I., Devonshire, A.L., Hollomon, D.W., eds). Elsevier: London **1992**, pp. 148-164.

Moss, S.R.; Cussans, G.W. in *Herbicide Resistance in Weeds and Crops. (*Caseley, J.C., Cussans, G.W., Atkin, R.K., eds) Butterworths:Oxford **1991,** pp. 45-56.

Moss, S. in *Resistance* '91-*Achievements and Developments in Combating Pesticide Resistance* (Denholm, I., Devonshire, A.L., Hollomon, D.W. ,eds). Elsevier: London **1992**. pp. 28-40.

Mouches, C.; Pasteur, N.; Berge, J.B.; Hyrien, O.; Raymond, M.; de Saint Vincent, B.R.; de Silvestri, M; Georghiou, G.P., *Science* **1986,** *233*, 778-780.

Nosticzius, A.; Muller, T.; Czimber, G. *Bot. Kozl.* **1979,** *66*, 299-305.

O'Keefe, D.P.; Romesser, J.A.; Leto, K.J. in *Phytochemical Effects of Environmental Compounds* (Saunders, J.A., Channing, L.K., Conn, E.E., eds.). Plenum: New-York, **1987,** pp. 151-173.

Peever, T.L.; Milgroom, M.G. *Phytopathology* **1993,** *83,* 1076-1081

Powles, S.B. in *Integrated Weed Management for Sustainable Agriculture, (*Malik, R.K., ed). Ind. Soc. Weed Sci.: Hisar, **1993**, Vol. 1 pp.189-194

Powles, S.B.; Matthews, J.M. in *Resistance '91. Achievements and Developments in Combating Pesticide Resistance* (Denholm, I., Devonshire, A.L., Hollomon, D.W. eds.) Elsevier: London, **1992** pp. 75-87

Raymond, M.; Marquinne, M.; Pasteur, N., in *Resistance '91-Achievements and Developments in Combating Pesticide Resistance* (Denholm, I., Devonshire, A.L., Hollomon, D.W., eds.) Elsevier: London, **1992** pp. 19-27.

Roush, R.T.; Daly, J.C., in *Pesticide Resistance . in Arthropods,(* Roush R.T.; Tabashnik, B.E., eds).; Chapman Hall: New York, **1990,** pp. 97-152.

Shaw, M.W., *Plant Pathol.* **1989,** *38*, 44-55.

Schimke, R.T., *Cell,* **1984**, *37,* 705-713.

Tabashnik, B.E. *J. Econ. Entomol.,* **1990,** *83*, 1170-1176.

Tardif, F.J.; Holtum, J.A.M.; Powles, S.B. *Planta,* **1993,** *190,* 176-181.

Taylor, C.E., in *Pest Resistance to Pesticides,* (Georghiou, G.P.; Saito, T., eds.) Plenum: New York. **1983,** pp. 163-173.

Via, S. in *Pesticide Resistance: Strategies and Tactics for Managment.* National Academy: Washington, D.C. **1986,**. pp. 222-235.

Wang, Y.; Jones, J.D.; Weller, S.C.; Goldsbrough, P.B. *Plant Mol. Biol.* **1991**, *17*, 1127-1138.

Williams, K., *Nature,* **1976,** *260,* 785-788.

Managing Resistance to Sterol Demethylation Inhibitors

Wolfram Köller, Department of Plant Pathology, Cornell University, New York State Agricultural Experiment Station, Geneva, NY 14456, USA

Resistance to sterol demethylation inhibitors with their specific mode of action evolved more slowly than with other site-specific fungicides such as the benzimidazoles. The difference in speed of resistance development is related to the sensitivity distribution found in wild-type pathogen populations. For sterol demethylation inhibitors, the distribution of isolate sensitivities prior to selective exposure is continuous in character and covers a wide range. This broad sensitivity distribution has impact on monitoring procedures, but also on the management of resistance. For the DMIs, monitoring of population sensitivities must be based on quantitative sensitivity data tested for a sufficiently large sample of individual isolates. Experimental evidence obtained for *Venturia inaequalis*, the causal agent of apple scab, indicated that resistance can be delayed by (a) minimizing disease pressure, (b) reducing the selection time, and (c) reducing the population size of resistant isolates. Reduction of population sizes of resistant isolates can be accomplished with high rates, because resistant phenotypes remain accessible to inhibition, albeit at higher doses.

Sterol demethylation inhibitors (DMIs) comprise a widely used class of modern systemic fungicides (Scheinpflug and Kuck, 1987). Although first representatives such as triadimefon were commercially introduced in the early 1970's, DMIs have remained subject of research and development, and new DMIs with improved properties are still in early stages of introduction or late stages of development. It can be expected that over 25 different commercial products will be available for the broad control of diseases on a wide range of crops.

Although DMIs are structurally diverse and contain either a pyridine, pyrimidine, imidazole and triazole as essential group, they all share a common mode of action in fungal sterol biosynthesis. The precise target site, a specific cytochrome P450 active in the oxidative demethylation of lanosterol or other precursors at the C14-position, was first identified for the pyrimidine triarimol and later confirmed for all other DMIs (Sisler and Ragsdale, 1984; Köller, 1992).

This highly specific mode of action implied a considerable risk of resistance. Field experience and research results, however, have indicated that resistance evoleved more slowly than for other site-specific fungicides such as the benzimidazoles (Köller and Scheinpflug, 1987; Köller, 1991). This difference among fungicides has impact on the monitoring of resistant populations, but also on tactics of anti-resistance strategies. Description of these differences will be the subject of this overview.

Origin and dynamics of DMI resistance

Widespread failures of satisfactory disease control due to the evolution of DMI resistance were first documented for barley powdery mildew in 1981 and later for powdery mildews of cucurbits and wheat (Brent and Hollomon, 1988; Köller and Scheinpflug, 1987). More recent reports suggest that the number of diseases affected by resistance has increased gradually over the course of prolonged DMI use. Widespread resistance was recently documented for leaf blotch of barley (Kendall et al, 1993). Strong indications of resistance have also been reported for powdery mildew of grapes and post-harvest control of *Penicillium digitatum* on citrus (Hollomon, 1993). Although several other diseases have been listed as potentially affected by DMI resistance (Hollomon, 1993), the magnitude of control problems remains to be evaluated.

Even though the incidence of DMI resistance is increasing, field problems have evolved more gradually for the DMIs than for other site-specific classes of fungicides such as the benzimidazoles and the phenylamides. Based on first experimental data describing the responses of powdery mildew populations to DMI treatment and considering the principles of dynamic responses of fungal populations to a changing environment, Wolfe (1982) proposed that population responses leading to the selection of resistant subpopulations were different for the different types of fungicides. It had been recognized that the frequencies of isolate sensitivities to DMIs in unexposed pathogen populations were broadly distributed and continuous in character, ranging from highly sensitive to resistant (Köller and Scheinpflug, 1987; Skylakakis and Hollomon, 1987). Although conclusive molecular or genetic explanations for these broad distributions in sensitivities have not yet been provided (Köller, 1988; Köller, 1992; Hollomon, 1993), the phenomenon appears to be related to the lower risk of resistance experienced with the DMIs.

In contrast to DMIs, distributions of population sensitivities to the high-risk benzimidazoles are discontinuous in character (Skylakakis and Hollomon, 1987; Köller, 1991). Strains resistant to benzimidazoles, which reside in low frequencies in wild-type populations, contain a mutated β-tubulin, the normally sensitive target site of benzimidazoles (Koenraadt et al., 1992). A single-base mutation leads to drastically reduced inhibitor binding affinity and a virtually insensitive subpopulation. Resistant strains are not controlled at any practical dose and rapidly increase in frequency in the presence of the fungicide. To the contrary, DMI-resistant strains have largely reduced sensitivities, but growth remains accessible to inhibition at higher doses. Regardless of these differences, under the selective force of a DMI, the proportion of DMI resistant phenotypes may reach a level where satisfactory disease control with recommended rates is no longer provided.

The continuous and broad sensitivity distribution of wild-type populations to DMIs has impact on the choice of monitoring procedures, the definition of resistance, and the prevention and management of resistance. Apple scab caused by *Venturia inaequalis* has become one of the more intensively studied diseases with regard to DMI resistance. A first report from an experimental apple orchard in Nova Scotia, Canada, had provided strong evidence for the evolution of DMI field resistance after 12 years of DMI testing (Hildebrand et al., 1988). A shift of the respective pathogen population toward resistance was confirmed more recently (Braun and McRae, 1992). The delayed introduction of DMIs for the control of apple scab in the United States provided an opportunity to initiate systematic studies on DMI resistance in New York State, a region that had first experienced resistance of *V. inaequalis* to dodine, followed by resistance to benzimidazoles (Gilpatrick, 1992).

The summary of results will therefore serve as an example for systematical studies dedicated to the development of DMI anti-resistance strategies.

Monitoring procedures.

Sensitivity monitoring serves the purpose of confirming practical resistance, of following population shifts toward resistance once a fungicide has been introduced, and of evaluating the relative success of anti-resistance strategies. The most appropriate monitoring technique is determined by the inhibitory characteristics of the fungicide under question, by the biology and epidemiology of the pathogen to be monitored, and by the variation and distribution of sensitivities residing in unexposed wild-type populations. Considering these multi-component requirements, it is immediately apparent that generalized monitoring techniques such as sensitivity tests of spore germination, germ tube elongation or mycelial growth are of limited value, even for a given disease.

The variable demands are exemplified by the monitoring techniques suitable for either benzimidazole or DMI resistance in populations of *V. inaequalis*. Benzimidazoles, as inhibitors of tubulin polymerization, inhibit cell division and thus curtail the elongation of germ tubes emerging from conidia. Consequently, germ tube elongation has become an appropriate and convenient developmental stage for sensitivity tests (Yoder et al., 1986). In contrast, germ tube elongation is relatively unaffected by DMIs as inhibitors of sterol biosynthesis. The initial steps of disease initiation, spore germination, appressoria formation and cuticle penetration, are undisturbed even in the presence of the inhibitor. Disease development is halted at the stage of mycelial development subsequent to the penetration into the leaf tissue (Scheinpflug, 1986). Therefore, spore germination is a questionable phase for DMI sensitivity tests, and mycelial growth was chosen as a more suitable stage for reliable and quantitative bioassays (Smith et al., 1991).

A second crucial consideration for monitoring techniques relates to sample size requirements. For benzimidazoles with their mechanism of resistance based on the point mutation of the target site and an almost insensitive subpopulation, which is clearly separated from the originally prevailing sensitive population, the sample size is a parameter of utmost importance. An example is shown in Fig 1. The response of mycelial colony growth to a single dose of 0.2 mg/L benomyl was evaluated for 280 single isolates obtained from apple orchards in New York State with past histories of benzimidazole resistance and stable resistant subpopulations. The combined data

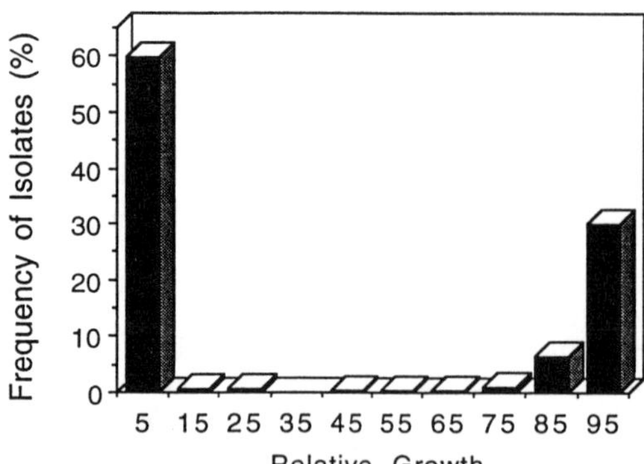

Fig. 1. Frequency distribution of *V. inaequalis* isolate sensitivities to benomyl in orchards with previous histories of benzimidazole resistance. The sensitivity of 280 isolates was tested by determining the relative growth (percent of mycelial growth) at a benomyl dose of 0.2 mg/L.

from 6 orchards demonstrate that sensitive and resistant isolates are easily discriminated, and that the frequency of resistant isoaltes can be determined unequivocally. For benzimidazole monitoring, the quantitative accuracy of sensitivity tests is therefore less important than sample size considerations.

Although never confirmed experimentally, the frequency of benzimidazole-resistant mutants residing in wild-type populations has been considered very low (Skylakakis and Hollomon, 1987). Consequently, any attempt to establish baseline sensitivities derived from a reasonable number of tested isolates would, with all likelihood, indicate the presence of an entirely sensitive population. Even if the shift of populations toward a frequency of 1% resistant isolates is the monitoring goal, statistical rigor would require the testing of 300 randomly sampled individual isolates (Leung et al., 1993). For fungicides, sample sizes tested have been frequently obscured by the presence of abundant conidia formed within a single lesion. However, these conidia were produced asexually and thus are of clonal character. Consequently, conclusions derived from sensitivity tests conducted with a large number of germinating conidia originating from only a few sporulating lesions can yield misleading results.

Monitoring requirements for the DMI are different from those established for the benzimidazoles. It had been observed that DMI sensitivities found in wild-type populations are broadly distributed and continuous in character, and that declining performances in powdery mildew control were accompanied by a continuous progression of population shifts toward higher frequencies of the less sensitive fraction of isolates present in wild-type populations (Köller and Scheinpflug, 1987; Skylakakis and Hollomon, 1987). Furthermore, DMI resistant mutants could be easily generated in the laboratory, but these mutants were, in contrast to benzimidazole-resistant mutants, not insensitive to the inhibitors (Hippe and Köller, 1984; Köller and Wubben, 1989). The DMI target sites remained accessible to tyipcal inhibitor action; only higher doses were required for equivalent effects. The accessibility of the target site in resistant isolates might be explained by the lack of mutational target site changes leading to drastically decreased inhibitor binding affinities.

The continuously decreasing efficacy of DMIs on the growth of less sensitive isolates residing in pathogen populations necessitated the precise determination of sensitivities of individual *V. inaequalis* isolates, before DMIs were used for the commercial control of apple scab. ED_{50} values of 300 isolates collected from three orchard populations in New York State were determined for the DMI flusilazole, based on the inhibition of mycelial colony expansion (Smith et al., 1991). As reported before for powdery mildews (Brent and Hollomon, 1988), ED_{50} values were log normally distributed. For flusilazole, they ranged from 0.0006 to 0.17 mg/L, with a mean sensitivity of 0.08 mg/L. Although the variation in sensitivities reflects the sum of test and biological variance, comparison of sensitivity data obtained by others and for several DMIs (Braun and McRae, 1992), the biological variance of tested isolate sensitivities exceeds a factor of 300.

Despite small differences between the orchard populations tested, variances of sensitivity distributions were homogenous. Based on sensitivity variations obtained for all 300 isolates, sample sizes necessary to detect differences in mean sensitivities were determined. Analysis showed that sample sizes of >50 did not greatly improve the precision of monitoring tests, whereas with sample sizes <15, detectability of population differences was insufficient (Smith et al., 1991).

Measurements of ED_{50} values for monoconidial isolates is sufficiently precise, but it is also relatively labor-intensive and cost-prohibitive if large numbers of sites are to be tested. Based on the data obtained for flusilazole sensitivities, a simplified yet sufficiently precise test was developed. It was proposed that the relative growth (= percent of growth compared to a nontreated control) of mycelial colonies at one discriminatory dose close to the mean baseline ED_{50} could serve as a simplified monitoring test. Under these test conditions, isolates less sensitive than typical baseline representatives would yield relative growth values <50, and those more resistant would yield values >50.

Because DMIs are not equally active as inhibitors, the discriminatory doses had to be determined for each DMI individually. This was accomplished by testing a small number of typical baseline isolates (Köller et al., 1991). For seven DMIs registered for the control of apple scab, mean baseline ED_{50} values ranged from 0.008 mg/L for flusilazole to 0.07 mg/L for myclobutanil. Comparison with published data and personal communication revealed that these data were similar for *V. inaequalis* populations in Europe, Canada and Japan. Extensive monitoring with flusilazole, fenarimol, myclobutanil and tebuconazole demonstrated that the proposed monitoring technique based on relative mycelial growth at one discriminatory dose and tested for 30-50 individual isolates per orchard site provided an adequate tool for the monitoring of DMI sensitivities. The baseline sensitivity distribution determined for fenarimol is illustrated in Fig. 2.

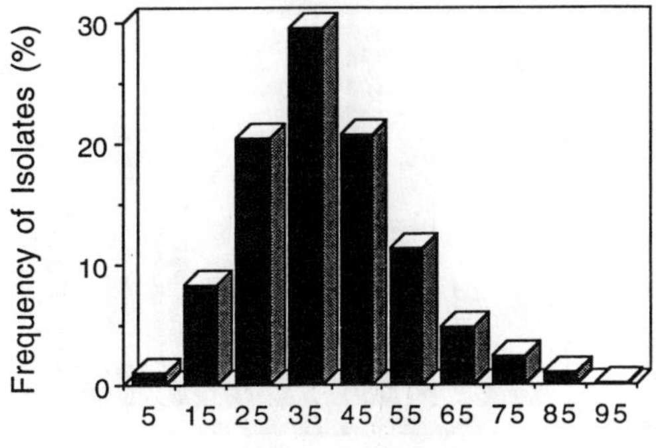

Fig. 2. Baseline frequency distribution of *V. inaequalis* isolate sensitivities to the DMI fenarimol in orchards never exposed to a DMI. The sensitivity of 530 isolates was tested by determining the relative growth (percent of mycelial growth) at a fenarimol dose of 0.05 mg/L.

Definition of DMI resistance.

The definition of what constitutes fungicide resistance responsible for control failures under field conditions is relatively simple (Köller, 1991). Resistant isolates have to be considerably less sensitive than typical baseline isolates, and they have to occur in considerably increased frequencies at sites with unsatisfactory disease control. If both criteria are met, the presence of practical resistance is established.

For fungicides such as the benzimidazoles with their clearly separated sensitive and resistant subpopulations (Fig. 1), the criteria are easily tested. The definition of resistance under practical conditions is more complicated with the DMIs and their broad distribution of sensitivities found in wild-type populations (Fig. 2). Ideally, the definition can be tested by comparison of baseline frequency distributions with distributions found in orchards with clear evidence of control failures. This goal was reached by testing the population of *V. inaequalis* at the experimental orchard at the Nova Scotia Research Station, where the DMI bitertanol had failed to control apple scab (Hildebrand et al., 1988). Analysis of myclobutanil sensitivities residing in this orchard, based on ED_{50} values of single isolates, revealed that the population had shifted toward higher frequencies of less sensitive isolates (Braun and McRae, 1992).

Surprisingly, two separate yet overlapping populations were identified, with mean ED_{50} values separated by a factor of approximately 10 (Fig. 3A). This bimodal distribution of sensitivities departed from the previous model of population dynamics derived from shifts of powdery mildew toward DMI resistance (Wolfe, 1982; Köller and Scheinpflug, 1987; Skylakakis and Hollomon, 1987). Here, the separation of two distinct populations had not been reported. Rather, continuous sensitivity distributions with increasing mean sensitivities had been considered as characteristic for population shifts toward resistance. However, a bimodal separation of DMI sensitive and resistant subpopulations similar to *V. inaequalis* was also reported as an intermediate stage of resistance evolution for *Rynchosporium secalis*, the causal agent of barley leaf blotch (Kendall et al, 1993). The genetic implications of this bimodal distribution remain untested at the present time. It might imply the presence of one major resistance gene, or it might merely reflect the decreasing levels of control achieved with isolates resisting increasingly higher DMI doses.

More relevant with regard to the definition of DMI sensitivities identifiable as "resistant" under practical conditions was the comparison of sensitivity frequencies based on relative growth at a single discriminatory dose, with fenarimol as a representative DMI (Fig. 3A). The bimodal separation of two distinct populations was also apparent under these test conditions. Analysis of frequencies demonstrated that isolates with relative growth values ranging from 0-60 had descreased in frequency, and that the frequency of isolates with relative growth values of 81-100 had increased. The latter isolates are therefore considered resistant, because they conform with the definition of practical resistance: increased

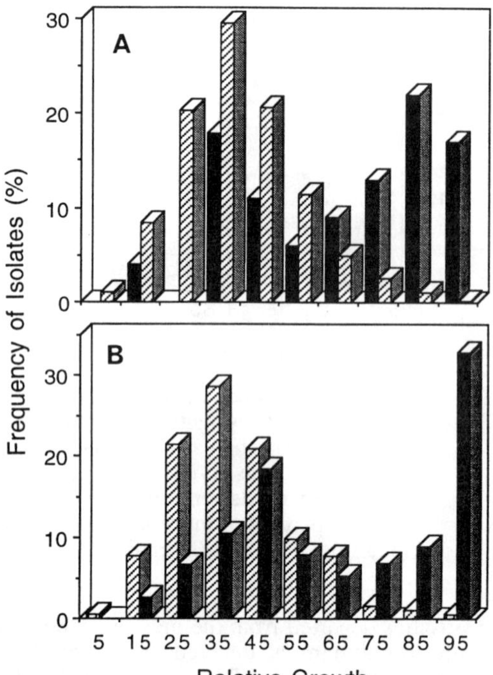

Fig. 3. Comparison of baseline frequency distributions of *V. inaequalis* populations with distributions determined in orchards with reported disease control failures. A: DMI baseline and threshold levels tested with fenarimol at a discriminatory dose of 0.05 mg/L. B: Dodine baseline and threshold levels tested with dodine at a discriminatory dose of 0.2 mg/L. Open bars represent baseline, closed bars represent threshold distributions.

frequencies at a site of unsatisfactory DMI performance.

Not surprisingly in view of overlapping distributions, the frequency of isolates with relative growth values of 61-80 had increased only by a factor of 2.8 and, therefore, were designated as reduced sensitive (Table 1). This small increase indicated that respective isolates were selected less rigorously than resistant isolates.

As discussed below in greater detail, the distribution of isolate sensitivities established for the particular orchard site with unsatisfactory control of scab with bitertanol defines the threshold of allowable shifts of *V. inaequalis* populations toward resistance. This threshold level is reached after resistant isolates reach a frequency of 39 % (Table 1). Threshold distributions and frequencies were very similar

Table 1. Baseline and threshold sensitivities of *V. inaequalis* to fenarimol at 0.05 mg/L.

		Frequency (%)		
Status [1]	RG	Baseline	Threshold	Increase
S	0-60	91.1	39	0.4
RS	61-80	8.9	22	2.8
R	81-100	1.1	39	35

[1] S, sensitive; RS, reduced sensitive; R, resistant.

with either flusilazole or myclobutanil employed as the diagnostic DMI (Braun and Köller, unpublished results).

Development of anti-resistance strategies.

The development of strategies coping with fungicide resistance has been dominated by mathematical computer models based on epidemiological principles, and on circumstantial evidence derived from field experiences (Staub, 1991; Köller, 1991). One of the more recent computer models and simulations was provided by Milgrom and Fry (1988). In this model, the time from a baseline situation with a ratio of disease caused by resistant (x_o) or sensitive (y_o) strains to a fixed value of x/y is determined by the equation:

$$t = [1/(r_R - r_S)] \times [\ln (x/y) - \ln (x_o/y_o)],$$

with r_R and r_S representing apparent infection rates (disease progression) of resistant and sensitive strains, respectively.

One basic tactic derived from this computer simulation and other models is that low infection rates (low disease pressure) of both the sensitive and resistant population (r_R and r_S) will delay resistance development (TACTIC 1). A second but almost redundant tactic is to minimize the selection time by a fungicide under risk, because r_R and r_S would be equal (TACTIC 2). A third tactic relates to decreasing the difference r_R - r_S during times of active selection (TACTIC 3). This can be accomplished by increasing r_S, the infection rate of sensitive isolates, by lowering fungicide doses and allowing the development of disease by the sensitive population. Under these conditions, resistant isolates would not be controlled, but their total size would increase more slowly, because the tissue area diseased by sensitive isolates would not be accessible to infection by resistant isolates. This strategy is of limited value for some diseases. With apple scab, for example, a high level of disease is economically unacceptable.

An alternative possibility of decreasing r_R - r_S would be the increased supression of epidemics caused by resistant isolates. This is not possible for fungicides with essentially insensitive subpopulations (e.g., benzimidazoles). With the DMIs, resistant subpopulations remain sensitive to inhibition at higher doses, so increased suppression appears feasible. Increased control of resistant isolates at higher doses would, in addition to decreasing r_R - r_S, decrease the frequency of resistant isolates, which also should delay resistance evolution (TACTIC 4).

Despite the existance of ground rules for anti-resistance strategies, their relative merits have rarely been assessed. For control of apple scab with DMIs, however, there is evidence that high disease pressure is perhaps the most important driving force of resistance selection. As outlined above, 12 years of continuous DMI testing at an experimental orchard site in Nova Scotia had shifted the *V. inaequalis* population to a threshold level of resistance (Table 1). In contrast, the sensitivity distribution had remained at baseline levels during 12 years of very similar DMI testing at an experimental orchard site in New York State (Table 2). Reconstruction of respective DMI use histories from reports published in "Pesticide Research Reports, Canada" and "Fungicide and Nematicide Tests" revealed that DMI use patterns over the 12 year period were almost identical at both sites (Braun and Köller, unpublished). Therefore, different DMI exposure histories could not account for the differences in orchard sensitivities.

The most pronounced difference between New York and Nova Scotia is a substantially higher disease pressure in Nova Scotia, as indicated by a a higher number and severity of infection cycles and a resulting higher incidence of disease on nontreated apple trees. Although higher scab severity in Nova Scotia mandates stricter control measures, the typical number of sprays per season over the 12 years period was only increased by one to two additional sprays. This increased selection time might have contributed to an increased speed of resistance selection, but the differences in disease severity appears more likely.

Table 2. Sensitivities of *V. inaequalis* at orchards with similar DMI histories (tested with fenarimol at 0.05 mg/L).

Status [1]	RG	Frequency (%) Baseline	NY[2]	NS[3]
S	0-60	91	94	39
RS	61-80	9	6	22
R	81-100	1	0	39

[1] S, sensitive; RS, reduced sensitive; R, resistant.
[2] Agricultural Experiment Station, Geneva, NY.
[3] Research Station, Agriculture Canada, Kentville, Nova Scotia.

The comparison of the two experimental orchards indicates the importance of disease severity as a serious driving force of resistance. Control of scab by means other than a DMI will substantially delay resistance development (TACTIC 1). For example, good disease control into the late season, which determines the inoculum level of ascospores formed in leaf litter and released during spring as primary apple scab inoculum, is an important measure of decreasing disease severity (MacHardy et al., 1993) and thus can be expected to delay DMI resistance.

Good control of scab into the late season, which allows a later start of scab control programs in the following spring (MacHardy et al., 1993), would also decrease the exposure and selection time of scab populations (TACTIC 2). This tactic was incorporated into a DMI use strategy for the control of scab in apple growing regions of the northeastern US (Wilcox et al., 1992). Taking full advantage of both the protective and after-infection activities of the DMIs, an integrated, reduced spray program was successfully developed. According to this program, the timing of sprays is independent of the occurrence of apple scab infections but coincides with applications of insecticides or acaricides at or near the four phenological stages: tight cluster, pink bud, petal fall and first cover approximately 10 days after petal fall. The program combines reduced fungicide use with simple integrated pesticide application events.

The relative merits of restricting the selection time by limited use of a fungicide has not been tested directly for the DMIs. However, the success of use restrictions was apparent for dodine, a scab fungicide that had been used extensively in the Eastern US and Canada for 10-15 years before resistance became a prohibitive problem (Gilpatrick, 1982). The pattern of slow resistance evolution (Gilpatrick, 1982) was similar to the current experience with the DMIs. When the DMI test procedure was adapted to this compound, and the similarity between dodine and the DMIs was indeed reflected by almost identical sensitivity distributions, both for baseline and threshold populations (Fig. 3B). This similarity was not recognized before, but further suggests that the former experience with dodine can guide anti-resistance measures for the DMIs.

A systematic survey of orchard sensitivities throughout New York State showed that in Hudson Valley orchards, where dodine had been restricted to two applications per season for over 20 years, scab populations had not yet shifted (Köller, unpublished). Because dodine and the DMIs are very similar with respect to the time frame of resistance evolution, in sensitivity distributions and the characteristics of shift patterns (Fig. 3B), the observation that limited dodine use greatly delayed resistance development is of relevance to the DMIs. Decreasing selection time through prudent use of DMIs can, therefore, be rated as a promising anti-resistance strategy (TACTIC 2).

Tactics applicable to the DMIs with their resistant but not insensitive subpopulations relate to a slower progression of disease caused by resistant isolates through use of high DMI rates (TACTICS 3). In contrast to benzimidazole-resistant subpopulations clearly separated from the sensitive population (Fig. 1), the DMI resistant subpopulation is defined as the least sensitive part of a continuous sensitivity distribution (Fig 2). For *V. inaequalis*, relative growth values of 80-100 were identified as resistant because respective isolates had increased in frequencies at a site of unsatisfactory disease control. These relative growth values translate into ED_{50} values of 0.3 to 4 mg/L for fenarimol, 0.5 to 6 mg/L for myclobutanil and 0.08 to 1 mg/L for flusilazole (Braun and Köller, unpublished).

Considering a 11-fold difference in sensitivities within the resistant population, it appears that a proportion of these isolates are not sufficiently controlled at low doses. They are allowed to sporulate between consecutive DMI applications and will therefore increase in frequency. Higher DMI doses would prevent these isolates from sporulation and, consequently, from selection. The character of these isolates under high dose conditions would thus change from "resistant" to "sensitive", and the size of the resistant population would become smaller (TACTIC 3). In addition, the frequency of resistant isolates would be lowered (TACTIC 4). High doses have not been considered before as anti-resistance strategy for the DMIs. Circumstantial evidence exists, however, that low rates of dodine with its similar sensitivity distribution (Fig. 3B) had increased the speed of resistance selection considerably (Gilpatrick, 1982).

High DMI rates as anti-resistance measure was tested in an orchard experiment. In addition, the previously proposed strategy of mixing a DMI with a protective fungicides such as mancozeb (Scheinpflug, 1988) was evaluated. Protective fungicides with their non-specific "biocidal" modes of action have never been affected by resistance (Buchenauer, 1990; Köller, 1991). They are confined to the plant surface and do not penetrate through the cuticle. At the surface, they inhibit spore germination and, therefore, prevent infection. The systemic DMIs are almost inactive as inhibitors of spore germination and stop infection subsequent to cuticle penetration (Scheinpflug, 1986). Consequently, the sites of action of the two partners in a DMI-protectant mixture are separated in space. The protectant will lower the number of infections eventually controlled by the DMI and, therefore, will lower the disease pressure exerted on the DMI. Consequently, the strategy lowers disease pressure (TACTIC 1).

The preliminary results of this orchard trial confirmed the predictions. After three consecutive DMI applications according to the delayed spray program developed by Wilcox et al. (1992), sensitivity distributions for isolates collected from trees treated with a high rate of fenarimol were not significantly different from non-treated control tree populations. In contrast, a significant shift toward a higher frequency of resistance was observed for isolates collected from trees treated with half of the fenarimol rate (Fig. 4). Shifts observed with a half-rate mixture of fenarimol with mancozeb were detectable but intermediate compared to those for full and half rates of

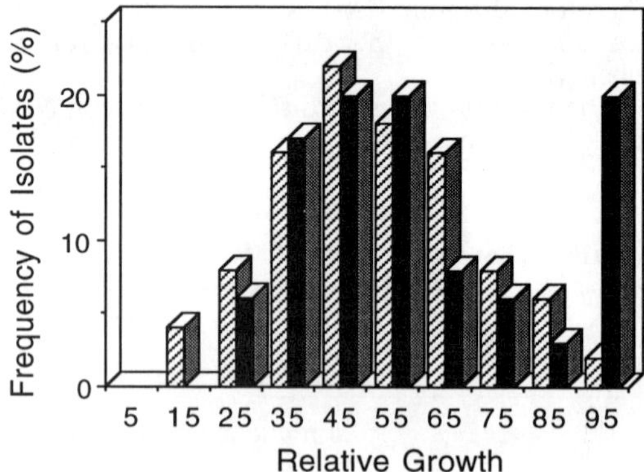

Fig. 4. Comparison of frequency distributions of *V. inaequalis* isolates collected from trees treated with a high dose of fenarimol with distributions determined for isolates collected from trees treated with half of that rate. DMI sensitivities were tested with fenarimol at a discriminatory dose of 0.05 mg/L. Lower part: Open bars represent high-rate and closed bars represent half rate treatments.

fenarimol. Although preliminary in character, the results indicate the following anti-resistance quality:

Fenarimol	> fenarimol + mancozeb	> fenarimol
(full rate)	(half rates)	(half rate)

Combination of both tactics, a high DMI rate in mixture with a reduced rate of mancozeb, may be expected to delay the development of resistance effectively.

Based on the results, relatively simple anti-resistance rules can be offered to apple growers in the northeastern US. These rules may also easily gain compliance, because they offer advantages to growers:

1. Use DMIs according to the reduced spray program developed by Wilcox et al. (1992).
   Anti-resistance tactics: Decrease of selection time (TACTIC 1).
   Compliance incentive: Tank-mix application with insecticides/acaricides.
2. Use DMIs at high rates in mixture with a protectant.
   Anti-resistance tactics: Decrease of disease pressure (TACTIC 1), size reduction of the resistant population (TACTIC 3), and favorable ratio of sensitive to resistant isolates (TACTIC 4).
   Compliance incentive: Under the conditions of the reduced spray program, the protective activity of DMIs is sometimes insufficient to warrant full control of the economically important fruit scab (Wilcox et al., 1992). This activity is improved with mixtures. Higher DMI rates are accepted, because the number of applications is reduced.
3. Start the scab program with an alternative after-infection compound such as dodine in orchards without prohibitive levels of dodine resistance.
   Anti-resistance tactics: Decrease of selection time (TACTIC 2).
   Compliance incentive: At the start of the season, simultaneous powdery mildew control provided by the DMIs is not required.
4. Continue a careful control program during the later parts of the season with fungicides other than DMIs, such as protectants or benzimidazoles in mixture with protectants.
   Anti-resistance tactics: Decrease of ascospore inoculum dose and thus disease pressure at the start of the following season (TACTIC 1) and decrease of selection time (TACTIC 2).
   Compliance incentive: DMIs have no activity against "summer diseases", which have to be controlled during the later part of the season.

These rules combine merits of experimentally tested anti-resistance tactics with good chances of grower compliance. The example also demonstrates that evaluative research has to consider adjustments of monitoring techniques to a particular disease and fungicide, incorporate use strategies according to the demands of a particular region and grower interests, and evaluate the relative merits of general anti-resistance strategies.

DMI characteristics and resistance.

Considering the large number of DMI products, it is not surprising that individual inhibitors vary greatly with regard to intrinsic activities but also levels of resistance. Although DMIs generally are cross-resistant to each other, it had been noted that resistance factors are widely different (Köller, 1988; Köller and Wubben, 1989). Both intrinsic activities and resistance factors can be expected to influence prevention and management of DMI resistance.

As discussed above, the prohibitive level of resistance of *V. inaequalis* populations was determined at an orchard site, where the DMI bitertanol had failed to control apple scab. It had been reported that bitertanol provided control levels superior to a control program with protective fungicides until 1983. From 1984 on and for four consecutive seasons, control levels were consistently inferior, culminating in a control failure in 1987 (Hildebrand et al., 1988). The analysis of data published in "Pesticide Research Reports, Canada" for the performance of other DMIs tested at the same site revealed that bitertanol had been used at a low rate, representing 70 % of the typical recommended rate. While bitertanol performances declined, control of scab with other DMIs applied at low rates also declined. However, at rates reflecting the high range of recommended field rates, scab control remained satisfactory, in spite of resistance (Table 3).

The example demonstrates that control levels achieved with DMIs remain dose-dependent, even under conditions of shifted populations. In view of the evidence that low DMI rates are increasing the speed of resistance selection, the subject relating to intrinsic activities, resistance factors and recommended field rates deserve increased attention. High rates of DMIs as an anti-resistance tactic are not clearly defined and are related to the intrinsic activity of a particular DMI in comparison to the application rate recommended by the manufacturer.

For example, sensitivities determined for the most resistant isolates of *V. inaequalis* and based on ED_{50} values were 1 mg/L for flusilazole, 3.5 mg/L for fenarimol and 6 mg/L for myclobutanil (Braun and Köller, unpublished). Typical recommended rates translate to dilute spray

Table 3. DMI performance at an experimental orchard in Nova Scotia.

Year	Control of fruit scab (%) PROT[1]	DMI	Compound[2]	Rate[3]
1984	89	41	BIT	250
		86	PYR	220
		83	MYC	250
1985	77	18	BIT	250
		67	MYC	250
		90	FLU	50
1986	90	37	BIT	250
		35	FLU	15
		88	FLU	35
1987	99	1	BIT	250
		45	FLU	20
		65	FLU	60

[1] Mancozeb or captan program.
[2] BIT, Bitertanol; FLU, Flusilazole; MYC, Myclobutanil; PYR, Pyrifenox.
[3] g (a.i.)/ha

concentrations of 20 mg/L for flusilazole, 30 mg/L for fenarimol and 65 mg/L for myclobutanil. The ratio between these two parameters, 20 for flusilazole, 8.5 for fenarimol and 11 for myclobutanil, implies that a typical fenarimol rate reflects only half of a typical flusilazole rate. Flusilazole should therefore be more effective in delaying the evolution of resistance.

Although other parameters such as systemic movement and persistence contribute to field performance, our experience from greenhouse and orchard experiments indicate that a correlation between intrinsic activities and levels of disease control exists for a number of DMIs evaluated (Köller and Wilcox, unpublished). For example, levels of scab control achieved with fenarimol and tebuconazole at 26 mg/L and 33 mg/L, respectively, were equivalent in an orchard trial (Table 4). The rate difference reflects exactly the difference of intrinsic activities, 0.04 mg/L for fenarimol and 0.05 mg/L for tebuconazole as mean baseline sensitivities (Köller et al., 1991). The definition of "high" rates recommended for the various DMIs deserves increased attention in view of accelerated selection of resistance at low rates.

A second consideration with potential impact on resistance management is related to different resistance factors. Disease control might be curtailed for a DMI with large factors of

Table 4. Dose dependent DMI performance in the control of apple scab.

Compound	Rate[1]	Control of scab (%) Leaf	Fruit
Fenarimol	13	18	65
Fenarimol	26	41	81
Tebuconazole	33	53	83
Tebuconazole	66	79	94

[1] mg/L in dilute spray application.

resistance, whereas control levels remain adequate with a second DMI characterized by small resistance factors. Such a case has been reported for the control of barley leaf blotch. Whereas disease control with triadimenol was severly affected, control achieved with tebuconazole remained adequate (Kendall et al., 1993). Resistant factors determined with the leaf blotch pathogen *Rhynchosporium secalis* were 64 for triadimenol, but only 9 for tebuconazole (Hollomon, 1993). As shown for the control of apple scab (Table 3) and leaf blotch (Kendall et al., 1993), certain DMIs with high intrinsic activities applied at high rates and DMIs with small resistance factors may still be used after the performance of other DMIs has already declined due to resistance.

Conclusions and outlook.

The past experience with DMI resistance development has demonstrated that two types of population responses to the selective force of a fungicide can occur. For fungicides such as the benzimidazoles with their almost insensitive and clearly separated subpopulation of resistant genotypes, the risk of resistance appears higher than for DMI fungicides with resistant subpopulations as part of a broad and continuous sensitivity distribution. Unfortunately, neither the number nor the nature of molecular mechanisms responsible for these broad sensitivity distributions has been clarified (Köller, 1992; Hollomon, 1993). However, for several benzimidazole-resistant pathogens it has been shown that resistance is based on a point mutation of the target site. Such a target site mutation in field isolates of pathogens has yet not been identified for the DMIs (Köller, 1992). If the lack of a target mutation in combination with a broad sensitivity distribution in wild-type populations is concomitant with a lowered risk of resistance, these characteristics could serve as guidance for the assessment of resistance risks of future specific fungicides.

References

Braun, P. G.; McRae, K. B. *Can. J. Plant Pathol.* **1992,** *14*, 215-220.

Brent, K. J.; Hollomon, D. W. In *Sterol Biosynthesis Inhibitors*; Berg, D.; Plempel, M., Ed.; Ellis Horwood: Chichester, 1988; pp 332-346.

Buchenauer, H. In *Chemistry of Plant Protection*, Haug, G.; Hoffmann, H., Eds.; Springer-Verlag: Berlin, Vol. 6, 1990; pp 217-292.

Gilpatrick, J. D. In *Fungicide Resistance in Crop Protection*; Decker, J.; Georgopoulos, S. G., Eds.; Pudoc: Wageningen, 1982; pp 195-206.

Hildebrand, P. D.; Lockhardt, C. L.; Newberry, R. J.; Ross, R. G. *Can. J. Plant Pathol.* **1989,** *10*, 311-314.

Hippe, S.; Köller, W. *Pestic. Biochem. Physiol.* **1986,** *26*, 209-219.

Hollomon, D. W. *Biochem. Soc. Trans.* **1993,** *21*, 1047-1051.

Kendall, S. J.; Hollomon, D. W.; Cooke, L. R.; Jones, D. R. *Crop Prot.* **1993,** *12*, 357-362.

Koenraadt, H.; Shauna, C., Somerville, S.C.; Jones, A.L. *Phytopathology* **1992,** *82*, 1348-1354.

Köller, W. In *Fungicide Resistance in North America*. Delp, C. J.,Ed.; APS-Press: St. Paul, 1988; pp 79-88.

Köller, W. In *CRC Handbook of Pest Management in Agriculture*; Pimentel, D., Ed.; CRC Press: Boca Raton, 1991; 2nd Ed., Vol. 2; pp 255-310.

Köller, W. In *Target Sites of Fungicide Action*; Köller, W., Ed.; CRC Press: Boca Raton, 1992; pp 255-310.

Köller, W.; Scheinpflug, H. *Plant Dis.* **1987,** *71,* 1066-1074.

Köller, W.; Wubben, J. P. 1989. *Pestic. Sci.* **1989,** *26*, 133-145.

Köller, W.; Parker, D. M.; Reynolds, K. L. *Plant Dis.* **1991,** *75*, 726-728.

Leung, H.; Nelson, R. J.; Leach, J. E. *Adv. Plant Pathol.* **1993,** *10*, 157-205.

MacHardy, W. E.; Gadoury, D. M.; Rosenberger, D. A.; *Plant Dis.* **1993,** *77*, 372-375.

Milgroom, M. G.; Fry, W. E. *Phytopathology* **1988,** *78*, 565-570.

Scheinpflug, H. In *Fungicde Chemistry - Advances and Practical Applications*; Green, M. B.; Spilker, D. A, Eds.; ACS: Washington, DC, 1986; pp 73-88.

Scheinpflug, H. In *Fungicide Resistance in North America*. Delp, C. J., Ed.; APS-Press: St. Paul, 1988, pp 93-95.

Scheinpflug, H.; Kuck, K.-H. In *Modern Selective Fungicides - Properties, Applications, Mechanism of Action*; Lyr, H., Ed.; Pitman Publishing: London, 1987; pp 173-204.

Sisler, H. D.; Ragsdale, N. N. In *Mode of Action of Antifungal Agents*; Trinci, A. P. J.; Ryley, J. F., Eds.; Cambridge University Press: Cambridge, 1984; pp 257-282.

Skylakakis, G.; Hollomon, D. W. In *Combating Resistance to Xenobiotics*; Ford, M. G.; Hollomon, D. W.; Khambay, B. P. S.; Sawicki, R. M., Eds.; Ellis Horwood: Chichester, 1987; pp 94-103.

Smith, F. D.; Parker, D. M.; Köller, W. *Phytopathology* **1991,** *81*, 392-398.

Staub, T. *Annu. Rev. Phytopathol.* **1991,** *29*, 421-442.

Wilcox, W. F.; Wasson, D. I.; Kovach, J. *Plant Dis.* **1992,** *76*, 669-677.

Wolfe, M. S. In *Fungicide Resistance in Crop Protection*; Decker, J.; Georgopoulos, S. G., Eds.; Pudoc: Wageningen, 1982; pp 139-148.

Yoder, K. S.; Davis, A. E.; Hadley, B. A. In *Methods for Evaluating Pesticides for Control of Plant Pathogen* ; Hickey, K. D., Ed.; APS Press: St. Paul, 1986; pp 73-75.

Status and Future of Synergists in Resistance Management

Gerald L. Lamoureux and Donald G. Rusness, Biosciences Research Laboratory, USDA, Agricultural Research Service, State University Station, Fargo, North Dakota 58105-5674

Synergists have been used to study and control resistant arthropods for ca. 45 years and more recently have been considered for use against resistant weed and plant pathogen populations. In spite of some success in the control of resistant arthropods with synergists, they are not widely used in the practical control of resistant agricultural pests and it is uncertain whether they will become an important factor in the near future. Difficulties associated with the use of synergists include environmental concerns, economic factors, the ability of resistant organisms to evolve additional resistance mechanisms, and the occurence of target site resistance which is difficult to synergize. Most pesticide synergists of current interest are inhibitors of pesticide detoxification enzymes (cytochrome P450s, glutathione transferases, or various hydrolytic enzymes). A number of these synergists are actually pesticides and pesticides may represent one of the best sources of synergists. Biological pesticides will become more important in the future and there has been some success in the synergism of biological pesticides. Currently, synergists are commonly used to study and monitor resistance and resistance mechanisms.

Pesticide Resistance

Pesticide resistance, which evolves from the repeated exposure of insect, weed, or plant pathogens to pesticides, began to emerge as a serious world-wide problem about 50 years ago. Today there are approximately 540 arthropod, 100 weed, and 100 plant pathogen species or biotypes that are resistant to pesticides (Metcalf, 1989; Green *et al.*, 1990; Roush and Tabashnik, 1990; Caseley *et al.*, 1991; Georghiou, 1990; LeBaron and McFarland, 1990; Powles and Holtum, 1994). Most serious cases of pesticide resistance are due to an increased rate of pesticide detoxification or changes in the pesticide target site (National Research Council, 1986). Pesticide resistance within a population is frequently expressed as a resistance ratio relative to an unexposed population (RR = LD_{50} resistant strain/LD_{50} normal strain) (Scott, 1990). In arthropods, resistance ratios of 10 to 2,000 are relatively common (Elzen *et al.*, 1992; Motoyama and Dauterman, 1992). Cross resistance, resistance to more than one class of pesticide, and multiple resistance, resistance by more than one mechanism, are also observed (National Research Council, 1986). Cross resistance and multiple resistance in insects such as the diamondback moth (*Plutella xylostella*) and the Colorado potato beetle (*Leptinotarsa decemlineata*) and in weeds such as annual rye grass (*Lolium rigidum*) and blackgrass (*Alopecurus myosuroides*) represent extremely difficult control problems (Christopher *et al.*, 1992; Powles and Matthews, 1991; Moss and Cussans, 1991; Powles and Holtum, 1994).

Synergism

Pesticide synergism is only one of several techniques that can be used to control or study pesticide resistance (Kemp and Caseley, 1991). In this discussion, synergism will be considered as the co-operative action of several components of a mixture such that the total effect is greater than the sum of the effects of the components used independently (Tammes, 1964).

Synergists have been used to study and combat pesticide resistance for ca. 45 years. Reviews on the synergism of insecticides include those by Casida, 1970; Wilkinson, 1976; Raffa and Priester, 1985 and B-Bernard

1054–7487/95/0350$12.00/0

and Philogene, 1993. Reviews that deal with herbicide synergism include Gressel, 1990; Kemp and Caseley, 1991; Gressel, 1993 and Gressel *et al.*, 1993. The most current review on fungicide synergism is by De Waard and Gisi, in press. Other discussions or reviews of fungicide synergism include DeWaard, 1985; Kataria and Gisi, 1990 and Kataria *et al.*, 1990.

The most commonly used synergists are inhibitors of pesticide detoxification reactions. Other synergists may alter the uptake or distribution of the pesticide (Smith and Vandenborn, 1992), act at or near the target site in a manner not completely understood (Liu and Plapp, Jr., 1992), or act at a different target site and produce an effect greater than the additive effect of the pesticides used singly (Hagimoto and Yoshikawa, 1972). The quantitative activity of synergists may be expressed as a synergistic ratio (SR = LD_{50} of the pesticide/LD_{50} of the pesticide + synergist) (Raffa and Priester, 1985).

Synergism of Target Site Resistance

Resistance due to target site alteration is common in arthropods, weeds, and plant pathogens and is one of the most difficult forms of resistance to overcome by synergism (Kemp and Caseley, 1991; Hammock and Soderlund, 1986). It is frequently due to reduced binding at the target site (Trebst, 1991; Davidse, 1990; Metcalf, 1989) and it is commonly associated with cross resistance to pesticides that function at the same target site.

Arthropods. Target site resistance in arthropods commonly involves changes in the sodium channel that regulates nerve membrane permeability to sodium, changes in acetylcholinesterase (AChE) involved in signal transmission (Metcalf, 1989), or changes in the γ-aminobutryic acid (GABA) gated chloride ion channel (Lee *et al.*, 1993). Changes in the sodium channel (Kdr resistance) is the basis for target site resistance to DDT, related chlorinated compounds, and the pyrethroids; altered AChE is the basis of target site resistance to organophosphorus (OP) and carbamate insecticides; and changes in the GABA gated chloride ion channel confers resistance to the cyclodiene insecticides. The synergism of pyrethroids by the formamidines (chlordimeform, BTS 27271, and amitraz) in Kdr-like target site resistant arthropods has been reported (Bagwell and Plapp, Jr., 1992). Based on binding studies, it was proposed that the formamidines function as synergists by acting at or near the target site (Liu and Plapp, Jr., 1992); however, a ligand that binds at a slightly different site than the pyrethroids was used in the binding studies (Liu and Plapp, Jr., 1991) and the results are equivocal. The formamidines are insecticidal and also synergize organophosphorus insecticides that function at a different target site; therefore, the possibility that the formamidines function by an alternate synergistic mechanism(s) should be investigated. A synergistic pyrethroid/chlordimeform combination was used commercially until chlordimeform was withdrawn from the market; however, other formamidines such as amitraz could be considered for use as pyrethroid synergists (Baggwell and Plapp, 1992). Two advantages of this synergism were observed. Synergism by the formamidines was greater in the target species than in a beneficial predator (Rajakulendran and Plapp, Jr., 1982) and this synergistic combination delayed the evolution of resistance in laboratory selection experiments (Crowder *et al.*, 1984; Plapp, Jr., 1986). It is not clear whether these synergistic combinations would give similar results in the field.

Weeds and Fungi. Target site resistance to herbicides and fungicides is very common and it has been observed with triazine, phenylurea, sulfonylurea, imidazolinone, and dinitroaniline herbicides and with benzimidazole and *N*-phenylcarbamate fungicides (Green, LeBaron, and Moberg, 1990; Caseley, Cussans, and Atkin, 1991; Powles and Holtum, 1994). Atrazine resistance is one of the most carefully studied forms of target site resistance. Amino acid changes at or near the active center result in a loss of binding with atrazine, yet the target site still binds plastoquinone Q_B and carries on its electron transport function at a near normal rate (Trebst, 1991). An excellent example of fungicide target site resistance is the resistance to the benzimidazole and the *N*-phenylcarbamate fungicides associated with decreased binding to β-tubulin due to changes in the amino acid sequence of β-tubulin (in Ziogas and Girgis, 1993). Target site resistance has not been commonly synergized in resistant weeds and plant pathogen species. However, when negative cross resistance is associated with target site resistance, this phenomena can potentially be used in resistance management.

Negative Cross Resistance

Negative cross resistance is resistance to one pesticide that is accompanied by increased susceptibility to a second pesticide. When negative cross resistance exists, the resistant organism can be controlled with the second pesticide and in some cases it has been be possible to delay the evolution of resistance by using these pesticides in combination (Gressel and Segel, 1990; Yamamoto *et al.*, 1993).

An excellent example of negative cross resistance is with a resistant green rice leafhopper (*Nephotettix cincticeps*) which is resistant to the *N*-methylcarbamates due to an altered AChE. This leaf hopper displays negative cross resistance to *N*-propylcarbamates and oxadiazolones (Yamamoto *et al.*, 1993). There appear to be two binding sites on the AChE from the resistant strain, one that binds *N*-methylcarbamates and oxadiazolones and a second that is highly sensitive to oxadiazolones and insensitive to *N*-methylcarbamates. The combination of an oxadiazolone and an *N*-methylcarbamate was synergistic in the green leafhopper and no further resistance developed when this combination was used on the resistant population in the laboratory. Since many pyrethroids are detoxified by cytochrome P-450 based mechanisms and many OP insecticides are activated by cytochrome P-450, negative cross resistance between pyrethroid and OP insecticides might be expected; however, this form of negative cross resistance is uncommon in arthropods (Forrester *et al.*, 1993; Hatano *et al.*, 1992).

Cross resistance and negative cross resistance are sometimes observed in herbicide-resistant weeds (in Gressel and Segel, 1990). Target site resistance to atrazine can be associated with cross resistance to other herbicides with a similar mode of action. Negative cross resistance between terbutryn and DCMU and between atrazine and bentazon or pyridate have been reported (Sinning *et al.*, 1989; Trebst, 1991; De Prado *et al.*, 1992). In atrazine resistant purple amaranth (*Amaranthus cruentus)*, the resistance ratio for atrazine was 500 while the resistance ratios for bentazon and pyridate were 0.4 and 0.5, respectively. Therefore, bentazon or pyridate could be used to aid in the control of atrazine resistant *A. cruentus*. In other cases of negative cross resistance involving triazine herbicides, sensitivity to the second herbicide was increased by 5- to 25-fold (in Gressel and Segel, 1990). In one chlorsulfuron-resistant mutant of *Datura innoxia*, a 33-fold increase in sensitivity to imazaquin was observed. Negative cross resistance has been used on a commercial scale to control benzimidazole resistant plant pathogens that display negative cross resistance to the *N*-phenylcarbamate fungicides, but fungi have evolved resistance to the *N*-phenylcarbamates as well (Katan *et al.*, 1989; Ziogas and Girgis, 1993).

Metabolic Resistance

Metabolic resistance, resistance due to an increased rate of pesticide detoxification, is common in insecticide resistance and has also been observed in cases of herbicide and fungicide resistance. Most pesticides are detoxified by a restricted group of enzymes that catalyze various oxidations, hydrolyses, and conjugation reactions (Gressel, 1993; B-Bernard and Philogene, 1993).

The control of metabolically resistant pests by synergists is conceptually simple. If resistance is due to a single detoxification reaction the toxicity of the pesticide can frequently be restored with inhibitors of the detoxification enzyme. Unfortunately, during the last 45 years this approach to managing resistant arthropods on a commercial scale has not been very successful (Raffa and Priester, 1985; Metcalf, 1989). This lack of success has been due to a variety of reasons including the ability of insects to evolve additional resistance mechanisms, unavailability of formulations that accommodate both the synergist and the pesticide, cost of the synergist, etc. (Raffa and Priester, 1985; Metcalf, 1989). However, with the increased pressure to reduce pesticide loads in the environment, this approach should still be further explored.

Glutathione (GSH) Conjugation

GSH conjugation is one of the most commonly utilized reactions in the detoxification of electrophilic pesticides in plants, insects and microorganisms (Lamoureux and Bakke, 1984; Lamoureux and Rusness, 1989a). These reactions are usually catalyzed by glutathione transferase enzymes (GSTs) frequently present as a multigene family of isozymes with a number of functions including the detoxification of xenobiotics and the metabolism of natural constituents (Lamoureux and Rusness, 1993). Although the substrate specificities of GSTs frequently overlap, they can be very substrate specific and their response to inhibitors can vary significantly (Yalcin *et al.*, 1983; Lamoureux *et al.*, 1991; Ploemen *et al.*, 1993). Effective inhibitors of GST include tridiphane, K-1 (2-phenyl-4*H*-1,3,2-benzodioxaphosphorin 2-oxide), K-2 (2-phenoxy-4*H*-1,3,2-benzodioxaphosphorin 2-oxide), triphenyl tin chloride, WARF (*N*,*N*-di-*n*-butyl-p-chlorobenzenesulfonamide), bromocresol green and various glutathione conjugates (Metcalf, 1989; Lamoureux and Rusness, 1989a; Shiotsuki *et al.*, 1992).

Arthropods. The evolution of insecticide resistance in arthropods is frequently associated with changes in the expression of GST isozymes (Clark *et al.*, 1984; Lamoureux and Rusness, 1989a; Cochrane *et al.*, 1992; Toung *et al.*, 1993). Organophosphorus compounds, halogenated hydrocarbons, and thiocyanates are among the insecticide classes detoxified by GSH conjugation in arthropods. However, the importance of GSH conjugation depends on both the insecticide and the insect species (Lamoureux and Rusness, 1989a), and many insects seem to have concurrently evolved other mechanisms so that the totality of resistance can include other metabolic mechanisms.

One of the earliest examples of the widespread evolution of resistance to a synthetic pesticide was to DDT in arthropods, as observed between 1946 and 1951 (in Metcalf, 1989). Resistance was due to an increased rate of metabolism of DDT to DDE, a reaction catalyzed by DDT-dehydrochlorinase (DDTase). Clark and Shamaan (1984) presented evidence that DDTase is GST. More recently, each of the GSTs from the house fly were purified by affinity chromatography and chromatofocusing and were found to contain DDTase activity (Dauterman, personal communication, 1994). A series of DDT analogues such as WARF, which produced synergistic ratios of 60 to 140 in resistant house fly, were developed to combat this resistance (in Metcalf, 1989). These synergists were temporarily effective in field trials, but Kdr target site resistance evolved quickly and these synergists became ineffective. WARF and related inhibitors compete with DDT for the active site on DDTase (Balabaskaran *et al.*, 1968). DDTase inhibitors competitive with GSH were also investigated, but they were not absorbed by the insects and were ineffective as synergists (Balabaskaran *et al.*, 1968; Blabaskaran and Smith, 1970).

Diazinon toxicity in a resistant strain of house fly was synergized by tridiphane, an inhibitor of GST (Lamoureux and Rusness, 1987). Diazinon is detoxified by GSH conjugation in the housefly (Hayaoka and Dauterman, 1983) and both tridiphane and a GSH conjugate of tridiphane are inhibitors of this reaction. The GSH conjugate of tridiphane was a more effective inhibitor than tridiphane and this conjugate was produced very rapidly *in vivo* where its concentration was sufficient to inhibit diazinon metabolism and synergize toxicity (Lamoureux and Rusness, 1987).

Fenitrothion insecticide is synergized in a resistant strain of house fly by IBP fungicide (Shiotuski and Eto, 1991) and related saligenin cyclic phosphates (SCPs) (Shiotuski *et al.*, 1992). Synergism was due to the inhibition of the GST catalyzed detoxification of fenitrothion. Two mechanisms of inhibition were observed. In one case a SCP (K-2) was metabolized to a GSH conjugate that inhibited fenitrothion:GST activity and in the second case K-2 reacted directly with the enzyme causing irreversible inhibition of GST activity. IBP and related compounds have also been reported to be esterase inhibitors responsible for the hydrolysis of OP insecticides (B-Bernard and Philogene,

1993). Therefore, caution should be exercised when these synergists are used to determine the mechanism of resistance.

Tetrachlorvinphos is detoxified in the house fly by GSH conjugation and it is synergized by diethyl maleate (Welling and De Vries, 1985). It is not clear whether synergism is due to simple inhibition of GST, depletion of GSH, or a combination of factors since GSH conjugation was inhibited only 50%, but a synergism ratio of 116 was obtained (Hollingworth, 1970; Mirer *et al.*, 1977; Welling and De Vries, 1985). Parathion toxicity in *Triatoma infestans* is synergized by *N*-ethylmaleimide (SR = 2.7) and synergism was attributed to inhibition of GST activity and increased parathion penetration (Wood *et al.*, 1986). Since *N*-ethylmaleimide reacts nonenzymatically with GSH and other sulfhydryl compounds, both the inhibition of GST activity and increased parathion penetration may have been due to reduced GSH levels.

Plants. Herbicide selectivity is frequently due to GST isozymes that catalyze the detoxification of herbicides. This includes most members of the triazine, chloroacetamide, thiocarbamate, and the diphenyl ether classes of herbicides as well as individual members of other classes of herbicides (Lamoureux and Rusness, 1989a). Although GSH conjugation is a common mechanism in herbicide selectivity between weeds and crops, it has not been documented as a common mechanism of herbicide resistance in weeds.

Atrazine was one of the first herbicides for which resistance became a significant problem. Field populations of atrazine resistant weeds were reported in 1968. There are ca. 57 atrazine resistant species/biotypes of weeds in over 1000 locations (Holt, Powles and Holtum, 1993). Target site resistance is the most common form of atrazine resistance (Caseley, Cussans, Atkin, 1991), but exceptions to this have been observed (Gressel *et al.*, 1982). In annual ryegrass, resistance was due to a cytochrome P450 mechanism (Burnet *et al.*, 1993). In velvetleaf (*Abutilon theophrasti*), resistance was due to the over-production of two existing GST isozymes and represents one of the few documented cases of herbicide resistance due to GST (Anderson and Gronwald, 1991). This velvet leaf had a resistance ratio of 10 and was capable of metabolizing atrazine at a 2- to 9-fold greater rate than a susceptible biotype (Gronwald *et al.*, 1989). It is likely that GST will be found as the basis of additional cases of herbicide resistance.

Tridiphane is a synergist of atrazine when used to control giant foxtail (*Setaria faberii*) and other grasses requiring high levels of atrazine for control in corn (Zorner and Olsen, 1981; Zorner and Sheppard, 1985). Atrazine is metabolized by GSH conjugation in both corn and giant foxtail and the selective synergism of atrazine by tridiphane in giant foxtail appears to be due to differences in the inhibition of the GST-catalyzed GSH conjugation of atrazine in these species (Lamoureux and Rusness, 1986; Boydston and Slife, 1986; Dionigi and Dekker, 1990). When giant foxtail was treated with 0.5 kg/ha tridiphane and 0.5 kg/ha atrazine, atrazine metabolism was inhibited 50%. The tridiphane GSH conjugate was present at a concentration well in excess of the Ki_{GSH} and thus would have significantly inhibited atrazine:GST activity (Lamoureux and Rusness, unpublished). Tridiphane also inhibits the *in vitro* GSH conjugation of PCNB, fluorodifen, propachlor, and diazinon (Lamoureux and Rusness, 1989) and it inhibits EPTC and alachlor metabolism *in vivo* (Ray *et al.*, 1990). Tridiphane might be effective in overcoming atrazine resistance in velvetleaf.

The detoxification of pesticides by GSH conjugation is dependent upon GST and GSH; therefore, pesticides can be synergized by manipulating either GST or GSH. In some plant species homoglutathione (hGSH, γ-glutamylcysteinyl-β-alanine) occurs in place of GSH and in these species hGSH conjugation occurs in a manner analogous to GSH conjugation (in Lamoureux and Rusness, 1989b). Resistance to a chloroacetanilide herbicide was positively correlated with levels of GSH or hGSH and to the rate of GSH or hGSH conjugation in 16 plant species (Breaux, 1987; Breaux *et al.*, 1987). When GSH and GST levels were elevated with a herbicide safener, resistance increased (Breaux *et al.*, 1987) and when GSH in corn seedlings was depleted with buthionine-*S,R*-sulfoxamine, an inhibitor of γ-glutamylcysteine synthetase, the toxicity of a chloroacetamide herbicide (metolachlor) increased (Farago *et al.*, 1993). Large-seeded crop plants such as corn and soybean tend to have higher levels of GSH or hGSH in the very early seedling stage than small-seeded weeds so synergism by depletion of GSH or GST is a possible method of pesticide synergism. If this approach to synergism is used, the age of the tissue should be considered since GSH and GST levels vary as a function of species and stage of development (in Lamoureux and Bakke, 1984). GST isozyme patterns vary depending upon chemical treatment and it may also be possible to reduce the level of a GST required for a particular detoxification by treatment with a second chemical agent. Both atrazine and the chloroacetamide herbicides are metabolized by GSH conjugation and there have been no reported cases of atrazine resistance when these compounds have been used in combination (Gressel *et al.*, 1993).

Cytochrome P-450 Monooxygenases

The cytochrome P450 microsomal monooxygenases are the most versatile of the enzymes utilized in the metabolism and detoxification of pesticides (Burden *et al.*, 1987; Durst *et al.*, 1992; Scott, 1993; Owen and deBoer, this volume). They usually exist as multigene families of membrane bound enzymes with varying and overlapping substrate specificity capable of catalyzing the oxidation of a diverse range of compounds. They are important in the metabolism and detoxification of xenobiotics and in the metabolism and regulation of endogenous substrates. The microsomal cytochrome P450 monooxygenases usually consist of two essential elements, a cytochrome P450 substrate binding component which acts as the terminal oxidase and the NADPH-cytochrome P450 reductase component that transfers electrons from NADPH to the cytochrome P450 (Scott, 1993).

Arthropods. The synergism of carbaryl by sesamax provided early evidence of the involvement of the cytochrome P450s in insecticide resistance (Eldefrawi *et al.*, 1960; Forrester *et al.*, 1993). It is now recognized that resistance to a broad range of insecticides is due to increased insecticide detoxification catalyzed by cytochrome P450s. These enzymes can confer high or low levels of resistance and they can also confer cross resistance to other insecticides (Scott, 1993). The cytochrome P450 monooxygenases from insects have been the subject of several reviews (Scott, 1993; Hodgson, 1985; Agosin, 1985; Hodgson, 1983). Insecticides metabolized by cytochrome P450 catalyzed reactions include carbaryl (aromatic epoxidation), aldrin (alicyclic epoxidation), pyrethrin I (aliphatic epoxidation), preocene I (heterocyclic epoxidation), carbaryl (hydroxylation), nicotine (alicyclic hydroxylation), DDT (aliphatic hydroxylation), pyrethrin I (aliphatic hydroxylation), carbaryl and monocrotophos (N-dealkylation), methoxychlor and tetrachlorvinphos (O-dealkylation), parathion and diazinon (desulfuration), and parathion and diazinon (phosphoester cleavage) (Brattsten *et al.*, 1986). Unfortunately, knowledge of the individual cytochrome P450 monooxygenases from insects that catalyze these reactions is limited.

Inhibitors of cytochrome P450 are effective insecticide synergists and have been the subject of a comprehensive review (Casida, 1970). Some of the most successful inhibitors of these enzymes include piperonyl butoxide (PBO), sesamex, sulfoxide, dillapiol, sesamine, MGK-264, and N-declyimidazole (B-Bernard and Philogene, 1993). PBO is the best known insecticide synergist and it is used commercially for this purpose. It has produced synergistic ratios of 300 to 923 with pyrethroids in resistant houseflies (Scott and Georghiou, 1986). In Australia, PBO is used with synthetic pyrethroids to improve the control of resistant cotton bollworm (*Helicoverpa armigera*). Unfortunately, PBO is in limited supply, may present some toxicological problems, and has a short half-life in the field (Fujitani *et al.*, 1993; Forrester *et al.*, 1993). Insects have overcome PBO synergism by various mechanisms including the evolution of pesticide target site resistance, altered pathways of insecticide metabolism, and metabolism of PBO (in Forrester *et al.*, 1993). PBO was one of the most effective of 65 compounds recently evaluated as pyrethroid synergists against a resistant cotton bollworm. A chemically related synthetic synergist was equally effective under field conditions. Although less effective, several registered insecticides were also synergists: pyrazophos, phosalone, fenthion, and azinphos ethyl (Forrester *et al.*, 1993).

Plants. The plant cytochrome P450 monooxygenases have been implicated in the metabolism of numerous herbicides, including the *N*-dealkylation of phenylurea herbicides in cotton (Frear *et al.*, 1968); diclofop, chlorsulfuron, and trisulfuron hydroxylation and chlortoluron *N*-demethylation and hydroxylation in wheat (Mougin *et al.*, 1990; Zimmerlin and Durst, 1990; Frear *et al.*, 1991); primisulfuron hydroxylation, chlortoluron ring-methyl hydroxylation, and bentazon hydroxylation in maize (Fonne-Pfister *et al.*, 1990; Fonne-Pfister and Kruez, 1990; McFadden *et al.*, 1990); metolachlor and bentazon hydroxylation in grain sorghum (Moreland *et al.*, 1990; Moreland *et al.*, 1993a); flumetsulam ring and ring-methyl hydroxylation in barley, wheat and maize (Frear *et al.*, 1993); and 2,4-D hydroxylation in tulip (*Tulipa gesneriana*) (Topal *et al.*, 1993). The plant cytochrome P450 monooxygenases have been the subject of several reviews (West, 1980; Higashi, 1985, Durst, 1991; Donaldson and Luster 1991; Durst *et al.*, 1992) and it is apparent that they play a key role in the metabolism and selectivity of herbicides in plants.

As discussed in a recent review, a number of cytochrome P450 inhibitors (tetcyclacis, PBO, ancymidol, BAS 110, BAS 111, DPP, menadione, metryapone, paclobutrazol, propiconazole, and tridiphane) inhibit herbicide metabolism and/or synergize the toxicity of some herbicides (Gressel, 1993). If cytochrome P450 inhibitors are to be used as herbicide synergists in the field, selectivity between crops and resistant weeds must be maintained; therefore, it may be necessary to meet one of the following conditions: the crop and weed must utilize different detoxification mechanisms, the crop must have a higher titer of the requisite cytochrome P450, the cytochrome P450 in the crop must be less susceptible to the inhibitor, the inhibitor can not reach the pesticide detoxification enzyme in the crop, or the inhibitor must be metabolized more rapidly in the crop. It appears that multiple cytochrome P450 enzymes are present in plants and they may respond differently to synergists and inducers; therefore, it may be possible to selectively synergize some herbicides (Gonneau, *et al.*, 1988; McFadden *et al.*, 1989; Frear *et al.*, 1993; Moreland *et al.*, 1993b; Owen and deBoer, this volume).

The importance of the cytochrome P450 enzymes in herbicide-resistant weeds is not clearly established, but many resistant annual ryegrass biotypes in Australia have an elevated capacity to metabolize sulfonylurea herbicides such as chlorsulfuron (Powles and Matthews, 1991; Christopher *et al.*, 1992); and in England there are ca. 60 biotypes of blackgrass with an elevated capacity to metabolize herbicides such as diclofop methyl or chlortoluron which are cross resistant to many other herbicides or families of herbicides (Moss and Cussans, 1991; Linda Hall, Univ. of Alberta, personal communication). Because of the potential for cross resistance, resistance due to cytochrome P450 can be extremely serious. It is likely that additional cases of herbicide resistance due to this mechanism will be discovered. The synergism of simazine and chlorotoluron toxicity with 1-aminobenzotriazole in resistant annual ryegrass are examples of synergism by the respective inhibition of cytochrome P450 catalyzed *N*-dealkylation and hydroxylation reactions (Burnet *et al.*, 1993a; Burnet *et al.*, 1993b). Also of importance is the finding that OP insecticides such as malathion and terbufos synergize the toxicity of a broad range of sulfonylurea and other herbicides by inhibition of the cytochrome P450 enzymes (Kreuz *et al.*, 1992; Christopher *et al.*, 1994; Diehl *et al.*, 1994; Baerg *et al.*, 1994).

Hydrolytic Reactions

Arthropods. Resistance to carbamate, OP, and synthetic pyrethroid insecticides is commonly due to hydrolytic enzymes (amidases, esterases, and epoxide hydrolysases) that catalyze the hydrolysis of CN, CO, CS, PO, PN, and PS bonds. Resistance can be due to either the over-production of existing esterases or the production of mutant ali-esterases as discussed in a recent review (Abdel-Aal *et al.*, 1993). When resistance is due to mutant ali-esterases, a negative correlation may be observed between esterase activity assayed with model substrates versus esterase activity assayed with the actual insecticide; however, with over-production of existing esterases a positive correlation may be observed (Abdel-Aal *et al.*, 1993). Examples of hydrolytic resistance include resistance to the pyrethroids, permethrin and cypermethrin, the acylurea, diflubenzuron (Ishaaya, 1993) and the OP insecticides such as malathion, dimethoate, parathion (Raffa and Priester, 1985; B-Bernard and Philogene, 1993), fenitrothion (Suzuki *et al.*, 1993), and chlorpyrifos (Dong and Scott, 1992). Resistance due to these hydrolytic enzymes can frequently be overcome with various inhibitors such as DEF (*S,S,S*-tributyl phosphotrithioate), TPP, IBP, K-1, and some insecticides that function as synergists (Raffa and Preister, 1985; in Ishaaya, 1993).

Resistance to cypermethrin is frequently due to oxidative metabolism, but in the cotton leafworm (*Trichoplusia ni*) and the cabbage looper (*Spodoptera littoralis*) it is due to hydrolytic metabolism that can be blocked by profenofos. As a result of inhibiting hydrolytic reactions, profenofos caused a 4-fold synergism of trans-permethrin and a 20-fold synergism of cis-permethrin (Ishaaya, 1993). In the whitefly (*Bemisia tabaci*), cypermethrin was synergized 50-fold (3 days post-treatment) by a 1:1 formulation with monocrotophos and cypermethrin and it was also syngerized in 1:8 formulation with methidathion (Ishaaya, 1993). The benzoylphenylureas, diflubenzuron and teflubenzuron, were synergized by subtoxic levels of the esterase inhibitors DEF and/or profenofos in cotton leafworm, beet armyworm (*Spodotera exigua)*, and flour bettle (*Tribolium castaneum*) (Ishaaya, 1993; Van Laecke and Degheele, 1993). Synergism was correlated with inhibition of hydrolysis (Ishaaya and Degheele, 1988; El Saidy *et al.*, 1989). Diflubenzuron was also synergized by dimethoate and diethylmaleate in the beet armyworm, but the mechanism of synergism is uncertain (Van Laecke and Degheele, 1993). The organophosphate insecticide azamethiphos was synergized 837-fold by DEF in a strain of house fly with a resistance factor of 1,967. Although more than one resistance mechanism was present, this appears to be an excellent example of synergism of an organophosphate insecticide by inhibition of the hydrolytic enzymes (Saito *et al.*, 1992). Chlorpyrifos was synergized up to 29-fold by DEF in a German cockroach (*Blattella germanica*). As with the house fly above, several resistance mechanisms may have been present in this German cockracoh since several P450 and GST inhibitors were also synergistic. DEF is a defoliant growth regulator; therefore, it is not suited for some agricultural applications.

Plants. Hydrolytic reactions have not been demonstrated to play a major role in herbicide resistance in weeds, but they are important in the activation, detoxification, and selectivity of some herbicides (Owen, 1989). Herbicides enzymatically metabolized by hydrolytic reactions include various phenoxy esters, phenoxy-phenoxy esters, benzoic acid esters (all activation reactions), some sulfonylureas (detoxification reaction), and some acylamides (detoxification reactions). The inhibition of propanil hydrolysis by carbamate and OP insecticides in rice is a well documented example of the synergism of a herbicide by inhibition of a hydrolytic detoxification reaction (in Lamoureux and Frear, 1979). Because of this interaction, rice treated with a combination of propanil and a carbamate or OP insecticide is readily injured. There are numerous examples of the interaction of insecticides with herbicides in plants and many of these synergist interactions are due to the inhibition of herbicide metabolism (Green and Bailey, 1989). Unfortunately, most of these interactions result in crop damage.

Sequestering

Arthropods. Resistance to OP, carbamate, and pyrethroid insecticides may be due to the presence of carboxylesterases that effectively bind these insecticides, but have very low catalytic rates for their hydrolysis (Devonshire and Moores, 1982; Brown, 1990). In the cotton aphid (*Myzus persicae*), resistance to these insecticides is due to the 67-fold overproduction a carboxylase (E4) that accounts for 3% of the total protein. Although E4 is irreversibly inhibited by DEF, synergists are unlikely to be effective in controlling this resistance due to the large amounts of E4 present and its low catalytic efficiency (Devonshire and Moores, 1982). In *Aphis gossypii*, resistance to fenitrothion was attributed to soluble carboxylesterase activities that both sequester and catalyze the hydrolysis of fenitrothion. Resistance appeared to be most closely related to sequestering activity and the Kd and Bmax for this activity in a resistant strain were 0.22 μM and 283 pmol/mg protein, respectively. The sequestering activity was ca. 10-fold higher in a resistant strain. The esterase inhibitor K-2 was both an effective inhibitor of the esterase activity $I_{50} = 0.8$ μM and sequestering activity $I_{50} = 1.9$ μM; however, other esterase inhibitors such as DEF and IBP were ineffective. Toxicity was synergized by K-2 (Suzuki *et al.*, 1993). Since K-2 is an inhibitor of GST activity (Shiotsuki *et al.*, 1992) and some GSTs have esterase activity (Lalah *et al.*, submitted 1994), it may be important to further document the nature of this enzyme since some GSTs have sequestering activity (Jakoby, 1978; Singh and Shaw, 1988).

Plants. Sequestering is not a common mechanism of herbicide selectivity between tolerant and susceptible plant species, but this mechanism has been proposed in some cases of paraquat resistance (in Coupland, 1991).

Synergism Due to Pesticide Interactions

Cost is a minor factor when synergists are used to study pesticide resistance mechanisms in the laboratory, but this is not the case when they are intended for the field control of agricultural pests. For the latter purpose, synergists that are also registered pesticides have a distinct advantage over single purpose synergists because costs associated with registration, construction of production facilities, marketing and market risk if synergism should fail, are all minimized.

Many synergistic pesticide interactions have been reported (Hatzios and Penner, 1985; Green and Bailey, 1989) and a few of these synergistic combinations are used in commercial control programs. Some synergistic combinations are shown in Table 1.

The OP and carbamate insecticides commonly synergize other insecticides by inhibiting the various cytochrome P450 and hydrolytic enzymes involved in insecticide detoxification. The OP insecticides methadithion and monocrotophos are used commercially to synergize cypermethrin and it was recently shown that pyrazophos and several other OP insecticides were nearly as effective as PBO in the synergism of fenvalerate (Forrester *et al.*, 1993; Table 1). It has long been known that the carbamate and OP insecticides synergize amide herbicides such as propanil by inhibiting amidase activity (Frear and Still, 1968); however, recent reports that the OP and some carbamate insecticides also synergize a broad range of herbicides by inhibiting cytochrome P450 enzymes is extremely interesting and potentially useful (Table 1; Kreuz *et al.*, 1992; Diehl *et al.*, 1994; Christopher *et al.*, 1994). Interestingly, in the synergism of nicosulfuron herbicide by the OP insecticide terbufos, an oxidative metabolite of terbufos was the most potent nicosulfuron synergist and cytochrome P450 inhibitor (Diehl *et al.*, 1994). Some modern fungicides and/or plant growth regulators that function by inhibiting critical cytochrome P450 enzymes also inhibit the cytochrome P450-catalyzed detoxification of various herbicides and synergize their toxicity (Table 1; Fonne-Pfister and Kreuz, 1990; Leah *et al.*, 1991). These same compounds might also serve as insecticide synergists.

The saligen cyclophosphates derived from fungicides such as IBP (Kitazin) synergize OP insecticides by inhibiting various GST and hydrolytic enzymes involved in their detoxification (Table 1) and the plant growth regulator DEF is an excellent synergist of insecticides metabolized by various hydrolytic enzymes (in B-Bernard and Philogene, 1993). The use of DEF may be limited because of its defoliant activity. Perhaps compounds such as IBP, K1 and K2 might also be synergistic with some herbicides.

The use of the herbicide tridiphane as a selective synergist of atrazine to control giant foxtail in corn is an example of a beneficial synergistic interaction between two herbicides. Tridiphane is also a synergist of diazinon insecticide in the resistant housefly. Inhibition of GST activity appeared to be the synergistic mechanism of action of tridiphane with both atrazine and diazinon (Lamoureux and Rusness, 1986; Lamoureux and Rusness, 1987); however, tridiphane is also an inhibitor of cytochrome P450 monooxygenase activity (Moreland *et al.*, 1993).

Some pesticides synergize herbicide activity

Table 1. Some Typical Synergistic Pesticide Interactions

PESTICIDE	SYNERGIST	MECHANISM	REFERENCE
	. . . . Insecticide Synergism		
Permethrin	IBP (F)	Esterase	Ishaaya & Cassida, 1981
Dimethoate	IBP (F)	Amidase	Chen & Dauterman, 1971
Parathion	IBP (F)	Esterase	Motoyama & Dauterman, 1974
Fenitrothion	IBP/SCPs(F)	GST	Shiotsuki & Eto, 1991
Permethrin	Simazine (H)	P450	Wilkins et al., 1993
Carbofuran	Atrazine (H)	P450	Chio & Sanborn, 1977
Pyrethroids	Prochloraz (F)		Colin & Belzunces, 1992
Pyrethroids	Chlordimeform (I)		Bagwell & Plapp, 1992
Fenvalerate	OP Insecticides (I)	P450	Forrester et al., 1993
Organophosphates	Kitazin/IBP (F)	Esterase	Forrester et al., 1993
Diflubenzuron	Profenofos (I)	Esterase	Ishaaya, 1993
Cypermethrin	Methadithion (I)	Esterase	" "
Cypermethrin	Monocrotophos (I)	Esterase	" "
Diazinon	Tridiphane (H)	GST	Lamoureux & Rusness, 1989
	. . . . Herbicide Synergism		
Alachlor	Atrazine (H)	GST/GSH(?)	Akobundu et al., 1975
Amitrole	Paraquat (H)		Putnam & Ries, 1967
Simazine	Paraquat (H)		" " "
Bentazon	BAS 110/111 (PGR)	P-450	Leah et al., 1991
Atrazine	Tridiphane (H)	GST	Lamoureux & Rusness, 1986
Simetryn	Thiobencarb (H)		Hagimoto & Yoshikawa, 1972
Paraquot	Diuron (H)		Headford, 1970
Imidazolinone	Glyphosate (H)		Boicion, 1986
MCPA	Dicamba (H)		in Gressel, 1990
Amitrole	Bromoxynil		" " "
Mefluidide	Acifluorfen (H)		Hook & Glen, 1984
Mefluidide	Chlorsulfuron (H)		Tautuydas, 1983
Nicosulfuron	Terbufos (I)	P-450	Diehl et al., 1994
Chlorsulfuron	Malathion (I)	P-450	Christopher et al., 1994
Primsulfuron	Malathion (I)	P-450	Kreuz et al., 1992
Sethoxydim	Fluazifop (H)		Harker & O'Sullivan, 1991
Imazethapyr	Imazaquin (H)		Riley & Shaw, 1993
Propanil	Carbaryl (I)	Amidase	Frear & Still, 1968
Propanil	Mobam (I)	Amidase	Reichel et al., 1991
Fluazifop-butyl	Ethephon (PGR)	Translocate	Lawrie & Clay, 1993
Dicamba	Ethephon (PGR)	Translocate	Binning et al., 1971
Naptalam	2,4-D, PCPA (H)	Uptake	Devlin & Yaklich, 1972
Glyphosate	Lactofen (H)	Translocate	Wells & Appleby, 1992
Dicamba	Chlorflurenol (PGR)	U/Translocate	Baradari et al., 1980
Diclofop	Propiconazole (F)	P-450	Fonne-Pfister & Kreuz, 1990
	. . . . Fungicide Synergism		
Propiconazole	Pyrazophos (F)		Zeun et al., 1992
Cymoxanil	Mancozeb (F)		Cohen & Gisi, 1993

Herbicide (H), Fungicide (F), Insecticide (I), and Plant Growth Regulator (PGR).

by stimulating uptake or translocation while in other cases the mechanism of synergism is not understood (Table 1). Some of these synergistic interactions may be rather complicated.

Many synergistic interactions between herbicides and other pesticides in plants have been reported; unfortunately, many of these result in the reduction of crop yield and are not useful (Hatzios and Penner, 1985; Green and Bailey, 1989). In some cases it may be difficult to develop satisfactory herbicide synergists because the selectivity between the weed and the crop may be lost if a synergist is used. However, it has been proposed that metabolic inhibitors could be used more successfully as herbicide synergists if crops were genetically engineered with target site resistance to the herbicide being synergized (Gressel, 1993). This approach might be useful in some serious cases of cross resistance where most practical herbicides are no longer effective because of monooxygenase activity in the resistant weed species, i.e. blackgrass and annual ryegrass.

The selection of pesticide combinations to be screened for synergism and the methods to screen for synergism are difficult problems because of the overwhelming number of potential combinations (Green and Bailey, 1989; Gressel, 1993). A consideration of the pesticide target site, the types of detoxification reactions, detailed information regarding the detoxification enzymes, and other known pesticide interactions may aid in that selection process.

Synergism of Biological Pesticides

Biological pesticides currently account for ca. $150 million of a world-wide expenditure of $25 billion for pesticide control and represent a very small portion of the pest control effort (Carlton, 1993). Biological pesticides frequently lack the effect needed or are expensive as compared to chemical control measures. Therefore, the synergism of biological pesticides may be appropriate.

Bacillus thuringiensis (Bt) and Bt toxins are among the most successful of the biological pesticides and are sometimes used to aid in the control of insecticide-resistant insects. There are ca. 30 subspecies of Bt and many of these produce multiple toxins which range 1000-fold in their toxicity to specific insect species. Some combinations of these Bt toxins appear to be synergistic, producing up to a 5-fold level of synergism; however, this synergism may be difficult to detect (Angsuthanasombat *et al.*, 1992; Tabashnik, 1992). The use of viruses as insecticides originated in Canada where a baculovirus was used for the control of the European spruce sawfly in pine forests (in Maeda and Hammock, 1993). Currently there are four viruses registered for use as insecticides in the United States (Vaughn, 1993) and viruses are also used as insecticides in Europe, Japan, and South America (Maeda and Hammock, 1993). The Baculoviruses are found only in arthropods and probably represent many of the best candidates for use as insecticides (Vaughn, 1993). Recent recombinant research to increase the insecticidal activity of baculoviruses has resulted in a significant increase (synergism) in the kill speed and yielded promising results (in Maeda and Hammock, 1993). The most successful recombinant baculoviruses have genes for the production of scorpion toxins, mite toxins, and juvenile hormone esterase, and produce effects comparable to classical insecticides (Maeda and Hammock, 1993).

Plant pathogens, primarily fungal pathogens such as *Colletotrichum gloeosporioides* and *Phytophthora palmivora*, are sometimes used as bioherbicides for the control of specific weed species, but their use is very restricted (Christy *et al.*, 1993). Problems associated with the use of these agents include cost and the requirements for specific environmental conditions for the infection to develop. Several studies with fungal plant pathogens have shown that a variety of herbicides with different modes of action (glyphosate, thidiazuron, bentazon, imazaquin, oryzalin, and acifluorfen) can facilitate the infection process and synergize the pathogen. It has also been shown that the phytotoxicity of bacterial plant pathogens can be synergized by herbicides such as glyphosate and sulfosate (Gressel *et al.*, 1993; Christy *et al.*, 1993). The basis of this synergistic interaction is not always clear, but in several cases the production of phytoalexins is reduced by the herbicide which then facilitates the infection process (Sharon *et al.*, 1992; Gressel *et al.*, 1993; Christy *et al.*, 1993).

Synergism Due to the Presence of Allelochemicals

Many plant species have well developed methods for limiting damage to various hazards, including insect pests. Among these defense mechanisms are the presence of both natural insecticides and synergists of these insecticides (B-Bernard and Philogene, 1993). It is recognized that some insecticides are more effective when used on specific plant biotypes that contain natural insecticide synergists. Therefore, one mechanism of utilizing pesticides to the greatest advantage is to consider the possibility of maintaining or building up the concentrations of these natural synergists in key crop plants. These naturally occurring synergists include compounds that are inhibitors of cytochrome P450 monooxygenases and GSTs (Manoharan and Uthamasamy, 1993; Das *et al.*, 1984; B-Bernard *et al.*, 1990; B-Bernard and Philogene, 1993).

Synergists Used to Study Resistance Mechanisms

Synergists are commonly used to study the evolution and the mechanisms of pesticide resistance (Raffa and Priester, 1985), but when used for this purpose, knowledge of the specificity of the synergists and enzymes is critical (Scott, 1990). Some GST inhibitors such as tridiphane also inhibit cytochrome P-450 activity (Moreland *et al.*, 1989) and some hydrolyase inhibitors such as the SCPs inhibit GST (Shiotsuki *et al.*, 1992). Some GST enzymes have esterase like activity (Lalah *et al.*, submitted), GST enzymes can bind ligands in a manner similar to some esterases with sequestering activity (Jackoby, 1978; Singh and Shaw, 1988), and monooxygenases catalyze reactions that yield insecticide metabolism products very similar to those produced by GST and some hydrolytic enzymes. Therefore, the interpretation of these studies can be very confusing and when alternate substrates are used in place of the pesticide of interest, these interpretations can become even more difficult (Brown, 1990).

Conclusions

Currently, pesticide synergists are commonly used to study and/or monitor pesticide resistance, but they are not widely used in commercial agriculture to directly control or to delay the evolution of pesticide resistant organisms. Most pesticide synergists inhibit the cytochrome P450 monooxygenase, GST, or hydrolytic enzymes responsible for the detoxification of pesticides and this is thought to be their primary mechanism of action. However, other mechanisms of synergism have been documented. Target site resistance, which is especially common in weeds and fungi and which also occurs in insects, is difficult to overcome with synergists.

The failure of synergists to achieve an important commercial role in the management of resistant pests is due to a variety of reasons, including the following: the ease with which some pests overcome synergists with alternate resistance mechanisms, the presence of multiple mechanisms of resistance within a population of pests, the high levels of synergist sometimes required to overcome resistance, cost, the toxicity of some synergists, and the lack of adequate synergists to overcome some forms of resistance. In spite of these problems, research on pesticide synergism should continue because the problem of pesticide resistance continues to grow, there may be few practical alternatives for some resistance problems, the rate of discovery of new classes of pesticides is very limited, there is a great deal of pressure to decrease pesticide use rates, and some registered pesticides can be used to synergize other pesticides.

To be successful in managing pesticide resistance, all available pest management tools should be considered and the most appropriate of these should be selected for each resistance problem. Pesticide synergists are only one of these tools.

Acknowledgments

Pesticide and synergist nomenclature used herein is described in the Pesticide Manual-9th Edition; Worthing, C.R.; Hance, R.J., Eds. The British Crop Protection Council; Unwin Brothers Limited: Old Woking, Surrey (1991) or in The Merck Index-11th Edition; Budavari, S.; O'Neil, M.J.; Smith, A., Eds. Merck and Co., Inc.: Rahway, NJ (1989) unless otherwise noted.

Mention of trademark or proprietary product does not constitute a guarantee or warranty of the product by the U. S. Department of Agriculture and does not imply its approval to the exclusion of other products that may also be suitable. U. S. Department of Agriculture, Agricultural Research Service, Northern Plains Area, is an equal opportunity/affirmative action employer and all agency services are available without discrimination.

References

Abdel-Aal, Y.A.I.; Ibrahm, S.A.; Lampert, E.P.; Rock, G.C. *Rev. Pestic. Toxicol.* **1993**, 2, 13-33.

Agosin, M. *Comprehensive Insect Physiology Biochemistry and Pharmacology*; Kerkut, G.A; Gilbert, L.I., Eds.; Pergamon Press: Oxford, **1985**; Vol. 12.

Akobundu, I.O.; Sweet, R.D.; Duke, W.B. *Weed Sci.* **1975**, 23, 20-25.

Anderson, M.P.; Gronwald, J.W. *Plant Physiol.* **1991**, 96, 104-109.

Angsuthanasombat, C.; Crickmore, N.; Ellar, D.J. *FEMS Microbiol. Lett.* **1992**, 94, 63-68.

Baerg, R.; Barrett, M.; Polge, N. *Pest. Biochem. Physiol.* (submitted).

Bagwell, R.D.; Plapp, F.W. Jr. *J. Econ. Entomol.* **1992**, 85, 658-663.

Balabaskaran, S.; Clark, A.G.; Cundell, A.; Smith, J.N. *Suppl. 66 The Australian Journal of Pharmacy* **1968**, 49(584).

Balabaskaran, S.; Smith, J.N. *Biochem. J.* **1970**, 117, 989-996.

Baradari, M.R.; Haderlie, L.C.; Wilson, R.G. *Weed Sci.* **1980**, 28, 197-200.

B-Bernard, C.; Arnason, J.T.; Philogene, B.J.R.; Lam, J.; Waddell, T. *Entomol. Exp. Appl.* **1990**, 57, 17-22.

B-Bernard, C.; Philogene, B.J.R. *J. Toxicol. Environ. Health* **1993**, 38, 199-223.

Binning, L.K.; Penner, D.; Meggitt, W.F. *Weed Sci.* **1971**, 19, 73-75.

Bocion, P. *European Patent EP 234*, **1986**, 379, 31.

Boydston, R.A.; Slife, F.W. *Weed Sci.* **1986**, 34, 850-858.

Brattsten, L.B.; Holyoke, C.W.; Leeper, J.R.; Raffa, K.F. *Science* **1986**, 231, 1255-1260.

Breaux, E.J. *Weed Sci.* **1987**, 35, 463-468.

Breaux, E.J.; Patanella, J.E.; Sanders, E.F. *J. Agric. Food Chem.* **1987**, 35, 474-478.

Brown, T.M. In *Managing Resistance to Agrochemicals*; Green, M.B.; LeBaron, H.M.; Moberg, W.K., Eds.; ACS Symposium Series 421; American Chemical Society: Washington, DC, **1990**; pp 61-76.

Bull, D.L.; Menn, J.J. In *Managing Resistance to Agrochemicals*; Green, M.B.; LeBaron, H.M.; Moberg, W.K., Eds.; ACS Symposium Series 421; American Chemical Society: Washington, DC, **1990**; pp 118-133.

Burden, R.S.; Carter, G.A.; Clark, T.; Cooke, D.T.; Croker, S.J.; Deas, A.H.B.; Hedden, P.; James, C.S.; Lenton, J.R. *Pestic. Sci.* **1987**, 21, 253-267.

Burnet, M.W.M.; Loveys, B.R.; Holtum, J.A.M.; Powles, S.B. *Planta* **1993b**, 190, 182-189.

Burnet, M.W.M.; Oveys, B.R.; Holtum, J.A.M.; Powles, S.B. *Pestic. Biochem. Physiol.* **1993a**, 46, 207-218.

Carlton, B.C. In *Pest Control with Enhanced Environmental Safety*; Duke, S.O.; Menn. J.J.; Plimmer, J.R., Eds.; ACS Symposium Series 524; American Chemical Society: Washington, DC, **1993**; pp 258-266.

Caseley, J.C.; Cussans, G.W.; Atkin, R.K., Eds. In *Herbicide Resistance in Weeds and Crops*; Butterworth-Heinemann, Ltd.: Oxford, **1991**; 513 pp.

Casida, J.E. *J. Agric. Food Chem.* **1970**, 18, 753-772.

Chen, P.; Dauterman, W.C. *J. Agric. Food Chem.* **1971**, 18, 753-772.

Chio, H.; Sanborn, J.P. *J. Econ. Entomol.* **1977**, 70, 544-546.

Christopher, J.T.; Powles, S.B.; Holtum, J.A.M. *Plant Physiol.* **1992**, 100, 1909-1913.

Christopher, J.T.; Preston, C.; Powles, S.B. *Pestic. Biochem. Physiol.* (submitted).

Christy, A.L.; Herbst, K.A.; Kostka, S.J.; Mullen, J.P.; Carlson, P.S. In *Pest Control with Enhanced Environmental Safety*; Duke, S.; Menn, J.; Plimmer, J., Eds.; ACS Symposium Series 524, American Chemical Society: Washington, DC, **1993**; pp 87-100.

Clark, A.G.; Shamaan, N.A. *Pestic. Biochem. Physiol.* **1984**, 22, 249-261.

Clark, A.G.; Shamaan, N.A.; Dauterman, W.C.; Hayaoka, T. *Pestic. Biochem. Physiol.* **1984**, 22, 51-59.

Cochrane, B.J.; Hargis, M.; Debelligny, P.C.; Holtsberg, F.; Coronella, J. In *Molecular Mechanisms of Insecticide Resistance*; Mullin, C.A.; Scott, J.G., Eds.; ACS Symposium Series 505, American Chemical Society: Washington, DC, **1992**; pp 53-70.

Cohen, Y.; Gisi, U. *Crop Protection* **1993**, 12, 284-289.

Colin, M.E.; Belzunces, L.P. *Pestic. Sci.* **1992**, 36, 115-199.

Coupland, D. In *Herbicide Resistance in Weeds and Crops*; Caseley, J.C.; Cussans, G.W.; Atkin R.K., Eds.; Long Ashton International Symposium, Butterworth-Heinemann, Ltd.: Oxford, **1991**; pp 263-278.

Crowder, L.A.; Jensen, M.P.; Watson, T.F. In *Proc. Beltwide Cotton Conf.* **1984**, 223-224.

Das, M.; Bickers, D.R.; Mukhtar, H. *Biochem. Biophys. Res. Commun.* **1984**, 120(2), 427-433.

Davidse, L.C. In *Managing Resistance to Agrochemicals*; Green; LeBaron; Moberg; Eds.; American Chemical Society: Washington, DC, **1990**; 215-223.

De Prado, R.; Sanchez, M.; Jorrin J.; Dominguez, C. *Pestic. Sci.* **1992**, 35, 131-136.

Devonshire, A.L.; Moores, G.D. *Pestic. Biochem. Physiol.* **1982**, 18, 235-246.

Devlin, R.M.; Yaklich, R.W. *Weed Sci.* **1971**, 19, 135-137.

De Waard M.A. *Fungicides for Crop Protection*; BCPC Monograph **1985**, 31, 89-95.

De Waard, M.A.; Gisi, U. In *Modern Selective Fungicides*, 2nd ed.; Lyr, H., Ed.; Gustav Fisher Verlag: Jena, Germany (in press).

Diehl, K.E.; Stoller, E.W.; Barrett, M. *Pestic. Biochem. Physiol.* (submitted).

Dionigi, C.P.; Dekker, J.H. *Pestic. Biochem. Physiol.* **1990**, 37, 287-292.

Dolphin, D.; Poulson, R.; Avramovic, O., Eds. In *Glutathione: Chemical, Biochemical, and Medical Aspects, Part B*; John Wiley & Sons, Inc.: New York, **1989**; 848 pp.

Donaldson, R.P.; Luster, D.G. *Plant Physiol.* **1991**, 96, 669-674.

Dong, K.; Scott, J.G. *Medical and Veterinary Entomology* **1992**, 6, 241-243.

Durst, F. *Frontiers in Biotransformation*; Taylor & Francis: New York, NY, **1991**; pp 191-232.

Durst, F.; Benveniste, I.; Salaun, J-P.; Werck-Reichhart, D. *Biochemical Soc. Trans.* **1992**, 20, 353-357.

Eldefrawi, M.E.; Miskus, R.; Sutcher, V.J. *J. Econ. Entomol.* **1960**, 53, 231-234.

El Saidy, M.F.; Auda, M.; Degheele, D. *Pestic. Biochem. Physiol.* **1989**, 35, 211-222.

Elzen, G.W.; Leonard, B.R.; Graves, J.B.; Burris, E.; Micinski, S. *J. Econ. Entomol.* **1992**, 85, 2064-2072.

Farago, S.; Kreuz, K.; Brunold, C. *Pestic. Biochem. Physiol.* **1993**, 47, 199-205.

Fonne-Pfister, R.; Gaudin, J.; Kreuz, K.; Ramsteiner, K.; Ebert, E. *Pestic. Biochem. Physiol.* **1990**, 37, 165-173.

Fonne-Pfister, R.; Kreuz, K. *Phytochemistry* **1990**, 29, 2793-2796.

Forrester, N.W.; Cahill, M.; Bird, L.J.; Layland, J.K. *Bulletin of Entomological Research Supplement Series* **1993**, Supplement No. 1, 62-82.

Frear, D.S.; Still, G.G. *Phytochemistry* **1968**, 7, 913-920.

Frear, D.S.; Swanson, H.R.; Tanaka, F.S. *Pestic. Biochem. Physiol.* **1993**, 45, 178-192.

Frear, D.S.; Swanson, H.R.; Thalacker, F.W. *Pestic. Biochem. Physiol.* **1991**, 41, 274-287.

Fujitani, T.; Tanaka, T.; Hashimoto, Y.; Yoneyama, M. *Toxicology* **1993**, 25;83, 93-100.

Georghiou, G.P. In *Managing Resistance to*

Agrochemicals: From Fundamental Research to Practical Strategies; Green, M.B.; LeBaron, H.M.; Moberg, W.E., Eds.; ACS Symposium Series 421; American Chemical Society: Washington, DC, **1990**; pp 18-41.

Gonneau, M.; Pasquestte, B.; Cabanne, F.; Scalla, R. *Weed Res.* **1988**, 28, 19-25.

Green, J.M.; Bailey, S.P. *Methods of Applying Herbicides*; McWhorter, C.G.; Gebhart, M.R., Eds.; Weed Science Society of America: Champaign, IL, **1989**; pp 37-61.

Green, M.B.; LeBaron, H.M.; Moberg, W.E., Eds.; *Managing Resistance to Agrochemicals: From Fundamental Research to Practical Strategies*; ACS Symposium Series 421, American Chemical Society, Washington, DC, **1990**; pp 1-496.

Gressel, J. *Rev. Weed Sci.* **1990**, 5, 49-82.

Gressel, J. In *Pest Control with Enhanced Environmental Safety*; Duke, S.O.; Menn. J.J.; Plimmer, J.R., Eds.; ACS Symposium Series 524; American Chemical Society: Washington, DC, **1993**; pp 48-61.

Gressel, J.; Ammon, H.U.; Fogelfors, H.; Gasquez, J.; Kay, Q.O.N.; Kees, H. In *Herbicide Resistance in Plants*; LeBaron, H.; Gressel, J., Eds.; J. Wiley Inc.: New York, **1982**; pp 31-55.

Gressel, J.; Segel, L.A. *Z. Naturforsch.* **1990**, 45c, 470-473.

Gressel, J.; Shaaltiel, Y.; Sharon, A.; Amsellem, Z. *Herbicide Bioassays* **1993**, 217-252.

Gronwald, J.W.; Andersen, R.N.; Yee, C. *Pestic. Biochem. Physiol.* **1989**, 34, 149-163.

Gronwald, J.W.; Connelly, J.A. *Pestic. Biochem. Physiol.* **1991**, 40, 284-294.

Hagimoto, H.; Yoshikawa, H. *Weed Res.* **1972**, 12, 21-30.

Hammock, B.D.; Soderlund, D.M. In *Pesticide Resistance: Strategies and Tactics for Management*; National Academy Press: Washington, DC, **1986**; pp 111-129.

Harker, K.N.; O'Sullivan, P.A. *Weed Technol.* **1991**, 5, 310-316.

Hatano, R.; Scott, J.G.; Dennehy, T.J. *J. Econ. Entomol.* **1992**, 85, 1088-1091.

Hatzios, K.K.; Penner, D. *Rev. Weed Sci.* **1985**, 1, 1-63.

Hayaoka, T.; Dauterman, W.C. *Pestic. Biochem. Physiol.* **1983**, 19, 344-349.

Headford, D.W.R. *Pestic. Sci.* **1970**, 1, 41- 42.

Higashi, K. *GANN Monograph on Cancer Research* **1985**, 30, 49-66.

Hodgson, E. *Insect Biochem.* **1983**, 13, 237-246.

Hodgson, E. *Comprehensive Insect Physiology Biochemistry and Pharmacology*; Kerkut, G.A; Gilbert, L.I., Eds.; Pergamon Press: Oxford, **1985**; Vol. 11.

Hollingworth, R.M. *Biochemical Toxicology of Insecticides*; O'Brien, R.D.; Yamamoto, I., Eds.; Academic Press: New York, **1970**; p 75.

Holt, J.S.; Powles, S.B.; Holtum, J.A.M. *Annu. Rev. Plant. Physiol.* **1993**, 44, 203-229.

Holtum, J.A.M.; Matthews, J.M.; Hausler, R.E.; Liljegren, D.R.; Powles, S.B. *Plant Physiol.* **1991**, 97, 1026-1034.

Hook, B.J.; Glenn, S. *Weed Sci.* **1984**, 32, 691-696.

Ishaaya, I. *Arch. Insect Biochem. Physiol.* **1993**, 22, 263-276.

Ishaaya, I.; Casida, J.E. *Pestic. Biochem. Physiol.* **1981**, 14, 178-184.

Ishaaya, I.; Degheele, D. *Pestic. Biochem. Physiol.* **1988**, 32, 180-187.

Jacoby, W.B. *Advances in Enzymology*; Meister, A., Ed.; John Wiley & Sons: New York, **1978**; Vol. 46, pp 383-414.

Kataria, H.R.; Gisi, U. *Crop Protection* **1990**, 9, 403-409.

Kataria, H.R.; Singh, H.; Gisi, U. *Crop Protection* **1989**, 8, 399-404.

Kemp, M.S.; Caseley, J.C. In *Herbicide Resistance in Weeds and Crops*; Caseley, J.C.; Cussans, G.W.; Atkin, R.K., Eds.; Long Ashton International Symposium, Butterworth-Heinemann, Ltd.: Oxford, **1991**; pp 279-292.

Kreuz, K.; Fonne'-Pfister, R. *Pestic. Biochem. Physiol.* **1992**, 43, 232-240.

Lalah, J.O.; Chien, C-I.; Motoyama, N.; Dauterman, W.C. *J. Econ. Entomol.*, (submitted).

Lamoureux, G.L.; Bakke, J.E. In *Foreign Compound Metabolism*; Caldwell, J.; Paulson, G.D., Eds.; Taylor and Francis: London, **1984**; pp 185-199.

Lamoureux, G.L.; Frear, D.S. In *Xenobiotic Metabolism: In Vitro Methods*; Paulson, G.D.; Frear, D.S.; Marks, E.P., Eds.; ACS

Symposium Series 97: Washington, DC, **1979**; pp 77-128.

Lamoureux, G.L.; Rusness, D.G. *Pestic. Biochem. Physiol.* **1986**, 26, 323-342.

Lamoureux, G.L.; Rusness, D.G. *Pestic. Biochem. Physiol.* **1987**, 27, 318-329.

Lamoureux, G.L.; Rusness, D.G. In *Glutathione: Chemical, Biochemical, and Medical Aspects, Part B*; Dolphin, D.; Poulson, R.; Avramovic, O., Eds.; John Wiley & Sons, Inc.: New York, **1989a**; pp 153-196.

Lamoureux, G.L.; Rusness, D.G. *Pestic. Biochem. Physiol.* **1989b**, 34, 187-204.

Lamoureux, G.L.; Rusness, D.G. In *Sulfur Nutrition and Assimilation in Higher Plants*; De Kok, L.J.; Stulen, I.; Rennenberg, H.; Brunold, C.; Rauser, W.E., Eds.; SPB Academic Publishing: The Hague, **1993**; pp 221-237.

Lamoureux, G.L.; Shimabukuro, R.H.; Frear, D.S. In *Herbicide Resistance in Weeds and Crops*; Caseley, J.C.; Cussans, G.W.; Atkin, R.K., Eds.; Butterworth Heinemann: Oxford, **1991**; pp 227-261.

Lawrie, J.; Clay, D.V. *Weed Res.* **1993**, 33, 375-382.

LeBaron, H.M.; McFarland, J. In *Managing Resistance to Agrochemicals: From fundamental Research to Practical Strategies*; Green, M.B.; LeBaron, H.M.; Moberg, W.E., Eds.; ACS Symposium Series 421; American Chemical Society: Washington, DC, **1990**; pp 336-352.

Lee, H.J.; Rocheleau, T.; Zhang, H.G.; Jackson, M.B.; ffrench-Constant, R.H. *FEBS Lett.* **1993**, 335, 315-318.

Liu, M-Y; Plapp, F.W. Jr. *Pestic. Biochem. Physiol.* **1991**, 41, 1-7.

Liu, M-Y.; Plapp, F.W. Jr. *Pestic. Biochem. Physiol.* **1992**, 43, 134-140.

Maeda, S.; Hammock, B.D. In *Pest Control with Enhanced Environmental Safety*; Duke, S.; Menn, J.; Plimmer, J., Eds.; ACS Symposium Series 524, American Chemical Society, Washington, DC, **1993**; pp 281-297.

Manoharan, T.; Uthamasamy, S. *Indian J. Agr. Sci.* **1993**, 63, 759-761.

Metcalf, R.L. *Pestic. Sci.* **1989**, 26, 333-358.

McFadden, J.J.; Frear, D.S.; Mansager, E.R. *Pestic. Biochem. Physiol.* **1989**, 34, 92-100.

McFadden, J.J.; Gronwald, J.W.; Eberlein, C.V. *Biochem. Biophys. Res. Commun.* **1990**, 168, 206-213.

Mirer, F.E.; Levine, B.S.; Murphy, Sh.D. *Chem. Biol. Interact.* **1977**, 17, 99.

Moreland, D.E.; Corbin F.T.; McFarland, J.E. *Pestic. Biochem. Physiol.* **1993a**, 45, 43-53.

Moreland, D.E.; Corbin, F.T.; McFarland, J.E. *Pestic. Biochem. Physiol.* **1993b**, 47, 206-214.

Moreland, D.E.; Corbin, F.T.; Novitzky, W.P.; Parker, C.E.; Tomer, K.B. *Z. Naturforsch., C:Biosci (ZNCBDA)*, **1990**, 45, 5584-64.

Moreland, D.E.; Novitzky, W.P.; Levi, P.E. *Pestic. Biochem. Physiol.* **1989**, 35, 42-50.

Moss, S.R.; Cussans, G.W. *Herbicide Resistance in Weeds and Crops*; Caseley, J.C.; Cussans, G.W.; Atkin, R.K., Eds.; Butterworth-Heinemann Ltd.: Oxford, **1991**; pp 45-55.

Motoyama, N.; Dauterman, W.L. *J. Agric. Food Chem.* **1974**, 22, 350-355.

Mougin, C.; Cabanne, F.; Canivenc, M.C.; Scalla, R. *Plant Sci.* **1990**, 195-203.

National Research Council In *Pesticide Resistance Strategies and Tactics for Management*; Natl. Acad. Press: Washington, DC, **1986**; 471 pp.

Owen, W.J. In *Herbicides and Plant Metabolism*; Dodge, A.D., Ed.; Cambridge University Press: New York, **1989**: pp 171-198.

Owen, W.J.; deBoer, G. In this volume.

Plapp, F.W., Jr. In *Pesticide Resistance Strategies and Tactics for Management*; Glass, E.H., Ed.; National Academy Press: Washington, DC, **1986**; pp 74-86.

Ploemen, J.H.T.M.; Vanommen, B.; Bogaards, J.J.P.; Vanbladeren, P.J. *Xenobiotica* **1993**, 23, 913-923.

Powles, S.B.; Holtum, J.A.M., Eds. *Herbicide Resistance in Plants Biology and Biochemistry;* Lewis Publishers: Boca Raton, Fl, **1994**; 360 pp.

Powles, S.B.; Matthews, J.M. *Resistance '91: Achievements and Developments in Combating Pesticide Resistance*; Denholm,

I; Devonshire, A.L.; Hollomon, D.W., Eds.; Elsevier Applied Science: London and New York, **1991**; pp 75-87.

Putnam, A.R.; Ries, S.K. *Weed Res.* **1967**, 7, 191-197.

Raffa, K.F.; Priester, T.M. *J. Agric. Entomol.* **1985**, 2, 27-45.

Rajakulendran, S.V.; Plapp, F.W., Jr. *J. Econ. Entomol.* **1982**, 75, 1089-1092.

Ray, P.G.; Markley, L.D.; Snelling J.; Coleman, S. *WSSA Abstracts* **1990**, 30, 209.

Reichel, H.; Sisler, H.D.; Kaufman, D.D. *Pestic. Biochem. Physiol.* **1991**, 39, 240-250.

Riley, D.G.; Shaw, D.R. *Weed Technol.* **1993**, 3, 95-98.

Roush, R.T.; Tabashnik, B.E., Eds. *Pesticide Resistance in Arthropods*; Chapman and Hall: New York, **1990**.

Saito, K.; Motoyama, N.; Dauterman, W.C. *J. Econ. Entomol.* **1992**, 85, 1041-1045.

Scott, J.G. In *Pesticide Resistance in Arthropods*; Roush, R.T.; Tabashnik, B.E. Eds.; Chapman and Hall: New York, **1990**; pp 39-57.

Scott, J.G. *Rev. Pestic. Toxicol., Vol. 2*; Roe, R.M.; Kuhr, R.J., Eds.; Toxicology Communications Inc.: Raleigh, NC, **1993**; 12 pp.

Scott, J.G.; Lee, S.S.T. *Arch. Insect Biochem. Physiol.* **1993**, 24, 1-19.

Sharon, A.; Amsellem, Z.; Gressel, J. *Plant Physiol.* **1992**, 98, 654-659.

Shiotsuki, T.; Eto, M. *J. Pestic. Sci.* **1991**, 16, 573-675.

Shiotsuki, T.; Kakimoto, T.; Eto, M. *Pestic. Biochem. Physiol.* **1992**, 42, 119-127.

Singh, B.R.; Shaw, R.W. *FEB* **1988**, 234, 379-382.

Sinning, I.; Michel, H.; Mathis, P.; Rutherford, W.A. *Biochemistry* **1989**, 256, 192-194.

Smith, A.M.; Vandenborn, W.H. *Weed Sci.* **1992**, 40, 351-358.

Suzuki, K.; Hama, H.; Konno, Y. *Appl. Entomol. Zool.* **1993**, 28, 439-450.

Tabashnik, B.E. *Appl. Environ. Microbiol.* **1992**, 58, 3343-3346.

Tammes, P.M.L. *Netherlands Journal of Plant Pathology* **1964**, 70, 73-80.

Tautuydas, K.J. *Proc. Plant Growth Regulator Soc. Amer.* **1983**, 10, 51-56.

Topal, A.; Adams, N.; Dauterman, W.C.; Hodgson, E.; Kelly, S.L. *Pestic. Sci.* **1993**, 38, 9-15.

Toung, Y.P.S.; Hsieh, T.S.; Tu, C.P.D. *J. Biol. Chem.* **1993**, 5;268, 9737-9746.

Trebst, A. *Herbicide Resistance in Weeds and Crops*; Caseley, J.C.; Cussans, G.W.; Atkin, R.K., Eds.; Butterworth-Heinemann Ltd.: Oxford, **1991**; pp 145-164.

Van Laecke, K.; Degheele, D. *Pestic. Sci.* **1993**, 37, 283-288.

Van Laecke, K.; Degheele, D. *Phytoparasitica* **1993**, 21, 9-21.

Vaughn, J.L. In *Pest Control With Enhanced Envionrmental Safety*; Duke, S.; Menn, J.; Plimmer, J., Eds.; ACS Symposium Series 524, American Chemical Society: Washington, DC, **1993**; pp 239-257.

Welling, W.; de Vries, J.W. *Pestic. Biochem. Physiol.* **1985**, 358-369.

Wells, B.H.; Appleby, A.P. *Weed Sci.* **1992**, 40, 171-173.

West, C.A. *The Biochemistry of Plants*; Stumph, P.K.; Conn, E.E., Eds.; Academic Press: New York, **1980**; Vol. 2, pp 317-364.

Wilkins, R.M.; Khalequzzaman, M. *Breighton Crop Protection Conf. Weeds*, **1993**, 3B, 157-162.

Wilkinson, C.F. In *Insecticides for the Future: Needs and Prospects*; Metcalf, R.L.; McKelvey, J.J., Eds.; John Wiley and Sons: New York, **1976**; pp 195-218.

Wood, E.J., de Villar, M.I.P.; Melgar, F.J.; Zerba, E.N. *Pestic. Biochem. Physiol.* **1986**, 26, 170-182.

Yalcin, S.; Jensson, H.; Mannervik, B. *Biochem. Biophys. Res. Commun.* **1983**, 114, 829-834.

Yamamoto, I.; Kyomura, N.; Takahashi, Y. *Arch. Insect Biochem. Physiol.* **1993**, 22, 277-288.

Zeun, R.; Sachse, B.; Buchenauer, H. *J. Plant Diseases and Protection* **1992**, 99, 273-285.

Zimmerlin, A.; Durst. F. *Phytochem.* **1990**, 29(6), 1729-1732.

Ziogas, B.N.; Girgis, S.M. *Pestic. Sci.* **1993**, 39, 199-205.

Zorner, P.S.; Olsen, G.L. *Proc. North Cent. Weed Control Conf.* **1981**, 36, 115.

Zorner, P.S.; Shappard, B.R. *Weed Sci. Soc. Am. Abstr.* **1985**, 25, 290.

REGULATION

Conversion of Hazard to Risk: A High-Cost Mistake

Sir Colin Berry, The London Hospital Medical College, London E1 2AD, UK

Imprecise risk assessment or the substitution of hazard identification for risk evaluation, may be costly. In a number of instances it has resulted in large expenditures with money spent unwisely in programmes apparently aimed at improving human health. Current science has produced techniques which enable us to be confident about disregarding hazards or to ensure that they are correctly evaluated.

Hazard may be defined as a situation that can in particular circumstances lead to harm. Mountains are a hazard; they become a **risk** only if we travel to them, climb them, ski down them, walk on them or dig through them and the risks that they then present will need to be defined before they can be evaluated. Thus, climbing in winter is more dangerous than in summer, the age of those involved will affect the outcome, travelling by motor vehicle will be a greater risk if alcohol is consumed and so on. The risk to a drunken driver of racing on a winters night to a ski village on snowy road is clearly a different one from a sober walker in mid-summer. This difference between hazard and risk is often blurred in the public assessment of factors which may cause harm, and it is worth considering why.

It should not be thought that the problem of risk assessment is a simple one. Although it is done in an informal way every time we cross a busy road we do not do it consciously - Hume said that rationality is a suit of clothes we put on for special occasions and this is certainly true for risk. Familiar risks are more easily understood than unfamiliar ones, voluntary risks are more readily tolerated than involuntary, shared risks are often disregarded and the risk of events occurring at a very low level are invariably overestimated by the uninformed. This pattern of thought adds greatly to regulatory costs, often without benefit. Data from Upton (1982) make this clear; in ranking events contributing to death, two groups questioned **(League of Women Voters, College Students)** had nuclear power (100 deaths) and pesticides (0 deaths) in the top 5 events and although both groups correctly included smoking (150,000 deaths), handguns (17,000 deaths) and The League of Women Voters included motor vehicles (50,000 deaths), neither mentioned electric power (14,000 deaths). **Business and Professional Club** members had motorcycles (3000 deaths) above motor vehicles but otherwise scored well. Whatever reservations one has about these figures, the differences in perception of the three groups of people are clear and show that we would be foolish to assume that those who do not know the data will ever reach a consensus.

In an often quoted paper Hill (1965) discussed the problems of sorting out association and causation in environmental associations. The critical factors are the strength of association (he used the example of the very large increase in scrotal cancer in sweeps) its consistency (the same effect should have been observed in different places by different people at different times); there should be specificity in the association of a specific type or site of disease limited to particular groups of workers, if there is no association with other modes of dying causation is implied. Dose repose relationships are most important, longer or greater exposures should show greater effects - Hill used the term biological gradients for this relationship. He emphasised that the effect should be biologically plausible but this cannot be demanded as what is plausible will depend on the state of knowledge at the time and

1054–7487/95/0368$12.00/0

also suggested that it should be coherent with the body of knowledge, the cause and effect proposed should not conflict with what is known about the natural history and biology of the disease in question. Experimental evidence can be powerfully supportive in a confirmatory sense and may help with analogy; it is fair to judge by this in some circumstances. The information that rubella and thalidomide produce malformations increases the suspicion that other drugs and viruses may act in the same way. Interestingly, he clearly indicated that if you need sophisticated statistics to make the point, the association was probably not biologically significant - we would, no doubt, have reservations about this today.

Nevertheless, these factors are important and have often been disregarded. To take an example where real concern underlay the process, the Boston Globe (1978) reported a leukaemia "epidemic" at the Portsmouth nuclear submarine facility (Kittery, Maine). Najarian and Colton (1978) examined 1450 death certificates of persons believed to have worked at the yard and who were "badged". They found six cases of leukaemia where they had expected one. This resulted in a survey of all white males who had ever worked at the shipyard (24,545 men) of whom 7,615 had received 2 rads of lifetime exposure at the naval facility. The group of exposed workers had slightly fewer deaths from leukaemia than those who had worked elsewhere in the yard and were unexposed (Rinsky et al 1981). Here argument by analogy (radiation may cause leukaemia) has been used to reach a premature conclusion, but this seems to me to have been a reasonable risk to evaluate; not a hazard. The Hill criteria, however, might have helped.

Adherence to some of them would certainly have helped in the ALAR controversy. In 1989 a "60 minutes" programme on television described daminozide (Alar) as the "most potent cancer causing agent in our food supply" - think of the other contenders! There followed a most extraordinary response to the supposed risk this compound presented to children in particular, supported in the main by ill-informed and often unqualified opinions from many who were clearly unaware of the fact that there was more hydrazine (the putative carcinogen) in a helping of mushrooms than anyone was likely to get from apples or apple juice. There was no effective public response and actresses became the preferred communicators on the issue. My own calculations of 28,000 apples/day as the danger level, if one assumed that the mouse splenic haemangiosarcomas were relevant to Man, was matched by a calculation by the *Washington Post* that 19,000 litres of apple juice would be the critical intake (Science 1991). The scientific issues were never discussed seriously in any forum before supermarket chains responded by withdrawing Alar treated apples from sale - a later WHO/FAO view that higher levels of Alar residues than those previously adopted would be safe did not save those apple growers who had been left with an unsaleable crop - one grower lost part of a farm his family had worked since 1912 due to consequent financial pressure (Whelan 1993).

Inappropriate response is not unimportant. It is estimated that the Water Industry in England and Wales will have invested £1 billion on capital works for advanced water treatment in the period 1989-1997, mainly in ozone and activated carbon plant. Severn Trent Water alone will have spent around £100 million on capital works and the estimate of their annual running costs is around 10% of the capital costs (Leahy 1993). The monitoring costs imposed by the Drinking water Regulations are around £5,000,000 annually, on a national basis. No-one appears convinced that this will confer a public health benefit commensurate with the cost. Part of the problem resides in the fact that we fail to consider benefit adequately in our calculations and there are many who do not find the value of disease free cheap food significant. The fall in gastric cancer rates has been related to better crop treatment and handling but we disregard these changes as we do in medicine. High immunisation rates are difficult to maintain; diphtheria is assumed to have disappeared as a result of immunisation procedures but recent events in the former USSR demonstrate that it is not only liberty that has constant vigilance as its price; this disease has returned promptly as collapsing central organisation results in the abandonment of appropriate childhood immunisation schedules.

However, we are developing a more sophisticated methodology to ascribe risk to particular exposures. It is perfectly possible to envisage a situation in which it will be possible to identify the particular causes of cancer within a particular sub-set of the diagnosis. As an example Taylor et al (1994) have found that radon-associated lung cancer has a quite specific mutation pattern; in investigating p53 mutations in uranium miners with lung cancer they found that the same AGG to ATG transversion was present at codon 249 in 31% of the large cell and squamous cancers they investigated including 3 of 5 cases in miners who had never smoked. This mutation had only been reported in 1 of 241 published p53 mutations from lung cancers. Similar molecular data exist which suggest a distinction between the types of carcinogen affecting colon and pancreatic carcinoma (similar patterns of mutation in p53 and Kras) and lung (a different pattern). This is not surprising, the ability to document it is of great potential value in discussions of causation.

Another important misuse of argument by analogy is the extrapolation "if a high dose causes harm, then a small dose must be potentially harmful". Concepts of thresholds are ignored and the idea (which most would reject) that they might have an operation under 1/100th the normal dose of anaesthetic is treated as the nonsense it is. What enables us to be confident about dose response relationships for chemicals is an understanding of mechanisms. Where the mechanism of a toxic effect has been identified, one can often reduce any arbitrarily set safety factor considerably but we continue to find them necessary in the handling of particular types of risk. This type perhaps best exemplified by genotoxic carcinogens, is often dealt with by prohibition of use which might leave a food residue in the United Kingdom (although not in the USA) but it is interesting to speculate that our rapidly increasing knowledge about the specificity of the type of genetic damage necessary for carcinogenesis in Man may alter this approach. Of course, the mode of action need not be understood at the molecular level to give an understanding of an effect, the data on cigarette smoking and chronic bronchitis are a good example of this.

Another difficulty is in deciding what we mean by safe. Siddall (1980) defined it in the following way - "Safety is the degree to which temporary ill health or injury, or chronic or permanent ill health or injury, or death are controlled, avoided, prevented, made less frequent or less probable in a group of people". The concept of a group is enormously important; as I have said, there is a tendency to ignore the consequences of 'negative' regulatory actions but a ban on the use of herbicides which resulted in the use of more labour and machinery would probably have an adverse outcome in terms of farm workers health; around 60 people were killed on farms in the United Kingdom last year.

It is easy to covert a perceived hazard to a real risk in this way. Critical evaluation of a number of studies relating phenoxy-herbicides to soft tissue sarcoma shows that estimates of the putative risk diminishes as confounding factors are identified. But let us look at some Hill criteria; the suggestion that soft tissue sarcomas are an entity in terms of causation is doubtful, there are no adequate exposure data let alone a dose response relationship, differing values are found in different studies of comparable design and no putative mechanism of causation exists. These factors should reassure us but it is foolish to pretend that an entirely rational consideration will prevail. Scientific progress and attention to new data will provide answers to many questions but not instantly; Hill criteria may be used to reassure those in this field but there is much patient explanation of the difficulties to the public to be done if inappropriate and sometimes damaging measures are not to be taken.

References

Hill, A.B; *Proc Roy Soc Med.* **1965**, *58*, 295-300.

Leahy, J.S.; **1993** Personal Communication.

Marshall, E.; *Science* **1991**, *254*, 20-22.

Najarian, T.; Colton, T.; *Lancet* **1981**, *i*, 1018-1020.

Rinsky, R.A.; Zumwalde, R.D.; Waxweiler, R.J.; Murray, W.E.; Bierbaum, P.J.; Landrigan, P.J.; Terpilak, M.; Cox, C.; *Lancet* **1981**, *i*, 231-235.

Sidall, E.; *Risk, Fear and Public Safety.* Atomic Energy of Canada Ltd, 1980

Taylor, J.A.; Watson, M.A.; Devereux T.R.; Michels, R.Y.; Saccomanno, G.; Anderson, M.; *Lancet* **1994**, *i*, 86-87.

Upton, A.C.; *Sci Am.* **1982**, *246(2)*, 41-49.

Whelan, E.N.; *Toxic Terror;* Prometheus Books: Buffalo, New York, **1993** pp194.

New Directions in Pesticide Risk Reduction: Report from North America

Lynn R. Goldman, M.D., Office of Prevention, Pesticides, and Toxic Substances (OPPTS), Environmental Protection Agency (MC 7101), 401 M Street, S.W., Washington, D.C. 20460
Kennan J. Garvey, Office of Pesticide Programs, OPPTS, Environmental Protection Agency (MC 7501C), 401 M Street, S.W., Washington, D.C. 20460
Christina Cortinas de Nava, National Institute of Ecology, Ministry of Social Development, Rio Elba 20, Piso 14, AP. Postal 06500, Mexico, DF
Karen Looye, Pest Management Secretariat, 10th Floor Trebla Building, 473 Albert Street, Ottawa, Ontario K1A 0C5 Canada

North American countries are taking new directions in pesticide regulation, including encouraging the development and use of reduced risk pesticides, alternative pest control methods, and improved regulatory requirements. This paper surveys major risk reduction activities that each North American country is taking independently, as well as some joint efforts begun in recent years.

North American countries are taking new directions in pesticide regulation. These new directions result from a growing awareness that chemicals, including pesticides, can pose a variety of serious, subtle, complex health and environmental risks, as seen in recent concerns about: (1) estrogenic chemicals in the environment, (2) methyl bromide's role in ozone depletion, and (3) special sensitivities of infants and children to pesticide risks, as described in a 1993 U.S. National Academy of Sciences (NAS) report. Past approaches to pesticide regulation focused on careful analysis and evaluation of discrete pesticides. These approaches are not sufficient to deal with new concerns, because they are slow, resource intensive, and often create new problems by not considering interrelated systems. The most effective, least costly solution is pollution prevention - preventing problems rather than trying to fix them retroactively.

Pest problems are real, but we need controls that both work and pose the least risk possible. Pest control cannot be purchased at the expense of our health and environmental quality. We need to look at risk concerns more comprehensively when evaluating ways to prevent risks in the first place, finding faster and more intelligent ways to reduce real risks and share our resources and expertise, so that national governments together can achieve more than any one government could individually.

All three North American countries - the U.S., Canada, and Mexico - are pursuing significant pesticide risk/use reduction initiatives, both individually and together. These countries are also active supporters and contributors to the efforts of the global community to achieve the same goals worldwide. This paper surveys major pesticide risk reduction activities that each North American country is taking independently, as well as joint efforts begun in recent years, including work being done through the Organization for Economic Cooperation and Development (OECD) and the newly established International Forum on Chemical Safety (IFCS).

I. United States' Risk Reduction Initiatives

In the United States, the new Administration has committed to move forward on pesticide regulation in a new way, by seeking a reduction in the use of higher risk pesticides, rather than placing primary emphasis on pesticide licensing. The Administration has the following basic goals regarding pesticide risk reduction:

- o Regulate pesticide use that poses unacceptable risks, but incorporate flexibility in applying restrictions. Adopt regional, pest, and crop specific approaches because one size does not fit all.

- o Consider pollution prevention in all actions.

- o Strive to understand how to provide for both a sustainable economy and a sustainable environment -- with the least harm to individuals.

In reaching these broad goals, the U.S. government has been guided by the following principles:

- o A firm commitment to reducing risks to people and the environment that may be associated with pesticides, and especially to providing greater assurance of protection for children, while ensuring the availability of cost-effective pest management techniques.

- o Recognition of the need to work with pesticide users to: (1) develop and introduce improved means of pest control; (2) reduce the use of high-risk pesticides; and (3) promote greater use of integrated pest management (IPM) techniques, including biological and cultural pest control systems and other sustainable agricultural practices.

- o Introduction of legislative and regulatory reforms and incentives for the development of pesticides that will eliminate or reduce risks.

Consistent with these goals and principles, the Administration announced in June 1993 its commitment to reduce pesticide use and promote sustainable agriculture through the development of legislative, regulatory, and administrative initiatives designed to achieve real risk reduction. The Administration has been working very closely with industry and public interest groups over the past year, and is proceeding on some major pesticide reform initiatives. Consistent with the approach of the President's National Performance Review, which is leading to widespread efforts to streamline and reinvent government processes, these changes will enable the pesticide program to work more effectively, by establishing a pesticide regulatory system that is based on sound science and is capable of acting promptly to reduce identified pesticide risks.

The specific risk reduction initiatives described for the U.S. are in the areas of: (a) proposed food safety legislative changes; (b) use/risk reduction initiative; (c) incentives for faster registration of reduced risk pesticides; and (d) other initiatives related to risk reduction.

A. Administration's Pesticide & Food Safety Legislative Reforms

The Clinton Administration's legislative initiatives are designed to: (1) maintain and enhance food safety for all Americans; (2) address recommendations of a 1993 National Academy of Sciences (NAS) report on ways to improve pesticide regulation to assure that children are fully protected from pesticide risks; and (3) strengthen regulatory agencies' ability to make and enforce sound, timely, science-based decisions to protect public health and the environment.

The legislative reform package includes changes to both the Federal Food, Drug, and Cosmetic Act (FFDCA) and the Federal Insecticide, Fungicide, and Rodenticide Act (FIFRA). The package builds on the recommendations of the NAS report and the input received from representatives of all interests concerned about pesticide use and regulation. These reforms will permit the U.S. to make real progress in enhancing public health and environmental protection.

The heart of the proposal is the establishment of a strong, health-based standard, defined for tolerance-setting purposes as "a reasonable certainty of no harm" to consumers, that would apply to all pesticide residues in food. Existing residue tolerances would have to be reviewed and brought into conformity with the new safety standard within fixed time frames. The proposals allow for a transition period under carefully prescribed conditions that will help avoid undue dislocations in agricultural production, but still ensure that all tolerances meet the new standard within fixed time frames. As recommended in the NAS report, EPA would be required to consider unique aspects of children's diets and potential sensitivities to pesticide risks. EPA would issue specific findings that tolerances are safe for infants and children.

In addition to the legislative proposals regarding tolerances, the U.S. recently developed a proposal to reinvent its tolerance-setting process, in response to recommendations from the National Academy of Sciences and many other quarters. The purpose of the proposal is to ensure that tolerances are more protective of public health and more closely reflect the actual residues that could be found in food. The proposal would allow EPA to use tolerances more directly in assessing dietary risks to the public, and EPA believes the proposal would also make the system for protecting the food supply more understandable and sensible. The legislative reform initiatives contain many complex elements. The ones relating particularly to risk reduction are summarized in the following paragraphs.

An important part of the legislative initiative would require EPA to establish criteria for designation of reduced risk pesticides. Registration applications that appear to meet the criteria would qualify for priority review, and, if approved, would be accorded two additional years of exclusive data use, beyond the ten years now provided in FIFRA.

Also, EPA would be authorized to grant special time-limited, conditional registrations for biologically-based pesticides posing low potential risks. As a class, biological pesticides share a set of desirable environmental characteristics. For example, biologicals include microbial pesticides (i.e. bacteria, fungi, and protozoa) which: (1) tend to be effective at controlling target pest organisms without adversely affecting others; (2) usually do not have toxic effects on animals and people; and (3) do not leave toxic or persistent chemical residues in the environment. Biologicals are usually very specific against target pests and exposure levels are often quite low. This class of pesticides comprises the single fastest-growing segment of registration activity within the Agency's pesticide programs. Over 200 microbials and biochemicals are now on the market. In order to encourage growth in the biological pesticide market, EPA has streamlined the data requirements for biologicals and reduced the target time for processing registration applications associated with biological pesticides. The legislative proposal for time-limited conditional registrations for these pesticides should further accelerate the registration and early market entry of these potentially safer pesticides. However, because EPA recognizes that biologicals can pose risk concerns, the proposal contains safeguards to ensure that only low potential risk products will qualify for this special conditional registration.

The statute would embody clear policy goals favoring reduced use and directing federal agencies to take a leadership role in promoting use reduction and IPM in their programs. The statute would also authorize regional ecosystem-based reduced use pilot projects designed to reduce aggregate pesticide risks and provide a mechanism to focus research priorities. EPA and USDA would be mandated to work together to develop and make available comparative information on the environmental and health effects of pesticides and to identify the research, education and extension activities that are most promising in terms of meeting pest management needs and reducing use of pesticides that raise risk concerns.

The statute would require EPA to identify pesticides of regulatory concern for which there are limited alternatives. USDA would be required to use this information to focus its research and education efforts. The statute would require USDA, in consultation with EPA, to develop a research and extension plan for each commodity determined to have an "alternatives problem."

EPA would be authorized to establish criteria for "prescription use" of pesticides. Such authority could permit retention of pesticides critical to IPM and pesticide resistance management programs. The current prohibition on requiring IPM training as part of certification and training programs would be repealed. In addition, following the model of the 1990 Farm Bill provisions, which applied only to restricted use pesticides, the legislative proposals would require record-keeping for all agricultural pesticide use.

No amount of legislative direction can succeed, however, unless it is accompanied by corresponding changes at the user level. Ultimately, it is the pesticide user who makes day-to-day pest management decisions that govern the introduction of pesticides into the environment. For legislative changes to have the desired effect at the field level, EPA needs to take a "systems" approach to the reduction of risk. Such an approach would provide users with the knowledge and technologies needed to reduce the risks associated with pesticide use. While EPA and FDA must retain a primary focus on regulatory action, they must also, in concert with USDA, work to create non-regulatory programs which encourage voluntary efforts to prevent or mitigate the human health and environmental impacts of pesticide use.

B. Use/Risk Reduction Initiative

The Administration's June 1993 initiative included a commitment to reduce the use of higher risk pesticides nationwide. For the first time ever, the U.S. government is committed to real reductions in pesticide use and risk. Reducing unnecessary use and risk prevents pollution and saves money. The initiative is designed to reduce the risk associated with the use of high risk pesticides, not simply to reduce the volume of use of all pesticides.

One remarkable feature of the initiative is the unprecedented solidarity of the three federal-agency players involved: EPA, USDA, and the Food and Drug Administration (FDA). These agencies have entered into a close partnership to develop and carry out new policies related to the initiative. They are cooperating as never before to make certain that things happen quickly, consistently and efficiently.

EPA is committed to making a concerted interagency effort to reduce pesticide risks and associated use, focusing its efforts on uses, commodities, and products that present the greatest opportunities for risk reduction. EPA will take a close look at geographical areas and commodities that currently account for the largest total volume of pesticide use, those requiring the greatest pounds per acre of pesticide application, and those using the highest risk pesticides.

As part of this effort, EPA and USDA will promote sustainable agriculture and IPM practices, including biological and cultural control systems, setting a goal of developing IPM programs for 75% of the total crop acreage in the U.S. by the year 2000. For agricultural uses of pesticides, EPA and USDA are going through a planning process with extensive public input, especially from growers. The goal is to develop a plan by October -- with commodity-specific goals for reducing pesticide use. The plan will consider local and regional conditions, as well as available and potential pest control strategies, both chemical and non-chemical. Growers, environmental groups, and other interested parties, will help develop these goals to achieve real risk reduction. Individual states are providing leadership by bringing interested parties together to discuss risk/use reduction opportunities.

At this time nothing has been decided. No commodities have been targeted, no mandatory use reductions have been proposed, and no chemicals have been targeted for removal solely because of this initiative.

The new U.S. initiatives raise significant implementation needs, including: (1) development of ecological and human health risk indices; (2) obtaining accurate baseline measurements; and (3) achieving voluntary cooperation of interested parties. Each of these is important to the success of the initiative. Accurate risk indices for individual pesticides are essential to determine risk trade-offs and desirable use/risk reduction strategies. Accurate baseline measurements are needed to measure progress. Voluntary cooperation of all interested parties is a prerequisite for successful implementation, since basing such a complex initiative principally on regulatory and enforcement tools is unworkable.

C. Registration of New, Reduced-Risk Pesticides and Restrictions on Older, Higher Risk Pesticides

The voluntary use/risk reduction initiative complements an earlier EPA initiative to encourage the development, registration and use of new, safer pesticides, and the removal over time of older, higher risk pesticides, which is consistent with the goal of bringing a pollution prevention focus to all of EPA's programs.

For a variety of reasons, older pesticides may often enjoy advantages over newer, and potentially safer, products that could replace them. For this reason, the EPA has initiated a voluntary pilot program using regulatory and non-regulatory incentives that would encourage the registration of lower-risk pesticides and ultimately would discourage the continued registration of older, higher risk pesticides. The long-term goals of this project are to: (1) get troublesome pesticides off the market faster when alternatives for the use exist; (2) substitute safer ones by giving registration applications in this category priority review within EPA; and (3) encourage the use of IPM and other alternative control methods which decrease the reliance on toxic and persistent chemicals.

A number of chemicals and use patterns have already been evaluated under the new incentive system. One product, hexaflumuron as a termiticide for post-construction uses, was registered in March as an alternative to chlorpyrifos. Four chemical uses have passed the reduced risk screen, and are being evaluated for registration on an expedited basis.

In addition to the registration of reduced risk pesticides, EPA is making efforts to determine what kinds of information people need, either on pesticide labels, advertising, or other educational resources, to allow them to make informed choices on pest management.

D. Other U.S. Initiatives Related to Risk Reduction

In addition to the three major initiatives described in the previous section, there are many other currents in the rapidly flowing waters of change in U.S. environmental policies affecting agriculture. In summary, other initiatives related to risk reduction include:

- Implementation of **new worker protection standards**, strengthening protection for the people most directly exposed to pesticides;
- Planned issuance of a final rule on **reporting of new adverse effects data.** Pesticide residues exceeding tolerance levels in food crops, and important ground or surface water contamination will be reportable.
- Rapid acceleration of the U.S. **reregistration program.** Over 10% of cases (54 of 408) have reregistration eligibility decisions, and all have data call-ins issued. There will be 40 to 50 decisions annually over the coming years.
- EPA Air program action to restrict and phase out use of **methyl bromide** as an ozone depleter. This is an indication that EPA will use the full range of its statutory authority to address serious environmental risks.
- Recent announcement of **Clean Water Act** legislative changes will also affect farmers, by strengthening regulation of non-point source pollution.
- Planned issuance of **ground water regulations** this year, requiring state management plans for certain pesticides, and establishing ground water risk criteria and naming specific pesticides that meet the criteria to be classified for Restricted Use.
- Working closely with the USDA/State IR-4 program to ensure that **minor uses of concern to growers** that lack only residue data are supported, using IR-4's expanded budget.
- Planned implementation this year of EPA's program for **protecting endangered species** from the effects of pesticides.
- Proposed regulations scheduled for later this year governing **pesticide containment structures and the design and residue removal for containers.**
- Efforts to exempt some controls, such as pheromones, from regulatory oversight,

so we can encourage safer control methods and focus our efforts in areas of greatest need.

The new directions described for the U.S. will benefit both the public and farmers. USDA has found that farmers benefit most when the public is assured of the safety of the food supply. The reality of food safety must be coupled with public confidence in the food supply. Increasing the public's confidence may mean the loss of some pesticides or uses, or reduced tolerances on others. However, these changes should occur without having as many of the abruptly implemented regulatory changes as seen in the past for individual pesticides. Over time, pesticide companies will begin to bring safer products on the market, and alternative control technologies will be developed. As public confidence increases, agriculture will benefit by having open markets, both domestically and internationally. The U.S. can continue to have an abundant and diverse food supply, while doing more to ensure food safety.

II. Canadian Risk Reduction Initiatives

All pest control products used in Canada must be registered by the federal government in accordance with the Pest Control Products Act. The products regulated are diverse, covering pesticides used in agriculture, forestry, industry, public health and household situations. Products are evaluated on the basis of data submitted by companies to prove that the product is safe and effective when used according to label directions. Maximum residue limits for residues of agricultural chemicals in food are set federally through the Food and Drugs Act. Provincial and territorial governments further regulate the sale, storage, use and disposal of pest control products.

In 1992, the Government of Canada committed new resources to improving the federal pesticide regulatory system with three stated objectives: (1) reduce the risks to people and the environment associated with pesticides; (2) ensure access to effective pest management tools; and (3) allow for a more open and transparent system. The specific improvements stem from the recommendations of an independent review by the Pesticide Registration Review Team. The Team included representatives of health, environmental, research and consumer groups as well as manufacturers and representatives from the agricultural and forestry sectors. Their final report - based on extensive negotiations and public consultation - reflected the need to design a comprehensive system to guide pesticide regulation into the future. The current government, elected in October 1993, is examining ways to accelerate these improvements.

A national policy on pesticide risk reduction is seen as part of the improvements to the pesticide regulatory system. This policy, which requires cooperation with the provinces and interested groups, is in the beginning stages of development. Many existing initiatives, both federal and provincial, already contribute to pesticide risk reduction. A national policy framework would serve to further clarify links between those existing initiatives and identify areas requiring further action.

Many of the measures announced in 1992 are aimed at reducing risks associated with pesticides. For example, policies such as the mandatory reporting of adverse effects of pesticides and a policy on the formulants permitted in pesticides will be developed.

Some new initiatives are also designed to ensure timely access to new and effective pest management tools by rendering the system more efficient. Access to a wider array of tools will allow the phasing out of older, more harmful products in favor of new, safer technologies. Measures to improve access to technologies include new or expanded programs for User Requested Minor Use Registration and User Requested Minor Use Label Expansion. International harmonization of data requirements for submissions for registrations of new products, as well as use of international submission reviews, are being explored through international fora in an effort to make the best use of resources and speed up the registration process.

Risks associated with pesticides will be reduced if it is ensured that pesticides registered at a time when requirements were less stringent are evaluated against today's standards. As part of its registration scheme, Canada undertakes the special review of pesticides in response to particular health or environmental concerns and reevaluations of older pesticides. As a result of these reviews, the registration may be confirmed, the conditions of registration such as permitted uses may be modified, or the registration suspended or canceled. Canada is looking increasingly to international cooperation as a means of speeding up its reevaluation process. Projects currently underway through international fora may facilitate the use of international reviews in determining risks to safety and the environment in order to make the best use of Canada's regulatory resources.

The consideration of alternative pest management strategies is taking on increased importance in Canada. As a result of the Pesticide Registration Review, the Pest Management Alternatives Office (PMAO) has been established. This government funded non-profit organization is mandated to coordinate the development of alternatives. The PMAO is responsible for promoting measures that encourage the judicious use of pesticides, promoting integrated pest management and encouraging the development of ecologically-sound pest management strategies. The PMAO also acts as a resource center for information on domestic and international research activities, preventative and control strategies and other issues relating to pest problems and their management.

In addition to new initiatives in pesticide regulation, many other federal policies and activities contribute to risk reduction. These include:

- Development and use of alternative approaches to pest management in the forest sector. The Canadian Forest Service is developing and encouraging alternative approaches to the control of forest pests through the investigation and development of environmentally acceptable products and forest management methods. Governments and academia are researching alternatives to chemical pesticides such as baculoviruses and pheromones. Public and private forest management agencies have committed to minimizing the use of chemicals in forest management and to expanding research into ecological approaches to resource management which will render forests less vulnerable to pests;

- Management of selected toxic substances in the wood preservation sector as part of initiatives under the Canadian Environmental Protection act;

- Development of environmental indicators for agriculture to provide information on key trends, evaluate environmental sustainability of the agricultural sector and facilitate the integration of environmental considerations in planning, policy and programs. Pesticide related indicators are planned;

- Regulation under the Canadian Environmental Protection Act to control and reduce the use of the ozone depleting pesticide methyl bromide.

Because of the split jurisdictions, federal, provincial and territorial cooperation is essential in the Canadian regulatory environment. Work is well underway on a "Standard for Pesticide Education, Training, and Certification in Canada." Work on establishing baseline environmental data through coordination of post registration monitoring and research, and the development of a national pesticide use database is also beginning as is work on a national compliance strategy.

A number of provinces have already developed pesticide reduction strategies involving use reduction targets or the promotion of integrated pest management. Provincial regulation and other initiatives in the area of pesticide use, such as user training, certification and extension work contribute to the objectives of pesticide risk reduction.

The chemical industry and user groups are also undertaking projects that contribute to risk reduction. Pesticide manufacturers are committed to product stewardship. Through their industry association, they have instituted a container recycling program for agricultural pesticides. Stringent new warehousing standards are also being implemented voluntarily by these companies. Agricultural user groups, in partnership with federal and provincial governments, are encouraging sustainable agriculture through projects such as the development of environmental farm plans and through promotion of integrated pest management.

Development of a national risk reduction policy will require close cooperation between all levels of government, pesticide users, manufacturers and public interest groups. Existing initiatives will form a solid basis for development of this policy.

III. Mexican Risk Reduction Initiatives

In Mexico, an organizational realignment in 1987 improved pesticide regulation, production control and use at the Federal level, through the coordination of the health, social development, agriculture and trade authorities. Most of the states have created committees in response to the Federal reorganization. In January 1994, Mexico issued a new Federal Phytosanitary Law strengthening the regulation of pesticide use and application. The new Law gives the Secretariat of Health the main authority for the pesticide Registry, cooperating with other Secretariats on the evaluation of applications. Mexico promotes IPM and related biological, legal, cultural and autocidal control methods.

The agriculture authorities have focused their attention on the control of pesticide use; the monitoring of pesticide residues on import and export products; the establishment of maximum residue levels for pesticides of interest for Mexico and not in use in the U.S. or Canada; the solution of differences in pesticide residue standards between Mexico and the U.S., and the evaluation of field biological effectiveness of pesticides. The health authorities have created a pesticide intoxication registry and develop regional training activities to improve the diagnosis and treatment of pesticide intoxications. The agrochemical industries association develops training programs for workers to avoid intoxications and promote product stewardship.

IV. Cooperation Among North American Countries

Budget constraints are forcing most governments to do more with less. One way to achieve this is by increasing the level of regional cooperation among North American countries, with the goal of achieving greater harmonization of risk reduction approaches. There are three main benefits of harmonization; (1) food safety, including greater compliance by reducing confusion over which standards to meet, (2) better science through cooperation, and (3) addressing trade issues.

A. CUSTA

A trade agreement between the U.S. and Canada was in place for a number of years. Although it has been superseded by the North American Free Trade Agreement (NAFTA), the Technical Working Groups established under CUSTA remain active as a result of letters exchanged between the U.S. and Canada. Harmonization efforts initiated by the Pesticides Technical Working Group have gained momentum over the past year and a number of projects are currently underway.

CUSTA originally established the Pesticides Technical Working Group to "work toward equivalent guidelines, technical regulations, standards, and test methods" for pesticides in the two countries. The working group has tackled a number of projects including the parallel review of an application for registration of an active ingredient, investigation of potential areas for sharing data, and several issues related to tolerances.

EPA and its Canadian counterparts have both received applications for the registration of tebufenozide, submitted by Rohm and Haas. This is a new pesticide which has passed the U.S. reduced risk screen and is receiving priority review. Through parallel and coordinated review of the application, each country hopes to gain insight into the other's system and decision making processes. The working group is also conducting a pilot project for sharing data on heavy duty wood preservatives that has been collected through the re-registration process in both countries. In both the registration and re-registration pilot projects, the two governments, with the cooperation of industry, are investigating means for facilitating the exchange of data which should prove beneficial in the harmonization of standards, while assuring food safety in both countries.

One of the tolerance projects involves selection of crop/pesticide combinations which have been trade irritants with the goal of potentially harmonizing tolerance levels between countries. Each country has chosen two combinations to review.

B. NAFTA

The North American Free Trade Agreement (NAFTA) entered into force on January 1, 1994. NAFTA includes provisions on agricultural and health issues such as pesticide standards. NAFTA recognizes the right of each country to set levels it feels are necessary to protect human health, provided these levels are based on science and are not arbitrary barriers to trade. NAFTA also allows countries to inspect commodities entering the country to determine compliance with their national levels. NAFTA includes a side agreement on environmental issues aimed at facilitating cooperation to enhance environmental standards in NAFTA countries. The side agreement established an Environmental Commission to provide a forum for public input on these issues.

NAFTA also contains provisions designed to enhance harmonization efforts between U.S., Canada, and Mexico. Under NAFTA each government is expected to consider existing standards - both from international sources and from other NAFTA countries when setting national health standards such as pesticide residue levels. In addition, NAFTA establishes a Sanitary and Phytosanitary Measures (SPS) Committee to consider means of achieving greater harmonization, including for pesticide residue tolerances, and a Standards-Related Measures (SRM) Committee to address technical standards harmonization issues, including pesticide registration requirements. Each country may still set national levels that differ from existing standards in other countries as long as the individual decisions to achieve these levels are based on sound science.

The U.S. has proposed the establishment of a pesticides subcommittee, under the SPS and SRM Committees, that would also work closely with the Environmental Commission established under NAFTA's environmental side agreement. Canada and Mexico have requested a U.S. proposal, outlining the relationship of the subcommittee to other groups established by NAFTA, and the U.S. has agreed to develop a formal proposal. Subject to all three countries' approval of the proposal, having a single focus on pesticide issues should conserve resources and avoid duplication of effort as each country gains experience under NAFTA.

Even before NAFTA, the U.S. had been assisting the Mexican government in understanding U.S. food safety laws and regulations, in order to reduce residue violation rates and the number of Mexican exports detained at the border. Similar to previous efforts between the U.S. and Canada, the U.S./EPA has been working with U.S./FDA and Mexico to identify crop-pesticide combinations which lack U.S. tolerances but are considered important for pest control on exported foods produced in Mexico. Then, staff will look for alternative means of pest control that already have, or do not require, U.S. tolerances. If satisfactory alternatives are not identified, our goal is to identify a few promising initial candidates for the establishment of additional U.S. tolerances. In addition, the U.S., Mexico and Canada are currently considering a number of options for instituting increased cooperation on risk reduction and harmonization activities.

C. Broader International Efforts

The U.S. and Canada have worked actively and cooperatively to establish a new pesticide program of work with the OECD's chemicals program. Mexico will become increasingly involved in the OECD pesticide activity, since it recently joined OECD as a full member. The OECD pesticide program includes a number of activities to achieve greater harmonization. The OECD risk reduction activity in particular will benefit North American risk reduction efforts. This OECD project has begun with a survey of current risk reduction activities in OECD member countries. This survey information will provide background for a workshop in 1995 to identify the most appropriate role for OECD in the area of pesticide risk reduction. The three North American countries participated a meeting in Stockholm last month, ending with a decision of the group to constitute itself as a continuing Intergovernmental Forum on Chemical Safety (IFCS). The first session of the Forum was the largest gathering ever of nations to discuss chemical safety (355 participants). The Forum considered the recommendations of the 1992 UN Conference on Environment and Development (UNCED) for improved chemical safety, and took steps to implement these recommendations. In the area of risk reduction, the Forum agreed that governmental experience and progress in national risk reduction programs shall be presented in a report by 1997, and serve as a basis of setting goals for the year 2000. In the area of risk assessment, the U.S. volunteered to host a workshop next year to establish a list of chemicals, including pesticides, for risk assessment by the year 1997. The workshop will also establish the type of risk assessment documents needed, and determine which risk assessment documents exist currently both nationally and internationally.

V. Conclusion

North American countries are working both individually and collectively on a wide variety of initiatives to enhance pesticide risk reduction activities. Due to the increasingly global nature of trade issues, pesticide regulatory policy must also expand beyond national borders to consider regional and global concerns. The close working relationships among North American countries has been strengthened by the CUSTA and NAFTA agreements, greatly increasing the likelihood of significant and complementary risk reduction efforts across their long borders. North American countries are finding ways to share their resources and expertise, to achieve the ultimate goal of pollution prevention.

These new directions in pesticide regulation create opportunities for increased public confidence in food safety, greater human health and environmental protection, a more predictable regulatory process, and more stable markets.

New Directions : Report from Europe

B Thomas, AgrEvo UK Limited, Chesterford Park, Saffron Walden, Essex CB10 1XL

The basic principles of the European Union's Registration Directive and its associated reregistration programme are described. The consequences of the Directive in terms of registration delays and resource implications are considered in the context of the impact on new and existing active ingredients. Practical consequences in relation to the adoption of a pan-European regulatory strategy and the need for a complete European data base are reviewed. Likely scenarios resulting from an extended European Union and the impact of the Directive on Central Europe are predicted.

Regulatory procedures and requirements have been developed within the 12 Member States of the European Union over a period of nearly 40 years. Although harmonisation throughout the Member States has been achieved in areas such as Maximum Residue Limits and Classification, Packaging and Labelling of pesticide products, attempts in the 1950's and more intensively in the 1970's to harmonise the registration of pesticides were unsuccessful. This failure was due to a number of factors, both political and scientific. In the late 1980's however a new initiative from the Commission led to a broader acceptance from the Member States possibly because of a greater environmental awareness within the European Community coupled with unilateral policy decisions in many Member States to either increase the regulatory control of pesticides or to reduce the overall use of pesticides or indeed both. Following some years of lengthy and complex discussion the first steps towards a harmonised registration system in the European Union were taken with the implementation of the so-called Registration Directive (Official Journal 1991) and with it fundamental changes in the way in which pesticide registrations will be dealt with in the future. These changes, which affects both Regulators and Industry alike are yet to be completely defined as the full requirements of the Directive continue to be developed. Additionally practical experience with the implementation of the Directive will lead to development of practical rules whilst the expansion of the European Union to include a further 4 Member States will have further repercussions. This paper discusses the likely impact of the Directive and its associated legislation not only in terms of its direct applicability in the European Union but also indirectly in areas such as Central Europe.

The European Registration 'Playing Field'

Conceptually the EU Registration Directive is simple and, some might argue, logical in that it firstly establishes a Community procedure for the 'approval' of active ingredients. Secondly, it allows Member States to register plant protection products containing these approved active ingredients within their own territories thus taking account of the varying climatic, environmental and agricultural conditions which pertain throughout the different countries in the European Union. Perhaps not unexpectedly however this conceptual simplicity is not reflected in the detailed procedures and requirements imposed on Member States and Industry by the implementation of the Directive. Thus complex procedures must be followed to achieve both Community approval of the active ingredient and the subsequent registration of the plant protection products. Accompanying these complex procedures is, perhaps the inevitable consequences of the harmonisation of regulatory data requirements, namely the increase in extent of these data requirements. Similarly the even

1054–7487/95/0382$12.00/0

more difficult objective of harmonising the evaluation of regulatory data has resulted in evaluation procedures (the so-called Uniform Principles) which are based on 'cut-off criteria' allowing less room for manoeuvre and directing the regulatory process away from the once traditional 'risk-benefit' analysis.

The Review Regulations

The comprehensive nature of the Directive and the significance of its future impact is further reflected in the fact that Regulations (Official Journal 1992 and Official Journal 1994) made under the Directive impose a demanding process for the 're-registration' of those active ingredients (and ultimately the plant protection products containing those active ingredients) on the Community market prior to the implementation of the Directive. Some 800 active ingredients are theoretically registered for use in the various, although not in all, Member States but in reality only about 350-400 of these are of any significance. The scheduled re-registration programme of reviewing all these 'existing' active ingredients is 10 years which is clearly over-ambitious but it nevertheless will have an immediate effect on those companies having compounds in the first list of 89 active ingredients to be reviewed.

Data Protection

A further major element of the Directive is the harmonisation of data protection provisions throughout the Union. Prior to the implementation of the Directive data protection laws or 'rules' varied significantly between Member States. One might therefore argue that harmonised data protection would bring some benefit to the Industry. Such benefits are however unlikely to be realised particularly with regard to existing (old) compounds which, because of the nature of the provisions, will undoubtedly make it easier for generic manufacturers to obtain registration for 'off-patent' products.

Implementation and Completion of the Directive

Two other key factors must be taken into account in assessing the impact of the Directive, at least in the short to medium-term. Firstly, very few Member States have, to date, implemented the Directive via National Legislation. This will inevitably prolong the transitional phase between National and Directive procedures, and with it the difficulties of predicting the attitudes and evaluative decisions of individual Member States. Secondly, many of the detailed Appendices, e.g. data requirements and the Uniform Principles remain to be finalised and subsequently adopted by Member States. This clearly poses problems in establishing data requirements for new compounds in various stages of development.

Consequences of the Directive

The Evaluation of 'existing' active ingredients has been allocated between Member States on a 'weighted' basis with the UK, France, Germany and Italy having 12 compounds each; Spain having 8; Portugal, Belgium, Greece and the Netherlands 6 each; Ireland and Denmark 4 each and finally Luxembourg having one. One of the major consequences of the Review Programme is the impact on the resources of Member States which will be faced with the problem of allocating their limited resources between the evaluation of the review compounds (an evaluation which according to the Regulation must be undertaken in 12 months) and the evaluation of new compounds (for which there is no time limit). Given the political influences on Agriculture within the European Union, there seems little or no pressure for either the Union or Member States to register new compounds with any degree of urgency - indeed some Member States would regard such delay as being politically desirable. By contrast, public and political pressures continue to stress the need for the review or even withdrawal of 'old' compounds in the mistaken belief that all such compounds are insufficiently supported by valid data bases.

Following completion of the review of an existing active ingredient and its inclusion in Annex I of the Directive the next phase of the Review Programme requires each Member State to re-evaluate all those products which contain the approved active ingredient and which are marketed in the territory of the Member State. When one considers that the UK Authorities have estimated that there are some 1,400 products on sale in the UK which are based on the 89 active ingredients in the First Review List, then it becomes readily apparent that certainly over the next decade, or even longer, Member State resources will be under considerable pressure. Although this Member State review of products will be phased in line with the inclusion of the active ingredients in Annex I, it should not be overlooked that the Second List of 90 compounds is expected to be published in late 1994 or early 1995, thus involving Member States in a further round of resource intensive activities.

One must therefore conclude that a 'negative' consequence of the Review Programme will be to delay the registration of new active ingredients and also of new uses or new formulations based on existing active ingredients.

Transition from a National to a Community approach

The transition from National to Community registration procedures is one which will not be achieved either easily or quickly. It should not be seen simply as the introduction of a new system but more as the need to adopt new attitudes, indeed a new philosophy. Given that National registration systems have been developed over some 40 years and that during this period regulatory decisions have been taken unilaterally, it will undoubtedly take some time to adapt to a system where major regulatory decisions will be taken on a Community basis and, under some circumstances, possibly by a weighted majority voting procedure. The extent and speed of these changes will vary from Member State to Member State which indicates the need to monitor these changes on a regular and effective basis.

Further changes

The European Registration 'Playing Field' is one which is not yet complete either in terms of its size or indeed the rules to which the 'game' must be played. The likely extension of the European Union in 1995 to include Sweden, Norway, Finland and Austria coupled with the increased power of the European Parliament will influence many ongoing and future discussions on a range of issues. The addition of these countries to the discussions on pesticide registration will result in a significant shift in the power base away from the more pragmatic countries such as France and the UK towards more 'environmentally active' countries such as Germany, Denmark and the Netherlands. For example, Sweden's pesticide use reduction programme is well established and much publicised whilst Austria's new registration requirements are amongst the most extreme in Europe.

The European Union's Common Agricultural Policy (CAP) and its associated Set Aside Schemes have resulted in the reduced production of major crops such as cereals and oil seed rape. Thus in the last few years the crop protection market in the European Union has decreased by at least 15%. Further changes in the CAP cannot be ruled out and may lead to even further decreases in production and in the sales of agrochemicals.

All those concerned with the Directive whether they be the Commission, Member States or Industry recognize that practical experience in applying the Directive will identify problems and that many 'rules' will have to be developed on the basis of this practical experience. One only has to consider that when the UK introduced its new legislative registration system in 1986 no new active ingredients were registered for 12 months and registrations of label extensions were considerably delayed. What then are the delays likely to be encountered when a new system is introduced throughout the European Union? Bearing in mind that problems, and their solutions, will have to be considered on an European basis it is difficult to believe that a 'steady state' will be achieved in less than 3 to 5 years.

Impact on New Compound Registration

From the comments made above it seems inevitable that the registration of new compounds will be delayed and that this delay will be most significant within the first few years of the implementation of the Directive. This in itself presents a problem but it is a problem which is exacerbated by the difficulty in predicting how extensive such delays will be. Prior to the implementation of the Directive it was possible to predict how long registration would take to achieve in individual Member States. Although there was always the possibility of an unexpected reaction depending on the Member State's specific evaluation of the compound in question, nevertheless the experience built up over a period of years enabled registration times to be predicted with a certain degree of 'standard error'. With the implementation of the Directive, Registration Personnel in Industry are starting from 'square one' with no experience on which to base an estimate of the registration time. This in turn has a direct consequence on planning the development of new compounds, product launches and investment in new production facilities. Furthermore such experience will need to be established over a period of time and based on the experience of all companies operating in all the Member States. Given the uncertainties involved it is perhaps a little dangerous to make predictions but it might be estimated that a product containing a new active ingredient is unlikely to appear on the European market until 1997 or more probably 1998.

Provisional Registration

To some extent this likely delay of registration under the Directive is recognized within the Directive itself in that provisions are made whereby 'provisional registration' in individual Member States is allowed whilst the Community evaluation is ongoing. Such a provisional registration may be granted by individual Member States for up to a period of 3 years. Simplistically this might be viewed as the possibility to operate much as is done presently, (i.e. applying for provisional registration in individual Member States), but at the same time and in parallel applying for Community approval and inclusion of the active ingredient in Annex I. Two factors however must be considered in assessing the full benefits of provisional registration and the speed with which it will be granted.

Firstly, and as discussed above, one of the many consequences of the Directive is a shift in the regulatory decision making process away from individual Member States to the Community. Member States may therefore be somewhat reluctant to take such decisions 'in isolation'. In other words Member States may not wish to grant provisional registration for an active ingredient which they feel might run into difficulties on a Community basis. In this respect it should not be overlooked that any provisional registrations granted for products containing an active ingredient which subsequently is not included in Annex I, must be withdrawn. Clearly under such circumstances Member States might be somewhat reluctant to act as independently as they might have done in the past. Instead they might prefer to wait and see how the Community evaluation is progressing or at least to solicit the views of other Member States who are considering an application for provisional registration.

Secondly, provisional registration under the Directive should not be confused with the types of provisional approvals previously granted by such countries as the UK and France. Typically, and on a case by case basis, these latter provisional registrations could be achieved on the basis of partially complete dossier provided the Authorities were satisfied with the submitted data and were assured regarding the submission of further data to complete the dossier and the likely outcome of these additional data. Classically, in France, it was thus possible to achieve a provisional registration (APV) on the basis of 12-months interim reports from the long-term toxicology studies. Under the Directive however provisional approval can only be granted by a Member State when a complete dossier has been submitted indeed the Directive lays down a procedure whereby the 'completeness' of the Dossier is 'checked' (ie formal approval by all Member States constituted as the Standing Committee on Plant Health) before the Member State can authorize provisional registration.

Traditionally many Companies have frequently targeted France as the first country in which to obtain registration. Whereas French 'pragmatism' is unlikely to be entirely submerged by the provisions of the Directive, it is clear that the 'old style' provisional registration will no longer be possible and that this traditional approach will have to be modified to the extent that a much wider, European approach will need to be adopted.

It has been argued that the fact that the Directive imposes a 'standardised' European dossier is one of the benefits to arise from the new system. Whereas there are undoubted benefits from such a 'standardised' dossier, these benefits do not really offset the financial disadvantages of delaying registration in a country with the market size of France.

Impact on Existing Compounds

Many of the consequences on products currently marketed in Europe have been discussed above. Some additional points worthy of consideration are as follows:

i) The deadline for the submission of dossiers for active ingredients in the First List of 89 is 30 April 1995. No flexibility is possible expect where delays in the preparation of the dossier is directly attributable to efforts made to submit a single dossier on behalf of more than one producer of the active ingredient (ie where a 'Task Force' has been established).

ii) Some companies have a significant number of compounds in the Review Programme -14 in the case of AgrEvo for example and this number of compounds and the limited time allowed for by the Regulation in which to submit the dossiers has obvious impact on Company resources - not only within Regulatory Affairs but also in those Departments such as Toxicology, Residues, Environmental Sciences, etc., which need to provide expert opinions on the data to be submitted.

iii) In some cases new data will have to be generated to fill data gaps which arise as a result of the new data requirements imposed by the Directive.

The Impact of Generic Manufacturers

In the short-term, i.e. before completion of the Review, there is little or no impact on these existing compounds and the sales of the wide range of products in the various Member States remains unaffected. One of the major impacts on existing compounds however results from the data protection provisions of the Directive.

Whereas the Directive addresses the issue of data protection it does so only in very broad terms and has left much of the interpretation of the complex details to be resolved by Member States. Despite considerable efforts by Industry to clarify these issues, doubts remain regarding the way in which these provisions will be applied in practise as administration of data protection will be undertaken by Member States and not the Commission.

It is quite clear however that as far as existing compounds are concerned registrations for generic manufacturers will become easier in the future. This is clearly of great importance to those companies which market products for which patent protection has expired or is due to expire over the next few years.

Management of the Registration Dossier

Traditionally the approach to the compilation of the registration dossiers has been to develop a 'core data base' and to add to this to meet the needs of individual European countries, eg local residues and environmental data, additional mutagenicity studies, etc. In theory the

Directive requires a dossier which to a large extent will have a pan-European applicability - even the residues data package will be representative of European regions rather than individual countries. These considerations therefore dictate the data necessary to complete the dossier and indeed the way in which the dossier is assembled in terms of individual study reports and summary documentation. An important principle arising from the Directive therefore is that a single dossier on the active ingredient will be submitted to all Member States and that the dossier on the formulation will also be essentially the same for all Member States.

This then requires a different approach to the management of the registration programme in the Community. As a first step the applicant must decide as to which Member State (the Rapporteur Member State) it should submit the dossier for evaluation and inclusion of the active ingredient in Annex I. A number of factors have to be taken into account in making such a decision, eg main use of the product, known workload of Member States, degree of contact with officials in Member States, etc.

Once the dossier is submitted and an initial evaluation undertaken the Rapporteur Member State it seems likely that various sections of the dossier, eg residues, toxicology, environmental fate and effects, will be considered by small teams (5 persons) of experts from different Member States before a final review of the whole dossier is undertaken by all Member States in plenary session. This evaluation procedures thus requires the management of the dossier on a truly European-wide basis to ensure the efficient progress of the evaluation procedure. Additionally, the management of the dossier for the Community evaluation must also be integrated with any activities associated with the possible application for provisional registration in some Member States.

Central Europe

This Paper has concentrated mainly on the European Union, although some account has been taken of the extended membership of the Union from January 1995. Whilst Central Europe requires special consideration there are some similarities with the regulatory developments in the European Union. Thus in the light the political changes which have occurred closer links between Central European and the European Union have been reflected in considerable interest being shown in the requirements and progress of the Registration Directive. There is therefore a clear indication that the requirements of the Directive are gradually being incorporated into the systems operating in many of the Central European countries.

This has some disadvantages in that the regulatory requirements in Central Europe are likely to increase over the next few years but to some extent this will be offset by the utilisation of the European Registration Dossier to meet the most, if not all, of these requirements. Any attempts however to introduce additional local requirements must be carefully monitored and resisted where not scientifically justified.

Clearly this is an area which is still developing and it is to be hoped that any efforts to extend the regulatory control of pesticides in Central Europe will draw on the Registration Directive as a model rather than to copy it directly.

Conclusions and Recommendations

This review of the regulatory scene in the European Union is of necessity a 'snapshot' and further changes can confidently be predicted. On the basis of current knowledge however it is readily apparent that the full implementation of the Registration Directive and its associated Regulations will represent the biggest single change ever to have occurred in terms of achieving registration of new products in Europe and in maintaining existing products on the market. Both Industry and Regulators must adapt to these changes and given the difficulties and confusion which will arise over the next few years, success in adapting to these changes will be reflected in the successful introduction of new products on to the market within a reasonable period. Failure to do so will result in the European crop protection industry being placed at a commercial disadvantage to its main competitors in the USA and Japan.

References

Official Journal of the European Communities No L230/1, 19 August **1991**, Council Directive of 15 July **1991** Concerning the Placing of Plant Protection Products on the Market : 91/414/EEC

Official Journal of the European Communities No L366/10, 15 December **1992**, Commission Regulation (EEC) No 3600/92

Official Journal of the European Communities No L107/8, 28 April **1994**, Commission Regulation (EC) No 933/94

Pesticide Registration and Establishment of Maximum Residue Limits in Latin America

Roberto H. González, Department of Plant Protection, University of Chile, Casilla 1004, Santiago, Chile.

Despite toxicological and environmental concerns about pesticide usage, the demand for these chemicals has continued increasing in the L. American region. The need to meet pesticide tolerances on international trade has prompted some governments to set up registration schemes which, for the most part, are mere authorizations for sale rather than control mechanisms. Nationally developed Maximum Residue Limits (MRLs) to suit local needs are practically non existing in the region, although Codex Alimentarius MRLs are being partly adopted at a greater pace. Analytical facilities are limited to very few countries, particularly to assist food exports rather than conducting domestic residue surveillances. The potentials and limitations of pesticide registration in the region are assessed.

For a vast agricultural region such as Latin America, having a wide array of socioeconomic and ecological conditions, the need to increase fiber and food production and share more strongly the demanding world food trade is currently urging governments to set up pesticide registration and control legislation schemes. However, many countries in the region still lack appropriate pesticide legislation and adequate mechanisms to enforce regulations effectively, notwithstanding the protection of human health and the environment are issues of great concern.

The primary objective in regulating chemical pesticides is to provide society with maximum protection from adverse effects while not denying it access to benefits (Snelson, 1978). In other words, the registration process aims at avoiding unnecessary dangerous dosages of approved pesticides and protects the environment from all possible risks. However, a number of other considerations are also included among the decisions towards registering a new chemical, considerations which vary widely among developed and developing countries. In fact, the registration process in some countries is merely a sale permit, which does not define specific usages and residue tolerances on target crops, whilst more advanced countries set up a registration scheme which implies the acceptance by statutory authority of extensive documented proof in support of all claims for efficacy and safety made for the proposed product. Hence, a pesticide assessment capacity is also part of a complex regulatory system.

Registration in most developed countries enables authorities to exercise control over use levels, labelling, packaging and advertising and thus to ensure that the interest of end-users are will protected. If every pesticide has to be registered the public will be aware that the product has satisfied the requirements of the law as to its effectiveness and safety when used according to good agricultural practices, the latter a component of the proper public's responsibility.

Governments of developed countries must establish legislations to regulate manufacture, sale and safe use pattern for each chemical, including the provision for safe legal limits for residues in food and feeds. On the contrary, many developing countries in the region are more concerned with efficacy and, consequently, the registration permit applies to a pest/disease problem rather than to a plant/product system on a crop by crop basis. In other words, once the registration is obtained, the chemical has no limitations in terms of candidate crops in which it can be used.

The directions on registered labels have been developed at great cost in time and scientific manpower, have been evaluated by experienced scientists and have finally been approved by government authorities. This information, in addition to the toxicological and

1054–7487/95/0389$12.00/0

environmental data, once the chemical is approved for use in a developed country, could be used for submission to the registration authorities of interested developing countries to avoid unnecessary duplication of efforts. In practice, it would only need the addition of use pattern data relevant to the registrant country. However, as a general rule, registration acceptances are granted on the basis of "a third country data" without pertinent, local inputs.

Very few pesticides just developed by the industry are first registered in L. America. They are rather second hand registrations already approved by developed countries and, therefore, with all the necessary basic credentials. However, triangulation from the industry to a developing country on its possible avenue to a developed nation, is not rare, particularly to draw field effectivity data. In this regard, the so-called experimental registration, which in a developed country requires very special authorizations, including the purchase and disposal of the target crop, it may sometimes skip some of these steps under particular developing countries legislations. In this regard, the region is currently noticing the advent of new chemicals produced by the Japanese industry which first land in L.America with a view to develop further efficacy information required by Western Hemisphere countries.

On the positive side, the implementation of the variety of registration schemes in the region has permitted the discard and suppression of tolerances of so far elsewhere cancelled pesticides, as a first step to harmonize pesticide usage within the region and avoid legal residue problems on the international trade front.

With respect to a possible harmonization of registration procedures, it would be unrealistic to expect that regulatory mechanisms in the region would ensure the adoption of a single and "developed country registration model", including safety testing, residue work and associated environmental issues. On the contrary, the adoption of these requisites would do nothing but delaying and perhaps precluding the registration of many valuable chemicals. In this concern, it should be noted that the region is nowadays witnessing the incorporation of environmentalists into the registration bodies in demand, among other things, of the inclusion of environmental impact studies as an additional tests requisite to approve new chemicals of possible concern to the wildlife forest and water resources. The bureaucracy of the impending registration schemes is still to be seen among countries in the region since no definite protocols are available to determine the pesticide effect once this has been released in particular environments. In the evaluation of safety of pesticides, as in other fields of human endeavor, some degree of risk must be considered acceptable to society. The alternative would be needless prohibition of important social and economic benefits.

All decisions regarding the adoption of new chemicals in the region have been made, nevertheless, on incomplete information to the real needs of the country. As a matter of fact, the most important element in this process has been the mature judgement of experienced professionals more than in a complete dossier such as that required by more stringent legislations. Registration, for the most part, has been granted when the authority is satisfied that the information made available at that time is sufficient to demonstrate that the product is effective against the target pest problem under practical condition of use.

International efforts promoting pesticide registration schemes

In view of the slow pace of government actions leading to the establishment of pesticide regulations, and following Resolution XII by the FAO Ad Hoc Consultation on Pesticides and Agriculture in 1975 that "FAO, in collaboration with WHO, call an International Consultation to analyze and discuss the basis for harmonizing the requirements for registration of pesticides in different countries ...", called for an Ad Hoc Government Consultation in 1975 to analyze the basis for harmonizing the requirements for registration of pesticides. This meeting was previously prepared by an Expert Group on Pesticide Registration to set up a uniform, basic registration scheme for its eventual adoption by member nations lacking this regulatory system (FAO, 1977a, 1977b).

The Group considered that, irrespective of the type of pesticides, its properties or applications, the information required for its evaluation and registration could conveniently be classified into six issues:

1) Chemical and physical properties
2) biological efficacy
3) toxicology
4) residues in foods and feeds
5) environmental considerations, and
6) labelling, packaging, transport and disposal.

All these six issues are generally represented among current national requisites to apply for a pesticide registration; however, their degree of fulfillment, is far from complete in all known national registration procedures.

A further meeting of this Group of Experts was convened in Rome in 1982 to issue relevant documentation for a second Government Consultation (FAO, 1982). The preparation of Guidelines and Model Schemes for the Registration Control of pesticides and the draft Code of Conduct in the Distribution and Use of Pesticide, were great achievements emerging from these meetings. With respect to the concern expressed about supplying pesticides to countries which did not have the infrastructure in place to control them, as well as in the absence of a registration scheme, the Consultation agreed that until a country had the capacity to manage the restriction of pesticides, the elaboration of an internationally agreed Code of Conduct in the Distribution and Use of Pesticides (the Code of Conduct) could offer a major contribution to the safe and efficient use of pesticides.

The International Code of Conduct and the Registration schemes

The Second FAO Government Consultation on International Harmonization of Pesticide Registration Requirements 1982, strongly supported the Code proposal and requested FAO to proceed with the preparation of a draft Code, in cooperation with other U.N. agencies, especially WHO, UNEP, ILO, UNIDO and with a number of international non-governmental organizations representing a wide spectrum of interests, including the pesticide industry (GIFAP), the International Organization of Consumers Union (IOCU) and the Environmental Liaison Centre (ELC). After extensive consultations over a period of more than three years, the Code was formally submitted to member governments for consideration by appropriate authorities. The Code was adopted by the 23rd Session of the FAO Conference, 1985, and addressed to member nations, particularly those lacking pesticide regulatory mechanisms.

The text of the Code was published by FAO in 1986 and includes 12 articles which make provisions for all pesticide related aspects such as pesticide management, testing of pesticides, health hazards, information exchange, labelling, storage and distribution. Of special interest to this paper is Article 6 dealing with Regulatory and Technical Requirements, which includes the setting up of a pesticide registration and control scheme at the national level, the acquisition and evaluation of supporting data for registration, and questions on quality verification.

Since the issuance of the Code in 1986, a number of important initiatives have been adopted in Latin America at the regional and subregional levels, to promote its implementation and assessing its impact and feasibility. Major issues emerging from meetings held in the 1987-1993 period refer to harmonization of registration procedures, strengthening of regulatory national services, survey missions on training capabilities, laboratory infrastructures for pesticide analysis, and a number of related initiatives.

In the meantime, the FAO Conference at its 1987 session directed the FAO to incorporate into Article 9 on Information Exchange of the Code of Conduct, the concept of Prior Informed Consent (PIC) with a view to assessing PIC options for interested countries.

At this point, it is worthwhile to defining Registration as per the International Code of Conduct, Article 1, as meaning "the process whereby the responsible national government authority approves the sale and use of a pesticide following the evaluation of comprehensive scientific data demonstrating that the product is effective for the purposes intended and not unduly hazardous to human or animal health or the environment".

Although, as a result of such lengthy definition, involving not always available data, it should be expected that each national Registration bureau is doing its best to come up, at least, with a roster of authorized chemicals in the hope that, in the foreseeable future, a more complete evaluation of candidate pesticides could be made.

Various sectors of the community have varying responsibilities under any pesticide registration scheme. Most of these are set out in the Code of Conduct and include manufacturers, governments, vendors and users. Therefore, the responsibility of an operational system must be necessarily shared among parties concerned, all efforts coordinated by the government authorities.

It has been suggested above, that developing countries do not need to introduce elaborate regulatory schemes in order to control pesticides effectively. They should design procedures suited to their own specific needs and should not attempt to adopt all the elements of regulatory schemes used in developed countries. For example, in the case of exporting countries, national registration mechanisms should provide for (a) making available pesticides needed for domestic crops, and (b) selecting very carefully the pesticides for export crops among those provided with tolerances by the importing country. Under the latter circumstances, the exporting country should also develop the necessary infrastructure to analyze pesticide residues and conduct supervised trials to establish preharvest intervals data and thus meeting import tolerances.

To sum up, when a country has decided that some measure of control of pesticides is desirable and feasible and once a commitment to that is reached, it will need to decide on the extent of the resource and effort to be put into that control.

Pesticide registration schemes in the region

Given the largely agricultural use of pesticides, the most appropriate responsible authority would normally be the Ministry of Agriculture. However, since human health and environmental issues are involved, a coordinated authority may deem necessary. For example, in Mexico the registration procedures depend from an Intersecretarial Commission (Agriculture, Health, Environment and Commerce), whilst in Chile, registration procedures are only authorized by the Ministry or Agriculture in coordination with Customs to permit registered pesticide imports.

Prior to the concern about pesticide registrations, a Pesticide Law was available in most countries of the region. Hence, many provisions still exist under the authority of such a law, covering imports, sale, use, application, labelling, storage and disposal. Consequently, the establishment of an effective, modern pesticide registration system may take several years if a new legal instrument is required under a coordinated authority.

Pesticide laws and registration procedures are not always harmonized in their definitions and meanings. An example of such legislation in which the definition and understanding of a registration process is the recent Pesticides Control Act issued in Belize in 1985, "an Act to provide for the control of the manufacture, importation, sale, storage and use of pesticides, and for matters connected therewith or incidental thereto". According to the Law a registered pesticide is defined as "a product declared as a registered pesticide and given in Schedule II to this Act, and is intended for general use". Under Schedule II, a few, selected examples of such "registered pesticides" include: atrazine, benomyl, dinoseb, alachlor, just to name a few.

A subregional approach for the setting up of a pesticide registration scheme has been proposed by the Central American countries during the Second Technical Meeting on Harmonization of Registration Procedures for Central America (El Salvador, October 1993). A similar approach is sought by Caribbean countries, except Dominican Republic.

The Cuban Registration procedure, approved in 1985, is an example of a sound but unrealistic legislation. The relevant norm establishes that prior to issuing a registration it will be mandatory to evaluate the biological effectivity and dosages, the analytical methods and the residue degradation curve. Once the product is evaluated the registration will be granted for a 5-year period and the approved chemicals will be published in the annual Official List of Authorized Pesticides. The 1993 Official List includes over 100 active ingredients and a number of mixtures and provides recommendations on target crops and authorized dosages. No provision for MRLs is made. Under our present state of information, candidate pesticides can not be evaluated as the norm indicates.

The Mexican Official Pesticide Catalog is an annual undertaking listing registered chemicals, approved used, dosages and legal residue tolerances. It comprises over 100 insecticides, 70 fungicides, 75 herbicides, 7 fumigants and 4 nematicides. However, the allowed uses in terms of target crops is very limited and we wonder if with such pesticide arsenal it is possible to restrict their uses only to the crops allowed in the Official List. On the other hand, the recent North American Free Trade Agreement (NAFTA) should also harmonize pesticide usages among participating countries, an action which will eventually require a whole revision of the Mexican registration procedures.

In South America, a complete pesticide legislation is found in Brazil. The basic law published in 1989 was issued under the authority of the Agriculture and Health sectors; however insofar registration concerns, the Environmental sector has added new, cumbersome inputs. Consequently, less tolerances than needed for the domestic production are provided and the single preharvest intervals (PHI) are not realistic for such a vast territory. For example, there are PHI data which are not compatible with the good agricultural practices, e.g. recommending a PHI of 30 days for a clear cut postharvest fungicide treatment.

In Argentina, the Secretary of Agriculture is responsible for pesticide registration and for the fixing of legal residue tolerances and preharvest intervals. However, the shortage of analytical facilities throughout the country makes these exercises a difficult, if not an impossible task.

Pesticide registration in Chile is carried out since 1984 by the Ministry of Agriculture. The process just requires the submission of the manufacturers pesticide profile data and a certification from Universities or National Agriculture Research Stations that the chemical has been field tested against local phytosanitary problems. Since Chile is the largest fresh fruit exporting country from the Southern Hemisphere, the private export sector has in addition contributed with analytical facilities to conduct residue degradation studies and establish updated PHIs for pesticide/crop combinations exported to over 45 world markets (Pesticide Agenda, Chilean Exporters Assoc., 1994).

Other large pesticide consuming countries in the region, namely Colombia, Ecuador, Perú and Venezuela, have been able to establish a so-called registration process, which is again a list of authorizations for sale of non quite scrutinized chemicals.

Conclusions

After 25 years of international efforts trying to implement at the country level the establishment of pesticide registration schemes, due to the great diversity in requirements and the great emphasis on data which is not available among developing countries, there has been a great delay and diversification of the original goals. Another fact which has to be understood is that the stringency in registration requirements mounted in recent years in the developed world, has impeded the adoption of newer chemicals to the detriment of developing countries where pesticide registration procedures are in an early stage of evolution.

The ultimate goal in regulating chemical pesticides is to provide users with maximum protection from adverse effects, a matter which should be easily feasible within a simple registration mechanism and information exchange, provided that a minimum of analytical skills are made available in the country.

Registration schemes in the region, can be easily mounted by submitting the necessary evaluations provided by manufacturers as they contain essential information on the chemical, physical and toxicological aspects of the chemical. To demonstrate that the product will provide effective control against the target pests, local field trials on efficacity can be carried out. Regarding toxicological data, no countries in the region have the expertise available to undertake a comprehensive toxicological evaluation, which is otherwise available as part of the registration package.

Maximum residue limits and the analytical methods used to regulate them are also properly considered as a part of registration requirements, even though they usually are not a direct part of the registration process itself. The registration authority has the obligation of developing efforts towards this aim. Some countries with registration laws do not have a national tolerance system and rely on the international MRLs recommended by the Codex Commission on Pesticide Residues. This choice is fine provided that the country develops analytical facilities to determine residues on food imports and exports.

The Registration schemes in most of the region is a licensing process which permits a pesticide product to be imported and marketed. This is an unsufficient action and countries should look forward to develop further evaluation methods in line with the domestic and foreign needs.

References

Chilean Exporters Association. Pesticide Agenda. 1988-1994.

FAO. Report of the First Session of the Group on Pesticide Registration Requirements, Rome, 28 June - 4 July, 1977. Meeting Rpt. AGP:1977/M/4

FAO. Report of the Ad Hoc Government Consultation on International Standardization of Pesticide Registration Requirements. Rome, 24-28 October, 1977. Meeting Rpt. AGP:1977/M/9.

FAO. Report of the Fourth Meeting of the Group on Pesticide Registration Requirements. Rome, 10-14 May, 1982. Meeting Rpt. AGP:1982/M/4.

FAO. Report of the Second Government Consultation on International Harmonization of Pesticide Registration Requirements. Rome, 11-15 October, 1982. Meeting Rpt. AGP:1982/M/5.

FAO. Código Internacional de Conducta para la Distribución y Utilización de Plaguicidas. Rome, 1986. 37 p.

FAO. Guidelines for the Registration and Control of Pesticides. Rome, 1985.

República de Cuba. Lista Oficial de Plaguicidas Autorizados. 1993, 260 p.

Snelson, J.T. The need for and principles of pesticide registration. FAO Plant Prot. 1978, 26:96-100.

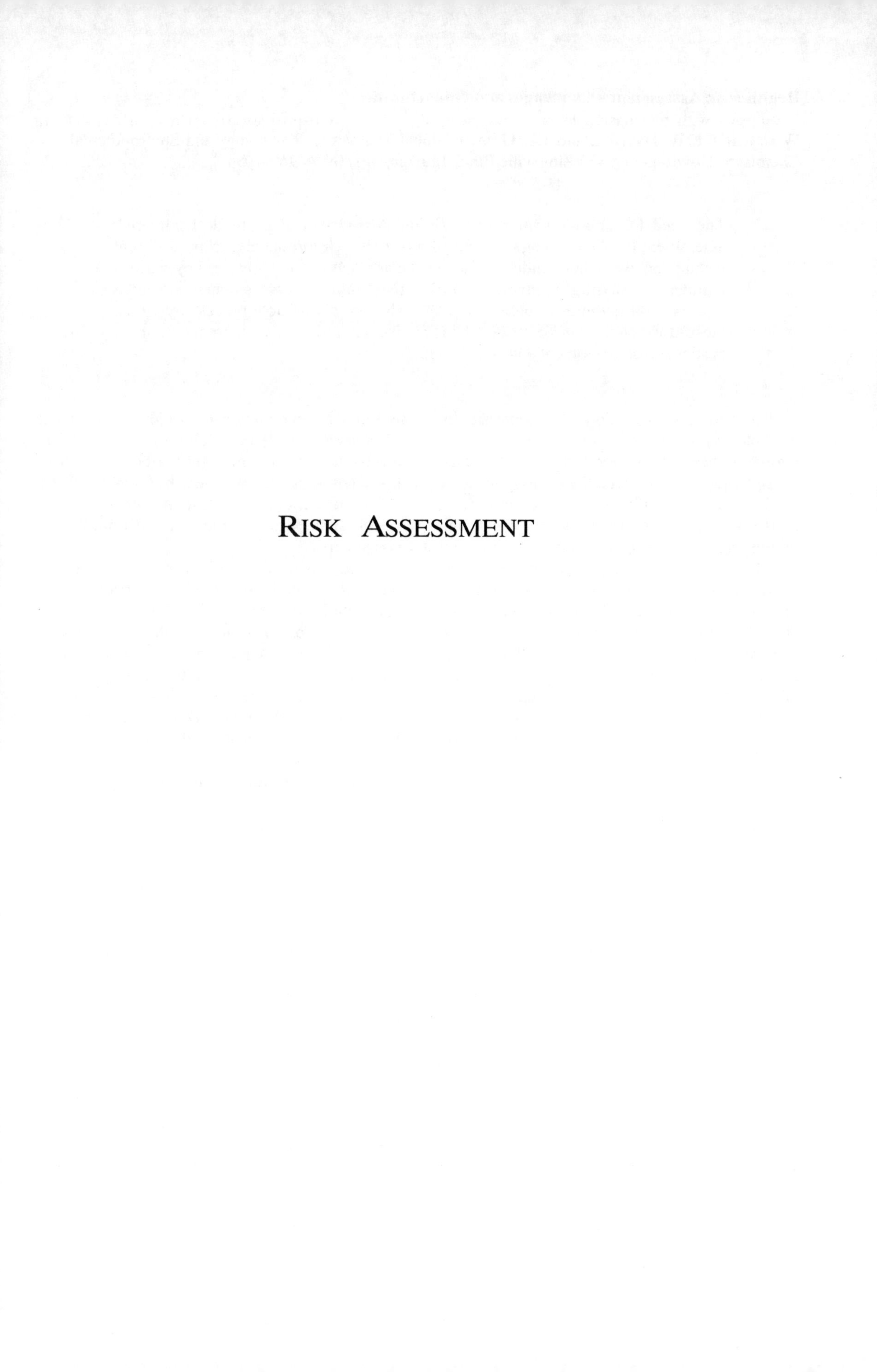

Risk Assessment

Health Risk Assessment - Challenges and Opportunities

W.L. Chen, E.W. Day, Jr., and J.E. Gibson, Global Regulatory, Toxicology and Environmental Chemistry, DowElanco, 9330 Zionsville Road, Indianapolis, IN 46268-1054

The need for international harmonization in the field of pesticide health risk assessment has been recognized for many years. Unfortunately, many different testing and evaluation guidelines have been adopted and implemented by various countries, creating significant differences in risk assessments. Recent international harmonization efforts have provided a workable process by which these differences can be resolved in moving towards global harmonization of pesticide risk assessment and registration.

The need for international harmonization in the field of pesticide health risk assessment and regulation has been recognized for many years. This recognition prompted the development, by the Food and Agriculture Organization (FAO), by the World Health Organization (WHO) and by the Organization for Economic Cooperation and Development (OECD), of recommendations for the harmonization of data requirements, and of guidelines as to methodologies for generating data. Unfortunately, many different testing and evaluation guidelines have been adopted and implemented by various countries, creating significant differences in risk assessments. This paper examines three key differences in: (1) Test Guidelines, (2) Maximum Tolerated Dose (MTD), and (3) Cancer Classification and Risk Assessment. The American, European and Japanese approaches in these important areas are compared in light of the current scientific knowledge.

Test Guidelines

There is no area of the chemicals field where the multiplicity of work is so obvious as in the pesticides sector. Many different data requirements and testing guidelines have been adopted and implemented by various countries, creating significant differences in studies required and the methodologies to be used in generating data.

Differences in the data requirements of various countries and in the methodologies accepted by them for the generation of data, result in very significant additional costs for industry in generating data to address issues that often have already been fully investigated. The additional data thus generated add little knowledge as to the health impact of the pesticide concerned, while the costs arising necessarily are passed on, ultimately to consumers.

The urgent need for international harmonization in the field of pesticides has prompted the recent development of the Pesticides Forum project by the Organization for Economic Cooperation and Development (OECD). In the formation meeting held in Saltsjobaden, Sweden on October 29-31, 1991, it was concluded that OECD would take the lead in developing internationally harmonized test guidelines which would be acceptable for pesticide registration by all member countries. With regard to data requirements, OECD will develop the common core data sets which are mutually acceptable for registration of new and old pesticide active ingredients.

This effort by OECD has provided a workable process by which the differences in test guidelines and data requirements can be resolved in moving towards global harmonization. The European Community (EC) countries have recently decided to stop developing their own EC test guidelines and to accept OECD guidelines as their study protocols. If this approach were to be adopted by the United States (U.S.) and Japan, then the goal of international harmonization of test guidelines is within reach.

Unfortunately, neither U.S. nor Japan have fully accepted studies performed according to OECD test guidelines when these differ from

1054–7487/95/0396$12.00/0

guidelines of the U.S. EPA FIFRA and Japanese Ministry of Agriculture, Fisheries and Food (MAFF). A comparison between different guidelines for acute toxicity studies is summarized in Table 1. It can be seen that, in comparison with OECD test guidelines, both U.S. EPA and Japanese MAFF require more animals to be tested in acute studies. Furthermore, Japanese MAFF requires pharmacokinetics and metabolism studies conducted in dogs in addition to rats. A comparison between different guidelines of metabolism studies is shown in Table 2.

The U.S. EPA has recently proposed to revise their reproductive and developmental toxicity test guidelines. In addition, a new test guideline for assessing immunotoxicity will be introduced as a required study. The new guidelines will require many major additions or changes. For example, sperm and cyclicity data will be required in reproduction studies. In developmental toxicity studies, dosing of pregnant animals throughout the entire period of gestation, rather than only during the major period of organogenesis. The number of pregnant rabbits will increase from 12 per group to 20 per group. These changes, if implemented, will increase the differences between U.S. test guidelines and the rest of the world.

In the interest of achieving global harmonization of test guidelines, it would be reasonable for the U.S. EPA not to implement these new guidelines at present but to submit these new test guidelines to OECD for development as internationally harmonized study protocols.

Maximum Tolerated Dose (MTD)

Much of the controversy on MTD has been a result of inconsistencies in its definition and interpretation. These inconsistencies exist between the U.S. and most of the OECD countries. It is a source of disharmony between EC countries and Japan on one side and U.S. EPA on the other, in the registration and regulation of pesticides. It is therefore important to examine the MTD as it is defined and interpreted globally.

The OECD guideline for carcinogenicity studies gives this description of the MTD (OECD, 1987):

> The highest dose level should be sufficiently high to elicit signs of minimal toxicity wthout substantially altering the normal life span due to effects other than tumors. Signs of toxicity are those that may be indicated by alterations in certain serum enzyme levels or slight depression of body weight gain (less than 10 percent).

The EC Registration Directive (Directive 91/414/EEC, 1993 draft Annex II) describes the dose-selection process as follows (EC, 1993a):

> The doses tested, including the highest dose tested, must be selected on the basis of results of short-term testing and the level of possible human exposure, and where available at the time of planning the studies concerned, on the basis of metabolism and toxicokinetics data, such that at the highest dose, definite but minimal signs of toxicity are elicited (viz, slight depression in body weight gain), without causing tissue necrosis or metabolic saturation, and without altering normal life span due to effects other than tumors. Higher doses causing excessive toxicity are not considered relevant to evaluations to be made.

The World Health Organization (WHO) in discussing the selection of dose levels and the validity of MTD concept, states (IPCS, 1990):

> Results of studies at dose levels many orders of magnitude above the level of human exposure . . . (are of) little relevance. Instead of using the MTD to select the top dose level, the use of properly designed biotransformation studies over a range of doses (including human exposure level) may provide a more rational basis for dose selection in long term animal studies.

All three guidelines define minimal toxicity for selection of the highest dose. In contrast, the EPA takes the position of eliciting significant toxicity as its requirement for MTD (EPA, 1987):

> The highest dose to be tested in the oncogenicity study should be selected

Table 1

Acute Toxicity Studies - Comparison Between Different Guidelines

Type of Study	OECD	EEC	EPA	JMAFF
Acute Oral Toxicity	**401**	**84/449 B.1**	**§81-1**	**-**
Species	not specified		rat	rat, dog or other animal
No of species	one (preferably rat)	=*	=	**two (including rat)**
No of animals in Test	**5 animals/group**	5 males, 5 females/group	5 males, 5 females/group	5 males, 5 females/group
No of dose groups	3	=	=	5
Limit test	2000 mg/kg[1]	5000 mg/kg	5000 mg/kg	5000 mg/kg
Observation period	14 days	=	=	=
Acute Dermal Toxicity	**402**	**84/49 B.3**	**§81-2**	**-**
No of species	one (rat, rabbit or **guinea pig**)	one (rat <u>or</u> rabbit)	=	=
No of animals in Test	5 animals/group	5 males, 5 females/group	5 males, 5 females/group	5 males, 5 females/group
No of dose groups	3	=	=	=
Limit test	2000 mg/kg[1]	=	=	=
Exposure period	24 hours	=	=	=
Observation period	14 days	=	=	=
Acute Inhalation Toxicity	**403**	**84/449 B.2**	**§83-1**	**-**
No of species	one (preferably rat)	=	=	=
No of animals in Test	5 males, 5 females/group	=	=	=
No of dose groups	3	=	=	not specified
Limit test	5000 mg/m^3	**20 g/m^3 of gas** or 5000 mg/m^3	=	=
Exposure period	4 hours	=	=	=
Observation period	14 days	=	=	=
Acute Intraperitoneal Toxicity	not requested	**proposed 91/414 A.5**	not requested	**not requested**

[1] when a limit test is done 5 males and 5 females per group are stipulated
* = similar to the previous square

Table 1

Acute Toxicity Studies - Comparison Between Different Guidelines (continued)

Type of Study	OECD	EEC	EPA	JMAFF
Acute Dermal Irritation	**404**	**84/449 B.4**	**§81-5**	**-**
No of species	one (preferably albino rabbit)	=	=	=
No of animals in Test	**3**	=	**6**	**6**
'Dose'	0.5 ml/0.5 mg	=	=	=
Exposure period	4 hours	=	=	=
Observation period	normally 72 hours if signs are still present up to 14 days	=	=	=
Acute Eye Irritation	**405**	**84/449 B.5**	**§81-4**	**-**
No of species	one (preferably albino rabbit)	=	=	=
No of animals in Test	3[2]	=	**6**	6[3]
'Dose'	0.1 ml, not more than 0.1g	=	=	=
Observation period	normally 72 hours, if signs of eye lesions are still present up to 21 days	=	=	=

[2] single animal, if marked effects are anticipated

3 If signs of irritation did not disappear after 72 hours, an additional test for the effect of eye washing shall be conducted with at least 3 animals. In this case, the treated eyes shall be washed out 2-3 minutes after instillation and subjected to observation and scoring as normal.

* = similar to the previous square

Table 2

Metabolism Studies - Comparison Between Different Guidelines

Type of Study	OECD	EEC	EPA	JMAFF
Metabolism in Animals	**417**		**§85-1**	
Species	not specified		rat	rat, dog or other animal
Application and dose levels	single oral, low dose (NOEL)[1] single oral, high dose (EL)[1]		**i.v. low dose (A)** single oral, low dose (B) **repeated oral, low dose (C)** single oral, high dose (D)	single oral, low dose single oral, high dose **(single oral, approx. dietary exposure[2])**
Parameters to determine:				
- Toxicokinetics - Absorption	determine, no method specified		determine, no method specified	**periodic measurement of the test compound in blood, plasma or serum**
- Excretion	determine label in urine, feces and expired air till more than **95%** of the label is recovered or for 7 days		determine label in urine, feces and expired air till more than **90%** of the label is recovered or for 7 days (Groups A-D)	determine label in urine, feces and expired air till more than **90%** of the label is recovered or for 7 days
- Distribution in tissues	**determine, no organs specified**		quantify residues in bone, brain, fat, gonads, heart, kidney, liver, lung, muscle, spleen, residual carcass and target organs (Groups B-D)	quantify residues in bone, brain, fat, gonads, heart, kidney, liver, lung, muscle, and spleen at different time points, determine distribution ratio
- Metabolism	identify metab., not specified in which excreta, organs or tissues		**identify metabolites in urine and feces (Groups A-D)**	determine, not specified in which excreta, organs or tissues
Dermal absorption study			**§85-3 (draft)**	
when required	not requested		required on individual basis (Zendzian protocol)	not requested

[1] applies to all guidelines (EL = (toxic) effect level), NOEL = No Effect-Level
[2] preferred

below a level which resulted in significant life-threatening toxicity in the subchronic study. The level should not be selected too far below a life-threatening level because the highest dose tested in the oncogenicity study should elicit significant toxicity without substantially altering the normal life span of the test species from effects other than tumor formation.

The consequences of selecting significant toxicity versus minimally toxic doses are enormous. As a result of differences in high dose selection, many rodent carcinogenicity studies were rejected by U.S. regulatory bodies, for not reaching MTD, even after they have been accepted by European and International agencies. The repeated studies add little knowledge in terms of real hazard identification or risk assessment of pesticides and hence are a waste of resources. The high doses themselves may produce tumors in animals which are not relevant to situations where humans are exposed to much lower concentrations. Differences in MTD may cause unwarranted concerns about residues allowed in food. This difference could give rise to potential trade barriers between nations and serve to increase the growing skepticism with which governments and scientists are regarded.

Recent efforts in harmonizing high dose selection or MTD have met with some success. The October 1993 draft consensus text of the International Conference on Harmonization (ICH) of Technical Requirements for the Registration of Pharmaceuticals for Human Use proposed five equally acceptable criteria for selecting the high dose (ICH, 1993): 1) dose-limiting pharmacodynamics, 2) a minimum of a 25-fold area-under-the-curve (AUC) ratio (rodent:human), 3) saturation of absorption, 4) maximum feasible dose, and 5) the MTD.

Another project underway is the international harmonization of risk assessment methodologies for carcinogens, mutagens, and reproductive toxins. Participants in this project include the WHO's International Program on Chemical Safety (IPCS), EC, OECD, and the U.S. EPA. Differences in approaches to high dose selection (MTD) hopefully will be harmonized in this process.

Cancer Classification and Risk Assessment

A recent survey of risk assessment methodologies practiced by OECD and selected non-OECD countries was conducted by IPCS (Dragula, 1994). While the linearized multistage extrapolation model is used by U.S. in assessing cancer risks, most countries employed the No Observed Adversed Effect Levels (NOAELs) with safety factors for non-genotoxic carcinogenic risk assessment.

The U.S. EPA (EPA, 1986) assumption for all substances showing carcinogenic activity in animal experiments is that no threshold exists (or at least none can be demonstrated), so there is some risk with every exposure. Thus, the dose-response curve derived based on this assumption shows zero risk only at zero exposure. All other exposures entail some risk. All substances identified as Categories A or B, and some Category C carcinogens (see Table 3 for details on carcinogen classification) are subjected to the same linearized multistage extrapolation procedure in risk assessment. This is a conservative procedure used to estimate the hypothetical upper bounds on the cancer risk and cannot be related to expected disease incidence in an actual population. In fact, EPA generally attaches the following description to its risk estimates:

> "The true value of the risk is unknown, and may be as low as zero."

Table 3

Human Classification of Carcinogens by EPA

A. Human Carcinogen
B. Probable Human Carcinogen
C. Possible
D. Inadequate Data
E. No Evidence (Negative)

A.
} Risk Assessment
B.

D.
} No Risk Assessment
E.

In EC and most OECD and non-OECD countries, carcinogens are divided into genotoxic and non-genotoxic categories. For genotoxic carcinogens, it is assumed that no threshold exists for these agents. Non-genotoxic carcinogens are treated as threshold toxicants. A NOAEL and safety factor are used to set allowable daily intakes (ADIs). The mechanisms of tumor induction are often investigated.

No linearized extrapolation models are used in cancer risk assessment. The U.K. described its rationale for not providing linearized mathematical model as follows (U.K., 1991): "The Committee does not support the routine use of quantitative (linearized model) risk assessment for chemical carcinogens. This is because the present models are not validated, are often based on incomplete or inappropriate data, are derived more from mathematical assumptions than from knowledge of biological mechanisms and, at least at present, demonstrate a disturbingly wide variation in risk estimates on the model adopted."

The linearized extrapolation model and the MTD used by U.S. EPA have also been severely criticized by others. Dr. Jay I. Goodman, in a paper in Molecular Carcinogenesis (1994), made the following statement (Ray, 1994):

> "In freshman chemistry laboratory we are taught not to extrapolate beyond a standard curve because one quickly ends up in ' never-never land.' All too often the low-dose extrapolations from carcinogen bioassay results are based upon one data point, i.e., the estimated line is drawn from a single point, a practice that would not be deemed acceptable in a geometry class . . . The implicit assumptions underlying extrapolation from the MTD . . . do not appear to be valid. Therefore, both the criteria for selection of the high dose used and the default criteria that are employed for extrapolation from high-dose must be reevaluated in a critical manner."

American and European regulatory agencies have adopted different cancer risk assessment methodologies, to a large degree due to their differences in the goal of risk assessment. The differences can be illustrated by comparing the goal described in the Risk Assessment Guidelines of the U.S. EPA (EPA, 1986) with that described in Commission Directive 93/67/EEC of the European Community (EC, 1993b).

U.S. EPA:

> It should be emphasized that the linearized multistage procedure leads to a plausible upper limit to the risk that is consistent with some proposed mechanisms of carcinogenesis. Such an estimate, however, does not necessarily give a realistic prediction of the risk. The true value of the risk is unknown, and may be as low as zero. . . . An established procedure does not yet exist for making "most likely" or "best" estimates of risk.

EC:

> 'Risk characterization' is the estimation of the incidence and severity of the adverse effects likely to occur in the human population or environmental compartment due to actual or predicted exposure to asubstance, and may include 'risk estimation', i.e. the quantification of that likelihood.

Two very different goals of risk assessment are apparent in these two statements. The U.S. EPA has established "a plausible upper limit to the risk," i.e. what is not likely to occur, as its goal, while the EC's goal is to describe "effects likely to occur in the human population . . . ." In the harmonization of cancer risk assessment methodology to be conducted by IPCS, these differences between U.S. EPA and EC need to be resolved.

A reasonable goal for risk assessment would be to present a combination of both approaches, providing an estimate of the upper bound to risk for genotoxic carcinogens, and an estimate of expected incidence of disease in the human population for non-genotoxic carcinogens. Such a broad perspective resulting from the evaluation of the scientific data would allow more informed judgments regarding the management of risk.

Conclusion

Although many of the recommendations and guidelines, developed by international organizations such as FAO, WHO and OECD to harmonize testing and evaluation of data, have been adopted by various countries, significant differences still remain in the national requirements of member countries.

Differences in three key areas: (1) Test Guidelines, (2) Maximum Tolerated Dose (MTD), and (3) Cancer Classification and Risk Assessment, have resulted in significant additional testing costs by requiring studies to address issues that have already been fully investigated. The repeated studies add little knowledge in terms of hazard identification or risk assessment of pesticides and hence are a waste of resources. Furthermore, differences in data interpretation and risk assessment may cause unwarranted concerns about residues allowed in food. These differences could give rise to potential trade barriers between nations and serve to increase the growing skepticism with which governments and scientists are regarded.

Three recent international harmonization efforts have provided a workable process by which the differences can be resolved in moving towards global harmonization of pesticide risk assessment and registration. These are: (1) The development of a harmonized European Community registration regime, (2) The International Conference on Harmonization (ICH) in pharmaceuticals, and (3) The OECD and IPCS pesticide harmonization programs.

References

Dragula, 1994. Dragula, C.; Burin, G.; International Harmonization for the Risk Assessment of Pesticides: Results of an IPCS Survey; Reg. Tox. Pharm., 1994 (in press).

EC, 1993a. EEC Directive 91/414. Council Directive Concerning the Placing of Plant Protection Products on the Market (regulating the registration of pesticides from July 1993 onwards), 1993.

EC, 1993b. Commission Directive of Laying Down the Principles for the Assessment of Risks to Man and the Environment of Substances Notified in Accordance with Council Directive 67/548/EEC, 1993.

EPA, 1986. U.S. Environmental Protection Agency (EPA), Guidelines for Cancer Risk Assessment. Fed. Reg. 51 (185): 33993 - 34003, 1986.

EPA, 1987. EPA Pesticide Assessment Guidelines, Subdivision F, Position Document - Selection of a Maximum Tolerated Dose (MTD) in Oncogenicity Studies. National Technical Information Service, Washington, DC, 1987.

ICH, 1993. International Conference on Harmonization (ICH) of Technical Requirements for the Registration of Pharmaceuticals for Human Use. Carcinogenicity: Guidance for Dose Selection for Carcinogenicity Studies of Therapeutics. Draft consensus text, ICH-2, Orlando, FL, 1993.

IPCS, 1990. Principles for the Toxicological Assessment of Pesticide Residues in Food. Environmental Health Criteria 104, WHO, Geneva, 1990.

OECD, 1987. Organization for Economic Cooperation and Development (OECD) Guidelines for Testing of Chemicals, "Carcinogenicity Studies", Guideline 459, 1987.

Ray, 1994. Ray, J.S.; Harbison, M.L.; McClain, R.M.; Goodman, J.I.; Molecular Carcinogenesis, 9: 155 - 166, 1994.

U.K., 1991. Guidelines for the Evaluation of Chemicals for Carcinogenicity. Committee on Carcinogenicity of Chemicals in Food, Consumer Products and the Environment, U.K. Department of Health, London, 1991.

Health Risk Assessment - Critical Issues for Improved Decision Making

Dr Andrew Cockburn, Head of Toxicology AgrEvo, AgrEvo UK Limited, Chesterford Park, Saffron Walden, Essex, UK

Health risk assessment is one of the most critical phases in pesticide registration. It not only provides the basis for product approval but also for risk communication. The latter in turn governs media reaction and public confidence. The current processes are unsatisfactory as they vary widely in theory and practice and lead to different assessments of risk for the same product in different countries. In important instances risks have been greatly exaggerated. This leads to confusion and an overall poor perception of pesticides. Radical reform is urgently required to reduce overregulation by developing a single science based global procedure which takes into account mechanistic data and also separates risk assessment from risk management. This would result in a faster flow of new compounds and an internationally agreed system of risk assessment, the transparency of which would increase public and industrial confidence alike.

And he gave it for his opinion, that whoever could make two ears of corn or two blades of grass to grow upon a spot of ground where only one grew before, would deserve better of mankind, and do more essential service to his country, than the whole race of politicians put together.

Jonathan Swift, 1667-1745

Pesticides have brought the world the most abundant, reliable, safe and cheap food in its history. Nevertheless, the industry faces an enormous challenge if it is to feed a population expanding at the unprecedented rate of 94 million people each year - another Britain every seven months (Harrison, 1994). The supply of food to meet world needs is now threatened by a number of external factors. Principal among these is the generally negative perception of agrochemicals by the public and politicians which has resulted in an uncoordinated escalation of the regulatory requirements of major international authorities thus retarding the supply of new products. The associated bureaucracy results in a huge waste of scarce public resources. This cannot be justified on a scientific basis as the perceived health risk of pesticides does not accord with reality. Notwithstanding, the charges persist that pesticide use can involve significant risks in terms of residues on food, workplace exposure and environmental damage.

Today, however, we are living longer and more healthily than ever before despite a meteoric expansion of industrial activities, and when the substantial effects of tobacco and increased exposure to sunshine are discounted there is no evidence that the overall incidence of cancer is rising. Moreover there is no evidence that potential toxic hazards such as pesticides, chemical waste and other forms of industrial pollution have had a major impact on overall rates of cancer (Coggon and Inskip, 1994). As stated by Hans Mohr in his seminal opening lecture at the Hamburg IUPAC meeting in 1990, "Despite all the evidence of our physical well-being beyond the dreams of all previous generations - our life expectancy is approaching the biological limits of the human life span - we seem to have become a nation of frightened people, the wealthiest and healthiest hypochondriacs that have ever existed in the world". The worry is nevertheless very real but it is quite irrational and often categorises the risk from pesticide residues far higher than that for more serious threats such as smoking or travelling to work each day. What has gone wrong? The fact is that pesticides make good "bad" news. They are designed to kill, they are not in the main under our individual control, the

1054–7487/95/0404$12.00/0

benefit to risk ratio looks small or is not seen and they even contaminate our food and water, at alarming levels of parts per million or even parts per billion! To balance this, industry has been slow to explain that "plants need medicines too"!

In response to such sensationalism the public have demanded action and governments have responded. In consequence, over the last three decades we have seen the construction of powerful National pesticide regulatory industries whose goals are to safeguard health and safety using the assessment of risk for the purposes of prioritisation. However these agencies being human ventures have developed self serving motivations and the degree and extent of regulation has inevitably expanded. In the context of an affluent society less motivated to take risks and lose coveted lifestyles, the perpetuation of chemophobia is in the interest of the regulatory industry. However the agencies' massive cost and inertia leads many to conclude that overregulation per se will ultimately prove adverse to the interests of the governments and public they serve when it leads to a decline of individual income and living standards (Gori, 1993). Put simply, the cost of overregulation is not commensurate with debatable benefits to the quality of public health. Indeed, it is likely that a number of substances which become delayed or fail to reach the market would have directly and positively influenced public health either as environmental health products or indirectly by providing improved food supply. There may even be a cost to public health where a critical product is held up by the regulatory system. Additionally there are huge budgetary costs, fees for litigation, unwarranted public anxiety and increased development costs leading to loss of jobs and contraction of the industry. We must therefore encourage the regulators to prioritise more effectively between major and minor risks to prevent an enormous waste of resources. To "err on the side of safety" may do more harm than good! Perhaps the "acceptable risk" does exist, we each make plenty of judgements regarding these in our day to day life, where shall we cross the road, shall I go out in a thunderstorm?

Regulatory reform of the entire pesticide health risk assessment process on a global scale is now long overdue. The current trend towards international harmonisation of guidelines for the conduct of mammalian toxicology should now be extended into a major initiative for harmonised interpretation of the data, and the theory and practice of the associated risk assessment. This is clearly a difficult but a feasible and essential goal to achieve: indeed the process has been started in Europe as part of the 91/414/EEC Pesticide Registration Directive concerning the placing of plant protection products on the market. The Directive lays down Uniform Principles aimed at providing member states with harmonised criteria and guidance in their assessment of data and subsequent authorisation of plant protection products within their own territories. Why should the same basic technical information lead to widely divergent regulatory decisions in different countries? (Nilsson et al, 1993). The present Congress with its many distinguished attendees of international reputation is in a strong position to promote the view that the decision process as it relates to agrochemicals must now be globally harmonised and made transparent and fully comprehensible outside the narrow circle of decision makers. To aid this major initiative, it is essential to segment the complex process of decision making on the acceptability of new substances into its component steps and to reiterate the need to isolate the exclusively scientific undertaking of risk assessment from risk management. Whereas the former can be the subject of scientific analysis and standardisation, the latter is subject to a variety of non-scientific pressures including local conditions, political considerations and media reaction which will vary from country to country.

To develop the argument it will be helpful to start by looking at the process stepwise and to consider various definitions.

THE DECISION MAKING PROCESS - WHAT IS RISK ASSESSMENT?

All pesticides, because of their intended use, are toxic to some form of life. Risk assessments are therefore necessary to estimate a level of human or environmental exposure which will not result in adverse human health effects in order to permit safe agricultural usage. However, because of marked differences in toxicity, this may require that preventative measures are put in place to manage the risk to within acceptable defined limits. The overall process is exceedingly complex and involves four major steps: (1) data collection (principally toxicity studies), (2) the assessment of risk, (3) the management of risk and (4) the communication of risk.

Step 1 - Data collection

For chemicals, or more specifically pesticides, toxicity data and physico-chemical information is accumulated in order to identify hazards upon which to base the risk assessment. The former depends primarily on animal and epidemiological models, the latter on a variety of laboratory based analytical measurements. Animals are used as surrogates for man and the principal purpose of toxicity testing is to provide data about the types of adverse effect, their incidence and relation to dose (exposure). There is now a reasonable degree of international agreement on the testing procedures although guidelines continue to evolve and a number of authorities remain extremely inflexible in accepting data which is technically sound but fails to meet exceedingly minor points of protocol detail. The need to repeat fully interpretable studies for an individual country is ethically and morally indefensible. Certain regulatory authorities have probably done more to advance the cause of animal rights than the campaigners themselves! It is therefore refreshing to see in the new EC "toxicology and metabolism" guidelines for the active substance referred to in Annex II of the Directive, that there is ample scope for judgement, flexibility and adaptation to scientific progress to ensure the most efficient and economic use of resources.

Different kinds of toxicity testing are required to estimate the totality of potential risks (adverse effects) to human health. A typical toxicology programme therefore includes animal tests to determine the immediate or longer term toxicity resulting from exposure by oral ingestion, inhalation, skin or eye contact, sensitisation, the potential to cause birth defects, effects on reproduction, specific tissue or organ effects, genetic effects, neurotoxicity, immunotoxicity or cancer. In addition metabolism and toxicokinetic studies are undertaken to determine the fate and significance of chemicals ingested as residues of the plant protection product left on crops. Innovations in the programme which demonstrate the new EU approach are exemplified in Table 1 below.

Animal testing is always limited by a degree of uncertainty when extrapolating data from animals to humans. This limitation may be overcome partially by human epidemiological studies which can provide an alternative source of data. These compare the health of a large group of persons that probably have been exposed to the substance to that of a similar group which probably has not. A disadvantage is the fact that such data may only be gained retrospectively in development terms, ie after marketing and hence exposure has already taken place. Epidemiological models are not very precise tools with which to establish degrees of risk partly because of the size of the sample required, the difficulty in obtaining adequate control populations, and accurate reporting of data.

There are many examples of difficulties in the estimation of risk, with different approaches taken by different authorities leading to bans in one country and full regulatory approval in another. Some insight into this apparently impossible situation can be gained by looking at the risk assessment process in more detail.

Table 1

COUNCIL DIRECTIVE 91/414/EEC

INNOVATIONS

- Flexible scientific approach
- Tiered approach
- Minimise or obviate animal use
- Limit information and studies to those appropriate for decision making
- Expert assessment of data

Step 2 - Risk assessment

Risk assessment is an analytic process based on scientific considerations. It involves four integrated steps, as identified by the European Commission Directive 93/67/EEC, the US Office of Science and Technology Policy (1986) and the National Academy of Sciences: (1) hazard identification for chemicals, this is actually identifying the critical toxicological end points for a given compound, (2) dose (concentration) response (effect) assessment, (3) exposure assessment, (4) risk characterisation.

The basic process in risk assessment is to derive and compare the estimated human exposure (dose) with a no-observed-adverse-effect-level (NOAEL) for the most critical effect normally derived from the most sensitive animal species/toxicity study conducted by a relevant dosing route. This can either result in no margin of safety (MOS) or a margin which then has to be considered for its acceptability. A major problem of the current system is that the outcome can vary greatly according to who is doing the interpretation, both within and between the following groups, regulator, industrialist or consumer and this apparent confusion only serves to increase uncertainty and lead to media exploitation.

"Risk assessment" is used as a generic term in this paper to describe the entire four stage process, and can be defined as the use of facts to predict the health effects on individuals or populations of hazardous materials and situations (National Research Council, 1983). For pesticides there are three target populations, manufacturers and farm workers, consumers and humans exposed indirectly via the environment. For consumers the concern is primarily directed towards cancer resulting from the consumption of food contaminated with pesticide residues. This unfounded worry is based on the fact that certain regulatory authorities and in particular the EPA have policy based on the outdated notion that one molecule of a rodent carcinogen in man's diet can lead to cancer. If true this would mean that cancer risks exist at any exposure level, however small. In contrast other chemical induced diseases, including teratogenesis, are considered to have exposure thresholds below which there is no risk. Therefore we have become concerned about carcinogens presenting a risk at exposures much lower than those associated with other diseases. However low levels of carcinogenic agents of natural origin are present throughout man's environment in the light (UVB), the air (eg radon) and the food and water that we consume. Many chemicals that we consume derive from the natural defence mechanisms of plants to ward off insects and other predators where it has been estimated that 99.99% of the pesticide we eat is the plants own (Ames, 1990a). Thus animals and man have to be and are extremely well defended by repair mechanisms against low doses of chemicals and physical agents. One does not expect, nor does one find, a general difference between synthetic and natural chemicals in their carcinogenic potential. Reducing our exposure to the remaining 0.01% of synthetic pesticides is both enormously expensive and will not reduce cancer rates. The probable causes of cancer which lead to the death of one in four people are becoming better known. Major causes are smoking, unbalanced diets (including a lack of vegetables), chronic infections, genetic factors and exposure to the sun. Occupational exposure and pollution seem to be a minor contribution accounting for no more than 1% or so and yet pollution in particular has become a major public preoccupation fuelled by media sound bites and political rhetoric.

"Hazard" is seen as the inherent properties of anything that in particular circumstances could lead to harm to a human or population (eg chemicals, electricity, etc).

"Risk" is the probability, great or small, that someone or a population will be harmed by the hazard.

The word risk probably originates from the Greek word ριψοκινδυνεΰω or the Italian word rischio. Originally it was related to the hazards of sailing near dangerous rocks and cliffs, particularly in swirling tides.

The first stage of the risk assessment process is "*hazard identification*". This is the identification of the adverse effects which a substance (or physical presence) has an inherent capacity to cause. It consists of a review and analysis of the toxicity and/or human epidemiological data generated in Step 1, Data

Collection. It involves identification of the critical adverse effect(s) which may be defined both qualitatively and quantitatively and may result in certain classifications.

"*Dose (concentration) - response (effect) assessment*" is really the heart of toxicology. It is the estimation of the relationship between dose, or level of exposure to a substance, and the incidence and severity of an adverse effect. It provides the basis for extrapolating from experimental animals to man.

The third stage, "*exposure assessment*" is defined as an estimation of the magnitude (concentration) and time to which human populations are, or may be, exposed from all possible sources. This information is generally developed either from direct monitoring or exposure modelling.

To date more attention has been focused on hazard identification and dose response than on exposure assessment. The latter can have just as great an effect on the outcome of the risk assessment as the former and we must now concentrate on improved techniques. This is because there remain a number of highly active substances which although of high hazard can have a very low risk to man due to negligible exposure as a consequence of training/ certification and the stringent use of personal protective equipment.

The final stage, "*risk characterisation*" combines information from the preceding three steps into estimation of the incidence and severity of adverse effects likely to occur in a human population due to actual or predicted exposure to a substance.

Risk assessment is by definition an iterative process which can and should be repeated as more information becomes available. Because of the uncertainty in risk assessment, flexibility and expert judgement are important elements in the overall process.

Step 3 - Risk management

Depending on the outcome of the risk characterisation, risk management may be necessary to control or reduce exposure. In other words, now that a problem has been identified, what is going to be done about it? This is decided by weighing policy alternatives and selecting the most appropriate regulatory action to deal with each risk, or in this case substance. Generally it involves integrating the results of risk assessment with risk reduction strategies and with social, economic and political concerns to reach a decision. This recognises that it is not possible to eliminate all risk associated with our present lifestyle. Risk reduction strategies provide measures which enable the risks for man and/or for the environment in connection with the marketing of the substance to be lessened. They may include

(i) modifications to the classification, packaging or labelling of the substance;
(ii) modifications to the safety data sheet;
(iii) various restrictions over the use proposed by the notifier.

The public's overall perception of the risk will depend on the transparency of the decision making procedure and the availability of the data even if they do not understand it fully. In turn industry's acceptance of the fairness of regulatory decisions depends upon the scientific integrity of the risk assessment, its consistency, and of the methodologies used. Equally important is effective communication of the risk information as well as the uncertainties in its derivation. As stated by Hans Mohr, "We may be experts for risk analysis, but we may not assume to be experts for acceptance. Acceptance is no longer a scientific problem."

Thus risk assessment is no use in isolation. It requires good management and good communication to achieve the most appropriate balance between risk and benefit. While differences between regulatory agencies occur at all stages of this process, they tend to become more and more pronounced towards the risk management, the non scientific, part of the process, see **Initiative 3**.

The existence of common principles for approval, restriction and cancellation of pesticide registrations would drastically reduce duplication of reviewer effort. The 1991 FAO/WHO Conference on Food Standards, Chemicals in Food and Food Trade, held in Rome, crystallised this concept by calling for the development of internationally recognised principles for risk assessment that are consistent and based on sound science.

Step 4 - Risk communication

There are obvious difficulties in explicitly stating and reporting risks due to the controversial nature of the message. However, given an open approach and the ready supply of information by industry or regulatory authorities, the public generally feel able to make up their own minds and will normally accept much higher levels of risk than when they are not involved. The purpose of the product label is to communicate risk in an informed way to the user. The chemical and regulatory industry has generally been perceived as too secretive in the past.

The risk communication process therefore involves the way in which the information is transmitted, the characteristics of the receiver and the nature of the risk itself. For example, invisible, involuntary risks, over which the public have no control and see no benefit, have a far higher alarm signal than visible voluntary risks, eg radiation v obesity. The invisible, involuntary risks are often associated with complex scientific and technical information, laced with a degree of actual uncertainty, compounded by "so called" experts arguing from opposing positions. Add to this the combination of political views and consumer protection groups and you have a recipe for concern and even outrage.

The net effect has been a breakdown of the communication process. This has led to public fear and confusion rather than an appreciation of the overall beneficial effects of agrochemicals.

In summary, the decision making process is complex. There are no absolutes, only varying degrees of uncertainty which are catered for by using relatively smaller or larger margins of safety. The data collection and risk assessment phases are scientific and therefore able to be harmonised and unified globally. The risk management and risk communication phases are essentially non-scientific and depend largely on local conditions and societal attitudes and can therefore be dealt with most expediently at the national level. Taking these factors into account, the following three initiatives are critical to the improvement of the decision making process.

INITIATIVE 1 - GLOBAL HARMONISATION OF HAZARD (TOXICITY) IDENTIFICATION AND DATA INTERPRETATION PROCEDURES

Evaluation of the same pesticide active ingredient at the National level for the purpose of regulation not only results in a tremendous duplication of effort but, with different outcomes, also fails to guarantee the same level of protection of man and the environment from country to country. This has now been addressed in the European Community or more correctly the European Union, where a single country, the rapporteur member state, undertakes assessment of the toxicology of the active ingredient using uniform principles for data interpretation. A system of mutual recognition then exists for acceptance of the review between community members (Council Directive, 1991). This process also avoids unnecessary trade barriers.

Historically, over the last 30 years the World Health Organisation and the Food and Agricultural Organisation (WHO/FAO) has convened a meeting of experts whose role was not only to study the health effects of pesticides but also to establish acceptable daily intakes (ADIs) and the maximum residue levels (MRLs) that could safely be retained on foods resulting from product use in accordance with good agricultural practice (GAP). This group is known as the JMPR (Joint Meeting on Pesticide Residues) and some 200 pesticides have now been evaluated. The decisions made by these two groups were used widely by both developed and developing nations who could not afford to have their own separate agencies - 30 years ago we had harmonisation! As each nation began to develop its own regulatory philosophy the unifying influence of WHO/FAO was lost. Not only do most countries now have their own system but some, such as the United States, create their own state regulations, not uncommonly in contradiction with their own National authority, in this case USEPA. We need to turn back from this isolated and independent approach to one of international cooperation and global harmonisation, a point stressed in the preceding paper. The cost of new product development is such that without harmonisation fewer and fewer products will survive, as well as reach the market, and food production will fail to meet the rapidly escalating demand.

Harmonisation of hazard identification

Chemical hazard for man is normally identified by animal toxicity testing carried out in conjunction with toxicokinetic studies designed to elucidate the metabolism and fate of the molecule. In vitro studies are employed increasingly where thorough validation has taken place. There already exists a reasonable degree of international agreement as to what constitutes an acceptable battery of toxicology protocols, for example those defined by the OECD. Such tests have proved very effective in "forcing" the appearance of adverse effects in experimental animals. For this reason standardised hazard identification is by no means an impossible task and has already been initiated within the pharmaceutical industry via the International Conferences on Harmonisation, ICH 1 and 2.

Harmonisation of interpretation

The ability to detect ever smaller quantities of chemicals in the environment, food or water causes problems of understanding when it comes to communicating the significance of vanishingly small quantities of residues or exposure. In the early 1900's analysts could detect milligram (10^{-3}) quantities of impurities. With a combination of gas chromatography and mass spectrometry (GCMS) it is now possible to detect picogram quantities (10^{-12}). The question is no longer is something present or not, but what does it mean. This ability has rendered the US Delaney clause, which postulates that any molecule of a carcinogenic substance can cause cancer and shall not be added to the food supply, hopelessly obsolete.

This law represents probably the major difference in the risk assessment process between the USA and the rest of the world. It does not permit consideration of the extent of human exposure, or the validity of the animal data for predicting effects to humans. It makes no allowance for how low the human exposure might be or the possibility that a threshold dose exists. It should now be repealed and replaced with a "de minimis" concept. This comes from the judicial doctrine of *de minimis non curat lex* which essentially means, the law does not concern itself with trifles. This would allow regulatory officials to disregard substances present in such concentrations that they are of no consequence. As long ago as the middle ages Paracelsus stated that it is only the dose which makes the poison and this fact provides a sound basis for utilising a de minimis or negligible risk approach based on the concept of threshold dose levels. The clash between old legislation and new science is inevitable. The sooner action takes place and the law is changed the quicker we achieve the goal of harmonisation of interpretation. The United States is the only country in the world with a Delaney clause and it is time to change. Zero risk is not an option and the public need to be educated to this concept. Despite this, a recent attempt by the EPA to challenge part of the Delaney Clause and exclude animal carcinogens with estimated lifetime upperbound cancer risks of 1 in a million or less failed in court. In contrast, the new EC guidelines 91/414 have been developed very recently and therefore take into account modern scientific thinking and a number of the innovations already introduced by the pharmaceutical industry.

Areas that by general consensus require repeal to pave the way to global harmonisation are discussed below:

Issue 1 - Maximum tolerated dose (MTD)

Most toxicity studies employ at least three dose levels, low, intermediate and high, as well as a control group, in order to define the shape of the dose response curve. The highest dose tested (HDT) should be chosen at a level which can reliably predict the possible occurrence of cancer at doses to which humans are likely to be exposed. Traditionally in the United States, dose level selection based on the maximally tolerated dose has been the only acceptable practice. It is chosen on the basis of 90 day subchronic testing but at best can only be viewed as an estimate, as the main study lasts some eight times longer during which period the animals age and their tolerance to the compound being fed may alter significantly. In consequence the MTD is not a fixed value and the main consideration has been, and is, to ensure an adequate margin of safety between the dose employed in the animal test and the maximum exposure likely to man. In this way any oncogenic potential will be displayed without the interference and potential complication of altered physiological response in the treated animals. For example, it should not upset normal body defenses, including metabolic, immune, tissue and genetic repair which operate at human exposure levels. All the major regulatory authorities worldwide accept this concept with the exception of the USA backed up by the US National Toxicology Programme and the IARC. Here there is a virtual religion in achieving heroic doses and many compounds have been sacrificed to the god MTD. EPA's definition states that the MTD "should not be selected too far below a life threatening level because the highest dose tested in the oncogenicity study must elicit significant toxicity". Thus, many fully interpretable studies where the top dose was not quite high enough, have had to be repeated purely to meet the EPA's MTD criteria, which is an enormous waste of scientific resources, and a dreadful and unjustifiable waste of laboratory animals. Ames and Gold (1990b) have reported that more than half the substances tested in rats and mice at the MTD, whether natural or synthetic, proved carcinogenic. This proportion seems too high in relation to what we already know about cancer causation and therefore implies problems with this type of bioassay technique. The probable aetiology is that excessive doses (the MTD and even half the MTD) lead to prolonged toxicity and tissue repair in various target tissues which results in an increased rate of transcription errors, thus increasing the probability of carcinogenesis. At human exposure levels this process will not occur.

Thus the outcome can be affected by the dose level chosen rather than the intrinsic potential of the compound and this is wrong. The use of the EPA's definition of MTD is not only inappropriate but results in a hyperconservative approach. This indicates the need for significant changes in the design, dose level selection, conduct and interpretation of animal cancer studies.

The EEC proposes to overcome this problem, see Annex II revision (May 1994) of the Pesticides Directive by referring only to the highest dose tested, not the MTD, "selected on the basis of the results of short-term testing and where available at the time of planning the studies concerned, on the basis of metabolism and toxicokinetic data." "The highest dose level in the carcinogenicity study should elicit signs of minimal toxicity such as slight depression in bodyweight gain without causing tissue necrosis or metabolic saturation and without substantially altering normal lifespan due to effects other than tumours." It is proposed that the HDT in the long term toxicity study should elicit definite signs of toxicity. In short a different dose level strategy is quite rightly proposed for oncogenicity and long term studies. This of course has a number of implications on the ubiquitous combined rat chronic and oncogenicity study and the acceptability of oncogenicity studies done to EU guidelines and then submitted in the USA.

Clearly, and as a minimum, we need international harmonisation on what constitutes an acceptable MTD in order to save time, scientific and most importantly animal resources.

Issue 2 - High dose/low dose quantitative risk assessment and threshold versus non-threshold dose response

Human health risk analysis has traditionally been based on expert judgement, usually involving safety factors. In the early 1970s quantitative risk assessment was introduced in the USA as a discipline to evaluate health risks using statistical models. The cancer dose-response model currently favoured by regulatory agencies in North America is the linearised multistage (LMS) model (Crump, 1994). Critically the model uses simplified assumptions and approximations and greatly exaggerates the human risk of cancer. The concept of extrapolating from an established small area of "certainty", usually two or three points of the animal dose response curve, to very low doses, many orders of magnitude below is of highly questionable value especially when the shape and nature of the dose response curve is unknown. Three patterns are most commonly seen. (1) The "flat" dose response curve where there is no observable or measurable effect over a significant dose range, spanning several orders of magnitude, eg 1 to 100 mg/kg bodyweight/day. (2) The opposite situation is where there is no tolerable dose and even a miniscule amount can cause an adverse effect. This line of argumentation has been extrapolated to the "one-hit" hypothesis where a single molecular event could in theory change DNA leading to an oncogenic response. However, pragmatism coupled with scientific knowledge and experience indicates that this is unlikely to be the case except for radiation and potent genotoxic carcinogens which are unlikely to be registered in today's world. This is because of the naturally occurring number of calculated DNA hits per day in man (Ames and Gold, 1990b) which approximate to 10^4, in which case if there were no other mitigating factors such as very powerful DNA repair mechanisms, cancer would be more prevalent. For these reasons it is questionable whether the so-called "no threshold" model actually exists. (3) The third scenario is the most common and is a threshold model. This relates to the situation where there is a "threshold" dosage below which there is no observable adverse effect, this is defined as the NOAEL, or the no observed effect level (NOEL), the absolute no effect level or the threshold level. A threshold exists because animals can detoxify, metabolise or excrete most if not all chemical substances, natural and synthetic alike. At higher doses these mechanisms can become saturated and a toxic effect(s) is observed. The typical dose response here has the shape of a hockey stick. The NOEL in the most sensitive species is normally divided by a safety factor of 100 to obtain the acceptable daily intake (ADI) or the more politically correct, reference dose (RfD) for man, which is expressed in mg/kg bodyweight/day. The ADI and RfD form the basis of assessing the safety/risk of the intake of a pesticide. Thus when human intake is at or below the ADI/RfD, the residue may be considered "safe", ie involving an acceptable risk.

In order to estimate hazard for man with a higher degree of validity more data below the MTD are required. When suitable studies are evaluated it is seen that the vast majority conform to a sigmoidal curve where decreasing the dose by a factor of say 5 would decrease the tumour burden by a factor of 25, not the 5 expected from linear response. Moreover, the current EPA linear extrapolation model which uses a virtually safe dose (VSD) concept of not increasing the cancer risk by more than one in a million, is highly conservative, dramatically overestimates risk and should not be used routinely. Even worse, the final risk estimates for pesticides normally appear as a figure like 2.46 in 10^{-6} when the true risk is unknown and could be as low as zero. The need for a more realistic risk assessment approach, using scientific biologically based information, ie improved bioassays backed up with mechanistic studies, is now urgent.

The EU neither utilises linear extrapolation modelling nor considers all animal carcinogens to be presumptive human carcinogens. As this is an exclusively scientific area estimations must be based on a weight of evidence approach, supplemented where necessary by quantitative approaches (GIFAP Technical Monograph, 1987).

Issue 3 - Extrapolation animal to man

Extrapolations from high to low dose and between species should preferably be made using physiologically based pharmacokinetic (PBPK) models. Such extrapolations should be performed on a case-by-case basis, taking all available mechanistic, metabolic and kinetic data into account. This allows the prediction of parent compound and metabolites reaching the toxicological target tissues in animals and aids the judgement of whether certain effective tissue doses and hence effects are likely to be species specific. Such models therefore have the benefit of reducing uncertainty and are now being recognised as a powerful analytical tool by regulatory agencies.

Issue 4 - Genotoxic v non-genotoxic carcinogenesis

An understanding of the fundamental mechanisms of carcinogenesis is rapidly evolving. Traditionally carcinogens have been classified into different groups according to their potential risk for man. Predictably agencies have arrived at different regulatory decisions in different countries. It is critical that regulatory decision making keeps up with the science: there is no need for such anomalies. Thus to determine with confidence the probability that a substance is only a rodent carcinogen or is also likely to be a human carcinogen it is critical to know the mechanism of action. This involves understanding the molecular and cellular processes leading to defined biological events. Such information viewed in conjunction with bioassay data and human epidemiology or molecular dosimetry data enables the entry of a compound into a given category with a sufficient weight and breadth of evidence for consensus agreement. Major classification schemes include those of the EEC, EPA and IARC, see Table 2 below.

Clearly the EU scheme has the flexibility to enable mechanistic studies to be taken into account which may prevent the classification of rodent specific carcinogens.

Assuming a given substance is found to cause tumours in a bioassay it is critical to establish whether these have arisen by "genotoxic" or "non-genotoxic" mechanisms, see Initiative 2, Mechanistic Studies. The former are substances or their metabolites which are capable of damaging DNA. The latter however increase cancer incidence in rodents without apparently interacting chemically with DNA or its associated maintenance functions (Ashby, 1992). In this case genotoxic events may occur as a secondary event to sustained biological change following prolonged dosing, eg disruption of normal hormonal balance or other physiological homeostasis.

It is helpful to provide working definitions for the two terms and the definitions presented by Butterworth (1990) have probably not been bettered:

"A genotoxic agent is one for which a primary biological activity of the chemical or a metabolite is alteration of the information

Table 2 Comparison of Approaches for Classifying Substances as Carcinogens

EEC	EPA		IARC	
Category 1 Known human carcinogens	GpA	Human carcinogen	Gp1	Carcinogenic to humans
Category 2 Regarded to be carcinogenic to man based on animal studies	GpB	Probable human carcinogen	Gp2a	Probably carcinogenic to humans
Category 3 Concern for man owing to possible carcinogenic effects	GpC	Possible human carcinogen	Gp2b	Possible carcinogenic to humans
	GpD	Not classifiable as to human carcinogenicity	Gp3	Not classifiable
	GpE	No evidence of carcinogenicity for humans	Gp4	Probably not carcinogenic to humans
Unlisted*				

* A compound should not be classified in any of the categories if the mechanism of experimental tumour formation is clearly identified, with good evidence that this process cannot be extrapolated to man

encoded in the DNA. These can be point mutations, insertions, deletions or changes in chromosome structure or number. Chemicals exhibiting such acitivity can usually be identified by assays that measure reactivity with the DNA, induction of mutations, induction of DNA repair or cytogenetic effects."

"Non-genotoxic chemicals are those that lack genotoxicity as a primary biological activity. While these agents may yield genotoxic events as a secondary result of other induced toxicity, such as forced cellular growth, their primary action does not involve reactivity with the DNA." They tend to produce cancer in a single organ, in a single sex and/or species of rodent.

Thus, agreement on what constitutes a genotoxic and a non genotoxic carcinogen and international acceptance of mechanistic studies designed to provide the necessary weight of evidence arguments should be part of the data base for each compound. Harmonisation of terminology and interpretation in this field would drmatically reduce the current uncertainty in hazard identification. At present however in the USA no exception is made for compounds that do not react with DNA. Compounds that cause cancer secondarily to tissue damage, altered hormonal status, proliferation of cells or subcellular organelles are treated as though they were genotoxic carcinogens themselves.

Issue 5 - Tumour diagnosis

The type of tumour, whether benign or malignant, reversible or non-reversible, eosinophilic or basophilic, the precise classification, within or outwith current or historical control and relevance to man, have all proved contentious in the context of carcinogen hazard assessment and hence classification. Moreover the use of strains such as the B6C3F1 mouse, with high background incidences of liver tumours, has also been a confounding influence. Initiatives to standardise the tumour diagnosis are proposed in the latest EC Annex II revision where standard terminology such as that used by the American Society of Toxicological Pathologists or the Hannover Tumour Registry (RENI) are recommended.

In conclusion, regulations that build in too many uncertainty factors, for example use of an exaggerated MTD, high to low dose extrapolation using unvalidated models, and classification of certain rodent specific carcinogens as significant for man, results in an ultra-conservative approach which is very costly for society. We now have the technology to reduce uncertainty in our decision making through mechanistic studies which can be factored into the risk assessment process. In turn this enables less conservative and more realistic interpretations of risk to be made.

As it is more comfortable to retain the status quo, international pressure is now required to bring about global standardisation by dealing with the above issues.

INITIATIVE 2 - REFINEMENT OF THE RISK ASSESSMENT PROCESS VIA MECHANISTIC STUDIES TO REDUCE UNCERTAINTY

Risk assessment is a very young developing science and many of the principles are evolving rapidly. This inevitably results in some uncertainty and as we have seen regulators fail safe and take the "worst-case" most conservative position by building in large margins of safety (MOS), safety factors (SF) or uncertainty factors (UF). However, as a better understanding of the nature of toxicological response becomes available due to rapid scientific advances and the inclusion of more mechanistic data by registrants, it is then possible to become more certain and reduce the degree of conservatism in the risk assessment models. To this end it is critical that regulatory authorities keep pace and recognise the benefits of taking mechanistic studies into account in their decision making. Similarly key assumptions and dogma must be revised where necessary in order to stand the tests of the courts, the scientific community and public opinion. The resources necessary for this would best be provided by a panel of international experts to ensure that the evolutionary process proceeds consistently and synchronously between different Nations.

In fact as scientific advances enable us to understand mechanisms of carcinogenesis it becomes increasingly possible to predict acceptable dose levels for each substance. Indeed a high proportion of rodent carcinogens are found to be "non-genotoxic" and therefore have a threshold dose below which no pathological effects are observed. The process of refinement requires that evaluation of human risk from rodent carcinogens should now take into account the weight of evidence for:

- non genotoxicity
- non-linear kinetics at high dose levels
- structure activity mechanisms
- human response

The grounds for considering certain rodent carcinogens with known mechanisms of action as acceptable products for human use are based on the following observations. Twenty of the 23 known human carcinogens are genotoxic (Shelby, 1988). A similar proportion of the expanded IARC list of human carcinogens is genotoxic (Shelby & Zeiger, 1990). In contrast about half of the known rodent carcinogens are non-genotoxic (Ashby & Tennant, 1991) and there is no evidence of untoward effects in man from these. This is probably because unlike their genotoxic counterparts they are often rodent specific, and normally only induce tumours in rodents at very high doses (MTD or above) and only after sustained periods of expsoure. There is therefore concern that much of the neoplasia in rodents is only an atefact.

Importance of non-genotoxic mechanisms

Non-genotoxic carcinogens, which will all normally be negative in standard batteries of mutagenicity and genotoxicity tests, exert different modes of action. Knowledge of the mode as well as the mechanism is essential for scientific decision making. The principal mechanism is proliferation. This leads to increased cell turnover, with less time for repair prior to DNA replication which in turn increases the chance of clonal expansion of DNA damaged cells and hence neoplasia. The mode of action varies between mitogenic agents, cytotoxicants and those operating through receptor based mechanisms, sometimes in combination. For example new information from CIIT (Butterworth et al, 1994) shows that the ability of chloroform to produce liver cancer in mice is secondary to a continual state of tissue damage and repair. Thus accurate science based prediction of risk for humans can only be undertaken when it is known if the compound is genotoxic or non-genotoxic and if the latter, by what mode and according to what shape of dose response it is inducing cell proliferation.

There are currently three well understood species specific non-genotoxic mechanisms of carcinogenesis: α-2μ globulin nephropathy and carcinogenesis, thyroid stimulating hormone (TSH) mediated thyroid tumours and bladder carcinogens associated with crystal and calculi formation. These are excellently reviewed by Swenberg (Swenberg et al, 1991). Further examples exist covering other organs and tissues including special problems associated with the interpretation of mouse liver carcinogenesis.

Importance of pharmacokinetic/toxicokinetic mechanisms

If the metabolic handling of a xenobiotic in man differs significantly either qualitatively or quantitatively from that in a rodent species which develops cancer it may be possible to show lack of relevance of the animal model. However it is necessary to be alert to the range of genetic polymorphism in man which may be a significant determinant of individual human susceptibility. Alternatively, if very high doses (MTD) are used in the rodent and neoplasia results it may be possible to show that normal metabolic and excretory pathways became saturated leading to excessive cell proliferation. Such non linear kinetics result in the previously mentioned "hockey stock" dose response, where tumours are only seen above the metabolic/ excretory threshold. Here there is no direct relationship between administered dose and target tissue dose. Comparative metabolic studies in vitro or in vivo or a combination of both can be extremely helpful in developing weight of evidence argumentation. Integration of such data with mechanistic toxicology is critical to achieve quality risk assessment.

Importance of structure activity mechanisms

Since the recognition of the electrophilic nature of genotoxins that damage DNA it has been possible to identify certain structural alerts in molecules possessing this capability. While the lack of such an identified moiety does not guarantee lack of oncogenicity it helps to add weight to the probability that a rodent carcinogen is more likely to be acting through a non-genotoxic mechanism.

Importance of human response

Where available, data and information on the extent and effects of human exposure can be of great value. Such data can be acquired through accidental, occupational or intentional clinical pharmacological human exposure. Exposure can be estimated by environmental monitoring (external dose), biological monitoring (internal dose), and biochemical effect monitoring (tissue dose). The tissue dose can be assessed from the amount of DNA adduct in target and non target tissues, and this is proportional to the external dose (Beland et al, 1988). To obtain direct evidence about the role of DNA adducts, follow-up studies are needed in animals and humans relating adducts to cancer. This could be done by investigating DNA adduct formation in toxicity studies. Thus, human molecular dosimetry data have the potential to significantly improve qualitative and quantitative risk assessment.

In conclusion, how can such diverse mechanistic information be factored into the assessment either qualitatively or quantitatively and what are the implications for policy development? The way forward must be for the process to accommodate the existing developments in research, which elucidate the mechanisms of carcinogenesis. This will permit greater accuracy in estimations of human risk and will thus ultimately lead to further improvement in human health. It is relevant that an EPA working group considering revisions to the 1986 Guidelines for Carcinogen Risk Assessment proposes a more flexible system taking into account mechanism of carcinogenesis and weight of evidence.

INITIATIVE 3 - SEPARATION OF RISK ASSESSMENT FROM RISK MANAGEMENT

As we have seen regulatory actions are based on two distinct elements, risk assessment and risk management. The former uses factual toxicity data to define the potential health effects, whereas the latter integrates this information with broad social, economic and political concerns to reach a decision.

We each take voluntary decisions on risk every day. However, for those areas we cannot control individually, society expects government agencies to weigh policy alternatives and then provide guidance or produce rules through regulatory action to manage the risk to acceptable levels. Both so-called voluntary and involuntary acceptance of risk varies according to cultural and socioeconomic conditions as well as the nature of the risk in question. For end use crop protection products the risk benefit considerations which are necessary to determine appropriate risk management options are normally best handled by authorities closest to the situation, or who at the very least are familiar with the typical circumstances of local use. This is why it makes sense to separate the entirely scientific risk assessment process, from the essentially administrative process of risk management and communication. The former should be undertaken centrally on the active ingredient by teams of international experts whereas the latter, which is by definition non-scientific and politically biased, should be performed locally. This concept would clear the way for a single global risk assessment process using standard hazard identification protocols with uniform principles for risk interpretation. This simply represents an expansion of the current European scheme to global dimensions.

Risk management of agricultural chemicals and biocides takes into account the different groups who may be exposed, namely manufacturers, farmworkers, bystanders and consumers. Whereas for the first two groups exposure occurs through voluntary handling by the individual, any exposure of the latter two groups is likely to be involuntary and to have the potential to affect larger populations. To protect the former workers the concept of acceptable operator exposure level (AOEL) is used whereas consumers are covered by the acceptable daily intake (ADI) approach. Overall, risk management leads to risk reduction by a variety of different measures, guidelines and rules imposed by the control agency.

CONCLUSION

Many of us are now fortunate to the extent that the biggest threat we face from food is having too much of it. We must not allow overregulation to jeopardise the supply of good value high quality food or environmental health products. Risks compete with risks and society must distinguish between significant risks and those which are trivial or even hypothetical. Pesticides are here to stay and it is imperative that the best science and registration systems are evolved to permit the most effective data collection, risk assessment, risk management and risk communication with the minimum use of resources. When we speak as one voice confidence will return and crop protection and environmental health products will be seen in their true perspective. The time is now right to bring together those groups, eg IPCS, WHO/FAO and OECD, already committed to the harmonisation of toxicology at the international level. We now need to establish a supranational task force whose aim would be to agree a global standard for risk assessment. If the problem is not tackled promptly, the bitter controversies that surround regulatory decision making will surely exact a very heavy price on those people least able to afford it.

REFERENCES

Ames, B N; Gold, L S. *Science* **1990**, 249,: 970-971.

Ames, B N; Gold L S. Chemical Carcinogenesis. *Proc Natl Acad Sci* **1990** USA, *87*,7777-7781.

Ashby J. In *Mechanisms of Carcinogenesis in Risk Identification* pp.135-164. IARC Scientific Publication Number 116, Lyon 1992.

Ashby J; Tennant R W. *Mutat.Res.* **1991** *257*, 209-227.

Beland F A; Fullerton N F; Kinouchi T; Poirier M C. In: *Methods for Detecting DNA Damaging Agents in Humans*: Bartich H; Hemminiki K; O'Neill I K, eds. *IARC Publications Number 89*, Lyon 1988; pp 175 -180.

Butterworth B E; *Mutat.Res.* **1990**, *239*, 117-132.

Butterworth B E; Larson J L; Conolly R B; Borghoff S J; Kedderis G L; Wolf D C. *Chemical Industry Institute of Toxicology (CIIT)* **1994**, *14*, No.2, 1-8.

Carcinogenic Risks of Assessment of Pesticides. GIFAP Technical monograph No 12, July 1987.

Coggon, D; Inskip, H. *British Medical Journal* **1994**, *308*, 705-708.

Commission Directive 93/67/EEC of 20 July 1993. *Official Journal of the European Communities* No.L.227/9.

Council Directive 91/414/EEC, *Official Journal of the European Communities*, No.L230, 19.8.1991, p.1 and OJ No.L170, 25.6.1992, p.40.

Council Directive 91/414/EEC, Annex VI draft.

Crump et al, *19*, 106-114, **1994**.

Gori, G B. *Regulatory Toxicology and Pharmacology* **1993**, *17*, 224-229.

Harrison, P. *The Third Revolution.* Penguin, 1994.

International Conference on Harmonisation of Technical Requirements for Registration of Pharmaceuticals for Human Use.

Mohr, H. *Pesticide Chemistry*; Frehse, H., Ed. VCH 1991; 31-33.

National Research Council: *Risk Assessment in the Federal Government: Managing the Process*. Washington, DC: National Academy Press, **1983**, 17-83.

Nilsson, R; Tasheva, M; Jaeger, B. *Regulatory Toxicology and Pharmacology*, **1993**, *17*, 292-332.

Shelby M D. *Mutat.Res.*, **1988,** *204*, 3-15.

Shelby M D; Zeiger E. *Mutat.Res.*, **1990** *257* 257-261.

Swenberg J A; Dietrich D A; McClain R M; Cohen S M. In *Mechanisms of Carcinogenesis in Risk Identification*, IARC Scientific Publication Number 116, Lyon 1992; pp 477-500.

Principles governing consumer safety in relation to pesticide residues. Report of a Joint FAO/WHO Meeting of Experts on Pesticide Residues. *WHO Tech.Rep.Ser.240.* 1962.

Ecotoxicological Risk Assessment of Pesticides in Soil

N.M. van Straalen, J.P. van Rijn & C.A.M. van Gestel, Vrije Universiteit, Department of Ecology and Ecotoxicology, De Boelelaan 1087, 1081 HV Amsterdam, The Netherlands

Ecological side-effects of pesticides may be evaluated on the basis of two different aspects, the magnitude and the duration of the effect. Since every pesticide is expected to cause some unavoidable negative effect, particular attention must be payed to the potential for recovery of the affected receptors, after the action of the product has faded. This paper proposes a theoretical framework that joins the magnitude and the duration of the effect into a single criterion, which may be useful when classifying pesticides for regulatory purposes.

From laboratory experiments, estimates may be obtained for the no-effect concentrations for potential ecological receptors and for the degradation time of the chemical. These data may be combined with information on the estimated initial concentration after application of the product, to estimate the time needed for achieving a "safe" concentration, the so-called ecotoxicological recovery time. The assumptions of the model underlying the estimation of ecotoxicological recovery time are discussed.

This newly developed methodology is applied to literature data for some pesticides applied to soil. Examples elaborated for carbofuran, benomyl, lindane and parathion show that the joint assessment approach may lead to a classification of pesticides that is different from the ones based on persistence or toxicity alone.

The intensity of application of plant protection products in The Netherlands and the detection of several active ingredients in ground water has urged the Ministry of Environment to start a re-evaluation of all registered products. As a basis for the evaluation, three generic criteria have been formulated which served to make up a list of products whose registration has to be reconsidered (MJP-G, 1991). The criteria are:

1. **Mobility**. If the active ingredient could leach from soil to contaminate the ground water in concentrations greater than 0.1 μL/L, the product was placed on the list.

2. **Aquatic toxicity**. If the application of a product could lead to concentrations in surface water exceeding one tenth of the $L(E)C_{50}$ of fish, crustaceans or algae, the product was placed on the list.

3. **Persistence in soil**. If the active ingredient had a half-life in soil greater than 60 days, the product was placed on the list.

When these criteria were applied to the set of registered products in the Netherlands, it appeared that about three quarters of them did not obey one or more of the criteria. The actual implementation of this policy has however not proceeded very far, due to continuing discussion on the validity of the criteria and their legal foundation in the Pesticide Act. One of the points for discussion concerned the use of persistence in soil; some have argued that persistence as such cannot be considered as an inherent negative property of a chemical, and that the third criterion of the above list does not find support in the Pesticide Act.

The present contribution aims to add a new argument to this discussion, that is, the coupling of ecotoxicity and persistence in soil. Following on some papers published earlier (Van Straalen *et al.*, 1992; Van Rijn *et al.*, 1994), a new methodology is proposed and applied to data for carbofuran, benomyl, parathion, and lindane.

Ecotoxicological Recovery Time

When a pesticide is introduced into an agro-ecosystem, its initial concentration will often be higher than the no-effect threshold, because the pesticide is assumed to suppress the target, and the target will rarely be the only organism affected. Hence, there is a fair chance that there will be ecological side-effects; these might be considered as acceptable if recovery from these effects follows within a certain time period. When evaluating the environmental risks of quickly degrading chemicals, the rate of recovery seems to be a more useful criterion than the magnitude of the effect itself.

There are three different aspects involved in the recovery process, and it may be useful to distinguish between these as follows:

1. **Chemical Recovery**. This is defined here as the disappearance of the chemical to a level where it cannot be detected anymore.

1054–7487/95/0418$12.00/0

2. **Ecotoxicological Recovery**. This is defined as the disappearance of a chemical to a level where it does not have any negative effects, or is not in any way biologically available. This level may be below or above the chemical detection limit.

3. **Ecological Recovery**. This implies the restoration of the original species composition and ecological functions to a level that is considered as "normal", usually referring to the situation before the application or to a synchronous control plot.

It may be assumed that ecological recovery will lag behind ecotoxicological recovery. Ecological recovery will however not only depend on the disappearance of the chemical, but also on site-specific conditions, such as availability of shelters, untreated field borders, life-histories of the recolonizing species, *etc.* (Jepson, 1988; Jepson & Thacker, 1990; Thomas *et al.*, 1990). Hence ecological recovery is not well suited to be laid down in generic criteria. Ecotoxicological recovery may be seen as a necessary (but not a sufficient) condition for ecological recovery and it is a concept that may be more suitable for use in a generic approach.

The approach laid down in this paper follows in the tradition of the statistical risk assessment methodologies as developed in Europe by Kooijman (1987), Van Straalen & Denneman (1989), Wagner & Løkke (1991) and Aldenberg & Slob (1993). These papers have all considered the concentration in the environment as a constant and have focused on estimating a maximal acceptable level for this concentration, based on effect data for a community of organisms. In the present contribution we extend this approach by focusing on the maximal acceptable recovery time, rather than on the maximum acceptable concentration.

The assumptions underlying the model proposed here may be listed as follows:

1. The concentration of the chemical undergoes simple first order degradation kinetics, leading to an exponential decay curve following a pulse on time zero. This assumption implies two model parameters, the initial concentration, C_0, and the half-life of the pesticide, DT_{50}.

2. The sensitivities of the species in the community that is exposed to the pesticide follow a log-logistic distribution, that is a symmetric, bell-shaped distribution on a logarithmic concentration axis. It is assumed that the sensitivity of each species is measured by its no-effect-concentration, that is the highest exposure concentration that still does not cause a negative effect on reproduction, growth or survival. This assumption implies two other parameters, the mean of the distribution, denoted here by x_m, and the standard deviation, denoted by s_m.

3. Ecotoxicological recovery is assumed to be complete when the concentration of the pesticide has decreased to a level such that the fraction of species with a no-effect-concentration which is still lower than this concentration is smaller than a predescribed, arbitrary small number. This assumption implies another parameter, the fraction of species that is potentially still unprotected after ecotoxicological recovery. This parameter is denoted by δ.

On the basis of the above listed assumptions, ecotoxicological recovery time, R (expressed in years) may be written as (Van Straalen *et al.*, 1992, Van Rijn *et al.*, 1994):

$$R = \frac{DT_{50}}{365 \ln 2}\left[\ln C_0 - x_m + \frac{s_m \sqrt{3}}{\pi} \ln \left\{\frac{1-\delta}{\delta}\right\}\right]$$

where:

DT_{50} = half-life for the degradation of the pesticide (days),
C_0 = initial concentration in soil (mg/kg),
x_m = mean of the no-effect concentrations (NEC) (mg/kg), each concentration being transformed to natural logarithms,
s_m = standard deviation of the ln (NEC),
δ = fraction of unprotected species.

The formula shows that duration and magnitude of the effect are integrated in a single quantity, R, and this quantity may be calculated in a straightforward way from other quantities that each have a clear physical meaning. The relevance of the assumptions underlying the model may be briefly addressed as follows.

1. Degradation of pesticides in soil will not follow linear kinetics if there is a lag time due to adaptation of the microflora (see, *e.g.* Mirgain *et al.*, 1993). Other processes invalidating simple linear kinetics are transport through the soil profile and uptake by plants (Leistra, 1986). Various soil factors influence the rate of degradation, such as soil organic matter, pH, temperature, etc. Considering these complicating factors, it may seem futile to even attempt using a single parameter (DT_{50}) to characterize the persistence of a pesticide in soil. Yet, the DT_{50} is one of best known properties of a pesticide (Worthing & Hance, 1991).

2. The initial concentration in soil, C_0 may be estimated from the dose applied (kg/ha), the density of the soil (kg/m^3), and the depth over which the pesticide is distributed (m). The dose can be taken as the recommended field application rate, while the distributional depth is often assumed to be 2.5 cm. Laboratory-field comparisons for earthworms, in which the initial field concentration of pesticides was estimated in this way have generally provided rather good results (Van Gestel, 1992; Heimbach, 1992). In the case of repeated applications of the pesticide it will of course be necessary to include residues from previous treatments in C_0.

3. The assumption on the statistical distribution for the sensitivities of species is questionable. For some chemicals, especially for the aquatic environment, the assumption has been verified (Kooijman, 1987; Volmer *et al.*, 1990; Niederlehner *et al.*, 1986). This can however only be done when there are about 20 different data on a great range of species. With fewer data, any test on the validity of an assumed distribution has a low power and will not be very informative.

The model will not be grossly in error when the logistic distribution is replaced by another symmetric, bell-shaped distribution, such as the normal distribution (*cf.* Aldenberg & Slob, 1993; Wagner & Løkke, 1991). However, in the case of pesticides one may expect a serious violation of the assumption due to the fact that some organisms will not be sensitive at all (*e.g.* plants to insecticides) while some are potentially sensitive (animals to insecticides). A distribution describing the sensitivities to a very selective pesticide would be double peaked, with the target species in the left peak, and most of the other organisms in the right peak. It seems appropriate not to apply the model to a very broad range of species when there is a suspicion that these will have heterogeneous sensitivities. Instead, one may assume that the logistic distribution will hold for a more or less homogeneous community (*e.g.* soil invertebrates, higher plants, soil microorganisms).

4. A further complication follows when the parameters of the assumed sensitivity distribution are estimated from a limited number of data. The formula for R given above assumes that the location and the shape parameter of the logistic distribution are estimated by their moment estimators (Johnson & Kotz, 1969). To achieve unbiased estimates for the mean and the standard deviation, the species that are used to estimate the parameters of the distribution should represent a random sample from the community which is exposed to the pesticide. There is however hardly any way to verify this assumption; usually the investigated species are not selected on the basis of their representativeness, but on their suitability for culture in the laboratory.

5. The final, much debated, assumption concerns the choice for the parameter δ. Any value, including $\delta = 0.05$, proposed in Van Straalen & Denneman (1989), will be arbitrary and not based on scientific arguments. The use of some kind of cut-off value is always necessary when using a bell-shaped distribution function, because the left tail of such a distribution extends to minus infinity (to zero on the concentration scale).

The formulation of distribution-based models for risk assessment, and especially their application for the derivation of environmental standards has received criticism from various authors (Hopkin, 1993; Forbes & Forbes, 1993; Smith & Cairns, 1993). The main argument concerns the ecological implications of the 95% protection level, for example when a key species in an ecosystem process, or a red list species happens to fall in the unprotected fraction of the community. The possible occurrence of situations like these requires a prudent use of the methods.

A clear advantage of the statistical approach is that it replaces the arbitrary safety factors that are used in more primitive assessments, and that it allows an objective, tractable procedure, that may be improved when ecotoxicological insights develop further. Van Leeuwen (1990) and Okkerman *et al.* (1992) have compared various versions of the statistical approach with other methods and with field data for aquatic ecosystems. The conclusion from these studies is that the distribution-based approach is usually somewhat more conservative but often not very much deviant from the more simple approaches using fixed assessment factors. It is also in good accordance with the lowest critical effect concentrations derived from field studies (Okkerman *et al.*, 1992).

Numerical examples

The methodology explained and discussed in the previous section was applied to four pesticides: carbofuran, benomyl, lindane and parathion.

Information on the terrestrial ecotoxicity of pesticides was collected from the literature. In the first place, we searched for studies in which soil organisms (invertebrates, micro-organisms) had been exposed experimentally to a range of concentrations of a pesticide mixed through soil. As a second step, we selected only those studies that showed a clear concentration-dependent response and that allowed a determination of sublethal endpoints. The sensitivity of the species was expressed as the highest concentration causing no significant decrease of reproduction, growth and survival. This so-called no-observed-effect concentration (NOEC) was taken as given by the author, or it was read from tables or figures provided in the paper. In a few cases where there was doubt on the interpretation of the data, confirmation was asked from the authors. LC_{50}-values were not used, because these data are of limited value when extrapolating ecotoxicity to higher levels of ecological organisation (Van Straalen, 1994).

In a similar way, information was collected on the rate of degradation of pesticides, expressed as DT_{50}-values. Only data referring to realistic conditions (using field soils, no synthetic media) were selected. When there were several DT_{50}s available, their harmonic mean was used in the calculation of R.

The initial concentration (C_0) of the pesticides was estimated from data on the recommended field application rate in various cropping systems. From these, the highest recommended field rate was used to estimate C_0. A distributional depth of 2.5 cm was assumed and a bulk density of the soil of 1061 kg/m^3. The latter figure corresponds to a "standard" soil with 25% clay and 10% organic matter, which are reference values in the Dutch soil protection policy.

The variability of the ecotoxicity data , being derived for various species (earthworms, springtails, beetles, *etc.*) is not only due to inter-species differences in sensitivity, but also due to variability in the conditions under which the species were tested. One of the most obvious conditions is the organic matter content of the soil, which will have a great influence on the bioavailability of the pesticide. An attempt was made to correct for these differences by normalizing the NOECs to a "standard" organic matter content of 10%. This procedure is explained in more detail in Van Rijn *et al.* (1994).

For both toxicity and degradation, the compounds benomyl and carbendazim were taken together, as the second is the degradation product of the first, and is mainly responsible for the toxicity. For the other compounds, it is assumed that degradation does not lead to products with an appreciable toxicity.

All original data, a full account of the model and the evaluation, plus a statistical elaboration of the uncertainty margins of R will be given elsewhere (J.P. van Rijn, unpublished). This paper provides only some indicative data for four selected pesticides. The original literature on which the evaluations were based is not reproduced in this paper. More compounds are being reviewed and will be considered in the final publication.

Table 1 provides a summary of the data starting with estimates for x_m and s_m. It appears that, among the four compounds considered, carbofuran is the most toxic product for soil invertebrates (lowest x_m), while parathion is the least toxic. The variability in sensitivity also differs between the compounds; especially for lindane the data are rather variable (high s_m), while the data for benomyl/carbendazim are not so widely dispersed.

The estimates obtained for degradation half-lives of the pesticides show that among the four compounds reviewed, benomyl/carbendazim is the most persistent, while parathion is the least persistent.

Finally, R-values were estimated assuming $\delta = 0.05$ (see the formula given above). The data listed in Table 1 indicate that ecotoxicological recovery from a single application at the maximal recommended field rate may last for a period of 77 days (parathion), one and a half year (carbofuran), two years (benomyl/carbendazim), or nearly three years (lindane). According to the classification scheme for R proposed by Van Rijn *et al.* (1994), parathion would be classified as a compound with "moderate risk", while the others would fall into the category of "high risk" pesticides.

Table 2 compares the classification of the four compounds on the basis of R with classifications based on persistence (DT_{50}) and toxicity (x_m) alone. This comparison demonstrates an interesting point: the joint assessment approach does not duplicate the information obtained from one of the other evaluation schemes. The value of R is very much influenced by the persistence

Table 1. Estimates for ecotoxicological recovery time for four pesticides in soil.

Parameter	Units	Symbol	Carbofuran	Benomyl/ Carbendazim	Lindane	Parathion
Number of species tested	–	m	6	6	4	8
Mean ln (NOEC)	ln(mg/kg)	x_m	−1.155	0.6520	0.254	1.844
Standard deviation of ln (NOEC)	ln(mg/kg)	s_m	1.269	1.214	3.323	1.764
Maximum initial concentration	mg/kg	C_o	37.7	2.07	55.4	30.2
Degradation half-life	days	DT_{50}	59	241	78	12
Ecotoxicological recovery time	years	R	1.6	2.0	2.8	0.21

Table 2. Ranking of four pesticides using three different criteria.

Pesticide	Ranking based on		
	Persistence	Toxicity	Joint Assessment
Carbofuran	3	1	3
Benomyl	1	3	2
Lindane	2	2	1
Parathion	4	4	4

Note: 1 = highest persistence, highest toxicity, longest ecotoxicological recovery time.

score, because it is directly proportional to DT_{50} (see the formula); toxicity and initial concentration have a smaller influence on R, because these parameters are under a logarithmic operator (see the formula). However, R does not seem to follow persistence only, as is demonstrated by the case of lindane: although this compound is neither the most toxic, nor the most persistent, it is still the least acceptable compound in terms of R. Benomyl/carbendazim is by far the most persistent compound, but due to its relatively low toxicity, it does not receive the highest R. Carbofuran is an example of a highly toxic compound that still receives a low score for R due to its relatively low persistence.

Discussion

The approach advocated in this paper evidently contains elements that are a gross over-simplification of the recovery process under real field conditions. It must be pointed out that the strength of the model is not to precisely predict degradation and recovery processes in the field; instead, it offers a generic system for the evaluation of non-persistent chemicals based on pieces of information that are usually available from the registration procedure. As such, it is an attempt to combine this information into a single score. It would be interesting to correlate this score with estimates for recovery times observed in field studies.

Evaluation of pesticides in terms of ecotoxicological recovery times may be a new argument in the discussion on the usefulness of persistence as an environmental criterion to be taken up in the "uniform principles" for the evaluation of plant protection products on the European market. From a scientific point of view, there is no reason to consider persistence as an inherent negative property of a pesticide; however, persistence coupled to ecotoxicological effects is clearly relevant. One way to combine these two is the parameter R, as proposed in this paper.

While the data reported in this paper are only preliminary, the outcomes must not be judged on their numerical details. A statistical analysis of the confidence that may be placed in estimates of R is being conducted at the moment. This involves the propagation of errors in the ecotoxicities, initial concentrations, and half-lives, using Monte Carlo simulations. Since the basic parameters may be assumed to vary independent from each other, the uncertainty in R will be considerable. Preliminary calculations show that the distribution of possible R-values is skewed to the left, and that the confidence range may span from 0.5 times the median in the lower range, to 5 times the median in the upper range.

References

Aldenberg, T.; Slob, W. *Ecotox. Environ. Saf.* **1993**, *25*, 48-63.

Heimbach, F. In *Ecotoxicology of Earthworms*, ed. P.W. Greig-Smith, H. Becker, P.J. Edwards & F. Heimbach. Intercept, Andover, **1992**, pp. 100-106.

Hopkin, S.P. *Oikos* **1993**, *66*, 137-141.

Jepson, P.C. *BCPC Monogr.* **1988**, *40*, 191-200.

Jepson, P.C.; Thacker, J.R. *Funct. Ecol.* **1990**, *4*, 349-358.

Johnson, N.L.; Kotz, S. *Distributions in Statistics: Continuous Distributions.* John Wiley & Sons, New York, **1969**.

Forbes, T.L.; Forbes, V.E. *Funct. Ecol.* **1993**, *7*, 249-254.

Kooijman, S.A.L.M. *Water Res.* **1987**, *21*, 269-276.

Leistra, M. *Pestic. Sci.* **1986**, *17*, 256-264.

MJP-G. (Meerjarenplan Gewasbescherming). *Proc. Netherlands Parliament*, **1991**, *21677*, 3-4.

Mirgain, I., Green, G.A.; Monteil, H. *Environ. Toxicol. Chem.* **1993**, *12*, 1627-1634.

Niederlehner, B.R.; Pratt; J.R., Buikema, Jr., J.; Cairns Jr., J. In *Community Toxicity Testing*, ed. J. Cairns Jr. ASTM STP 920, Philadelphia, **1986**, pp. 30-48.

Okkerman, P.C.; Van de Plassche, E.J.; Emans,

H.J.B.; Canton, J.H. *Ecotox. Environ. Saf.* **1992**, *25*, 341-359.

Smith, E.P.; Cairns, J.Jr. *Ecotoxicology* **1993**, *2*, 203-219.

Thomas, C.F.G.; Hol, E.H.A.; Everts, J.W. *Funct. Ecol.* **1990**, *4*, 357-368.

Van Gestel, C.A.M. *Ecotox. Environ. Saf.* **1992**, *23*, 221-236.

Van Leeuwen, C.J. *Environ. Manage.* **1990**, *14*, 779-792.

Van Rijn, J.P.; Hermans, M.; Van Gestel, C.A.M.; Van Straalen, N.M. In *Environmental Toxicology in South East Asia*, ed. B. Widianarko, K. Vink & N.M. van Straalen. VU University Press, Amsterdam, **1994**, pp. 289-300.

Van Straalen, N.M. In *Environmental Toxicology in South East Asia*, ed. B. Widianarko, K. Vink & N.M. van Straalen. VU University Press, Amsterdam, **1994**, pp. 33-47.

Van Straalen, N.M.; Denneman, C.A.J. *Ecotox. Environ. Saf.* **1989**, *18*, 269-251.

Van Straalen, N.M.; Schobben, J.H.M.; Traas, T.P. *Pestic. Sci.* **1992**, *34*, 227-231.

Volmer, J.; Kördel, W.; Klein, W. *Chemosphere* **1990**, *17*, 1493-1500.

Wagner, C.; Løkke, H. *Water Res.* **1991**, *25*, 1237-1242.

Worthing, C.R.; Hance, R. (eds.) *The Pesticides Manual*. British Crop Protection Council, **1991**.

Ecological Risk Assessment of Pesticides: The Role of Field Data

James F. Hobson, Technology Sciences Group, Inc., 1101 17th St. N.W. Washington, D.C. 20036

Aquatic field and simulated field studies, and terrestrial vertebrate field studies, have been a central, although controversial, component of the regulatory process for agrochemicals in the United States for over ten years. In the "New Paradigm" for Environmental Fate and Ecological Effects being implemented by the EPA's Office of Pesticide Programs, prospective field or simulated field studies are now falling out of favor. These studies are being replaced in large part by mitigation and regulatory decisions based on lower tiered laboratory ecotoxicology studies and simplistic (modelled) estimates of exposure. In addition, post-registration monitoring may play an increased role as a regulatory tool to gather the necessary information for risk assessment and ultimately risk management. However, there are some issues that can only be demonstrated through use of field or simulated field evaluations, during either pre- or post-registration monitoring. This paper will examine the positive contribution that well designed field or simulated field programs with clear objectives can make to an ecological risk assessment. Several case studies from the literature will be discussed to illustrate this point including: an aquatic microcosm, an aquatic residue monitoring study, a small scale avian field study and an evaluation of incorporation efficiency.

In the United States field and simulated field studies in support of pesticide registrations have been an important regulatory tool for the Ecological Effects Branch (EEB) in the Office of Pesticide Programs (OPP) and a major area of concern for the agrochemical industry. Field and simulated field studies were designed to provide a definitive, unequivocal demonstration of safety. However, in practice these studies were rarely unequivocal in the regulatory context (AEDG 1994). Furthermore, the conduct and review of these field programs have often resulted in significant delays in the registration process. In part, because of this, the emphasis on field studies has now changed.

In October 1992 the OPP's *Ecological and Environmental Fate Data Requirements Task Force* issued a report after six months of work (EPA 1992). This report has become known as the "*New Paradigm*" and emphasizes risk mitigation early in the regulatory process and reduced use of field and simulated field studies. Mitigation (i.e., management) of risk incorporates the concept of ecological risk assessment (ERA) before registration or reregistration. However, field and simulated field efforts to examine environmental fate and effects will likely continue to play an important role in the ERA of many pesticides for both industry and the Environmental Fate and Ecological Effects Division (EF&ED) of OPP. For example, such studies may be required as part of post-registration monitoring (which is yet to be defined). In addition, some registrants may not be willing to accept an unfavorable risk assessment based on lower tiered studies when field studies could potentially clarify issues and result in a positive, or favorable, risk assessment.

This paper will discuss the role of field data in ERA emphasizing how well designed studies, can make clear and substantive contributions to an ERA and the regulatory process for a given pesticide. Instead of attempting to prove unequivocal "safety" for a compound using field studies as these well focused, often smaller scale, field studies can provide significant contributions to the overall weight of evidence. In the past the agrochemical industry has resisted large scale field studies in the area of ecological effects; however, industry may, in the future, find significant advantages in smaller scale ecotoxicology and environmental fate studies.

Since the New Paradigm stresses making regulatory decisions earlier without higher tiered studies, such decisions will by nature be more conservative. Figure 1 demonstrates that when an iterative tiered testing scheme, such as the one used by the EF&ED to evaluate agrochemicals, is employed the confidence in the risk assessment increases with each progressive tier. However, when risk management decisions are implemented based on risk assessments using only laboratory ecotoxicology data (i.e., Tiers I and II) and preliminary environmental fate modelling, the ERA must allow for increased uncertainty (i.e., less confidence). The estimates of the level of concern (LOC) (the level of exposure expected to result in adverse effects based on toxicology studies) and the estimated environmental concentration (EEC) (the concentration predicted to occur in the environment) must then be farther apart to obtain a given level of confidence in an ERA which predicts in adequate margins of safety. LOC and the EEC are close together the risk assessment could potentially be refined possibly using field data to achieve an acceptable margin of safety. By increasing the available data, the use of higher tiered studies serve to decrease the CIs.

Lower tiered studies are designed to differentiate inoquous or very non-hazardous chemicals from compounds which require further examination and possibly testing to manage potential risk. Thus lower tiered risk assessments are by nature very conservative.

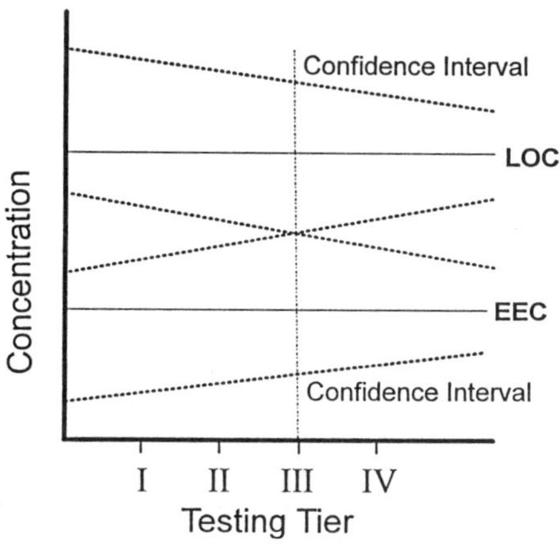

Fig. 1 Decreasing confidence intervals around estimates of effect and exposure levels with higher tiers of testing. The vertical line at Tier III is a hypothetically where the EPA will chose to stop testing. Adapted from Sutter (1993).

When risk assessments based on lower tiered studies are not favorable to the compound in question, or if mitigation measures proposed by EPA are unacceptable, the registrant may look to carefully designed field or simulated field studies to refine the risk assessment. The objective is to reduce uncertainty and increase confidence levels in an effort to potentially achieve registration without mitigation. If field studies or simulated field studies have a clear focused design and the results are clearly presented, they may still effectively contribute to the overall ERA for many pesticides.

Current Developments in Ecological Risk Assessment

Ecological Risk Assessment is a rapidly expanding field which is applied to many areas in the regulation of chemicals. In the United States ERA, is used under a variety of statutes including the Federal Insecticide, Fungicide and Rodenticide Act (FIFRA), and the Toxic Substances Control Act (TSCA), Superfund (CERCLA), the Cleanwater Act (CWA), the Resource Conservation and Recovery Act (RCRA), and others. The data upon which ERAs are based can range from laboratory fate and effects programs for a single chemical under a Section 4 test rule (Hobson 1992) to an *in situ* assessment of an ecosystem contaminated with a suite of chemicals associated with a Superfund site (Jenkins *et al. inpress*). However, the most extensive field programs, especially for single chemicals, have been required by the Ecological Effects Branch (EEB) of the OPP. Such programs are required because the underlying statute (FIFRA) mandates that to register a pesticide the EPA must determine that the chemical will not "generally cause unreasonable effects on the environment" (P.L.95693, Sec. 3(c)(5)(D)). Other statutes such as TSCA require the Agency to have reason to believe that a chemical is causing unreasonable adverse effects on the environment before any higher tiered laboratory or field testing is required. Furthermore, the approaches to ERA differ for these different applications.

In February 1992, EPA's Risk Assessment Forum published a report titled "*Framework for*

Ecological Risk Assessment", the first step in a long-term effort to develop risk assessment guidelines for ecological effects (Risk Assessment Forum 1992). This document is provides a general framework that enables a user to apply a common structure and approach to various types of ecological risk assessment.

The *"Framework"* uses broadly accepted fundamental concepts which are presented in the flow diagram in Figure 2. There are two basic components of the risk assessment, the exposure and effects information, that are analyzed and then these are integrated in the risk characterization. To these basic components the "Framework" adds a Problem Formulation Phase and adjunct activities involving interactions between Risk Assessors and Risk Managers (which will not be discussed in detail in this paper) and Data Acquisition, Verification and Monitoring. The latter activities are central to the theme of the current paper. Risk assessment is often an iterative process and it is through these latter data acquisition activities that information enters the risk assessment process. If additional information is needed to refine the risk assessment at any phase of the risk assessment process (or even the risk management process), additional data acquisition can be conducted and the risk assessment process can then continue or be reinitiated at an earlier phase.

Under the New Paradigm, EF&ED is

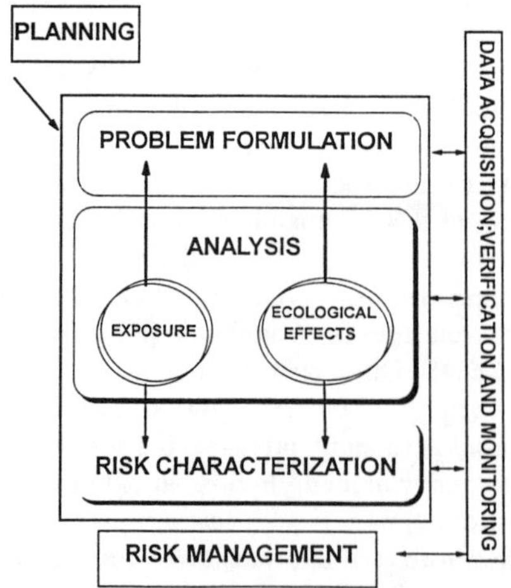

Fig. 2 Framework for Ecological Risk Assessment. Adapted from EPA (1992).

attempting to truncate the data acquisition process after the lower tiered studies when ever possible. This is being done to facilitate meeting Congressionally mandated timelines for reregistration and because "... field studies do not provide risk managers with the kind of information that greatly enhances risk management decisions ...." (EPA 1993). When possible regulatory decisions are to be made in the absence of higher tiered testing. Although this is expedient for the EF&ED this is often likely to be unfavorable to the registrant. As mentioned earlier, ERAs based on lower tiered studies are generally more conservative due to lower confidence in the assessment. It is likely that in many cases registrants may choose not to truncate the data acquisition phase, but to conduct additional data acquisition, verification, and monitoring as shown in Figure 2. The additional data should be the basis to refine the risk assessment. This is especially true if the study is well designed and the results clearly presented.

Field and simulated field programs are often important in refining the risk assessment of chemical, because they offer a more realistic assessment of fate and effects under actual or simulated field conditions and they allow evaluation of the integrated effects of numerous natural processes on the fate and effects of a chemical (eg., photolysis, hydrolysis, adsorption/desorption, etc.). For these reasons, and others, field and simulated field studies are likely to continue to play a key role in the refinement of ERAs for pesticides either through pre-registration studies or post-registration monitoring.

The purpose of this paper is to examine the additional information and insight that is generally unavailable from laboratory studies, or modelling, but can potentially be provided by field or simulated field studies. However, this paper will not discuss the large scale Tier IV studies (eg., mesocosms, or avian field studies) which are not often well focused. The latter studies are designed to address concerns of aquatic or avian risk under worst-case exposure conditions. Instead this paper will discuss four generally small-scale field efforts which address specific questions, that when answered, contributed to the overall weight of evidence of the risk assessment for the given compound.

These case studies represent two effects studies (one aquatic and one terrestrial) and two exposure studies (one aquatic and one terrestrial). These studies are an aquatic microcosm (Drenner *et al.* 1993), a small scale avian field study (Hawkes *et al.* 1991), an aquatic residue monitoring study (Bievers et al. 1991), and a granular incorporation study which was conducted as part of a Level I avian field study (Hobson et al. 1988).

Effects Assessment

Aquatic Effects Study (Case Study #1)

In the past, aquatic field studies have usually involved mesocosms (simulated field studies) or pond studies (actual field studies). Microcosms are smaller scale experimental systems designed to evaluate aquatic ecosystems and have been evaluated and proposed as an intermediate level study for use in the evaluation of pesticide ecotoxicity (SETAC 1991). Microcosms are simulated aquatic systems that can be used to evaluate a suite of ecological, ecotoxicological or environmental fate processes. For example, microcosms can be used simply to evaluate the fate of a compound in an aquatic system (i.e., an integrated aquatic dissipation) incorporating all of the natural processes that influence the dissipation of given compound (eg., photolysis, hydrolysis, biolysis, adsorption/desorption, etc.). Or microcosms can be used to identify compound-specific effects for specific taxa or life stages of organisms that are particularly sensitive to a given compound (eg., a juvenile hormone mimic). Microcosms have also been proposed to replace current mesocosm designs (SETAC 1991) and have been used as adjunct studies in combination with aquatic field studies and mesocosms (Drenner *et al.* 1993).

As part of any data gathering exercise such as the registration of pesticides, data is often generated that warrants a more refined evaluation, particularly when the data is to be used in a risk assessment. In such a situation a focused study design to elucidate the specific issue of concern may be appropriate, and such a study may require conduct under field or simulated field conditions. For example, Drenner and co-workers (1993) present a microcosm evaluation of the toxicity of pyrethroid (bifenthrin) to a filter-feeding fish (*Dorosoma cepedianum*).

In a three year program to evaluate an Alabama farm pond adjacent to cotton acreage treated with a pyrethroid insecticide (bifenthrin), a substantial winter die-off of gizzard shad was observed. Although it was likely that this was simply a relatively typical winter die-off of this species, questions were raised by the Agency about the role of the pyrethroid in the observed mortality. Although the pond water residues were extremely low (ng/L), the sensitivity of this compound to shad was not known. In fact, little is known about the toxicity of any pesticides to filter feeding fish species such as shad. This study was designed to evaluate the toxicity of sediment (soil) bound bifenthrin to this species.

In order to assess the sensitivity of shad to bifenthrin, a large-scale microcosm study was conducted using 30 5.5-M^3 circular fiberglass tanks according to methods described in Drenner *et al.* (1993). This scale of test system was necessary, because shad can not be held in small square tanks, but had been successfully maintained in these larger round tanks (Drenner pers. comm). In addition, capture, transport, and maintenance of shad is problematic. Shad are unusually sensitive to temperature and dissolved oxygen (DO) changes and can easily be stressed in transport, often resulting in significant transfer mortality. Furthermore, long-term maintenance in a confined test system, even in large-scale microcosms, is difficult. This work was conducted at Texas Christian University by Dr. Ray W. Drenner and co-workers who have extensive experience with shad and other filter-feeding fish species, microcosm test systems.

The results of the pond study showed that the compound entered the pond primarily as soil bound residues on solids during runoff events; furthermore the mortalities observed were well after treatment and any potential drift-related exposures. In this adjunct study the microcosms were treated with soil slurries prepared with bifenthrin treated, aged soil. Initial measured concentrations in whole-water samples were 0, 90, 185, 250, 1550, and 7750 ng/L bifenthrin for a control and five exposure levels, respectively. Two additional controls, one with no soil slurry, and one with control sediment and no fish, was also included. Water column concentrations declined steadily over an eight-day exposure period to approximately 21% of nominal. Effects on fish survival followed a typical dose response (Fig. 3) with 8-d LC50 values calculated as 521 ng/L and 207 ng/L for 1-h and 8-d averaged concentrations, respectively.

These 8-d LC50 values are in the same order of magnitude as the results of laboratory studies with the results of the 96-h LC50s, 100 ng/L and 260 ng/L, for rainbow trout and bluegill, respectively (Drenner *et al.* 1993). Furthermore a histopathological component to this study showed epithelial loss and secondary lamellae clubbing in a dose response fashion consistent with observations by other workers. These values were much higher than the mean maximum concentration of this compound observed in the Alabama pond (17.9 ng/L) following a rainfall event (Drenner *et al.* 1993). The results of this study provide a measured toxicity value for soil-bound bifenthrin to shad.

This focused microcosm study was used to effectively address a key issue, toxicity of shad, that arose during the data generation phase of the effort to support registration of this pyrethroid. This result contributed to the overall weight of evidence that the mortality of shad in the Alabama pond was likely not related to bifenthrin.

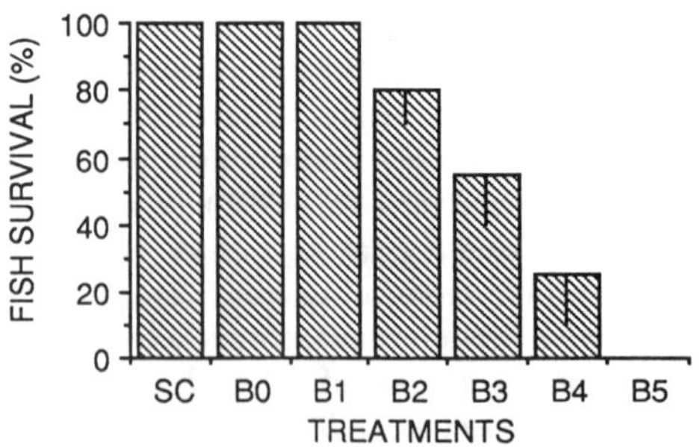

Fig. 3 Fish survival across increasing concentrations of bifenthrin after seven days of exposure. Bars represent means of two replicate tanks for each treatment. Reproduced with permission from *Env. Tox. Chem.*

Terrestrial Vertebrate Effects Study (Case Study #2)

Tier III avian studies such as pen studies can add substantial information beyond the laboratory evaluations because they allow the natural behavior of the test species to be evaluated as it relates to the ecotoxicology and pharmacology of an active ingredient or product formulation (i.e., granular insecticides and nematicides). On the other hand, terrestrial vertebrate field studies can raise questions which can only be addressed by returning to more focused evaluations.

Level one field studies are designed to determine if avian mortality occurs associated with use of a pesticide product (Fite et. al. 1988). These studies have relied heavily on carcass searching as the principle means of determining impact. One criticism of these studies for both organophosphates (OPs) and carbamates has been that birds which have ingested a lethal dose of a compound can possibly fly away from the site of ingestion to die, or birds might seek cover when becoming intoxicated and not be easily observed during subsequent carcass searching activities. A discussion of this topic is provided in *Assessing Pesticide Impacts on Birds, the Final Report of the Avian Effects Dialogue Group*, 1988 to 1993 (AEDG 1994) and Mineau and Collins (1988). The latter authors conclude that offsite migration may be a problem with OPs, but may not be a problem with carbamates, based on the pharmacology of the latter class of compounds.

An extensive Level I field study was conducted in 1987 with granular aldicarb (TEMIK) (Hobson *et al.* 1988). Multiple sites (28 plots in five states) in three crops including citrus, cotton, and potatoes were evaluated. On a combined study acreage of 1,366 acres only 11 birds with aldicarb residues were found dead. Many Agency staff believe that a significant portion of birds that ingest a lethal dose of an OP or carbamate fly off-site or into the edge habitat to die. Hypothetically, these birds are not observed during subsequent carcass searches (AEDG 1994). However, with a carbamate like aldicarb, this behavior pattern would not be expected based on toxicokinetics of the chemical which allows for both reversal of binding (inhibition) and rapid metabolism (detoxification) of the compound (Klaassen et al. 1986). If a bird ingests a lethal dose the onset of symptoms of toxicity and death are extremely rapid (i.e., generally within 5 to 10 minutes, Hawkes et al. 1991). If death does not occur quickly the compound is readily metabolized, the cholinesterase inhibition reversed, and the animal fully recovers. Based on these facts and due to criticism of large-scale studies with other carbamates an additional focused study was conducted to refute the concerns related to off-site migration with aldicarb.

Morning doves (*Zenaida macroura*) and Northern bobwhite (*Colinus virginianus*) were trapped, tagged with radio transmitters, dosed with aldicarb, and quickly ($\leq$ 10 minutes) released in the same locality along the edge cotton fields (Hawkes *et al.* 1991). These species were chosen because doves were the species most frequently observed as mortalities in a recent extensive field study and this species is considered highly mobile and could have a high potential to move off-site. Similarly, bobwhite have a tendency to look for dense cover when threatened or stressed. Both species are seed eaters and therefore assumed to be more susceptible to ingestion of granules. Birds were dosed via oral gavage with aldicarb in corn oil or corn oil vehicle at, below, or slightly above the LD50 as determined for the local populations of the same species. The LD50 values for doves and bobwhite were determined to be 0.82 and 1.48 mg/L. Birds were released in less than 10 minutes of dosing using controlled conditions that did not stimulate flight. Most birds quickly began feeding in the field near the edge.

Birds were monitored visually for the first 1 to 2 hours and using radio-telemetry for seven days post-dosing. None of the doves dosed and released were observed to seek dense cover. Of the bobwhite mortalities observed, 67% died on the field within 1 meter of the release cage and 30% sought cover within 2 meters of the cage, while untreated bobwhite sought cover within 25 meters of the release site. In both species birds generally expressed symptoms of toxicity within 15 minutes and died within one hour or recovered. Recovery was generally complete within 2.0 hours. Birds intoxicated with a lethal dose of aldicarb remained on the field or in the edge immediately adjacent to the open field. Where according to the authors most if not all of these carcasses would easily be found by using standard carcass searching techniques. Birds are apparently limited in their cover seeking ability by the rapid onset of the toxicity of aldicarb.

This off-site migration study clearly demonstrated that birds intoxicated with carbamates such as aldicarb do not migrate off-site to die. In a field study a significant proportion of bird carcasses would likely be found within the perimeter of the agricultural field. Furthermore, this work supports the conclusion that the carcasses found in the 1987 study were a relatively accurate measure of the mortality that may have occurred associated with the use of aldicarb within the confines of the carcass efficiency determined for the larger field study. The results of this study served to refute a standard criticism of avian field studies which was inappropriate to carbamates such as aldicarb. This contributed to the overall weight of evidence for the risk assessment of aldicarb and served to increase the confidence in the results of the previous avian field study where carcass searching was a key part of the evaluation (Hobson *et al.* 1988).

Environmental Fate and Exposure Assessment (Case Study #3)

Aquatic Exposure Assessment

Discussions of the Aquatic Effects Dialogue Group facilitated by RESOLVE[1] (AEDG 1992) identified three types of field studies to assess aquatic exposure. These were small-plot simulated runoff studies, large-scale field run-off studies and large-scale monitoring studies. The latter are also called Aquatic Residue Monitoring Studies (40 CFR 158; Pesticide Assessment Guidelines, Subdivision E, Guideline 72-7) and, although larger in scale than the other types of studies discussed in this paper, are an example of focused field evaluations. These studies were designed to measure residues in aquatic systems associated with normal labeled use of an agricultural chemical. These studies were required for a number of compounds under reregistration especially to establish or verify the

---

[1] RESOLVE: Center for Environmental Dispute Resolution; 1250 24th St. N.W., #500, Washington, D.C. 20037

EECs for given use pattern. Aquatic residue monitoring studies measure the magnitude, frequency and duration of chemical residues in streams, drainages or waterbodies associated with specific agricultural use patterns. The studies usually include multiple sites (as many as five or more) and regions (as many as three or more).

Although the New Paradigm deemphasizes pre-registration field studies, mitigation and post-registration monitoring are discussed. Aquatic residue monitoring studies are an excellent example of what may be required for post-registration activities. While these studies have numerous disadvantages (eg., they are labor-intensive, expensive, do not provide probablistic or predictive information and are not appropriate for some compounds), aquatic residue monitoring studies do measure real-world exposure potential including the spray drift component, and can verify exposure model calculations. The following discussion describes a successful application of an aquatic monitoring study to estimate the aquatic residues of a rice fungicide outside the rice paddy as described by Biever and co-workers (1991).

Residues of iprodione, a rice fungicide, and its two major metabolites were measured in water and sediments from below rice paddy outfalls down stream to the first major water source (i.e., river or estuary) and one site upstream from the outfall (Biever *et al.* 1991). Five sites chosen in three states (Texas, Louisiana, and Arkansas) included three freshwater and two saltwater sites. Paddies were chosen which also had iprodione use on other fields within the same drainage in order to assess potential additive exposure. The product was applied at the maximum label rate of 0.5 lbs. a.i. twice (the maximum frequency) at 14 days apart.

The impetus for this requirement was a risk assessment based on tier II studies and simplistic modelling (Biever *et al.* 1991). The EECs (maximum receiving water concentrations) were calculated based on paddy water concentration to be 734 µg/L assuming direct application to water. The EECs were measured in an aquatic dissipation study to be 700 µg/L. Following a worst-case (2 in.) rainfall event and a receiving water dilution factor of 2.0, the resulting concentration in the adjacent stream was estimated by the EPA to be 220 µg/L. This calculated exposure level is higher than either of the two lowest NOEC (no observed effect concentration) values of 170 µg/L and 7.5 µg/L for *Daphnia* and mysid shrimp, respectively. Because the EEC value exceeded the LOCs, resulting in a presumption of hazard, the registrant was required to measure the exposure levels under field conditions, i.e., an Aquatic Residue Monitoring Study, a condition of reregistration.

Based on these NOEC levels low limits of detection were required. In order to measure the parent compound and its two major degradates in water the detection limits were required to be at or below 3.5 µg/L (one twentieth the lowest NOEC) in water. This low detection limit was a challenge to meet and maintain.

The results of this study showed that maximum water concentrations of the parent molecule to range from <MDL (approximately 0.35 µg/L) to 121.3 µg/L in drainage ditches and in adjacent streams following off-target deposition immediately after some applications, major runoff events and draining. However, iprodione dissipated quickly following rain events or drainage to < 2.0 µg/L usually in less than seven days (Bievers *et al.* 1991). Maximum concentrations of iprodione were well below the predicted EEC of 220 µg/L and acute laboratory studies. Similarly the maximum 21-day average water concentrations were <MDL to 54.2 µg/L in freshwater and 3.3 to 7.1 µg/L in brackish water. These values were also below the chronic NOEC values for fresh and saltwater species, respectively. The major metabolite was observed at lower concentrations, but residues were somewhat more persistent than the parent moiety. The minor metabolite was only observed infrequently at low concentrations.

The results of this study provided realistic exposure based on measured concentrations in water under different agronomic practices, geographic regions, timing and weather conditions and include drainages associated with both fresh and salt water. These results allowed the risk assessment to be based on the actual measured (i.e., real) residues which were lower than the model predicted EEC. Such studies could be used effectively as part of the New Paradigm, especially as post-registration monitoring.

Terrestrial Exposure Assessment (Case Study #4)

Exposure assessment in terrestrial ecotoxicology studies is generally much more difficult to assess than in aquatic systems. This is linked to number of characteristics of the compound(s) being evaluated (eg., formulation, use pattern and environmental fate characteristics) and the organisms of concern (eg., mobility, behavior, etc.). A number of new, as well as standard techniques have been applied to assessing avian exposure in the past few years. Exposure estimates are generally compared to estimates of toxicity reported in more controlled laboratory studies for the same compound and if possible the same organism. In some cases, classic environmental residue techniques have been applied to avian tissues and feed items in the field and then compared to the results of laboratory exposures to the same compound and organism (Lincer 1975).

In addition, some innovative techniques have been applied to the assessment of residues. Kendall *et al.* (1989) have proposed to establish starling population on study plots by providing nestboxes. This species was chosen partially because it can be manipulated in the field. As part of this study design the throat of young birds can be temporarily, partially restricted and food items provided by adults can subsequently be removed and analyzed, allowing a direct measure of food residues actually being consumed by young. This alleviates the need for assumptions about dietary content or what proportion of food comes from the study site or what portion of the site. These latter issues are potential sources of error in extrapolating from residues to actual dietary exposure.

The level of concern for acute avian mortality related to granular insecticides (eg., granular formulations of organophosphate and carbamate compounds) has been based on the LD50/Sq.Ft. as determined by the application rate (lb/A.I.) and the number of granules per pound (i.e., the number granules equivalent to the LD50). As originally proposed by Felthousen (1977, as referenced in EPA 1989) the number of LD50's per square foot is used by the EEB as a measure of "risk". This approach to avian risk has played a prominent role in such EPA documents as the *Carbofuran Special Review Technical Support Document* (EPA 1989). The now famous 29 October 1992 ("Linda Fisher") memo from Linda Fisher, then Assistant Administrator (Fisher 1992), to Doug Campt, Director of *OPP,* which outlined the New Paradigm, continues to use this approach to establish an unacceptable level of exposure as compared to the LOC for acute avian risk. However, this evaluation ignores differences in incorporation efficiency for different compounds, formulations, and use patterns.

As discussed above, in 1987, an extensive avian field effort was conducted to assess the potential effects of TEMIK Brand pesticides on birds associated with citrus, cotton, and potato fields (Hobson *et al.* 1988)[2]. As described by these authors an "incorporation study" was included in this program to evaluate exposure. This study was designed to establish the placement of granules in soil using the same application methods and equipment as employed in the avian field study. Incorporation efficiency was evaluated using granules coated with fluorescent dye. Blacklight photography was used to monitor the placement of granules in the soil bed and on the soil surface. These techniques were also compared to a surface band application typical of corn root worm insecticides/nematicides which are less efficient at incorporation.

The granule incorporation study utilized each of seven application techniques used in the avian field study. The results demonstrated that all of these application techniques were highly efficient and allowed no significant surface exposure of granules to occur. Spills at the end of rows and subsequent discing was simulated. Photographs (blacklight) of the same area of soil surface were taken before and after end row discing, per label instructions. This technique essentially eliminated the presence of granules on the surface. Proper application techniques applied to the incorporation of TEMIK reduced the exposure and the potential toxicity (hazard) to avian populations.

---

[2] The incorporation study was part of the overall presentation (Hobson et al. 1988) and was designed and conducted by Mr. J.D. Fish of Rhone-Poulenc Ag Co., Research Triangle Park, NC.

Current EPA calculations (LD50/sq. ft.) fail to differentiate between different application techniques. For example the incorporation study described here included a comparison of the typical TEMIK applications with corn root worm "band" applications. The latter techniques generally involve application in the top 2 inches of the soil surface which may result in substantial surface exposure of granules. This granular-incorporation study showed that these latter application techniques were far less efficient than those used with TEMIK supporting the results of the avian field study which showed only 11 avian mortalities attributed to TEMIK use on a combined study area of 1,366 acres (28 fields/groves in 5 States) (Hobson *et al.* 1988).

Summary

The role of field studies in the ERA of pesticides will continue to change as the New Paradigm is implemented and pre- and post-registration testing and monitoring requirements are better defined. Whether or not the Agency requires the field or simulated field efforts to support registrations, registrants may find added value in conducting such studies. This is especially true when risk assessments based on lower tiered studies, therefore a more conservative ERA, are used in making regulatory decisions that are not favorable to a given product. For example if strict mitigation measures are required based on such a lower tiered ERA, the impact of the mitigation measures may be unacceptable to the registrant. In such a case refinement of the risk assessment using a focused field or simulated field study or other higher tiered study may be advantageous to the registrant.

The type of study discussed above, although experimental in nature, is now generally conducted according to GLPs and with full analytical support. However, such studies are not standardized and the relative weight that may be given to such a study by the Agency in consideration of registration is uncertain. The clarity and completeness of the presentation to the Agency may be very important and oral presentations and discussions with Agency reviewers is recommended.

The types of studies discussed above can serve to address specific questions raised in the context of larger studies. For example, the results of the two adjunct terrestrial studies discussed above (Hobson *et al.* 1991, Hawkes *et al.* 1991), provide strong support for the results and conclusions of avian field studies conducted in support of TEMIK Brand granular pesticides.

The studies discussed above are conducted in a manner consistent with the ERA process outlined in the "Framework for Ecological Risk Assessment" (Risk Assessment Forum, 1992). All four studies were designed after thorough definition of the problem and with clearly defined assessment and measurement endpoints. As discussed in the Framework document, additional data acquisition, verification, or monitoring can take place on an iterative basis throughout the ERA process and even during risk management. In addition the four studies discussed included two studies representing each of the basic components of the ERA analysis phase, effects characterization and exposure characterization (Fig. 2). Thus these examples of well-defined, well-focused field studies illustrate how field data can provide valuable information for the evaluation of the potential environmental risks related to pesticides use.

References

Aquatic Effects Dialogue Group. *Improving Aquatic Risk Assessment Under FIFRA*; **RESOLVE**: **1992.**

Avian Effects Dialogue Group. Assessing Pesticide Impacts on Birds; **RESOLVE: 1994.**

Biever, R.C.; Kendall, T.Z.; Hobson, J.F. *Environmental Concentrations of Iprodione and Two of its Degradates in Waters and Sediments Associated with Rice Production*; Presented at the 12th Annual Meetings of the Society of Environmental Toxicology and Chemistry, Seattle, WA. Nov. **1991**

Drenner, R.W.; Hoagland, K.D.; Smith, J.D.; Barcellona, W.J.; Johnson, P.C.; Palmieri, M.A.; Hobson, J.F. *Env. Tox. Chem.* **1993,** 12, 1297-1306.

EPA. Carbofuran: Special Review Technical Support Document; EPA: Washington, **1989.**

EPA. *Implementation of the Policy Decisions of the Ecological Effects and Environmental Fate Task Force,* August 25, **1993**

Fite, E.C.; Turner, L.W.; Cook, N.J.; Stunkard, C. *Terrestrial Field Studies;* EPA, **1988**

Hawkes, A.W.; Brewer, L.W.; Hobson; *Survival and Cover Seeking Response of Northern Bobwhites and Mourning Doves with Aldicarb*; Presented at the 12th Annual Meetings of the Society of Environmental Toxicology and Chemistry, Seattle, WA. Nov. **1991**.

Jenkins, K.D.; Lee, C.R.; Hobson, J.F. *Ecologial Risk Assemssment of a Hazardous Waste Site at the Naval Weapons Station, Concord, CA;* In: G.M. Rand (Ed.) *Environmental Toxicology*; Hemisphere: *in press*

Klaassen, C.D.; Amdur, M.O.; Duoll, J.D. Toxicology: The Basic Science of Poisons, Third Edition; Macmillian, New York, **1986**.

Risk Assessment Forum. *Frmework for Ecological Risk Assessment;* EPA: **1992**.

SETAC. *Workshop on Aquatic Microcosms for Ecological Assessment of Pesticides;* **RESOLVE: 1992.**

INDEX

Index

N

Indexing: Janet S. Dodd
Production: Janet S. Dodd and Charlotte McNaughton
Acquisition: Anne Wilson
Cover design: Alan Kahan

Printed by United Book Press, Inc., Baltimore, MD
Bound by American Trade Bindery, Baltimore, MD

Highlights from ACS Books

Good Laboratory Practice Standards: Applications for Field and Laboratory Studies
Edited by Willa Y. Garner, Maureen S. Barge, and James P. Ussary
ACS Professional Reference Book; 572 pp; clothbound ISBN 0–8412–2192–8

Silent Spring Revisited
Edited by Gino J. Marco, Robert M. Hollingworth, and William Durham
214 pp; clothbound ISBN 0–8412–0980–4; paperback ISBN 0–8412–0981–2

The Microkinetics of Heterogeneous Catalysis
By James A. Dumesic, Dale F. Rudd, Luis M. Aparicio, James E. Rekoske, and Andrés A. Treviño
ACS Professional Reference Book; 316 pp; clothbound ISBN 0–8412–2214–2

Helping Your Child Learn Science
By Nancy Paulu with Margery Martin; Illustrated by Margaret Scott
58 pp; paperback ISBN 0–8412–2626–1

Handbook of Chemical Property Estimation Methods
By Warren J. Lyman, William F. Reehl, and David H. Rosenblatt
960 pp; clothbound ISBN 0–8412–1761–0

Understanding Chemical Patents: A Guide for the Inventor
By John T. Maynard and Howard M. Peters
184 pp; clothbound ISBN 0–8412–1997–4; paperback ISBN 0–8412–1998–2

Spectroscopy of Polymers
By Jack L. Koenig
ACS Professional Reference Book; 328 pp;
clothbound ISBN 0–8412–1904–4; paperback ISBN 0–8412–1924–9

Harnessing Biotechnology for the 21st Century
Edited by Michael R. Ladisch and Arindam Bose
Conference Proceedings Series; 612 pp;
clothbound ISBN 0–8412–2477–3

From Caveman to Chemist: Circumstances and Achievements
By Hugh W. Salzberg
300 pp; clothbound ISBN 0–8412–1786–6; paperback ISBN 0–8412–1787–4

The Green Flame: Surviving Government Secrecy
By Andrew Dequasie
300 pp; clothbound ISBN 0–8412–1857–9

For further information and a free catalog of ACS books, contact:
American Chemical Society
Distribution Office, Department 225
1155 16th Street, NW, Washington, DC 20036
Telephone 800–227–5558